T0211469

LONDON MATHEMATICAL SOCIETY LECTURE NOTE SERIES

Managing Editor: Professor M. Reid, Mathematics Institute, University of Warwick, Coventry CV4 7AL, United Kingdom

The titles below are available from booksellers, or from Cambridge University Press at
http://www.cambridge.org/mathematics

London Mathematical Society Lecture Note Series: 407

The Maximal Subgroups of the Low-Dimensional Finite Classical Groups

JOHN N. BRAY

Queen Mary, University of London

DEREK F. HOLT

University of Warwick

COLVA M. RONEY-DOUGAL

University of St Andrews

CAMBRIDGE
UNIVERSITY PRESS

CAMBRIDGE
UNIVERSITY PRESS

Shaftesbury Road, Cambridge CB2 8EA, United Kingdom

One Liberty Plaza, 20th Floor, New York, NY 10006, USA

477 Williamstown Road, Port Melbourne, VIC 3207, Australia

314–321, 3rd Floor, Plot 3, Splendor Forum, Jasola District Centre, New Delhi – 110025, India

103 Penang Road, #05–06/07, Visioncrest Commercial, Singapore 238467

Cambridge University Press is part of Cambridge University Press & Assessment, a department of the University of Cambridge.

We share the University's mission to contribute to society through the pursuit of education, learning and research at the highest international levels of excellence.

www.cambridge.org
Information on this title: www.cambridge.org/9780521138604

First published 2013

A catalogue record for this publication is available from the British Library

ISBN   978-0-521-13860-4   Paperback

Additional resources for this publication at www.cambridge.org/9780521138604

Contents

v

Contents   vii

Foreword

In this book the authors determine the maximal subgroups of all the finite classical groups of dimension 12 or less. This work fills a long-standing gap in the literature. Behind this gap there is a story which I am pleased to have the opportunity to tell.

The completion of the classification of finite simple groups was first announced in the early 1980s. It was clear then (and before) that for many applications of the classification one would need detailed knowledge of the maximal subgroups of the simple groups and of their automorphism groups. Around that time, I gave a Part III course at Cambridge about the classification and its impact. Full of enthusiasm, I set a fearsome exam – I remember giving it to John Conway to check, and him saying that he couldn't do any of the questions, but he thought it was probably OK. The second highest mark was 18%, scored by a rather strong student. The top mark was 97%, scored by Peter Kleidman, a young American.

Soon afterwards, Kleidman started as my first research student. Michael Aschbacher had just published his fundamental theorem on maximal subgroups of the finite classical groups. The time seemed right to attempt to use this to determine all the maximal subgroups of the classical groups of low dimensions (up to 20, say, I thought optimistically). This was Kleidman's initial project. As it turned out, in his thesis he solved many other maximal subgroup problems, and this project occupied just one chapter. Nevertheless it was rather an interesting chapter, consisting of tables of all the maximal subgroups of finite simple classical groups of dimension up to 12. No proofs were given, just an outline of the strategy and a few examples of how the calculations were performed.

After he had graduated, Kleidman and I wrote a book on the subgroups of the finite classical groups, which was an analysis of the structure, conjugacy and maximality of the subgroups arising in Aschbacher's theorem now known as geometric subgroups. For the maximality questions we assumed that the dimension was more than 12, and it was the intention that Kleidman would extend and write up his thesis work on the low dimensions as a separate book. Indeed, this project was accepted as a volume to appear in the Longman Research Notes series, and has been referred to as such in many articles. Unfortunately, he did not write this book, and left mathematics at the age of about thirty to pursue other interests such as working on Wall Street and producing Hollywood movies.

The non-appearance of Kleidman's book left a yawning gap in the literature for over twenty years. We are fortunate indeed that a number of years ago the authors of this volume took it upon themselves to fill this gap. They have done this in marvellously complete fashion, presenting the material with great clarity and attention to detail. Full proofs and comprehensive background material are given, making the book easily accessible to graduate students. It should also be said that their results go quite a way beyond Kleidman's thesis, in that they handle almost simple classical groups rather than just simple ones, which is important for applications.

It is marvellous to have this volume on the bookshelf where previously there was such an evident space, and I congratulate the authors on their achievement.

Martin Liebeck
Imperial College London

Preface

The aim of this book is to classify the maximal subgroups of the almost simple finite classical groups of dimension at most 12. We also include tables describing the maximal subgroups of the almost simple finite exceptional groups that have faithful representations of degree at most 12.

A group G is *simple* if it has order greater than 1 and has no normal subgroups other than the trivial subgroup and G itself, and is *almost simple* if $S \leqslant G \leqslant \operatorname{Aut} S$ for some non-abelian simple group S. A group is *perfect* if it is equal to its derived group. A group G is called *quasisimple* if G is perfect and G modulo its centre is a non-abelian simple group.

The study and classification of the (maximal) subgroups of the finite simple groups and their variations has a long history, and the completion of the classification of the finite simple groups provided further motivation.

The term 'classical group' is used frequently in the literature, but it is rarely, if ever, defined precisely. We shall not attempt a formal definition here, and we shall avoid using it in a precise sense. Our general intention is to use it very inclusively. We shall certainly regard all of the named groups (like $\operatorname{GL}_n(q)$, $\operatorname{O}_n^\epsilon(q)$, $\operatorname{PCSp}_n(q)$, etc.) in Table 1.2 as classical groups, but we also include among the classical groups arbitrary subgroups between the Ω-groups and the A-groups in the table, and also quotients of the groups in the first of each of the paired rows in the table by arbitrary subgroups of the scalars. Furthermore, we include all almost simple extensions of the simple classical groups.

Maximal subgroups of classical groups. In [1], Aschbacher proved a fundamental theorem that describes the subgroups of almost all of the finite almost simple classical groups (the only exceptions are certain extensions of $S_4(2^e)$ and $\operatorname{O}_8^+(q)$). This theorem divides these subgroups into nine classes. The first eight of these consist roughly of groups that preserve some kind of geometric structure; for example the first class consists (roughly) of the reducible groups, which fix a proper non-zero subspace of the vector space on which the group acts naturally. Subgroups of classical groups that lie in the first eight classes are of *geometric type*. The ninth class, denoted by $\mathscr{C}_9$ or $\mathscr{S}$, consists (roughly) of those subgroups that are not of geometric type and which, modulo the subgroup of scalar matrices, are almost simple. An alternative proof of Aschbacher's theorem, as a corollary to a version of the theorem for algebraic groups, can be found in [82]. We present a detailed version of Aschbacher's theorem, based on the treatment in [66], in Section 2.2. An interesting version of Aschbacher's theorem is presented in [91], which emphasises the links between the subgroup structure of the finite classical groups and of the algebraic groups of Lie type.

In [66], Kleidman and Liebeck provide an impressively detailed enumeration of the maximal subgroups of geometric type of the almost simple finite classical groups of dimension greater than 12. In this book, we shall extend the work of Kleidman and Liebeck to handle dimensions at most 12, and also classify the maximal subgroups of these groups that are in Class $\mathscr{S}$.

With the exception of Kleidman's work [62] on $\Omega_8^+(q)$, and classifications of Kleidman [62, 64, 63] and Cooperstein [14] of maximal subgroups of some of the exceptional groups of Lie type, which we simply cite and reproduce in our tables, our approach in this book has been to use these previous classifications for checking purposes only: our proofs make no use of the references below, although we have compared our results with them and, where there are differences, verified that our tables are correct.

The most complete previous work on low-dimensional classical groups is undoubtedly Peter Kleidman's PhD thesis [61], where he presents a classification, without proof, of the maximal subgroups of the simple classical groups in dimensions up to 12. This is a remarkable achievement. Kleidman intended to publish a subsequent book, with the same goal as ours: the classification of maximal subgroups of the almost simple classical groups in dimension up to 12. Unfortunately, this has not been published, and the present work is an attempt to carry out Kleidman's plan.

We base the following historical summary on the surveys by King [60], and Kleidman and Liebeck [65]. The complete description of the subgroup structure of the groups $L_2(q)$ is usually attributed to Dickson [22], but this topic was also investigated by E. H. Moore [94] and Wiman [115]. The subgroups of $L_3(q)$ were described by Mitchell for q odd [92] and then by Hartley for q even [40]. In both cases the subgroups that lie in $U_3(\sqrt{q})$ were identified. A more modern treatment of subgroups of $L_3(q)$ is provided by Bloom in [3]. The maximal subgroups of $L_4(q)$ for even q were listed independently by Mwene [95] and Zalesskii [116]. A partial classification for odd q can be found in [96] and, independently, for $p > 5$ in [117]. Further results on this case can be found in [59, Section 5]. Mitchell classified the maximal subgroups of $S_4(q)$ for odd q in [93]. Flesner [26, 27] partially classified the maximal subgroups of $S_4(2^e)$.

The maximal subgroups of $L_5(q)$ were determined by Di Martino and Wagner for q odd [23], and independently by Wagner [110] and Zalesskii [116] for q even. Kondrat'ev classified the quasisimple absolutely irreducible subgroups of $GL_6(q)$ [71]. There is a brief survey of many of these results in [72].

For higher dimensions, many people have concentrated on the case $q = 2$. In 1984, Darafsheh classified the maximal subgroups of $GL_6(2)$ [21], building on Harada and Yamaki's paper [39], which classified the insolvable irreducible subgroups of $GL_n(2)$ with $n \leqslant 6$. The subgroups of $GL_n(2)$ for $n \leqslant 10$ were studied extensively by Kondrat'ev: see [67, 68, 69, 70].

In the work of King and others (see [60] and the references therein) a different approach is taken: rather than concentrating on a family of almost simple groups, such as those with socle $L_6(q)$, one concentrates on a family of potentially maximal subgroups, such as those of type $\Gamma L_{n/2}(q^2)$ in $SL_n(q)$, and tries to determine maximality. A great deal is known in this direction: see the results cited in [60], together with [15, 16, 17, 99], amongst others. Again, we have not used these works in our proofs, but have compared our results with them: we mention them in the relevant sections of Chapters 2, 3 and 6 of this book.

An approach which is slightly orthogonal to our present purposes, but which has been used as a tool in the proof of many deep theorems, is to classify subgroups of classical groups containing elements of specified orders. We shall not use these results, so will not provide an extensive list of papers, but the interested reader could start by looking at [35, 36] and the references therein.

Maximal subgroups of non-classical simple groups. See [83, 114] for an excellent survey and introduction, respectively, on the whole of this field. The ATLAS [12] is also an essential reference in this area.

For the alternating groups, the O'Nan–Scott Theorem (see, for example, Chapter 4 of [8]) provides a subgroup classification similar to, but much simpler than, the Aschbacher classification of matrix groups over finite fields, and results of Liebeck, Praeger and Saxl [78] enable us to determine maximality. As in the case of finite matrix groups, we have a final class of almost simple primitive permutation groups which need to be listed individually by degree, and these lists are currently complete up to degree 4095 [18, 98].

The maximal subgroups of the almost simple exceptional Lie type groups have not yet been fully classified, although a great deal is known about them. A discussion of the overall strategy for their classification, and a brief summary of the main theorems in this area, appear in [91, Chapter 29]; whilst further references to the literature can be found in [66, Table 1.3.B]. In particular, the maximal subgroups of all of the simple, and most of the almost simple groups of this type with representations of degree up to 12 have been classified: see [106, Section 15] for the Suzuki groups $^2B_2(q)$, [76, 64] for the Ree groups

$^2G_2(q)$, [2, 14, 64] for $G_2(q)$ and [63] for $^3D_4(q)$. In this book we extend these classifications to the remaining almost simple groups with representations of degree at most 12.

The sporadic groups can be handled on a case-by-case basis. Most of the necessary information is available, including references to the literature, in [12] or, more usefully for computational purposes, in [111]. At the time of writing, all of the maximal subgroups of the almost simple groups with socle a sporadic group are known except for almost simple maximal subgroups of the Monster whose socle is one of a small list of groups. See [111] for a description of the state of play: note that it has recently been shown by R.A. Wilson that $L_2(41)$ is a maximal subgroup of the Monster, correcting an earlier error in the literature.

Computational applications. We were partly motivated to carry out this classification by its applications to computational group theory. In [9], results of Kovács, Aschbacher and Scott dating from the mid 1980s are used to reduce the computation of the maximal subgroups of a general finite group G to the case when G is almost simple. Polynomial-time algorithms for constructing the geometric-type maximal subgroups of the classical groups (in all dimensions) are presented in [45] for the linear, unitary and symplectic groups, and in [46] for the orthogonal groups, and they have been implemented in MAGMA [5].

The Class $\mathscr{S}$ subgroups arising from representations of almost simple groups in their defining characteristic are generally moderately straightforward to construct using standard functionality for computing with modules over groups. Most of the quasisimple Class $\mathscr{S}$ subgroups that are not in defining characteristic can be constructed by restricting a representation of a quasisimple group in characteristic 0 to the required finite field. The associated almost simple groups can be constructed over the finite field from a knowledge of the relevant group automorphisms and the computation of module isomorphisms.

Various databases of characteristic 0 representations are available either directly on the web, or via computer algebra systems such as GAP [29] and MAGMA. A facility of this type [111] has been under construction and continuous development for several years now. More recently, Steel [103] has constructed almost all of the characteristic 0 representations in [42] of quasisimple groups in dimensions up to 250. We used data from these databases to carry out some of the calculations needed to complete the classification.

Although the bulk of the arguments used in our classification theorems are theoretical, a substantial number of them make use of computer calculations. These calculations require only small amounts of computer time (generally at most a few seconds) and could be easily carried out using existing functionality and databases in either GAP or MAGMA. The MAGMA commands for each of these individual calculations are given in the files on the webpage http://www.cambridge.org/9780521138604, to enable the user to verify them easily. Whenever we use such a calculation in a proof in this book, we refer to it as a "computer calculation" in the text, and direct the user to the individual online file that contains the commands to carry it out. The files have names like 4134d8calc, which contains calculations with the 8-dimensional characteristic 0 representation of the group $4^{\cdot}L_3(4)$. The matrices defining the images of these representations are stored in data files, which are also on the website, and are accessed by the commands that carry out the calculations.

We shall assume that the reader has a general knowledge of group theory and of group representation theory as might be acquired from advanced undergraduate courses on these topics. Some knowledge of the general theory of classical groups over finite fields and of their associated bilinear, sesquilinear and quadratic forms would also be helpful, because we shall only briefly summarise what we need for this book. Good sources are the books by Rob Wilson [114] (which also includes a great deal of information about maximal subgroups of simple groups), Don Taylor [108] or Chapter 2 of [66]. We do not require any familiarity with algebraic groups, but the interested reader should consult [91] for an introduction which is especially well-suited to our current purposes.

Finally, in any classification project of this scale, it is inevitable that some mistakes will have slipped into our tables. At the time of publication, we know of no such errors, but an errata list has been created at

http://www.cambridge.org/9780521138604,

and we shall keep this up to date. We would be extremely grateful to be informed of any errata.

Acknowledgements. Writing this book has been a long-running project, and a great many people have helped us at various stages of it. We would like to thank Martin Liebeck for kindly agreeing to write the foreword to this volume. Many thanks go to Gunter Malle for his heroic proof-reading of this entire volume, mostly over the Christmas holidays preceding its publication; the book has been vastly improved by his efforts. Special thanks go also to Rob Wilson for proof-reading large portions of the book, and his thoughtful and extremely helpful comments. In addition we would like to thank Jeff Burdges, Peter Cameron, John Cannon, Ian Gent, Steve Linton, Frank Lübeck, Chris Parker and Anna Schroeder. Needless to say, any and all remaining mistakes and infelicities in this volume are entirely the responsibility of the authors! We gratefully acknowlege the financial support that we have received from EPSRC grants GR/S30580/01 and EP/C523229/1, the Mathematisches Forschungsinstitut Oberwolfach, and the seminar budgets of our respective universities. Finally, Derek thanks Cathy for systems support, and Colva thanks Struan for his encouragement, tolerance and sense of humour during the many years it has taken to write this book.

1

Introduction

1.1 Background

Given a group G, we write Soc G for the *socle* of G: the subgroup of G generated by its minimal normal subgroups. A group G is *almost simple* if $S \leqslant G \leqslant \operatorname{Aut} S$ for some non-abelian simple group S. Note that $S = \operatorname{Soc} G$. A group G is *perfect* if $G = G'$. A group G is *quasisimple* if G is perfect and $G/Z(G)$ is a non-abelian simple group.

Aschbacher [1] proves a classification theorem, which subdivides the subgroups of the finite classical groups into nine classes. The first eight of these consist roughly of groups that preserve some kind of geometric structure; for example the first class, $\mathscr{C}_1$, consists (roughly) of the reducible groups, which fix a proper non-zero subspace of the vector space on which the group acts naturally. Subgroups of classical groups that lie in the first eight classes are of *geometric type*. The ninth class, denoted by $\mathscr{C}_9$ or $\mathscr{S}$, consists (roughly) of those absolutely irreducible subgroups that are not of geometric type and which, modulo the central subgroup of scalar matrices, are almost simple.

In [66], Kleidman and Liebeck provide an impressively detailed enumeration of the maximal subgroups of geometric type of the finite classical groups of dimension greater than 12. More precisely, they classify the conjugacy classes of maximal subgroups $\bar{H}$ of those almost simple groups $\bar{G}$ for which $\bar{\Omega} := \operatorname{Soc} \bar{G} = \Omega/Z(\Omega)$ for some classical quasisimple group Ω, with $\bar{H} \cap \bar{\Omega} = K/Z(\Omega)$ for a subgroup K of Ω of geometric type.

In this book, we determine the maximal subgroups of all such almost simple groups $\bar{G}$ with dimension at most 12. For the subgroups of geometric type, Kleidman and Liebeck proved that their lists contain all such maximal subgroups even in dimensions at most 12. But their determination of when these subgroups are actually maximal applies only to dimensions greater than 12. It turns out that they are nearly all maximal, with just a few exceptions in small dimensions: all of the exceptions are in dimension at most 8.

We do not, however, restrict ourselves to the subgroups of geometric type, and include those subgroups in Aschbacher Class $\mathscr{S}$ in our classification. It is a feature of the groups in this class that they are not, as far as we know, susceptible to a uniform description across all dimensions, but can only be listed for each individual dimension and type of classical group. Fortunately, lists are available of all irreducible representations of degree up to 250 of all finite quasisimple groups G. These have been compiled by Lübeck [84] for representations of G in defining characteristic (when G is a group of Lie type), and by Hiß and Malle [42] for all other representations. These lists provide us with a complete set of candidates for the quasisimple normal subgroups S of maximal subgroups in Class $\mathscr{S}$ of the finite classical groups of dimension up to 250.

We are, however, left with two major problems. Firstly, in order to find the *almost* simple maximal subgroups of the almost simple classical groups $\bar{G} = G/\mathrm{Z}(\Omega)$, we need to determine which of the automorphisms of the simple groups $S/\mathrm{Z}(S)$ in the lists of candidates can be adjoined within $\bar{G}$. Secondly, we need to determine which of the candidates that we construct are actually maximal subgroups of the almost simple groups. Indeed, our approach to the project as a whole follows the same general pattern as [66]: first we find the candidates for the maximal subgroups within each of the nine Aschbacher classes, then we determine which are maximal within their own class, and finally we decide maximality itself by identifying cases in which maximal groups in one class are properly contained in a subgroup in another class.

The $\mathrm{O}_8^+(q)$ case is handled in detail in [62], so we shall not repeat that work here: we will simply reproduce the table of maximal subgroups from [62], but in the format we are using for the remainder of our tables.

Structure of this book. In the remainder of this chapter we present basic results on the structure and representations of simple groups; this material will be required both for the study of geometric type groups and of groups in Class $\mathscr{S}$. Topics covered include: novelty maximal subgroups; finite fields; sesquilinear and quadratic forms, including the specification of our standard forms; introduction to the classical groups, including the specification of our standard outer automorphisms; some relevant representation theory; tensor products; exceptional properties of various small classical groups; permutation and matrix representations of the classical groups; properties of the natural matrix representations of the classical groups; Zsigmondy primes; quadratic reciprocity.

In Chapter 2 we first state our main theorem, Theorem 2.1.1. Then in Section 2.2 we introduce the *types* of geometric subgroups: these are families of subgroups with the property that if H is a geometric maximal subgroup of a quasisisimple classical group, then H is a member of one of these families. For each geometric Aschbacher class, we define the corresponding types, give the structure of the maximal groups of each type, and prove some elementary properties. This detailed definition of the types enables us to state our version of Aschbacher's theorem in Subsection 2.2.9, which is essentially the refined version given in [66]. In Section 2.3 we establish some results concerning maximality of the geometric subgroups that can be proved simultaneously for more than one dimension at a time.

In Chapter 3 we consider each dimension from 2 to 12 in turn, and determine which subgroups are maximal amongst the geometric subgroups of the almost simple classical groups of that dimension. Thus by the end of this chapter we have produced a list of candidate geometric maximals, which only need to be compared with the Class $\mathscr{S}$ groups to determine their maximality.

Section 4.1 presents our overall strategy for classifying the maximal subgroups in Class $\mathscr{S}$, and a description of the subdivision between Class $\mathscr{S}_1$ (cross characteristic) and Class $\mathscr{S}_2$ (defining characteristic). The remainder of Chapter 4 is devoted to finding the maximal subgroups in Class $\mathscr{S}_1$. There is a section (4.2) on working with algebraic irrationalities. After this we present in Section 4.3 our list of $\mathscr{S}_1$-candidates (taken from [42]). The next few sections concern dimensions 2 to 6, and alternate between theory and practice. In Sections 4.4 and 4.5, we first describe how to calculate the stabiliser of a quasisimple group in cross characteristic in a conformal classical group (the group of all linear mappings that multiply the form by a scalar), and hence to determine the exact structure and conjugacy behaviour of a candidate maximal $\mathscr{S}_1$-subgroup of a quasisimple group, and then carry out these calculations in some detail in dimensions up to 6. Next, in Sections 4.6 and 4.7 we present methods to determine the actions of the duality and field automorphisms of a quasisimple classical group on its quasisimple subgroups, and apply these methods in dimensions up to 6. In Section 4.8 we then determine whether any of these $\mathscr{S}_1$-subgroups contain one another, and hence are non-maximal. After carrying out all of these calculations in detail in dimensions up to 6, in Section 4.9 we perform the same calculations for dimensions 7 to 12. This means that we have determined those subgroups of the almost simple classical groups in dimension at most 12 that are maximal amongst the $\mathscr{S}_1$-subgroups: this list is summarised in Section 4.10.

In Chapter 5 we move on to considering the maximal subgroups arising from representations of groups of Lie type in defining characteristic. Note that we in fact define a class $\mathscr{S}_2^*$ of subgroups (see Definition 5.1.15) and work only with them. In Section 5.1 we present as much of the general theory of representations of groups of Lie type in defining characteristic as we will need to perform our calculations. In Section 5.2 we briefly present some information about symmetric and anti-symmetric powers of modules. We then consider the possible families of groups of Lie type in turn, basing our lists of candidates on

[84]. For each possible candidate maximal subgroup we determine the structure of the maximal such subgroup of the quasisimple group, the module on which it acts, the number of conjugacy classes in the quasisimple group, and the stabiliser of one such class in the conformal classical group. Then in Section 5.3 we consider groups with non-abelian composition factor $L_2(q)$, in Section 5.4 the groups $L_n(q)$ and $U_n(q)$ with $n \geqslant 3$, in Section 5.5 the groups $S_n(q)$, in Section 5.6 the groups $O_n^\varepsilon(q)$ and $^3D_4(q)$, and finally in Section 5.7 the remaining groups of Lie type. We summarise our findings to this point in Section 5.8. Next, in Section 5.9 we determine the action of duality and field automorphisms on $\mathscr{S}_2^*$-subgroups. In Section 5.10 we determine all containments between the $\mathscr{S}_2^*$-subgroups, and then finally in Section 5.11 we summarise the results of this chapter. Thus Chapter 5 determines all subgroups that are maximal amongst the $\mathscr{S}_2^*$-subgroups in dimension at most 12.

By the end of Chapter 5 we have produced a list of subgroups of the almost simple classical groups such that all maximal subgroups lie in this list. We then proceed in Chapter 6 to determine the containments between these subgroups, and hence to prove the main theorem of the book, Theorem 2.1.1. In Section 6.2 we determine all containments between $\mathscr{S}_1$-maximals and $\mathscr{S}_2^*$-maximals, to produce a set of $\mathscr{S}^*$-maximals (where $\mathscr{S}^* = \mathscr{S}_1 \cup \mathscr{S}_2^*$), and then in Section 6.3 we determine all containments between geometric and $\mathscr{S}^*$-maximals.

Aschbacher's theorem does not apply to certain extensions of $S_4(2^i)$ and $O_8^+(q)$ (that is, those that involve the exceptional graph automorphism and the triality graph automorphism, respectively): note, however, that Aschbacher's paper [1] includes a variant of his theorem for the relevant extensions of $S_4(2^i)$, but we shall not deem this variant to be part of "Aschbacher's theorem". Since the $O_8^+(q)$ case is fully handled in [62], we do not concern ourselves with that. In Chapter 7 we calculate the maximal subgroups of those almost simple extensions of $S_4(2^i)$ to which Aschbacher's theorem does not apply, as well as the maximal subgroups of the finite almost simple exceptional groups that have a faithful projective representation in defining characteristic of degree at most 12, namely those with socles $^2B_2(q) = Sz(q)$, $G_2(q)$, $^2G_2(q) = R(q)$ and $^3D_4(q)$. For many of these groups such classifications are already known, and we merely provide references to the original calculations, however we occasionally include our own proofs if we feel this may be helpful for the reader. Finally, in Chapter 8, we present tables of our results.

1.2 Notation

Here we list some general notation used in the book. More specialised notation will be introduced as it arises, and in particular our notation for the classical groups is presented in Subsection 1.6.3 and for the outer automorphisms of the classical groups in Subsection 1.7.1.

By $[a, b]$ we denote the least common multiple of two positive integers a and b, and by (a, b) we denote their greatest common divisor. If $a \in \mathbb{N}$ and p is a prime, we write $(a)_p$ for the highest power of p that divides a. In Section 1.13 we define the notion of a *Zsigmondy prime* for $q^n - 1$, where q is a prime power and $n \geqslant 3$ is an integer. We shall denote such a prime by $z_{q,n}$.

If we write δ_{ij} (with two subscripts), we will mean the Kronecker delta: $\delta_{ii} = 1$ for all i and $\delta_{ij} = 0$ whenever $i \neq j$.

For group elements g and h, we write g^h for $h^{-1}gh$ and $[g, h]$ for $g^{-1}h^{-1}gh$. As usual, $Z(G)$ is the centre of the group G, and $N_G(H)$ and $C_G(H)$ are the normaliser and centraliser of H in G. The derived subgroup of G is written as $[G, G]$ or G', and we define $G^{(n)} = [G^{(n-1)}, G^{(n-1)}]$ for $n > 1$, and use G^∞ to denote $\bigcap_{i \geqslant 0} G^{(i)}$. We write $\operatorname{Aut} G$ for the automorphism group of G, and $\operatorname{Inn} G$ and $\operatorname{Out} G$ respectively for its inner and outer automorphism group.

Our notation for group structures is based on that in the ATLAS [12]. Note in particular that this means that we generally use ATLAS notation for simple groups. Thus, for example, $A \times B$ is the direct product of groups A and B, we write $A{:}B$ to denote a split extension of A by B, we write $A{\cdot}B$ to denote a non-split extension (or possibly one in which A is trivial), and we write $A.B$ when we do not know or do not wish to specify whether the extension splits. The symbol $A \wr B$ is defined for an arbitrary group A and a permutation group B, and denotes the wreath product of A by B. If G is a group with a unique index 2 subgroup H, then we sometimes write $\frac{1}{2}G$ to denote H.

The cyclic group of order n is denoted by C_n or (particularly when as a component of a group structure) just by n. An elementary abelian group of order p^n is denoted by E_{p^n} or just by p^n. By $[n]$ we denote a group of order n, of unspecified structure. For elementary abelian groups A we write A^{m+n} to mean a group with an elementary abelian normal subgroup A^m such that the quotient is isomorphic to A^n. The group A^{m+n} is usually, but not always, special. For n even, D_n denotes the dihedral group of order n, and for n a power of 2, we write Q_n for the quaternion group of order n. For r an odd prime, we write r_+^{1+2n} for an extraspecial group of order r^{1+2n} and exponent r, and r_-^{1+2n} for an extraspecial group of the same order, but exponent r^2. We write 2_+^{1+2m} for an extraspecial group of order 2^{1+2m} that is isomorphic to a central product of m copies of D_8, and we write 2_-^{1+2m} for an extraspecial group of the same order, but that is isomorphic to a central product of $m - 1$ copies of D_8 and one of Q_8.

For L an arbitrary finite group, $P(L)$ denotes the minimum degree of a non-trivial permutation representation of L.

We write $\mathbb{F}_q$ for a finite field of order $q = p^e$, with a fixed primitive element

$\omega = \omega_q$, and Frobenius automorphism $\phi : x \mapsto x^p$. For a field F, we write $F^\times$ to denote the multiplicative group of F, char F to denote the characteristic of F, and $(F, +)$ to denote the additive group of F.

As usual, I_n is the $n \times n$ identity matrix and J_n is the $n \times n$ matrix with all entries 1. We write $\mathrm{diag}(\alpha_1, \alpha_2, \ldots, \alpha_n)$ for the matrix $A = (a_{ij})_{n \times n}$ with $a_{ii} = \alpha_i$ for all i and $a_{ij} = 0$ otherwise. We write $\mathrm{antidiag}(\alpha_1, \ldots, \alpha_n)$ for the matrix $A = (a_{ij})_{n \times n}$ with $a_{i,n-i+1} = \alpha_i$ for all i and $a_{ij} = 0$ otherwise. The transpose of A is denoted by A^T, and its trace by $\mathrm{tr}(A)$. We denote the elementary matrix with a 1 in position (i, j) and 0 in all other positions by $\mathrm{E}_{i,j}$.

The set of all $(m \times n)$-matrices with entries in the field F is denoted by $\mathrm{M}_{m \times n}(F)$, or by $\mathrm{M}_{m \times n}(q)$ when $F = \mathbb{F}_q$. For a vector space V over a field F, we write $\mathrm{GL}(V)$ for the *general linear group* of V, which is the set of all invertible F-linear maps from V to itself. For a given basis of V, we can identify $\mathrm{GL}(V)$ with $\mathrm{GL}_n(F)$ (or $\mathrm{GL}_n(q)$ when $F = \mathbb{F}_q$), the set of all invertible $n \times n$ matrices over F. Our convention is that linear maps act on the right, with corresponding action of matrices on row vectors by right multiplication.

We write $\mathrm{GL}_n^\pm(q)$, and related notation such as $\mathrm{SL}_n^\pm(q)$, to denote the linear and unitary groups: the $+$ sign corresponds to the linear groups, and the $-$ sign to the unitary groups. (We may also of course denote the unitary groups by the more usual $\mathrm{GU}_n(q)$ and $\mathrm{SU}_n(q)$.)

If U and W are subspaces of V, and $G \leqslant \mathrm{GL}(V)$, then we write $\mathrm{N}_G(W)$ for the stabiliser in G of W, and $\mathrm{N}_G(W, U)$ for $\mathrm{N}_G(W) \cap \mathrm{N}_G(U)$. If $G \leqslant S_n$ stabilises a set $W \subseteq \{1, \ldots, n\}$, or $G \leqslant \mathrm{GL}(V)$ stabilises a subspace $W \leqslant V$, then by G^W we mean the image of the induced action of G on W.

For a vector space V, we write V^* for the dual space of V. If V is equipped with a reflexive form β (see Section 1.5), then we write $V = A \perp B$ to mean that V decomposes as a direct sum of A and B, such that $\beta(a, b) = 0$ for all $a \in A$, $b \in B$. For $v \in V$, we write $v^\perp$ for the subspace $\{ w \in V \mid \beta(v, w) = 0 \}$, and similarly $W^\perp = \{ v \in V \mid \beta(w, v) = 0 \text{ for all } w \in W \}$. If Q is a quadratic form on V, and W is a non-degenerate subspace of V, then $\mathrm{sgn}(W)$ denotes the sign of the restriction of Q to W (so $\mathrm{sgn}(W)$ can be $\circ$, $+$ or $-$).

1.3 Some basic group theory

An *automorphism* of a group G is a bijective homomorphism from G to itself: the set of all automorphisms of G forms a group, $\mathrm{Aut}\,G$. For a fixed $g \in G$, we denote the automorphism $x \mapsto x^g = g^{-1}xg$ by c_g. An automorphism ϕ of G is *inner* if there exists a g in G such that $\phi = c_g$. We denote the group of all inner automorphisms of G by $\mathrm{Inn}\,G$. Note that $\mathrm{Inn}\,G \cong G/\mathrm{Z}(G)$. It is an easy exercise to prove that $\mathrm{Inn}\,G \trianglelefteq \mathrm{Aut}\,G$; the quotient $\mathrm{Aut}\,G/\mathrm{Inn}\,G$ is

Out G, the *outer automorphism group* of G. An *outer automorphism* of G is often defined to be an element of Aut $G \setminus$ Inn G rather than an element of Out G. In this book, however, despite the risk of causing confusion, we find it convenient to use "outer automorphism" to denote either a non-trivial element α of Out G (which is consequently only defined modulo Inn G, and hence not itself an automorphism of G), or a representative α of a non-trivial coset of Inn G in Aut G, depending on the context.

Let $\alpha \in$ Aut G, let C be a conjugacy class of G, and let $C^\alpha = \{x^\alpha : x \in C\}$. We say that α *stabilises* C if $C^\alpha = C$. The following lemma is elementary.

Lemma 1.3.1 *Let $\alpha \in$ Aut G and let C be a conjugacy class of G. Then C^α is also a conjugacy class of G. Furthermore, if α stabilises C and $x \in C$, then there exists $g \in G$ with $x^{\alpha c_g} = x$.*

So, for a given C, the class C^α depends only on the coset of Inn G in which α lies, and hence there is an induced action of Out G on the set of conjugacy classes of G. So when we write C^α with α an outer automorphism of G or talk about an outer automorphism stabilising C, then it does not matter whether we are thinking of α as an element of Aut G or of Out G.

The following theorem is a straightforward consequence of the classification of finite simple groups, and was known as the *Schreier Conjecture* before the completion of the classification.

Theorem 1.3.2 *Let S be a finite non-abelian simple group. Then Out S is soluble.*

We will occasionally need the concept of isoclinism. The commutator $[x, y]$ of two elements x and y of a group G is unchanged if we multiply x and y by central elements of G. Thus we can think of the commutator map not as a map from $G \times G$ to G, but instead as a map from $G/Z(G) \times G/Z(G)$ to G.

Definition 1.3.3 Two groups G and H are *isoclinic* if there are isomorphisms $\rho : G/Z(G) \to H/Z(H)$ and $\theta : G' \to H'$ which form a commutative diagram with the commutator maps from $G/Z(G) \times G/Z(G)$ to G' and from $H/Z(H) \times H/Z(H)$ to H'.

$$
\begin{array}{ccc}
\dfrac{G}{Z(G)} \times \dfrac{G}{Z(G)} & \xrightarrow{\ (\rho,\rho)\ } & \dfrac{H}{Z(H)} \times \dfrac{H}{Z(H)} \\[2ex]
{\scriptstyle [\,,\,]}\Big\downarrow & & \Big\downarrow{\scriptstyle [\,,\,]} \\[2ex]
G' & \xrightarrow{\ \theta\ } & H'
\end{array}
$$

The dihedral group of order 8 and the quaternion group of order 8 are isoclinic; see just before Subsection 1.3.1 for another example.

If G is finite, then $G^\infty = \cap_{i=1}^\infty G^{(i)}$ is the first perfect group in the derived series of G. If S is non-abelian and simple, and $S \trianglelefteq G \leqslant \operatorname{Aut} S$, then Theorem 1.3.2 implies that $G^\infty = S$.

Recall from Section 1.1 that a group G is quasisimple if G is perfect with $G/\mathrm{Z}(G)$ a non-abelian simple group. We shall use the following lemma implicitly throughout much of the book, without further citation.

Lemma 1.3.4   *Let $G = Z \dot{} S$ be quasisimple, with Z central and S non-abelian simple. Then $\operatorname{Aut} G$ can be naturally regarded as a subgroup of $\operatorname{Aut} S$.*

Proof   Let α be a non-trivial element of $\operatorname{Aut} G$. If α induces the identity map on G/Z, then α acts as the identity on all commutators $[g, h]$, and so induces the identity on all of G. Thus $\operatorname{Aut} G$ acts faithfully on G/Z.   □

A *stem extension* of a group G by a group K is a group $C = K \dot{} G$ such that $K \leqslant \mathrm{Z}(C) \cap C'$. In particular, a quasisimple group is a stem extension of a non-abelian simple group.

Definition 1.3.5   The *Schur multiplier* $M(G)$ of a group G is the largest K such that there exists a stem extension of G by K.

It is not immediately apparent that $M(G)$ is well-defined, but it turns out that the corresponding stem extension is determined up to isoclinism by G, (see Definition 1.3.3 of isoclinism), and up to isomorphism if G is perfect.

As an example, the symmetric groups S_n for $n \geqslant 4$ have two isoclinic double covers, $2 \dot{} \mathrm{A}_n.2^+$ and $2 \dot{} \mathrm{A}_n.2^-$, whilst by the above remark for $n \geqslant 5$ and $n \neq 6$ the groups A_n have a unique double cover (which is true also for A_4.) This notation comes from [12], and $2 \dot{} \mathrm{A}_5.2^+$ is the group of which the character table is printed there. The inverse images of transpositions in S_5 have orders 2 and 4 in $2 \dot{} \mathrm{A}_5.2^+$ and $2 \dot{} \mathrm{A}_5.2^-$, respectively.

1.3.1 Maximal subgroups of almost simple groups

Let G be an almost simple group with socle S, and let M be a maximal subgroup of G. The group M is a *triviality* if $S \trianglelefteq M$. The trivialities correspond to the maximal subgroups of the soluble group G/S and are very easy to determine. They are consequently generally omitted from tables of maximal subgroups of almost simple groups (for example, in the ATLAS [12]), and are excluded from the statements of Aschbacher's theorem and the O'Nan–Scott Theorem.

The following result is fundamental in the analysis of maximal subgroups of almost simple groups. It was first proved by Wilson [113], though we follow the proof of Liebeck, Praeger and Saxl [79, end of paper].

Theorem 1.3.6 *Let G be a finite almost simple group with socle S. Suppose that M is a maximal subgroup of G. Then $S \cap M \neq 1$.*

Proof If $G = S$ then the result is trivial, so suppose not, and assume that $S \cap M = 1$. Then $M < \langle S, M \rangle \leqslant G$, whence $G = SM = S{:}M$, by the maximality of M. Let N be a minimal normal subgroup of M. Then N is characteristically simple. One of the following cases must arise:

(i) $N \cong E_{p^r}$, where p is prime, $r \geqslant 1$ and $p \mid |S|$;

(ii) $N \cong E_{p^r}$, where p is prime, $r \geqslant 1$ and $p \nmid |S|$;

(iii) $N \cong T \times \cdots \times T \cong T^m$, where $m \geqslant 1$ and T is non-abelian simple.

The quotient G/S is soluble by Theorem 1.3.2, so Case (iii) does not arise.

In all cases, $M \leqslant N_G(N)$. Moreover $N_G(N) \cap S = C_S(N)$, and of course $C_S(N) \neq S$, since N is a subgroup of $\mathrm{Aut}\, S$. Thus $N_G(N) \neq G$ and hence the maximality of M implies that $N_G(N) = M$ and so $C_S(N) = 1$.

In Case (i), the conjugation action of N on S centralises 1_S, so must centralise some non-trivial elements of S (since the orbits of N have p-power order and $p \mid |S|$). This contradicts $C_S(N) = 1$.

In Case (ii), we let $q \mid |S|$, with q prime (so $q \neq p$), and we let Q be a Sylow q-subgroup of S normalised by N, which exists as the number of Sylow q-subgroups of S is a divisor of $|S|$, and therefore not divisible by p. Suppose that N also normalises a Sylow q-subgroup Q_1 of S. Then $Q_1 = Q^{x^{-1}}$ for some $x \in S$. So now N and N^x normalise Q and so are Sylow p-subgroups of $N_{SN}(Q) = N_S(Q)N$. Hence there exist $y \in N_S(Q)$ and $z \in N$ such that $N^{xyz} = N = N^z$, and so there exists $y \in N_S(Q)$ such that $N^{xy} = N$. Now $[g, xy] \in N \cap S = 1$ for all $g \in N$, and so $xy \in C_S(N) = 1$. Hence $x = y^{-1} \in N_S(Q)$, and so $Q_1 = Q$. Therefore $M = N_G(N) \leqslant N_G(Q)$, because N normalises a unique Sylow q-subgroup of S, and so $M < QM < SM = G$, contradicting the maximality of M. $\square$

In Chapter 7, we shall need the following immediate corollary.

Corollary 1.3.7 *Let G be a finite almost simple group with socle S, and let M be a maximal subgroup of G such that $S \not\leqslant M$. Then there exists a characteristically simple group N with $1 < N < S$ such that $M = N_G(N)$.*

Proof By the previous result, we may assume that $M \cap S \neq 1$. Notice that $M \cap S \trianglelefteq M$, so we may choose N to be minimal subject to $1 < N \leqslant M \cap S$ and $N \trianglelefteq M$. Thus N is a minimal normal subgroup of M, and so is characteristically simple. Clearly $M \leqslant N_G(N)$, and from $N \leqslant M \cap S < S$ we deduce that $S \not\leqslant N_G(N)$. Maximality of M then gives $M = N_G(N)$. $\square$

Definition 1.3.8 A maximal subgroup M of an almost simple group G is called an *ordinary* maximal subgroup if $S \cap M$ is a maximal subgroup of S. We say that M is a *novel* maximal subgroup (or, simply, *a novelty*) if $S \cap M$ is non-maximal in S.

Suppose that $H < S$ and we are considering $M = N_G(H)$ as a candidate for being a novel maximal subgroup of G. This is only possible if $N_G(H)S = G$, so we shall assume that to be the case. Then M fails to be maximal in G if and only if $M < N_G(K)$ for some K such that $H < K < S$. By replacing H by $N_S(H)$, we may assume that $N_S(H) = H$, and similarly we may restrict attention to groups K with $N_S(K) = K$.

If $N_G(K)S \neq G$ for some such K then M is not a proper subgroup of $N_G(K)$. This motivates the following definition.

Definition 1.3.9 Let G be almost simple with socle S. If $H = N_S(H) < K = N_S(K) < S < G$, but $N_G(K)S \neq G$, then $M = N_G(H)$ is called a *type 1 novelty* with respect to K.

An example is $G = \mathrm{PGL}_2(7)$, $S = \mathrm{L}_2(7)$, $H = \mathrm{D}_6$, $M = \mathrm{D}_{12}$. The only possibility for K is S_4, but $N_G(K) = K$ in that case.

The following is essentially equivalent to [113, Proposition 2.3 (e),(f)], but in a slightly different context.

Proposition 1.3.10 *Let G be almost simple with socle S. Suppose that G has subgroups $H < K < S < G$, with $N_S(H) = H$, $N_S(K) = K$, and $N_G(H)S = N_G(K)S = G$. Then $N_G(H) \nleq N_G(K)$ if and only if there exists $H_0 < K$ with H and H_0 conjugate in $N_G(K)$ but not in K. In this situation H and H_0 are also conjugate in S.*

Proof Let M denote $N_G(H)$. Suppose first that $M \nleq N_G(K)$, and let m be an element of $M \setminus N_G(K)$. Then $H = H^m < K^m \neq K$, and $N_G(K)S = G$ implies that $m = ns$ for some $n \in N_G(K)$ and $s \in S$, and hence that $K^m = K^s$. So $H_0 := H^{s^{-1}} < K$, and $H^n = H^{ms^{-1}} = H_0$, so H and H_0 are conjugate in $N_G(K)$ and in S. But if $H_0^k = H$ with $k \in K$, then $s^{-1}k \in N_S(H) = H$, so $s \in K$ and hence $K^m = K^s = K$, a contradiction.

Suppose conversely that $H_0 < K$, where H_0 is such that H and H_0 are conjugate in $N_G(K)$ but not in K, and let $n \in N_G(K)$ with $H^n = H_0$. Since $MS = G$, we can write $n = ms$ with $m \in M$ and $s \in S$. If $m \in N_G(K)$, then $s \in N_S(K) = K$, so $H_0 = H^n = H^{ms} = H^s$, contradicting the assumption that H and H_0 are not K-conjugate. So $m \notin N_G(K)$ and hence $M \nleq N_G(K)$. $\square$

Definition 1.3.11 Let G be almost simple with socle S. If $H < K < S < G$, with $N_S(H) = H$, $N_S(K) = K$, $N_G(H)S = N_G(K)S = G$, and such that $M = N_G(H) \nleq N_G(K)$, then M is a *type 2 novelty* (or *Wilson novelty*) with respect to K.

An example of this in [12] is $S = \text{He}$ (the sporadic Held group), $G = \text{He:2}$, $H = (A_5 \times A_5){:}2^2$, $M = (S_5 \times S_5){:}2$. The only possible K is $S_4(4){:}2$. The two conjugacy classes of groups isomorphic to H in K are fused in $N_G(K) = S_4(4){:}4$. Another example, from Table 8.13, is $G = S_4(7){:}2$, $H = L_2(7)$, $M = L_2(7){:}2$. The only possible K is A_7. The two conjugacy classes of groups isomorphic to H in K are fused in $N_G(K) = S_7$.

1.4 Finite fields and perfect fields

The reader unfamiliar with the theory of finite fields should consult [11, Chapter 7], which contains all of the basic concepts that we will need, and much more. The following dates back to E.H. Moore in 1893.

Theorem 1.4.1 *For each prime p and each $e \geqslant 1$, there is exactly one finite field of p^e elements, up to isomorphism, and these are the only finite fields.*

Although it is only defined up to isomorphism, and there is no satisfactory canonical way to define it, it is customary to regard 'the' finite field of order q as a fixed object, and to denote it by $\mathbb{F}_q$. The following result is also standard, and we will use it implicitly throughout the rest of the book.

Proposition 1.4.2 *Any finite subgroup of the multiplicative group of a field is cyclic. In particular, the multiplicative group of a finite field is cyclic.*

In the light of Proposition 1.4.2, the field $\mathbb{F}_q$ always contains elements of multiplicative order $q - 1$, and such an element ω is called a *primitive* element of $\mathbb{F}_q$. Clearly $\mathbb{F}_q = \mathbb{F}_p(\omega)$, and so the minimal polynomial f of ω over $\mathbb{F}_p$ must be of degree e, where $q = p^e$. An irreducible polynomial of degree e over $\mathbb{F}_p$ of which the roots are primitive elements of $\mathbb{F}_q$ is called a *primitive* polynomial.

Definition 1.4.3 Let f be a polynomial with coefficients in a field F. If f can be written as a product of linear factors over an extension field K of F and no proper subfield of K that contains F has this property, then K is a *splitting field* for f over F.

For any field of characteristic p, the map $\phi : x \mapsto x^p$ defines an injective endomorphism of $\mathbb{F}_q$ of order e, called the *Frobenius endomorphism*.

Definition 1.4.4 A field F is *perfect* if either char $F = 0$, or if char $F = p > 0$ and the Frobenius endomorphism is an automorphism, in which case it is called the *Frobenius automorphism*.

In this book, we will mostly work with finite fields, but some results also hold for arbitrary fields, and some for perfect fields. In particular, all finite fields are perfect, and the Frobenius automorphism of the finite field $\mathbb{F}_q$ with $q = p^e$ has order e, and generates the complete automorphism group of $\mathbb{F}_q$. So, if f is a primitive polynomial of degree e over $\mathbb{F}_p$ with root $\omega \in \mathbb{F}_q$, then the e elements in the set $\{\omega^{p^i} \mid 0 \leqslant i < e\}$ are all roots of f in $\mathbb{F}_q$. Hence $\mathbb{F}_q$ is a splitting field of f over $\mathbb{F}_p$.

The following lemma is technical, but will make several appearances in Chapter 3 when analysing $\mathscr{C}_5$-subgroups (see Definition 2.2.11).

Lemma 1.4.5   *Let $q = p^e$, and let r be such that e/r is prime.*

(i) *If $\alpha + \alpha^{-1} \in \mathbb{F}_{p^r}$ for all non-zero $\alpha \in \mathbb{F}_{p^e}$, then $p = 2$, $e = 2$ and $r = 1$.*
(ii) *The prime e/r is 2 if and only if $\alpha^{q-1} + \alpha^{1-q} \in \mathbb{F}_{p^{2r}}$ for all $\alpha \in \mathbb{F}_{p^{2e}}^{\times}$.*

Proof (i) For each $\lambda \in \mathbb{F}_{p^r}$ there are at most two solutions to the equation $\alpha^2 - \alpha\lambda + 1 = 0$ (that is, to $\alpha + \alpha^{-1} = \lambda$), so $p^e - 1 \leqslant 2p^r$, forcing $p^e = 4$.
(ii) For each $\lambda \in \mathbb{F}_{p^{2r}}$ there are at most two solutions for α^{q-1} to the quadratic equation $(\alpha^{q-1})^2 - \alpha^{q-1}\lambda + 1 = 0$, and neither solution is 0. Given a value for α^{q-1}, there are at most $q - 1$ values for α. Thus we account for at most $2p^{2r}(p^e - 1)$ non-zero field elements as solutions. Therefore, if we are to account for all non-zero elements of $\mathbb{F}_{p^{2e}}$ as solutions, then $2p^{2r}(p^e - 1) \geqslant p^{2e} - 1$, so $2p^{2r} \geqslant p^e + 1$, and hence $p^{2r+1} \geqslant p^e + 1$. If $r < e/2$ then $2r + 1 < e$, because r divides e, and we have a contradiction.

Conversely, $\alpha^{q-1} + \alpha^{1-q}$ is fixed by the automorphism $x \mapsto x^q$ of $\mathbb{F}_{q^2}$. $\square$

1.4.1 Conway polynomials

Although $\mathbb{F}_q$ is unique up to isomorphism, it arises as the splitting field of $\Phi(q-1)/e$ different primitive polynomials, where Φ is the Euler Phi-function and $q = p^e$. For computational purposes, it is useful to agree on a standard primitive polynomial, so that different computer algebra systems can use the same representation of the elements of $\mathbb{F}_q$. Unfortunately, there appears to be no natural or canonical way of choosing such a standard polynomial.

The standard that has been generally agreed upon is known as the *Conway polynomial* for $\mathbb{F}_q$. (This is an unfortunate choice of name, because there is another meaning of Conway polynomial in knot theory!) They were originally introduced by Richard Parker, who computed many examples. To define them, we first define an ordering on the set of all polynomials of degree n over $F = \mathbb{F}_p$, and it is here that an apparently arbitrary choice had to be made.

We order $\mathbb{F}_p$ itself by $0 < 1 < 2 < \cdots < p - 1$. Then the polynomial

$$x^n - \alpha_{n-1}x^{n-1} + \alpha_{n-2}x^{n-2} - \cdots + (-1)^n\alpha_0$$

is mapped onto the word $\alpha_{n-1}\alpha_{n-2}\cdots\alpha_1\alpha_0$, and the resulting words are ordered lexicographically using the above ordering of $\mathbb{F}_p$.

The Conway polynomial for $\mathbb{F}_p$ is defined to be the least primitive polynomial of degree 1 under this ordering. In other words, it is $x - \alpha$, where α is the smallest primitive element in $\mathbb{F}_p$.

For non-prime fields, there is an extra condition. It is a standard result that $\mathbb{F}_{p^m}$ is a subfield of $\mathbb{F}_{p^n}$ if and only if m divides n. For compatibility between the Conway polynomial f of the field $\mathbb{F}_{p^n}$ and its subfields, it is required that if α is a root of the Conway polynomial f of $\mathbb{F}_{p^n}$ then, for all proper divisors m of n, α^t with $t := (p^n - 1)/(p^m - 1)$ should be a root of the Conway polynomial of $\mathbb{F}_{p^m}$. We can then define the Conway polynomial f of $\mathbb{F}_{p^n}$ to be the least primitive polynomial under the ordering defined above that satisfies this compatibility condition for all divisors of n.

It is a slightly tricky exercise to show that a primitive polynomial f exists that satisfies this property. A proof can be found in the thesis of W. Nickel [97]. One disadvantage of this definition is that Conway polynomials are extremely difficult to calculate for large q, though a substantial number are known and available on [86], and for certain values of p and e they are easily calculated.

1.5 Classical forms

In this section we present basic material on the theory of classical forms. The reader unfamiliar with this material may wish to consult Taylor's textbook [108], where these topics are covered in much greater detail. We shall start with a brief introduction to sesquilinear and quadratic forms, culminating in the Birkhoff–von Neumann Theorem, which classifies such forms, subject to certain additional symmetry conditions. After a short digression into representing forms via matrices, we then present the definitions of our standard unitary, symplectic, symmetric bilinear and quadratic forms.

Throughout this section, we let V be a vector space of dimension $n > 0$ over the field F.

Definition 1.5.1 Let $\sigma \in \operatorname{Aut} F$. A map $\beta : V \times V \to F$ is a *σ-sesquilinear form* if, for all $u, v, w \in V$ and all $\lambda, \mu \in F$:

(i) $\beta(u, v + w) = \beta(u, v) + \beta(u, w)$;

(ii) $\beta(u + v, w) = \beta(u, w) + \beta(v, w)$; and

(iii) $\beta(\lambda u, \mu v) = \lambda \mu^\sigma \beta(u, v)$.

The form β is *bilinear* if $\sigma = 1$, and *symmetric* if $\beta(u, v) = \beta(v, u)$ for all $u, v \in V$.

Definition 1.5.2 The map $Q : V \to F$ is a *quadratic form* if:

(i) $Q(\lambda v) = \lambda^2 Q(v)$ for all $\lambda \in F, v \in V$; and

(ii) the map β, defined by $\beta(u, v) := Q(u + v) - Q(u) - Q(v)$ for all $u, v \in V$, is a symmetric bilinear form on V.

We call β the *polar form* of Q.

Definition 1.5.3   Let β be a σ-sesquilinear form on V, let Q be a quadratic form on V and let $g \in \mathrm{GL}(V)$. Then g is an *isometry* of β (or of Q) if $\beta(ug, vg) = \beta(u, v)$ (respectively, $Q(vg) = Q(v)$) for all $u, v \in V$. The element g is a *similarity* of β (or of Q) if there is a $\lambda \in F \setminus \{0\}$ such that $\beta(ug, vg) = \lambda\beta(u, v)$ (respectively, $Q(vg) = \lambda Q(v)$) for all $u, v \in V$.

Definition 1.5.4   Let κ be a σ-sesquilinear or quadratic form. The *isometry group* of κ is the group of all isometries of κ, and the *similarity group* of κ is the group of all similarities of κ.

Note that the isometry group of κ is normal in the similarity group of κ: if κ is identically zero then both groups are equal to $\mathrm{GL}(V)$, and otherwise this follows from the fact that the map $g \mapsto \lambda_g$ defined by $\kappa(ug, vg) = \lambda_g \kappa(u, v)$ (or, if κ is quadratic, $\kappa(vg) = \lambda_g \kappa(v)$) is a homomorphism.

Definition 1.5.5   Two σ-sesquilinear forms β and γ on V (respectively, two quadratic forms Q_1 and Q_2 on V) are *isometric* or *equivalent* if there is a $g \in \mathrm{GL}(V)$ such that $\gamma(u, v) = \beta(ug, vg)$ (respectively $Q_1(v) = Q_2(vg)$) for all $u, v \in V$. The forms are *similar* if there is a $g \in \mathrm{GL}(V)$ and $\lambda \in F \setminus \{0\}$ such that $\gamma(u, v) = \lambda\beta(ug, vg)$ (respectively, $Q_1(v) = \lambda Q_2(vg)$) for all $u, v \in V$.

Definition 1.5.6   The σ-sesquilinear form β is *non-degenerate* if $\beta(u, v) = 0$, for a fixed $v \in V$ and all $u \in V$, implies that $v = 0$.

Let Q be a quadratic form on V with polar form β. Then Q is *non-degenerate* if β is a non-degenerate bilinear form. The form Q is *non-singular* if $Q(v) \neq 0$ for all $v \in V$ such that $\beta(w, v) = 0$ for all $w \in V$: so, a non-degenerate quadratic form is non-singular.

(It will follow from Proposition 1.5.25 that a σ-sesquilinear form β is non-degenerate if and only if $\beta(u, v) = 0$, for a fixed $u \in V$ and all $v \in V$, implies that $u = 0$.)

Definition 1.5.7   Let β be a σ-sesquilinear form on V. A subspace W of V is *non-degenerate* if the restriction of β to W is non-degenerate. The subspace W is *totally singular* or *totally isotropic* if β restricted to W is identically 0. Similarly, a vector v is *singular* if $\beta(v, v) = 0$, and *non-singular* ortherwise.

Let Q be a quadratic form on V with polar form β. A subspace W of V is *totally singular* if $Q(w) = 0$ for all $w \in W$, and is *totally isotropic* if $\beta(v, w) = 0$

for all $v, w \in W$. A vector w is *singular* if $Q(w) = 0$; *isotropic* if $\beta(w, w) = 0$; and *non-singular* if $Q(w) \neq 0$.

By considering $\beta(w, w) = Q(2w) - 2Q(w) = 2Q(w)$, we see that all singular vectors are isotropic, but the converse need not be true in characteristic 2.

We are interested in certain σ-sesquilinear forms that have more symmetry than in the general case above.

Definition 1.5.8 Let β be a σ-sesquilinear form. If there exist $\lambda \in F^{\times}$ and $\tau \in \operatorname{Aut} F$ such that $\beta(v, u) = \lambda \beta(u, v)^{\tau}$ for all $u, v \in V$ then β is *quasi-symmetric*. If the form β satisfies the property that $\beta(u, v) = 0$ if and only if $\beta(v, u) = 0$ then β is *reflexive*.

The proof of the following is left as an easy exercise.

Lemma 1.5.9 *Every quasi-symmetric σ-sesquilinear form is reflexive.*

Definition 1.5.10 Let β be a reflexive σ-sesquilinear form, and let W be a subspace of V. The *orthogonal complement* of W, denoted by $W^{\perp}$, is:

$$W^{\perp} := \{ v \in V \mid \beta(w, v) = 0 \ \forall w \in W \} = \{ v \in V \mid \beta(v, w) = 0 \ \forall w \in W \}.$$

So, by Lemma 1.5.9, for any quasi-symmetric σ-sesquilinear form, a subspace W is non-degenerate if and only if $W \cap W^{\perp} = \{0\}$.

Definition 1.5.11 The *Witt index* of a σ-sesquilinear or quadratic form is the maximum dimension of a totally singular subspace of V.

Lemma 1.5.12 *Let β be a non-zero quasi-symmetric σ-sesquilinear form in characteristic $p \geqslant 0$, and let λ and τ be as in Definition 1.5.8. Then:*

(i) $\sigma^2 = 1$, $\lambda \lambda^{\sigma} = 1$, *and* $\tau = \sigma$.
(ii) *If $\sigma \neq 1$ then β is similar to a σ-sesquilinear form β' such that $\beta'(u, v) = \beta'(v, u)^{\sigma}$.*
(iii) *If $\sigma = 1$, $\lambda = -1$, and $p \neq 2$, then $\beta(v, v) = 0$ for all $v \in V$.*

Proof (i) We calculate that $\beta(u, v) = \lambda \beta(v, u)^{\tau} = \lambda \lambda^{\tau} \beta(u, v)^{\tau^2}$. By choosing v, w with $\beta(v, w) \neq 0$, and considering the non-zero scalar multiples of u, we see that $\lambda \lambda^{\tau} = 1$ and $\tau^2 = 1$. Now, notice that

$$\mu^{\sigma} \beta(u, v) = \beta(u, \mu v) = \lambda \beta(\mu v, u)^{\tau} = \lambda \lambda^{\tau} \mu^{\tau} \beta(u, v)^{\tau^2}.$$

Therefore $\mu^{\tau} = \mu^{\sigma}$ for all $\mu \in F$, so $\sigma = \tau$.
(ii) Suppose that $\sigma \neq 1$ and $\nu^{\sigma} + \lambda^{\sigma} \nu = 0$ for all $\nu \in F$. Setting $\nu = 1$, we deduce that $\lambda = -1$. Thus $\nu^{\sigma} - \nu = 0$ for all $\nu \in F$, a contradiction. Thus there exists $\nu \in F$ with $\nu^{\sigma} + \lambda^{\sigma} \nu = \mu \neq 0$. A short calculation shows that $\mu^{\sigma} = \mu \lambda$. Using this, and setting $\beta'(u, v) = \mu \beta(u, v)$, gives the result.
(iii) The assumptions imply that $\beta(v, v) = -\beta(v, v)$. $\qquad \square$

The following is essentially the Birkhoff–von Neumann Theorem (see [108, Theorem 7.1]), and follows straightforwardly from Lemma 1.5.12.

Theorem 1.5.13 *Let β be a quasi-symmetric σ-sesquilinear form on a vector space V over a field F, and let λ be as in Definition 1.5.8. Then up to similarity one of the following holds, for all $u, v \in V$.*

(i) $\beta(u, v) = 0$.

(ii) $\sigma = 1$, $\lambda = -1$ and $\beta(v, v) = 0$, so $\beta(v, u) = -\beta(u, v)$.

(iii) $\sigma^2 = 1 \neq \sigma$ and $\lambda = 1$, so $\beta(v, u) = \beta(u, v)^\sigma$.

(iv) $\sigma = 1$ and $\lambda = 1$, so $\beta(v, u) = \beta(u, v)$.

These cases are mutually exclusive, except that Case (ii) in characteristic 2 also satisfies Case (iv).

Definition 1.5.14 In Case (ii) we say that β is *alternating* or *symplectic*, in Case (iii) β is *σ-Hermitian* or *unitary*, and in Case (iv) β is *symmetric bilinear*, or sometimes *orthogonal* when char F is odd.

Note that for the above definitions we do not assume that the forms concerned are non-degenerate or non-singular.

The Birkhoff–von Neumann Theorem justifies restricting our study of quasi-symmetric sesquilinear forms to the symplectic, unitary and orthogonal forms, except when the form is symmetric bilinear and char $F = 2$. The following proposition, which follows from $\beta(v, v) = 2Q(v)$, shows that this is precisely when quadratic forms will be useful.

Proposition 1.5.15 *Let V be a vector space over F, equipped with a quadratic form Q with polar form β, and suppose that char $F \neq 2$. Then Q and β determine one another, and $g \in \mathrm{GL}(V)$ is an isometry (or a similarity) of Q if and only if g is an isometry (or a similarity) of β. Furthermore, the form Q is non-singular if and only if Q is non-degenerate, which is true if and only if β is non-degenerate. In addition, the Witt index of β is the same as that of Q.*

When studying quadratic forms in odd characteristic, we therefore can prove results via either the quadratic or the symmetric bilinear form. Combining this result with the Birkhoff–von Neumann Theorem, we see that there are essentially five types of non-degenerate forms to study: the zero form, symplectic forms, unitary forms, symmetric forms, and quadratic forms. Furthermore, it suffices to study quadratic forms only for q even, since when q is odd they are determined by their polar forms.

1.5.1 The matrix formulation

In this subsection we study a practical way of representing and calculating with forms. Let $\mathscr{B} = (e_1, \ldots, e_n)$ be a basis of V. Recall our notation for finite fields and matrices from Section 1.2.

Definition 1.5.16 The matrix of the sesquilinear form β with respect to $\mathscr{B}$ is $B = (b_{ij})_{n \times n}$ where $b_{ij} = \beta(e_i, e_j)$ for all i and j.

Let Q be a quadratic form with polar form β. We can calculate that

$$Q(\sum_{i=1}^n \lambda_i e_i) = \sum_{i=1}^n \lambda_i^2 Q(e_i) + \sum_{i<j} \lambda_i \lambda_j \beta(e_i, e_j).$$

Definition 1.5.17 The matrix of the quadratic form Q with polar form β with respect to $\mathscr{B}$ is the upper-triangular matrix $A = (\alpha_{ij})$, where $\alpha_{ij} = \beta(e_i, e_j)$ if $i < j$, $\alpha_{ii} = Q(e_i)$ and $\alpha_{ij} = 0$ if $i > j$.

The matrix of the polar form of Q is then $B = A + A^{\mathsf{T}}$. When κ is a form with matrix C, we shall often assume that the basis $\mathscr{B}$ is present, and (abusing language) refer to C as being the form.

Lemma 1.5.18 *Let $v, w \in V$, let β be a σ-sesquilinear form on V with matrix B, and let Q be a quadratic form on V with matrix A. Then $\beta(v, w)$ is the single entry of the matrix $vBw^{\sigma\mathsf{T}}$, and $Q(v)$ is the single entry of the matrix vAv^{T}.*

Proof This is a straightforward calculation. □

Definition 1.5.19 A square matrix B is called *symmetric, anti-symmetric* or *σ-Hermitian*, if $B^{\mathsf{T}} = B$, $B^{\mathsf{T}} = -B$, or $B^{\mathsf{T}} = B^{\sigma}$, respectively. It is called *alternating* if it is anti-symmetric and all of its diagonal entries are 0.

Lemma 1.5.20 *The matrix of the form β is symmetric, alternating or σ-Hermitian if and only if β is.*

Proof This is a straightforward calculation. □

Recall Definition 1.5.3 of an isometry of a form.

Lemma 1.5.21 *Let β be a σ-sesquilinear form on F^n with matrix B, and let Q be a quadratic form on F^n with matrix A, whose polar form has matrix C. A matrix $M \in \mathrm{GL}_n(F)$ is an isometry of β if and only if $MBM^{\sigma\mathsf{T}} = B$. The matrix M is an isometry of Q if and only if $MCM^{\mathsf{T}} = C$, and the diagonal entries of A and MAM^{T} coincide.*

Proof Suppose first that $MBM^{\sigma\mathsf{T}} = B$. Then

$$\beta(vM, wM) = vMB(wM)^{\sigma\mathsf{T}} = v(MBM^{\sigma\mathsf{T}})w^{\sigma\mathsf{T}} = \beta(v, w),$$

so the result follows from Lemma 1.5.18. The converse, and the arguments for quadratic forms, are equally easy. □

For quadratic forms there is an alternative way of checking whether a matrix is an isometry which is useful when doing calculations.

Definition 1.5.22   For an $n \times n$ matrix $A = (\alpha_{ij})$, define $A^{\mathrm{UT}} = (\beta_{ij})$, where $\beta_{ii} = \alpha_{ii}$ for $1 \leqslant i \leqslant n$, $\beta_{ij} = \alpha_{ij} + \alpha_{ji}$ for $1 \leqslant i < j \leqslant n$, and $\beta_{ij} = 0$ if $i > j$.

Note that $(BA^{\mathrm{UT}}B^{\mathsf{T}})^{\mathrm{UT}} = (BAB^{\mathsf{T}})^{\mathrm{UT}}$ for all $n \times n$ matrices A, B. Using this notation, the following is straightforward.

Lemma 1.5.23   *If Q is a quadratic form on F^n with form matrix A, and $M \in \mathrm{GL}_n(F)$, then M is an isometry of Q if and only if $(MAM^{\mathsf{T}})^{\mathrm{UT}} = A$.*

Recall Definition 1.5.5 of an isometry between forms. The following fundamental lemma will be used without reference throughout much of the book.

Lemma 1.5.24   *Two σ-sesquilinear forms on F^n with matrices B and B' are isometric if and only if there exists $L \in \mathrm{GL}_n(F)$ such that $B' = LBL^{\sigma\mathsf{T}}$. Similarly, two quadratic forms on F^n with matrices A and A' are isometric if and only if there exists $L \in \mathrm{GL}_n(F)$ such that $A' = (LAL^{\sigma\mathsf{T}})^{\mathrm{UT}}$.*

Proof   Suppose that $C = LBL^{\sigma\mathsf{T}}$. Then $\beta(vL, wL) = vLBL^{\sigma\mathsf{T}}w^{\sigma\mathsf{T}} = \gamma(v, w)$. Conversely, if B and C are isometric then there exists an L in $\mathrm{GL}_n(F)$ such that $\beta(vL, wL) = \gamma(v, w)$. The quadratic case is similar. □

Recall Definition 1.5.6 of a non-degenerate sesquilinear form.

Proposition 1.5.25   *Let β be a σ-sesquilinear form with matrix B. Then β is non-degenerate if and only if B has non-zero determinant.*

Proof   If $u \in V^{\perp}$ then $uBv^{\sigma\mathsf{T}} = 0$ for all v, in particular if v is one of the e_i. Thus $uB = 0$. Conversely, if $uB = 0$ then $uBv^{\sigma\mathsf{T}} = 0$ for all v, whence $\beta(u, v) = 0$ for all v, and so $u \in V^{\perp}$. Therefore $V^{\perp} \neq \{0\}$ if and only if B is not invertible. □

We now proceed to consider the types of non-zero forms individually, and introduce some notation for their isometry groups. When the field F is finite, we have preferred choices of the basis of V that result in specific form matrices, which we call our *standard form matrices*. We call their isometry groups (as subgroups of $\mathrm{GL}_n(F)$) our *standard copies* of these groups. We shall summarise these definitions later in Table 1.1.

1.5.2 Alternating forms

Recall the definition of an alternating form from Definition 1.5.14. The following is standard; see for instance [108, p69].

Proposition 1.5.26 *If V admits a non-degenerate alternating form then V has a basis $(e_1, f_1, e_2, f_2, \ldots, e_m, f_m)$ such that:*

$$\beta(e_i, e_j) = \beta(f_i, f_j) = 0, \quad \beta(e_i, f_j) = \delta_{ij} \quad for \ all \quad i, j,$$

where δ_{ij} is the Kronecker delta. Thus $\dim V = n = 2m$ is even, and there is a unique isometry class of such forms on V.

It follows that the isometry groups of any two non-degenerate alternating forms on V are isomorphic.

The matrix $\mathrm{antidiag}(1, \ldots, 1, -1, \ldots, -1)$, with $\frac{n}{2}$ 1's and $\frac{n}{2}$ -1's, defines our standard symplectic form, which is the result of ordering the above basis as $(e_1, \ldots, e_m, f_m, \ldots, f_1)$. MAGMA uses this standard form.

Definition 1.5.27 The notation $\mathrm{Sp}_n(F)$ denotes the isometry group of our standard alternating form of dimension n over F. If $F = \mathbb{F}_q$ is finite, we usually write $\mathrm{Sp}_{2m}(q)$ instead of $\mathrm{Sp}_{2m}(F)$. This is the *symplectic group* on F^n.

1.5.3 Hermitian forms

Recall the definition of a σ-Hermitian form from Definition 1.5.14. To define such a form, we require an automorphism σ of F of order 2, and we define F_0 by $F_0 = \mathrm{C}_F(\sigma) = \{\, x \in F \mid x^\sigma = x \,\}$. Then F has dimension 2 over F_0.

Proposition 1.5.28 *Let β be a non-degenerate σ-Hermitian form over a finite field F. Then $F = \mathbb{F}_{q^2}$ for some prime power q, the automorphism σ is the map $\sigma : x \mapsto x^q$, and $F_0 = \mathbb{F}_q$.*

Throughout the rest of the book, we will use the previous result without further citation to deduce that if $H \leqslant \mathrm{GL}_n(q)$ for a non-square q then H does not preserve a non-degenerate unitary form on $\mathbb{F}_q$.

Proposition 1.5.29 *If V admits a non-degenerate σ-Hermitian form β, then V has a basis with respect to which β has matrix $\mathrm{diag}(a_1, \ldots, a_n)$, where $a_i \in F_0^\times$ for all i. When F is finite, V has a basis for which $a_i = 1$ for all i, and so all non-degenerate σ-Hermitian forms are isometric.*

Proof Since β is non-degenerate, we may choose a non-singular vector $e_1 \in V$. Then $V = \langle e_1 \rangle \perp \langle e_1 \rangle^\perp$, so by induction we may choose a basis for V, as required. Now suppose that $F = \mathbb{F}_{q^2}$ is finite and $\sigma : x \mapsto x^q$, as in Proposition 1.5.28, and that β has matrix $\text{diag}(a_1, \dots, a_n)$. The fact that β is σ-Hermitian implies that $a_i = a_i^q$ for all i; that is, $a_i \in \mathbb{F}_q$. Since $\mathbb{F}_{q^2}^\times$ is cyclic, there exists $b_i \in \mathbb{F}_{q^2}^\times$ with $b_i b_i^\sigma = b_i^{1+q} = a_i^{-1}$, and then $\beta(b_i e_i, b_i e_i) = 1$. □

The above proposition shows that for finite F all isometry groups of non-degenerate σ-Hermitian forms are isomorphic; this explains why for many purposes one may suppress the precise description of the form when discussing unitary groups.

Our standard σ-Hermitian form over any field has matrix I_n. Its isometry group is the set of $n \times n$ matrices A over F such that $A A^{\sigma\mathsf{T}} = I_n$, by Lemma 1.5.21. The MAGMA standard σ-Hermitian form is different, and has matrix $\text{antidiag}(1, \dots, 1)$.

Definition 1.5.30 We write $\text{GU}_n(F)$ to denote the isometry group of our standard σ-Hermitian form in dimension n over a field with an automorphism of order 2. If $F = \mathbb{F}_{q^2}$, we usually write $\text{GU}_n(q)$ rather than $\text{GU}_n(F)$. The groups $\text{GU}_n(F)$ are *general unitary groups*.

1.5.4 Symmetric bilinear forms in characteristic not 2

Recall the definition of a symmetric bilinear form from Definition 1.5.14. Recall also from Proposition 1.5.15 that the study of symmetric bilinear forms in characteristic not 2 is equivalent to the study of quadratic forms.

It can be shown that over an arbitrary field F of characteristic not 2, for each non-degenerate symmetric bilinear form there exists a basis such that the form matrix is diagonal. However, in the finite case we can say far more; for a proof of the following, see [108, p138].

Theorem 1.5.31 *Let $F = \mathbb{F}_{p^e}$ with p odd. Then up to isometry there are precisely two non-degenerate forms on V, corresponding to the cases when the determinant of the form matrix is a square or non-square of $F^\times$. If n is odd then there is a unique similarity class, and if λ is non-square then a form with matrix B is not isometric to a form with matrix λB. If n is even then there are two similarity classes.*

By Proposition 1.5.15 the same is true for quadratic forms: over a finite field of odd characteristic there are always two isometry classes of forms, and these consist of two similarity classes when the dimension is even and one when the dimension is odd.

In odd dimension, we take I_n as the matrix of our standard symmetric bilinear form, so that our standard quadratic form has matrix $I_n/2$. If μ is any non-zero non-square, then μI_n is in the other isometry class of non-degenerate symmetric forms. Our standard symmetric form differs from the MAGMA standard form which is antidiag$(1, \ldots, 1, \frac{1}{2}, 1, \ldots, 1)$, with the $\frac{1}{2}$ in position $\frac{n+1}{2}$.

Definition 1.5.32 If $F = \mathbb{F}_q$ has odd order, and n is odd, then the isometry group of our standard symmetric (or quadratic) form is denoted by $\mathrm{GO}_n(q)$, or sometimes as $\mathrm{GO}_n^\circ(q)$, the *general orthogonal group*.

As we saw in Theorem 1.5.31, in even dimension, if F is finite and of odd order, then any two non-degenerate symmetric bilinear forms are similar if and only if they are isometric. Thus there are up to two isomorphism classes of isometry groups of such forms, and it turns out that there are precisely two.

Definition 1.5.33 Let β be a non-degenerate symmetric bilinear form in even dimension over a finite field of odd characteristic. Then β has *plus type* if it is isometric to the form antidiag$(1, \ldots, 1)$, otherwise it has *minus type*.

It turns out that a non-degenerate symmetric bilinear form has Witt index $n/2$ or $n/2 - 1$ when the form has plus or minus type, respectively. We shall state this result for the corresponding quadratic forms (which is valid also in even characteristic) in Proposition 1.5.39.

Our standard plus type form is antidiag$(1, \ldots, 1)$, with quadratic form antidiag$(1, \ldots, 1, 0, \ldots, 0)$ with $n/2$ 1's. For minus type, we use the form matrix I_n when this is not isometric to the previous form: we will see in Proposition 1.5.42 that this is the case if and only if $n \equiv 2 \pmod 4$ and $q \equiv 3 \pmod 4$. Otherwise, our standard symmetric bilinear form of minus type has matrix diag$(\omega_q, 1, \ldots, 1)$, where ω_q is a fixed primitive element of $\mathbb{F}_q^\times$, with corresponding quadratic form diag$(\omega_q/2, 1/2, \ldots, 1/2)$. Our standard form of plus type agrees with the one in MAGMA, but our form of minus type does not.

Definition 1.5.34 The isometry group of our standard plus type form is denoted $\mathrm{GO}_n^+(q)$, and the isometry group of our standard minus type form is denoted $\mathrm{GO}_n^-(q)$. These groups are also called the *general orthogonal groups*.

Definition 1.5.35 We say that a non-degenerate symmetric bilinear form with q odd has *square discriminant* if the determinant of its matrix is a square in $\mathbb{F}_q^\times$ and *non-square discriminant* otherwise. Similarly, the discriminant of a quadratic form with q odd is the discriminant of its polar form.

1.5.5 Symmetric bilinear forms in characteristic 2

Recall Definition 1.5.14 of a symmetric bilinear form and Definition 1.4.4 of a perfect field. The following proposition shows that we need not consider isometry groups of symmetric but not alternating forms in characteristic 2.

Proposition 1.5.36   *Let F have characteristic 2, and let V be an irreducible FG-module of dimension at least 2. If G is a group of isometries of a symmetric bilinear form β on V then $\beta(v,v) = 0$ for all $v \in V$, so β is alternating.*

Proof   Pick any non-zero $w \in V$. Then since G is irreducible, w^G spans V and, since $\dim V \geqslant 2$, there exists $g \in G$ such that w and wg are linearly independent. Define $e_1 := w + wg$, which is non-zero. As before, e_1^G spans V, so we can choose a basis $(e_1, e_2, \ldots, e_n)$ for V from the orbit e_1^G. Then

$$\beta(e_1, e_1) = \beta(w, w) + \beta(wg, wg) + \beta(w, wg) + \beta(wg, w)$$
$$= 2\beta(w, w) + 2\beta(w, wg)$$
$$= 0.$$

Therefore $\beta(e_i, e_i) = 0$ for all i. Taking $v = \sum_{i=1}^{n} a_i e_i$ to be an arbitrary element of V, we get $\beta(v, v) = \sum_{i=1}^{n} a_i^2 \beta(e_i, e_i) + 2\sum_{i<j} a_i a_j \beta(e_i, e_j) = 0 + 0 = 0$, as required. $\square$

We may generalise this to the following result; since we will not be using it, we omit the proof.

Proposition 1.5.37   *Let β be a non-degenerate symmetric bilinear form on V, over a perfect field F of characteristic 2. Then V has a basis $(e_1, f_1, \ldots, e_m, f_m, d_1, \ldots, d_r)$ such that $0 \leqslant r \leqslant 2$ and:*

$$\beta(e_i, e_j) = \beta(f_i, f_j) = 0, \quad \beta(e_i, f_j) = \delta_{ij} \quad \textit{for all} \quad i, j,$$
$$\beta(d_i, e_j) = \beta(d_i, f_j) = 0, \quad \beta(d_i, d_j) = \delta_{ij} \quad \textit{for all} \quad i, j.$$

It follows from the above Proposition that in even dimension $n = 2m + r \geqslant 2$ there are always two isometry classes of non-degenerate forms, corresponding to $r = 0$ and $r = 2$, whilst in odd dimension there is only one. It is not too hard to see that the isometry groups of these forms for $r = 0, 1, 2$ are respectively $\mathrm{Sp}_{2m}(F)$, $\mathrm{Sp}_{2m}(F)$ and $(F, +)^{1+2m} : \mathrm{Sp}_{2m}(F)$, so we get nothing new.

1.5.6 Quadratic forms in characteristic 2

Recall the definition of a quadratic form and its polar form from Definition 1.5.2. We have seen that in characteristic not 2 a quadratic form and its polar form determine one another, and share the same groups of isometries and similarities. Thus it remains only to study quadratic forms in characteristic 2. Recall Definition 1.4.4 of a perfect field.

We will assume throughout this section that Q is a quadratic form in characteristic 2 with polar form β. Then $\beta(v, v) = 2Q(v) = 0$ for all $v \in V$, and so β is a symplectic form.

Lemma 1.5.38 *Let $n = 2m$ be even, and let F be perfect. Then a quadratic form Q is non-singular if and only if Q is non-degenerate.*

Proof If Q is non-degenerate then $V^\perp = \{0\}$, so Q is non-singular.

Suppose that Q is non-singular. Assume, by way of contradiction, that there exists a non-zero $v \in V^\perp$. Then $Q(v) = \lambda$ for some $\lambda \in F^\times$, and since F is perfect there exists $\mu \in F$ with $\mu^2 = \lambda$. Thus $Q(\mu^{-1}v) = 1$, so without loss of generality we may assume that $Q(v) = 1$. If $\dim V^\perp > 1$ then there exists $w \in V^\perp \setminus \langle v \rangle$, and without loss of generality $Q(w) = 1$. Then $Q(v + w) = Q(v) + Q(w) = 0$, a contradiction. Therefore $\dim V^\perp = 1$. Now, $V/V^\perp$ is non-degenerate, so the polar form of Q on $V/V^\perp$ is a non-degenerate symplectic form, and thus by Proposition 1.5.26 the dimension of $V/V^\perp$ is even, a contradiction. $\square$

The following result is standard; see for instance [108, p139]. Recall from Definition 1.5.11 that the Witt index of a form is the maximum dimension of a totally singular subspace.

Proposition 1.5.39 *Let Q be a non-degenerate quadratic form on F^{2m}, with $F = \mathbb{F}_{2^e}$. Then there exists a basis $(e_1, \dots, e_m, f_m, \dots, f_1)$ of V such that*

$$Q(e_i) = Q(f_i) = 0 \text{ for all } i \text{ with } 1 \leqslant i \leqslant m - 1,$$
$$\beta(e_i, e_j) = \beta(f_i, f_j) = 0 \text{ for all } i, j, \quad \beta(e_i, f_j) = \delta_{ij} \text{ for all } i, j.$$

Up to both similarity and isometry, there are exactly two choices for the values of $Q(e_m)$ and $Q(f_m)$. One possibility is $Q(e_m) = Q(f_m) = 0$, giving Witt index $n/2$. The other is $Q(e_m) = 1$, $Q(f_m) = \mu$, where the polynomial $x^2 + x + \mu$ is irreducible over F, giving Witt index $n/2 - 1$.

We take the forms in Proposition 1.5.39 as our standard quadratic forms in characteristic 2.

Definition 1.5.40 Let Q be a non-degenerate quadratic form in even dimension over a finite field of even characteristic. Then Q has *plus* type if it is isometric to the form in Proposition 1.5.39 with $Q(e_m) = Q(f_m) = 0$, and *minus* type if it is isometric to the other form in Proposition 1.5.39. The isometry group is called the *general orthogonal group* in each case. In plus type it is denoted by $\mathrm{GO}_n^+(F)$, and in minus type it is denoted by $\mathrm{GO}_n^-(F)$.

The case when char $F = 2$ and $n = 2m + 1$ is odd is less interesting to us.

Theorem 1.5.41 *Let char $F = 2$, with F perfect, and let $n = 2m + 1 > 1$. If Q is any quadratic form then the isometry group of V is reducible.*

Proof If Q is singular then the isometry group of Q fixes the subspace of V

on which the form is identically zero. If Q is non-singular then reason as in the proof of Lemma 1.5.38 to see that $\dim(V^\perp) = 1$, so the isometry group of Q must fix $V^\perp$. $\square$

It turns out (see [108, p139]) that all non-singular quadratic forms in odd dimension over a perfect field of even characteristic are isometric, and there exists a basis $(e_1, \ldots, e_m, d, f_m, \ldots, f_1)$ such that:

$$Q(e_i) = Q(f_i) = 0 \text{ for all } i, \quad Q(d) = 1,$$
$$\beta(d, e_i) = \beta(d, f_i) = \beta(e_i, e_j) = \beta(f_i, f_j) = 0 \text{ for all } i, j,$$
$$\beta(e_i, f_j) = \delta_{ij} \text{ for all } i, j.$$

The isometry group is $\mathrm{Sp}_{2m}(F)$, acting naturally on $V/\langle d \rangle$ (see [108, Theorem 11.9]) and so we will assume that orthogonal groups in odd dimension are defined over fields of characteristic not 2.

We finish this subsection with a collection of results that will enable us to calculate the sign of a quadratic form or its polar form in various situations.

Proposition 1.5.42 *Let V be a vector space of even dimension n over $\mathbb{F}_q$ equipped with a non-singular quadratic form Q with polar form β. Let A and B be the form matrices of Q and β with respect to some fixed basis of V. Then*

(i) *The Witt index of Q is equal either to $n/2$ (plus type) or to $n/2 - 1$ (minus type), and all forms with the same Witt index are isometric.*

(ii) *If q is odd and n is even, then the form is of plus type if and only if either the discriminant ($\det B$) is square and $n(q-1)/4$ is even, or the discriminant is non-square and $n(q-1)/4$ is odd.*

(iii) *If q is even, $n = 2$, and V has a basis (e_1, f_1) such that $\beta(e_1, f_1) = 1$, $\beta(e_1, e_1) = \beta(f_1, f_1) = 0$, $Q(e_1) = 1$ and $Q(f_1) = \mu$, then Q is of minus type if and only if the polynomial $x^2 + x + \mu$ is irreducible over F.*

(iv) *If $V = W \oplus W^\perp$ and the restrictions of Q to W and $W^\perp$ have types t_1, t_2 with $t_i = 1$ or -1 for plus type or minus type, then Q has type $t_1 t_2$.*

(v) *The form over $\mathbb{F}_{q^k}$ defined by A is of plus type when k is even.*

Proof (i) In even characteristic, this is Proposition 1.5.39. In odd characteristic, see [108, p138].

(ii) This is standard; see for instance [66, Proposition 2.5.10].

(iii) If $p(x) = x^2 + x + \mu$ is irreducible over $\mathbb{F}_q$, then Q has minus type by definition. If $p(x)$ has a root $a \in \mathbb{F}_q$, then $Q(ae_1 + f_1) = 0$, so $\langle ae_1 + f_1 \rangle$ is totally singular, and hence Q has non-zero Witt index, and has plus type.

(iv) follows from (ii) when q is odd. When q is even, (iv) follows directly from our standard forms except when $t_1 = t_2 = -1$ since if at least one of the forms is of plus type, then one may re-order a basis for W_1 and W_2 to yield either

Table 1.1 *Our standard classical forms*

Case	conditions	form type	isom. gp	form
L	—	zero	$\mathrm{GL}_n(q)$	$0_{n \times n}$
S	—	alternating	$\mathrm{Sp}_n(q)$	$\mathrm{antidiag}(1, \ldots, 1, -1, \ldots, -1)$
U	—	σ-Hermitian	$\mathrm{GU}_n(q)$	I_n
O°	qn odd	symmetric	$\mathrm{GO}_n(q)$	I_n
O⁺	q odd, n even	symmetric	$\mathrm{GO}_n^+(q)$	$\mathrm{antidiag}(1, \ldots, 1)$
O⁻	q odd, n even	symmetric	$\mathrm{GO}_n^-(q)$	I_n if $n(q-1)/4$ is odd
				$\mathrm{diag}(\omega_q, 1, \ldots, 1)$ otherwise
O⁺	q, n even	quadratic	$\mathrm{GO}_n^+(q)$	$\mathrm{antidiag}(1, \ldots, 1, 0, \ldots, 0)$
O⁻	q, n even	quadratic	$\mathrm{GO}_n^-(q)$	$\mathrm{antidiag}(1, \ldots, 1, 0, \ldots, 0)$
				$+\mathrm{E}_{m,m} + \mu \mathrm{E}_{m+1,m+1}$

our standard plus type form (if $t_1 = t_2 = +1$) or our standard minus type form (if $t_1 t_2 = -1$). If $t_1 = t_2 = -1$, then is sufficient to deal with the case when $\dim W = \dim W^\perp = 2$ and the restrictions of Q to W_1 with basis (e_1, f_1) and $W_1^\perp$ with basis (e_2, f_2) have the same standard form matrix $\begin{pmatrix} 1 & 1 \\ 0 & \mu \end{pmatrix}$. But then the subspace of V spanned by $e_1 + e_2, f_1 + f_2$ is totally singular, so the Witt index of V is 2, and Q is of plus type.

(v) follows from (ii) and (iii) and the fact that irreducible equations of degree 2 over $\mathbb{F}_q$ become reducible over $\mathbb{F}_{q^2}$. □

1.5.7 Summary of standard forms

In Table 1.1, we summarise the standard forms associated with the classical groups in their standard representations, as described in this section. The form is given as a matrix, which, in the final two cases only, is the matrix of the quadratic form. A form specified in the table as $\mathrm{antidiag}(a, \ldots, a, b, \ldots, b)$ has equal numbers of a's and b's, but a form specified as $\mathrm{diag}(\lambda, 1, \ldots, 1)$ has just one λ. We define m by $n = 2m$ or $2m + 1$, when n is even or odd respectively. In the last line in the table, the polynomial $x^2 + x + \mu$ is irreducible over $\mathbb{F}_q$.

When q is odd, one may recover our standard quadratic form Q from our standard symmetric bilinear form β via $\beta(v, v) = 2Q(v)$.

1.6 The classical groups and their orders

In the previous section, we defined our standard classical forms, and their associated isometry groups $\mathrm{GL}_n(q)$, $\mathrm{GU}_n(q)$, $\mathrm{Sp}_n(q)$ and $\mathrm{GO}_n^\varepsilon(q)$ (where $\varepsilon \in$

$\{\circ, +, -\}$). In this section we define various other associated classical groups, and explore some of their basic properties, such as their order.

1.6.1 Semilinear maps

Definition 1.6.1 Let V and W be vector spaces over a common field F, and let $\theta \in \operatorname{Aut} F$. A *$\theta$-semilinear map* $f : V \to W$ is any map satisfying $(v + w)f = vf + wf$ and $(\lambda v)f = \lambda^{\theta}(vf)$ for all $v, w \in V$, $\lambda \in F$. The θ-semilinear map f is *non-singular* if it satisfies $vf = 0$ if and only if $v = 0$. The map f is called *semilinear* if f is θ-semilinear for some θ.

So a linear map is just a θ-semilinear map with $\theta = 1$. It is an easy exercise to show that the set

$$\{f : V \to V \mid f \text{ non-singular } \theta\text{-semilinear map for some } \theta \in \operatorname{Aut} F\}$$

forms a group under composition.

Definition 1.6.2 The group of all non-singular semilinear maps from V to itself is denoted by $\Gamma L(V)$. We may also denote this by $\Gamma L_n(F)$, or by $\Gamma L_n(q)$ when $F = \mathbb{F}_q$, where $n = \dim V$.

One may check that the map from $\Gamma L_n(F)$ to $\operatorname{Aut} F$ which sends the θ-semilinear map f to θ is a homomorphism with kernel $\operatorname{GL}_n(F)$.

For a fixed basis $(e_1, \ldots, e_n)$ of V and $\theta \in \operatorname{Aut} F$, define $\bar{\theta} : V \to V$ by $(\sum_{i=1}^{n} \lambda_i e_i)\bar{\theta} = \sum_{i=1}^{n} \lambda_i^{\theta} e_i$. Then $\{\bar{\theta} : \theta \in \operatorname{Aut} F\}$ is a complement to $\operatorname{GL}_n(F)$ in $\Gamma L_n(F)$. If A is the matrix with respect to this basis of $f \in \operatorname{GL}_n(F)$, then the effect of the conjugation action of $\bar{\theta}$ on A is to replace the matrix entries by their images under θ.

Definition 1.6.3 Let $F = \mathbb{F}_q$ with $q = p^e$. We denote the generating field automorphism $x \mapsto x^p$ by ϕ. We also write ϕ rather than $\bar{\phi}$ for the element of $\Gamma L_n(q)$ corresponding to ϕ. Thirdly, we denote by ϕ the automorphism of $\operatorname{GL}_n(q)$ induced by conjugation by $\bar{\phi}$: that is, replace all matrix entries by their p-th powers.

So ϕ has three different meanings: it is a field automorphism, a semilinear map, and an automorphism of $\operatorname{GL}_n(q)$. Since these meanings are generally compatible, we hope that this practice will not cause confusion.

Definition 1.6.4 A θ-semilinear map f is a *semi-isometry* of a form β (or a quadratic form Q), if $\beta(vf, wf) = \beta(v, w)^{\theta}$ (respectively, $Q(vf) = Q(v)^{\theta}$) for all $v, w \in V$. It is a *semi-similarity* if there exists $0 \neq \lambda \in F$ such that $\beta(vf, wf) = \lambda\beta(v, w)^{\theta}$ (respectively, $Q(vf) = \lambda Q(v)^{\theta}$) for all $v, w \in V$.

Notice that the set of all semi-isometries is a group containing the isometry group of the form as a normal subgroup, and that the group of all semi-similarities contains the similarity group as a normal subgroup.

1.6.2 Definitions of the classical groups

Throughout this subsection, let F be a finite field, and let $V = F^n$ be equipped with a form β, which is one of: the zero form, our standard unitary form, our standard symplectic form, or our standard symmetric form from Table 1.1. If β is non-zero symmetric, then V may also be equipped with our standard non-degenerate quadratic form Q. For each of these possible forms we define a series of groups, which we will denote by

$$\Omega \leqslant S \leqslant G \leqslant C \leqslant \Gamma \leqslant A. \tag{1.1}$$

Our notation for each of these groups for each form is given in Section 1.6.3.

Recall Definition 1.5.4 of the *isometry group* of V, and the definitions of $\mathrm{Sp}_n(F)$, $\mathrm{GU}_n(F)$ and $\mathrm{GO}_n^\varepsilon(F)$ from Definitions 1.5.27, 1.5.30, 1.5.32, 1.5.34 and 1.5.40. We define $u := 2$ if β is unitary, and $u := 1$ in all other cases. For the remainder of this subsection we let $F = \mathbb{F}_{q^u}$.

Definition 1.6.5 The groups $\mathrm{GL}_n(F)$, $\mathrm{Sp}_n(F)$, $\mathrm{GU}_n(F)$ and $\mathrm{GO}_n^\varepsilon(F)$ are the *general* groups of V. The general group is denoted by G in Series 1.1.

For groups preserving non-degenerate or non-singular forms other than our standard forms, we use similar notation. For example, a group preserving a non-degenerate symplectic form on $\mathbb{F}_q^n$ with form matrix B would be denoted by $\mathrm{Sp}_n(q, B)$, and if W is a space carrying a non-standard form then we may also write $\mathrm{GU}(W)$, for example, if the form is understood. However, if no form or module is specified then the standard form is always assumed.

We shall always assume that q is odd for $\mathrm{GO}_n^\circ(q)$, since if $n \geqslant 1$ and q is even, then $\mathrm{GO}_n^\circ(q)$ is reducible by Theorem 1.5.41.

Definition 1.6.6 The *special group* of V is the subgroup of the general group consisting of all matrices of determinant 1. Thus we refer to the *special linear group*, denoted by $\mathrm{SL}_n(q)$; the *special unitary group*, denoted by $\mathrm{SU}_n(q)$; and the *special orthogonal group*, denoted by $\mathrm{SO}_n^\varepsilon(q)$, with $\varepsilon \in \{\circ, +, -\}$. The special group is denoted by S in Series 1.1.

The following can be proved by showing that $\mathrm{Sp}_{2m}(F)$ is generated by symplectic transvections: see [108, Corollary 8.6].

Theorem 1.6.7 *All elements of* $\mathrm{Sp}_{2m}(F)$ *have determinant 1.*

We will use this result frequently without reference: rather than referring to a general or special symplectic group we will just refer to a symplectic group.

In addition, in characteristic 2 it is a straightforward exercise using Proposition 1.6.9 to show that all isometries of a quadratic form have determinant 1, and hence the special orthogonal group coincides with the general orthogonal group. We shall use whichever term is more convenient in this case. Note that in characteristic 2, some authors write $SO_n^{\pm}(q)$ to denote a certain subgroup of index 2 in $GO_n^{\pm}(q)$: we shall define this subgroup shortly, and denote it $\Omega_n^{\pm}(q)$.

For linear, symplectic and unitary groups, the special group is quasisimple except for a few small dimensions and prime powers: see Proposition 1.10.3 for the exceptions. However, if $n \geqslant 2$ then there is an epimorphism from the special orthogonal group to $\{\pm 1\}$, whose kernel is generally quasisimple.

Definition 1.6.8   Let Q be a quadratic form on V, of sign ε, and let β be its polar form. Let $v \in V$ be non-singular. We define the *reflection* $r_v : V \to V$ by $(x)r_v = x - \beta(v,x)v/Q(v)$.

The following result is well known; see for example [108, Corollary 11.42].

Proposition 1.6.9   *The group* $GO_n^{\varepsilon}(q)$ *is generated by the set of reflections in non-singular vectors, provided that* $(n, q, \varepsilon) \neq (4, 2, +)$.

Definition 1.6.10   Assume that $(n, q, \varepsilon) \neq (4, 2, +)$. Let $g = \prod_{i=1}^{k} r_{v_i}$ be an element of $GO_n^{\varepsilon}(q)$.

If q is odd then the *spinor norm* of g is $+1$ if $\prod_{i=1}^{k} \beta(v_i, v_i)$ is a square in $\mathbb{F}_q^{\times}$ and -1 if it is a non-square. If q is even then the *quasideterminant* of g is $+1$ if k is even and -1 if k is odd.

The additive version of the quasideterminant for even q is known as the *Dickson determinant* [108, p160]. It can be shown that the spinor norm and quasideterminant are well-defined homomorphisms, and the following proposition (see [108, Theorems 11.43, 11.50]) provides a way to calculate them.

Proposition 1.6.11   *Let* $g \in GO_n^{\varepsilon}(q)$, *let* $A := I_n - g$ *and suppose that* A *has rank* k. *If* q *is odd then let* F *be the matrix of the invariant symmetric bilinear form of* $SO_n^{\varepsilon}(q)$, *and let* B *be a* $k \times n$ *matrix over* $\mathbb{F}_q$ *whose rows form a basis of a complement of the nullspace of* A. *Then:*

  (i) *If* q *is even and* $(n, q, \varepsilon) \neq (4, 2, +)$, *then the quasideterminant of* g *is* 1 *if* k *is even and* -1 *otherwise.*
  (ii) *If* q *is odd, then the spinor norm of* g *is* 1 *if* $\det(BAFB^{\mathsf{T}})$ *is a square in* $\mathbb{F}_q^{\times}$ *and* -1 *otherwise.*

We record some additional information about $GO_4^+(2)$.

Lemma 1.6.12 ([66, Proposition 2.5.9]) *The group* $GO_4^+(2)$ *has exactly three subgroups of index two. One of these is the subgroup generated by reflections, and a second is the subgroup of all isometries that induce even permutations on the set W of totally singular 2-spaces. There is an equivalence relation on W given by $U_1 \sim U_2$ if and only if $\dim(U_1 \cap U_2)$ is even, and this partitions W into two equivalence classes. The third subgroup of index two is the group of all isometries that fix each equivalence class setwise.*

Definition 1.6.13 The Ω group in Series 1.1 is defined to be equal to the special group if the form on V is linear, symplectic or unitary. If the form on V is quadratic of sign ε, and $(n, q, \varepsilon) \neq (4, 2, +)$, then the Ω group, denoted by $\Omega_n^\varepsilon(q)$, is defined to be the kernel of the spinor norm or the quasideterminant map on $SO_n^\varepsilon(q)$, when q is odd or even, respectively. The group $\Omega_4^+(2)$ is the third subgroup described in Lemma 1.6.12. We will call Ω the *generally quasisimple classical group*.

Note that this definition of $\Omega_4^+(2)$ agrees with that in MAGMA, and that in all other cases it can be shown that if $n \geqslant 2$ then $SO_n^\varepsilon(q)$ has a unique subgroup of index 2 and so that, although there are different definitions in the literature of the spinor norm map, $\Omega_n^\varepsilon(q)$ is well-defined. Note however that when q is even, the kernel of the quasideterminant map, which we denote $\Omega_n^\pm(q)$, is denoted by $SO_n^\pm(q)$ by some authors.

We now consider the larger terms in Series 1.1. Recall Definition 1.5.4 of the similarity group of a form.

Definition 1.6.14 The *conformal group* C in Series 1.1 is the similarity group of V. Thus we refer to the *conformal symplectic group*, $CSp_n(q)$; the *conformal unitary group*, $CGU_n(q)$; and the *conformal orthogonal group*, $CGO_n^\varepsilon(q)$.

The conformal linear group is equal to the general linear group, and we will prefer the term general linear group. It will follow from Lemma 1.8.9 that if $n \geqslant 2$, or $n \geqslant 3$ when the form is bilinear, then the conformal group is equal to the normaliser in the corresponding linear group of the quasisimple group.

Definition 1.6.15 We shall call the subgroup of the outer automorphism group of Ω that is induced by conjugation by elements of C the *group of conformal automorphisms* of Ω.

Now we consider the next group up in the series.

Definition 1.6.16 The *conformal semilinear group* of V is the group of all semi-similarities of V. These groups are denoted by $\Gamma L_n(q)$, $C\Gamma Sp_n(q)$, $C\Gamma U_n(q)$, and $C\Gamma O_n^\varepsilon(q)$. The conformal semilinear group is denoted by Γ whenever the context makes the meaning of this clear, and in particular in Series 1.1.

Let $\phi : V \to V$ be as in Definition 1.6.3. If the general group preserves a form with matrix B, and all entries of B are in $\mathbb{F}_p$, then $B^\phi = B$, and so ϕ normalises the conformal group C. In that case, one may prove that the group Γ is a semidirect product of C and $\langle \phi \rangle$. This applies to all of our standard forms except for some bilinear and quadratic forms of minus type.

It was shown in [6] that the element of Out Ω determined by ϕ can sometimes depend on the choice of fixed form. We specified our standard forms in Table 1.1, and are using them throughout this section. Recall from Subsections 1.5.2 to 1.5.6 that our standard forms are not always the same as the forms fixed by the groups returned by the corresponding MAGMA functions. For example, the group that we shall denote $U_6(3).\langle \phi \rangle$ (which is sometimes denoted $P\Sigma U_6(3)$) is not isomorphic to the group returned by the MAGMA function PSigmaU(6,3).

Definition 1.6.17 The *semilinear group* of V is the group of all of semi-isometries of the form. For our standard forms, these groups are denoted by $\Gamma L_n(q)$, $\Gamma Sp_n(q)$, $\Gamma U_n(q)$, and $\Gamma O_n^\varepsilon(q)$.

Note that, although the names of these groups begin with Γ, they are not the groups Γ in the Series 1.1, which are the groups defined in Definition 1.6.16. For the linear groups, there is one further distinct group in Series 1.1.

Definition 1.6.18 If β is identically zero and $n \geqslant 3$, then the group A in Series 1.1 is the split extension of $\Gamma L_n(q)$ by the inverse-transpose map $\gamma := -\mathsf{T}$, where γ commutes with the field automorphisms. For all other forms and dimensions, $A := \Gamma$.

When $n = 2$, the inverse-transpose map is induced by an inner automorphism of $L_2(q)$ and of $SL_2(q)$.

We complete this section with a brief discussion of projective groups.

Definition 1.6.19 For each of the groups that we have defined in this section, we also define a projective version, which is the quotient by scalar matrices. We denote this either with a prefix P, as in $PGL_n(q)$, or with an overbar, as in $\overline{\Omega}$. The simple classical groups also have ATLAS-style notation: see Table 1.2.

Definition 1.6.19 yields a second chain of subgroups:

$$\overline{\Omega} \leqslant \overline{S} \leqslant \overline{G} \leqslant \overline{C} \leqslant \overline{\Gamma} \leqslant \overline{A}. \tag{1.2}$$

We will see in Section 1.7 that if the form is unitary, symplectic over $\mathbb{F}_{2^e}$, or orthogonal in odd dimension, then the conformal group is obtained by adjoining scalars to the general group, in which case the projective versions of these two groups are the same. We shall defer further discussion of the indices of these groups in one another until Table 1.3.

The following notation will be used extensively throughout the book.

Definition 1.6.20 Let κ be a standard form from Table 1.1. Let H be any group such that $\Omega \leqslant H \leqslant A$, or $\overline{\Omega} \leqslant H \leqslant \overline{A}$. If κ is identically zero and $n \geqslant 2$, then H lies in *Case* **L**. If κ is unitary and $n \geqslant 3$, then H lies in *Case* **U**. If κ is symplectic and $n \geqslant 4$, then H lies in *Case* **S**. If κ is symmetric bilinear or quadratic and $n \geqslant 7$, then H lies in *Case* **$\mathbf{O}^\varepsilon$**.

But note that we sometimes consider $\mathrm{SL}_2(q)$ to lie in Case **S**, since $\mathrm{SL}_2(q) = \mathrm{Sp}_2(q)$ by Lemma 1.12.1. We will always state when we are doing this.

The following is classical; for a textbook reference see [10, Chapter 12].

Theorem 1.6.21 *Assume that $\overline{\Omega}$ is simple. Then $\overline{A} = \mathrm{Aut}\,\overline{\Omega}$, except when $\Omega = \mathrm{Sp}_4(2^e)$ or $\Omega = \Omega_8^+(q)$.*

As we shall see in Section 1.7, there is a graph automorphism that squares to a generating field automorphism in Case **S** when $n = 4$ and q is even (note also that $S_4(2) \cong S_6$ is not simple). There is a graph automorphism of order 3 in Case $\mathbf{O}^+$ when $n = 8$. When q is odd, this is an automorphism of the projective group $\mathrm{O}_8^+(q)$, but not of $\Omega_8^+(q)$.

1.6.3 Notation for the classical groups

Our notation for the classical groups is summarised in Table 1.2. (Although our usage of 'Case **L**', 'Case **S**' normally implies the restrictions on the dimensions described in the previous subsection, we use the notation in this table for all $n \geqslant 1$.) Unfortunately there is a lack of consistency in the literature for this notation. For example, $\mathrm{GO}_n^\varepsilon(q)$ is used with different meanings in [12] and in [66]. Our notation is closer to that in [12], but we introduce some new symbols such as $\mathrm{CGO}_n^\varepsilon(q)$.

Table 1.2 *Notation for the classical groups*

Case	Ω	S	G	C	Γ	A
L	$\mathrm{SL}_n(q)$	$\mathrm{SL}_n(q)$	$\mathrm{GL}_n(q)$	$\mathrm{GL}_n(q)$	$\Gamma\mathrm{L}_n(q)$	$\Gamma\mathrm{L}_n(q){:}\langle\gamma\rangle$ [†]
	$\mathrm{L}_n(q)$	$\mathrm{L}_n(q)$	$\mathrm{PGL}_n(q)$	$\mathrm{PGL}_n(q)$	$\mathrm{P}\Gamma\mathrm{L}_n(q)$	$\mathrm{P}\Gamma\mathrm{L}_n(q){:}\langle\gamma\rangle$ [†]
U	$\mathrm{SU}_n(q)$	$\mathrm{SU}_n(q)$	$\mathrm{GU}_n(q)$	$\mathrm{CGU}_n(q)$	$\mathrm{C}\Gamma\mathrm{U}_n(q)$	$\mathrm{C}\Gamma\mathrm{U}_n(q)$
	$\mathrm{U}_n(q)$	$\mathrm{U}_n(q)$	$\mathrm{PGU}_n(q)$	$\mathrm{PGU}_n(q)$	$\mathrm{P}\Gamma\mathrm{U}_n(q)$	$\mathrm{P}\Gamma\mathrm{U}_n(q)$
S	$\mathrm{Sp}_n(q)$	$\mathrm{Sp}_n(q)$	$\mathrm{Sp}_n(q)$	$\mathrm{CSp}_n(q)$	$\mathrm{C}\Gamma\mathrm{Sp}_n(q)$	$\mathrm{C}\Gamma\mathrm{Sp}_n(q)$
	$\mathrm{S}_n(q)$	$\mathrm{S}_n(q)$	$\mathrm{S}_n(q)$	$\mathrm{PCSp}_n(q)$	$\mathrm{PC}\Gamma\mathrm{Sp}_n(q)$	$\mathrm{PC}\Gamma\mathrm{Sp}_n(q)$
$\mathbf{O}^\varepsilon$	$\Omega_n^\varepsilon(q)$	$\mathrm{SO}_n^\varepsilon(q)$	$\mathrm{GO}_n^\varepsilon(q)$	$\mathrm{CGO}_n^\varepsilon(q)$	$\mathrm{C}\Gamma\mathrm{O}_n^\varepsilon(q)$	$\mathrm{C}\Gamma\mathrm{O}_n^\varepsilon(q)$
	$\mathrm{O}_n^\varepsilon(q)$	$\mathrm{PSO}_n^\varepsilon(q)$	$\mathrm{PGO}_n^\varepsilon(q)$	$\mathrm{PCGO}_n^\varepsilon(q)$	$\mathrm{PC}\Gamma\mathrm{O}_n^\varepsilon(q)$	$\mathrm{PC}\Gamma\mathrm{O}_n^\varepsilon(q)$

† The automorphism γ is only defined when $n \geqslant 3$.

Notice that each of the four cases in the table has two lines. The top line gives the notation for the groups in Series 1.1, and the second gives the notation for the groups in Series 1.2. Thus the top group in Column Ω is the generally quasisimple group, as in Definition 1.6.13. The top group in Column S is the special classical group, as in Definition 1.6.6. The top group in Column G is the general classical group, as defined in Definition 1.6.5. The top group in Column C is the conformal classical group, as in Definition 1.6.14. The top group in Column Γ is the conformal semilinear group, as in Definition 1.6.16. The top group in Column A is as in Definition 1.6.18.

1.6.4 Orders of classical groups

The orders of the classical groups are well known; see for example [108, p19, p118, p70, p141] for derivations of the orders of the general groups, and [66, Tables 2.1.C, 2.1.D] (reproduced here as Table 1.3) for a convenient summary of the indices of groups in Series 1.1 and 1.2.

Theorem 1.6.22 *Let $q = p^e$, where p is a prime. The order of $\mathrm{GL}_n(q)$ is*

$$q^{n(n-1)/2} \prod_{i=1}^{n} (q^i - 1).$$

The order of $\mathrm{GU}_n(q)$ is

$$q^{n(n-1)/2} \prod_{i=1}^{n} (q^i - (-1)^i).$$

When n is even, the order of $\mathrm{Sp}_n(q)$ is

$$q^{n^2/4} \prod_{i=1}^{n/2} (q^{2i} - 1).$$

When nq is odd, the order of $\mathrm{GO}_n^\circ(q)$ is

$$2q^{(n-1)^2/4} \prod_{i=1}^{(n-1)/2} (q^{2i} - 1).$$

When n is even, the order of $\mathrm{GO}_n^\pm(q)$ is

$$2q^{n(n-2)/4}(q^{n/2} \mp 1) \prod_{i=1}^{n/2-1} (q^{2i} - 1).$$

For any group H in Table 1.2, the order of H can be calculated from Theorem 1.6.22 and the information in Table 1.3.

Table 1.3 *Indices of classical groups*

Case	$\|S : \Omega\|$	$\|G : S\|$	$\|C : G\|$	$\|\Gamma : C\|$	$\|A : \Gamma\|$
L	1	$q - 1$	1	e	2 *
U	1	$q + 1$	$q - 1$	$2e$	1
S	1	1	$q - 1$	e	1
O°	2 †	2	$(q - 1)/2$	e	1
O$^\pm$	2	$(2, q - 1)$	$q - 1$	e	1

Case	$\|G \cap Z(\mathrm{GL}_n(q^u))\|$	$\|\overline{S} : \overline{\Omega}\|$	$\|\overline{G} : \overline{S}\|$	$\|\overline{C} : \overline{G}\|$	$\|\overline{\Gamma} : \overline{C}\|$	$\|\overline{A} : \overline{\Gamma}\|$
L	$q - 1$	1	$(q - 1, n)$	1	e	2 *
U	$q + 1$	1	$(q + 1, n)$	1	$2e$	1
S	$(2, q - 1)$	1	1	$(2, q - 1)$	e	1
O°	2	2 †	1	1	e	1
O$^\pm$	$(2, q - 1)$	$a_\pm$ ‡	$(2, q - 1)$	$(2, q - 1)$	e	1

Note the restrictions on n and q in Theorem 1.6.22.

∗ If $n \in \{1, 2\}$ then $A = \Gamma$.
† If $n = 1$ then $S = \Omega$.
‡ The entries a_+ and a_- are defined by: $a_\pm \in \{1, 2\}$; $a_+ a_- = 2^{(2,q)}$; if q is odd then $a_+ = 2$ if and only if $n(q - 1)/4$ is even.

1.7 Outer automorphisms of classical groups

In this section we will introduce our standard notation for outer automorphisms of their classical groups Ω and $\overline{\Omega}$, and give presentations for $\mathrm{Out}\,\overline{\Omega}$. See Section 1.6 for an introduction to the classical groups.

1.7.1 Standard outer automorphisms

Let Ω be as in Definition 1.6.13, and let $\overline{\Omega}$ be the generally simple group $\Omega/Z(\Omega)$ (see Proposition 1.10.3 for exactly when $\overline{\Omega}$ is simple). Let B be the matrix of one of the standard forms in Table 1.1.

We shall use the symbols δ, δ', γ, τ, ϕ and φ to denote generators of the outer automorphism group of $\overline{\Omega}$. We abuse notation and use the same symbols to denote both their inverse images in $\mathrm{Aut}\,\overline{\Omega}$ and, when they exist, specific matrices in $\mathrm{GL}_n(q)$ that induce them by conjugation. Presentations of $\mathrm{Out}\,\overline{\Omega}$ on these generators will be listed in the following subsection.

We shall refer to δ and δ' as *diagonal* automorphisms, γ and τ as *graph* automorphisms, and ϕ or φ as a *field* automorphism. These names correspond to the terminology used in the theory of algebraic groups.

We now proceed to define these automorphisms as specific elements of

Aut $\overline{\Omega}$. Recall Definition 1.6.3 of the automorphism ϕ of $\mathrm{GL}_n(q)$. If $B^\phi = B$, then ϕ induces an automorphism of Ω of order e, which we shall also denote by ϕ. Consulting Table 1.1, we see that $B^\phi \neq B$ if and only if $\Omega = \Omega_n^-(q)$, n is even, and either q is even or q is odd and the form has non-square discriminant. In these exceptional cases, the corresponding automorphism of Ω will be denoted by φ rather than ϕ, where φ is defined to be ϕ followed by conjugation by a fixed element $c \in \mathrm{GL}_n(q)$ with $cBc^\mathsf{T} = B^\phi$.

Case L. Here δ is the diagonal automorphism of $\Omega = \mathrm{SL}_n(q)$ induced by $\mathrm{diag}(\omega, 1, 1, \ldots, 1) \in \mathrm{GL}_n(q)$. Its order in $\mathrm{Out}\,\overline{\Omega}$ is $(q-1, n)$. Projectively, δ extends $\mathrm{L}_n(q)$ to $\mathrm{PGL}_n(q)$. The field automorphism ϕ is as in Definition 1.6.3, and projectively ϕ extends $\mathrm{PGL}_n(q)$ to $\mathrm{P\Gamma L}_n(q)$. For $n \geqslant 3$, γ is the duality (or graph) automorphism $g \mapsto g^{-\mathsf{T}}$ of Ω, as in Definition 1.6.18, and projectively γ extends $\mathrm{P\Gamma L}_n(q)$ to $\mathrm{Aut}\,\mathrm{L}_n(q)$. Note that γ is undefined when $n = 2$, and that the inverse-transpose map is induced by conjugation by an element of $\mathrm{SL}_2(q)$.

Case U. Here δ is the diagonal automorphism of $\Omega = \mathrm{SU}_n(q)$ induced by $\mathrm{diag}(\omega^{q-1}, 1, 1, \ldots, 1) \in \mathrm{GU}_n(q)$, where $\omega = \omega_{q^2}$ is a primitive element of $\mathbb{F}_{q^2}^\times$. Its order in $\mathrm{Out}\,\overline{\Omega}$ is $(q+1, n)$, and projectively δ extends $\mathrm{U}_n(q)$ to $\mathrm{PGU}_n(q)$. For $n \geqslant 3$, the duality or graph automorphism is $\gamma : g \mapsto g^{-\mathsf{T}}$ of Ω. The field automorphism ϕ is as in Definition 1.6.3, of order $2e$ in $\mathrm{Out}\,\overline{\Omega}$, and projectively ϕ extends $\mathrm{PGU}_n(q)$ to $\mathrm{P\Gamma U}_n(q)$. A consequence of our choice of standard form I_n is that $\phi^e = \gamma$.

Case S. When q is odd, δ is the diagonal automorphism of $\Omega = \mathrm{Sp}_n(q)$ induced by $\delta = \mathrm{diag}(\omega, \ldots, \omega, 1, \ldots, 1) \in \mathrm{CSp}_n(q) \setminus \mathrm{Sp}_n(q)$, with $n/2$ ω's and $n/2$ 1's. Its order in $\mathrm{Out}\,\overline{\Omega}$ is 2, and projectively it extends $\mathrm{S}_n(q)$ to $\mathrm{PCSp}_n(q)$. Observe that $\delta B \delta^\mathsf{T} = \omega B$. When q is even, δ is trivial. The field automorphism ϕ is as in Definition 1.6.3, has order e in $\mathrm{Out}\,\overline{\Omega}$, and projectively extends $\mathrm{PCSp}_n(q)$ to $\mathrm{PC\Gamma Sp}_n(q)$, which is equal to $\mathrm{Aut}\,\mathrm{S}_n(q)$ except when $n = 4$ and q is even.

When $n = 4$ and q is even, there is a graph automorphism γ of Ω with $\gamma^2 = \phi$ in $\mathrm{Out}\,\overline{\Omega}$, which projectively extends $\mathrm{PC\Gamma Sp}_4(2^e)$ to $\mathrm{Aut}\,\mathrm{S}_4(2^e)$. We shall define γ precisely in Section 7.2. Note that we exclude $\mathrm{Sp}_4(2) \cong \mathrm{S}_6$, as it is not quasisimple.

Case O°. Recall that nq is odd. Here δ denotes a diagonal automorphism of $\Omega = \Omega_n^\circ(q)$, of order 2 in $\mathrm{Out}\,\overline{\Omega}$, induced by an element of $\mathrm{SO}_n^\circ(q) \setminus \Omega_n^\circ(q)$, which projectively extends $\mathrm{O}_n^\circ(q)$ to $\mathrm{PSO}_n^\circ(q)$. (We can choose $\delta = r_\square r_\boxtimes$ using the notation defined in [66, §2.6].) The field automorphism ϕ is as in Definition 1.6.3, has order e in $\mathrm{Out}\,\overline{\Omega}$, and projectively extends $\mathrm{PCGO}_n^\circ(q)$ to $\mathrm{PC\Gamma O}_n^\circ(q) = \mathrm{Aut}\,\mathrm{O}_n^\circ(q)$.

Case O$^\pm$. Here γ denotes a graph automorphism of $\Omega = \Omega_n^\pm(q)$, of order 2 in $\mathrm{Out}\,\overline{\Omega}$, induced by an element of $\mathrm{GO}_n^\pm(q) \setminus \mathrm{SO}_n^\pm(q)$ when q is odd and by an element of $\mathrm{SO}_n^\pm(q) \setminus \Omega_n^\pm(q)$ when q is even. Projectively, γ extends $\mathrm{PSO}_n^\pm(q)$ to

$\mathrm{PGO}_n^\pm(q)$ when q is odd, and $\mathrm{O}_n^\pm(q)$ to $\mathrm{PSO}_n^\pm(q) = \mathrm{PGO}_n^\pm(q)$ when q is even. (For all q we can choose $\gamma = r_\square$, as in [66, §2.7–2.8].)

When q is odd and B has square discriminant, δ' denotes a diagonal auto-morphism of Ω, of order 2 in $\mathrm{Out}\,\bar\Omega$, induced by an element of $\mathrm{SO}_n^\pm(q) \setminus \Omega_n^\pm(q)$. Projectively, δ' extends $\mathrm{O}_n^\pm(q)$ to $\mathrm{PSO}_n^\pm(q)$. (We can choose $\delta' = r_\square r_\boxtimes$ as in [66, §2.6].) If B has non-square discriminant or q is even then δ' is trivial.

When q is odd, Ω has a diagonal outer automorphism δ. Projectively, δ extends $\mathrm{PGO}_n^\pm(q)$ to $\mathrm{PCGO}_n^\pm(q)$. If $\Omega = \Omega_n^+(q)$, then δ is the automorphism induced by $\delta = \mathrm{diag}(\omega, \ldots, \omega, 1, \ldots, 1) \in \mathrm{CGO}_n^+(q) \setminus \mathrm{GO}_n^+(q)$, with $n/2$ ω's and $n/2$ 1's. Observe that $\delta B \delta^\mathsf{T} = \omega B$ and $\det \delta = \omega^{n/2}$. Therefore $\det \omega^{-1}\delta^2 = 1$, and by Definition 1.6.10 the matrix $\omega^{-1}\delta^2$ has spinor norm 1 when $4 \mid n$ and -1 when $n \equiv 2 \pmod 4$. So δ has order 2 in $\mathrm{Out}\,\mathrm{O}_n^+(q)$ when $4 \mid n$, or when B has non-square discriminant. But δ has order 4 when $n \equiv 2 \pmod 4$ and B has square discriminant. When q is even, δ is trivial.

When q is odd and $\Omega = \Omega_n^-(q)$, choose two elements $a, b \in \mathbb{F}_q^\times$ with $a^2 + b^2 = \omega$, the primitive element of $\mathbb{F}_q^\times$, and let

$$X = \begin{pmatrix} a & b \\ -b & a \end{pmatrix} \quad \text{and} \quad Y = \begin{pmatrix} 0 & \omega \\ -1 & 0 \end{pmatrix}.$$

Then δ is the diagonal automorphism of Ω induced by $\delta = \mathrm{diag}(X, \ldots, X)$, (with $n/2$ X's) when B has square discriminant, and by $\delta = \mathrm{diag}(Y, X, \ldots, X)$ (with one Y and $n/2 - 1$ X's) when B has non-square discriminant. In both cases, $\delta \in \mathrm{CGO}_n^-(q) \setminus \mathrm{GO}_n^-(q)$ with $\delta B \delta^\mathsf{T} = \omega B$ and $\det \delta = \omega^{n/2}$, which was also true for the δ defined for $\Omega_n^+(q)$.

Observe also that, if we define $C = \mathrm{diag}(1, -1, 1, -1, \ldots, 1, -1) \in \mathrm{GO}_n^-(q)$, then $(\delta C)^2 = \omega I_n$, so δC has order 2 in $\mathrm{Out}\,\bar\Omega$. If $4 \mid n$ then $C \in \Omega_n^-(q)$ by Definition 1.6.10, so δ also has order 2 in $\mathrm{Out}\,\bar\Omega$. If $n \equiv 2 \pmod 4$ then $C \in \mathrm{GO}_n^-(q) \setminus \mathrm{SO}_n^-(q)$. From the fact [66, Proposition 2.8.2] that $\mathrm{PCGO}_n^\pm(q)/\Omega_n^\pm(q)$ is isomorphic to D_8 and $C_2 \times C_2$ when the the discriminant of B is square and non-square, respectively, it follows that δ has order 4 in $\mathrm{Out}\,\Omega$ when B has square discriminant and order 2 when B has non-square discriminant. This again corresponds to the order of $\delta \in \mathrm{Out}\,\Omega_n^+(q)$. Note that our definition of δ for $\Omega_n^-(q)$ is different from that of [66]; in fact our δ is equal to their δC.

The field automorphism ϕ of Ω is induced by the map ϕ as in Definition 1.6.3 when $\Omega = \Omega_n^+(q)$, or when $\Omega = \Omega_n^-(q)$, q is odd, and B has square discriminant; it is undefined otherwise. Projectively, ϕ extends $\mathrm{PCGO}_n^\pm(q)$ to $\mathrm{PC\Gamma O}_n^\pm(q)$. For the cases when the ϕ is undefined as an automorphism of Ω, we proceed to define a field automorphism φ of Ω.

When $\Omega = \Omega_n^-(q)$, q is odd, and B has non-square discriminant, the matrix $c := \mathrm{diag}(\omega^{(p-1)/2}, 1, 1, \ldots, 1)$ satisfies $c B c^\mathsf{T} = B^\phi$, and so c conjugates Ω^ϕ back to Ω. We define the field automorphism φ of Ω to be ϕ followed by conjugation by c, so φ can also be regarded as the ϕ-semilinear map ϕc and, as such, it is a semi-isometry of B. It can be checked that φ^e is induced by a matrix in $\mathrm{GO}_n^-(q)$ with determinant -1, so $\varphi^e = \gamma$ in Out $\overline{\Omega}$. Projectively, φ extends $\mathrm{PCGO}_n^-(q)$ to $\mathrm{P}\Gamma\mathrm{O}_n^-(q)$, but it is not always a split extension.

Similarly, when $\Omega = \Omega_n^-(q)$ with q even, we define φ to be ϕ followed by conjugation by a certain matrix c. We shall not need to carry out calculations with c in this case, so we shall not define it precisely, and refer the reader to [66, §2.8] for a precise definition of φ. Again, $\varphi^e = \gamma$ in Out $\overline{\Omega}$, and projectively φ extends $\mathrm{PCGO}_n^-(q) = \mathrm{PGO}_n^-(q)$ to $\mathrm{P}\Gamma\mathrm{O}_n^-(q)$.

Finally, if $\Omega = \Omega_8^+(q)$ then τ denotes a graph automorphism of order 3 in Out $\overline{\Omega}$ that is inverted by γ, extending $\mathrm{P}\Gamma\mathrm{O}_8^+(q)$ to Aut $\mathrm{O}_8^+(q)$. Note that τ does not lift to an automorphism of Ω when q is odd, and is undefined except for $\Omega = \Omega_8^+(q)$.

1.7.2 Presentations of outer automorphism groups of $\overline{\Omega}$

In each of these presentations, $q = p^e$ with p prime. All of the relations in these presentations can be readily derived from the information presented in the previous subsection, with the possible exception of the relation $[\delta, \phi] = 1$ in $\mathrm{O}_n^-(q)$ when q is odd and the discriminant is square. But in that case e must be odd, so ϕ has odd order. It normalises the subgroup $\langle \gamma, \delta \rangle$, which is dihedral of order 8. Since D_8 has no non-trivial odd order automorphisms, ϕ must centralise $\langle \gamma, \delta \rangle$.

$\mathrm{L}_2(q) = \mathrm{S}_2(q)$:

$$\langle \delta, \phi \mid \delta^{(q-1,2)} = \phi^e = [\delta, \phi] = 1, \rangle.$$

$\mathrm{L}_n(q)$, $n \geqslant 3$:

$$\langle \delta, \gamma, \phi \mid \delta^{(q-1,n)} = \gamma^2 = \phi^e = [\gamma, \phi] = 1, \, \delta^\gamma = \delta^{-1}, \, \delta^\phi = \delta^p \rangle.$$

$\mathrm{U}_n(q)$, $n \geqslant 3$:

$$\langle \delta, \gamma, \phi \mid \delta^{(q+1,n)} = \gamma^2 = 1, \, \phi^e = \gamma, \, \delta^\gamma = \delta^{-1}, \, \delta^\phi = \delta^p \rangle.$$

$\mathrm{S}_n(q)$, $n \geqslant 2$, $(n,p) \neq (4,2)$:

$$\langle \delta, \phi \mid \delta^{(q-1,2)} = \phi^e = [\delta, \phi] = 1 \rangle.$$

$\mathrm{S}_4(p^e)$, $p = 2$:

$$\langle \gamma, \phi \mid \gamma^2 = \phi, \, \phi^e = 1 \rangle.$$

$\mathrm{O}_n^\circ(q)$, $n \geqslant 3$ odd:

$$\langle \delta, \phi \mid \delta^2 = \phi^e = [\delta, \phi] = 1 \rangle.$$

$\mathrm{O}_n^+(q)$, $n \geqslant 6$ even, $n \neq 8$, q even:

$$\langle \gamma, \phi \mid \gamma^2 = \phi^e = [\gamma, \phi] = 1 \rangle.$$

$\mathrm{O}_8^+(q)$, q even:

$$\langle \tau, \gamma, \phi \mid \tau^3 = \gamma^2 = (\gamma\tau)^2 = \phi^e = [\tau, \phi] = [\gamma, \phi] = 1 \rangle.$$

$\mathrm{O}_n^-(q)$, $n \geqslant 4$ even, q even:

$$\langle \gamma, \varphi \mid \gamma^2 = 1, \ \varphi^e = \gamma \rangle.$$

$\mathrm{O}_8^+(q)$, q odd:

$$\langle \delta', \tau, \gamma, \delta, \phi \mid \delta'^2 = \tau^3 = \gamma^2 = (\gamma\tau)^2 = \delta^2 = 1, \ \delta^\tau = \delta', \delta'^\tau = \delta\delta',$$
$$(\delta\gamma)^2 = \delta', \phi^e = [\delta, \phi] = [\tau, \phi] = [\gamma, \phi] = 1 \rangle.$$

$\mathrm{O}_n^+(q)$, $n \geqslant 12$, $4 \mid n$, q odd:

$$\langle \delta', \gamma, \delta, \phi \mid \delta'^2 = \gamma^2 = \delta^2 = 1, \ (\delta\gamma)^2 = \delta', \phi^e = [\delta, \phi] = [\gamma, \phi] = 1 \rangle.$$

$\mathrm{O}_n^+(q)$, $n \geqslant 6$, $n \equiv 2 \pmod 4$, $q \equiv 1 \pmod 4$:

$$\langle \delta', \gamma, \delta, \phi \mid \delta'^2 = \gamma^2 = 1, \ \delta^2 = \delta', \ \delta^\gamma = \delta^{-1}, \phi^e = [\gamma, \phi] = 1, \ \delta^\phi = \delta^p \rangle.$$

$\mathrm{O}_n^+(q)$, $n \geqslant 6$, $n \equiv 2 \pmod 4$, $q \equiv 3 \pmod 4$:

$$\langle \gamma, \delta, \phi \mid \gamma^2 = \delta^2 = [\delta, \gamma] = \phi^e = [\gamma, \phi] = [\delta, \phi] = 1 \rangle.$$

$\mathrm{O}_n^-(q)$, $n \geqslant 4$, $4 \mid n$ or $q \equiv 1 \pmod 4$, q odd:

$$\langle \gamma, \delta, \varphi \mid \gamma^2 = \delta^2 = [\delta, \gamma] = [\delta, \varphi] = 1, \ \varphi^e = \gamma \rangle.$$

$\mathrm{O}_n^-(q)$, $n \geqslant 4$, $n \equiv 2 \pmod 4$, $q \equiv 3 \pmod 4$:

$$\langle \delta', \gamma, \delta, \phi \mid \delta'^2 = \gamma^2 = 1, \ \delta^2 = \delta', \delta^\gamma = \delta^{-1}, \phi^e = [\gamma, \phi] = [\delta, \phi] = 1 \rangle.$$

1.8 Representation theory

This section contains an assortment of results from representation theory that we shall need. We remind the reader that our default assumption throughout the book is that groups are finite. We are mainly interested in representations over finite fields, but we shall at times need to consider characteristic 0 representations. The reader unfamiliar with representation theory should consult Isaac's excellent textbook [56].

A representation of G over a field F is by definition a homomorphism $\rho : G \to \mathrm{GL}_n(F)$ for some n. We shall generally refer to n as the *dimension* of ρ (although it is more usually called the *degree* of ρ).

Definition 1.8.1   Representations $\rho, \rho' : G \to \mathrm{GL}_n(F)$ are *equivalent* if there exists $x \in \mathrm{GL}_n(F)$ with $x^{-1}(g\rho')x = g\rho$ for all $g \in G$.

Representations have associated (right) FG-modules, where equivalent representations correspond to isomorphic modules.

A representation $\rho : G \to \mathrm{GL}_n(F)$ is *faithful* if its kernel is trivial. It is *irreducible* if $\rho(G)$ stabilises no proper non-zero subspace of F^n, and is *reducible* otherwise. A representation ρ is *absolutely irreducible* if the natural action of $\rho(G)$ on K^n is irreducible for every extension K of F. The same terms are used for the corresponding FG-module. Schur's Lemma states that if $\rho : G \to \mathrm{GL}_n(F)$ is absolutely irreducible, then the centraliser of $\rho(G)$ in $\mathrm{GL}_n(F)$ consists just of the scalar matrices in $\mathrm{GL}_n(F)$. If ρ is irreducible but not absolutely irreducible and F is finite, then the centraliser is isomorphic to $K^\times$, for some proper extension K of F.

A *projective representation* is a homomorphism $\rho : G \to \mathrm{PGL}_n(F)$. A projective representation lifts to an ordinary representation $\hat{\rho} : \hat{G} \to \mathrm{GL}_n(F)$ for some central extension $\hat{G}$ of G, and we say that ρ is irreducible, etc. if $\hat{\rho}$ is.

1.8.1 Dual modules

Let F be a field, A an associative unital F-algebra, and V a right A-module. We make the dual vector space V^* of V into a left A-module, by defining $v(af) := (va)f$ for $v \in V$, $a \in A$ and $f \in V^*$. The *opposite algebra* A^{op} has the same underlying set, addition and F-multiplication as A, but the product $a \circ b$ in A^{op} is the same as the product ba in A. We make V^* into a right A^{op}-module by defining fa to be af, where $f \in V^*$ in both cases, a is regarded as an element of A^{op} in the first case, and as an element of A in the second.

We now let V be finite dimensional with basis $(e_1, \ldots, e_n)$, and let V^* have dual basis $(e_1^*, \ldots, e_n^*)$ where $e_i\, e_j^* = \delta_{ij}$. It is a routine exercise to show that, if $a \in A$ has matrix M in its action on V with respect to the basis $(e_1, \ldots, e_n)$, then $a \in A^{\mathrm{op}}$ has matrix M^T with respect to the basis $(e_1^*, \ldots, e_n^*)$.

If $A = FG$ is a group algebra, then the map $g \mapsto g^{-1}$ extended F-linearly gives an isomorphism between A and A^{op}. We use this isomorphism to make V^* into a right FG-module which, for the purposes this book, we take as the definition of the dual module and its associated representation (although it would be more standard to call it the *contragredient module*).

Definition 1.8.2   Let $\rho : G \to \mathrm{GL}_n(F)$ be a representation with associated FG-module V. Then the dual right FG-module V^*, with associated representation ρ^*, is defined by

$$v(f(g\rho^*)) := (v(g^{-1}\rho))f$$

for $v \in V$, $g \in G$ and $f \in V^*$. When there is no possibility for confusion, we shall suppress the representations and write simply

$$v(fg) = (vg^{-1})f.$$

We shall make frequent use of the following well-known result.

Proposition 1.8.3 *Let $\rho : G \to \mathrm{GL}_n(F)$ be a representation with associated FG-module V. Then $g\rho^* = (g\rho)^{-\mathsf{T}}$, with respect to the dual basis of the natural basis of V.*

Proof This follows from Definition 1.8.2, and the fact that matrices are transposed in their action on the dual space under the opposite algebra. $\qquad\square$

1.8.2 Actions of group and field automorphisms on representations

Let $\rho : G \to \mathrm{GL}_n(q)$ be a representation of G. For $\alpha \in \mathrm{Aut}\, G$, we define the representation $^\alpha\rho : G \to \mathrm{GL}_n(q)$ by $g(^\alpha\rho) = (g^\alpha)\rho$ for $g \in G$. (So ρ and $^\alpha\rho$ have the same image.) If $\alpha \in \mathrm{Inn}\, G$ then $^\alpha\rho$ is equivalent to ρ, so the action of $\mathrm{Aut}\, G$ on the representations of G induces a left action of $\mathrm{Out}\, G$ on the set of equivalence classes of representations of G.

Definition 1.8.4 Two representations $\rho, \rho' : G \to \mathrm{GL}_n(F)$ are said to be *quasi-equivalent* if there exists $\alpha \in \mathrm{Aut}\, G$ such that $^\alpha\rho$ is equivalent to ρ'.

Analogously, for an automorphism θ of $\mathrm{GL}_n(q)$ (or, more generally, for an automorphism θ of a classical group containing $\mathrm{Im}(\rho)$), we define $\rho^\theta : G \to \mathrm{GL}_n(q)$ by $g(\rho^\theta) = (g\rho)^\theta$. This induces a right action of $\mathrm{Out}\,\mathrm{GL}_n(q)$ on the set of equivalence classes of representations of G. We are primarily concerned with the cases when θ is the duality (inverse-transpose) automorphism or a field automorphism ϕ of $\mathrm{GL}_n(q)$ (or of a classical group containing $\mathrm{Im}(\rho)$).

Definition 1.8.5 Let ρ be a representation of G, with $\mathrm{Im}\,\rho$ a subgroup of a classical group C. Let $\alpha \in \mathrm{Aut}\, G$ and let $\theta \in \mathrm{Aut}\, C$. We say that α or θ *stabilises* ρ if $^\alpha\rho$ or ρ^θ is equivalent to ρ.

Lemma 1.8.6 *Let $\rho, \rho' : G \to \mathrm{GL}_n(F)$ be faithful and let $\alpha \in \mathrm{Aut}\, G$. Then:*

(i) *ρ and ρ' are quasi-equivalent if and only if they have conjugate images in $\mathrm{GL}_n(F)$.*

(ii) *α stabilises ρ if and only if there exists $g \in \mathrm{GL}_n(F)$ that normalises, and induces α on, $\mathrm{Im}(\rho)$.*

Proof (i) Suppose that ρ and ρ' are quasi-equivalent, and hence ρ' and $^\theta\rho$ are equivalent for some $\theta \in \text{Aut } G$. Then $\text{Im}(\rho')$ is conjugate in $\text{GL}_n(F)$ to $\text{Im}(^\theta\rho) = \text{Im}(\rho)$.

Conversely, suppose that $x^{-1}\text{Im}(\rho')x = \text{Im}(\rho)$ with $x \in \text{GL}_n(F)$. Since ρ is faithful, it has an inverse on its image. Then the map $\mu : G \to G$ defined by $g^\mu = (x^{-1}(g\rho')x)\rho^{-1}$ lies in $\text{Aut } G$, so ρ' is equivalent to $^\mu\rho$.

(ii) The map α stabilises ρ if and only if there exists $x \in \text{GL}_n(F)$ with $x^{-1}(g\rho)x = (g^\alpha)\rho$, which says that x normalises and induces α on $\text{Im}(\rho)$. □

We shall apply the same terminology to the associated FG-modules of representations. So we have induced a left action of $\text{Out } G$ and a right action of $\text{Out } \text{GL}_n(F)$ on the isomorphism classes of FG-modules of dimension n over F, and we can talk of an automorphism stabilising a module as well as a representation.

1.8.3 Representations that preserve forms

Recall Definition 1.5.4 of the isometry and similarity groups of forms. Recall also Definition 1.4.4 of a perfect field.

Definition 1.8.7 Let G be a subgroup of $\text{GL}_n(F)$, and let β be a non-degenerate symplectic, unitary, orthogonal or quadratic form. We say that G *preserves β up to scalars* if G is a subgroup of the similarity group of β, and that G *preserves β* (or *preserves β absolutely*, for emphasis) if G is a subgroup of the isometry group of β.

Lemma 1.8.8 *Let F be a field, and let G be an absolutely irreducible subgroup of $\text{GL}_n(F)$.*

(i) *If G preserves a non-degenerate quadratic, bilinear or σ-Hermitian form up to scalars, and G is perfect, then G preserves that form absolutely.*

(ii) *Up to multiplication of the form by a scalar, G preserves at most one bilinear form, at most one σ-Hermitian form for a given σ, at most one quadratic form when $\text{char } F$ is not 2, and at most one quadratic form when F is perfect.*

(iii) *If $F = \mathbb{F}_q$ is finite, and G simultaneously preserves a σ-Hermitian and a bilinear form, then G is conjugate to a subgroup of $\text{GL}_n(q_0)$ for some proper subfield $\mathbb{F}_{q_0}$ of $\mathbb{F}_q$.*

Proof (i) If $g \in G$ scales a given form by $\lambda_g \in F^\times$, the multiplicative group of F, then the map $g \mapsto \lambda_g$ is a homomorphism from G into $F^\times$, so G perfect implies that $\lambda_g = 1$ for all g.

(ii) Let B_1, B_2 be the matrices of two bilinear or two σ-Hermitian forms of which G is a group of isometries, as in Definition 1.5.16. Since G is absolutely irreducible, each form is non-degenerate, so by Proposition 1.5.25 each B_i has non-zero determinant. Then $B_1 B_2^{-1}$ is a matrix commuting with G, whence by Schur's lemma it is a scalar, as G is absolutely irreducible.

Let Q_1, Q_2 be two quadratic forms of which G is a group of isometries: as before, since G is absolutely irreducible each form is non-degenerate. By Proposition 1.5.15, quadratic forms correspond to symmetric bilinear forms when char $F \neq 2$, so we may assume that char $F = 2$. By the preceding paragraph, the bilinear forms associated with Q_1 and Q_2 differ by a scalar multiple, and hence, for some scalar λ, the quadratic form $Q := Q_1 + \lambda Q_2$ satisfies $Q(v_1 + v_2) = Q(v_1) + Q(v_2)$ for all vectors v_1, v_2. Let v_1, v_2 be linearly independent vectors (the result is trivially true for $n = 1$). If $Q(v_1) \neq 0$ then, since F is assumed perfect, all of its elements have square roots, and we can multiply v_1 by a scalar to get $Q(v_1) = 1$. Similarly, if $Q(v_2) \neq 0$ we may assume $Q(v_2) = 1$ and then $Q(v_1 + v_2) = 0$. So in any case there exists a non-zero vector v with $Q(v) = 0$. But the set of all vectors with $Q(v) = 0$ is a G-invariant subspace, so $f = 0$, which completes the proof of (ii).

(iii) Suppose that G preserves both a non-degenerate bilinear form with matrix B_1 and a non-degenerate σ-Hermitian form with matrix B_2. Then B_2 is invertible, and so the associated representation ρ is equivalent under $B_1 B_2^{-1}$ to ρ^σ, where σ is the involutory automorphism of $\mathbb{F}_q$. Then, by Corollary 1.8.14, G can be written over a proper subfield of $\mathbb{F}_q$. $\square$

Lemma 1.8.9 *Let $G \leqslant \mathrm{GL}_n(F)$ be an absolutely irreducible group consisting of isometries of a non-degenerate quadratic, bilinear or σ-Hermitian form β. If β is quadratic and char $F = 2$, assume also that F is perfect. Then $\mathrm{N}_{\mathrm{GL}_n(F)}(G)$ consists of similarities of β.*

Proof Let B be the matrix of β, as in Definition 1.5.16 or 1.5.17. If $h \in \mathrm{N}_{\mathrm{GL}_n(F)}(G)$ then $G = G^h$ is a group of isometries of $hBh^{\sigma\mathsf{T}}$ (or $(hBh^\mathsf{T})^{\mathrm{UT}}$ for a quadratic form), which must be a scalar multiple of B by Lemma 1.8.8 (ii), so h is a similarity of B. $\square$

Recall the definitions of the classical groups from Subsection 1.6.2.

Lemma 1.8.10   (i) *Let G and H be two absolutely irreducible subgroups of $\mathrm{GU}_n(q)$ that are conjugate in $\mathrm{GL}_n(q^2)$. Then G and H are conjugate in $\mathrm{GU}_n(q)$.*

(ii) *Let G and H be two absolutely irreducible subgroups of $\mathrm{Sp}_n(q)$ or $\mathrm{GO}_n^\varepsilon(q)$ that are conjugate in $\mathrm{GL}_n(q)$. Then G and H are conjugate in $\mathrm{CSp}_n(q)$ or $\mathrm{CGO}_n^\varepsilon(q)$, respectively.*

Proof   (i) Let $\sigma : x \mapsto x^q$ be the involutory field automorphism of $\mathbb{F}_{q^2}$. Let B

be the matrix of the non-degenerate σ-Hermitian form for which G is a group of isometries and suppose that $a^{-1}Ga = H$ with $a \in \mathrm{GL}_n(q^2)$. Then H is a group of isometries of $aBa^{\sigma\mathsf{T}}$ and thus, by Lemma 1.8.8 (ii), $aBa^{\sigma\mathsf{T}} = \lambda B$ for some $\lambda \in \mathbb{F}_{q^2}$. In fact, by Lemma 1.5.20 the matrix $aBa^{\sigma\mathsf{T}}$ is σ-Hermitian, so $\lambda \in \mathbb{F}_q$. Therefore $a \in \mathrm{CGU}_n(q)$. But $\mathrm{CGU}_n(q) = \langle Z, \mathrm{GU}_n(q) \rangle$ where Z is the group of scalars of $\mathrm{GL}_n(q^2)$, and so conjugacy in $\mathrm{GU}_n(q)$ follows.

(ii) Let B be a matrix of a non-degenerate symmetric or anti-symmetric bilinear form for which G and H are groups of isometries, so that B is unique up to multiplication by non-zero scalars. Now $H = aGa^{-1}$ for some $a \in \mathrm{GL}_n(q)$. So H is a group of isometries of aBa^{T} and thus $aBa^{\mathsf{T}} = \lambda B$ for some $\lambda \in \mathbb{F}_q$. Therefore $a \in \mathrm{CSp}_n(q)$ (respectively $\mathrm{CGO}_n^{\varepsilon}(q)$).   □

1.8.4 Other results

The following result is well-known, but we could not find a convenient reference.

Lemma 1.8.11   *Let V be a G-module, and suppose that V decomposes as a direct sum $V = V_1 \oplus \cdots \oplus V_t$ of irreducible G-submodules. Assume further that the V_i are pairwise nonisomorphic. Then the V_i are the only non-zero irreducible G-submodules of V.*

Proof   If $t = 1$ then V is irreducible and the result is clear, so let $U_i = \oplus_{j \neq i} V_j$ and let W be an irreducible G-submodule of V.

If $W \nleq U_i$ for some i, then since $V/U_i \cong V_i$ is irreducible it follows that $\langle U_i, W \rangle = V$. Also, $W \cap U_i$ is a G-submodule of W, so $W \cap U_i = \{0\}$. Thus $W \cong V/U_i \cong V_i$. Thus if $W \not\cong V_i$ for all i then $W \leq U_i$ for all i, so $W \leqslant \cap_i U_i = \{0\}$, a contradiction.

So $W \cong V_i$ for some i. If $W \cap V_i = \{0\}$, then W is isomorphic to an irreducible submodule of $V/V_i \cong U_i$, which is a contradiction, because by hypothesis none of the summands of U_i are isomorphic to V_i. So $W = V_i$.   □

Proposition 1.8.12 ([19, Theorem 29.7])   *Let $F \leqslant E$ be fields and $\rho, \rho' : G \to \mathrm{GL}_n(F)$ be representations. If ρ and ρ' are equivalent as representations over E, then they are equivalent over F.*

Proposition 1.8.13 ([20, Theorem 74.9])   *Let $\rho : G \to \mathrm{GL}_n(E)$ be an absolutely irreducible representation of G, where $\mathrm{char}\, E = p > 0$. Let $F = \langle \mathbb{F}_p, \mathrm{tr}(g\rho) : g \in G \rangle \leqslant E$ be the field of traces. Then ρ is equivalent (over E) to a representation with image in $\mathrm{GL}_n(F)$.*

Corollary 1.8.14   *Let ρ and E be as in Proposition 1.8.13. If ρ is equivalent to ρ^{ϕ} for an automorphism ϕ of F, then ρ is equivalent to a representation with image in $\mathrm{GL}_n(K)$, where K is the fixed field of ϕ.*

Proof Since $\mathrm{tr}(g\rho^\phi) = (\mathrm{tr}(g\rho))^\phi$, this follows from Proposition 1.8.13. $\qquad\square$

1.9 Tensor products

In this section we collect some elementary properties of tensor products of linear maps, matrices, modules and groups. Throughout the section, let V_1 and V_2 be vector spaces over a field F.

Let $\tau_i : V_i \to V_i$ $(i = 1, 2)$ be linear maps. Then the linear map $\tau_1 \otimes \tau_2 :$ $V_1 \otimes V_2 \to V_1 \otimes V_2$ is defined by putting $(v_1 \otimes v_2)(\tau_1 \otimes \tau_2) = v_1\tau_1 \otimes v_2\tau_2$ and extending additively. If the τ_i are both σ-semilinear maps (Definition 1.6.1) for the same σ, then the same definition yields a well-defined σ-semilinear map $\tau_1 \otimes \tau_2$.

A semilinear map $\tau_1 \otimes \tau_2$ defined in this way *preserves* the tensor decomposition $V_1 \otimes V_2$. If $\dim V_1 = \dim V_2$, and $\tau_1 : V_1 \to V_2$, $\tau_2 : V_2 \to V_1$ are σ-semilinear maps, then we can also define a σ-semilinear map on $V_1 \otimes V_2$ by $v_1 \otimes v_2 \mapsto v_2\tau_2 \otimes v_1\tau_1$, and such a map *interchanges* the tensor factors. We can extend this in the obvious way to define σ-semilinear maps on the n-th tensor power $V^{\otimes n}$ that permute the tensor factors.

Definition 1.9.1 Given two matrices $A = (\alpha_{ij})_{a \times b}$ and $B = (\beta_{ij})_{c \times d}$ over a common field, their *Kronecker product* $A \otimes B$ is an $ac \times bd$ block matrix, with blocks of size $c \times d$, where the (k, l)th block (for $1 \leqslant k \leqslant a$, $1 \leqslant l \leqslant b$) is $\alpha_{kl}B$.

For $i = 1, 2$, let $(v_{i1}, \ldots, v_{id_i})$ be bases of V_i, and let τ_i be linear maps on V_i with corresponding matrices A_i. Then the matrix of $\tau_1 \otimes \tau_2$ with respect to the basis $(v_{11} \otimes v_{21}, v_{11} \otimes v_{22}, \ldots, v_{11} \otimes v_{2d_2}, v_{12} \otimes v_{21}, \ldots, v_{1d_1} \otimes v_{2d_2})$ of $V_1 \otimes V_2$ is $A_1 \otimes A_2$. The following result is standard, and is left as an exercise.

Proposition 1.9.2 *Let A_1, A_2 and A_3 be matrices over a common field F, and let $\sigma \in \mathrm{Aut}\, F$. Then $(A_1 \otimes A_2) \otimes A_3 = A_1 \otimes (A_2 \otimes A_3)$, $(A_1 \otimes A_2)^{\mathsf{T}} = A_1^{\mathsf{T}} \otimes A_2^{\mathsf{T}}$, and $(A_1 \otimes A_2)^\sigma = A_1^\sigma \otimes A_2^\sigma$.*

For $i = 1, 2$, if A_i and C_i are square matrices of degrees d_i, then $(A_1 \otimes A_2)(C_1 \otimes C_2) = (A_1C_1 \otimes A_2C_2)$ and $\det(A_1 \otimes A_2) = (\det A_1)^{d_2}(\det A_2)^{d_1}$.

Assume in addition that V_1 and V_2 are (finite-dimensional) FG-modules, corresponding to representations ρ_1 and ρ_2. The tensor product $\rho_1 \otimes \rho_2$ acting on $V_1 \otimes V_2$ is defined by $g(\rho_1 \otimes \rho_2) = g\rho_1 \otimes g\rho_2$. The next result follows from Proposition 1.9.2 and the fact that the identity maps on V_1 and V_2 induce an FG-isomorphism $V_1 \otimes V_2 \to V_2 \otimes V_1$ by interchanging the tensor factors, and will be used implicitly in Chapter 5.

Lemma 1.9.3 *Let V_1 and V_2 be FG-modules. Then, as FG-modules,*

(i) $(V_1 \otimes V_2)^\sigma \cong V_1^\sigma \otimes V_2^\sigma$ *for any field automorphism σ of F;*

(ii) $(V_1 \otimes V_2)^* \cong V_1^* \otimes V_2^*$, *where V^* denotes the dual module of V;*

(iii) $V_1 \otimes V_2 \cong V_2 \otimes V_1$.

Proposition 1.9.4 *For $i = 1, 2$, suppose that the action of G on V_i preserves a bilinear form β_i with matrix B_i. Then:*

(i) *G preserves a bilinear form with matrix $B_1 \otimes B_2$ on $V_1 \otimes V_2$. Moreover $B_1 \otimes B_2$ is non-singular if and only if both B_1 and B_2 are. Abstractly, the form with matrix $B_1 \otimes B_2$ is $\beta_1 \otimes \beta_2$ defined by $(\beta_1 \otimes \beta_2)(u_1 \otimes v_1, u_2 \otimes v_2) = \beta_1(u_1, u_2)\beta_2(v_1, v_2)$ and extended F-bilinearly.*

(ii) *If B_1 and B_2 are both symmetric or both anti-symmetric then $B_1 \otimes B_2$ is symmetric, and if one of B_1 and B_2 is symmetric and the other anti-symmetric then $B_1 \otimes B_2$ is anti-symmetric.*

(iii) *If char $F = 2$ and the forms are alternating, then $B_1 \otimes B_2$ is also alternating, and G preserves a quadratic form Q such that $Q(u \otimes v) = 0$ for all $u \in V_1$ and $v \in V_2$. If, in addition, F is finite and β_1 and β_2 are non-degenerate, then Q is of plus type.*

Proof All of these assertions except for those pertaining to Q follow directly from Proposition 1.9.2. So assume that char $F = 2$ and the β_i are alternating forms. We define a quadratic form Q on $V_1 \otimes V_2$ with polar form $\beta = \beta_1 \otimes \beta_2$ by specifying that $Q(e_i \otimes f_j) = 0$ for all i and j, where $(e_1, \ldots, e_n)$ is a basis for V_1 and $(f_1, \ldots, f_m)$ is a basis for V_2. For all $\lambda_i, \mu_j \in F$, vectors $u, u_i \in V_1$ and $v, v_j \in V_2$ we have:

$$
\begin{aligned}
Q((\textstyle\sum_{i=1}^r \lambda_i u_i) \otimes v) &= Q(\textstyle\sum_{i=1}^r \lambda_i (u_i \otimes v)) \\
&= \textstyle\sum_{i=1}^r \lambda_i^2 Q(u_i \otimes v) + \sum_{1 \leqslant i < j \leqslant r} \lambda_i \lambda_j \beta_1(u_i, u_j)\beta_2(v, v) \\
&= \textstyle\sum_{i=1}^r \lambda_i^2 Q(u_i \otimes v).
\end{aligned}
$$

Similarly,

$$
Q(u \otimes (\sum_{j=1}^s \mu_j v_j)) = \sum_{j=1}^r \mu_j^2 Q(u \otimes v_j).
$$

Combining all of the above gives $Q(u \otimes v) = 0$ for all $u \in V_1$ and $v \in V_2$. So $Q((e_i \otimes f_j).g) = Q(e_i.g \otimes f_j.g) = 0 = Q(e_i \otimes f_j)$ for all $g \in G$ and for all i, j. Therefore G preserves the quadratic form Q.

If the β_i are non-degenerate, then V_1 has a maximal totally singular subspace W_1 of dimension $(\dim V_1)/2$ under β_1. Then β_1 and Q are both identically zero on $W_1 \times V_2$, so Q is of plus type when F is finite, by Proposition 1.5.42. $\square$

It is possible that $V_1 \otimes V_2$ is isomorphic as FG-module to an $F_0 G$-module, for a proper subfield F_0 of F. In characteristic 2 we can make no assertions about the type of the corresponding quadratic form in such cases.

The following result is a consequence of Proposition 1.9.2.

Proposition 1.9.5 *If G is a group of isometries of sesquilinear forms on V_1 and V_2 with corresponding form matrices B_1 and B_2 then G is also a group of isometries of the sesquilinear form with matrix $B_1 \otimes B_2$ on $V_1 \otimes V_2$, and $B_1 \otimes B_2$ is non-degenerate if and only if B_1 and B_2 are.*

Propositions 1.9.4 and 1.9.5 extend in an obvious way to tensor products with more than two tensor factors.

Definition 1.9.6 For $i = 1, \ldots, k$, with $k \geqslant 2$, if G preserves forms β_i on modules V_i which are either all bilinear or all sesquilinear, then we call the form $\beta_1 \otimes \cdots \otimes \beta_k$ the *induced form*.

Definition 1.9.7 Let $G \leqslant \mathrm{GL}_{n_1}(F)$ and $H \leqslant \mathrm{GL}_{n_2}(F)$. Then we define $G \otimes H = \{g \otimes h \mid g \in G, h \in H\} \leqslant \mathrm{GL}_{n_1 n_2}(F)$.

It follows from Proposition 1.9.2 that this operation is associative. If $G = \langle X \rangle$ and $H = \langle Y \rangle$, then $G \otimes H = \langle \{x \otimes 1 : x \in X\} \cup \{1 \otimes y : y \in Y\} \rangle$.

For $i = 1, 2$, define representations $\rho_i : G \times H \to \mathrm{GL}_{n_i}(F)$ by $(g, h)\rho_1 = g$ and $(g, h)\rho_2 = h$. Then $G \otimes H = \mathrm{Im}(\rho_1 \otimes \rho_2)$.

Proposition 1.9.8 *Let G and H be matrix groups over the same field. Then*

$$G \otimes H \cong \frac{G \times H}{\{(\lambda \mathrm{I}_{n_1}, \lambda^{-1} \mathrm{I}_{n_2}) : \lambda \mathrm{I}_{n_1} \in G, \lambda^{-1} \mathrm{I}_{n_2} \in H\}}.$$

Proof Note that $(g, h)(\rho_1 \otimes \rho_2) = g \otimes h = 1 = \mathrm{I}_{n_1 n_2}$ if and only if $g = \lambda \mathrm{I}_{n_1}$ and $h = \lambda^{-1} \mathrm{I}_{n_2}$ for some $0 \neq \lambda \in F$. $\square$

1.10 Small dimensions and exceptional isomorphisms

The results in this section are all standard. We have taken them from [66, Proposition 2.9.1], but more information is available in [108]. The given isomorphisms will be used without further reference.

Proposition 1.10.1 *Let $\mathcal{C}$ be the following collection of groups: $\mathrm{L}_n(q)$ with $n \geqslant 2$, $\mathrm{U}_n(q)$ with $n \geqslant 2$, $\mathrm{S}_n(q)$ with $n \geqslant 2$, $\mathrm{O}_n^\varepsilon(q)$ with $n \geqslant 3$ and q odd if n is odd. Then the following is a complete list of the isomorphisms between pairs of* *elements of C.*

$$L_2(q) \cong S_2(q) \cong U_2(q) \cong O_3(q),$$
$$O_4^+(q) \cong L_2(q) \times L_2(q), \quad O_4^-(q) \cong L_2(q^2),$$
$$O_5(q) \cong S_4(q), \quad O_6^+(q) \cong L_4(q), \quad O_6^-(q) \cong U_4(q),$$
$$L_2(4) \cong L_2(5), \quad L_2(7) \cong L_3(2), \quad S_4(3) \cong U_4(2).$$

Proposition 1.10.2 *The only alternating or symmetric groups that are isomorphic to almost simple classical groups are* A_5, A_6, A_8 *and* S_6, *and*

$$L_2(4) \cong L_2(5) \cong A_5, \quad L_2(9) \cong A_6, \quad L_4(2) \cong A_8, \quad S_4(2) \cong S_6.$$

Proposition 1.10.3 *Let G be one of $L_n(q)$, $U_n(q)$, $S_n(q)$ or $O_n^\varepsilon(q)$, with $n \geqslant 2$, and q odd when G is $O_n^\circ(q)$.*

(i) *If G is soluble, then G is isomorphic to one of the following:*

$$L_2(2) \cong S_3, \quad L_2(3) \cong A_4, \quad U_3(2) \cong 3^2{:}Q_8,$$
$$O_2^+(q) \cong (q-1)/(2, q-1), \quad O_4^+(2) \cong S_3 \times S_3, \quad O_4^+(3) \cong A_4 \times A_4,$$
$$O_2^-(q) \cong (q+1)/(2, q-1).$$

(ii) *If G is not simple and not soluble, then G is isomorphic to $O_4^+(q)$ for $q \geqslant 4$, or to $S_4(2)$.*

(iii) *If G is simple, then the corresponding group $SL_n(q)$, $SU_n(q)$, $Sp_n(q)$ or $\Omega_n^\varepsilon(q)$ is quasisimple.*

Isomorphisms between the almost simple exceptional groups will be discussed in Section 4.1 and Chapter 7.

1.11 Representations of simple groups

In this section we collect some results about representations of simple groups as permutation and matrix groups, and about the classical groups in their natural representation.

Definition 1.11.1 For an arbitrary finite group G, let

$$P(G) = \min\{\, n : G \text{ has a non-trivial permutation representation of degree } n \,\},$$

noting that if G is simple then this is the same as the minimum degree of a faithful permutation representation. Let $\overline{\mathbb{F}_p}$ be the algebraic closure of $\mathbb{F}_p$ for a prime p,

$$R_p(G) = \min\{\, n : G \text{ is isomorphic to a subgroup of } \mathrm{PGL}_n(\overline{\mathbb{F}_p}) \,\},$$
$$R(G) = \min\{\, R_p(G) : \text{ all primes } p \,\}.$$

Table 1.4 *Simple classical groups with low degree projective representations*

S	$R_p(S)$
$O_6^{\pm}(p^e)$	4
$O_5(p^e)$, p odd	4
$O_4^-(p^e)$	2
$O_3(p^e)$, p odd	2
$S_4(2)'$	3

We now present a series of results giving values of these functions on various simple groups S.

Theorem 1.11.2 *Let S be a non-abelian simple group with $P(S) \leqslant 12$. Then either S is an alternating group in its natural action, or S has a primitive action on n points as given below:*

n	6	7	8	9	10	11	12
S	A_5	$L_2(7)$	$L_2(7)$	$L_2(8)$	A_5, A_6	$L_2(11), M_{11}$	$L_2(11), M_{11}, M_{12}$

Proof We consult Sims's classification [102] of primitive permutation groups of degree at most 20 to get the result. □

Proposition 1.11.3 ([66, Proposition 5.3.3]) *Let G be quasisimple. Then $R_p(G) \geqslant R_p(G/Z(G))$ for all primes p.*

Lemma 1.11.4 ([66, Lemma 5.5.3]) *Let G be a finite perfect group with a unique minimal normal subgroup N. If $N \cong E_{p^t}$ with $t \geqslant 2$, and p' is a prime other than p, then $R_{p'}(G) \geqslant \min\{P(G/N), p^{t/2}\}$.*

For a recent reference for the following result, see for instance [84]. It roughly states that the lowest degree representation in defining characteristic is the natural representation – all exceptions in Table 1.4 are due to isomorphisms.

Theorem 1.11.5 *Let S be a non-abelian simple classical group in dimension d over $\mathbb{F}_{p^e}$. Then $R_p(S) = d$ except for the groups occurring in Table 1.4.*

Proposition 1.11.6 ([66, Proposition 5.3.7]) *If $5 \leqslant n \leqslant 8$ then $R_p(A_n)$ is as given in Table 1.5. If $n \geqslant 9$ then $R(A_n) = n - 2$.*

Theorem 1.11.7 ([75]) *Let S be a simple linear, symplectic or unitary group defined over $\mathbb{F}_{p^e}$, and let p' be a prime other than p. Then $R_{p'}(S) \geqslant e(S)$, where $e(S)$ is as in Table 1.6. In particular, if S is a simple non-orthogonal classical group in dimension d and $d \geqslant 9$ then $R_{p'}(S) > d^2$.*

Table 1.5 *Values of $R_p(A_n)$ for $5 \leqslant n \leqslant 8$*

n	$R_2(A_n)$	$R_3(A_n)$	$R_5(A_n)$	$R_p(A_n),\ p \geqslant 7$
5	2	2	2	2
6	3	2	3	3
7	4	4	3	4
8	4	7	7	7

Table 1.6 *Selected values of $e(S)$*

$L_2(4)$, $L_2(9)$	2, 3
$L_2(q)$ otherwise	$(q-1)/(2, q-1)$
$L_3(2)$, $L_3(4)$	2, 4
$L_n(q)$, $n \geqslant 3$ otherwise	$q^{n-1} - 1$
$S_4(2)'$, $S_6(2)$	2, 7
$S_{2m}(q)$, q odd, otherwise	$(q^m - 1)/2$
$S_{2m}(q)$, q even, otherwise	$q^{m-1}(q^{m-1} - 1)(q - 1)/2$
$U_4(2)$, $U_4(3)$	4, 6
$U_n(q)$, n odd	$q(q^{n-1} - 1)/(q + 1)$
$U_n(q)$, n even, otherwise	$(q^n - 1)/(q + 1)$

The preceding results can easily be generalised to direct products of simple classical groups using the following lemma.

Lemma 1.11.8 ([66, Proposition 5.5.7]) *Let $S_1, \ldots, S_t$ be non-abelian simple groups, let $G = S_1 \times \cdots \times S_t$, let p be a prime, and let $n_i = R_p(S_i)$. Then $R_p(G) \geqslant \sum_{i=1}^{t} n_i$.*

1.12 The natural representations of the classical groups

In this section we collect some basic results concerning the natural representations of the classical groups.

Lemma 1.12.1 *The group $\mathrm{Sp}_2(q)$ is equal to $\mathrm{SL}_2(q)$.*

Proof Use Lemma 1.5.21 and Table 1.1 to see that any 2×2 matrix of determinant 1 is an isometry of our standard symplectic form. Therefore $\mathrm{SL}_2(q) \leqslant \mathrm{Sp}_2(q)$, and hence these two groups are equal. □

Proposition 1.12.2 (i) *The groups $\mathrm{SL}_n(q)$, $\mathrm{SU}_n(q)$ and $\mathrm{Sp}_n(q)$ are absolutely irreducible on $\mathbb{F}_{q^u}^n$ for all n and q, where $u = 2$ for $\mathrm{SU}_n(q)$ and $u = 1$ otherwise.*

(ii) *The group $\Omega_n^\varepsilon(q)$ is absolutely irreducible on $\mathbb{F}_q^n$ if and only if $n > 2$. The group $SO_n^\varepsilon(q)$ is absolutely irreducible on $\mathbb{F}_q^n$ if and only if one of the following holds:*

   (a) *$n > 2$;*
   (b) *$n = 2$, $\varepsilon = -$ and q is even;*
   (c) *$n = 2$, $\varepsilon = +$, q is even and $q > 2$.*

   The group $GO_n^\varepsilon(q)$ is absolutely irreducible if and only if $(n, q, \varepsilon) \notin \{(2, 2, +), (2, 3, +)\}$.

(iii) *The group $Sp_4(2)'$ is absolutely irreducible on $\mathbb{F}_2^4$.*

Proof Part (i) in all dimensions, Part (ii) in dimension at least 5, and Part (iii) follow from Theorem 1.11.5 since a non-absolutely irreducible representation would give rise to a representation of degree properly dividing n in the same characteristic. For the low-dimensional orthogonal groups see for instance [66, Proposition 2.10.6]: this reference is incorrect regarding $SO_2^+(2)$. □

In the next lemma we study the 4-dimensional orthogonal groups. Recall the definitions of the automorphisms of $O_n^\pm(q)$ from Subsection 1.7.1: in dimension 4 the outer automorphisms ϕ, δ and δ' have the same interpretation as there. Also, recall the isomorphisms given in Proposition 1.10.1, and that ω denotes a primitive element of $\mathbb{F}_q^\times$.

Lemma 1.12.3 (i) *The natural module for $\Omega_4^+(q)$ is isomorphic to the tensor product of two copies of the natural module for $SL_2(q)$, one for each direct factor of the preimage group $SL_2(q) \times SL_2(q)$. Let the subgroup S of $P\Gamma O_4^+(q)/O_4^+(q)$ be $\langle \phi \rangle$ if q is even, and $\langle \delta', \delta, \phi \rangle$ if q is odd. If $O_4^+(q) \trianglelefteq \bar{G} \leqslant O_4^+(q).S$ then G preserves the tensor product, whilst if $\bar{G} \leqslant P\Gamma O_4^+(q)$ but $\bar{G} \nleqslant O_4^+(q).S$ then G interchanges the two factors of the tensor product.*

(ii) *The natural module for $\Omega_4^-(q)$ is isomorphic to the tensor product of a copy of the natural module M for $SL_2(q^2)$ and the image of M under the automorphism $\sigma : x \mapsto x^q$ of $\mathbb{F}_{q^2}$.*

Proof (i) Let $W = \mathbb{F}_q^2$, and consider $SL_2(q)$ acting naturally on W. Then, by Lemma 1.12.1, the module W admits a non-degenerate symplectic form f. The tensor product representation of $SL_2(q) \times SL_2(q)$ on $W \otimes W$ has image isomorphic to $SL_2(q) \otimes SL_2(q)$, as in Definition 1.9.7, which is isomorphic to a central product of the two copies of $SL_2(q)$, by Proposition 1.9.8. By Proposition 1.9.4, $SL_2(q) \times SL_2(q)$ is a group of isometries of a bilinear form β when q is odd, and a quadratic form Q of plus type when q is even. In fact the matrix of β is antidiag$(1, -1, -1, 1)$, so by Proposition 1.5.42 the form β is also of plus type when q is odd. Hence $SL_2(q) \otimes SL_2(q) \leqslant GO_4^+(q)$. If $q > 2$ then the smallest normal subgroup of $\mathrm{SL}_2(q)$ with a quotient of 2-power order is $\mathrm{SL}_2(q)$ itself. Thus $\mathrm{SL}_2(q) \otimes \mathrm{SL}_2(q)$ is a subgroup of $\Omega_4^+(q)$, and hence by Proposition 1.10.1 is equal to $\Omega_4^+(q)$. The result can be checked by direct calculation using Lemma 1.6.12 when $q = 2$.

The permutation matrix corresponding to $(2, 3)$ is an isometry of Q over $\mathbb{F}_q$ which interchanges the two tensor factors, and hence lies in $\mathrm{GO}_4^+(q) \setminus \mathrm{SO}_4^+(q)$ when q is odd, and in $\mathrm{GO}_4^+(q) \setminus \Omega_4^+(q)$ when q is even. The field automorphism ϕ acting as a ϕ-semilinear map on both copies of W induces a ϕ-semilinear map on $W \otimes W$, which preserves the tensor factors, and acts as the field automorphism ϕ on both of the $\mathrm{SL}_2(q)$ factors. So, if q is even, then there is a non-trivial action $\tau :$ $\Gamma\mathrm{O}_4^+(q) = \langle \mathrm{SO}_4^+(q), \phi \rangle \to S_2$ on the two tensor factors with $\ker \tau = \langle \Omega_4^+(q), \phi \rangle$. If q is odd, then the outer automorphism $\delta = \mathrm{diag}(\omega, \omega, 1, 1)$ preserves the two tensor factors, and, since $\mathrm{C}\Gamma\mathrm{O}_4^+(q)$ is generated by $\Omega_4^+(q)$ together with δ, ϕ and the automorphism induced by the permutation matrix corresponding to $(1, 2)$, there is an action $\tau : \mathrm{C}\Gamma\mathrm{O}_4^+(q) \to S_2$ with $\ker \tau = \langle \Omega_4^+(q), \delta', \delta, \phi \rangle$.

(ii) See [108, pp199–201], where explicit isomorphisms are constructed. □

Lemma 1.12.4 *Let q be odd, and let $n \geqslant 4$ be even. Then our standard copy of $\mathrm{Sp}_n(q)$ contains no $\mathrm{GL}_n(q)$-conjugate of our standard copy of $\Omega_n^{\pm}(q)$.*

Proof Since, by Proposition 1.12.2, $\Omega_n^{\pm}(q)$ is absolutely irreducible on $\mathbb{F}_q^n$, it follows from Lemma 1.8.8 (ii) that it cannot preserve both a symmetric bilinear and an anti-symmetric bilinear form, and the result follows. □

Lemma 1.12.5 *Let q be odd, and let $n \geqslant 4$ be even. The no $\mathrm{GL}_n(q^r)$-conjugate of our standard copy of $\mathrm{Sp}_n(q)$ is contained in the standard copy of $\Omega_n^+(q^r)$ for any r.*

Proof The proof of this is almost identical to that of Lemma 1.12.4 □

We now prove some results on the traces of elements of classical groups in their natural representation.

Lemma 1.12.6   (i) *Let $g \in \mathrm{Sp}_n(q)$ or $\mathrm{GO}_n^{\varepsilon}(q)$. Then $\mathrm{tr}(g^{-1}) = \mathrm{tr}(g)$.*
  (ii) *Let $g \in \mathrm{GU}_n(q)$. Then $\mathrm{tr}(g^{-1}) = \mathrm{tr}(g)^{\sigma}$ where σ is the automorphism $x \mapsto x^q$ of $\mathbb{F}_{q^2}$.*

Proof (i) Let A be the matrix of our standard alternating or symmetric bilinear form, or the polar form of our standard quadratic form, as in Table 1.1. Then $gAg^{\mathsf{T}} = A$ by Lemmas 1.5.21 and 1.5.23, so $g^A = A^{-1}gA = g^{-\mathsf{T}}$. Transposition and conjugation preserve traces, so the result follows.
(ii) Let A be the matrix of our standard σ-Hermitian form. Then $gAg^{\sigma\mathsf{T}} = A$, so $g^A = A^{-1}gA = g^{-\sigma\mathsf{T}}$ and $\mathrm{tr}(g^{-\sigma\mathsf{T}}) = \mathrm{tr}(g^{-1})^{\sigma}$. □

Proposition 1.12.7 (i) *Let G be $\mathrm{SL}_n(q)$ with $n \geqslant 2$, $\mathrm{Sp}_n(q)$ with $n \geqslant 2$, or $\Omega_n^\varepsilon(q)$ with $n \geqslant 4$. Then, for any $\alpha \in \mathbb{F}_q$, there exists $g \in G$ with $\mathrm{tr}(g) = \alpha$.*

(ii) *Let G be $\mathrm{SU}_n(q)$ with $n \geqslant 3$. Then, for any $\alpha \in \mathbb{F}_{q^2}$, there exists $g \in G$ with $\mathrm{tr}(g) = \alpha$.*

Proof (i) By considering natural embeddings $\mathrm{SL}_n(q) < \mathrm{SL}_{n+1}(q)$, $\mathrm{Sp}_n(q) < \mathrm{Sp}_{n+2}(q)$, $\Omega_n^+(q) < \Omega_{n+1}(q)$ and $\Omega_n^\pm(q) < \Omega_{n+2}^\pm(q)$, it can be seen that we only need to prove this result for the smallest value of n in each case. For $\mathrm{SL}_2(q) = \mathrm{Sp}_2(q)$, choose $g = \begin{pmatrix} \alpha & 1 \\ -1 & 0 \end{pmatrix}$.

By Lemma 1.12.3 the natural module for $\Omega_4^+(q)$ is isomorphic to the tensor product of the natural module for two copies of $\mathrm{SL}_2(q)$. The trace of the Kronecker product of two matrices is equal to the product of their traces, so we can choose

$$g = \begin{pmatrix} \alpha & 1 \\ -1 & 0 \end{pmatrix} \otimes \begin{pmatrix} 1 & 1 \\ -1 & 0 \end{pmatrix}.$$

By Lemma 1.12.3 the natural module for $\Omega_4^-(q) \cong \mathrm{L}_2(q^2)$ is isomorphic over $\mathbb{F}_{q^2}$ to $M \otimes M^\sigma$, where M is the natural module for $\mathrm{SL}_2(q^2)$, and $\sigma : x \mapsto x^q$ is an automorphism of $\mathbb{F}_{q^2}$. So, for an element of $\mathrm{SL}_2(q^2)$ of trace β, the trace of the corresponding element of $\Omega_4^-(q)$ is β^{1+q}. A counting argument, and the fact that $\mathbb{F}_q^\times$ is cyclic, shows that for all $\alpha \in \mathbb{F}_q$ there exists $\beta \in \mathbb{F}_{q^2}$ with $\beta^{1+q} = \alpha$.

(ii) Again we only need to prove this for the smallest value of n, that is $n = 3$. We consider a matrix of the form

$$g = \begin{pmatrix} \alpha & 0 & \beta \\ \gamma & 0 & \alpha \\ 0 & \delta & 0 \end{pmatrix}.$$

For g to have determinant 1, we require $(\beta\gamma - \alpha^2)\delta = 1$, and for g to be an isometry of the unitary form I_3, we require $\alpha^{q+1} + \beta^{q+1} = \alpha^{q+1} + \gamma^{q+1} = 1$, $\alpha\gamma^q + \alpha^q\beta = 0$, and $\delta^{q+1} = 1$. Let $\alpha \in \mathbb{F}_{q^2}$ be given. If $\alpha = 0$, then choose $\beta = \gamma = \delta = 1$ to get $g \in \mathrm{SU}_3(q)$ with trace α.

Otherwise, $1 - \alpha^{q+1} \in \mathbb{F}_q$, so we can find $\gamma \in \mathbb{F}_{q^2}$ with $\alpha^{q+1} + \gamma^{q+1} = 1$, and then we choose $\beta = -\gamma^q/\alpha^{q-1}$ to get $\alpha\gamma^q + \alpha^q\beta = 0$. Then $\beta^q = -\alpha^{q-1}\gamma$, so $\beta^{1+q} = \gamma^{1+q}$, and hence $\alpha^{q+1} + \beta^{q+1} = 1$. Therefore

$$(\beta\gamma - \alpha^2)(\beta\gamma - \alpha^2)^q = \beta^{1+q}\gamma^{1+q} + \alpha^{2+2q} - \alpha^2\beta^q\gamma^q - \alpha^{2q}\beta\gamma =$$
$$\beta^{2+2q} + \alpha^{2+2q} + 2\alpha^{1+q}\beta^{1+q} = (\alpha^{1+q} + \beta^{1+q})^2 = 1.$$

Let $\delta = (\beta\gamma - \alpha^2)^{-1}$, then $\delta^{1+q} = \delta^{-1-q} = 1$, so $g \in \mathrm{SU}_3(q)$ and $\mathrm{tr}(g) = \alpha$. $\square$

Lemma 1.12.8   *Let q be odd. Then the set of traces of elements of $\Omega_3(q)$ does not lie in any proper subfield of $\mathbb{F}_q$.*

Proof   The group $\Omega_3(q)$ is the symmetric square representation of $\mathrm{SL}_2(q)$, acting on the basis $(v_1 \otimes v_1, 1/2(v_1 \otimes v_2 + v_2 \otimes v_1), v_2 \otimes v_2)$ (see Proposition 5.3.6 for more details). If $\lambda \in \mathbb{F}_q^\times$, then $\mathrm{SL}_2(q)$ contains $\mathrm{diag}(\lambda, \lambda^{-1})$. With respect to the above basis this corresponds in $\Omega_3(q)$ to $\mathrm{diag}(\lambda^2, 1, \lambda^{-2})$, of trace $\mu := \lambda^2 + \lambda^{-2} + 1$.

Since the equation for μ is of degree 4, each possible trace μ can be produced by at most 4 non-zero elements λ. Thus μ can take at least $\lceil (q-1)/4 \rceil$ values. If $q \neq 9$ then $\lceil (q-1)/4 \rceil$ is greater than any proper factor of q, and we are done. Let ω be a primitive element of $\mathbb{F}_9^\times$. Then the symmetric square of

$$\begin{pmatrix} \omega & 1 \\ 2 & 0 \end{pmatrix} \in \mathrm{SL}_2(9)$$

has trace $\omega^2 + 2$, which does not lie in $\mathbb{F}_3$. $\square$

1.13 Some results from number theory

In this section we collect some facts about Zsigmondy primes, an identity concerning least common multiples, and an introduction to quadratic reciprocity.

Theorem 1.13.1 (Zsigmondy [118])   *Let $q \geqslant 2$ be a prime power and $n \geqslant 3$, with $(q, n) \neq (2, 6)$. Then there exists at least one prime q_n such that q_n divides $q^n - 1$ but does not divide $q^i - 1$ for $i < n$.*

Definition 1.13.2   We call such primes q_n *Zsigmondy primes*, and denote a Zsigmondy prime for $q^n - 1$ by $z_{q,n}$.

Lemma 1.13.3   (i) *Let $q \geqslant 2$ be a prime power and $n \geqslant 3$. Then $q^n + 1$ is divisible by $z_{q,2n}$ if and only if $(q, n) \neq (2, 3)$.*
  (ii) *If $z_{q,n}$ divides $q^m - 1$ then n divides m.*
  (iii) *The prime $z_{q,n} \equiv 1 \pmod{n}$, so in particular, $q_n > n$.*

The following result is an immediate corollary of Theorem 1.6.22, and will be used frequently when discussing groups in Aschbacher's Class $\mathscr{C}_3$. Of course p is assumed to be prime throughout.

Proposition 1.13.4   (i) *Let $q = p^e$ with $e > 1$ and $(p, e) \neq (2, 3)$. Then $z_{p,2e}$ divides $|\mathrm{SL}_2(q)|$, and if $z_{p,i}$ divides $|\mathrm{SL}_2(q)|$ then $i \leqslant 2e$.*
  (ii) *Let $n \geqslant 3$, and assume that $(q, n) \neq (2, 6)$. Then $z_{q,n}$ divides $|\mathrm{SL}_n(q)|$, and if $z_{q,i}$ divides $|\mathrm{SL}_n(q)|$ then $i \leqslant n$.*

(iii) *Let* $n \geqslant 3$ *be odd, and assume that* $(q, n) \neq (2, 3)$. *Then* $z_{q,2n}$ *divides* $|\mathrm{SU}_n(q)|$, *and if* $z_{q,i}$ *divides* $|\mathrm{SU}_n(q)|$ *then* $i \in \{2n, 2n - 4\}$ *or* $i \leqslant 2n - 6$.

(iv) *Let* $n \geqslant 4$ *be even, and assume that* $(q, n) \neq (2, 4)$. *Then* $z_{q,2n-2}$ *and* $z_{q,n}$ *divide* $|\mathrm{SU}_n(q)|$, *and if* $z_{q,i}$ *divides* $|\mathrm{SU}_n(q)|$ *then* $i \in \{2n-2, 2n-6, n, n-2\}$ *or* $i \leqslant 2n - 10$.

(v) *Let* $n \geqslant 4$ *be even, and assume that* $(q, n) \neq (2, 6)$. *Then* $z_{q,n}$ *divides* $|\mathrm{Sp}_n(q)|$, *and if* $z_{q,i}$ *divides* $|\mathrm{Sp}_n(q)|$ *then* $i \in \{n, n-2, \ldots, 2\}$ *or* $i \leqslant n/2$.

(vi) *Let* $n \geqslant 5$ *be odd, and let* q *be odd. Then* $z_{q,n-1}$ *divides* $|\Omega_n^\circ(q)|$, *and if* $z_{q,i}$ *divides* $|\Omega_n^\circ(q)|$ *then* $i \in \{n-1, n-3, \ldots, 2\}$ *or* $i \leqslant (n-1)/2$.

(vii) *Let* $n \geqslant 6$ *be even, and assume that* $(q, n) \neq (2, 8)$. *Then* $z_{q,n-2}$ *divides* $|\Omega_n^+(q)|$, *and if* $z_{q,i}$ *divides* $|\Omega_n^+(q)|$ *then* $i \in \{2, 4, \ldots, n-2\}$ *or* $i \leqslant n/2$.

(viii) *Let* $n \geqslant 4$ *be even, and assume that* $(q, n) \neq (2, 6)$. *Then* $z_{q,n}$ *divides* $|\Omega_n^-(q)|$, *and if* $z_{q,i}$ *divides* $|\Omega_n^-(q)|$ *then* $i \in \{2, 4, \ldots, n\}$ *or* $i \leqslant n/2$.

In the following lemma we derive simpler expressions for the number of conjugacy classes of subfield and unitary groups in $\mathrm{SL}_n(q)$ than those given in [66, Table 3.5.A]. Let $(a)_p$ denote the highest power of the prime p dividing a. Recall that we write (a, b) for the greatest common divisor of positive integers a and b, and $[a, b]$ for their least common multiple.

Lemma 1.13.5 *Let* $q = p^e$ *be a prime power, let* f *be a divisor of* e, *let* $q_0 = p^f$ *and let* $n > 1$ *be a positive integer. Then*

(i)

$$\frac{q - 1}{[q_0 - 1, (q - 1)/(q - 1, n)]} = \left(\frac{q - 1}{q_0 - 1}, n\right).$$

(ii)

$$\frac{q + 1}{[q_0 + 1, (q + 1)/(q + 1, n)]} = \left(\frac{q + 1}{q_0 + 1}, n\right).$$

(iii) *If* e *is even then*

$$\frac{q - 1}{[q^{e/2} + 1, (q - 1)/(q - 1, n)]} = (q^{e/2} - 1, n).$$

Proof (i) Let r be a prime dividing $q - 1$, and let $a = (q - 1)_r$, $b = (q_0 - 1)_r$ and $c = (n)_r$. Now $a \geqslant b$, since $(q_0 - 1) \mid (q - 1)$. Also $(q - 1, n)_r = \min\{a, c\}$, so

$$((q - 1)/(q - 1, n))_r = a - \min\{a, c\} = \max\{a - c, 0\}.$$

Thus

$$([q_0 - 1, (q - 1)/(q - 1, n)])_r = \max\{b, \max\{a - c, 0\}\} = \max\{b, a - c\}.$$

Hence we conclude that

$$((q-1)/[q_0-1, (q-1)/(q-1, n)])_r = a - \max\{b, a-c\} = \min\{a-b, c\}.$$

Conversely, $((q-1)/(q_0-1))_r = a - b$, so $((q-1)/(q_0-1), n)_r = \min\{a-b, c\}$. The proofs of (ii) and (iii) are similar. $\qquad\square$

When determining minimal fields of representations, we will frequently need to determine for which finite fields certain integers are squares. To do so, we use the law of quadratic reciprocity.

Definition 1.13.6   Let $m \in \mathbb{Z}$ and $p \in \mathbb{N}$, such that p is prime. The *Legendre symbol* $\left(\frac{m}{p}\right)$ takes value 0 if $p \mid m$, value 1 if $m \bmod p$ is a square in $\mathbb{F}_p^\times$, and value -1 if $m \bmod p$ is a non-square in $\mathbb{F}_p^\times$.

Part (ii) of the following is proved in [74, p78]. The other parts are straightforward, and their proofs are left as an exercise.

Proposition 1.13.7   *Let $p \in \mathbb{N}$ be an odd prime.*

(i) *If $p \equiv 1 \bmod 4$ then $\left(\frac{-1}{p}\right) = 1$, and if $p \equiv 3 \bmod 4$ then $\left(\frac{-1}{p}\right) = -1$.*

(ii) *If $p \equiv \pm 1 \bmod 8$ then $\left(\frac{2}{p}\right) = 1$ and if $p \equiv \pm 3 \bmod 8$ then $\left(\frac{2}{p}\right) = -1$.*

(iii) *For all integers m and n and all primes p, $\left(\frac{mn}{p}\right) = \left(\frac{m}{p}\right)\left(\frac{n}{p}\right)$.*

The following was conjectured by Legendre and proved by Gauss. For a proof, see almost any textbook on number theory, for example [74, p78].

Proposition 1.13.8 (Law of Quadratic Reciprocity)   *Let $p, q \in \mathbb{N}$ be odd primes. Then*

$$\left(\frac{p}{q}\right) = \left(\frac{q}{p}\right)(-1)^{\frac{p-1}{2}\frac{q-1}{2}}.$$

Thus, for example, to calculate in which finite fields there exists an element $\sqrt{5}$, we find the values of q for which $\left(\frac{5}{q}\right) = 1$. For these values of q, we have $1 = \left(\frac{q}{5}\right)(-1)^{\frac{4}{2} \times \frac{q-1}{2}} = \left(\frac{q}{5}\right)$, and so q is a square modulo 5. So 5 is a square in $\mathbb{F}_q^\times$ when $q \equiv \pm 1 \bmod 5$, and 5 is a square in $\mathbb{F}_{q^2}^\times$ but not in $\mathbb{F}_q$ when $q \equiv \pm 2 \bmod 5$.

2

The main theorem and the types of geometric subgroups

2.1 The main theorem

Chapters 2 to 7 of this book are devoted to the proof of our main result:

Main Theorem 2.1.1 *Let q be a prime power, let $n \leqslant 12$, and let Ω be quasisimple and equal to one of $\mathrm{SL}_n(q)$, $\mathrm{SU}_n(q)$, $\mathrm{Sp}_n(q)$, $\Omega_n^\varepsilon(q)$, $\mathrm{Sz}(q) = {}^2\mathrm{B}_2(q)$, $\mathrm{G}_2(q)$, $\mathrm{R}(q) = {}^2\mathrm{G}_2(q)$ or ${}^3\mathrm{D}_4(q)$. Let $\overline{G}$ be an almost simple extension of $\overline{\Omega} := \Omega/\mathrm{Z}(\Omega)$. Then representatives of the conjugacy classes of maximal subgroups of $\overline{G}$ that do not contain $\overline{\Omega}$ are as specified in the appropriate table in Chapter 8.*

The maximal subgroups $\overline{H}$ of $\overline{G}$ in Tables 8.1 to 8.85 are defined by describing the structure of the inverse images H in Ω of $\overline{H} \cap \overline{\Omega}$. We refer the reader to Section 8.1 for further information on how to read the tables.

In Table 2.1 we give a rough description of eight classes, $\mathscr{C}_1$–$\mathscr{C}_8$, of subgroups of the classical groups, based on [66, 1.2.A]. Recall that we define $u = 2$ in Case **U** and $u = 1$ otherwise. See Definitions 2.2.8 and 2.2.14 for the (differing)

Table 2.1 *Rough descriptions of Aschbacher classes*

$\mathscr{C}_i$	Rough description
$\mathscr{C}_1$	stabilisers of totally singular or non-singular subspaces
$\mathscr{C}_2$	stabilisers of decompositions $V = \oplus_{i=1}^t V_i$, $\dim(V_i) = n/t$
$\mathscr{C}_3$	stabilisers of extension fields of $\mathbb{F}_{q^u}$ of prime index dividing n
$\mathscr{C}_4$	stabilisers of tensor product decompositions $V = V_1 \otimes V_2$
$\mathscr{C}_5$	stabilisers of subfields of $\mathbb{F}_{q^u}$ of prime index
$\mathscr{C}_6$	normalisers of symplectic-type or extraspecial groups in absolutely irreducible representations
$\mathscr{C}_7$	stabilisers of decompositions $V = \otimes_{i=1}^t V_i$, $\dim(V_i) = a$, $n = a^t$
$\mathscr{C}_8$	groups of similarities of non-degenerate classical forms

meanings of *preserving* these decompositions. We shall give considerably more information about these classes of subgroups in Section 2.2.

Definition 2.1.2   Let H be a subgroup of G, where $\Omega \leqslant G \leqslant A$ as in Series 1.1, with Ω one of $\mathrm{SL}_n(q)$, $\mathrm{SU}_n(q)$, $\mathrm{Sp}_n(q)$ or $\Omega_n^\varepsilon(q)$ and dimension restrictions as in Definition 1.6.20. If H is a subgroup of a member of Class $\mathscr{C}_i$ for some i with $1 \leqslant i \leqslant 8$, then H is a *geometric* group.

Recall (Theorem 1.3.2) that, if a group S is non-abelian simple, then $\mathrm{Out}\, S$ is soluble, and (Definition 1.8.7) what it means for a group to preserve a form.

Definition 2.1.3   Let H be a subgroup of G, where $\Omega \leqslant G \leqslant A$ as in Series 1.1, with Ω one of $\mathrm{SL}_n(q)$, $\mathrm{SU}_n(q)$, $\mathrm{Sp}_n(q)$ or $\Omega_n^\varepsilon(q)$. Then H lies in *Class $\mathscr{S}$* of G if $H/(H \cap \mathrm{Z}(\mathrm{GL}_n(q^u)))$ is almost simple and the following all hold:

(i) H does not contain Ω;
(ii) H^∞ acts absolutely irreducibly;
(iii)   there does not exist a $g \in \mathrm{GL}_n(q^u)$ such that $(H^\infty)^g$ is defined over a proper subfield of $\mathbb{F}_{q^u}$;
(iv) H^∞ preserves a non-zero unitary form if and only if $\Omega = \mathrm{SU}_n(q)$;
(v) H^∞ preserves a non-zero quadratic form if and only if $\Omega = \Omega_n^\varepsilon(q)$;
(vi) H^∞ preserves a non-zero symplectic form and no non-zero quadratic form if and only if $\Omega = \mathrm{Sp}_n(q)$;
(vii) H^∞ preserves no non-zero classical form if and only if $\Omega = \mathrm{SL}_n(q)$.

Definition 2.1.4   Let H and K be subgroups of a generally quasisimple classical group Ω, and suppose that H and K have been specified up to conjugacy in $\mathrm{Aut}\,\Omega$. Then by a *containment* of H in K we mean not just that K has a subgroup isomorphic to H, but also that there exists an $\mathrm{Aut}\,\bar{\Omega}$-conjugate of K that contains an $\mathrm{Aut}\,\bar{\Omega}$-conjugate of H.

Part (ii) of the definition ensures that an $\mathscr{S}$-subgroup H is not contained in a member of $\mathscr{C}_1$ or $\mathscr{C}_3$, since (as we shall see in Section 2.2), all members of $\mathscr{C}_1$ are reducible and no member K of $\mathscr{C}_3$ has K^∞ absolutely irreducible. Part (iii) of the definition ensures that H is not contained in a member of $\mathscr{C}_5$. Parts (iv), (v) and (vi) ensure that H is not contained in a member of $\mathscr{C}_8$. Note that H^∞ may be imprimitive, have a normal absolutely irreducible sympectic-type of extraspecial subgroup, or preserve a tensor decomposition: that is, a member of Class $\mathscr{S}$ may be contained in a member of $\mathscr{C}_2 \cup \mathscr{C}_4 \cup \mathscr{C}_6 \cup \mathscr{C}_7$.

Theorem 2.1.5 (Aschbacher's theorem: approximate version)   *Let H be a subgroup of a group G in Column A of Table 1.2. Then H is either a geometric subgroup of G, or a member of Class $\mathscr{S}$.*

Chapters 2–6 are devoted to the proof of Theorem 2.1.1 for the classical groups, and Chapter 7 to the proofs for extensions of $\mathrm{Sp}_4(2^e)$ with $e > 1$ that are not contained in $\mathsf{A} = \Gamma\mathrm{Sp}_4(2^e)$, and for almost simple extensions of $\mathrm{Sz}(q)$, $\mathrm{G}_2(q)$, $\mathrm{R}(q)$ and $^3\mathrm{D}_4(q)$. (Recall that $\mathrm{S}_4(2)$ is not simple.) For the classical groups, we do not prove Theorem 2.1.1 for $\Omega = \Omega_8^+(q)$, and our table in this case is taken from [62]. For the exceptional groups, we do not include proofs for $\mathrm{G}_2(q)$ (q odd), $\mathrm{R}(q) = {}^2\mathrm{G}_2(q)$ or $^3\mathrm{D}_4(q)$, and the tables in these cases are taken respectively from [64], [64] and [63]. The maximal subgroups of $\mathrm{Sz}(q)$ are determined in [106, Theorems 9 and 10], but not of their almost simple extensions. The maximal subgroups of $\mathrm{G}_2(q)$ (q even) are determined in [14], whilst those of their almost simple extensions are covered by [2, (17.3)].

The geometric maximal subgroups of the classical groups of dimension greater than 12 are described in [66, Main Theorem], which also states that [66, Tables 3.5.A–F] include all of the geometric maximals in all dimensions. So, for dimensions up to 12, we only need to determine which of the subgroups in these tables are maximal.

Chapters 2 and 3 in this book are devoted to determining those subgroups that are maximal among the geometric subgroups. Chapter 2 contains general arguments concerning maximality, whereas Chapter 3 resolves the remaining cases dimension by dimension.

Class $\mathscr{S}$ is divided into Classes $\mathscr{S}_1$ (cross characteristic) and $\mathscr{S}_2$ (defining characteristic). Some $\mathscr{S}_2$-subgroups are naturally contained in members of $\mathscr{C}_4$ or $\mathscr{C}_7$, so we define a Class $\mathscr{S}_2^*$ which excludes these subgroups, and let Class $\mathscr{S}^* = \mathscr{S}_1 \cup \mathscr{S}_2^*$. In Chapters 4 and 5, we determine the subgroups that are maximal among those in Classes $\mathscr{S}_1$ and $\mathscr{S}_2^*$, respectively. For $\mathscr{S}$-subgroups, we need to start by determinining the candidate subgroups of Ω. The lists of irreducible representations of quasisimple groups in [42] (cross characteristic) and [84] (defining characteristic) are used for this, but there is additional work to be done to determine the outer automorphisms present in the $\mathscr{S}^*$-maximals.

In Chapter 6, the determination of the maximal subgroups of the almost simple classical groups is completed by finding the containments between those subgroups that have been found to be maximal among those of geometric type, the $\mathscr{S}_1$-maximal subgroups, and the $\mathscr{S}_2^*$-maximal subgroups.

Conjugacy of geometric subgroups The conjugacy of all subgroups of the classical groups that are maximal amongst the geometric subgroups is described in [66, Main Theorem], so we will not include any conjugacy calculations for geometric subgroups. However, the forms that their groups preserve, and the elements chosen to generate the outer automorphism groups, sometimes differ from ours. We therefore finish this section with a brief description of the "translation" that is necessary to produce the conjugacy information that appears in the tables of geometric subgroups in Chapter 8. Recall the definitions of our standard outer automorphisms in Section 1.7. It is shown in [6] that if one form can be transformed to another by a matrix that is centralised by the standard Frobenius automorphism, then the standard Frobenius automorphism acts in the same way on both groups of isometries.

For the linear groups, the form preserved is the zero form and our standard outer automorphisms agree with those in [66], except that we write γ for the duality automorphism that [66] denotes ι.

The standard unitary form in [66] is equal to ours. Our representatives for the outer automorphisms are identical to those in [66], so although it is proved in [6] that the action of the Frobenius automorphism depends on the choice of form, our conjugacy class stabilisers are identical to those in [66].

Our standard symplectic basis is just a reordering of the symplectic basis in [66]. We choose the same matrix for the automorphism δ, and the change of basis does not change its action. It is proved in [6] that when $\Omega = \mathrm{Sp}_n(q)$ the action of the Frobenius automorphism does not depend on the choice of form. Therefore our conjugacy class stabilisers are identical to those in [66].

For the orthogonal groups in odd dimension we use the same standard form as [66]. Our automorphism δ is equal to the automorphism $r_\square r_\boxtimes$ in [66]. We use the same automorphism ϕ. Therefore it is straightforward to write the stabilisers from [66] in our notation.

For the orthogonal groups of plus type, we use the same standard form as [66] in even characteristic. In odd characteristic our standard basis is a reordering of the basis in [66]. Thus in all cases the standard forms agree, up to the order of basis vectors. Our automorphism γ is the automorphism $r_\square$ in [66]; when q is odd our automorphism δ' is equal to $r_\square r_\boxtimes$ in [66]; when q is odd our δ is equal to the δ in [66]; and for all q our ϕ is equal to the ϕ in [66].

For the orthogonal groups of minus type, we use the same standard form as [66] in odd characteristic, and in even characteristic our standard basis is just a reordering of the standard basis in [66]. Our automorphism γ is the automorphism $r_\square$ in [66]. If q is odd and $n(q-1)/4$ is odd then our automorphism δ' is $r_\square r_\boxtimes$ in [66], and our automorphism ϕ is equal to the ϕ in [66]. However, if q is odd and $n(q-1)/4$ is even then our automorphism δ is equal to the automorphism δ in [66] if $n \equiv 0 \bmod 4$, and to $\delta r_\square$ if $n \equiv 2 \bmod 4$. Finally, if q is even, or if q is odd and $n(q-1)/4$ is even, then our automorphism φ is the automorphism ϕ in [66].

2.2 Introducing the geometric types

In this section we introduce the members of Classes $\mathscr{C}_1$ to $\mathscr{C}_8$. In each class we define various *types* of geometric subgroup, and define a group to be a member of the class only when it is of one of the types. This will enable a more detailed statement of Aschbacher's theorem in Theorem 2.2.19. For each type, we give the structure of the corresponding subgroup of the generally quasisimple group. We assume throughout this section that the dimension restrictions of Definition 1.6.20 apply, with the addition of $\Omega_8^+(q)$ to our list of possibilities. Thus although we print tables for $\Omega_6^+(q)$ (for example), and include Aschbacher Class information for these, these classes should be understood as corresponding to the descriptions in [1]: we are *not* formally defining Class $\mathscr{C}_1$ through to Class $\mathscr{S}$ for these groups.

2.2.1 Class $\mathscr{C}_1$

In this subsection we introduce Class $\mathscr{C}_1$, and describe some elementary properties of its members. Roughly speaking, these groups are the stabilisers of subspaces.

We shall refer to the Class $\mathscr{C}_1$ groups by their *type*. Recall Definition 1.5.7. A group of type P_k is a *maximal parabolic*: in Cases **S**, **U** and **O**$^\varepsilon$ this is the stabiliser of a totally singular subspace of dimension k, whilst in Case **L** it is the stabiliser of any k-space. A group of type $P_{k,n-k}$ is the stabiliser of two subspaces, one of dimension k, and the other of dimenson $n - k$, such that the $(n - k)$-space contains the k-space. A group of type $A \oplus B$ is the stabiliser of a pair of subspaces with trivial intersection which span the space. A group of type $A \perp B$ is the stabiliser of a pair of non-degenerate subspaces which are mutually orthogonal and span the space. We write t.s. as an abbreviation for *totally singular*, n.d. for *non-degenerate*, and n.s. for *non-singular*. In Case **O**$^\varepsilon$, let W be a non-degenerate k-space. Then $\text{sgn}(W)$ is the sign of the restriction of the quadratic form to W.

Definition 2.2.1 A subgroup H of $\Gamma\text{L}_n(q)$ is *reducible* if H stabilises a proper non-zero subspace of $\mathbb{F}_q^n$. Let G be a group such that $\Omega \trianglelefteq G \leqslant A$, as in Series 1.1, and let $K \leqslant G$. If $G \leqslant \Gamma$ then K lies in *Class $\mathscr{C}_1$* if $K = \text{N}_G(W)$ or $\text{N}_G(W, U)$, where W (or U and W) appear in Table 2.2. Otherwise, K lies in Class $\mathscr{C}_1$ if $K = \text{N}_A(H) \cap G$, where H is a $\mathscr{C}_1$-subgroup of Γ.

In Case **O**$^+$, there are two conjugacy classes in $\Omega_n^+(q)$ of groups of type $P_{n/2}$, which are conjugate under γ. In Case **O**$^\pm$ if k is odd (so that q is odd) then there are two classes of groups of type $\text{GO}_k^\circ(q) \perp \text{GO}_{n-k}^\circ(q)$, which are conjugate under δ, the automorphism that multiplies the form by a non-square.

Table 2.2 *Types of $\mathscr{C}_1$-subgroups*

Case	Type	Description	Conditions
All	P_k	$N_G(W)$, W t.s. k-space	$k < n$ in Case **L**
			$k < n/2$ in Case **O**$^-$
			$k \leqslant n/2$ otherwise
L	$P_{k,n-k}$	$N_G(W,U)$ with $W < U$,	$k < n/2$
		$\dim(W) = k$ and	
		$\dim(U) = n - k$	
L	$\mathrm{GL}_k(q) \oplus \mathrm{GL}_{n-k}(q)$	$N_G(W,U)$ with	$k < n/2$
		$W \cap U = 0$, $\dim(W) = k$	
		and $\dim(U) = n - k$	
U	$\mathrm{GU}_k(q) \perp \mathrm{GU}_{n-k}(q)$	$N_G(W)$, W n.d. k-space	$k < n/2$
S	$\mathrm{Sp}_k(q) \perp \mathrm{Sp}_{n-k}(q)$	$N_G(W)$, W n.d. k-space	$k < n/2$
O$^\varepsilon$	$\mathrm{GO}_k^{\varepsilon_1}(q) \perp \mathrm{GO}_{n-k}^{\varepsilon_2}(q)$	$N_G(W)$, W n.d. k-space,	$k \leqslant n/2$
		$\varepsilon_1 = \mathrm{sgn}(W)$ and	$\varepsilon = + \Rightarrow \varepsilon_1 = \varepsilon_2$
		$\varepsilon_2 = \mathrm{sgn}(W^\perp)$	$\varepsilon = - \Rightarrow \varepsilon_1 = -\varepsilon_2$
			$(k, \varepsilon_1) \neq (n - k, \varepsilon_2)$
			q even $\Rightarrow k$ even
O$^\pm$	$\mathrm{Sp}_{n-2}(q)$	$N_G(W)$, W n.s. 1-space	q and n even

In all other cases and for all other types there is a single conjugacy class in Ω of each type of group for each k in Table 2.2. Note that our definition of types P_k in $\mathrm{SL}_n(q)$ is very slightly different from that in [66]: since the groups of type P_k are conjugate under duality to those of type P_{n-k}, these two types are identified in [66] where we have preferred to leave them separate.

In Table 2.3 we describe the structure of the $\mathscr{C}_1$-subgroups in the quasisimple group Ω. For odd q, the group $\mathrm{GL}_n(q)$ has a unique subgroup of index 2, which we denote $\frac{1}{2}\mathrm{GL}_n(q)$. With two exceptions, these results are all straightforward generalisations of those in [66, Propositions 4.1.3, 4.1.4, 4.1.6, 4.1.7, 4.1.17–4.1.20], where the corresponding subgroups of $\bar{\Omega}$ are described. The first difference is that we correct an error in the statement of [66, Proposition 4.1.18], regarding the structure of the parabolic subgroups of $\mathrm{U}_{2n}(q)$. The second is that we use the matrices given in [46, Lemma 4.3] to deduce that the extension of order 4 of $\Omega_k^{\varepsilon_1}(q) \times \Omega_{n-k}^{\varepsilon_2}(q)$ in the penultimate row of Table 2.3 is of shape 2^2.

We collect some facts about the stabilisers of totally singular subspaces. Let r denote the dimension of a maximal totally singular subspace. Thus $r = n$ in Case **L**, $r = \lfloor n/2 \rfloor$ in Cases **U**, **S**. **O**$^+$ and **O**$^\circ$, and $r = n/2 - 1$ in Case **O**$^-$. A *maximal flag* in V is a chain of totally singular i-spaces V_i, with

$$\{0\} = V_0 < V_1 < \cdots < V_r.$$

Table 2.3 *Structures of $\mathscr{C}_1$-subgroups in Ω*

Case	Type	Shape of $H < \Omega$	Notes
L	P_k	$[q^a]{:}(\mathrm{SL}_k(q) \times \mathrm{SL}_{n-k}(q)){:}(q-1)$	
L	$P_{k,n-k}$	$[q^b]{:}(\mathrm{SL}_k(q)^2 \times \mathrm{SL}_{n-2k}(q)){:}(q-1)^2$	
L	$\mathrm{GL}_k(q) \oplus \mathrm{GL}_{n-k}(q)$	$(\mathrm{SL}_k(q) \times \mathrm{SL}_{n-k}(q)){:}(q-1)$	
U	P_k	$[q^b]{:}(\mathrm{SL}_k(q^2) \times \mathrm{SU}_{n-2k}(q)).(q^2-1)$	$k < n/2$
		$[q^b]{:}\mathrm{SL}_k(q^2).(q-1)$	$k = n/2$
U	$\mathrm{GU}_k(q) \perp \mathrm{GU}_{n-k}(q)$	$(\mathrm{SU}_k(q) \times \mathrm{SU}_{n-k}(q)).(q+1)$	
S	P_k	$[q^c]{:}(\mathrm{GL}_k(q) \times \mathrm{Sp}_{n-2k}(q))$	
S	$\mathrm{Sp}_k(q) \perp \mathrm{Sp}_{n-k}(q)$	$\mathrm{Sp}_k(q) \times \mathrm{Sp}_{n-k}(q)$	
O$^\varepsilon$	P_k	$[q^d]{:}(\mathrm{GL}_k(q) \times \Omega^\varepsilon_{n-2k}(q))$	q even
		$[q^d]{:}\frac{1}{2}\mathrm{GL}_k(q)$	$k = \lfloor n/2 \rfloor$, q odd
		$[q^d]{:}(\frac{1}{2}\mathrm{GL}_k(q) \times \Omega^\varepsilon_{n-2k}(q)).2$	otherwise
O$^\varepsilon$	$\mathrm{GO}^{\varepsilon_1}_k(q) \perp \mathrm{GO}^{\varepsilon_2}_{n-k}(q)$	$(\Omega^{\varepsilon_1}_k(q) \times \Omega^{\varepsilon_2}_{n-k}(q)).2$	$k = 1$ or q even
		$(\Omega^{\varepsilon_1}_k(q) \times \Omega^{\varepsilon_2}_{n-k}(q)).2^2$	otherwise
O$^\pm$	$\mathrm{Sp}_{n-2}(q)$	$\mathrm{Sp}_{n-2}(q)$	q even

In the table, $a = k(n-k)$, $b = k(2n-3k)$, $c = k(n + \frac{1-3k}{2})$ and $d = k(n - \frac{1+3k}{2})$.

The following results can all be found in [66, §4.1]; by $\mathrm{N}_G(W)^W$ we mean the restriction of the stabiliser in G of the subspace W to its action on W.

Lemma 2.2.2   *Let $P \in \mathrm{Syl}_p(\Omega)$.*

(i) *In Cases **L**, **U**, **S**, **O**$^\circ$ and **O**$^-$, the group P stabilises a unique maximal flag and a unique totally singular i-space for $1 \leqslant i \leqslant r$.*

(ii) *In Case **O**$^+$, the group P stabilises a unique totally singular i-space for $1 \leqslant i \leqslant n/2 - 1$, precisely two totally singular $n/2$-spaces, and precisely two maximal flags. The maximal flags are interchanged by $\mathrm{N}_{\mathrm{GO}^+_n(q)}(P)$.*

(iii) *If $W < V$ is totally singular then $\mathrm{N}_\Omega(W)$ contains a conjugate of P.*

(iv) *If $W < V$ is totally singular, then $\mathrm{SL}(W) \leqslant \mathrm{N}_\Omega(W)^W$.*

(v) *In Cases **U**, **S** and **O**$^\varepsilon$, if $W < V$ is a totally singular k-space, then there exist spaces X and Y such that $V = (W \oplus Y) \perp X$, where Y is a totally singular k-space, $W \oplus Y$ is non-degenerate and $X = (W \oplus Y)^\perp$.*

(vi) *In Cases **S**, **U** or **O**$^+$, let $n = 2r$ be even, let $W_1 = \langle e_1, \ldots, e_r \rangle$ and let $W_2 = \langle f_1, \ldots, f_r \rangle$. Let K be the stabiliser in $\mathrm{Sp}_n(q)$, $\mathrm{GU}_n(q)$ or $\mathrm{GO}^+_n(q)$ (respectively) of both W_1 and W_2. Then $K \cong \mathrm{GL}_r(q^u)$, the restricted representation $\rho : K \to \mathrm{GL}(W_1)$ is the natural representation of $\mathrm{GL}_r(q^u)$, and as K-modules $W_2 = W_1^{\gamma *}$, where $*$ is duality and γ is the power of the Frobenius automorphism sending $x \mapsto x^q$.*

Table 2.4 *Types of $\mathscr{C}_2$-subgroups*

Case	Type	Description of $\mathcal{D}$	Conditions
L	$\mathrm{GL}_m(q) \wr \mathrm{S}_t$	any decomposition	
U	$\mathrm{GU}_m(q) \wr \mathrm{S}_t$	V_i n.d.	
U	$\mathrm{GL}_{n/2}(q^2).2$	V_i t.s., $t = 2$	
S	$\mathrm{Sp}_m(q) \wr \mathrm{S}_t$	V_i n.d.	
S	$\mathrm{GL}_{n/2}(q).2$	V_i t.s.; $t = 2$	q odd
$\mathbf{O}^\varepsilon$	$\mathrm{GO}_1(p) \wr \mathrm{S}_n$	V_i n.d.; $t = n$	$q = p > 2$; $\varepsilon = \circ \Leftrightarrow n$ odd;
			$\varepsilon = - \Leftrightarrow (n \equiv 2 \bmod 4$
			and $q \equiv 3 \bmod 4)$;
			$\varepsilon = +$ otherwise
$\mathbf{O}^\varepsilon$	$\mathrm{GO}_m^{\varepsilon_1}(q) \wr \mathrm{S}_t$	V_i n.d.; $\varepsilon_1 = \mathrm{sgn}(V_i)$	$m > 1$; m even $\Rightarrow \varepsilon = \varepsilon_1^t$;
			m odd and t even $\Rightarrow$
			$\varepsilon = (-1)^{(q-1)n/4}$
$\mathbf{O}^+$	$\mathrm{GL}_{n/2}(q).2$	V_i t.s.; $t = 2$	
$\mathbf{O}^\pm$	$\mathrm{GO}_{n/2}^\circ(q)^2$	V_i n.d.; $t = 2$;	$\varepsilon = (-1)^{(q+1)/2}$; q odd
		V_i nonisometric	

2.2.2 Class $\mathscr{C}_2$

In this subsection we introduce the members of Class $\mathscr{C}_2$ and define certain of their subgroups that will be useful to us later.

We shall refer to the families of $\mathscr{C}_2$-groups by their *type*. The types of $\mathscr{C}_2$-group are listed in Table 2.4, taken from [66, Table 4.2.A]. In each type we require that the group is the stabiliser of a decomposition $\mathcal{D}$ of V into t subspaces, each of dimension $m = n/t$:

$$\mathcal{D}: \quad V = V_1 \oplus V_2 \oplus \cdots \oplus V_t. \tag{2.1}$$

We use the abbreviations n.d. for non-degenerate and t.s. for totally singular.

Definition 2.2.3 A subgroup H of $\Gamma\mathrm{L}_n(q)$ is *imprimitive* if H preserves a direct sum decomposition $\mathcal{D}$ of $V = \mathbb{F}_q^n$, as in Equation 2.1. Let G be a group such that $\Omega \lhd G \leqslant A$, as in Series 1.1, and let $K \leqslant G$. If $G \leqslant \Gamma$ then K lies in *Class $\mathscr{C}_2$* if K is the stabiliser in G of an imprimitive decomposition described in Table 2.4. Otherwise, K lies in Class $\mathscr{C}_2$ if $K = \mathrm{N}_A(H) \cap G$, where H is a $\mathscr{C}_2$-subgroup of Γ.

In Cases **L**, **U** and **S**, the $\mathscr{C}_2$-subgroups of $G \leqslant \Gamma$ are all irreducible, and so it is common for the definition of imprimitivity to include a requirement that the group H acts irreducibly on $\mathbb{F}_q^n$, and hence a requirement that H permutes the V_i transitively. However, there exist $\mathscr{C}_2$-groups in Case $\mathbf{O}^\varepsilon$ where the subgroup

Table 2.5 Shapes of $\mathscr{C}_2$-subgroups in Ω

Case	Type	Structure of $H < \Omega$
L	$\mathrm{GL}_m(q) \wr \mathrm{S}_t$	$\mathrm{SL}_m(q)^t.(q-1)^{t-1}.\mathrm{S}_t$
U	$\mathrm{GU}_m(q) \wr \mathrm{S}_t$	$\mathrm{SU}_m(q)^t.(q+1)^{t-1}.\mathrm{S}_t$
U	$\mathrm{GL}_{n/2}(q^2).2$	$\mathrm{SL}_{n/2}(q^2).(q-1).2$
S	$\mathrm{Sp}_m(q) \wr \mathrm{S}_t$	$\mathrm{Sp}_m(q)^t{:}\mathrm{S}_t$
S	$\mathrm{GL}_{n/2}(q).2$	$\mathrm{GL}_{n/2}(q).2$
O$^\varepsilon$	$\mathrm{GO}_1(p) \wr \mathrm{S}_n$	$2^{n-1}.\mathrm{A}_n$ if $q \equiv \pm 3 \bmod 8$
		$2^{n-1}.\mathrm{S}_n$ if $q \equiv \pm 1 \bmod 8$
O$^\varepsilon$	$\mathrm{GO}_m^{\varepsilon_1}(q) \wr \mathrm{S}_t$	$\Omega_m^{\varepsilon_1}(q)^t.2^{(2,q-1).(t-1)}.\mathrm{S}_t$
O$^+$	$\mathrm{GL}_{n/2}(q).2$	$\mathrm{SL}_{n/2}(q).\frac{(q-1)}{(q-1,2)}.(n/2,2)$
O$^\pm$	$\mathrm{GO}_{n/2}^\circ(q)^2$	$\mathrm{SO}_{n/2}^\circ(q)^2$

H of $\Omega_n^\varepsilon(q)$ is in fact reducible. With our weak definition of imprimitivity, all groups in $\mathscr{C}_2$ are imprimitive.

In Table 2.5 we describe the structure of the $\mathscr{C}_2$-subgroups of Ω. The information about the shape of H can be deduced from [66, Propositions 4.2.4, 4.2.5, 4.2.7, 4.2.9–4.2.11, 4.2.14–4.2.16]. We have not attempted here to specify the centre of H precisely, but instead to exhibit certain useful subgroups of H.

Let H be of one of the types given in Table 2.4, let Ω be the quasisimple classical group containing H and let G be the corresponding general group, as in Definition 1.6.5. Write $G_{\mathcal{D}}$ for the stabiliser in G of the decomposition $\mathcal{D}$. Write $H_{(\mathcal{D})}$ for the kernel of the action of H on the decomposition $\mathcal{D}$, namely those elements of H that map V_i to V_i for $1 \leqslant i \leqslant t$, and define $G_{(\mathcal{D})}$ similarly. Let $G^{\mathcal{D}} := G_{\mathcal{D}}/G_{(\mathcal{D})}$, and let G_i be the restriction of $G_{(\mathcal{D})}$ to V_i, so that G_i is the general group on V_i. If the type is not $\mathrm{GO}_{n/2}^\circ(q)^2$ then $G_{\mathcal{D}} = G_{(\mathcal{D})}{:}J = G_1 \wr J$, where $J \cong \mathrm{S}_t$. Note in particular that $J \leqslant G$.

Our proofs of maximality of $\mathscr{C}_2$-subgroups will have two main structures, depending on whether $t = 2$ or $t \geqslant 3$. We first make some definitions and basic observations for $t \geqslant 3$. If $t \geqslant 3$, then define $L := \langle J' \rangle^H$ (the normal closure of J' under H) so that $L \leqslant H$ since $\Omega = G'$. Define the symbols $L_{(\mathcal{D})}$ and $L^{\mathcal{D}}$ analogously to $G_{(\mathcal{D})}$ and $G^{\mathcal{D}}$.

Lemma 2.2.4 Let $t \geqslant 3$.

(i) $\mathrm{A}_t \cong L^{\mathcal{D}} \cong J' \leqslant H$, and hence $L^{\mathcal{D}}$ acts primitively on $\mathcal{D}$.

(ii) L is perfect provided $t \geqslant 5$.

(iii) The restriction of $L_{(\mathcal{D})}$ to V_i is equal to G_i, for $1 \leqslant i \leqslant t$.

(iv) If $t \geqslant 4$ and $G_1 \neq \mathrm{GL}_1(2)$, or if $m \geqslant 2$, then $V_1, \ldots, V_t$ are pairwise nonisomorphic as $L_{(\mathcal{D})}$-modules.

Proof (i) This is clear from Table 2.5.

(ii) The normal closure of a perfect subgroup is perfect.

(iii) Since $G_{\mathcal{D}}$ is a wreath product, we may use wreath product notation for elements of H. The element $x_g := (g, g^{-1}, I_m, \ldots, I_m) \in G_{(\mathcal{D})}$ has determinant 1 for all $g \in G_1$. If $\Omega = \Omega_n^\varepsilon(q)$ then it is easily seen that x_g has spinor norm (q odd) or quasideterminant (q even) 1. Thus $x_g \in H$ for all $g \in G_1$. The group J' is a subgroup of H, so $a := (I_m, I_m, \ldots, I_m)(1\ 2\ 3) \in L$. Therefore $y_g := (a^{-1})^{x_g^{-1}} a = (g, g^{-2}, g, 1 \ldots, 1) \in L$ for all $g \in G_1$, so $L_{(\mathcal{D})}^{V_1} = G_1$. The result for $i \geqslant 1$ now follows from Part (i).

(iv) If $G_1 \neq \mathrm{GL}_1(2)$ then G_1 is non-trivial. The element y_g from the proof of Part (iii) shows that if $t \geqslant 4$ then V_1 and V_4 are nonisomorphic as $L_{(\mathcal{D})}$-modules. The same element y_g shows that if $t = 3$ and $m \geqslant 2$ then V_1 and V_2 are nonisomorphic, since G_1 contains involutions. The result now follows from the primitivity of A_t. $\qquad\square$

When $t = 2$ we will require a different style of argument, so we make some extra definitions. Let $H < \Omega$ be an imprimitive group, of one of the types given in Table 2.4, and assume that H preserves an imprimitive decomposition into two subspaces, V_1 and V_2. Recall the dimension restrictions in Definition 1.6.20. Let Ω_i be the generally quasisimple group on V_i, as in Definition 1.6.13. Then H contains a subgroup isomorphic to Ω_1' when the decomposition is into totally singular subspaces in Cases **U**, **S** and $\mathbf{O}^\varepsilon$, or to $\Omega_1' \times \Omega_2'$ otherwise, and this subgroup is perfect if and only if none of the following occurs:

(i) $n = 2$;

(ii) $n = 4$ and $q \leqslant 3$;

(iii) $\Omega = \mathrm{SU}_6(2)$ and $\mathcal{D}$ is a decomposition into non-degenerate subspaces;

Assume that $\Omega_1' \times \Omega_2'$ is perfect, and let $N = H^\infty$, so that N is a subdirect product of $\Omega_1' \times \Omega_2'$. Let X be a non-abelian composition factor of N. In Case **L** the group $N \cong \mathrm{SL}_{n/2}(q)^2$ and $X = \mathrm{L}_{n/2}(q)$. If H is of type $\mathrm{GU}_{n/2}(q) \wr S_2$ then $N \cong \mathrm{SU}_{n/2}(q)^2$ and $X = \mathrm{U}_{n/2}(q)$. If H is of type $\mathrm{Sp}_{n/2}(q) \wr S_2$ then $N \cong \mathrm{Sp}_{n/2}(q)'^2$ and $X = \mathrm{S}_{n/2}(q)'$. In the non-degenerate types in Case $\mathbf{O}^\varepsilon$ the group $N \cong \Omega_{n/2}^{\varepsilon_1}(q)^2$ and $X = \mathrm{O}_{n/2}^{\varepsilon_1}(q)$. For the totally singular decompositions in Cases **U**, **S** and $\mathbf{O}^+$ the group N is isomorphic to $\mathrm{SL}_{n/2}(q^u)$ (recall that $u = 2$ in Case **U**, and 1 otherwise) and $X = \mathrm{L}_{n/2}(q^u)$.

Note that it is immediate from Lemma 2.2.2 (vi) that if $t = 2$, and either the Case is **L** or the Case is **S**, **U** or $\mathbf{O}^+$ and the decomposition is into totally singular subspaces, then H_{V_1} contains $\mathrm{SL}_{n/2}(q^u)$ in its natural action, and as an H_{V_1} module $V_2 = V_1^{\sigma*}$, where $*$ is duality and σ is the power of the Frobenius automorphism mapping $x \mapsto x^q$.

2.2.3 Class $\mathscr{C}_3$

In this subsection we introduce the members of $\mathscr{C}_3$, and prove some preliminary results about their structure and actions.

Recall Definition 1.6.1 of a semilinear map. When discussing $\mathscr{C}_3$-subgroups, we write $\Gamma L_{n/s}(q^s)$ for the group consisting of all semilinear maps f on $\mathbb{F}_{q^s}^{n/s}$ such that $(\lambda v)f = \lambda(vf)$ for all $\lambda \in \mathbb{F}_q$. (This notation is convenient but dangerous because, for example, $\Gamma L_{6/3}(4^3)$ is not the same as $\Gamma L_{4/2}(8^2)$.)

Let s be a divisor of n, and let $m = n/s$. There is an $\mathbb{F}_q$-vector space isomorphism from $\mathbb{F}_{q^{us}}$ to $\mathbb{F}_{q^u}^s$, and this induces an $\mathbb{F}_{q^u}$-vector space isomorphism $\alpha : V_s = \mathbb{F}_{q^{us}}^{n/s} \to V = \mathbb{F}_{q^u}^n$. In turn, this induces an embedding of $\mathbb{F}_{q^{us}}^\times$ in $\mathrm{GL}_n(q^u)$, and also of $\Gamma L_{n/s}(q^{us}) = N_{\mathrm{GL}_n(q^u)}(\mathbb{F}_{q^{us}})$ in $\mathrm{GL}_n(q^u)$. See, for example, [45] for an explicit embedding.

Recall Definition 1.6.4 of a semi-similarity. If β_s is a σ-sesquilinear or quadratic form on V_s then one may consider the set of all elements of $\Gamma L_n(q^u)$ that act as semi-similarities on $(V_s, \beta_s)\alpha$. The type of β_s can be deduced from the type of the $\mathscr{C}_3$-subgroup, as in Table 2.6, but see [66, Section 4.3] for information about how to construct the form $\beta_s\alpha$ on V: the construction depends on the type of β_s. Further information on the groups in this class can be found in [30].

Definition 2.2.5 A subgroup H of $\mathrm{GL}_n(q^u)$ is *semilinear* if there exists a divisor s of n, and an $\mathbb{F}_{q^u}$-vector space isomorphism from V_s to V, such that all elements of H act semilinearly on V_s.

Let G be a group such that $\Omega \leqslant G \leqslant A$, as in Series 1.1, and let $K \leqslant G$. If $G \leqslant \Gamma$ then K lies in *Class $\mathscr{C}_3$* if K is the set of all semi-similarities f of the image under α of (V_s, β_s) such that there exists $\lambda \in \mathbb{F}_{q^u}$ and $\theta \in \mathrm{Aut}\,\mathbb{F}_{q^{us}}$ with

$$\beta_s(v\alpha f\alpha^{-1}, w\alpha f\alpha^{-1}) = \lambda\beta_s(v, w)^\theta$$

$$\text{or, if } \beta \text{ is quadratic, } \beta_s(v\alpha f\alpha^{-1}) = \lambda\beta_s(v)^\theta$$

for all $v, w \in V_s$, where β_s is as in Table 2.6. Otherwise, K lies in Class $\mathscr{C}_3$ if $K = N_A(H) \cap G$, where H is a $\mathscr{C}_3$-subgroup of Γ. [In the rest of our discussion, we will omit the isomorphism α, for ease of reading.]

Note that some authors include irreducibility, or absolute irreducibility, as a requirement for a group to be semilinear. We will prove later on (see Lemma 2.3.14 and Proposition 3.3.4) that all $\mathscr{C}_3$-subgroups *are* irreducible, but we do not require irreducibility as part of the definition of the class. Not all Class $\mathscr{C}_3$ groups are absolutely irreducible.

We refer to the families of $\mathscr{C}_3$-subgroups by their *type*. The types of $\mathscr{C}_3$-subgroup are listed in Table 2.6, taken from [66, Table 4.3.A]. The information about the structure of $H < \Omega$ is a straightforward consequence of [66, Propositions 4.3.6, 4.3.7, 4.3.10, 4.3.14, 4.3.16, 4.3.17, 4.3.18, 4.3.20], where the shape of

Table 2.6 *Types of $\mathscr{C}_3$-subgroups*

Case	Type	Structure of $H < \Omega$	Conditions
L	$\mathrm{GL}_m(q^s)$	$\left(\frac{(q-1,m)(q^s-1)}{q-1} \circ \mathrm{SL}_m(q^s)\right) \cdot \frac{(q^s-1,m)}{(q-1,m)} . s$	s prime
U	$\mathrm{GU}_m(q^s)$	$\left(\frac{(q+1,m)(q^s+1)}{q+1} \circ \mathrm{SU}_m(q^s)\right) \cdot \frac{(q^s+1,m)}{(q+1,m)} . s$	s odd prime
S	$\mathrm{Sp}_m(q^s)$	$\mathrm{Sp}_m(q^s) . s$	s prime
S	$\mathrm{GU}_{n/2}(q)$	$\mathrm{GU}_{n/2}(q) . 2$	q odd, $s=2$
O$^\varepsilon$	$\mathrm{GO}^\varepsilon_m(q^s)$	$\Omega^\varepsilon_m(q^s) . [cs]$	s prime, $m \geqslant 3$
			$\varepsilon = + \Rightarrow c = (s,2)$
			$\varepsilon \in \{\circ, -\} \Rightarrow c = 1$
O$^\pm$	$\mathrm{GO}^\circ_{n/2}(q^2)$	$(q \mp 1, 4)/2 \times \Omega^\circ_{n/2}(q^2) . 2$	$qn/2$ odd, $s=2$
O$^+$	$\mathrm{GU}_{n/2}(q)$	$((q+1) \circ \mathrm{SU}_{n/2}(q)) \cdot [(q,2)(q+1,\frac{n}{2})]$	$n \equiv 0 \bmod 4$, $s=2$
O$^-$	$\mathrm{GU}_{n/2}(q)$	$\left(\frac{q+1}{(q+1,2)} \circ \mathrm{SU}_{n/2}(q)\right) \cdot (q+1,\frac{n}{2})$	$n \equiv 2 \bmod 4$, $s=2$

$\bar{H} < \bar{\Omega}$ is given. In each type we require that the degree s of the field extension is a prime divisor of n, and we set $m = n/s$.

The groups of type $\mathrm{GO}^\circ_{n/2}(q^2)$ and type $\mathrm{GO}^\pm_n(q^2)$ have been studied in [15].

Let H be a $\mathscr{C}_3$-subgroup of Ω. In Case **L**, let Ω_1 be the subgroup $\mathrm{SL}_m(q^s)$ of H, and in Case **U**, let Ω_1 be $\mathrm{SU}_m(q^s)$. If H is of type $\mathrm{Sp}_m(q^s)$ then let Ω_1 be the subgroup $\mathrm{Sp}_m(q^s)$. If H is of type $\mathrm{GU}_{n/2}(q)$ then let Ω_1 be the subgroup $\mathrm{SU}_{n/2}(q)$. If H is of type $\mathrm{GO}^\varepsilon_m(q^s)$ then let Ω_1 be $\Omega^\varepsilon_m(q^s)$. If H is of type $\mathrm{GO}^\circ_{n/2}(q^2)$ then let Ω_1 be the subgroup $\Omega^\circ_{n/2}(q^2)$. Except for some small values of m, s and q, the group Ω_1 is quasisimple and equal to H^∞.

Lemma 2.2.6 ([66, Lemma 4.3.2])   *If $m \geqslant 2$, then either Ω_1 is an irreducible subgroup of Ω or H is of type $\mathrm{GU}_2(q)$ in $\mathrm{Sp}_4(q)$.*

Recall the dimension assumptions in Definition 1.6.20.

Lemma 2.2.7   *Let H be a $\mathscr{C}_3$-subgroup of Ω. Then H is insoluble if and only if one of the following holds.*

(i)   *The Case is **L** or **U**, and $s \neq n$.*
(ii)   *The Case is **S**, and if H is of type $\mathrm{GU}_{n/2}(q)$ then $(n,q) \neq (4,3)$.*
(iii)   *The Case is **O$^\circ$**.*
(iv)   *The Case is **O$^\pm$** and $m \geqslant 4$.*

If H is insoluble then $H^\infty \cong \Omega_1$. Thus, if $\Omega_1 \neq \Omega^+_4(q)$ and H is insoluble then H^∞ is quasisimple.

Proof   Apply Proposition 1.10.3 to the shape of H, as given in Table 2.6.   □

Table 2.7 *Types of $\mathscr{C}_4$-subgroups*

Case	Type	Conditions
L	$\mathrm{GL}_{n_1}(q) \otimes \mathrm{GL}_{n_2}(q)$	$1 < n_1 < \sqrt{n}$
U	$\mathrm{GU}_{n_1}(q) \otimes \mathrm{GU}_{n_2}(q)$	$1 < n_1 < \sqrt{n}$
S	$\mathrm{Sp}_{n_1}(q) \otimes \mathrm{GO}_{n_2}^{\varepsilon}(q)$	q odd, $n_2 \geqslant 3$
O$^+$	$\mathrm{Sp}_{n_1}(q) \otimes \mathrm{Sp}_{n_2}(q)$	$n_1 < \sqrt{n}$
O$^\pm$	$\mathrm{GO}_{n_1}^{\circ}(q) \otimes \mathrm{GO}_{n_2}^{\pm}(q)$	q odd, $n_1 \geqslant 3$, $n_2 \geqslant 4$
O$^\circ$	$\mathrm{GO}_{n_1}^{\circ}(q) \otimes \mathrm{GO}_{n_2}^{\circ}(q)$	$3 \leqslant n_1 < \sqrt{n}$
O$^+$	$\mathrm{GO}_{n_1}^{\varepsilon_1}(q) \otimes \mathrm{GO}_{n_2}^{\varepsilon_2}(q)$	$n_1, n_2 \geqslant 4$, n_i even,
		q odd, $\varepsilon_1 = \varepsilon_2 \Rightarrow n_1 < \sqrt{n}$

Case	Shape of $H < \Omega$
L	$(\mathrm{SL}_{n_1}(q) \circ \mathrm{SL}_{n_2}(q)).[(q-1, n_1, n_2)^2]$
U	$(\mathrm{SU}_{n_1}(q) \circ \mathrm{SU}_{n_2}(q)).[(q+1, n_1, n_2)^2]$
S	$(\mathrm{Sp}_{n_1}(q) \circ \mathrm{GO}_{n_2}^{\varepsilon}(q)).(n_2, 2)$
O$^+$	$(\mathrm{Sp}_{n_1}(q) \circ \mathrm{Sp}_{n_2}(q)).(2, q-1, n/4)$
O$^\pm$	$\mathrm{SO}_{n_1}^{\circ}(q) \times \Omega_{n_2}^{\pm}(q)$
O$^\circ$	$(\Omega_{n_1}(q) \times \Omega_{n_2}(q)).2$
O$^+$	$(\mathrm{SO}_{n_1}^{\varepsilon_1}(q) \circ \mathrm{SO}_{n_2}^{\varepsilon_2}(q)).[c]$

2.2.4 Class $\mathscr{C}_4$

In this subsection we briefly introduce the members of Class $\mathscr{C}_4$.

Definition 2.2.8 A group $G \leqslant \Gamma\mathrm{L}(V)$ *preserves* a tensor product decomposition $V = V_1 \otimes V_2$ if for all $g \in G$ there exist $g_1 \in \Gamma\mathrm{L}(V_1)$ and $g_2 \in \Gamma\mathrm{L}(V_2)$ such that for all $v_1 \in V_1$ and $v_2 \in V_2$

$$(v_1 \otimes v_2)g = v_1 g_1 \otimes v_2 g_2.$$

Notice in particular that the group above does not interchange the two tensor factors. Recall Definition 1.9.7, of the tensor product of two groups defined over a common field.

We shall refer to the families of $\mathscr{C}_4$-subgroups by their *type*. The first part of Table 2.7 is taken from [66, Table 4.4.A]: in each type we require that $n_1 n_2 = n$. In Cases **U**, **S** and **O**$^\varepsilon$, the type of the induced form (as in Definition 1.9.6) can be deduced from Propositions 1.9.4 and 1.9.5. The groups occuring below the horizontal line only occur when $n > 12$, and will generally be excluded from our discussions.

Definition 2.2.9 A subgroup H of $\mathrm{GL}_n(q)$ is a *tensor product group* if H preserves a tensor product decomposition $\mathbb{F}_q^n = V_1 \otimes V_2$. Let G be a group such that $\Omega \trianglelefteq G \leqslant A$, as in Series 1.1, and let $K \leqslant G$. If $G \leqslant \Gamma$ then K lies in

Class $\mathscr{C}_4$ if K is the stabiliser in G of a tensor product decomposition with V_1 of dimension n_1 and V_2 of dimension n_2, where V_1 and V_2 are equipped with a zero or non-degenerate form, as described in Table 2.7. Otherwise, K lies in Class $\mathscr{C}_4$ if $K = \mathrm{N}_A(H) \cap G$, where H is a $\mathscr{C}_4$-subgroup of Γ.

The information about the shape of H in the second part of the table can easily be deduced from [66, Propositions 4.4.10, 4.4.11, 4.4.12, 4.4.14–4.4.17]. Recall Definition 1.5.35 of the discriminant of a form. In the final type of the second table, let d_i be the discriminant of the form on V_i, for $i = 1, 2$. Then the term c is 4 if any of the following hold: $\varepsilon_1 = \varepsilon_2 = -$; $\varepsilon_1 = \varepsilon_2 = +$ and at at least one of d_1 or d_2 is non-square; $\varepsilon_1 = \varepsilon_2 = +$ and $n \equiv 4 \bmod 8$; $\varepsilon_1 = +$ and $\varepsilon_2 = -$ with at least one of d_1 or d_2 non-square. Otherwise, c is 8.

The following lemma is standard (see [66, 4.4.3] for example), and will be useful for both Class $\mathscr{C}_4$ and Class $\mathscr{C}_7$.

Lemma 2.2.10 *Suppose that* $G = G_1 \otimes \cdots \otimes G_t$ *preserves a decomposition* $V = V_1 \otimes \cdots \otimes V_t$, *with* $G_i \leqslant \mathrm{GL}(V_i)$ *for* $1 \leqslant i \leqslant t$. *If* G_1 *is irreducible on* V_1 *and* G_i *is absolutely irreducible on* V_i *for* $i \geqslant 2$ *then* G *is irreducible on* V. *If each* G_i *is absolutely irreducible on* V_i *then* G *is absolutely irreducible on* V.

2.2.5 Class $\mathscr{C}_5$

In this subsection we briefly introduce the members of $\mathscr{C}_5$. Let $\mathbb{F}_{q^{u/r}}$ be a subfield of index r in $\mathbb{F}_{q^u}$, and let V_r be the $\mathbb{F}_{q^{u/r}}$-span of an $\mathbb{F}_{q^u}$-basis B of $V = \mathbb{F}_{q^u}^n$. Then as an $\mathbb{F}_{q^{u/r}}$-space, $V_r \cong \mathbb{F}_{q^{u/r}}^n$.

If $g \in \mathrm{GL}(V_r)$ then g acts naturally on V, so there is a natural embedding $\mathrm{GL}_n(q^{u/r}) \cong \mathrm{GL}(V_r) \leqslant \mathrm{GL}(V) \cong \mathrm{GL}_n(q^u)$, which extends to $\Gamma\mathrm{L}_n(q^{u/r}) \leqslant \Gamma\mathrm{L}_n(q^u)$. Thinking of working with respect to the basis B, notice that we can also characterise $\mathrm{GL}(V_r)$ as being the centraliser in $\mathrm{GL}(V)$ of a representative for the Galois group $\mathrm{Gal}(\mathbb{F}_{q^u} : \mathbb{F}_{q^{u/r}})$.

Recall Definition 1.6.4 of a semi-similarity. If V is equipped with a non-degenerate σ-Hermitian, symplectic or quadratic form β, and if β_r is a non-degenerate σ-Hermitian, symplectic, or quadratic form on V_r, then the embedding of V_r in V may induce an embedding of the semi-similarity group of V_r into the semi-similarity group of V. See Table 2.8 and [66, §4.5] for more information about the possible embeddings that can give rise to maximal subgroups. Except for the embedding of $\mathrm{Sp}_n(q)$ in $\mathrm{SU}_n(q)$, in Table 2.8 the form β_r on V_r is simply the restriction of the form β to the elements of V_r, viewed as an $\mathbb{F}_{q^{u/r}}$-vector space: note that β_r may be of a different type from β. As regards $\mathrm{Sp}_n(q)$ in $\mathrm{SU}_n(q)$, the symplectic form β_2 satisfies $\beta_2(v_1, v_2) = \lambda\beta(v_1, v_2)$ for all $v_1, v_2 \in V_2$, for some fixed $\lambda \in \mathbb{F}_{q^2}^\times$ such that $\lambda + \lambda^q = 0$.

In the following definition, for $G \leqslant \Gamma$ we write $N_G(V_r)$ to denote the stabiliser in G of $V_r \subseteq V$, that is, the elements of G that send V_r to itself, and in addition, if V_r is a space equipped with a form, act as semi-similarities of V_r.

Definition 2.2.11 A subgroup H of $\mathrm{GL}_n(q)$ is *subfield* if H is absolutely irreducible and there exists a proper subfield $\mathbb{F}_{q_0}$ of $\mathbb{F}_q$ and an element $g \in \mathrm{GL}_n(q)$ such that

$$H^g \leqslant \langle \mathrm{Z}(\mathrm{GL}_n(q)), \mathrm{GL}_n(q_0) \rangle.$$

That is, up to scalars, H is conjugate to a group over a proper subfield of $\mathbb{F}_q$.

Let G be a group such that $\Omega \trianglelefteq G \leqslant A$, as in Series 1.1, and let $K \leqslant G$. If $G \leqslant \Gamma$ then K lies in *Class $\mathscr{C}_5$* if $K = N_G(V_r)(\mathrm{Z}(\mathrm{GL}_n(q^u)) \cap G)$, for some formed space V_r as in Table 2.8. Otherwise, K lies in Class $\mathscr{C}_5$ if $K = N_A(H) \cap G$, where H is a $\mathscr{C}_5$-subgroup of Γ.

All columns of Table 2.8 except for that which describes the shape of $H <$ Ω are taken from [66, Table 4.5.A]. The information about the shape of H is largely from [66, Propositions 4.5.3–4.5.6, 4.5.8, 4.5.10]. However, we use Lemma 1.13.5 to simplify the description of the structure of the groups of type $\mathrm{GL}_n(q_0)$ and type $\mathrm{GU}_n(q_0)$. For the groups of type $\mathrm{GO}_n^\circ(q^{1/2})$, to determine the precise structure of the $\mathscr{C}_5$-subgroup we consider writing an element $g \in$ $\mathrm{SO}_n^\circ(q^{1/2}) \setminus \Omega_n^\circ(q^{1/2})$ as a product of reflections $g = r_{v_1} \ldots r_{v_k}$. To calculate the spinor norm of g in $\mathrm{GO}_n^\circ(q)$ (so q is odd), evaluate $\prod_{i=1}^{k} \beta(v_i, v_i) = \lambda$. Since $\lambda \in$ $\mathbb{F}_{q^{1/2}}$, this must be a square in $\mathbb{F}_q^\times$ and so g lies in $\Omega_n^\circ(q)$, by Definition 1.6.10. Thus the group $\Omega_n^\circ(q^{1/2}).2$ given in [66, Proposition 4.5.8] is in fact $\mathrm{SO}_n^\circ(q^{1/2})$. In Case $\mathbf{O}^+$, type $\mathrm{GO}_n^+(q_0)$ with $q = q_0^2$, the integer b is defined as follows:

$$b = \begin{cases} 1 & \text{if } n \equiv 2 \bmod 4 \text{ and } q_0 \equiv 1 \bmod 4 \\ 2 & \text{otherwise.} \end{cases}$$

In Case $\mathbf{O}^+$, type $\mathrm{GO}_n^-(q_0)$, the integer b is defined as follows:

$$b = \begin{cases} 2 & \text{if } n \equiv 2 \bmod 4 \text{ and } q_0 \equiv 1 \bmod 4 \\ 1 & \text{otherwise.} \end{cases}$$

All members of Class $\mathscr{C}_5$ are subfield groups. Given our restrictions on the dimension in Cases $\mathbf{U}$ and $\mathbf{O}^\varepsilon$, all groups in Class $\mathscr{C}_5$ are absolutely irreducible by Proposition 1.12.2. The groups of type $\mathrm{GL}_n(q_0)$, $\mathrm{GU}_n(q_0)$ and $\mathrm{Sp}_n(q_0)$ have been studied in [17].

The following result is clear from Definition 2.2.11 and Table 2.8.

Lemma 2.2.12 *Let H be a $\mathscr{C}_5$-subgroup of Ω. Then all elements of H^∞ have trace in some proper subfield of $\mathbb{F}_{q^u}$.*

Table 2.8 *Types of $\mathscr{C}_5$-subgroups*

Case	Type	Shape of $H < \Omega$	Conditions
L	$GL_n(q_0)$	$SL_n(q_0).\left[\left(\frac{q-1}{q_0-1},n\right)\right]$	$q = q_0^r$, r prime
U	$GU_n(q_0)$	$SU_n(q_0).\left[\left(\frac{q+1}{q_0+1},n\right)\right]$	$q = q_0^r$, r odd prime
U	$Sp_n(q)$	$Sp_n(q).[(q+1,n/2)]$	n even
U	$GO_n^\varepsilon(q)$	$SO_n^\varepsilon(q).[(q+1,n)]$	q odd
S	$Sp_n(q_0)$	$Sp_n(q_0).(2,q-1,r)$	$q = q_0^r$, r prime
O°	$GO_n^\circ(q_0)$	$\Omega_n^\circ(q_0)$	$q = q_0^r$, r odd prime
		$SO_n^\circ(q_0)$	$q = q_0^2$
O⁺	$GO_n^+(q_0)$	$\Omega_n^+(q_0)$	$q = q_0^r$, r prime, r odd or q even
		$SO_n^+(q_0).b$	$q = q_0^2$, q odd
O⁺	$GO_n^-(q_0)$	$\Omega_n^-(q_0)$	$q = q_0^2$, q even
		$SO_n^-(q_0).b$	$q = q_0^2$, q odd
O⁻	$GO_n^-(q_0)$	$\Omega_n^-(q_0)$	$q = q_0^r$, r odd prime

2.2.6 Class $\mathscr{C}_6$

In this subsection we present some basic information about the $\mathscr{C}_6$-subgroups. For any prime r and any integer $m \geqslant 1$, there are two isomorphism types of extraspecial groups of order r^{1+2m}; see, for example, [31, Theorem 5.2]. If r is odd then we are only concerned with the groups of exponent r, denoted r_+^{1+2m}, since the normaliser in $GL_{r^m}(q^u)$ of an extraspecial group of the other isomorphism type (denoted r_-^{1+2m}) is a proper subgroup of the normaliser of an extraspecial group of exponent r. If $r = 2$ then the extraspecial group of minus type is a central product of a quaternion group of order 8 with zero or more dihedral groups of order 8, whilst the group of plus type is a central product of dihedral groups of order 8. By taking a central product of either type of extraspecial 2-group with a cyclic group of order 4, we obtain a 2-group of *symplectic type*. The extraspecial and symplectic type groups of order r^{1+2m} or $2^{2+2m} = 4 \circ 2_+^{1+2m} = 4 \circ 2_-^{1+2m}$ act on $\mathbb{F}_{q^u}^{r^m}$ whenever $q^u - 1$ is divisible by r or 4 respectively, and this action is absolutely irreducible.

We shall refer to the $\mathscr{C}_6$-subgroups by their *type*, as in Table 2.9. In Table 2.9 $n = r^m$ with r prime. If type $\mathscr{C}_6$-subgroups occur in Cases **L** and **U**, then when $n = 2^m \geqslant 4$ the power q^u is the minimal power of p such that $q^u \equiv 1 \pmod 4$, and otherwise q^u is the minimal power of p such that $q^u \equiv 1 \pmod r$. Thus if $\Omega = SL_n(p^e)$ contains $\mathscr{C}_6$-subgroups then e is odd, and in addition if n is even then $e = 1$. If $Sp_n(p^e)$ or $\Omega_n^+(p^e)$ contains $\mathscr{C}_6$-subgroups then $e = 1$. Table 2.9 is derived from [66, Propositions 4.6.5–4.6.9]. If $n \leqslant 12$, then recalling our

Table 2.9 *Types of $\mathscr{C}_6$-subgroups*

Case	Type	Shape of $H < \Omega$	Conditions
L	$r^{1+2m}.\mathrm{Sp}_{2m}(r)$	$3^{1+2}{:}Q_8$	$n = 3, q \equiv 4, 7 \bmod 9$
		$((q-1,n) \circ r^{1+2m}).\mathrm{Sp}_{2m}(r)$	n odd, otherwise
L	$2^{2+2m}.\mathrm{Sp}_{2m}(2)$	$2^{1+2}_-{:}3 \cong 2{\cdot}A_4$	$n = 2, q \equiv \pm 3 \bmod 8$
		$2^{1+2}_-{:}S_3 \cong 2{\cdot}S_4$	$n = 2, q \equiv \pm 1 \bmod 8$
		$(4 \circ 2^{1+4}).A_6$	$n = 4, q \equiv 5 \bmod 8$
		$((q-1,n) \circ 2^{1+2m}).\mathrm{Sp}_{2m}(2)$	n even, otherwise
U	$r^{1+2m}.\mathrm{Sp}_{2m}(r)$	$3^{1+2}{:}Q_8$	$n = 3, q \equiv 2, 5 \bmod 9$
		$((q+1,n) \circ r^{1+2m}).\mathrm{Sp}_{2m}(r)$	n odd, otherwise
	$2^{2+2m}.\mathrm{Sp}_{2m}(2)$	$(4 \circ 2^{1+4}).A_6$	$n = 4, q \equiv 3 \bmod 8$
		$((q+1,n) \circ 2^{1+2m}).\mathrm{Sp}_{2m}(2)$	n even, otherwise
S	$2^{1+2m}_-.\Omega^-_{2m}(2)$	$2^{1+2m}_-.\mathrm{SO}^-_{2m}(2)$	$q \equiv \pm 1 \bmod 8$
		$2^{1+2m}_-.\Omega^-_{2m}(2)$	$q \equiv \pm 3 \bmod 8$
O$^+$	$2^{1+2m}_+.\Omega^+_{2m}(2)$	$2^{1+2m}_+.\mathrm{SO}^+_{2m}(2)$	$q \equiv \pm 1 \bmod 8$
		$2^{1+2m}_+.\Omega^+_{2m}(2)$	$q \equiv \pm 3 \bmod 8$

dimension restrictions in Definition 1.6.20, we see that $\mathscr{C}_6 = \varnothing$ in Cases $\mathbf{O}^\varepsilon$, and $n = 4$ or 8 in Case **S**.

Definition 2.2.13 A subgroup H of $\mathrm{GL}_n(q)$ is an *extraspecial normaliser group* if $n = r^m$ for some prime r and H has an extraspecial normal subgroup of order r^{1+2m} and exponent $r(2, r)$.

Let G be a group such that $\Omega \trianglelefteq G \leqslant A$, as in Series 1.1, and let $K \leqslant G$. Then K lies in *Class $\mathscr{C}_6$* if $K = N_G(R)$ where R is an absolutely irreducible r-group as in Table 2.9.

2.2.7 Class $\mathscr{C}_7$

In this subsection we present some basic information about Class $\mathscr{C}_7$.

Definition 2.2.14 A group $G \leqslant \Gamma\mathrm{L}(V)$ *preserves* a tensor induced decomposition $V = V_1 \otimes V_2 \otimes \cdots \otimes V_t$ if for all $g \in G$ there exist $g_i \in \Gamma\mathrm{L}(V_i)$ and $\sigma \in S_t$ such that for all $v_i \in V_i$

$$(v_1 \otimes \cdots \otimes v_t)g = v_{1\sigma}g_{1\sigma} \otimes \cdots \otimes v_{t\sigma}g_{t\sigma}.$$

If non-degenerate forms β_i have been defined on the V_i, then we require in addition that the g_i are elements of the Γ-group for that form, as in Series 1.1.

Notice in particular that G can permute the t tensor factors.

Definition 2.2.15 A subgroup H of $\mathrm{GL}_n(q)$ is a *tensor induced group* if

H preserves a tensor induced decomposition $\mathbb{F}_q^n = V_1 \otimes V_2 \otimes \cdots \otimes V_t$ with $\dim V_i = m$ for all i and $n = m^t$.

Let G be a group such that $\Omega \trianglelefteq G \leqslant A$, as in Series 1.1, and let $K \leqslant G$. If $G \leqslant \Gamma$ then K lies in *Class* $\mathscr{C}_7$ if K is the stabiliser in G of a tensor induced decomposition, as in Table 2.10. Otherwise, K lies in Class $\mathscr{C}_7$ if $K = \mathrm{N}_A(H) \cap G$, where H is a $\mathscr{C}_7$-subgroup of Γ.

Our dimension restrictions in Definition 1.6.20 mean that the groups described below the horizontal line in Table 2.10 do not occur in any of our Cases in dimension at most 12.

All members of Class $\mathscr{C}_7$ are tensor induced. Note that some authors require a tensor induced group to permute the tensor factors transitively: as can be seen from Table 2.10 all groups in Class $\mathscr{C}_7$ of dimension at most 12 in fact act on the t tensor factors as S_t, although this is not true for all dimensions.

The first part of Table 2.10 is taken from [66, Table 4.7.A], where we have imposed the additional condition that the derived group of the Ω-group on V_1 is quasisimple from Part (b) of their definition of Class $\mathscr{C}_7$. The second part of Table 2.10 can be deduced from [66, Propositions 4.7.3–4.7.8].

We briefly explain the shapes of the groups in Table 2.10. The normaliser in $\mathrm{GL}_n(q)$ of a $\mathscr{C}_7$-subgroup of type $\mathrm{GL}_m(q) \wr \mathrm{S}_t$ is a *tensor wreath product* of $\mathrm{GL}_m(q)$ and S_t. It is a quotient of the standard wreath product, where all t copies of the central subgroup of $\mathrm{GL}_m(q)$ have been identified. Recall Definition 1.9.7 of the tensor product of two groups: here, the base group of the standard wreath product has been replaced by the tensor product of the t copies of $\mathrm{GL}_m(q)$.

Lemma 2.2.16 *Let H be a $\mathscr{C}_7$-subgroup of Ω. Then H is insoluble and H^∞ is absolutely irreducible.*

Proof Here H contains a central product T of t copies of Ω_m, where Ω_m is the generally quasisimple group on V_1, as in Definition 1.6.13. The restrictions on m and q given in Table 2.10 ensure that Ω_m is in fact quasisimple (or $\mathrm{Sp}_4(2)$), so that T^∞ is perfect and H is insoluble. Proposition 1.12.2 implies that Ω_m is absolutely irreducible, so T is absolutely irreducible by Lemma 2.2.10. □

2.2.8 Class $\mathscr{C}_8$

In this subsection we present some basic information about the $\mathscr{C}_8$-subgroups. We refer to the families of $\mathscr{C}_8$-subgroups by their *type*, as listed in Table 2.11. Each type arises from the fact that it may be possible to define a second form on an already-formed space.

Definition 2.2.17 Let G be a group such that $\Omega \trianglelefteq G \leqslant A$, as in Series 1.1,

Table 2.10 *Types of $\mathscr{C}_7$-subgroups*

Case	Type	Conditions
L	$\mathrm{GL}_m(q) \wr \mathrm{S}_t$	$m \geqslant 3$
U	$\mathrm{GU}_m(q) \wr \mathrm{S}_t$	$m \geqslant 3$, $(m, q) \neq (3, 2)$
S	$\mathrm{Sp}_m(q) \wr \mathrm{S}_t$	qt odd, $(m, q) \neq (2, 3)$
O°	$\mathrm{GO}_m(q) \wr \mathrm{S}_t$	$m \geqslant 3$, $(m, q) \neq (3, 3)$
O⁺	$\mathrm{Sp}_m(q) \wr \mathrm{S}_t$	qt even, $(m, q) \notin \{(2, 2), (2, 3)\}$
O⁺	$\mathrm{GO}_m^\varepsilon(q) \wr \mathrm{S}_t$	q odd, $\varepsilon = + \Rightarrow m \geqslant 6$, $\varepsilon = - \Rightarrow m \geqslant 4$

Case	Shape of $H < \Omega$	Conditions
L	$(q - 1, m).\mathrm{L}_m(q)^2.[(q - 1, m)^2]$	$t = 2$, $m \equiv 2 \bmod 4$, $q \equiv 3 \bmod 4$
	$(q - 1, m).\mathrm{L}_m(q)^t.[(q-1, \frac{n}{m})(q-1, m)^{t-1}].\mathrm{S}_t$	otherwise
U	$(q + 1, m).\mathrm{U}_m(q)^2.[(q + 1, m)^2]$	$t = 2$, $m \equiv 2 \bmod 4$, $q \equiv 1 \bmod 4$
	$(q + 1, m).\mathrm{U}_m(q)^t.[(q+1, \frac{n}{m})(q+1, m)^{t-1}].\mathrm{S}_t$	otherwise
S	$2.\mathrm{S}_m(q)^t.2^{t-1}.\mathrm{S}_t$	
O°	$\Omega_m^\circ(q)^t.2^{t-1}.\mathrm{S}_t$	
O⁺	$(q - 1, 2).\mathrm{S}_m(q)^2$	$t = 2$, $m \equiv 2 \bmod 4$
	$(q - 1, 2).\mathrm{S}_m(q)^t.(q - 1, 2)^{t-1}.\mathrm{S}_t$	otherwise
O⁺	$2.\mathrm{PSO}_m^\varepsilon(q)^2.[4]$	$t = 2$, $m \equiv 2 \bmod 4$
	$2.\mathrm{PSO}_m^-(q)^2.[8]$	$t = 2$, $m \equiv 0 \bmod 4$
	$2.\mathrm{PSO}_m^\varepsilon(q)^3.[2^5].3$	$t = 3$, $m \equiv 2 \bmod 4$, V_1 non-square discriminant
	$2.\mathrm{PSO}_m^\varepsilon(q)^t.[2^{2t-1}].\mathrm{S}_t$	otherwise

Table 2.11 *Types of $\mathscr{C}_8$-subgroups*

Case	Type	Shape of $H < \Omega$	Conditions
L	$\mathrm{Sp}_n(q)$	$(q - 1, n)\dot{\ }\mathrm{S}_n(q)$	n even, $(q - 1, n/2) = (q - 1, n)/2$
		$(q - 1, n)\dot{\ }\mathrm{PCSp}_n(q)$	n even, $(q - 1, n/2) = (q - 1, n)$
L	$\mathrm{GU}_n(q^{1/2})$	$\mathrm{SU}_n(q^{1/2}).(q^{1/2} - 1, n)$	q square
L	$\mathrm{GO}_n^\varepsilon(q)$	$\mathrm{SO}_n^\varepsilon(q).(q - 1, n)$	q odd
S	$\mathrm{GO}_n^\pm(q)$	$\mathrm{GO}_n^\pm(q)$	q even

and let $K \leqslant G$. If $G \leqslant \Gamma$ then K lies in *Class $\mathscr{C}_8$* if K is the intersection with G of the Γ-group of a classical group given Table 2.11. Otherwise, K lies in Class $\mathscr{C}_8$ if $K = \mathrm{N}_A(H) \cap G$, where H is a $\mathscr{C}_8$-subgroup of Γ.

All columns of Table 2.11 except for the one describing the shape of $H < \Omega$ are taken from [66, Table 4.8.A]. The shape of $H < \Omega$ can generally be deduced from [66, Propositions 4.8.3–6], together with Lemma 1.13.5: the one exception is our division of the symplectic subgroups of $\mathrm{SL}_n(q)$ into two families, which is not mentioned in [66, Proposition 4.8.3], but is immediate from the generating matrices constructed in [45].

Note that in [66, Table 4.8.A] Class $\mathscr{C}_8$ is defined to be empty when $n = 2$. However, we feel it is more uniform to include these groups at first: we will prove in Lemma 3.1.1 that they are equal to other subgroups when $n = 2$.

In the next result, let G be the general group corresponding to Ω (see Definition 1.6.5) and let Γ be the conformal semilinear group, as in Definition 1.6.16: recall the dimension restrictions in Definition 1.6.20.

Proposition 2.2.18   *Let J be a subgroup of G, and assume that J is irreducible but not absolutely irreducible on $V := \mathbb{F}_{q^u}^n$. If $G = \mathrm{Sp}_n(q)$ with $n \geqslant 4$ and q is even then $\mathrm{N}_\Gamma(J)$ is contained in a member of $\mathscr{C}_3 \cup \mathscr{C}_8$. If $G = \mathrm{GO}_n^+(q)$ and $n \geqslant 8$ then $\mathrm{N}_\Gamma(J)$ is contained in a member of $\mathscr{C}_2 \cup \mathscr{C}_3$. If $G = \mathrm{GL}_n(q)$ and $n \geqslant 2$, or $G = \mathrm{GU}_n(q)$ and $n \geqslant 3$, or $G = \mathrm{Sp}_n(q)$ with $n \geqslant 4$ and q odd, or $G = \mathrm{GO}_n^-(q)$ and $n \geqslant 8$, then $\mathrm{N}_\Gamma(J)$ is contained in a member of $\mathscr{C}_3$.*

Proof   This is essentially [66, Lemma 4.3.12], but we prove a more detailed version. Since J is irreducible but not absolutely irreducible, $E := \mathrm{End}_{\mathbb{F}_{q^u} J}(V)$ is a field that properly contains $\mathbb{F}_{q^u}$. Let F be a subfield of E such that $\mathbb{F}_{q^u}$ has prime index in F. Then $\mathrm{N}_\Gamma(J) \leqslant \mathrm{N}_\Gamma(E) \leqslant K := \mathrm{N}_\Gamma(F)$. Now, since the non-zero elements of F are the non-zero elements of a field embedded in $\mathrm{GL}_n(q^u)$ and K is irreducible, K is by definition a member of Aschbacher's class of semilinear groups (see [1, p472]). Thus either K is contained in a member of $\mathscr{C}_3$, or K is one of the groups that lie in Aschbacher's Class $\mathscr{C}_3$ but not in ours. Consulting [66, p112] we see that the only possibilities are: (i) $K \leqslant \mathrm{C}\Gamma\mathrm{Sp}_n(q)$ with q even, and $K \cap \mathrm{Sp}_n(q) \cong \mathrm{GU}_{n/2}(q).2$ with $F = \mathbb{F}_{q^2}$; or (ii) $K \leqslant \mathrm{C}\Gamma\mathrm{O}_{2s}^\pm(q)$, with $K \cap \Omega_{2s}^\pm(q)$ of shape $\Omega_2^\pm(q^s).[cs]$ for some $1 \leqslant c \leqslant 2$, with $F = \mathbb{F}_{q^s}$. In either case, fix an isomorphism α from $\mathbb{F}_{q^s}^{n/s}$ to $\mathbb{F}_q^n$ (where $s = 2$ in Case (i)).

First assume that K is of shape $\mathrm{GU}_{n/2}(q).2.(q-1).e$ in $\Gamma = \mathrm{C}\Gamma\mathrm{Sp}_n(q)$ with q even. Let β be the unitary form on $\mathbb{F}_{q^2}^{n/2}$ for which K is a group of semi-similarities, and for $v \in \mathbb{F}_q^n$ define $Q(v) = \beta(v\alpha, v\alpha)$. Then it may be checked (or see [66, p117–118]) that Q is a non-degenerate quadratic form, so K is contained in a $\mathscr{C}_8$-subgroup of Γ of type $\mathrm{GO}_n^\pm(q)$.

Next assume that K is of shape $\Omega_2^+(q^s).[cs].[(2, q-1)2(q-1)].e$ in $\mathrm{C}\Gamma\mathrm{O}_{2s}^+(q)$. The group $\Omega_2^+(q^s)$ is reducible, stabilising an imprimitive decomposition into two totally singular subspaces. These correspond under α in $\mathbb{F}_q^{2s}$ to two totally singular s-spaces (see [66, p120]) that are stabilised by the (normal) subgroup of K isomorphic to $\Omega_2^+(q^s)$, so K is contained in a $\mathscr{C}_2$-subgroup of type $\mathrm{GL}_{n/2}(q).2$.

Finally assume that K is of shape $\Omega_2^-(q^s).[cs][(2, q-1)2(q-1)].e$ in $\Gamma = \mathrm{C}\Gamma\mathrm{O}_{2s}^-(q)$, and notice that s is odd. The group $\mathrm{GO}_2^-(q^s) \trianglelefteq K$ contains a characteristic cyclic subgroup $S \cong q^s + 1$, which is $\mathrm{SO}_2^-(q^s)$ when q is odd and $\Omega_2^-(q^s)$ when q is even. By [66, Lemma 4.3.11], after identifying $\mathbb{F}_{q^s}^2$ with $\mathbb{F}_{q^{2s}}$, the group S preserves a unitary form; so the characteristic subgroup S of K may be identified with $\mathrm{GU}_1(q^s)$. Therefore, $\mathrm{N}_\Gamma(J) \leqslant K \leqslant \mathrm{N}_\Gamma(\mathrm{GU}_1(q^s))$. Now $\mathrm{GU}_1(q^s)$ embeds in a $\mathscr{C}_3$-subgroup of $\mathrm{GU}_s(q)$, which is in a $\mathscr{C}_3$-subgroup of $\mathrm{GO}_{2s}^-(q)$. We claim that $\mathrm{N}_\Gamma(\mathrm{GU}_1(q^s)) < \mathrm{N}_\Gamma(\mathrm{GU}_s(q))$, which will complete the proof.

To see this, observe first that the normaliser in $\mathrm{GL}_{2s}(q)$ of $\mathrm{GU}_1(q^s)$ is a $\mathscr{C}_3$-group of type $\mathrm{GL}_1(q^{2s})$ and shape $(q^{2s} - 1).2s$. By [52, Satz 3], the group $\mathrm{GU}_1(q^s)$ is self-centralising in $\mathrm{GO}_{2s}^-(q)$. Since the index of $\mathrm{GO}_{2s}^-(q)$ in $\mathrm{C}\Gamma\mathrm{O}_{2s}^-(q)$ is $(q-1)e$ (where $q = p^e$), we deduce that $|\mathrm{N}_\Gamma(\mathrm{GU}_1(q^s))| \leqslant (q^s + 1)2s(q-1)e$. But there is a unique class of subgroups of $\Omega_{2s}^-(q)$ of type $\mathrm{GU}_s(q)$, and a unique class of subgroups of $\mathrm{SU}_s(q)$ of type $\mathrm{GU}_1(q^s)$ (see [66, Propositions 4.3.6, 4.3.18]), so a straightforward order calculation shows that $\mathrm{N}_\Gamma(\mathrm{GU}_1(q^s))$ has a subgroup of order at least $2es(q-1)(q^s+1)$ that lies in $\mathrm{N}_\Gamma(\mathrm{GU}_s(q))$. $\square$

2.2.9 Aschbacher's theorem, revisited

We are now in a position to state a more accurate version of Aschbacher's theorem. The theorem given below is essentially [66, Main Theorem, p 57], and so is slightly different from the version in [2]: the differences are justified in [66].

Recall Table 1.2 of our notation for groups, Definition 2.1.2 of a *geometric subgroup*, and Definition 2.1.3 of *Class $\mathscr{S}$*.

Theorem 2.2.19 *Let Ω be a quasisimple classical group, and let G be any group such that $\Omega \trianglelefteq G \leqslant A$, where A is the corresponding group in Column A of Table 1.2.*

(i) *Let H be a geometric subgroup of G that is maximal in G. Then:*

   (a) *the group H is a member of $\mathscr{C}_i$ for some $1 \leqslant i \leqslant 8$;*

   (b) *the shape of $H \cap \Omega$ is as given in Tables 2.3, 2.5, 2.6, 2.7, 2.8, 2.9, 2.10 or 2.11;*

   (c) *the number of conjugacy classes in Ω of groups of the same type as H, and their stabilisers in G, are as given in [66, Tables 3.5.A–F, Column V], except that in $\mathrm{SL}_n(q)$ with $n \geqslant 3$ our groups of type P_k and P_{n-k} are both conjugate to their groups of type P_k (where $k < n/2$).*

(ii) *If K is any other maximal subgroup of G, and K does not contain Ω, then K lies in Class $\mathscr{S}$.*

2.3 Preliminary arguments concerning maximality

In this section we shall work through Classes $\mathscr{C}_1$ to $\mathscr{C}_8$, and establish some generic results concerning maximality in dimension up to 12. In particular, if it is possible to prove that groups of one type are or are not contained in groups of some other type, without needing to use arguments that depend on specific dimensions, then we shall do so here. This will greatly shorten our arguments in the next chapter, where we consider one dimension at a time. Recall throughout our dimension restrictions in Definition 1.6.20: although the classical groups can arise in dimensions not covered by our Cases (e.g. $\Omega_5(q)$) they are either isomorphic to classical groups which we *are* considering or, in the case of $\Omega_8^+(q)$ only, the maximal subgroups of all almost simple extensions of $\overline{\Omega}$ have already been classified (in [62]).

Recall Definition 1.6.19 of $\overline{\Omega}$, and the dimension restrictions in Definition 1.6.20. We also remind the reader that the files of MAGMA calculations that we refer to are available on the webpage

http://www.cambridge.org/9780521138604.

2.3.1 Reducible groups

Recall Definition 2.2.1 of the $\mathscr{C}_1$-subgroups, their types as given in Table 2.2, and their structures in the quasisimple group Ω as in Table 2.3.

Proposition 2.3.1   *Let $n \leqslant 12$, let $H \leqslant \Omega$ be of type P_k, where $1 \leqslant k \leqslant n/2$ and Ω is the quasisimple group in Case **L**, **U**, **S** or $\mathbf{O}^\varepsilon$. Let G be almost simple with socle $\overline{\Omega}$, and let H_G be the $\mathscr{C}_1$-subgroup of G of type P_k. Then H is the stabiliser in Ω of a totally singular k-space W and the following hold:*

(i) *The group H is maximal amongst the geometric subgroups of Ω unless $\Omega = \Omega_n^+(q)$ and $k = n/2 - 1$.*

(ii) *If $\Omega = \Omega_n^+(q)$ and $k = n/2 - 1$, then H_G is non-maximal in every G such that $G \leqslant \mathrm{O}_n^+(q).K$, where $K = \langle \phi \rangle$ if q is even, $K = \langle \delta, \phi \rangle$ if q is odd and $n(q-1)/4$ is odd, and $K = \langle \delta', \delta, \phi \rangle$ if q is odd and $n(q-1)/4$ is even. If $\mathrm{O}_n^+(q) \trianglelefteq G \leqslant \mathrm{P\Gamma O}_n^+(q)$ but $G \not\leqslant \mathrm{O}_n^+(q).K$ then H_G is maximal amongst the geometric subgroups of G.*

Proof   By definition of type P_k, the group H is the stabiliser of a totally singular k-space, and so contains a Sylow p-subgroup P of Ω by Lemma 2.2.2 (iii). By Theorem 2.2.19, no member of $\mathscr{C}_2 \cup \cdots \cup \mathscr{C}_8$ contains such a subgroup, as none of them has order divisible by $|P|$. Thus if $H < K < \Omega$ for some geometric subgroup K, then without loss of generality $K \in \mathscr{C}_1$.

Groups of type $G_k \oplus G_{n-k}$ or $G_k \perp G_{n-k}$ do not have order divisible by $|P|$.

The natural module for groups of type $P_{m,n-m}$ (which only occur in Case **L**) has composition factors of dimension $m, m, n - 2m$, and so such groups cannot contain H, whose natural module has composition factors of dimension k and $n - k$. So $K = \mathrm{N}_\Omega(W_1)$ for some totally singular m-space $\{0\} \neq W_1 < V$.

Since $n \geqslant 2$, and $n \geqslant 10$ in Case $\mathbf{O}^+$, by Lemma 2.2.2 (i),(ii) the subgroup P of H fixes a unique 1-dimensional totally singular subspace $\langle v \rangle$, which is a term in the unique maximal flag that P stabilises if $\Omega \neq \Omega_n^+(q)$, and in both such flags in Case $\mathbf{O}^+$. Therefore $\langle v \rangle \leqslant W$ and $\langle v \rangle \leqslant W_1$, so $W_1 \cap W \neq \{0\}$. The group $\mathrm{SL}(W)$ is a subgroup of H^W by Lemma 2.2.2 (iv), so H^W is irreducible on W, and so $W < W_1$. In Case **L** the group H acts as $\mathrm{GL}(V/W)$ on V/W, and hence acts irreducibly on V/W by Proposition 1.12.2, a contradiction. This completes Case **L**.

In Cases **S**, **U** and $\mathbf{O}^\varepsilon$, the fact that W_1 is totally singular and $W < W_1$ implies that $W_1 < W^\perp$. Therefore H stabilises $W_1/W < W^\perp/W$, and hence acts reducibly on $W^\perp/W$. By Lemma 2.2.2 (v) there exist two subspaces X and Y of V such that Y is a totally singular k-space, $W \oplus Y$ is non-degenerate and $X = (W \oplus Y)^\perp$. Hence $X \cap W = \{0\}$ and $X \leqslant W^\perp$, and so $W^\perp = W \perp X$. Then $\Omega(X) \leqslant H^X$ and $H^{W^\perp/W}$ is reducible so, by Proposition 1.12.2, $\dim(X) = 2$ and $\Omega = \Omega_n^\varepsilon(q)$.

Suppose finally that $\dim(X) = 2$ and $\Omega = \Omega_n^\varepsilon(q)$. Then n is even and $\dim(W) = n/2 - 1$. Since $W < W_1$ and W_1 is totally singular, the only possibility is $\dim(W_1) = n/2$, and so $\varepsilon = +$. The proper containment of H in two copies of $P_{n/2}$ in Case $\mathbf{O}^+$ is discussed in [66, Prop 6.1.1]. Let H stabilise $W := \langle e_1, \ldots, e_{n/2-1} \rangle$. Then H is contained in two $\Omega_n^+(q)$-classes of groups of type $P_{n/2}$, namely the stabiliser of $W_1 = \langle e_1, \ldots, e_{n/2} \rangle$ and $W_2 = \langle e_1, \ldots, e_{n/2-1}, f_{n/2} \rangle$. Each class has stabiliser K, as defined in the statement, and outer automorphisms not in K interchange W_1 and W_2. The space W is stabilised by all of $\mathrm{PC\Gamma O}_n^+(q)/\mathrm{O}_n^+(q)$, so the result follows. □

Recall Definition 1.5.35 of the discriminant of a form.

Proposition 2.3.2 *Let $n \leqslant 12$. In Case **L**, let $H \leqslant \Omega$ be of type $\mathrm{GL}_k(q) \oplus \mathrm{GL}_{n-k}(q)$. In Cases **S**, **U** and $\mathbf{O}^\varepsilon$, let $H \leqslant \Omega$ be the stabiliser of a k-dimensional non-degenerate subspace W. Assume additionally that $n \geqslant 4$ in Cases **L** and **U**, and that $n - k \geqslant 7$ in Case $\mathbf{O}^\pm$. Let G be an almost simple group with socle $\bar{\Omega}$, and let H_G be the subgroup of G of the same type as H.*

(i) *In Case **L**, if $G \leqslant \mathrm{P\Gamma L}_n(q)$ then H_G is non-maximal in G, but otherwise H_G is maximal amongst the geometric subgroups of G.*

(ii) *In Cases **S** and **U**, H is maximal amongst the geometric subgroups of Ω.*

(iii) *In Case $\mathbf{O}^\varepsilon$, if $(k, q, \mathrm{sgn}(W)) \notin \{(2, 2, +), (2, 3, +)\}$ then H is maximal amongst the geometric subgroups of Ω. If $(k, q, \mathrm{sgn}(W)) = (2, 2, +)$ then*

H_G is non-maximal. If $(k, q, \mathrm{sgn}(W)) = (2, 3, +)$ then H_G is maximal amongst the geometric subgroups of G if and only if $G \not\leqslant \mathrm{PGO}_n^\varepsilon(q)$ and n is even.

Proof Note that $k < n/2$ in Cases **L**, **U**, **S**, **O**° and **O**$^+$, and $k \leqslant n/2$ in Case **O**$^-$. If $k = 3$ and $\Omega = \Omega_7^\circ(q)$, let $n_1 = 4$ and $n_2 = 3$, otherwise let $n_1 = k$ and $n_2 = n - k$. Let V_1 be the n_1-dimensional space stabilised by H and let V_2 be the n_2-dimensional space stabilised by H, so that $V = V_1 \perp V_2$. Then H contains a subgroup $L = \Omega_1 \times \Omega_2 := \Omega(V_1) \times \Omega(V_2)$.

Our assumptions on n and k imply that Ω_2 is quasisimple, with the exceptions of $(n, n_2, q, \text{Case}) \in \{(6, 4, 2, \mathbf{S}), (4, 3, 2, \mathbf{U}), (5, 3, 2, \mathbf{U}), (7, 3, 3, \mathbf{O}°)\}$. We start by considering these four exceptions.

In $\mathrm{Sp}_6(2)$, by Theorem 2.2.19 the only possibly maximal geometric subgroup with order divisible by $|H|$ is $\mathrm{SO}_6^-(2) \in \mathscr{C}_8$. However, H would have index 12 in $\mathrm{SO}_6^-(2) \cong \mathrm{U}_4(2){:}2$, whereas by Theorem 1.11.2 the two conjugacy classes of lowest index subgroups of $\mathrm{U}_4(2){:}2$ have indices 2 and greater than 12, respectively.

In $\mathrm{SU}_4(2)$ the group H has order 648, which by Theorem 2.2.19 is not a divisor of the order of any other geometric subgroup.

In $\mathrm{SU}_5(2)$ with $k = 2$, the order of H is 3888, and by Theorem 2.2.19 the only potentially maximal geometric subgroup of $\mathrm{SU}_5(2)$ to have order a multiple of $|H|$ is of type $\mathrm{GU}_1(2) \perp \mathrm{GU}_4(2)$. However, H acts irreducibly on V_1 and V_2 by Proposition 1.12.2, and so does not stabilise a non-degenerate 1-space.

Now consider $\Omega_7(3)$ with $n_2 = 3$. By Theorem 2.2.19, all geometric subgroups of $\Omega_7(3)$ lie in $\mathscr{C}_1 \cup \mathscr{C}_2$. The group H acts irreducibly on both V_1 and V_2 by Proposition 1.12.2, and so does not stabilise any other subspace, so is not contained in a member of $\mathscr{C}_1$. The only $\mathscr{C}_2$-subgroup of $\Omega_7(3)$ is isomorphic to $2^6{:}A_7$, which does not have order divisible by $|H|$. Thus H is maximal amongst the geometric subgroups of $\Omega_7(3)$.

Having dealt with these exceptional cases, Ω_2 is quasisimple. Suppose, by way of contradiction, that $H \leqslant K < \Omega$, where K is maximal amongst the geometric subgroups of Ω and is not of the same type as H. Note if $\Omega \neq \Omega_7(q)$ or $n_2 \neq 3$ then $n_2 > n/2$.

First we prove that $K \notin \mathscr{C}_2 \cup \mathscr{C}_4 \cup \mathscr{C}_7$. Otherwise, $K = \Omega_{\mathcal{D}}$, where $\mathcal{D}$ is either a direct sum decomposition $V = W_1 \oplus \cdots \oplus W_t$ or a tensor decomposition $W_1 \otimes \cdots \otimes W_t$. Here $1 < t \leqslant n \leqslant 12$, whilst $\dim(W_i)$ is a proper divisor of n for $1 \leqslant i \leqslant t$. By Theorem 1.11.2, since Ω_2 is quasisimple $P(\Omega_2) > n$ unless $\Omega_2 \cong \Omega_3(q) \leqslant \Omega_7(q)$ for $q \in \{5, 7, 9\}$.

Considering these three exceptions, note that $\mathscr{C}_4$ and $\mathscr{C}_7$ are void in $\Omega_7(q)$. If V_1 is of $+$ type then $\Omega_1 \times \Omega_2 \cong 2.\mathrm{L}_2(q)^2 \times \mathrm{L}_2(q) \leqslant H$, which is not a subgroup of $2^6{:}S_7$ for $q \in \{5,7,9\}$. If V_1 is of $-$ type then $L_2(q^2) \times L_2(q) \leqslant H$, which is not a subgroup of $2^6{:}S_7$.

Having dealt with these exceptions, we deduce that Ω_2 and $\overline{\Omega}_2$ are not subgroups of S_t. Since Ω_2 is quasisimple, $\Omega_2 \leqslant \Omega_{(\mathcal{D})}$, the pointwise stabiliser of $\{W_1,\ldots,W_t\}$. Hence $\dim(W_i) \geqslant R_p(\overline{\Omega}_2)$ for some i with $1 \leqslant i \leqslant t$. If $\Omega \neq \Omega_n^{\circ}(q)$ then $R_p(\overline{\Omega}_2) = n_2$ by Theorem 1.11.5. Therefore $\dim(W_i) \mid n$ and $\dim(W_i) \geqslant n_2 > n/2$, a contradiction. So $\Omega = \Omega_n^{\circ}(q)$. If $n = 7$ or 11 then the only imprimitive decompositions are 1-dimensional, so $\Omega_{(\mathcal{D})}$ cannot contain Ω_2. Otherwise, $n = 9$ and $R_p(\overline{\Omega}_2) \geqslant 4 > 3$, a contradiction.

Next we show that $K \notin \mathscr{C}_3$. Otherwise, $\Omega_2 \leqslant \Gamma$, where Γ is a $\mathscr{C}_3$-subgroup of degree n/s for some prime divisor s of n. Since Ω_2 is quasisimple, $\Omega_2 \leqslant K' \leqslant \mathrm{GL}_{n/s}(q^s)$. In Cases **L**, **U**, **S** and $\mathbf{O}^{\pm}$, since $R_p(\Omega_2) = n_2$ by Theorem 1.11.5, it follows that $n_2 \leqslant n/s$, contradicting $n_2 > n/2$. In Case $\mathbf{O}^{\circ}$ if $n = 7$ or 11 then $\mathscr{C}_3 = \varnothing$. Otherwise $n = 9$, and we get a contradiction as before.

Next we show that $K \notin \mathscr{C}_5$. Otherwise $\Omega_2 \leqslant \Omega_{q_0}^{\infty}$, where $\Omega_{q_0}^{\infty}$ is a quasisimple subfield group over $\mathbb{F}_{q_0}$, and in $\Omega_7(q)$ if $n_1 = 4$ then $\Omega_1 \leqslant \Omega_{q_0}^{\infty}$ as well. Now, Ω_1 and Ω_2 are both subgroups of H, acting naturally on $V_1 < V$ and $V_2 < V$. Furthermore at least one of Ω_1 and Ω_2 has dimension at least 4 in Case $\mathbf{O}^{\varepsilon}$. Thus in all cases the traces of elements of H^{∞} are all elements of $\mathbb{F}_{q^u}$ by Proposition 1.12.7 ($u = 2$ in Case **U**, 1 otherwise), contradicting the fact that all elements of $\Omega_{q_0}^{\infty}$ have traces in $\mathbb{F}_{q_0}$.

Suppose next that $K \in \mathscr{C}_6$. Then $n = r^b \in \{4,5,7,8,9,11\}$ with r a prime divisor of $q^u - 1$, the group $\Omega \neq \Omega_n^{\varepsilon}(q)$, and if $\Omega = \mathrm{Sp}_n(q)$ then $n = 8$ (since $n \geqslant 6$ for H to be defined), all from Table 2.9. Since Ω_2 is quasisimple and the only non-abelian composition factor of K is a subgroup of $\mathrm{PGL}_{2b}(r)$, there is a non-trivial representation ρ such that $\Omega_2\rho \leqslant \mathrm{PGL}_{2b}(r)$, and the possibilities for n imply that $2b \leqslant 6$. Given $n_2 \geqslant 3$, with $n_2 > n/2$ in Cases **L** and **U** and $n_2 = 6$ in Case **S**, Theorem 1.11.7 shows that the only possibilities are $\Omega_2 = \mathrm{SL}_3(2)$, $\mathrm{SL}_3(4)$, $\mathrm{SU}_3(3)$, $\mathrm{SU}_4(2)$ or $\mathrm{SU}_4(3)$. If $\Omega_2 = \mathrm{SL}_3(2)$ then there are no $\mathscr{C}_6$-subgroups as they require $q \equiv 1 \bmod r$. If $\Omega_2 = \mathrm{SL}_3(4)$ then $r = 3$ but $n \in \{4,5\}$ is not a power of 3, a contradiction. If $\Omega_2 = \mathrm{SU}_3(3)$ or $\mathrm{SU}_4(3)$ then Theorem 1.11.7 states that $R_{p'}(\overline{\Omega}_2) = 6$, forcing $b \geqslant 3$, so $n = r^b \geqslant 8 \geqslant 2n_2$, a contradiction. Finally, if $\Omega_2 = \mathrm{SU}_4(2)$ then $r = 3$, and hence n is a power of 3. However, $5 \leqslant n \leqslant 7$, a contradiction.

Next we prove that $K \notin \mathscr{C}_8$. Suppose otherwise, so $\Omega = \mathrm{SL}_n(q)$ or $\mathrm{Sp}_n(q)$. In Case **L**, K consists of isometries of a non-degenerate unitary, symplectic or quadratic form κ. Since $n_2 > 2$, the only $\mathrm{SL}(V_2)$-invariant form on V_2 is the zero form, so V_2 must be totally singular with respect to κ, but $\dim(V_2) > n/2$, a contradiction. If $\mathscr{C}_8 \neq \varnothing$ in Case **S** then q is even and $K = N_\Omega(Q)$ for some quadratic form Q whose bilinear form is equal to the symplectic form f. The space (V_2, f) is non-degenerate, so by definition the space (V_2, Q) is also non-degenerate. Therefore $\Omega_2 = \mathrm{Sp}_{n_2}(q) \leqslant \mathrm{GO}_{n_2}^{\pm}(q)$, a contradiction as $|\mathrm{Sp}_{n_2}(q)| > |\mathrm{GO}_{n_2}^{\pm}(q)|$ for $n_2 \geqslant 4$.

Finally, consider $K \in \mathscr{C}_1$. In Case **L**, the group K is not of type $P_{j,n-j}$ for any j, as then the natural module for K has composition factors of dimension $j, j, n-2j$, whereas the natural module for H has composition factors of dimension k and $n-k$. Similarly K is not of type $\mathrm{GL}_j(q) \oplus \mathrm{GL}_{n-j}(q)$ for any $j \neq k$. However, H *is* contained in the parabolic subgroup P_k that stabilises V_1, and is also contained in the conjugate P_k^{γ} that stabilises V_2 in $\mathrm{SL}_n(q)$. Therefore H is indeed non-maximal in $\mathrm{SL}_n(q)$. We note that γ interchanges the stabilisers of V_1 and V_2, so the extension of H by γ is not contained in the normaliser of P_k or of P_k^{γ}, and hence is maximal amongst the geometric subgroups.

In Cases **S** and **U** if $K \in \mathscr{C}_1$ then H fixes some non-zero proper subspace of V other than V_1 and V_2. However, by Proposition 1.12.2 the groups Ω_i act irreducibly on V_i for $i = 1, 2$, a contradiction.

In Case **O**$^{\varepsilon}$ the group Ω_2 acts irreducibly on V_2 by Proposition 1.12.2, so if $K \in \mathscr{C}_1$ then H^{V_1} must act reducibly on V_1. Now, $H^{V_1} \cong \mathrm{GO}(V_1)$, so by Proposition 1.12.2 the group $H^{V_1} \cong \mathrm{GO}_2^+(q)$ for $q \leqslant 3$. These two exceptions are discussed in [66, Proposition 6.1.2], where it is shown that if $q = 2$ then K is the stabiliser of a non-singular vector in V_1, whilst if $q = 3$ then K is the stabiliser of a non-singular vector of either square or non-square norm (two groups). In both cases, the containment of H in K is easily shown to be proper for $n \geqslant 10$. In the latter case, if n is even then a novel maximal subgroup may arise under the diagonal automorphism which multiplies the form by a non-square, as this interchanges the two choices for K whilst normalising H.   □

Recall that the $\mathscr{C}_1$-subgroup of type $\mathrm{Sp}_{n-2}(q)$ only occurs when $\Omega = \Omega_n^{\pm}(q)$ and q is even, and that Case **O**$^+$ requires $n \geqslant 10$.

Proposition 2.3.3   *If $H < \Omega$ is of type $\mathrm{Sp}_{n-2}(q)$ in Case $\mathbf{O}^{\pm}$ with $n \leqslant 12$, then H is maximal amongst the geometric subgroups of Ω.*

Proof   The proof is similar to that of Proposition 2.3.2. Now, q is even and

$$\Omega \in \{\Omega_8^-(q), \Omega_{10}^{\pm}(q), \Omega_{12}^{\pm}(q)\},$$

so $\mathscr{C}_4 \cup \mathscr{C}_6 \cup \mathscr{C}_7 \cup \mathscr{C}_8 = \varnothing$, by Theorem 2.2.19. Since $n - 2 \geqslant 6$ the group $H \cong \mathrm{Sp}_{n-2}(q)$ is simple. Assume by way of contradiction that $H < K < \Omega$, where K is maximal amongst the geometric subgroups of Ω.

We show first that $K \notin \mathscr{C}_1$. The group H fixes a unique non-singular 1-space, W, and acts as $\mathrm{Sp}_{n-2}(q)$ on $W^{\perp}/W$. Suppose $K = \mathrm{N}_{\Omega}(U)$ for some $\{0\} < U < V$. If U is non-degenerate, then the largest possible non-abelian composition factor of K is $\Omega_{n-2}^{\pm}(q)$, which is smaller than $\mathrm{Sp}_{n-2}(q)$. Thus U is totally singular, of dimension k say, and as a K-module, V has composition factors of dimension k (twice), and $n - 2k$ (or possibly $1, 1$ if $n - 2k = 2$). Thus U is a totally singular 1-space, but then again the largest composition factor of K is $\Omega_{n-2}^{\pm}(q)$.

We show next that $K \notin \mathscr{C}_2$. Note that $n-2 \in \{6, 8, 10\}$, so $P(\mathrm{Sp}_{n-2}(q)) > n$ by Theorem 1.11.2. Thus, if $K \in \mathscr{C}_2$ then $\mathrm{Sp}_{n-2}(q)$ lies in the kernel of the block action. This contradicts Theorem 1.11.5, which states that $R_2(\mathrm{Sp}_{n-2}(q)) = n - 2 > n/2$. Similarly, $K \notin \mathscr{C}_3$, for otherwise $R_2(K') \leqslant n/2$.

Finally, $K \notin \mathscr{C}_5$, for otherwise $K^{\infty} \cong \Omega_n^{\pm}(q_0)$ for some proper divisor q_0 of q. A short calculation shows that the 2-part of $|K^{\infty}|$ is less than that of $|H|$, a contradiction. $\qquad\square$

Proposition 2.3.4 *Let $3 \leqslant n \leqslant 12$, let G be almost simple with socle $L_n(q)$, and let $H < \mathrm{SL}_n(q)$ and $H_G < G$ be of type $P_{k,n-k}$. If $G \leqslant \mathrm{P\Gamma L}_n(q)$ then H_G is non-maximal in G. However, if $G \nleqslant \mathrm{P\Gamma L}_n(q)$ then H_G is maximal amongst the geometric subgroups of G.*

Proof Here $H = \mathrm{N}_{\mathrm{SL}_n(q)}(V_1, V_2)$, where $\dim(V_1) = k$ for some $k < n/2$, $\dim(V_2) = n - k$, and $V_1 < V_2$. Now H contains a Sylow p-subgroup of $\mathrm{SL}_n(q)$, and as in Proposition 2.3.1 we deduce that if there exists a geometric group K with $H < K < \mathrm{SL}_n(q)$ then K is of type P_j for some j. Since the only proper non-trivial subspaces stabilised by H are V_1 and V_2, the group K is the stabiliser in $\mathrm{SL}_n(q)$ of V_1 or V_2. The group H is normalised by the duality automorphism, whereas the two choices for K are conjugate under γ, so H extends to a group that is not contained in any geometric subgroup. $\qquad\square$

Since the $\mathscr{C}_1$-subgroups occur in every dimension, we make the following definition for the sake of brevity in the next chapter. We shall in fact prove that in dimension up to 12 all such groups have standard reducible behaviour.

Definition 2.3.5 Let Ω be a quasisimple group in Case **L**, **U**, **S** or $\mathbf{O}^{\varepsilon}$, and let G be almost simple with socle $\bar{\Omega}$. The group Ω has *standard reducible behaviour* if the following hold:

(i) In Case **L**, if $H < \mathrm{SL}_n(q)$ is of type P_k, then H is maximal amongst the geometric subgroups of $\mathrm{SL}_n(q)$. In Case **L**, if $H_G < G$ is of type $P_{k,n-k}$ or type $\mathrm{GL}_k(q) \oplus \mathrm{GL}_{n-k}(q)$, then H_G is maximal amongst the geometric subgroups of G if and only if $G \nleqslant \mathrm{P\Gamma L}_n(q)$.

(ii) In Cases **U** and **S**, if $H < \Omega$ is of one of the types listed in Table 2.2, then H is maximal amongst the geometric subgroups of Ω.

(iii) In Cases $\mathbf{O}^{\circ}$ and $\mathbf{O}^-$, if $H < \Omega$ is of type P_k, then H is maximal amongst the geometric subgroups of Ω.

(iv) In Case $\mathbf{O}^+$, if $H < \Omega$ is of type P_k, and $k \neq n/2 - 1$, then H is maximal amongst the geometric subgroups of Ω. Let K be as given in Proposition 2.3.1 (ii). (That is, $K = \langle \phi \rangle$ if q is even, $K = \langle \delta, \phi \rangle$ if q is odd and $n(q-1)/4$ is odd, and $K = \langle \delta', \delta, \phi \rangle$ if q is odd and $n(q-1)/4$ is even.) In Case $\mathbf{O}^+$ if $H_G < G$ is of type $P_{n/2-1}$, then H_G is maximal amongst the geometric subgroups of G if and only if $G \nleq \mathrm{O}_n^+(q).K$.

(v) In Case $\mathbf{O}^\varepsilon$, if $H < \Omega$ is of type $\mathrm{GO}_k^{\varepsilon_1}(q) \perp \mathrm{GO}_{n-k}^{\varepsilon_2}(q)$, and $(k, q, \varepsilon_1) \notin \{(2, 2, +), (2, 3, +)\}$, then H is maximal amongst the geometric subgroups of $\Omega_n^\varepsilon(q)$. In Case $\mathbf{O}^\varepsilon$, if $H_G < G$ is of type $\mathrm{GO}_2^+(2) \perp \mathrm{GO}_{n-2}^{\varepsilon_2}(2)$, then H_G is not maximal. In Case $\mathbf{O}^\varepsilon$, if $H_G < G$ is of type $\mathrm{GO}_2^+(3) \perp \mathrm{GO}_{n-2}^{\varepsilon_2}(3)$ then H_G is maximal amongst the geometric subgroups of G if and only if n is even and $G \nleq \mathrm{PGO}_n^\varepsilon(3) = \mathrm{O}_n^\varepsilon(3).\langle \delta', \gamma \rangle$.

(vi) In Case $\mathbf{O}^\pm$, if $H < \Omega$ is of type $\mathrm{Sp}_{n-2}(q)$, then H is maximal amongst the geometric subgroups of $\Omega_n^\pm(q)$.

2.3.2 Imprimitive groups

In this section we prove various preliminary results about the maximality of groups in Class $\mathscr{C}_2$. Recall the types of $\mathscr{C}_2$-subgroups, as given in Table 2.4, and their structures as given in Table 2.5.

Proposition 2.3.6 *Let H be a $\mathscr{C}_2$-subgroup of Ω, of one of the following types:*

(i) $\mathrm{GL}_1(2) \wr \mathrm{S}_n$, *in Case* $\mathbf{L}$;

(ii) $\mathrm{GL}_1(3) \wr \mathrm{S}_n$, *in Case* $\mathbf{L}$;

(iii) $\mathrm{GL}_1(4) \wr \mathrm{S}_n$, *in Case* $\mathbf{L}$;

(iv) $\mathrm{GL}_2(2) \wr \mathrm{S}_{n/2}$, *in Case* $\mathbf{L}$;

(v) $\mathrm{GU}_2(2) \wr \mathrm{S}_{n/2}$, *in Case* $\mathbf{U}$;

(vi) $\mathrm{Sp}_2(2) \wr \mathrm{S}_{n/2}$, *in Case* $\mathbf{S}$;

(vii) $\mathrm{GO}_2^+(2) \wr \mathrm{S}_{n/2}$, *in Case* $\mathbf{O}^+$;

(viii) $\mathrm{GO}_2^+(3) \wr \mathrm{S}_{n/2}$, *in Case* $\mathbf{O}^+$;

(ix) $\mathrm{GO}_2^+(4) \wr \mathrm{S}_{n/2}$, *in Case* $\mathbf{O}^+$;

(x) $\mathrm{GO}_3(3) \wr \mathrm{S}_{n/3}$, *in Case* $\mathbf{O}^\varepsilon$;

(xi) $\mathrm{GO}_4^+(2) \wr \mathrm{S}_{n/4}$, *in Case* $\mathbf{O}^+$.

The group of type $\mathrm{GL}_1(2) \wr \mathrm{S}_3$ is equal to the $\mathscr{C}_1$-subgroup of type $\mathrm{GL}_1(2) \oplus \mathrm{GL}_2(2)$ of $\mathrm{SL}_3(2)$, and will be considered as such. The group of type $\mathrm{GL}_1(3) \wr \mathrm{S}_3$ is equal to the $\mathscr{C}_8$-subgroup $\mathrm{SO}_3(3)$ of $\mathrm{SL}_3(3)$, and will be considered as such. The group of type $\mathrm{GL}_1(4) \wr \mathrm{S}_2$ is equal to the $\mathscr{C}_8$-subgroup $\mathrm{GU}_2(2)$ of $\mathrm{SL}_2(4)$, but will be considered as a $\mathscr{C}_2$-subgroup. For all other choices of H from the above list, the group H is non-maximal in Ω and does not extend to a novel maximal subgroup in any extension of Ω.

Note: There is an additional equality, $\mathrm{Sp}_2(2) \wr \mathrm{S}_2 = \mathrm{GO}_4^+(2)$, but we assume that $\Omega \neq \mathrm{Sp}_4(2)$ since $\mathrm{Sp}_4(2)$ is not quasisimple.

Proof We consider each possible H in turn. In each part, let H_G be the subgroup of the general group that is of the same type as H.

(i) The group $\mathrm{GL}_1(2)$ is trivial, so H is equal to S_n acting via permutation matrices. Thus H fixes $v = (1, 1, \ldots, 1)$, and also the $(n-1)$-dimensional space consisting of all even weight vectors. Thus if n is even then $H \leqslant P_{1,n-1} \leqslant P_1$, as v has even weight, and if n is odd then $H \leqslant \mathrm{GL}_1(2) \oplus \mathrm{GL}_{n-1}(2) < P_1$. The order of H is $n!$, whilst $|P_{1,n-1}| = 2^{2n-3} \cdot |\mathrm{SL}_{n-2}(2)|$ and $|\mathrm{GL}_1(2) \oplus \mathrm{GL}_{n-1}(2)| = |\mathrm{GL}_{n-1}(2)|$, so it is straightforward to check that H is contained in $P_{1,n-1}$ or $\mathrm{GL}_1(2) \oplus \mathrm{GL}_{n-1}(2)$, and that this containment is proper if and only if $n > 3$. Here H, $P_{1,n-1}$ and $\mathrm{GL}_1(2) \oplus \mathrm{GL}_{n-1}(2)$ are all normalised by the inverse-transpose automorphism, so if $n > 3$ then H does not extend to a novelty.

(ii) The group H_G is isomorphic to $2 \wr \mathrm{S}_n$, and if H_G is constructed as a standard wreath product then H_G consists of isometries of the bilinear form with matrix I_n, and hence is contained in $K := \mathrm{GO}_n(3, I_n)$. A straightforward calculation using Theorem 1.6.22 shows that this containment is proper if and only if $n > 3$, so when $n = 3$ we shall consider H as a $\mathscr{C}_8$-subgroup. Since $g \in K$ if and only if $gg^\mathsf{T} = I_n$, the inverse-transpose automorphism centralises K. Thus if $n > 3$ then H does not extend to a novelty.

(iii) The group H_G is isomorphic to $3 \wr \mathrm{S}_n$, so if H_G is the standard wreath product then H_G consists of isometries of the unitary form with matrix I_n (our standard form), and hence is contained in $\mathrm{GU}_n(2)$. A straightforward calculation using Theorem 1.6.22 shows that this containment is proper if and only if $n > 2$. A short calculation shows that all elements of $\langle \phi, \gamma \rangle$ normalise both this copy of $\mathrm{GU}_n(2)$ and H_G, since if $g \in \mathrm{GU}_n(2)$ then $gg^{\phi\mathsf{T}} = I_n$, and H_G consists of the set of all monomial matrices. Thus if $n > 2$ then H does not extend to a novelty.

(iv) The groups $\mathrm{GL}_2(2)$ and $\mathrm{Sp}_2(2)$ are equal, so the standard copy of H consists of isometries of a symplectic form with matrix J, where J is a direct sum of $n/2$ copies of $\mathrm{antidiag}(1, 1)$. One may easily check that H is properly contained in $\mathrm{Sp}_n(2, J)$ for all n, using Theorem 1.6.22. The only non-trivial outer automorphism of $\mathrm{L}_n(2)$ is the duality automorphism. One may check that the inverse-transpose automorphism normalises $\mathrm{Sp}_n(2, J)$, since $J = J^{-1}$, and that this automorphism also normalises this copy of H, since H consists of all matrices with one non-zero 2×2 block in each pair $\{2i-1, 2i\}$ of rows and columns. Thus H does not extend to a novelty.

(v) The group H_G is isomorphic to $\mathrm{GU}_2(2) \wr \mathrm{S}_{n/2}$. A straightforward MAGMA calculation (file `Chap2calc`) shows that the group $\mathrm{GU}_2(2, \mathrm{antidiag}(1, 1))$ (which is MAGMA's copy of $\mathrm{GU}_2(2)$) fixes an imprimitive decomposition, into a sum of two non-degenerate 1-spaces $X \perp Y$. Our standard copy of $\mathrm{GU}_2(2)$ has form I_2, and is generated by antidiag$(1, \omega^2)$ and antidiag$(1, 1)$ (where ω is a primitive element of $\mathbb{F}_4^\times$), so that H_G is a subgroup of the group of type $\mathrm{GU}_1(2) \wr S_n$, and it is easy to check that this containment is proper. It is clear that $\mathrm{GU}_1(2) \wr S_n$ and $\mathrm{GU}_2(2, I_2)$ are normalised by ϕ, so H does not extend to a novelty.

(vi) The groups $\mathrm{Sp}_2(2)$ and $\mathrm{GO}_2^-(2)$ are equal, so H consists of isometries of a quadratic form (whose sign, ε, depends on whether $n/2$ is even or odd). It is straightforward to check that H is properly contained in $\mathrm{GO}_n^\varepsilon(2)$ if and only if $n > 4$ (but recall that we exclude $\mathrm{Sp}_4(2)$ from our calculations since it is not quasisimple). Since $\mathrm{Sp}_n(2)$ has trivial outer automorphism group, H does not extend to a novelty.

(vii) The group H_G is equal to $\mathrm{GO}_2^+(2) \wr S_{n/2}$. The standard copy of $\mathrm{GO}_2^+(2)$ is generated by a reflection in $e_1 + f_1$, and so is reducible. Thus H_G is reducible, and a relatively straightfoward calculation shows that H_G is properly contained in a $\mathscr{C}_1$-subgroup of $\mathrm{GO}_n^+(2)$. Furthermore, since $n \geqslant 10$, Aut $\Omega_n^+(2) = \mathrm{GO}_n^+(2)$, so H does not extend to a novelty.

(viii) The group H_G is equal to $\mathrm{GO}_2^+(3) \wr S_{n/2}$. A short MAGMA calculation (file Chap2calc) shows that our standard copy of $\mathrm{GO}_2^+(3)$ is completely reducible, and fixes non-degenerate 1-spaces $E_1 := \langle e_1 + f_1 \rangle$ and $E_2 := \langle e_1 - f_1 \rangle$. It follows that H_G is contained in a $\mathscr{C}_2$-subgroup K that is the stabiliser in $\mathrm{GO}_n^+(3)$ of a decomposition $V = V_1 \oplus V_2$ with $V_1 = \langle e_1 + f_1, e_2 + f_2, \ldots, e_{n/2} + f_{n/2} \rangle$ and $V_2 = \langle e_1 - f_1, e_2 - f_2, \ldots, e_{n/2} - f_{n/2} \rangle$. It is straightforward to check that the V_i are both of type $-$ if $n \equiv 4 \bmod 8$, both of type $+$ if $n \equiv 0 \bmod 8$, and are similar but non-isometric if $n \equiv 2 \bmod 4$, and that all of these containments are proper, since $n \geqslant 10$. The automorphism δ of $\mathrm{GO}_2^+(3)$ has standard representative $D_2 = \mathrm{diag}(-1, 1)$, and extends $\mathrm{GO}_2^+(3)$ to an irreducible but imprimitive group, interchanging E_1 and E_2. The matrix $D_n := \mathrm{diag}(-1, \ldots, -1, 1, \ldots, 1)$ induces the δ automorphism of $\mathrm{GO}_n^+(3)$ and acts on each 2-space $\langle e_i, f_i \rangle$ as D_2. Therefore D_n preserves the imprimitive decomposition $V_1 \oplus V_2$, so H does not extend to a novelty.

(ix) Let $\mathcal{D}$ be the decomposition preserved by H_G. Consulting Table 2.8, we see that our standard copy of $\mathrm{GO}_n^+(4)$ contains a $\mathscr{C}_5$-subgroup naturally isomorphic to our standard copy of $\mathrm{GO}_n^+(2)$, and that there exists a nonstandard copy of $\mathrm{GO}_n^+(4)$ which contains a $\mathscr{C}_5$-subgroup naturally isomorphic to our standard copy of $\mathrm{GO}_n^-(2)$. Fix these two versions of $\mathrm{GO}_n^+(4)$. If $n/2$ is odd then let $K = \mathrm{GO}_n^-(2)$, and otherwise let $K = \mathrm{GO}_n^+(2)$. Now, K contains a $\mathscr{C}_2$-subgroup L of type $\mathrm{GO}_2^-(2) \wr S_{n/2}$, preserving a decomposition $\mathbb{F}_2^n = W = V_1 \oplus \cdots \oplus V_{n/2}$, with each V_i of minus type. By Proposition 1.5.42, the form with the same matrix on V_i is of $+$-type when V_i is extended to an $\mathbb{F}_4$-space, so without loss of generality, $V = V_1 \mathbb{F}_4 \oplus \cdots \oplus V_{n/2} \mathbb{F}_4$ is the same decomposition $\mathcal{D}$. Therefore $L \leqslant H$. However, $\mathrm{GO}_2^+(4) \cong \mathrm{GO}_2^-(2)$, so in fact $L = H_G$, and hence H_G and

$H_\Omega = \frac{1}{2}H_G$ (the unique index 2 subgroup of H_G) are non-maximal. The only remaining automorphism of $\Omega_n^+(4)$ is ϕ (note that since the nonstandard form of $+$ type is written over $\mathbb{F}_2$ we can still use the standard copy of ϕ). Now, ϕ centralises K and hence centralises H, so no extension of H is maximal.

(x) The group H_G is equal to $\mathrm{GO}_3(3) \wr \mathrm{S}_{n/3}$. A straightforward MAGMA calculation (file `Chap2calc`) shows that our standard copy of $\mathrm{GO}_3(3)$ consists of the set of all monomial matrices and, since our standard form is the identity matrix, the standard basis vectors span non-degenerate 1-spaces. Therefore, H_G is (properly) contained in $\mathrm{GO}_1(3) \wr \mathrm{S}_n$, again preserving the identity form (note that if n is even we require the discriminant of Ω to be square). If n is odd then $\mathrm{PGO}_n(3) = \mathrm{Aut}\,\Omega_n(3)$, and if n is even then there are two $\mathrm{GO}_n^\pm(3)$-classes of groups of type H, interchanged by all elements of $\mathrm{CGO}_n^\pm(3) \setminus \mathrm{GO}_n^\pm(3)$, so in neither case does H extend to a novelty.

(xi) The group H_G is equal to $\mathrm{GO}_4^+(2) \wr \mathrm{S}_{n/4}$. A MAGMA calculation (file `Chap2calc`) shows that $\mathrm{GO}_4^+(2)$ is imprimitive, preserving a decomposition into two non-degenerate 2-spaces of minus type. Thus H_G is properly contained in $\mathrm{GO}_2^-(2) \wr \mathrm{S}_{n/2}$. Since $\mathrm{Aut}\,\Omega_n^+(2) = \mathrm{GO}_n^+(2)$, no extension of H is maximal. □

Recall the definitions of T, L, N, G_i and Ω_i, given in Subsection 2.2.2.

Lemma 2.3.7 *Let $4 \leqslant n \leqslant 12$, and let H be a $\mathscr{C}_2$-subgroup of Ω, preserving a decomposition $\mathcal{D} : V = V_1 \oplus \cdots \oplus V_t$ with $\dim(V_i) = m$, with dimension restrictions as in Definition 1.6.20.*

(i) *Assume that $t = 2$ and Ω_1 is perfect and, if $n = 4$ and $\Omega = \mathrm{SU}_n(q)$ or $\mathrm{Sp}_n(q)$, then assume that the decomposition is into non-degenerate subspaces. Then V_1 and V_2 are the only irreducible N-submodules of V.*

(ii) *If $t \geqslant 3$ and $G_1 \notin \{\mathrm{GL}_1(2), \mathrm{GO}_2^+(2), \mathrm{GO}_2^+(3)\}$, then $V_1, \ldots, V_t$ are the only irreducible $L_{(\mathcal{D})}$-submodules of V.*

(iii) *Assume that $t = 2$ and $\Omega_1 \times \Omega_2$ is perfect and, if $\Omega = \Omega_n^+(q)$ and $n \equiv 2 \bmod 4$, then H is not of type $\mathrm{GL}_{n/2}(q).2$. If there exists $K < \Omega$ with $H \leqslant K$ then $K \notin \mathscr{C}_1$. If, in addition, H is not of type $\mathrm{GO}_{n/2}^\circ(q)^2$, then H is irreducible.*

(iv) *If $t \geqslant 3$ and $G_1 \notin \{\mathrm{GL}_1(2), \mathrm{GO}_2^+(2), \mathrm{GO}_2^+(3)\}$, then L, and hence H, is absolutely irreducible.*

(v) *If $t \geqslant 5$, $G_1 \notin \{\mathrm{GL}_1(2), \mathrm{GO}_2^+(2), \mathrm{GO}_2^+(3)\}$, and there exists $K < \Omega$ such that $H \leqslant K$, then $K \notin \mathscr{C}_3$.*

(vi) *If $t = 4$ and $\Omega \notin \{\mathrm{SL}_4(2), \mathrm{SL}_4(3)\}$ then H' is absolutely irreducible. If in addition there exists $K < \Omega$ with $H \leqslant K$ then $K \notin \mathscr{C}_3$.*

(vii) *Assume that $t = 2$, the decomposition is into two totally singular subspaces, and H is insoluble. In Cases $\mathbf{U}$, $\mathbf{S}$ or $\mathbf{O}^\varepsilon$, if there exists $K < \Omega$ with $H \leqslant K$, then $K \notin \mathscr{C}_3$.*

Proof (i) By assumption, Ω_1 and Ω_2 are perfect, and they are irreducible by Proposition 1.12.2. Therefore V_1 and V_2 are irreducible. If $N = \Omega_1 \times \Omega_2$ then V_1 and V_2 are non-isomorphic N-modules. If $N \cong \Omega_1$ then the decomposition is into totally singular subspaces, so N acts on V_1 as $\mathrm{SL}_{n/2}(q^u)$. As an N-module $V_2 \cong V_1^{\sigma*}$ by Lemma 2.2.2 (vi). Since $n > 4$, as N-modules $V_1 \not\cong V_1^{\sigma*}$, and hence by Lemma 1.8.11 they are the only N-submodules of V.

(ii) Lemma 2.2.4 (iv) states that the V_i are pairwise non-isomorphic as $L_{(\mathcal{D})}$-submodules whenever $G_1 \neq \mathrm{GL}_1(2)$. If also $G_1 \neq \mathrm{GO}_2^+(2), \mathrm{GO}_2^+(3)$ then the V_i are irreducible $L_{(\mathcal{D})}$-submodules. As in Part (i) this implies the required result.

(iii) Suppose that H acts reducibly on V. Assume first that either $n > 4$, or $\Omega = \mathrm{SL}_n(q)$, or the decomposition is into non-degenerate subspaces. Then by Part (i), any N-invariant non-trivial proper subspace of V is equal to V_1 or V_2, so without loss of generality H is contained in the stabiliser in Ω of V_1. If H is not of type $\mathrm{GL}_{n/2}(q).2$ in Case $\mathbf{O}^+$ with $n \equiv 2 \bmod 4$, or of type $\mathrm{GO}_{n/2}^\circ(q)^2$, then H contains an element interchanging V_1 and V_2, a contradiction. If H is of type $\mathrm{GO}_{n/2}^\circ(q)^2$, then V_1 and V_2 are non-degenerate and non-isometric. Looking at the types of reducible groups in Table 2.2, we see that there is no group K stabilising such a pair of subspaces, so if $H \leqslant K < \Omega$ then $K \notin \mathscr{C}_1$.

If $\Omega = \mathrm{Sp}_4(q)$ or $\mathrm{SU}_4(q)$, and the decomposition is into totally singular subspaces, we need a more detailed argument, which we present only for the symplectic case: the unitary case is similar. Note that $q > 2$ by assumption that N is perfect. The group H contains subgroups $T \cong \mathrm{SL}_2(q)$ and $S \cong \mathrm{GL}_2(q)$. With respect to our standard symplectic basis, the standard block decomposition may be chosen to be $V_1 \oplus V_2 = \langle e_1, e_2 \rangle \oplus \langle f_2, f_1 \rangle$. For all $x \in \mathrm{GL}_2(q)$, the group S contains elements which act as x on V_1 and as a conjugate of $x^{-\mathsf{T}}$ on V_2. Thus V_1 and V_2 are non-isomorphic as S-submodules, and hence are the only irreducible T-submodules of V. Since H contains an element interchanging V_1 and V_2, it follows that H is irreducible.

(iv) By Lemma 2.2.4 (iv), the $L_{(\mathcal{D})}$-modules V_i are pairwise nonisomorphic, since if $t = 3$ then $m \geqslant 2$. The group $L_{(\mathcal{D})}^{V_i}$ is equal to G_i for all i, by Lemma 2.2.4 (iii). It follows from Part (ii) and the transitivity of A_t that L is irreducible when $G_i \notin \{\mathrm{GL}_1(2), \Omega_2^+(2), \Omega_2^+(3)\}$. To see that L is, in addition, absolutely irreducible in these cases, let $g \in \mathrm{C}_{\mathrm{GL}_n(q^u)}(L)$. Since g centralises the absolutely irreducible group $L_{V_i}^{V_i}$ for $1 \leqslant i \leqslant t$, the restriction of g to each V_i is a scalar from $\mathrm{GL}_m(q^u)$. Since $L^{\mathcal{D}}$ is transitive on $\{V_1, \ldots, V_t\}$, these scalars must all be equal and so $g \in \mathrm{Z}(\mathrm{GL}_n(q^u))$. Hence L is absolutely irreducible.

(v) If K is a $\mathscr{C}_3$-subgroup of Ω, then K' is not absolutely irreducible. If $H \leqslant K$ then $L' \leqslant K'$. However, $n \geqslant t \geqslant 5$, so L is perfect and absolutely irreducible by Part (iv).

(vi) In Case $\mathbf{O}^\pm$ we deduce from Table 2.4 that $m \geqslant 3$. The derived group of H acts on $\mathcal{D}$ as A_4, so if $g \in \Omega_1$, then $\mathrm{diag}(g, g^{-1}, g^{-1}, g), \mathrm{diag}(g, g, g^{-1}, g^{-1}) \in H'$, where Ω_1 is one of $\mathrm{SL}_m(q)$, $\mathrm{SU}_m(q)$, $\mathrm{Sp}_m(q)$ or $\Omega_m^\varepsilon(q)$. We are assuming that $(m, q) \notin \{(1, 2), (1, 3)\}$ in Case **L**, so Ω_1 contains elements of order greater than 2. Thus the V_i are pairwise nonisomorphic as $(H')_{\mathcal{D}}$ modules, so as in Part (iv) we conclude that H' is absolutely irreducible on V, and hence H is not contained in a $\mathscr{C}_3$-subgroup.

(vii) Suppose that K is semilinear, preserving a field extension of degree s. The group H^∞ is isomorphic to $\mathrm{SL}_{n/2}(q)$ or $\mathrm{SL}_{n/2}(q^2)$, so it follows from Proposition 1.11.3 and Theorem 1.11.5 that $s = 2$, and in particular that $\Omega \neq \mathrm{SU}_n(q)$. If $\Omega = \mathrm{Sp}_n(q)$ then $K^\infty \cong \mathrm{SU}_{n/2}(q)$ or $\mathrm{Sp}_{n/2}(q^2)$, contradicting Theorem 1.6.22. Similarly, if $\Omega = \Omega_n^\varepsilon(q)$ then $K^\infty \cong \mathrm{SU}_{n/2}(q)$ or $\Omega_{n/2}^{\varepsilon_1}(q^2)$, and again we contradict Theorem 1.6.22. □

The next lemma is concerned with whether one type of $\mathscr{C}_2$-subgroup can contain another.

Lemma 2.3.8   *Let $n \leqslant 12$ be as in Definition 1.6.20, and let H be a $\mathscr{C}_2$-subgroup of Ω, preserving a decomposition $\mathcal{D}$ into t subspaces of dimension m.*

(i) *Assume that $t \geqslant 5$ and $G_1 \notin \{\mathrm{GL}_1(2), \mathrm{GO}_2^+(2), \mathrm{GO}_2^+(3)\}$. If there exists $K \leqslant \Omega$ with $H \leqslant K$, and $K \in \mathscr{C}_2$ preserves a decomposition $\mathcal{D}_1$ into t_1 subspaces, then $t_1 \geqslant 5$.*

(ii) *Assume that $t = 2$, $n \geqslant 6$, and $\Omega \neq \mathrm{SU}_6(2)$. If there exists $K \leqslant \Omega$ with $H \leqslant K$, then K is not a $\mathscr{C}_2$-subgroup of a different type from H.*

Proof   (i) Suppose that $H \leqslant K$, and that $2 \leqslant t_1 \leqslant 4$. The group L is perfect by Lemma 2.2.4 (ii), so $L \leqslant K^\infty \leqslant K_{(\mathcal{D}_1)}$. However, $K_{(\mathcal{D}_1)}$ is reducible, whereas L is irreducible by Lemma 2.3.7 (ii), a contradiction.

(ii) Suppose otherwise, and let K preserve a decomposition $\mathcal{D}_1$ into t_1 subspaces.

We deal first with $\mathrm{SL}_8(2)$. If $t_1 = 8$ then K is reducible, but L is irreducible by Lemma 2.3.7 (iii). If $t_1 = 4$ then K is soluble, but H is insoluble.

Now consider the general case. Let X and N be as in Subsection 2.2.2. If $N \nleqslant K_{(\mathcal{D}_1)}$ then there is a non-trivial homomorphism from N into S_{t_1}. The group X is either $\mathrm{L}_{n/2}(q^u)$ for $u \in \{1, 2\}$ or $\mathrm{S}_{n/2}(q)'$ or $\mathrm{U}_{n/2}(q)$, all with $n/2 \geqslant 3$, or $\mathrm{O}_{n/2}^{\varepsilon_1}(q)$ with $n/2 \geqslant 5$. By Theorem 1.11.2, if $P(X) \leqslant n \leqslant 12$ and $\Omega \neq \mathrm{SL}_8(2)$ then $X = \mathrm{S}_4(2)' \cong \mathrm{A}_6$ in $\mathrm{Sp}_8(2)$, or $X = \Omega_6^+(2) \cong \mathrm{L}_4(2)$ in $\Omega_{12}^+(2)$. In $\mathrm{Sp}_8(2)$ the only other $\mathscr{C}_2$-subgroup is soluble. If $X = \Omega_6^+(2)$ then $P(X) = 8$ by Theorem 1.11.2, but in $\Omega_{12}^+(2)$ there is no $\mathscr{C}_2$-decomposition into more than six blocks. Thus in all remaining cases $N \leqslant K_{(\mathcal{D}_1)}$, contradicting Lemma 2.3.7 (i). □

Next we consider whether $\mathscr{C}_2$-subgroups can be contained in $\mathscr{C}_4$-subgroups.

Lemma 2.3.9   *Let $n \leqslant 12$ be as in Definition 1.6.20, and let H be a $\mathscr{C}_2$-subgroup of Ω, preserving a decomposition $\mathcal{D}$ into t subspaces of dimension m.*

Assume that $H_{(\mathcal{D})}$ is insoluble. If there exists a subgroup K of Ω such that $H \leqslant K$, where K is a $\mathscr{C}_4$-subgroup preserving a decomposition into spaces of dimension m_1 and m_2, then $m_1, m_2 > m$.

Proof The group K^∞ is isomorphic to $\Omega_1 \circ \Omega_2$, where Ω_1 acts on the tensor factor of dimension m_1 and Ω_2 acts on the tensor factor of dimension m_2, and we assume without loss of generality that $m_1 > m_2$. We assume at least one of m_1 and m_2 are less than or equal to m and derive a contradiction.

First assume that H is not of type $\mathrm{GL}_{n/2}(q^2).2$ (Case **U**) or $\mathrm{GL}_{n/2}(q).2$ (Cases **S** and **O**$^+$). Then $T := H_{\mathcal{D}}^\infty \cong \Omega_3^t$, where Ω_3 is one of $\mathrm{SL}_m(q)$, $\mathrm{SU}_m(q)$, $\mathrm{Sp}_m(q)'$ or $\Omega_m^\varepsilon(q)$. Assume first that every direct factor of T projects non-trivially onto Ω_1. Since Ω_3 is quasisisimple, this implies that $|\overline{\Omega_3}|^t$ divides $|\overline{\Omega_1}|$. Calculating the p-part of the order of Ω_3^t and Ω_1 shows that this is impossible. Thus one direct factor, say S, of T has trivial image in Ω_1, and so is a subgroup of Ω_2. Now, Ω_2 acts homogeneously, with V splitting as a direct sum of m_1 isomorphic irreducible Ω_2-submodules each of dimension m_2. Thus V must split into a direct sum of $m_1 > 1$ isomorphic S-submodules, and so the only possibility is that H is of type $\mathrm{GO}_4^+(q) \wr S_{n/4}$ and K is of type $\mathrm{Sp}_2(q) \otimes \mathrm{Sp}_{n/2}(q)$. At most one copy of $\mathrm{SL}_2(q)$ can project non-trivially onto the factor $\mathrm{Sp}_2(q)$, which implies that a covering group of $\mathrm{L}_2(q)^{n/2-1}$ must embed in $\mathrm{Sp}_{n/2}(q)$, contradicting Proposition 1.11.3 and Lemma 1.11.8.

Assume instead that H preserves a decomposition into two totally singular subspaces, so that $H^\infty \cong \mathrm{SL}_{n/2}(q^2)$ or $\mathrm{SL}_{n/2}(q)$. The unique non-abelian composition factor of H is larger than either of the non-abelian composition factors of K, a contradiction.   □

Lemma 2.3.10   *Let $n \leqslant 12$ be as in Definition 1.6.20, and let H be a $\mathscr{C}_2$-subgroup of Ω, preserving a decomposition $\mathcal{D}$ into t subspaces of dimension m. Assume that there exists a subgroup K of Ω with $H \leqslant K$, and assume also that H is not one of the non-maximal groups listed in Proposition 2.3.6.*

  (i) *If $t \geqslant 5$ then $K \notin \mathscr{C}_5$.*
 (ii) *Suppose that one of the following holds: $\Omega = \mathrm{SL}_n(q)$ and $m \geqslant 2$; $\Omega = \mathrm{SU}_n(q)$, $m \geqslant 2$, $\mathcal{D}$ is a decomposition into non-degenerate subspaces, and $(m, q) \notin \{(2, 2), (2, 3), (3, 2)\}$; $\Omega = \mathrm{Sp}_n(q)$ and $\mathcal{D}$ is a decomposition into non-degenerate subspaces; $\Omega = \Omega_n^\varepsilon(q)$ and $m \geqslant 3$. Then $K \notin \mathscr{C}_5$.*
(iii) *Suppose that $\mathcal{D}$ is a decomposition into 2 totally singular subspaces. Suppose also that $\Omega \neq \mathrm{SL}_n(q)$ or $\mathrm{SU}_4(q)$. Then $K \notin \mathscr{C}_5$.*

Proof Note that our assumptions on t and m ensure that $n \geqslant 4$ throughout, that $\Omega \notin \{\mathrm{SU}_4(2), \mathrm{SU}_4(3)\}$, and that $\Omega \neq \mathrm{SU}_6(2)$ in Part (ii).

Recall Definition 2.2.11 of the $\mathscr{C}_5$-subgroups. Let $K \in \mathscr{C}_5$. Then in Case **L**, $K^\infty \cong \mathrm{SL}_n(q_0)$ for some subfield $\mathbb{F}_{q_0} \subset \mathbb{F}_q$ of prime index. In Case **U**, either

$K^\infty \cong \mathrm{SU}_n(q_0)$ for $\mathbb{F}_{q_0}$ a subfield of $\mathbb{F}_q$ of odd prime index, or $K^\infty \cong \Omega_n^\varepsilon(q)$ (q odd), or $K^\infty \cong \mathrm{Sp}_n(q)$ or $\mathrm{Sp}_4(2)'$. In Case $\mathbf{S}$, $K^\infty \cong \mathrm{Sp}_n(q_0)$ or $K^\infty \cong \mathrm{S}_4(2)'$, for some subfield $\mathbb{F}_{q_0} \subset \mathbb{F}_q$ of prime index. In Case $\mathbf{O}^\varepsilon$, if $K \in \mathscr{C}_5$ then $K^\infty \cong \Omega_n^{\varepsilon_1}(q_0)$ for some prime index subfield $\mathbb{F}_{q_0}$ of $\mathbb{F}_q$.

(i) Recall the notation y_g, G_1 and L from Lemma 2.2.4 (iii). For all $g \in G_1$ the element $z_g := y_g(2,3,4) = (g, g^{-2}, g, 1, \ldots, 1)(2,3,4) \in L$; we calculate that $\mathrm{tr}(z_g) = \mathrm{tr}(g) + (n - 4m)$. The group L is perfect so $L \leqslant K^\infty$.

In Cases $\mathbf{L}$ and $\mathbf{U}$ there exist elements of $G_1 = \mathrm{GL}_m^\pm(q)$ whose trace lies in no proper subfield of $\mathbb{F}_{q^u}$ (note in Case $\mathbf{U}$ that $\mathrm{GU}_1(q)$ contains elements of trace any $(q + 1)$-th root of unity) . In Case $\mathbf{S}$, $m \geqslant 2$ so the result follows from Proposition 1.12.7 (i). In Case $\mathbf{O}^\varepsilon$ if $m = 1$ then $q = p$, so $\mathscr{C}_5 = \varnothing$. Since $t \geqslant 5$ and $n \leqslant 12$, we may assume that $m = 2$. If $G_1 = \mathrm{GO}_2^+(q)$ then, with respect to our standard form, G_1 contains $d_\lambda = \mathrm{diag}(\lambda, \lambda^{-1})$ for all $\lambda \in \mathbb{F}_q^\times$, of trace $\lambda + \lambda^{-1}$. Thus each possible trace is produced by up to two d_λ, so at least $\lceil (q - 1)/2 \rceil$ traces occur. This is larger than the order of any proper subfield of $\mathbb{F}_q$ unless $q = 4$, in which case H is listed in Proposition 2.3.6. Now consider $G_1 = \mathrm{GO}_2^-(q)$, and let $\Lambda = \{\lambda \in \mathbb{F}_{q^2} \; : \; \lambda^{(q+1)/(q+1,2)} = 1\}$. Then for each such λ, the group G_1 contains elements of trace $\lambda + \lambda^{-1}$, so at least $\lceil (q+1)/(2(q-1,2)) \rceil$ traces occur. This is larger than the order of any proper subfield of $\mathbb{F}_q$ unless $q = 9$. We check that in $\mathbb{F}_{81}$ the element $\lambda + \lambda^{-1}$, where λ is a primitive fifth root of unity, does not lie in $\mathbb{F}_3$.

(ii) The restrictions on m and q imply that if $\mathscr{C}_5 \neq \varnothing$ then $\Omega(V_1)$ is perfect, so that H^∞ contains the perfect group $\Omega(V_1)^t$. If $\Omega = \mathrm{SU}_n(q)$, assume for now that in addition that $m \geqslant 3$. Then, by Proposition 1.12.7 and Lemma 1.12.8, for a suitable choice of basis, H^∞ contains elements $d := \mathrm{diag}(a, \mathrm{I}_m, \ldots, \mathrm{I}_m)$ where the set of possible traces of $a \in \Omega(V_1)$, and hence the set of possible traces of d, does not lie in any proper subfield of $\mathbb{F}_{q^u}$.

We now consider $m = 2$ in Case $\mathbf{U}$. The group H^∞ contains elements $\mathrm{diag}(\alpha, \alpha^{-1})$ for all non-zero $(q + 1)$th roots of unity, α, so at least $(q + 1)/2$ traces arise. Hence K is of type $\mathrm{GO}_n^\varepsilon(q)$ or type $\mathrm{Sp}_n(q)$. Let A_1 be the first direct factor of $(H_{(\mathcal{D})})^\infty$, so that $A_1 \cong \mathrm{SU}_2(q)$. Since A_1 is irreducible on V_1, the subspace V_1 is either totally singular or non-degenerate under the symplectic or quadratic form f for which K is a group of isometries. The group A_1 acts non-trivially on V_1 whilst centralising V/V_1, so V_1 is non-degenerate under f. If f is quadratic then $\mathrm{SU}_2(q) \leqslant \mathrm{GO}_2^\varepsilon(q)$, a contradiction since $\mathrm{GO}_2^\varepsilon(q)$ is soluble whilst $\mathrm{SU}_2(q)$ is insoluble for $q > 3$. As for the symplectic case, note that K_{V_1} must act as $\langle \mathrm{Sp}_2(q), Z(\mathrm{GU}_2(q)) \rangle$. If $t \geqslant 3$ then H contains elements that act on V_1 via elements of $\mathrm{GU}_2(q) \setminus \mathrm{SU}_2(q)$, whilst centralising V_3, a contradiction. For $t = 2$, so that $n = 4$, we note that $|H| = 2q^2(q^2 - 1)^2(q + 1)$, whereas $|K| = (q + 1, 2)q^4(q^2 - 1)(q^4 - 1)$, which contradicts Lagrange's Theorem.

(iii) First consider Cases $\mathbf{S}$ and $\mathbf{O}^+$. If $q \leqslant 3$ then $\mathscr{C}_5 = \varnothing$, so assume without loss of generality that $q > 3$. The group H is of type $\mathrm{GL}_{n/2}(q).2$ and $N \cong \mathrm{SL}_{n/2}(q)$. With respect to our standard forms, elements of N are block diagonal, with blocks $A, FA^\gamma F$, where $F = \mathrm{antidiag}(1, 1, \ldots, 1)$, γ is the inverse-transpose map, and $A \in \mathrm{SL}_{n/2}(q)$. Let $x_\alpha := \mathrm{diag}(\alpha, \alpha^{-1}, 1, \ldots, 1) \in \mathrm{SL}_{n/2}(q)$. Then $\lambda_\alpha := \mathrm{tr}(\mathrm{diag}(x_\alpha, Fx_\alpha^{-\mathsf{T}}F)) = 2\alpha + 2\alpha^{-1} + n - 4$. If $K \in \mathscr{C}_5$ then there exists a proper subfield $\mathbb{F}_{q_0}$ of $\mathbb{F}_q$ such that $\lambda_\alpha \in \mathbb{F}_{q_0}$ for all α. For each $\lambda_\alpha \in \mathbb{F}_{q_0}$ there are at most two solutions for $\alpha \in \mathbb{F}_q$, giving at most $2q_0$ traces in $\mathbb{F}_{q_0}$ as α varies. Thus if $q \neq 4$ then it is not possible for all traces of elements of H^∞ to lie in $\mathbb{F}_{q_0}$. If $q = 4$ then $\Omega = \Omega_n^+(4)$, and $K \in \{\Omega_n^+(2), \Omega_n^-(2)\}$. However, $H \cong \mathrm{GL}_{n/2}(4).2$, which contradicts Lagrange's theorem.

In Case **U**, the group H is of type $\mathrm{GL}_{n/2}(q^2).2$ and $N \cong \mathrm{SL}_{n/2}(q^2)$. The group $K^\infty \in \{\mathrm{SU}_n(q_0), \mathrm{Sp}_n(q), \Omega_n^\varepsilon(q)\}$, where $\mathbb{F}_{q_0}$ is any subfield of $\mathbb{F}_{q^2}$ of odd prime index. With an appropriate choice of form, elements of N are block diagonal, with blocks $A, FA^{\gamma\sigma}F$, where $F = \mathrm{antidiag}(1, 1, \ldots, 1)$ and $A \in \mathrm{SL}_{n/2}(q^2)$.

We deal first with $K^\infty = \mathrm{Sp}_n(q)$. Since K is a $\mathscr{C}_5$-subgroup, the group K^∞ is a $\mathrm{GL}_n(q^2)$-conjugate of our standard copy of $\mathrm{Sp}_n(q)$. If $N = H^\infty$ is a subgroup of K^∞, then $\mathrm{Sp}_n(q)$ contains a reducible subgroup J isomorphic to N. Since V splits as a direct sum of two non-isomorphic N-submodules of dimension $n/2$, on each of which N acts faithfully, we deduce that the same must be true for the natural $\mathbb{F}_q$-module for J. However, $|\mathrm{SL}_{n/2}(q^2)| > |\mathrm{GL}_{n/2}(q)|$, a contradiction.

Assume now that $K^\infty = \mathrm{SU}_n(q_0)$ or $\Omega_n^\varepsilon(q)$, so that $q > 2$. Let

$$x_\alpha := \begin{pmatrix} \alpha & 0 & 0 \\ 0 & 0 & -1 \\ 0 & \alpha^{-1} & 0 \end{pmatrix} \oplus I_{n/2-3} \in \mathrm{SL}_{n/2}(q^2).$$

Then $\mathrm{tr}(\mathrm{diag}(x_\alpha, Fx_\alpha^* F)) = \alpha + \alpha^{-q} + (n - 6)$. If $K \in \mathscr{C}_5$ then there exists a proper subfield $\mathbb{F}_{q_0}$ of $\mathbb{F}_{q^2}$ such that for all $\alpha \in \mathbb{F}_{q^2}^\times$ there exists $\beta_\alpha \in \mathbb{F}_{q_0}$ with $\alpha + \alpha^{-q} = \beta_\alpha$. Let α be a primitive element of $\mathbb{F}_{q^2}^\times$, then since $q > 2$ the sum $\alpha + \alpha^{-q}$ is not centralised by $x \mapsto x^q$, so H^∞ contains elements whose trace does not lie in $\mathbb{F}_q$. Thus K is not of type $\Omega_n^\varepsilon(q)$. If $q \neq 4$ then the set $\{\beta_\alpha : \alpha \in \mathbb{F}_q^\times\}$ lies in no proper subfield of $\mathbb{F}_q$, contradicting the fact that $\mathbb{F}_{q_0}$ must have odd index in $\mathbb{F}_{q^2}$ and hence not contain $\mathbb{F}_q$. Thus $q = 4$. But then $q^2 = 2^4$, and $\mathbb{F}_{q^2}$ has no proper subfields of odd index, a contradiction. $\quad\square$

Lemma 2.3.11 *Let $5 \leqslant n \leqslant 12$, with n as in Definition 1.6.20, and let H be a $\mathscr{C}_2$-subgroup of Ω, stabilising a decomposition into n subspaces. If there exists a subgroup K of Ω with $H \leqslant K$, then $K \notin \mathscr{C}_6$.*

Proof Suppose, by way of contradiction, that $H \leqslant K$ for some $K \in \mathscr{C}_6$. Then $\Omega = \mathrm{SL}_n(q)$ or $\mathrm{SU}_n(q)$. We see from Table 2.5 that $H^\infty \cong (q-1)^{n-1}{:}\mathrm{A}_n$ in Case **L** and that $H^\infty \cong (q+1)^{n-1}{:}\mathrm{A}_n$ in Case **U**.

If $n = 5$ then $K^\infty \cong 5^{1+2}.\mathrm{Sp}_2(5)$. If $\mathscr{C}_6 \neq \varnothing$ and $\Omega = \mathrm{SL}_5(q)$ then $q \geqslant 11$. If $\mathscr{C}_6 \neq \varnothing$ and $\Omega = \mathrm{SU}_5(q)$ then $q \geqslant 4$. Therefore $|K^\infty| < |H^\infty|$, a contradiction. If $n \in \{7, 9, 11\}$ then $|K^\infty| < |H^\infty|$ for all q.

Suppose finally that $n = 8$, so that $K^\infty \cong (4 \circ 2^{1+6}).\mathrm{Sp}_6(2)$. If $q \geqslant 7$ then $|(q-1)^7.A_8|$ is larger than $|K^\infty|$, so without loss of generality $q \leqslant 5$. If $\mathscr{C}_6 \neq \varnothing$ then $\Omega = \mathrm{SL}_8(5)$ or $\mathrm{SU}_8(3)$, so $H^\infty \cong 4^7.A_8$, and one may check that $|H^\infty|$ does not divide $|K^\infty|$, a contradiction. $\square$

Lemma 2.3.12 *Let $n \leqslant 12$ be as in Definition 1.6.20 and let H be a $\mathscr{C}_2$-subgroup of Ω, preserving a decomposition $\mathcal{D} : V = V_1 \oplus \cdots \oplus V_t$ with $\dim V_i = m$. Assume that there exists a subgroup K of Ω, with $H \leqslant K$.*

(i) *Suppose that one of the following holds: $\Omega = \mathrm{SL}_n(q)$ with $m \geqslant 2$ and $(m, q) \notin \{(2, 2), (2, 3)\}$; $\Omega = \mathrm{SU}_n(q)$; $\Omega = \mathrm{Sp}_n(q)$ with $(m, q) \neq (2, 2)$; $\Omega = \Omega_n^\varepsilon(q)$. Then $K \notin \mathscr{C}_8$.*

(ii) *If $t = n \geqslant 4$ and $q \geqslant 5$, then $K \notin \mathscr{C}_8$.*

Proof (i) Suppose otherwise, and let f be the form for which K is a group of similarities. If $\mathscr{C}_8 \neq \varnothing$ then $\Omega = \mathrm{SL}_n(q)$ or $\mathrm{Sp}_n(2^i)$. In Case **S**, the V_i are non-degenerate since q is even.

Let A be the subgroup of $H_\mathcal{D}$ consisting of elements acting non-trivially on V_1 but as scalars on $V_2 \oplus \cdots \oplus V_t$, so that A contains $\mathrm{SL}(V_1)$ or $\mathrm{Sp}(V_1)$. Since V_1 is an irreducible A-module, V_1 is either non-degenerate or totally singular under f.

Suppose first that V_1 is totally singular under f. By an easy generalisation of Lemma 2.2.2 (vi) (with V_1 in place of W_1), the irreducibility of the action of A on V_1 implies that A is also irreducible on a totally singular subspace W_2, such that $W_1 \cap W_2 = \{0\}$. This contradicts the fact that A acts as scalars on a complement to V_1.

Suppose instead that V_1 is non-degenerate under f. If f is quadratic then $\mathrm{Sp}_m(q)$ or $\mathrm{GL}_m(q) \leqslant \mathrm{CGO}_m^\varepsilon(q)$, so by Theorem 1.6.22 $m = q = 2$, which we have excluded. Thus f is unitary or symplectic, and so $\Omega = \mathrm{SL}_n(q)$. Dropping down to $H^\infty \leqslant K^\infty$ we get $\mathrm{SL}_m(q) \leqslant \mathrm{SU}_m(q)$ or $\mathrm{SL}_m(q) \leqslant \mathrm{Sp}_m(q)$, so that $m = 2$. Consulting Table 2.5 we see that if $t > 2$ then H has a subgroup which acts as $\mathrm{GL}_2(q)$ on one block whilst centralising at least one other. Thus H consist of isometries of f, and so $\mathrm{GL}_2(q)$ is a subgroup of $\mathrm{Sp}_2(q)$ or $\mathrm{GU}_2(q)$, a contradiction since $q \neq 2$.

If $t = 2$, so that $n = 4$, then $|H| = 2q^2(q^2 - 1)^2(q - 1)$. If f is symplectic then $|K| = (q - 1, 2)q^4(q^2 - 1)(q^4 - 1)$, forcing $q \in \{2, 3\}$. If f is unitary then $|K| = (4, q^{1/2} - 1)q^3(q - 1)(q^{3/2} + 1)(q^2 - 1)$, a contradiction.

(ii) If $t = n$ and $\mathscr{C}_8 \neq \varnothing$ then $\Omega = \mathrm{SL}_n(q)$. For $\alpha \in \mathbb{F}_q^\times$, define

$$x_\alpha = (1,3,2)^{\mathrm{diag}(\alpha^{-1},\alpha,1,\dots,1)}(1,2,3)(2,3,4)$$

$$= \begin{pmatrix} \alpha & 0 & 0 & 0 \\ 0 & 0 & \alpha^{-2} & 0 \\ 0 & 0 & 0 & \alpha \\ 0 & 1 & 0 & 0 \end{pmatrix} \oplus I_{n-4},$$

and note that $x_\alpha \in L$. Now $\mathrm{tr}(x_\alpha) = \alpha + (n-4)$, and $\mathrm{tr}(x_\alpha^{-1}) = \alpha^{-1} + (n-4)$. Thus, by Lemma 1.12.6, if there exists $\alpha \in \mathbb{F}_q^\times$ with $\alpha \neq \alpha^{-1}$, or $\alpha^{-1} \neq \alpha^{\sqrt{q}}$ for q square, then L is not a group of isometries of a symplectic, quadratic or unitary form. The first condition only fails when $q \leqslant 3$, and the second only when $q = 4$. □

The preceding results allow us to prove a more general result concerning the maximality of $\mathscr{C}_2$-subgroups.

Proposition 2.3.13 *Let $n \in \{5,7,11\}$ be as in Definition 1.6.20, and let H be a $\mathscr{C}_2$-subgroup of Ω. Then H is maximal amongst the geometric subgroups of Ω if and only if one of the following holds: $\Omega = \mathrm{SL}_n(q)$ with $q \geqslant 5$; $\Omega = \mathrm{SU}_n(q)$; $\Omega = \Omega_n(q)$. If H is non-maximal then H does not extend to a novel maximal subgroup.*

Proof Note that H preserves a decomposition into n subspaces. The result for $q \leqslant 4$ in Case **L** follows immediately from Proposition 2.3.6.

Assume, by way of contradiction, that there exists a geometric subgroup $K \neq H$ of Ω with $H \leqslant K$. Without loss of generality, $K \in \mathscr{C}_i$ for some $1 \leqslant i \leqslant 8$. It follows from Lemma 2.3.7 (iv),(v) that $K \notin \mathscr{C}_1 \cup \mathscr{C}_3$. There is a unique type of $\mathscr{C}_2$-subgroup, so $K \notin \mathscr{C}_2$. Since n is prime, $\mathscr{C}_4 = \mathscr{C}_7 = \varnothing$. It follows from Lemma 2.3.10 (i) that $K \notin \mathscr{C}_5$. It follows from Lemma 2.3.11 that $K \notin \mathscr{C}_6$. For $\mathscr{C}_8$ the result follows from Lemma 2.3.12 (ii). □

2.3.3 Semilinear groups

Recall Definition 2.2.5 of Class $\mathscr{C}_3$. In this subsection we prove various preliminary results about the maximality of $\mathscr{C}_3$-subgroups. Recall Definition 1.13.2 of a Zsigmondy prime $z_{q,n}$. In this subsection, we let s denote the degree of the field extension preserved by a $\mathscr{C}_3$-subgroup H of Ω, let $m = n/s$, and write β_s for the form preserved on $\mathbb{F}_{q^{us}}^{n/s}$ (note that this is only relevant to distinguish H in Cases **S** and **O**$^\varepsilon$ when $s = 2$). For maximality, we require s to be prime.

Lemma 2.3.14 *Let $3 \leqslant n \leqslant 12$ be as in Definition 1.6.20, and let H be a*

$\mathscr{C}_3$-*subgroup of* Ω. *Assume that* H *is not of type* $\mathrm{GU}_2(q)$ *in* $\mathrm{Sp}_4(q)$. *Then* H *is irreducible.*

Proof Assume first that $s = n$, so that n is an odd prime and $\Omega = \mathrm{SL}_n^{\pm}(q)$. In Case **L**, $|H| = (q^n - 1)n/(q - 1)$, so $|H|$ is divisible by a Zsigmondy prime $z_{q,n}$. If H is reducible, then H must stabilise a k-space for some k. This implies that H is contained in a parabolic subgroup, P_k. However, by Table 2.3 and Proposition 1.13.4 the prime $z_{q,n}$ does not divide $|P_k|$ for any k. In Case **U**, $|H| = (q^n + 1)n/(q + 1)$ so $|H|$ is divisible by some $z_{q,2n}$ (recall that $\mathrm{SU}_3(2)$ is soluble). Similarly to Case **L**, we first deduce that H is not contained in a parabolic subgroup, and from this conclude that H must stabilise a non-degenerate k-space for some k. Thus either H is contained in a group of type $\mathrm{GU}_k(q) \perp \mathrm{GU}_{n-k}(q)$, or a group of type $\mathrm{GU}_{n/2}(q) \wr S_2$. However, none of these groups have order divisible by $z_{q,2n}$.

If $s \neq n$, then the result follows from Lemma 2.2.6. □

Lemma 2.3.15 *Let* $n \in \{3, 5, 7, 11\}$ *be as in Definition 1.6.20, and let* H *be a* $\mathscr{C}_3$-*subgroup of* Ω. *If there exists a subgroup* K *of* Ω *such that* $H \leqslant K$ *then* $K \notin \mathscr{C}_2 \cup \mathscr{C}_6$.

Proof Since $\mathscr{C}_3 \neq \varnothing$ and n is an odd prime, $\Omega = \mathrm{SL}_n(q)$ or $\mathrm{SU}_n(q)$.

Since n is prime, $|H|$ is divisible by some $z_{q,n}$ in Case **L**, and by some $z_{q,2n}$ in Case **U** (recall that $\Omega \neq \mathrm{SU}_3(2)$). By Lemma 1.13.3 (iii) both $z_{q,n}$ and $z_{q,2n}$ are greater than n.

There is exactly one type of $\mathscr{C}_2$-subgroup, namely $\mathrm{GL}_1(q) \wr S_n$ or $\mathrm{GU}_1(q) \wr S_n$. If K is of one of these types, then the prime divisors of $|K|$ are the primes dividing $q \pm 1$, and primes less than or equal to n. In particular, $z_{q,n}$ and $z_{q,2n}$ do not divide $|K|$.

There is at most one type of $\mathscr{C}_6$-subgroup, namely $n^{1+2}.\mathrm{Sp}_2(n)$. If L is of this type, then $|L|$ is divisible by n and the prime divisors of $n^2 - 1$ and of $q \pm 1$, but by no other primes. Therefore $z_{q,n}$ and $z_{q,2n}$ do not divide $|L|$. □

Lemma 2.3.16 *Let* $H = \mathrm{N}_\Omega(\mathbb{F}_{q^{us}}, \beta_s)$ *be a* $\mathscr{C}_3$-*subgroup of* Ω, *with* n *as in Definition 1.6.20.*

(i) *Suppose that* $4 \leqslant n \leqslant 12$ *and* $s = 2$. *Assume that if* H *is of type* $\mathrm{GU}_2(q)$ *in* $\mathrm{Sp}_4(q)$ *then* $q \neq 3$. *If there exists a subgroup* K *of* Ω *such that* $H \leqslant K$ *then* $K \notin \mathscr{C}_2$.

(ii) *Suppose that* $6 \leqslant n \leqslant 12$ *with* $s = 3$, *and assume that* $\Omega \neq \Omega_n^+(q)$. *If there exists a subgroup* K *of* Ω *such that* $H \leqslant K$ *then* $K \notin \mathscr{C}_2$.

Proof We prove both parts at once. By Lemma 2.2.7, our assumptions on n, q, s and H ensure that H is insoluble and the quasisimple group H^∞ is one of $\mathrm{SL}_{n/s}(q^s)$, $\mathrm{SU}_{n/s}(q^s)$, $\mathrm{Sp}_{n/s}(q^s)$, $\mathrm{SU}_{n/2}(q)$ or $\Omega_{n/s}^{\varepsilon_1}(q^s)$. By Theorem 1.11.2,

$P(\bar{H}^\infty) > n$ for each of these types. Thus if $K \in \mathcal{C}_2$ then it follows that $H^\infty \leqslant K_{(\mathcal{D})}$, the kernel of the action on blocks. This contradicts the fact that H^∞ is irreducible, by Lemma 2.2.6.   □

Lemma 2.3.17 *Let $n \leqslant 12$ be as in Definition 1.6.20 and let H be a $\mathcal{C}_3$-subgroup of Ω. Assume that H is not of type $\mathrm{GU}_2(q)$ in $\mathrm{Sp}_4(q)$. If there exists a subgroup K of Ω such that $H \leqslant K$, and K is not of the same type as H, then $K \notin \mathcal{C}_3$.*

Proof Assume otherwise, by way of contradiction. Let $H = \mathrm{N}_\Omega(\mathbb{F}_{q^{us}})$, and let $K = \mathrm{N}_\Omega(\mathbb{F}_{q^{ut}})$ be a $\mathcal{C}_3$-subgroup of Ω.

If there is more than one type of $\mathcal{C}_3$-subgroup, then n is not prime. Since n is not prime, it follows from Lemmas 2.2.6 and 2.2.7 that H^∞ is the group Ω_1, as defined just before Lemma 2.2.6.

Assume first that $s \neq t$, so $\mathbb{F}_{q^t} \neq \mathbb{F}_{q^s}$, and $\mathbb{F}_{q^t}, \mathbb{F}_{q^s} \subset \mathrm{End}_{\mathbb{F}_{q^u} H^\infty}(\mathbb{F}_{q^u}^n)$. This implies that the centraliser of H^∞ in $\mathrm{GL}_n(q^u)$ has order greater than $q^s - 1$ and hence H^∞ is not absolutely irreducible on $\mathbb{F}_{q^s}^{n/s}$, contradicting Proposition 1.12.2.

Thus $s = t$, and so $\Omega = \mathrm{Sp}_n(q)$ or $\Omega_n^\pm(q)$ and $s = t = 2$. Assume first that $\Omega = \mathrm{Sp}_n(q)$. Then one of H and K is of shape $\mathrm{Sp}_{n/2}(q^2).2$, the other is of shape $\mathrm{GU}_{n/2}(q).2$, and q is odd. By Proposition 1.13.4, there exists a $z_{q,n}$ dividing $|\mathrm{Sp}_{n/2}(q^2).2|$ but not $|\mathrm{GU}_{n/2}(q).2|$ (since $n/2$ is even), whereas if H is of type $\mathrm{GU}_{n/2}(q)$ then $n \neq 4$, so some $z_{q,n-2}$ divides $|H|$ but not $|\mathrm{Sp}_{n/2}(q^2).2|$.

In Case $\mathbf{O}^+$, H and K must be of two of the following types: $\mathrm{GO}_{n/2}^+(q^2)$ (so $n/2$ is even), $\mathrm{GO}_{n/2}^\circ(q^2)$ (so $n/2$ is odd), $\mathrm{GU}_{n/2}(q)$ (so $n/2$ is even). Therefore, H and K are of types $\mathrm{GO}_{n/2}^+(q^2)$ and $\mathrm{GU}_{n/2}(q)$ (in some order), and $n/2$ is even, and so $n = 12$. By Proposition 1.13.4, some prime $z_{q,8}$ divides $|\Omega_6^+(q^2)|$ but not $|\mathrm{SU}_6(q)|$. Conversely, $z_{q,10}$ divides $|\mathrm{SU}_6(q)|$ but not $|\Omega_6^+(q^2)|$.

In Case $\mathbf{O}^-$, H and K must be of the following types: $\mathrm{GO}_{n/2}^-(q^2)$ (so $n/2$ is even), $\mathrm{GO}_{n/2}^\circ(q^2)$ (so $n/2$ is odd), $\mathrm{GU}_{n/2}(q)$ (so $n/2$ is odd). Therefore, H and K are of types $\mathrm{GO}_{n/2}^\circ(q^2)$ and $\mathrm{GU}_{n/2}(q)$ (in some order), and q and $n/2$ are odd, so $n = 10$. By Proposition 1.13.4, a prime $z_{q,8}$ divide $|\Omega_5(q^2)|$ but not $|\mathrm{GU}_5(q)|$. Conversely, a prime $z_{q,10}$ divides $|\mathrm{GU}_5(q)|$ but not $|\Omega_5(q^2)|$.   □

Lemma 2.3.18 *Let $n \in \{6, 8, 10, 12\}$, and let $\Omega = \mathrm{SL}_n(q)$ or $\mathrm{Sp}_n(q)$. Let $H = \mathrm{N}_\Omega(\mathbb{F}_{q^2}, \beta_2)$ be a $\mathcal{C}_3$-subgroup of Ω. If there exists a subgroup K of Ω with $H \leqslant K$, then $K \notin \mathcal{C}_4$.*

Proof Let K be a $\mathcal{C}_4$-subgroup of Ω. The composition factors of K lie in

$$\{\mathrm{L}_{n_1}(q), \mathrm{L}_{n_2}(q), \mathrm{Sp}_{n_1}(q), \mathrm{O}_{n_2}^\varepsilon(q) : n_1, n_2 > 1, \ n_1 n_2 = n, \ n_1 \text{ is even}\}.$$

Since H has a composition factor isomorphic to one of $\mathrm{L}_{n/2}(q^2)$, $\mathrm{Sp}_{n/2}(q^2)$ or $\mathrm{U}_{n/2}(q)$ (with q odd), it follows that $H \not\leqslant K$.   □

Lemma 2.3.19 *Let $3 \leqslant n \leqslant 12$ be as in Definition 1.6.20 and let H be a $\mathscr{C}_3$-subgroup of Ω. Assume that $\Omega \neq \mathrm{SL}_3(4), \mathrm{SU}_6(2)$. If there exists a subgroup K of Ω with $H \leqslant K$, then $K \notin \mathscr{C}_5$.*

Proof Let H preserve a field extension of degree s, let K be a $\mathscr{C}_5$-subgroup over $\mathbb{F}_{q_0^u}$, where $q_0^r = q$ for some r, and assume by way of contradiction that $H \leqslant K$.

Case L. Here $K < \mathrm{Z}(\mathrm{SL}_n(q))\mathrm{GL}_n(q_0)$, and r is prime. Note that $(q_0, rn) \neq (2, 6)$ since we are assuming that $n \geqslant 3$ and that $\Omega \neq \mathrm{SL}_3(4)$. If $n = s$ is prime then $|H|$ is a multiple of $(q^s - 1)/(q - 1)$, so $z_{q_0, rn}$ divides $|H|$. Otherwise, $\mathrm{SL}_{n/s}(q_0^{rs}) \leqslant H$, with $n/s \geqslant 2$ and $r > 1$, so by Proposition 1.13.4 a prime $z_{q_0, rn}$ divides $|H|$. If $z_{q_0, i}$ divides $|K|$ then $i \leqslant \max\{r, n\}$, a contradiction.

Case U. Here H is of type $\mathrm{GU}_{n/s}(q^s)$, s is odd, and $r = 2$ if and only if K is of type $\mathrm{Sp}_n(q)$ or $\mathrm{GO}_n^\varepsilon(q)$. Note that $(q_0, 2nr) \neq (2, 6)$, that $(q_0, nr) \neq (2, 6)$ if n is even, since $n \geqslant 3$ and $\mathrm{SU}_3(2)$ is solvable, and that $(q_0, 2r(n - s)) \neq (2, 6)$. If $s = n$ then $|H|$ is divisible by $z_{q_0, 2nr}$. Otherwise, $\mathrm{SU}_{n/s}(q^s) \leqslant H$, $n/s \geqslant 2$, and $s > 1$. By Proposition 1.13.4, if n is odd then $|H|$ is divisible by $z_{q_0, 2nr}$, whilst if n is even then $|H|$ is divisible by both $z_{q_0, nr}$ and $z_{q_0, 2r(n-s)}$.

If K is of type $\mathrm{GU}_n(q_0)$ then $r \geqslant 3$. If $z_{q_0, i} \mid |K|$ then from Proposition 1.13.4 we see that $i \leqslant \max\{2r, 2n\} < nr$, so $z_{q_0, 2nr}, z_{q_0, nr}$ do not divide $|K|$.

If K is of type $\mathrm{Sp}_n(q)$ then $q_0 = q$, $r = 2$ and $n > 2$ is even. By Table 2.8 and Proposition 1.13.4, if $z_{q, i}$ divides $|K|$ then $i \leqslant n$, so if $n \neq 2s$ then $z_{q_0, 2r(n-s)}$ does not divide $|K|$. So assume $n = 2s$. Then $|H| = |\mathrm{SU}_2(q^s)| \frac{(q^s + 1)}{(q+1)} . s$, whereas $|K| = |\mathrm{Sp}_{2s}(q)| . (q + 1, s)$, so a higher power of $z_{q_0, nr}$ divides $|H|$ than $|K|$.

If K is of type $\mathrm{GO}_n^\varepsilon(q)$, then $q_0 = q$ and $r = 2$. If $n = 3$ then we get a contradiction from $|H| = 3(q^2 - q + 1)$ whilst $|K| = q(q^2 - 1)(q + 1, 3)$. If $n = 4$ then $\mathscr{C}_3 = \varnothing$. If $n \geqslant 5$ is odd and $z_{q, i}$ divides $|K|$ then $i \leqslant n - 1$ by Proposition 1.13.4, so $z_{q, 2nr}$ does not divide $|K|$, a contradiction, so $n \geqslant 6$. Thus the group K is of type $\mathrm{GO}_n^+(q)$, and $z_{q, i}$ divides $|K|$, then $i \leqslant n - 2$ by Proposition 1.13.4, so $z_{q_0, nr}$ does not divide $|K|$. If K is of type $\mathrm{GO}_n^-(q)$ and $z_{q, i}$ divides $|K|$ then $i \leqslant n$, so if $n \neq 2s$ then $z_{q, 2r(n-s)}$ does not divide K. If $n = 2s$ then a higher power of $z_{q_0, nr}$ divides $|H|$ than $|K|$.

Case S. Here H is of type $\mathrm{Sp}_{n/s}(q^s)$ or type $\mathrm{GU}_{n/2}(q)$, and $K \cong \mathrm{Sp}_n(q_0).(2, q - 1, r)$, with $q_0^r = q$ and r prime. If $z_{q_0, i}$ divides $|K|$ then $i \leqslant n$ by Proposition 1.13.4. Note that $(q_0, nr) \neq (2, 6)$, and that $r(n - 2) > 2$. A straightforward argument rules out the containment of type $\mathrm{Sp}_{n/s}(q^s)$ in K. If H is of type $\mathrm{GU}_{n/2}(q)$ then q is odd and $H \cong \mathrm{GU}_{n/2}(q).2$. The order of H is divisible by $z_{q_0, rn}$ if $n/2$ is odd, and by $z_{q_0, r(n-2)}$ if $n/2$ is even (since q_0 is odd). No prime $z_{q_0, rn}$ divides $|K|$, whilst if $z_{q_0, r(n-2)}$ divides $|K|$ then $n = 4$ and $r = 2$, in which case $z_{q_0, r(n-2)}^2$ divides $|H|$ but not $|K|$.

Case O°. Here $\Omega^\circ_{n/s}(q^s) \leqslant H$, with s prime, $n/s \geqslant 3$, and $K \cong \Omega^\circ_n(q_0).(2, r)$, with r prime. Note that $n/s \geqslant 3$ and q_0 is odd, so $r(n - s) > n$ and $z_{q_0, r(n-s)}$ exists. Then, by Proposition 1.13.4 (and a direct calculation when $n/s = 3$) a prime $z_{q_0, r(n-s)}$ divides $|H|$ and if $z_{q_0, i}$ divides $|K|$ then $i \leqslant n - 1$.

Case O⁺. Here H is of type $\mathrm{GO}^+_{n/s}(q^s)$, type $\mathrm{GO}^\circ_{n/2}(q^2)$, or type $\mathrm{GU}_{n/2}(q)$, whilst K is of type $\mathrm{GO}^\pm_n(q_0)$, and r is prime. Now $n/s \geqslant 4$ and $n \geqslant 10$, so if n/s is even then $z_{q_0, r(n-2s)}$ exists, and $r(n - 2s) \geqslant n$. In addition, $z_{q_0, r(n-2)}$ exists, and $r(n - 2) > n$. We now consult Proposition 1.13.4. If H is of type $\mathrm{GO}^+_{n/s}(q^s)$ then $z_{q_0, r(n-2s)}$ divides $|H|$ (even when $n = n/s = 4$). If H is of type $\mathrm{GO}^\circ_{n/2}(q^2)$ then $z_{q_0, r(n-2)}$ divides $|H|$. If H is of type $\mathrm{GU}_{n/2}(q)$ then $n/2$ is even, and $z_{q_0, r(n-2)}$ divides $|H|$.

We now consider the possibilites for K, again using Proposition 1.13.4. If $|K|$ is of type $\mathrm{GO}^+_n(q_0)$ and $z_{q_0, i}$ divides $|K|$ then $i \leqslant n-2$, so neither $z_{q_0, r(n-2s)}$ nor $z_{q_0, r(n-2)}$ divide $|K|$, a contradiction. If $|K|$ is of type $\mathrm{GO}^-_n(q_0)$ and $z_{q_0, i}$ divides $|K|$, then $i \leqslant n$. Thus $z_{q_0, r(n-2)}$ does not divide $|K|$, and if $z_{q_0, r(n-2s)}$ divides $|K|$ then $n/s = 4$ and $r = 2$. If $n/s = 4$, $r = 2$ and H is of type $\mathrm{GO}^+_{n/s}(q^s)$ then a higher power of $z_{q_0, r(n-2s)}$ divides $|H|$ than $|K|$.

Case O⁻. Here H is of type $\mathrm{GO}^-_{n/s}(q^s)$ with $n/s \geqslant 4$, type $\mathrm{GO}^\circ_{n/2}(q^2)$, or type $\mathrm{GU}_{n/2}(q)$ with $n/2$ odd. Since $n \geqslant 8$ and $r \geqslant 2$, primes $z_{q_0, rn}$ and $z_{q_0, r(n-2)}$ exist, and both rn and $r(n - 2)$ are greater than n. If H is of type $\mathrm{GO}^-_{n/s}(q^s)$, type $\mathrm{GO}^\circ_{n/2}(q^2)$, or type $\mathrm{GU}_{n/2}(q)$ then $|H|$ is divisible by $z_{q_0, rn}$, $z_{q_0, r(n-2)}$ or $z_{q_0, rn}$ respectively, by Proposition 1.13.4. The group K is of type $\mathrm{GO}^-_n(q_0)$ and if $z_{q_0, i}$ divides K then $i \leqslant n$, so neither $z_{q_0, rn}$ nor $z_{q_0, r(n-2)}$ divide $|K|$. $\qquad\square$

We show now that the semilinear groups do not, in general, preserve classical forms.

Lemma 2.3.20   *Let $3 \leqslant n \leqslant 12$ be as in Definition 1.6.20 and let H be a $\mathscr{C}_3$-subgroup of Ω. If there exists a subgroup K of Ω with $H \leqslant K$, then $K \notin \mathscr{C}_8$.*

Proof   Let H preserve a field extension of degree s, let K be a $\mathscr{C}_8$-subgroup of Ω, and assume by way of contradiction that $H \leqslant K$. If $\mathscr{C}_8 \neq \varnothing$, then $\Omega = \mathrm{SL}_n(q)$ or $\mathrm{Sp}_n(2^e)$, so H is not of type $\mathrm{GU}_{n/2}(q)$.

Assume first that K is of one of the following types: $\mathrm{GU}_n(q^{1/2})$ with n even, $\mathrm{GO}^\circ_n(q)$, or $\mathrm{GO}^+_n(q)$. If $(n, q) = (6, 2)$ then $\Omega = \mathrm{Sp}_6(2)$ and K is of type $\mathrm{GO}^+_6(2)$, whilst $H \cong \mathrm{Sp}_2(8).3$. Thus $|H|$ does not divide $|K|$, a contradiction. We therefore assume that $(n, q) \neq (6, 2)$. Then $z_{q,n}$ divides $|H|$, by Proposition 1.13.4 if $s \neq n$ and directly from Table 2.6 if $s = n$. However, $z_{q,n}$ does not divide $|K|$, a contradiction.

Suppose next that $\Omega = \mathrm{SL}_n(q)$, and that K is of type $\mathrm{GU}_n(q^{1/2})$ with n odd, so that s is odd. Let $q = q_0^2$. Then $|H|$ is divisible by $z_{q_0, n}$ (since $q_0^{2n} - 1$

divides $|H|$), whilst $|K|_{p'}$ divides

$$|\mathrm{SU}_n(q_0)|_{p'}(q-1) = (q_0^2-1)^2(q_0^3+1)(q_0^4-1)\cdots(q_0^n+1).$$

Now, $(q_0^n+1, q_0^n-1) = 2$ and there is no term $q_0^{n/2}+1$ in the expression for $|K|$, as n is odd. Therefore, $z_{q_0,n}$ does not divide $|K|$.

Suppose next that $\Omega = \mathrm{SL}_n(q)$, and that K is of type $\mathrm{Sp}_n(q)$ with q odd, so that $|K| = |\mathrm{Sp}_n(q)|(q-1, n/2)$. By Lemma 2.2.7 the group H is insoluble, with $H^\infty \cong \mathrm{SL}_{n/s}(q^s)$ being irreducible but not absolutely irreducible, and $|H| = |\mathrm{SL}_{n/s}(q^s)|(q^s-1)s/(q-1)$. We now apply Proposition 2.2.18 to deduce that $H \leqslant \mathrm{N}_K(H') \leqslant L$, for some $\mathscr{C}_3$-subgroup L of K. Thus L is an extension by scalars of some $\mathscr{C}_3$-subgroup L_1 of $\mathrm{PCSp}_n(q)$. Arguing as in the proof of Lemma 2.3.17, we see that H and L_1 preserve field extensions of the same degree, but then $|H| > |L|$, a contradiction.

Suppose next that $\Omega = \mathrm{SL}_n(q)$ or $\mathrm{Sp}_n(q)$, and that K is of type $\mathrm{GO}_n^-(q)$. If $n = 4$ then $H^\infty \cong K^\infty$, but H^∞ is not absolutely irreducible whilst K^∞ is. If $n = 6$ then H is divisible by $z_{q,3}$, whilst K is not. If $n \geqslant 8$ then by Proposition 2.2.18 we deduce that H is contained in a $\mathscr{C}_3$-subgroup L of K, where L is contained in a $\mathscr{C}_3$-subgroup of $\mathrm{CGO}_n^-(q)$. As before we deduce that H and L preserve field extensions of the same degree, and hence that $|H| > |L|$, a contradiction.

Suppose finally that $\Omega = \mathrm{SL}_n(q)$, and that K is of type $\mathrm{Sp}_n(q)$ with q even. By Proposition 2.2.18 we deduce that H is contained in a subgroup L of K that lies in $\mathscr{C}_3 \cup \mathscr{C}_8$. If $L \in \mathscr{C}_3$ then we get a contradiction as for q odd, so $L \in \mathscr{C}_8$. Thus L is a subgroup of K of type $\mathrm{GO}_n^\pm(q)$. For type $\mathrm{GO}_n^+(q)$ if $n \geqslant 6$ then we get an easy contradiction using Zsigmondy primes, as in the second paragraph of this proof (for $(n,q) = (6,2)$ this is a direct calculation). For type $\mathrm{GO}_n^-(q)$ we argue just as in the preceding paragraph. $\qquad\square$

We finish this subsection with a more general result about semilinear groups.

Proposition 2.3.21 *Let $n \in \{3,5,7,11\}$ be as in Definition 1.6.20, and let H be a $\mathscr{C}_3$-subgroup of $\Omega \neq \mathrm{SL}_3(4)$. Then H is maximal amongst the geometric subgroups of Ω.*

Proof In Case $\mathbf{O}^\circ$, if n is prime then $\mathscr{C}_3 = \varnothing$, so $\Omega = \mathrm{SL}_n(q)$ or $\mathrm{SU}_n(q)$.

Suppose, by way of contradiction, that there exists $K < \Omega$ such that $H \leqslant K$, where K is geometric and is not of the same type as H. By Theorem 2.2.19, we may assume that $K \in \mathscr{C}_i$ for some $1 \leqslant i \leqslant 8$. It follows from Lemma 2.3.14 that $K \notin \mathscr{C}_1$. It is immediate from Lemma 2.3.15 that $K \notin \mathscr{C}_2 \cup \mathscr{C}_6$. Since n is prime, there is a unique type of $\mathscr{C}_3$-subgroup, and $\mathscr{C}_4 = \mathscr{C}_7 = \varnothing$. By Lemma 2.3.19, if $\Omega \neq \mathrm{SL}_3(4)$ and n is odd then $K \notin \mathscr{C}_5$. It is immediate from Lemma 2.3.20 that $K \notin \mathscr{C}_8$, so we are done. $\qquad\square$

2.3.4 Tensor product groups

Recall the types of $\mathscr{C}_4$-subgroup in Table 2.7: note that the conditions in that table imply that $n \geqslant 6$. We first collect some information regarding $\mathscr{C}_4$-subgroups which, despite their inclusion in Class $\mathscr{C}_4$, are never maximal.

Proposition 2.3.22   *Let G be an almost simple group with socle $\bar{\Omega}$. Let H_G be a $\mathscr{C}_4$-subgroup of G, of one of the following types:*

  (i) $\mathrm{GL}_2(2) \otimes \mathrm{GL}_{n/2}(2)$ *(Case* **L***)*,
 (ii) $\mathrm{GU}_2(2) \otimes \mathrm{GU}_{n/2}(2)$ *(Case* **U***)*,
(iii) $\mathrm{Sp}_{n/3}(3) \otimes \mathrm{GO}_3(3)$ *(Case* **S***)*,
 (iv) $\mathrm{Sp}_2(2) \otimes \mathrm{Sp}_{n/2}(2)$ *(Case* **O**$^+$*)*,
  (v) $\mathrm{GO}_3(3) \otimes \mathrm{GO}_{n/3}^{\pm}(3)$ *(Case* **O**$^{\pm}$*)*.

Then H_G is not maximal in G.

Proof   Let H be the corresponding subgroup of Ω. First suppose that H is of type $\mathrm{GL}_2(2) \otimes \mathrm{GL}_{n/2}(2)$, so $H \cong \mathrm{SL}_2(2) \times \mathrm{SL}_{n/2}(2)$. We copy the proof of [66, Proposition 6.3.1(i)] that H is (properly) contained in a $\mathscr{C}_3$-subgroup K of type $\Gamma\mathrm{L}_{n/2}(4)$. Now $\mathrm{SL}_2(2) \cong \mathrm{S}_3$, so $H' = 3 \times \mathrm{SL}_{n/2}(2) = L \times \mathrm{SL}_{n/2}(2)$. We may identify L with the non-zero scalars of $\mathbb{F}_4$, thus setting up an isomorphism between $\mathbb{F}_4^1$ and $\mathbb{F}_2^2$. Since L acts irreducibly on $\mathbb{F}_2^2$, the group H' is irreducible but not absolutely irreducible, by Lemma 2.2.10, and we may identify tensors $v \otimes w$ with elements λw, where $\lambda \in \mathbb{F}_4$, so that $L \times 1$ is the subgroup corresponding to the scalars of $\Gamma\mathrm{L}_{n/2}(4)$. Thus H is contained in a member K of type $\Gamma\mathrm{L}_{n/2}(4)$ of our Class $\mathscr{C}_3$. The only non-trivial outer automorphism of $\mathrm{SL}_n(2)$ is the duality automorphism γ, which is induced by the inverse-transpose map. The inverse-transpose automorphism normalises the standard constructions of H, and also normalises the subgroup $L \times 1$ of H and hence preserves the isomorphism from $\mathbb{F}_4^{n/2}$ to $\mathbb{F}_2^n$, so the extension of H by duality is contained in the extension of K by duality.

Next suppose that H is of type $\mathrm{GU}_2(2) \otimes \mathrm{GU}_{n/2}(2)$, preserving a decomposition $V = W_1 \otimes W_2$, and let H_G be the subgroup of $\mathrm{GU}_n(2)$ of the same type as H. The group $\mathrm{GU}_2(2)$ is imprimitive on $W_1 = \mathbb{F}_4^2$, and is equal to $\mathrm{GU}_1(2) \wr \mathrm{S}_2$, preserving a decomposition of $W_1 := \mathbb{F}_4^2$ into $W_{11} \oplus W_{12} = \langle v_1 \rangle \oplus \langle v_2 \rangle$. Thus H_G preserves an imprimitive decomposition of V as $(W_{11} \otimes W_2) \oplus (W_{12} \otimes W_2)$, and is therefore properly contained in a $\mathscr{C}_2$-subgroup of type $\mathrm{GU}_{n/2}(2) \wr \mathrm{S}_2$. The final automorphism to consider is ϕ, which (setwise) stabilises both $W_{11} \otimes W_2$ and $W_{12} \otimes W_2$, so H does not extend to a novelty.

Next suppose that $H \cong \mathrm{Sp}_{n/3}(3) \times \mathrm{GO}_3(3)$, so $\Omega = \mathrm{Sp}_n(3)$. As shown in the proof of Proposition 2.3.6 (x) our standard copy of $\mathrm{GO}_3(3)$ is imprimitive on $\mathbb{F}_3^3$, and consists of monomial matrices, with blocks being non-degenerate

1-spaces. Forming a tensor product of monomial matrices with our standard copy of $\mathrm{Sp}_{n/3}(3)$ will result in matrices with three (non-degenerate) blocks, so H is (properly) contained in a $\mathscr{C}_2$-subgroup K of type $\mathrm{Sp}_{n/3}(3) \wr S_3$ inside a copy of the symplectic group whose form matrix B is the Kronecker product of I_3 with our standard symplectic form. The only non-trivial outer automorphism of $\mathrm{Sp}_n(3)$ is the diagonal automorphism δ, which can be chosen to have basis vectors as eigenvalues, and so preserves this imprimitive decomposition as well as H. Thus H does not extend to a novel maximal subgroup.

Next suppose that $H \cong \mathrm{Sp}_2(2) \times \mathrm{Sp}_{n/2}(2)$, so $\Omega = \Omega_n^+(q)$. By [66, Proposition 6.3.1(ii)], the group H is properly contained in a $\mathscr{C}_3$-subgroup of type $\Gamma\mathrm{U}_{n/2}(2)$. The class stabiliser of H is trivial, so H cannot extend to a novelty.

Next suppose that $H \cong \mathrm{SO}_3(3) \times \Omega_{n/3}^\pm(3)$, so $\Omega = \Omega_n^\pm(3)$. As previously noted, $\mathrm{GO}_3(3)$ is imprimitive, so H is properly contained in a $\mathscr{C}_2$-subgroup K of type $\mathrm{GO}_{n/3}^\pm(3) \wr S_3$. See [66, Proposition 6.3.2] for more details. The normaliser N of H in $\mathrm{CGO}_n^\pm(3)$ is $\mathrm{CGO}_3(3) \otimes \mathrm{CGO}_{n/3}^\pm(3) = \mathrm{GO}_3(3) \otimes \mathrm{CGO}_{n/3}^\pm(3)$. So N is imprimitive, and hence H does not extend to a novelty. $\square$

Let H be a $\mathscr{C}_4$-subgroup of Ω. Then we write $L = \Omega_1 \circ \Omega_2 \leqslant H$, where Ω_1 is the generally quasisimple group on V_1 and Ω_2 is the generally quasisimple group on V_2. Considering the restrictions on n_1 and n_2 in Table 2.7, and our assumption that $n \leqslant 12$, we note that Ω_1 is quasisimple unless $(n_1, q) \in \{(2,2)$ (Cases **L**, **U** and **O**$^+$), $(2,3)$ (Cases **L**, **U** and **O**$^+$), $(3,2)$ (Case **U** only), $(3,3)$ (Cases **O**$^\pm$ only)$\}$, and is soluble for these exceptional values. In Cases **L** and **O**$^-$ the group Ω_2 is quasisimple. In Case **U** the group Ω_2 is quasisimple if and only if $(n_2, q) \neq (3,2)$, and is soluble otherwise. In Case **S** the group Ω_2 is quasisimple if and only if $(n_2, q, \varepsilon) \notin \{(3,3,\circ), (4,3,+)\}$, and is soluble for these exceptional values. In Case **O**$^+$, if $\Omega_2 = \mathrm{Sp}_{n_2}(q)$ then Ω_2 is always quasisimple (recall that $n \geqslant 10$ in Case **O**$^+$), whilst if $\Omega_2 = \Omega_{n_2}^+(q)$ then Ω_2 is perfect if and only if $(n_2, q) \neq (4,3)$, and is soluble otherwise. Note in particular that $\Omega_1 \otimes \Omega_2$ is perfect when $q > 3$.

Lemma 2.3.23 *Let $n \leqslant 12$ be as in Definition 1.6.20, and let H be a $\mathscr{C}_4$-subgroup of Ω. Assume that H is not one of the non-maximal groups listed in Proposition 2.3.22. If there exists a subgroup K of Ω such that $H \leqslant K$, then $K \notin \mathscr{C}_1 \cup \mathscr{C}_3$.*

Proof If L is perfect then this follows from Lemma 2.2.10, so assume otherwise. Since Ω_1 and Ω_2 are absolutely irreducible on V_1 and V_2 respectively, it is immediate from Lemma 2.2.10 that $K \notin \mathscr{C}_1$. We first consider the types where Ω_1 is not perfect. If $(n_1, q) = (2,2)$ then H is listed in Proposition 2.3.22, contrary to assumption.

Assume that $(n_1, q) = (2,3)$. Then $\Omega_1 = \mathrm{SL}_2(3)$, so the derived group of

Ω_1 is absolutely irreducible. In Cases **L**, **U** and **O**$^+$, the group Ω_2 is perfect so the result is immediate. In Case **S** we note that the group Ω_2 is quasisimple unless $\Omega_2 = \Omega_3(q) \cong L_2(3)$ or $\Omega_2 = \Omega_4^+(3)$. The first possibility is excluded by our assumption that H is not listed in Proposition 2.3.22, whilst for the second we note that it is easy to check in MAGMA (file `Chap2calc`) that the derived group of $\Omega_4^+(3)$ is also absolutely irreducible.

Next assume that $n_1 = 3$ and $\Omega = \mathrm{SU}_n(2)$. It is easy to check in MAGMA (file `Chap2calc`) that $\mathrm{SU}_3(2)'$ is absolutely irreducible on V_1. Also, $\mathrm{SU}_{n_2}(2)$ is both absolutely irreducible on V_2 by Proposition 1.12.2 and perfect. Therefore H' is absolutely irreducible on V, and the result follows.

The possibility that $(n_1, q) = (3, 3)$ in Cases **O**$^\pm$ is excluded by our assumption that H is not one of the groups listed in Proposition 2.3.22.

Thus we may assume without loss of generality that Ω_1 is perfect and Ω_2 is not perfect. Thus $\Omega = \mathrm{Sp}_n(q)$ or $\Omega_n^+(q)$, and since $n \leqslant 12$ the only possibility is that H is of type $\mathrm{Sp}_4(3) \otimes \mathrm{GO}_3(3)$ in $\mathrm{Sp}_{12}(3)$. However, this H is listed in Proposition 2.3.22. $\square$

Lemma 2.3.24 *Let $n \leqslant 12$ be as in Definition 1.6.20, and let H be a $\mathscr{C}_4$-subgroup of Ω, with $q > 3$. If there exists a subgroup K of Ω with $H \leqslant K$ and $K \in \mathscr{C}_2$, then K preserves a decomposition into at least five subspaces.*

Proof Suppose otherwise, and let K preserve an imprimitive decomposition $\mathcal{D}$ into two, three or four subspaces. If $H \leqslant K$ then $L = L^\infty \leqslant K^\infty$. However, our assumption that $q > 3$ implies that L^∞ is irreducible by Lemma 2.2.10, whereas $K^\infty = K_{(\mathcal{D})}$ is reducible. $\square$

Recall Definition 2.2.15 of the $\mathscr{C}_7$-subgroups.

Lemma 2.3.25 *Let $n \leqslant 12$ be as in Definition 1.6.20, and let H be a $\mathscr{C}_4$- or $\mathscr{C}_7$-subgroup of Ω. Assume that H is not one of the groups listed in Proposition 2.3.22. If there exists a subgroup K of Ω such that $H \leqslant K$, then $K \notin \mathscr{C}_5$.*

Proof Here H contains the subgroup $S := \Omega_1 \circ \cdots \circ \Omega_t$. Considering Tables 2.7 and 2.10, and using the fact that $n \leqslant 12$, we see that one of the following holds: Case **L**, with $\Omega_i = \mathrm{SL}_{n_i}(q)$ and $t = 2$; Case **U**, with $\Omega_i = \mathrm{SU}_{n_i}(q)$ and $t = 2$; Case **S**, with $\Omega_i = \mathrm{Sp}_{n_i}(q)$ and $t \leqslant 3$; Case **S**, with $t = 2$, $\Omega_1 = \mathrm{Sp}_{n_1}(q)$ and $\Omega_2 = \Omega_{n_2}^\varepsilon(q)$; Case **O**$^\circ$, with $\Omega_i = \Omega_m^\circ(q)$; Case **O**$^+$, with $\Omega_i = \mathrm{Sp}_{n_i}(q)$; Case **O**$^\pm$, with $\Omega_i = \Omega_{n_i}^{\varepsilon_i}(q)$. We will show that S^∞, and hence H^∞, are not contained in a member of $\mathscr{C}_5$, using the fact that $\mathrm{tr}(a \otimes b) = \mathrm{tr}(a)\mathrm{tr}(b)$. Assume, by way of contradiction, that $H \leqslant K$ for some $K \in \mathscr{C}_5$.

Consider first Cases **L** and **S**. If $\mathscr{C}_5 \neq \varnothing$ then $q > 3$, and so S is perfect. By Proposition 1.12.7, for any $\alpha \in \mathbb{F}_q$ there exists an element $a \in \Omega_1$ of trace α, and by the same result, together with Lemma 1.12.8, there exist elements

$b_i \in \Omega_i$ for $2 \leqslant i \leqslant t$ of non-zero trace. Then elements of S have q different traces, contradicting Lemma 2.2.12.

Consider next Case **U**, so that $t = 2$ and $n_2 \geqslant 3$. Proposition 2.3.22 excludes $\mathrm{SU}_6(2)$, so Ω_2 is perfect, and by Proposition 1.12.7 each element of $\mathbb{F}_{q^2}$ is the trace of at least one of its elements. For all n and q, the group Ω_1 contains an element of non-zero trace, so if $\alpha \in \mathbb{F}_{q^2}$ then there exists an element of $S^\infty \leqslant H^\infty$ of trace α.

Consider next Case **O**°, so that $H \in \mathscr{C}_7$, $t = 2$, $m = 3$ and $q > 3$. Hence S is perfect. By Lemma 1.12.8 the set of traces of elements of $\Omega_3(q)$ does not lie in any proper subfield of $\mathbb{F}_q$. Recall that $\Omega_3(q)$ is $\mathrm{SL}_2(q)$ acting on the symmetric square of its natural module (see Section 5.2). The symmetric square of antidiag$(1, -1)$ has trace -1, so the traces of elements of $S^\infty \leqslant H^\infty$ do not lie in any proper subfield of $\mathbb{F}_q$.

Consider Cases **O**$^\pm$, so $n = 12$ and $H \in \mathscr{C}_4$. If $\mathscr{C}_5 \neq \varnothing$ then $q \geqslant 4$, so each Ω_i is perfect. If the type is $\mathrm{Sp}_2(q) \times \mathrm{Sp}_6(q)$ then each Ω_i contains elements of all traces in $\mathbb{F}_q$, so $S^\infty \leqslant H^\infty$ is not contained in a member of $\mathscr{C}_5$. Thus H is of type $\mathrm{GO}_3(q) \times \mathrm{GO}_4^\pm(q)$. By Lemma 1.12.8 the set of traces of elements of $\Omega_3(q)$ lie in no proper subfield of $\mathbb{F}_q$. By Proposition 1.12.7 the group $\Omega_4^\pm(q)$ contains elements of all traces in $\mathbb{F}_q$, a contradiction. $\qquad\square$

Finally, we consider containments of $\mathscr{C}_4$-subgroups in $\mathscr{C}_8$-subgroups.

Lemma 2.3.26 *Let $n \leqslant 12$ be as in Definition 1.6.20 and let H be a $\mathscr{C}_4$-subgroup of Ω. If there exists a subgroup K of Ω with $H \leqslant K$, then $K \notin \mathscr{C}_8$.*

Proof Suppose otherwise. By Tables 2.7 and 2.11, the group $\Omega = \mathrm{SL}_n(q)$. Let W be a non-trivial irreducible Ω_2-submodule of V. Then since Ω_2 acts homogeneously on V, the space W has dimension $n_2 > 2$ and Ω_2 acts as $\mathrm{SL}_{n_2}(q)$ on W. Therefore W is totally singular with respect to the classical form for which K is a group of similarities, and so by a slight generalisation of Lemma 2.2.2 (vi), as an Ω_2-module $V/W^\perp \cong W^*$ or $W^{*\gamma}$. Since $n_2 > 2$, this contradicts the homogeneity of the action of Ω_2. $\qquad\square$

2.3.5 Subfield groups

In this section we prove various results about the maximality of $\mathscr{C}_5$-subgroups, and in particular determine their maximality for $5 \leqslant n \leqslant 12$.

Recall Definition 2.2.11 of the $\mathscr{C}_5$-subgroups, and our dimension assumptions in Definition 1.6.20. Recall that $u = 2$ in Case **U** and $u = 1$ otherwise. First we show that, with only a small number of possible exceptions, the $\mathscr{C}_5$-subgroups do not preserve an imprimitive, tensor or tensor induced decomposition.

Lemma 2.3.27   *Let H be a $\mathscr{C}_5$-subgroup of Ω, with $n \leqslant 12$, as in Definition 1.6.20. If $\Omega = \mathrm{SL}_n(q)$ then assume that $n \geqslant 3$. If $\Omega = \mathrm{SU}_n(q)$ then assume that $n \geqslant 4$, and that if $n = 4$ then H is of type $\mathrm{GU}_4(q_0)$ or $\mathrm{Sp}_4(q)$. If there exists a subgroup K of Ω with $H \leqslant K$, then $K \notin \mathscr{C}_2 \cup \mathscr{C}_4 \cup \mathscr{C}_7$.*

Proof   Suppose, by way of contradiction, that $H \leqslant K < \Omega$, where the group $K \in \mathscr{C}_2 \cup \mathscr{C}_4 \cup \mathscr{C}_7$. Then $K = \Omega_{\mathcal{D}}$, where $\mathcal{D}$ is either a direct sum decomposition $V = W_1 \oplus \cdots \oplus W_t$ or a $\mathscr{C}_4$- or $\mathscr{C}_7$-decomposition $V = W_1 \otimes \cdots \otimes W_t$. Our assumptions imply that H is insoluble and H^∞ is quasisimple and absolutely irreducible.

By Theorem 1.11.2, the group H^∞ has no non-trivial permutation representations of degree less than or equal to $n \leqslant 12$, so H^∞ is contained in the pointwise stabiliser of $\{W_1, \ldots, W_t\}$. By Theorem 1.11.5, the quasisimple group H^∞ has no non-trivial representation in defining characteristic in dimension properly dividing n, a contradiction.   □

Lemma 2.3.28   *Let $n \leqslant 12$ be as in Definition 1.6.20, and let H be a $\mathscr{C}_5$-subgroup of Ω. If there exists a subgroup K of Ω with $H \leqslant K$ and K of a different type than H, then $K \notin \mathscr{C}_5$.*

Proof   We first consider the types $\mathrm{GL}_n(q_0)$, $\mathrm{GU}_n(q_0)$, $\mathrm{Sp}_n(q_0)$ in Case **S** and $\mathrm{GO}_n^\varepsilon(q_0)$ in Case **O**$^\varepsilon$. Assume that H and K are of two of these types, so that q_0 is not prime. It follows that H is insoluble, and $H^\infty = \Omega_{q_0}$, the Ω-group over $\mathbb{F}_{q_0^u}$ where $q_0^r = q$. Furthermore, K^∞ contains the Ω-group over $\mathbb{F}_{q_1^u}$ where $q_1^s = q$. Here, by assumption, r and s are distinct primes. Thus $\mathbb{F}_{q_0^u}$ is not contained in $\mathbb{F}_{q_1^u}$. By Proposition 1.12.7 and Lemma 1.12.8, the group H^∞ contains elements of all traces in $\mathbb{F}_{q_0^u}$, whilst all traces of elements of K^∞ lie in $\mathbb{F}_{q_1^u}$, contradicting Lemma 2.2.12.

The remaining possible types for $\{H, K\}$ are $\{\mathrm{GU}_n(q_0), \mathrm{Sp}_n(q)\}$ (Case **U**), $\{\mathrm{GU}_n(q_0), \mathrm{GO}_n^\varepsilon(q)\}$ (Case **U**, with q odd), $\{\mathrm{Sp}_n(q), \mathrm{GO}_n^\pm(q)\}$ (Case **U**, with q odd) and $\{\mathrm{GO}_n^+(q_0), \mathrm{GO}_n^-(q_1)\}$ (Cases **U** and **O**$^\pm$, with $q_1^2 = q$). The only soluble possibility for H is type $\mathrm{GO}_3(3)$ (in which case there is no $\mathscr{C}_5$-subgroup of type $\mathrm{GU}_n(q_0)$) or type $\mathrm{GO}_4^+(3)$ in $\mathrm{SU}_4(3)$, which we will consider at the end of the proof. Otherwise, $H^\infty = \Omega_{q_0}$.

Proposition 1.12.7 rules out $\{\mathrm{GU}_n(q_0), \mathrm{Sp}_n(q)\}$, since $\mathbb{F}_{q_0^2}$ is an odd degree subfield of $\mathbb{F}_{q^2}$, and $\{\mathrm{GU}_n(q_0), \mathrm{GO}_n^\varepsilon(q)\}$ if $n \geqslant 4$. Lemma 1.12.8 and Proposition 1.12.7 give a contradiction if $\{H, K\}$ are of types $\{\mathrm{GU}_3(q_0), \mathrm{GO}_3(q)\}$. Lemmas 1.12.4 and 1.12.5 rule out types $\{\mathrm{Sp}_n(q), \mathrm{GO}_n^\pm(q)\}$. If the types are $\{\mathrm{GO}_n^+(q_0), \mathrm{GO}_n^-(q_1)\}$ and $q_0 = q_1$ then Lagrange's theorem gives a contradiction, if $q_0 \neq q_1$ then we apply Proposition 1.12.7, since $|\mathbb{F}_q : \mathbb{F}_{q_0}|$ is odd.

Finally consider H of type $\mathrm{GO}_4^+(3)$ and K of type $\mathrm{Sp}_4(3)$. Then $\Omega_4^+(3) \leqslant H' \leqslant K'$, which contradicts Lemma 1.12.4.   □

Proposition 2.3.29 *Let $n \leqslant 12$ be as in Definition 1.6.20, and let H be a $\mathscr{C}_5$-subgroup of Ω. If $\Omega = \mathrm{SL}_n(q)$ then assume that $n \geqslant 3$. If $\Omega = \mathrm{SU}_n(q)$ then assume that $n \geqslant 4$ and that, if $n = 4$, then H is of type $\mathrm{GU}_4(q_0)$ or $\mathrm{Sp}_4(q)$. Then H is maximal amongst the geometric subgroups of Ω.*

Proof Suppose, by way of contradiction, that $H \leqslant K < \Omega$, where K is not of the same type as H. By Theorem 2.2.19, without loss of generality $K \in \mathscr{C}_i$ for some $1 \leqslant i \leqslant 8$. Our assumptions on the type and dimension of H imply that H^∞ is quasisimple.

The group H^∞ is absolutely irreducible by Proposition 1.12.2, so the group $K \notin \mathscr{C}_1 \cup \mathscr{C}_3$. It follows from Lemma 2.3.27 that $K \notin \mathscr{C}_2 \cup \mathscr{C}_4 \cup \mathscr{C}_7$, and from Lemma 2.3.28 that $K \notin \mathscr{C}_5$.

Suppose next that $K \in \mathscr{C}_6$, then, since H^∞ is quasisimple, there exists a non-trivial representation ρ mapping H^∞ to $\mathrm{S}_{2m}(r)$ or $\mathrm{O}_{2m}^\varepsilon(r)$, where $n = r^m$ and r divides $q - 1$. However, both of these groups are smaller than the simple group $\overline{H^\infty}$, a contradiction.

Finally, suppose that $K \in \mathscr{C}_8$, so that $\Omega = \mathrm{SL}_n(q)$ or $\mathrm{Sp}_n(2^i)$. The group $\mathrm{SL}_n(q_0)$ contains all transvections, so a straightforward calculation shows that since $n > 2$ the group $\mathrm{SL}_n(q_0)$ preserves no non-zero bilinear or unitary form on $\mathbb{F}_q^n$. Similarly, considering the transvections in $\mathrm{Sp}_n(q_0)$ shows that $\mathrm{Sp}_n(q_0)$ fixes no quadratic form. $\qquad\square$

2.3.6 Extraspecial normaliser groups

Recall the definition of the $\mathscr{C}_6$-subgroups from Table 2.9. In this subsection we show that $\mathscr{C}_6$-subgroups are maximal if $4 \leqslant n \leqslant 12$. If $\mathscr{C}_6 \neq \varnothing$ then $\Omega \neq \Omega_n^\varepsilon(q)$.

Let H be a $\mathscr{C}_6$-subgroup of Ω. If $n = r^m > 3$ then $\mathrm{Sp}_{2m}(r)$ is perfect unless $m = r = 2$, and $\Omega_{2m}^-(2)$ is perfect when $r = 2$. Let $R = O_r(H^\infty)$ be the extraspecial group, or the 2-group of symplectic type. Recall Definition 1.11.1 of $R_k(G)$.

Lemma 2.3.30 *Let $4 \leqslant n \leqslant 12$ be as in Definition 1.6.20 and let H be a $\mathscr{C}_6$-subgroup of Ω. Then the following all hold.*

(i) *The group H is insoluble.*
(ii) *If r is odd then $R \cong r^{1+2m}$. If $r = 2$ then $R \cong 4 \circ 2^{1+2m}$ in Cases $\mathbf{L}$ and $\mathbf{U}$, and $R \cong 2_-^{1+2m}$ in Case $\mathbf{S}$.*
(iii) *The group H^∞ is absolutely irreducible.*
(iv) *If $n \in \{4, 8, 9\}$ then $P(\overline{H^\infty}) > n$, and if $n \in \{5, 7, 11\}$ then $P(\overline{H^\infty}) \geqslant n$.*
(v) *Let r' be a prime other than r. Then $R_{r'}(\overline{H^\infty}) \geqslant n$.*

Proof (i) and (ii) This is clear from the structures given in Table 2.9, noting that the quotient H/R acts irreducibly on R.

(iii) As mentioned in the introduction to the $\mathscr{C}_6$-subgroups, $R < H^\infty$ acts absolutely irreducibly on V.

(iv) Let $L = H^\infty$, and let $\overline{X} < \overline{L}$. We need to show that $|\overline{L} : \overline{X}| > n$. If $\overline{RX} \neq \overline{L}$ then

$$|\overline{L} : \overline{X}| \geqslant |\overline{L} : \overline{RX}| \geqslant P(L/R).$$

The group L/R is one of $\mathrm{Sp}_4(2)$, $\Omega_4^-(2)$, $\mathrm{Sp}_2(p)$ for $p \in \{5, 7, 11\}$, $\mathrm{Sp}_8(2)$ or $\Omega_8^-(2)$ or $\mathrm{Sp}_4(3)$, so the result is clear.

Suppose instead that $\overline{RX} = \overline{L}$ and let $\overline{S} = \overline{R} \cap \overline{X}$. Then $\overline{S} \trianglelefteq \overline{R}$, since $\overline{R}$ is elementary abelian. Also $\overline{S} \trianglelefteq \overline{X}$ since $\overline{R} \trianglelefteq \overline{L}$. Hence $\overline{S} \trianglelefteq \overline{RX} = \overline{L} = \overline{H^\infty}$. Since $\overline{R}$ is a minimal normal subgroup of $\overline{L}$, and $\overline{X}$ is a proper subgroup of $\overline{L}$, the group $\overline{S}$ is trivial. Therefore $|\overline{L} : \overline{X}| = |\overline{R}| = n^2 > n$.

(v) Our assumptions on n ensure that $\overline{H^\infty}$ is a perfect group, with a unique minimal normal subgroup $\overline{R}$ (as H^∞/R acts irreducibly on $\overline{R}$). The group $\overline{R}$ is elementary abelian, of shape r^{2m}, where $2m \geqslant 2$. Therefore by Lemma 1.11.4, $R_{r'}(\overline{H^\infty}) \geqslant \min\{P(H^\infty/R), r^m\} = n$. $\square$

Proposition 2.3.31 *Let $4 \leqslant n \leqslant 12$, as in Definition 1.6.20, and let H be a $\mathscr{C}_6$-subgroup of Ω. Then H is maximal amongst the geometric subgroups of Ω.*

Proof Suppose, by way of contradiction, that there exists a geometric subgroup $K < \Omega$ with $H \leqslant K$, where K is not of the same type as H. By Theorem 2.2.19 we may assume without loss of generality that $K \in \mathscr{C}_i$ for some i. Let R be the extraspecial or symplectic-type normal subgroup of H^∞.

By Lemma 2.3.30 (iii) H^∞ is absolutely irreducible, so $K \notin \mathscr{C}_1 \cup \mathscr{C}_3$.

Suppose $K \in \mathscr{C}_2 \cup \mathscr{C}_4 \cup \mathscr{C}_7$, and denote the decomposition of V preserved by K by $\mathcal{D}$. By Lemma 2.3.30 (iv) if $n \notin \{5, 7, 11\}$ then $P(\overline{H^\infty}) > n$, so the group $H^\infty \leqslant \Omega_{(\mathcal{D})}$. Therefore, if $n \notin \{5, 7, 11\}$, then $K \in \mathscr{C}_4 \cup \mathscr{C}_7$, as if $K \in \mathscr{C}_2$ then $\Omega_{(\mathcal{D})}$ is reducible. Therefore $\overline{H^\infty} \leqslant \mathrm{L}(V_1) \times \cdots \times \mathrm{L}(V_t)$ for $t \geqslant 2$. The image of $\overline{R}$ in $\mathrm{L}(V_i)$ must be non-trivial for some i. The group $\overline{H^\infty}$ acts faithfully and irreducibly on $\overline{R}$, so the whole of $\overline{H^\infty}$ must embed in $\mathrm{L}(V_i)$. But $\dim(V_i)$ is a proper divisor of n, contradicting Lemma 2.3.30 (v).

We deal now with $n = 5, 7, 11$, so that $\Omega = \mathrm{SL}_n(q)$ or $\mathrm{SU}_n(q)$. Since n is prime, Classes $\mathscr{C}_4$ and $\mathscr{C}_7$ are empty, so $K \in \mathscr{C}_2$. If $n = 5$ then $H^\infty \cong 5^{1+2}.\mathrm{A}_5$, whereas $K^\infty \cong (q \pm 1)^4{:}\mathrm{A}_5$ by Table 2.4. The only non-abelian composition factor of K is A_5, so if $H^\infty \leqslant K^\infty$ then $R \leqslant (q \pm 1)^4$. Since R is non-abelian this is a contradiction. If $n = 7, 11$, then $H^\infty/R \cong \mathrm{SL}_2(n)$, and the only non-abelian composition factor of K is A_n. The only subgroup of A_n with composition factor isomorphic to $\mathrm{L}_2(n)$ is isomorphic to $\mathrm{L}_2(n)$, so if $H \leqslant K$ then $R \leqslant (q \pm 1)^{n-1}$, a contradiction.

The field size q^u is the minimal power of p such that $p^{ue} \equiv 1 \bmod r$ or

$p^{ue} \equiv 1 \bmod 4$. Therefore R, and hence H^∞, cannot be represented over a proper subfield of $\mathbb{F}_{q^u}$. So $K \notin \mathscr{C}_5$.

There is a unique type of $\mathscr{C}_6$-subgroup, when one exists, so $K \notin \mathscr{C}_6$.

So suppose $K \in \mathscr{C}_8$. Then both $\mathscr{C}_6$ and $\mathscr{C}_8$ are nonempty, so $\Omega = \mathrm{SL}_n(q)$. The field size q is an odd power of a prime, and hence K is not of type $\mathrm{GU}_n(q^{1/2})$. Therefore, $H^\infty \leqslant \mathrm{Sp}_n(q)$ or $\Omega_n^\varepsilon(q)$. However, $H^\infty \cap \mathrm{Z}(\mathrm{GL}_n(q)) \cong C_4$ or C_r, whereas $\mathrm{Sp}_n(q) \cap \mathrm{Z}(\mathrm{GL}_n(q))$ and $\Omega_n^\varepsilon(q) \cap \mathrm{Z}(\mathrm{GL}_n(q))$ have order at most 2. $\quad\square$

2.3.7 Classical groups

Recall the types of $\mathscr{C}_8$-subgroup from Table 2.11.

Proposition 2.3.32 *Let $5 \leqslant n \leqslant 12$, with n as in Definition 1.6.20, and let H be a $\mathscr{C}_8$-subgroup of Ω. Then H is maximal amongst the geometric subgroups of Ω.*

Proof Suppose, by way of contradiction, that $H \leqslant K < \Omega$, where K is maximal amongst the geometric subgroups of Ω and is not of the same type as H. The group H^∞ is quasisimple because $n \geqslant 5$, and is absolutely irreducible by Proposition 1.12.2, so $K \notin \mathscr{C}_1 \cup \mathscr{C}_3$.

We show next that $K \notin \mathscr{C}_2 \cup \mathscr{C}_4 \cup \mathscr{C}_7$. Otherwise, $K = \Omega_\mathcal{D}$, where $\mathcal{D}$ is either a direct sum decomposition $V = W_1 \oplus \cdots \oplus W_t$ or a tensor decomposition $W_1 \otimes \cdots \otimes W_t$. Here $1 < t \leqslant n$, whilst $\dim(W_i)$ is a proper divisor of n for $1 \leqslant i \leqslant t$. For all types, $P(H^\infty) > n$ by Theorem 1.11.2, so if $K \in \mathscr{C}_2 \cup \mathscr{C}_7$ then H^∞ is a subgroup of the kernel of the action on the set of W_i. Theorem 1.11.5 states that $R_p(\overline{H^\infty}) > n/2$, so $K \notin \mathscr{C}_2 \cup \mathscr{C}_4 \cup \mathscr{C}_7$.

By Proposition 1.12.7, each element of $\mathbb{F}_q$ arises as the trace of an element of H^∞, so $K \notin \mathscr{C}_5$ by Lemma 2.2.12.

Assume next that $K \in \mathscr{C}_6$. Then $n = r^b$ for some prime r that divides $q - 1$, and the only non-abelian composition factor of K is $\mathrm{S}_{2b}(r)$ or $\mathrm{O}_{2b}^\pm(r)$. However, both of these groups are smaller than $\overline{H^\infty}$, a contradiction.

Finally, let $K \in \mathscr{C}_8$. In Case **L** the groups H and K are of different types from $\mathrm{Sp}_n(q)$, $\mathrm{SU}_n(q^{1/2})$ and $\mathrm{GO}_n^\varepsilon(q)$ (with q odd for this final type). The groups of type $\mathrm{Sp}_n(q)$ and $\mathrm{GO}_n^\varepsilon(q)$ do not contain any of the groups of the other types, by Theorem 1.6.22 and Lemmas 1.12.4, 1.12.5. The group $\mathrm{SU}_n(q^{1/2})$ does not contain $\mathrm{Sp}_n(q)$ or $\Omega_n^\varepsilon(q)$, by Theorem 1.6.22 and Proposition 1.13.4. In Case **S**, the groups H and K are of types $\mathrm{GO}_n^+(q)$ and $\mathrm{GO}_n^-(q)$, which contradicts Theorem 1.6.22. $\quad\square$

3

Geometric maximal subgroups

The maximal subgroups of the finite classical groups are divided into two broad classes by Aschbacher's theorem (see Theorem 2.1.5 for a rough statement): the geometric subgroups and those in Class $\mathscr{S}$. In this chapter we shall classify those subgroups that are maximal amongst the geometric subgroups of the finite classical groups in dimension up to 12.

For a more precise statement, first recall Definition 2.1.2 of the geometric subgroups, our dimension assumptions from Definition 1.6.20, and the more precise version of Aschbacher's theorem given in Theorem 2.2.19. Let G be an almost simple with socle S, where S is simple and one of:

$$\mathrm{L}_n(q), 2 \leqslant n \leqslant 12; \ \mathrm{U}_n(q), 3 \leqslant n \leqslant 12; \ \mathrm{S}_n(q), 4 \leqslant n \leqslant 12; \ \mathrm{O}_n^\varepsilon(q), 7 \leqslant n \leqslant 12.$$

In this chapter we shall classify those subgroups of G that are maximal amongst the set of all geometric subgroups of G. Later, in Chapter 6, we shall determine all containments between those subgroups of G that are maximal amongst the geometric subgroups and the $\mathscr{S}^*$-maximal subgroups of G (see Definition 6.1.1 for the meaning of $\mathscr{S}^*$-maximal).

The structure of this chapter is straightforward: we consider each dimension in turn. We remind the reader that the files of MAGMA calculations that we refer to are available on the webpage http://www.cambridge.org/9780521138604.

3.1 Dimension 2

Let $q = p^e$ be a prime power. We note that by Definition 1.6.20 the group $\Omega = \mathrm{SL}_2(q)$, and that Classes $\mathscr{C}_4$ and $\mathscr{C}_7$ are empty. We assume that $q \geqslant 4$ as $\mathrm{SL}_2(2)$ and $\mathrm{SL}_2(3)$ are soluble.

We start with a result that allows us to treat Class $\mathscr{C}_8$ as empty in dimension 2: recall the $\mathscr{C}_8$-subgroups from Definition 2.2.17.

Lemma 3.1.1 (i) *The group* $\mathrm{Sp}_2(q)$ *is equal to* $\mathrm{SL}_2(q)$, *and hence is not a maximal subgroup of* $\mathrm{SL}_2(q)$.

(ii) *If q is a square, then the group* $\mathrm{SU}_2(q^{1/2})$ *is conjugate to* $\mathrm{SL}_2(q^{1/2})$ *in* $\mathrm{SL}_2(q)$, *and hence may be considered as a member of* $\mathscr{C}_5$.

(iii) *The* $\mathscr{C}_8$-*subgroup of type* $\mathrm{GO}_2^+(q)$ *is equal to the* $\mathscr{C}_2$-*subgroup of* $\mathrm{SL}_2(q)$.

(iv) *The* $\mathscr{C}_8$-*subgroup of type* $\mathrm{GO}_2^-(q)$ *is equal to the* $\mathscr{C}_3$-*subgroup of* $\mathrm{SL}_2(q)$.

Proof (i) This is Lemma 1.12.1.

(ii) Let ω be a primitive element of $\mathbb{F}_q^\times$. One may check that a 2×2 matrix of determinant one with entries over $\mathbb{F}_{q^{1/2}}$ is an isometry of the antidiagonal unitary form with entries $\pm\omega^{(q+1)/2}$ (in odd characteristic), or both entries 1 (in even characteristic). Thus $\mathrm{SL}_2(q^{1/2})$ is conjugate to a subgroup of $\mathrm{SU}_2(q^{1/2})$, and since by Theorem 1.6.22 these groups have the same order, they are equal.

(iii) Let C_1 be the $\mathscr{C}_8$-subgroup of Ω of type $\mathrm{GO}_2^+(q)$, so that by Table 2.11 $C_1 \cong \mathrm{SO}_2^+(q).2$, of order $2(q-1)$ by Theorem 1.6.22. We check that the group $K = \langle \mathrm{diag}(\omega, \omega^{-1}), \mathrm{antidiag}(1, -1) \rangle$ of all monomial matrices of determinant 1 is the $\mathscr{C}_2$-subgroup of $\mathrm{SL}_2(q)$, and preserves our standard form of plus type. Since $|K| = |C_1|$ these groups are equal.

(iv) Let C_2 be the $\mathscr{C}_8$-subgroup of Ω of type $\mathrm{GO}_2^-(q)$, so that by Table 2.11 $C_2 \cong \mathrm{SO}_2^-(q).2$. The group $\Omega_2^-(q)$ is cyclic, and if q is odd then $\mathrm{SO}_2^-(q)$ is also cyclic, of order $q+1$. Thus either $\Omega_2^-(q)$ or $\mathrm{SO}_2^-(q)$ has order $q+1$, and hence is a Singer cycle of $\mathrm{SL}_2(q)$. All Singer cycles of $\mathrm{SL}_2(q)$ are conjugate by [53, 7.3], so either $\Omega_2^-(q)$ or $\mathrm{SO}_2^-(q)$ are conjugate to the characteristic subgroup of the $\mathscr{C}_3$-subgroup K of $\mathrm{SL}_2(q)$. Now, both C_2 and K have order $2(q+1)$, which by [53, 7.3] is the order of the normaliser of an element of order $q+1$ in $\mathrm{SL}_2(q)$, so $C_1 = K$ as required. $\square$

Recall Definition 2.2.1 of the $\mathscr{C}_1$-subgroups.

Proposition 3.1.2 *Let $n = 2$ and let H be a* $\mathscr{C}_1$-*subgroup of Ω. Then H is maximal amongst the geometric subgroups of Ω.*

Proof This follows immediately from Proposition 2.3.1. $\square$

Recall Definition 2.2.3 of the $\mathscr{C}_2$-subgroups. If $H < \mathrm{SL}_2(q)$ is a $\mathscr{C}_2$-subgroup, then H is of type $\mathrm{GL}_1(q) \wr S_2$.

Lemma 3.1.3 *Let H be a* $\mathscr{C}_2$-*subgroup of* $\mathrm{SL}_2(q)$. *Then H is maximal amongst the geometric subgroups of* $\mathrm{SL}_2(q)$ *if and only if* $q \notin \{5, 7, 9\}$. *If $q = 4$ then H is equal to the* $\mathscr{C}_5$-*subgroup of* $\mathrm{SL}_2(4)$, *but will be considered as a* $\mathscr{C}_2$-*subgroup. If G is* $\mathrm{PGL}_2(7)$, $\mathrm{PGL}_2(9)$, M_{10} *or* $\mathrm{Aut}\,\mathrm{L}_2(9)$ *then the* $\mathscr{C}_2$-*subgroup of an almost simple group G with socle* $\mathrm{L}_2(q)$ *is maximal amongst the geometric subgroups of G. For all other almost simple G with socle* $\mathrm{L}_2(q)$, *where* $q \in \{5, 7, 9\}$, *the* $\mathscr{C}_2$-*subgroup of G is not maximal.*

Proof The group H is of shape $(q-1)\cdot 2$, and is dihedral if and only if q is even. Suppose that $H \leqslant K < \mathrm{SL}_2(q)$, where K is maximal amongst the geometric subgroups of $\mathrm{SL}_2(q)$ and is not of the same type as H. The standard copy of H is equal to $\langle \mathrm{diag}(\omega, \omega^{-1}), \mathrm{antidiag}(1, -1) \rangle$, where ω is a primitive element of $\mathbb{F}_q^\times$. There is no 1-dimensional subspace that is fixed by both of these matrices, so $K \notin \mathscr{C}_1$.

There is a unique type of imprimitive group when $n = 2$, so $K \notin \mathscr{C}_2$. Since $q > 3$ it follows that $|H|$ does not divide $2(q+1)$ so $K \notin \mathscr{C}_3$.

Assume next that $K \in \mathscr{C}_5$, so that by Table 2.8 the group $K = \mathrm{SL}_2(q_0)$ or $K = \mathrm{SL}_2(q^{1/2}).2$, and in the latter case q is odd. Notice that $\mathrm{tr}(\mathrm{diag}(\alpha, \alpha^{-1})) = \alpha + \alpha^{-1}$. If $q > 4$ then not all of these traces lie in a proper subfield of $\mathbb{F}_q$, by Lemma 1.4.5, so if $K = \mathrm{SL}_2(q_0)$ we are done. If $q = 4$ then a computer calculation (file `Chap3calc`) shows that H is conjugate to the $\mathscr{C}_5$-subgroup $\mathrm{SL}_2(2)$ of $\mathrm{SL}_2(4)$. If $K \neq \mathrm{SL}_2(q_0)$ then $K = \mathrm{SL}_2(q^{1/2}).2$, and q is both odd and a square. Any index 2 subgroup H_1 of H contains elements of trace $\alpha^2 + \alpha^{-2}$, for all non-zero $\alpha \in \mathbb{F}_q$. This can be written as a degree 4 polynomial in α, so the set of traces of elements of H_1 has size at least $(q-1)/4$. This is bigger than $q^{1/2}$ if and only if $q > 9$. When $q = 9$, a computer calculation (file `Chap3calc`) shows that H is properly contained in $\mathrm{SL}_2(3).2$. There are two classes in $\mathrm{SL}_2(9)$ of groups of type $\mathrm{SL}_2(3).2$, which are normalised by $\langle \phi \rangle$ and interchanged by δ and $\delta\phi$. There is a single class of groups of type H, and the extension of H by ϕ is contained in the extension of K by ϕ. Thus $\bar{H}$ is not maximal in $\mathrm{L}_2(9)$ and $\mathrm{P\Sigma L}_2(9)$, but could give rise to a novelty in $\mathrm{PGL}_2(9)$, M_{10} or $\mathrm{Aut}\,\mathrm{L}_2(9) = \mathrm{P\Gamma L}_2(9)$.

The largest cyclic subgroup of a $\mathscr{C}_6$-subgroup has order at most 8, so if $q > 9$ then there is a contradiction. If $q = 4, 8, 9$ then there are no $\mathscr{C}_6$-subgroups, by Definition 2.2.13. If $q = 5$ then $H \cong \mathrm{Q}_8$, and so the $\mathscr{C}_2$-subgroup of $\mathrm{GL}_2(5)$ normalises Q_8, and hence is properly contained in a $\mathscr{C}_6$-subgroup. If $q = 7$ then H is the unique order 12 subgroup of the $\mathscr{C}_6$-subgroup. The group H is normalised by the diagonal automorphism, whereas the two classes of $\mathscr{C}_6$-subgroups are interchanged, which could give rise to a novelty in $\mathrm{PGL}_2(7)$. Groups in Class $\mathscr{C}_8$ have already been considered under other classes, by Lemma 3.1.1.

When $q = 7, 9$ all groups which could be novel maximal subgroups are maximal amongst the geometric subgroups of the corresponding overgroups of Ω, as we have now considered all possible geometric overgroups. $\qquad\square$

Recall Definition 2.2.5 of the $\mathscr{C}_3$-subgroups, and Definition 1.13.2 of Zsigmondy primes $z_{q,n}$. If $H < \mathrm{SL}_2(q)$ is a $\mathscr{C}_3$-subgroup, then H is of type $\mathrm{GL}_1(q^2)$.

Lemma 3.1.4 *Let $H < \mathrm{SL}_2(q)$ be a $\mathscr{C}_3$-subgroup. Then H is maximal amongst the geometric subgroups of $\mathrm{SL}_2(q)$ if and only if $q \neq 7$. The $\mathscr{C}_3$-subgroup of $\mathrm{PGL}_2(7)$ is maximal amongst the geometric subgroups of $\mathrm{PGL}_2(7)$.*

Proof The group H is of shape $(q + 1).2 = (p^e + 1).2$, by Table 2.6. Suppose that $H \leqslant K < \mathrm{SL}_2(q)$, where K is maximal amongst the geometric subgroups of $\mathrm{SL}_2(q)$ and is not of the same type as H.

The reducible groups have order $q(q - 1)$, so it follows from Lagrange's theorem that $K \notin \mathscr{C}_1$. If $K \in \mathscr{C}_2$ then $|K| = 2(q - 1)$, by Table 2.5, so $K \notin \mathscr{C}_2$. There is a unique type of semilinear group in $\mathrm{SL}_2(q)$, so $K \notin \mathscr{C}_3$. Classes $\mathscr{C}_4$, $\mathscr{C}_7$ and $\mathscr{C}_8$ are empty in dimension 2.

If $K \in \mathscr{C}_5$ then $e > 1$ and $|K| = (2, q - 1, r)q_0(q_0^2 - 1)$, where $q_0^r = q$, by Table 2.8, so that $e = rf$ for some f. By Theorem 1.13.1 $|H|$ is divisible by some $z_{p,2e}$ unless $p = 2$ and $e = 3$, whilst $z_{p,2e}$ does not divide $|K|$, a contradiction. If $(p, e) = (2, 3)$ then $|K|$ is smaller than $|H|$, a contradiction.

If $K \in \mathscr{C}_6$ then $K \leqslant 2 \cdot S_4$ by Table 2.9, so the largest cyclic subgroup of K has order at most 8. Therefore, if $q > 7$ then $K \notin \mathscr{C}_6$. Class $\mathscr{C}_6 = \varnothing$ when $q = 4$, and the $\mathscr{C}_6$-subgroup $2 \cdot A_4 < \mathrm{SL}_2(5)$ has no subgroups of order 12, so assume that $q = 7$. Then H is a Sylow 2-subgroup of $\mathrm{SL}_2(7)$, so $H < K$ as K has order divisible by 16. There are two classes of $\mathscr{C}_6$-subgroups in $\mathrm{SL}_2(7)$, interchanged by the diagonal automorphism, but only one class of $\mathscr{C}_3$-subgroups. Therefore the $\mathscr{C}_3$-subgroup of $\mathrm{PGL}_2(7)$ is not contained in the $\mathscr{C}_6$-subgroup.

We are considering Class $\mathscr{C}_8$ as empty, by Lemma 3.1.1. No other geometric subgroups contain the $\mathscr{C}_3$-subgroup of $\mathrm{PGL}_2(7)$. □

Recall Definition 2.2.11 of the $\mathscr{C}_5$-subgroups.

Lemma 3.1.5 *Let $n = 2$, and let $H < \mathrm{SL}_2(q)$ be a $\mathscr{C}_5$-subgroup of type $\mathrm{GL}_2(q_0)$, where $q_0^r = q$ for some prime r, let G be almost simple with socle $L_2(q)$, and let H_G be the corresponding $\mathscr{C}_5$-subgroup of G. Then H is maximal amongst the geometric subgroups of $\mathrm{SL}_2(q)$ if and only if $q_0 \neq 2$ or $q = 4$. If $q_0 = 2$ and $q \neq 4$, then H_G is not maximal in G. If $q = 4$, then H_G is equal to the $\mathscr{C}_2$-subgroup of G, and was considered in Class $\mathscr{C}_2$.*

Proof Suppose that $H \leqslant K < \mathrm{SL}_2(q)$, where K is maximal amongst the geometric subgroups of $\mathrm{SL}_2(q)$ and is not of the same type as H. Consulting Table 2.8, we see that $H \cong \mathrm{SL}_2(q_0).(2, q - 1, r)$, so $|H| = q_0(q_0^2 - 1)(2, q - 1, r)$.

The group $\mathrm{SL}_2(q_0)$ is absolutely irreducible for all q_0 by Proposition 1.12.2, so $K \notin \mathscr{C}_1$.

If $K \in \mathscr{C}_2 \cup \mathscr{C}_3$ then the p-part of $|K|$ is $(2, q)$, so if $q_0 \neq 2$ then $K \notin \mathscr{C}_2 \cup \mathscr{C}_3$. If $q = 4$ then H is conjugate to the imprimitive group of type $\mathrm{GL}_1(4) \wr S_2$: note that $\mathrm{SL}_2(4) \cong A_5$ has a single class of subgroups isomorphic to S_3. If $q_0 = 2$ (so that $r = e$ is an odd prime) then $H \cong \mathrm{SL}_2(2) \cong S_3$ is properly contained in a $\mathscr{C}_3$-subgroup $K \cong (q^e + 1):2$. To see this, we check using MAGMA (file Chap3calc) that H is semilinear, and so preserves an $\mathbb{F}_2$-vector space isomorphism from $\mathbb{F}_2^2$ to $\mathbb{F}_4^1$, which can clearly be extended to map $\mathbb{F}_{2^e}^2$ to $\mathbb{F}_{2^{2e}}^1$.

The outer automorphism δ is trivial, since q is even, and in $\mathrm{SL}_2(q)$ there is a single conjugacy class of groups of type H, and a single class of groups of type K. Since ϕ normalises the standard copy of K, and up to conjugacy K has a unique subgroup of order 6, the group H does not extend to a novelty.

Classes $\mathscr{C}_4$ and $\mathscr{C}_7$ are empty for $n = 2$. The group K is not a $\mathscr{C}_5$-subgroup, by Lemma 2.3.28. Class $\mathscr{C}_6$ is empty unless $q = p$, so $K \not\in \mathscr{C}_6$. The groups in Class $\mathscr{C}_8$ have already been considered, by Lemma 3.1.1. $\qquad\square$

Recall Definition 2.2.13 of the $\mathscr{C}_6$-subgroups.

Lemma 3.1.6 *Let* $H < \mathrm{SL}_2(q)$ *be a* $\mathscr{C}_6$*-subgroup. Then* H *is maximal amongst the geometric subgroups of* $\mathrm{SL}_2(q)$.

Proof Suppose, by way of contradiction, that $H \leqslant K < \mathrm{SL}_2(q)$, where K is maximal amongst the geometric subgroups of $\mathrm{SL}_2(q)$ and is not of the same type as H.

The order of H is 24 or 48. If $\mathscr{C}_6 \neq \varnothing$ then q is prime (and by assumption is greater than 3), so $(q, |H|) = 1$. The group H is not cyclic, so cannot be contained in a group of shape $p{:}(p - 1)$, and hence $K \not\in \mathscr{C}_1$. The group H does not have a non-absolutely-irreducible subgroup of index at most 2, so $K \not\in \mathscr{C}_2 \cup \mathscr{C}_3$. Since q is prime, $K \not\in \mathscr{C}_5$. Since there is a unique type of $\mathscr{C}_6$-subgroup, $K \not\in \mathscr{C}_6$. Classes $\mathscr{C}_4$ and $\mathscr{C}_7$ are empty. We are considering Class $\mathscr{C}_8$ as empty by Lemma 3.1.1. $\qquad\square$

3.2 Dimension 3

In dimension 3, by Definition 1.6.20, we assume that the group Ω is $\mathrm{SL}_3(q)$ or $\mathrm{SU}_3(q)$. We assume that $q > 2$ in Case **U** as $\mathrm{U}_3(2)$ is soluble. Although $\mathrm{L}_3(2)$ is isomorphic to $\mathrm{L}_2(7)$, we do analyse $\mathrm{L}_3(2)$, since its subgroups belong to different Aschbacher classes in the two representations. Note that Classes $\mathscr{C}_4$ and $\mathscr{C}_7$ are empty.

Recall Definition 2.2.1 of the $\mathscr{C}_1$-subgroups, and Definition 2.3.5 of *standard reducible behaviour*.

Proposition 3.2.1 *Let* $n = 3$. *Then* Ω *has standard reducible behaviour.*

Proof Let H be a $\mathscr{C}_1$-subgroup of Ω. If H is of type P_i or $P_{i,j}$ then the result follows from Propositions 2.3.1 and 2.3.4. Therefore, assume that H is of type $\mathrm{GL}_1(q) \oplus \mathrm{GL}_2(q)$ or $\mathrm{GU}_1(q) \perp \mathrm{GU}_2(q)$, so that $H \cong \mathrm{GL}_2^{\pm}(q)$, and hence $|H| = q(q \mp 1)(q^2 - 1)$.

Suppose that $H \leqslant K < \Omega$, where K is maximal amongst the geometric subgroups of Ω and is not of the same type as H. For $\mathscr{C}_1$ we note in Case **L** that $H \not\leqslant P_{1,2}$, since H acts irreducibly on a 2-space, and in Case **U** that H stabilises no non-zero totally singular subspace, and so is not contained in P_1 (the only other $\mathscr{C}_1$-subgroup). In Case **L** the group H stabilises both a 1-space and a complementary 2-space, so the subgroup H_G of $\mathrm{GL}_3(q)$ of the same type as H is a subgroup of both a parabolic subgroup P_1 of $\mathrm{GL}_3(q)$ and its conjugate under duality, P_1^γ. However H_G is normalised by γ, whilst $P_1 \neq P_1^\gamma$.

If $q = 2$ then $\Omega = \mathrm{SL}_3(2)$, and Classes $\mathscr{C}_5$, $\mathscr{C}_6$ and $\mathscr{C}_8$ are all empty. The only $\mathscr{C}_2$-subgroup K is reducible, coincides with H, and is considered as a $\mathscr{C}_1$-subgroup, by Proposition 2.3.6 (i). If $K \in \mathscr{C}_3$ then K has odd order, but H has even order.

If $\Omega = \mathrm{SL}_3(3)$ then $\mathscr{C}_5 = \mathscr{C}_6 = \mathscr{C}_8 = \emptyset$. The $\mathscr{C}_2$- and $\mathscr{C}_3$-subgroups are smaller than H. If $\Omega = \mathrm{SU}_3(3)$ then $\mathscr{C}_6 = \mathscr{C}_8 = \emptyset$. The $\mathscr{C}_2$-subgroups are the same size as H but are irreducible, whilst the groups in $\mathscr{C}_3 \cup \mathscr{C}_5$ are smaller than H. We now assume $q > 3$.

Since H is insoluble for $q > 3$ we see that $K \notin \mathscr{C}_2 \cup \mathscr{C}_3 \cup \mathscr{C}_6$. For $\mathscr{C}_5$ we note from Proposition 1.12.7 that H^∞ contains elements of all traces in $\mathbb{F}_q$. Therefore if $K \in \mathscr{C}_5$ then $\Omega = \mathrm{SU}_3(q)$ with $q > 3$ odd, and K is of type $\mathrm{GO}_3(q)$. Then $|K| = (q+1,3)q(q^2-1)$, but $q+1 > (q+1,3)$, a contradiction.

If $K \in \mathscr{C}_8$ then $\Omega = \mathrm{SL}_3(q)$ with q odd, and K consists of similarities of a unitary or orthogonal form κ. The group H is transitive on the non-zero vectors of the fixed 2-dimensional subspace U. If κ is unitary then any 2-dimensional subspace contains singular vectors (see [108, Corollary 10.3]), so U is totally singular under κ. However, any maximal totally singular subspace has dimension 1. For κ orthogonal, $|H^\infty| > |\mathrm{CGO}_2^\pm(q)|$, but H^∞ acts irreducibly on U, a contradiction. $\square$

Recall Definition 2.2.3 of the $\mathscr{C}_2$-subgroups. Consulting Table 2.4 we see that when $n = 3$, the types of $\mathscr{C}_2$-subgroup are $\mathrm{GL}_1(q) \wr S_3$ and $\mathrm{GU}_1(q) \wr S_3$. Recall that $\mathrm{SU}_3(2)$ is soluble.

Proposition 3.2.2 *Let $n = 3$ and let H be a $\mathscr{C}_2$-subgroup of Ω. If $q \geqslant 5$ or $\Omega = \mathrm{SU}_3(q)$ then H is maximal amongst the geometric subgroups of Ω. If $\Omega = \mathrm{SL}_3(2)$ or $\mathrm{SL}_3(3)$ then H is equal to a $\mathscr{C}_1$- or $\mathscr{C}_8$-subgroup, respectively (and is considered as such). If $\Omega = \mathrm{SL}_3(4)$ then H is not maximal, and does not extend to a novel maximal subgroup.*

Proof The claims for $\mathrm{SL}_3(q)$ with $q \leqslant 4$ are immediate from Proposition 2.3.6, so assume that $q \geqslant 5$ when $\Omega = \mathrm{SL}_3(q)$.

Suppose, by way of contradiction, that $H \leqslant K$, where K is maximal amongst the geometric subgroups of Ω and is not of the same type as H. Consulting Table 2.5 we see that $H \cong (q-1)^2.S_3$ in Case **L** and $H \cong (q+1)^2.S_3$

in Case **U**. Without loss of generality, H contains the matrices

$$x := \begin{pmatrix} 0 & 1 & 0 \\ 0 & 0 & 1 \\ 1 & 0 & 0 \end{pmatrix}, \quad y := \begin{pmatrix} 0 & 1 & 0 \\ 1 & 0 & 0 \\ 0 & 0 & -1 \end{pmatrix}$$

and $z_\alpha := \mathrm{diag}(\alpha, 1, \alpha^{-1})$, where α is any element of $\mathbb{F}_q^\times$ in Case **L**, and α is any $(q-1)$th power in $\mathbb{F}_{q^2}^\times$ in Case **U**.

Consider first Class $\mathscr{C}_1$. The eigenvalues of $\langle x \rangle$ show that if $p \neq 3$ then $\mathbb{F}_{q^u}^2$ is a direct sum of two $\mathbb{F}_{q^u}\langle x \rangle$-submodules: $U = \langle (1,1,1) \rangle$ and a 2-dimensional space W consisting of all vectors of coordinate sum 0. If $q^u \equiv 1 \pmod 3$ then W contains two 1-dimensional $\mathbb{F}_{q^u}\langle x \rangle$-submodules $\langle (1, \beta, \beta^2) \rangle$ and $\langle (1, \beta^2, \beta) \rangle$, where β is a cube root of unity, whilst if $q^u \equiv 2 \pmod 3$ then W is irreducible. None of these submodules are preserved by y, so H is irreducible. If $p = 3$ then U is the only 1-space preserved by $\langle x \rangle$, and $U < W$. Once again, y preserves neither U nor W so H is irreducible. Therefore $K \notin \mathscr{C}_1$.

In each case there is a unique type of imprimitive group when $n = 3$ so $K \notin \mathscr{C}_2$. If $K \in \mathscr{C}_3$ then consulting Table 2.6 we see that $K \cong C.3$, where C is cyclic. The group H does not contain a cyclic subgroup of index dividing 3, so $K \notin \mathscr{C}_3$. Class $\mathscr{C}_4$ is empty, as n is prime.

Assume next that $K \in \mathscr{C}_5$. In Case **L** the group H' contains z_α for all $\alpha \in \mathbb{F}_q^\times$. There exists a subfield $\mathbb{F}_{q_0}$ of $\mathbb{F}_q$ of prime index such that all traces of elements of K' lie in $\mathbb{F}_{q_0}$. Therefore, $\alpha + \alpha^{-1} \in \mathbb{F}_{q_0}$ for all $\alpha \in \mathbb{F}_q^\times$, contradicting Lemma 1.4.5 as we assumed that $q \geqslant 5$. Consider next Case **U**. If K is of type $\mathrm{GO}_3(q)$ then q is odd and $|K| = (q+1,3)q(q^2-1)2/(2,q-1)$, contradicting Lagrange's theorem. The group H' contains z_α for all $\alpha \in \mathbb{F}_{q^2}^\times$ of order $q+1$, so if K is of type $\mathrm{GU}_n(q_0)$ then $\lambda^{q-1} + \lambda^{1-q} \in \mathbb{F}_{q_0^2}^\times$ for all $\lambda \in \mathbb{F}_{q^2}^\times$. This contradicts Lemma 1.4.5, as $\mathbb{F}_{q_0^2}$ is of odd index in $\mathbb{F}_{q^2}$.

Next suppose that $K \in \mathscr{C}_6$. Then q is prime, $q > 3$ and, consulting Table 2.9, we see that $|K|$ divides $2^3 3^4$. If $q > 11$ then $(q-1)^2.6 > 2^3 3^4$, a contradiction. If $\Omega = \mathrm{SL}_3(5)$, $\mathrm{SU}_3(7)$ or $\mathrm{SL}_3(11)$ then $\mathscr{C}_6 = \varnothing$. In both $\mathrm{SU}_3(5)$ and $\mathrm{SL}_3(7)$ the group H has the same order as K, but H has derived length 3 whilst K has derived length 4, a contradiction. In $\mathrm{SU}_3(11)$ the group H is bigger than K.

Class $\mathscr{C}_7$ is empty, as n is prime, so suppose that $K \in \mathscr{C}_8$. Then $\Omega = \mathrm{SL}_3(q)$ and either $K = \mathrm{SU}_3(q^{1/2}) \times (q^{1/2}-1,3)$ or $K = \mathrm{SO}_3(q) \times (q-1,3)$. We are assuming that $q \geqslant 5$, so both are ruled out by Lagrange's theorem. $\qquad \square$

Recall Definition 2.2.5 of the $\mathscr{C}_3$-subgroups. If $n = 3$ and $H < \Omega$ is a $\mathscr{C}_3$-subgroup, then H is of type $\mathrm{GL}_1(q^3)$ or type $\mathrm{GU}_1(q^3)$.

Proposition 3.2.3 *Let $n = 3$ and let H be a $\mathscr{C}_3$-subgroup of Ω, let G be almost simple with socle $\bar{\Omega}$, and let H_G be the corresponding $\mathscr{C}_3$-subgroup of*

G. Then H is maximal amongst the geometric subgroups of Ω if and only if $\Omega \neq SL_3(4)$. If H is not maximal, then H_G is maximal amongst the geometric subgroups of G if and only if G is not a subgroup of a conjugate of $L_3(4).\langle\phi,\gamma\rangle$.

Proof If $\Omega \neq SL_3(4)$ then this follows immediately from Proposition 2.3.21, so assume that $\Omega = SL_3(4)$ and hence $H \cong 3 \times 7{:}3$. Note that H is the Sylow 7-normaliser of $SL_3(4)$, and there is a unique $SL_3(4)$-class of $\mathscr{C}_3$-subgroups.

Assume that $H \leqslant K < SL_3(4)$, where K is a $\mathscr{C}_i$-subgroup for some $1 \leqslant i \leqslant 8$ and is not of the same type as H. It follows from Lemma 2.3.14 that $K \notin \mathscr{C}_1$. It is immediate from Lemma 2.3.15 that $K \notin \mathscr{C}_2 \cup \mathscr{C}_6$. Since n is prime, there is a unique type of $\mathscr{C}_3$-subgroup, and $\mathscr{C}_4 = \mathscr{C}_7 = \varnothing$. It is immediate from Lemma 2.3.20 that $K \notin \mathscr{C}_8$.

Therefore $K \in \mathscr{C}_5$, and by Table 2.8, $K \cong SL_3(2) \times 3$, and H is a proper subgroup of K since H is also the Sylow 7-normaliser of K. The natural copy of K is stabilised by $T := \langle\phi,\gamma\rangle$, since ϕ centralises elements of $SL_3(2)$ and inverts the scalars from $\mathbb{F}_4$, whilst γ normalises $SL_3(2)$ and inverts the scalars. Since K contains a unique conjugacy class of subgroups of order $|H|$, we deduce that $H.T < K.T$ by Proposition 1.3.10. However, there are three $SL_3(4)$-classes of groups of the same type as K (permuted by δ), and only one class of groups H_G, so H_G is maximal amongst the geometric subgroups of those almost simple G with socle $L_3(4)$ that are not contained in (a conjugate of) $L_3(4).T$. $\square$

Recall Definition 2.2.11 of the $\mathscr{C}_5$-subgroups.

Proposition 3.2.4 *Let $n = 3$ and let H be a $\mathscr{C}_5$-subgroup of Ω. If $\Omega \neq SU_3(3)$ then H is maximal amongst the geometric subgroups of Ω. Otherwise, H is not maximal and does not extend to a novel maximal subgroup.*

Proof If $\Omega = SL_3(q)$ then this is Proposition 2.3.29, so assume that Ω is $SU_3(q)$. Thus $SU_3(q_0) \leqslant H$, where $q_0^r = q$ for some odd prime r, or $\Omega_3(q) \leqslant H$.

In $SU_3(3)$ the only $\mathscr{C}_5$-subgroup H is isomorphic to $SO_3(3) \cong S_4$. Let $H_G = \langle GO_3(3), \zeta I_4\rangle$ be the corresponding subgroup of $GU_3(3)$, where ζ is a primitive fourth root of unity in $\mathbb{F}_9^\times$. In dimension 3, our standard symmetric bilinear form has the same matrix as our standard unitary form, namely a basis of orthonormal vectors. It is straightforward to check that the standard imprimitive wreath product $GL_1(3) \wr S_3 \cong GO_3(3)$, and preserves the same form as H. The blocks of imprimitivity are spanned by the basis vectors, and so are non-degenerate subspaces with respect to both the symmetric bilinear form on $\mathbb{F}_3^3$ and the unitary form on $\mathbb{F}_{3^2}^3$, so H_G naturally embeds as a proper subgroup of the $\mathscr{C}_2$-subgroup K_G of type $GU_1(3) \wr S_3$ in $GU_3(3)$, and hence is not maximal. There is a single class of groups K in $SU_3(3)$, and it is straightforward to check that K contains a single class of subgroups isomorphic to H, so by Proposition 1.3.10 H does not extend to a novel maximal subgroup.

We may assume from now that $q > 3$. Assume, by way of contradiction, that $H \leqslant K < \mathrm{SU}_3(q)$, where K is maximal amongst the geometric subgroups of $\mathrm{SU}_3(q)$ and is not of the same type as H. By Proposition 1.12.2, the groups $\mathrm{SU}_3(q_0)$ and $\Omega_3(q)$ are absolutely irreducible, so $K \notin \mathscr{C}_1$. Furthermore, if $q > 3$ then $\mathrm{SU}_3(q_0)$ and $\Omega_3(q)$ have no non-absolutely irreducible normal subgroups of index 3, so $K \notin \mathscr{C}_3$.

If q is not an odd prime power of 2 then H is insoluble, and $H^\infty \cong \mathrm{SU}_3(q_0)$ or $\Omega_3(q)$. For these q it follows immediately that $K \notin \mathscr{C}_2 \cup \mathscr{C}_6$. If q is a proper power of 2 then $\mathscr{C}_6 = \varnothing$, so assume that $K \in \mathscr{C}_2$. Then K is of shape $(q+1)^2.\mathrm{S}_3$, and so $|K|$ is not divisible by 4, unlike $|\mathrm{SU}_3(2)|$, a contradiction.

It follows from Lemma 2.3.28 that $K \notin \mathscr{C}_5$, whilst $\mathscr{C}_8 = \varnothing$. $\qquad\square$

Recall Definition 2.2.13 of the $\mathscr{C}_6$-subgroups.

Proposition 3.2.5 *Let $n = 3$ and let H be a $\mathscr{C}_6$-subgroup of Ω. Then H is maximal amongst the geometric subgroups of Ω.*

Proof Suppose, by way of contradiction, that $H \leqslant K < \Omega$, where the group K is maximal amongst the geometric subgroups of Ω and is not of the same type as H.

If $\mathscr{C}_6 \neq \varnothing$, then either $q = p \equiv 1 \bmod 3$ and $\Omega = \mathrm{SL}_3(p)$, or $q = p \equiv 2 \bmod 3$ and $\Omega = \mathrm{SU}_3(p)$. By Table 2.9, $3^{1+2}.Q_8 \leqslant H \leqslant 3^{1+2}.\mathrm{SL}_2(3)$. Let R be the extraspecial normal subgroup of H, so that $R \cong 3^{1+2}$ and R is absolutely irreducible.

Working in $\mathrm{SL}_3(7)$, it is straightforward using MAGMA (file `Chap3calc`) to check that the second derived group of $3^{1+2}.Q_8$ contains R, so $K \notin \mathscr{C}_1 \cup \mathscr{C}_3$. If $K \in \mathscr{C}_2$ then K preserves a decomposition into three subspaces, so the second derived group of K is reducible, a contradiction. Recall from Definition 2.2.13 that q^u is the smallest power of p for which there exist cube roots of unity, so $K \notin \mathscr{C}_5$. There is a unique type of $\mathscr{C}_6$-subgroup, so $K \notin \mathscr{C}_6$. If $\mathscr{C}_8 \neq \varnothing$ then $\Omega = \mathrm{SL}_3(p)$, and so by Table 2.11 the only $\mathscr{C}_8$-subgroup K is $3 \times \mathrm{SO}_3(p)$. Now, H' contains $Z(\mathrm{SL}_3(p)) \cong 3$, whilst K' has trivial center, so $K \notin \mathscr{C}_8$. $\qquad\square$

Recall Definition 2.2.17 of the $\mathscr{C}_8$-subgroups.

Proposition 3.2.6 *Let $n = 3$ and let H be a $\mathscr{C}_8$-subgroup of Ω. Then H is maximal amongst the geometric subgroups of Ω.*

Proof If $\mathscr{C}_8 \neq \varnothing$, then $\Omega = \mathrm{SL}_3(q)$ with $q > 2$. Assume, by way of contradiction, that $H \leqslant K < \mathrm{SL}_3(q)$, where K is maximal amongst the geometric subgroups of $\mathrm{SL}_3(q)$ and is not of the same type as H.

We deal first with $q \in \{3, 4\}$, where H is soluble. If $q = 3$, then the group $H \cong \mathrm{SO}_3(3) \cong \mathrm{S}_4$ (and recall from Proposition 2.3.6 that H is equal to the unique $\mathscr{C}_2$-subgroup, but is considered under $\mathscr{C}_8$). If $q = 4$ then $H \cong \mathrm{SU}_3(2)$.

Since $q \leqslant 4$ and $n = 3$, the group $K \in \mathscr{C}_1 \cup \mathscr{C}_3 \cup \mathscr{C}_5$. Now, H is absolutely irreducible by Proposition 1.12.2, so $K \notin \mathscr{C}_1$. Furthermore $|H|$ does not divide $|\Gamma\mathrm{L}_1(q^3)|$, so $K \notin \mathscr{C}_3$. Finally, if $K \in \mathscr{C}_5$ then $q = 4$, but $|3 \times \mathrm{SL}_3(2)|$ is not divisible by $|H|$.

We therefore assume that $q \geqslant 5$, so that $H^\infty \cong \mathrm{SU}_3(q^{1/2})$ or $\Omega_3(q)$. The group H^∞ is absolutely irreducible by Proposition 1.12.2 so $K \notin \mathscr{C}_1 \cup \mathscr{C}_3$. Furthermore, $K \notin \mathscr{C}_2 \cup \mathscr{C}_6$ because H is insoluble.

Consider next Class $\mathscr{C}_5$. If H is unitary then, by Proposition 1.12.7, all elements of $\mathbb{F}_q$ occur as traces of elements of H^∞, so $K \notin \mathscr{C}_5$. The group $\Omega_3(q)$ contains elements whose trace lies in no proper subfield of $\mathbb{F}_q$ by Lemma 1.12.8, so again $K \notin \mathscr{C}_5$.

Finally, since we have $|\Omega_3(q)| = \frac{1}{(q-1,2)}q^2(q-1)(q+1)$, whilst $|\mathrm{SU}_3(q^{1/2})| = q^{3/2}(q-1)(q^{3/2}+1)$, Lagrange's theorem shows that neither quasisimple classical group can contain the other, so $K \notin \mathscr{C}_8$. $\qquad\square$

3.3 Dimension 4

In dimension 4 we find Cases **L**, **S** and **U**, by Definition 1.6.20. We assume throughout this section that the groups in Case **S** do *not* involve a graph automorphism; that is, they are subgroups of $\mathrm{PC}\Gamma\mathrm{Sp}_4(q)$ or $\mathrm{C}\Gamma\mathrm{Sp}_4(q)$, and so Aschbacher's Theorem applies: we will consider the graph automorphism in Chapter 7. In Case **S** we also assume that $q > 2$ as $\mathrm{Sp}_4(2)$ is not quasisimple. Classes $\mathscr{C}_4$ and $\mathscr{C}_7$ are empty.

Recall Definition 2.2.1 of the $\mathscr{C}_1$-subgroups, and Definition 2.3.5, of *standard reducible behaviour*.

Proposition 3.3.1 *Let $n = 4$. Then Ω has standard reducible behaviour.*

Proof This is immediate from Propositions 2.3.1, 2.3.2 and 2.3.4: note that there are no stabilisers of non-degenerate subspaces in Case **S**. $\qquad\square$

Recall Definition 2.2.3 of the $\mathscr{C}_2$-subgroups. If H is a $\mathscr{C}_2$-subgroup in dimension 4 then one of the following holds: $\Omega = \mathrm{SL}_4(q)$ and H is of type $\mathrm{GL}_1(q) \wr \mathrm{S}_4$ or $\mathrm{GL}_2(q) \wr \mathrm{S}_2$; $\Omega = \mathrm{SU}_4(q)$ and H is of type $\mathrm{GU}_1(q) \wr \mathrm{S}_4$ or $\mathrm{GU}_2(q) \wr \mathrm{S}_2$ or $\mathrm{GL}_2(q^2).2$; $\Omega = \mathrm{Sp}_4(q)$ and H is of type $\mathrm{Sp}_2(q) \wr \mathrm{S}_2$ or $\mathrm{GL}_2(q).2$, with q odd in the latter case.

Proposition 3.3.2 *Let $n = 4$, let H be a $\mathscr{C}_2$-subgroup of Ω, preserving a decomposition into four subspaces, let G be almost simple with socle $\bar{\Omega}$, and let H_G be the corresponding $\mathscr{C}_2$-subgroup of G. Then H is maximal amongst the geometric subgroups of Ω if and only if either $\Omega = \mathrm{SL}_4(q)$ and $q \geqslant 7$, or $\Omega = \mathrm{SU}_4(q)$ and $q \neq 3$. If $\Omega = \mathrm{SL}_4(q)$ and $q \leqslant 4$ then H_G is not maximal in*

G. If $\Omega = \mathrm{SL}_4(5)$ *then* H_G *is maximal amongst the geometric subgroups of* G *if and only if* $G \nleqslant \mathrm{L}_4(5).\langle \delta^2, \gamma \rangle$. *If* $\Omega = \mathrm{SU}_4(3)$ *then* H_G *is maximal amongst the geometric subgroups of* G *if and only if* $G \nleqslant \mathrm{U}_4(3).\langle \delta^2, \phi \rangle$.

Proof The exceptions in Case **L** when $q \leqslant 4$ follow immediately from Proposition 2.3.6, so we will assume that $q \geqslant 5$ in Case **L**. Assume that $H \leqslant K < \Omega$, where initially we will assume K is maximal amongst the geometric subgroups of Ω and is not of the same type as H.

It follows from Lemma 2.3.7 (vi) that $K \notin \mathscr{C}_1 \cup \mathscr{C}_3$ and H' is irreducible. If $K \in \mathscr{C}_2$ then K is of type $\mathrm{GL}_2^{\pm}(q) \wr S_2$ or $\mathrm{GL}_2(q^2).2$, so K' is reducible, a contradiction. Class $\mathscr{C}_4$ is empty as $n = 4$.

Consider Class $\mathscr{C}_5$. In Case **L**, let $\alpha \in \mathbb{F}_q^{\times}$. In Case **U**, let α be any $(q-1)$th power in $\mathbb{F}_{q^2}^{\times}$. It is straightforward to write down a matrix in H' with trace α as a product of commutators: the first commutator is a diagonal matrix with one entry equal to α, and the second a 3-cycle which moves all other non-zero entries off the diagonal. Thus in Case **L** H' contains elements of all traces in $\mathbb{F}_q^{\times}$, so $K \notin \mathscr{C}_5$, and in Case **U** the traces of elements of H' form a set of size at least $q + 1$ (since H' also contains elements of trace 0), and hence do not lie in any proper subfield of $\mathbb{F}_{q^2}$. Thus $K \notin \mathscr{C}_5$.

If $K \in \mathscr{C}_6$ then q is an odd prime, $|K| = 23040$ if $q \equiv \pm 3 \bmod 8$, and $|K| = 46080$ if $q \equiv \pm 1 \bmod 8$. In Case **L**, $|H| = 24(q-1)^3$, so $|H| > |K|$ for $q > 11$, and if $q \in \{3, 7, 11\}$ then $\mathscr{C}_6 = \varnothing$. If $q = 5$ then one may check, using MAGMA (file Chap3calc) that there is a (proper) containment $H < K$ and that K contains a unique conjugacy class of subgroups of type H. So the containment extends to the normaliser of the $\mathrm{SL}_4(5)$-class of K in $\Gamma\mathrm{L}_4(5)$. That is, $H.\langle \delta^2, \gamma \rangle < K.\langle \delta^2, \gamma \rangle < L := \mathrm{SL}_4(5).\langle \delta^2, \gamma \rangle$. There are two classes of $\mathscr{C}_6$-subgroups in $\mathrm{SL}_4(5)$, interchanged by δ, so if G is almost simple with socle $\mathrm{L}_4(5)$, and is not a subgroup of L, then G_H is not contained in G_K. In Case **U**, $|H| = 24(q+1)^3$, so $|H| > |K|$ for $q > 7$. If $q = 3$ then, as for $q = 5$ in Case **L**, one may check directly (file Chap3calc) that H is properly contained in K, and that the group L_H in $L := \mathrm{U}_4(3).\langle \delta^2, \phi \rangle$ is contained in L_K. There are two classes of $\mathscr{C}_6$-subgroups in $\mathrm{SU}_4(3)$, interchanged by δ, so if G is almost simple with socle $\mathrm{U}_4(3)$, and is not a subgroup of L, then $G_H \nleqslant G_K$. If $q = 5$ then $\mathscr{C}_6 = \varnothing$, and if $q = 7$ then $|H|$ does not divide $|K|$.

Class $\mathscr{C}_7$ is empty as $n = 4$, and $K \notin \mathscr{C}_8$ by Lemma 2.3.12 (ii).

In $\mathrm{SL}_4(5)$ and $\mathrm{SU}_4(3)$ we have only found one type of group in Class $\mathscr{C}_i$ for $1 \leqslant i \leqslant 8$ that (properly) contains H, so in extensions where this containment does not extend the group of type H is maximal amongst the geometric subgroups. $\square$

Proposition 3.3.3 *Let* $n = 4$, *let* H *be a* $\mathscr{C}_2$-subgroup of Ω *preserving a decomposition into two subspaces, let* G *be almost simple with socle* $\overline{\Omega}$, *and let*

H_G *be the subgroup of* G *of the same type as* H. *In Case* **S**, *assume that* $q > 2$.
Then H *is maximal amongst the geometric subgroups of* Ω *if and only if* $q > 3$
or one of the following holds:

(i) $\Omega = \mathrm{SU}_4(3)$ *and* H *is of type* $\mathrm{GU}_2(3) \wr S_2$;
(ii) $\Omega = \mathrm{Sp}_4(3)$ *and* H *is of type* $\mathrm{Sp}_2(3) \wr S_2$;

If H *is not maximal in* Ω *then* H_G *is maximal amongst the geometric subgroups
of* G *if and only if one of the following holds:* $\Omega = \mathrm{SL}_4(3)$ *and* $G \nleqslant \mathrm{L}_4(3).\langle\gamma\rangle$;
or $\Omega = \mathrm{SU}_4(3)$, *the group* H *is of type* $\mathrm{GL}_2(9).2$ *and* $G \nleqslant \mathrm{U}_4(3).\langle\delta^2, \phi\rangle$.

Proof Assume that $H \leqslant K < \Omega$, where we assume in the first instance that
$K \in \mathscr{C}_i$ for $1 \leqslant i \leqslant 8$ and is not of the same type as H. It is immediate from
Lemma 2.3.7 (iii) that $K \notin \mathscr{C}_1$.

We deal first with $q \leqslant 3$. The $\mathscr{C}_2$-subgroup of $\mathrm{SL}_4(2)$ is shown in Proposition 2.3.6 to be non-maximal, and not to extend to a novelty.

We next consider $\mathrm{SL}_4(3)$, so that $|H| = 2304$. Order considerations show
that $K \notin \mathscr{C}_2 \cup \mathscr{C}_3$. Classes $\mathscr{C}_4$, $\mathscr{C}_5$ and $\mathscr{C}_6$ are empty. A MAGMA calculation
(file **Chap3calc**) shows a copy of H is a proper subgroup of the standard copy
K of the group of type $\mathrm{Sp}_4(3)$, so H is not maximal amongst the geometric
subgroups of $\mathrm{SL}_4(3)$. Modulo an inner automorphism of $\mathrm{Sp}_4(3)$, the automorphism γ centralises the standard copy of $\mathrm{Sp}_4(3)$ (and hence H). However, there
are two classes of groups K, interchanged by δ, whilst there is a single class of
groups H. Since we have now considered all groups in Class $\mathscr{C}_i$ for $1 \leqslant i \leqslant 8$,
we conclude that $H_{\mathrm{PGL}_4(3)}$ is maximal amongst the geometric subgroups of
$\mathrm{PGL}_4(3)$.

We next consider $\mathrm{SU}_4(2)$. The group of type $\mathrm{GU}_2(2) \wr S_2$ is considered in
Proposition 2.3.6, so let H be of type $\mathrm{GL}_2(4).2$, so that $H \cong A_5{:}2$. A straightforward MAGMA calculation (file **Chap3calc**) shows that H is properly contained
in the subfield group K of type $\mathrm{Sp}_4(2)$. Both H and K are centralised by ϕ so
H does not extend to a novelty.

Next consider $\mathrm{SU}_4(3)$, so that H is of type $\mathrm{GU}_2(3) \wr S_2$ and order 4608,
or type $\mathrm{GL}_2(9).2$ and order 2880: since the larger group is soluble and the
smaller is insoluble, neither contains the other. Order considerations show that
K is not of type $\mathrm{GU}_1(4) \wr S_4$, so $K \notin \mathscr{C}_2$. Classes $\mathscr{C}_3$ and $\mathscr{C}_4$ are empty. If
$K \in \mathscr{C}_5$ then, consulting Table 2.8, order considerations imply that $H \cong (4 \circ
\mathrm{SL}_2(9)).2$ and $K \cong 4 \circ \mathrm{Sp}_4(3)$. A MAGMA calculation (file **Chap3calc**) shows
that there is a proper containment $H < K$ and that K has a unique conjugacy
class of subgroups of the same type as H. So the containment extends to the
normaliser of the $\mathrm{SU}_4(3)$-class of K in $\mathrm{C\Gamma U}_4(3)$. That is, $H.T < K.T$ for
$T := \mathrm{SU}_4(3).\langle\delta^2, \phi\rangle$. There are two classes in $\mathrm{SU}_4(3)$ of groups of type $\mathrm{Sp}_4(3)$
and one of groups of type $\mathrm{GL}_2(9).2$, so the containment does not extend beyond

T. If $K \in \mathscr{C}_6$ and H is of type $\mathrm{GU}_2(3) \wr \mathrm{S}_2$ then H has index 5 in K, by Table 2.9, which is impossible. Thus H is of type $\mathrm{GL}_2(9).2$, and has index 8 in K, with $K \cong (4 \circ 2^{1+4}).\mathrm{A}_6$. The A_6 quotient of K acts irreducibly on the 2^4 layer, so K has no subgroups of index 8, a contradiction. Classes $\mathscr{C}_7$ and $\mathscr{C}_8$ are empty.

Finally, consider $\mathrm{Sp}_4(3)$, so that H is of type $\mathrm{Sp}_2(3) \wr \mathrm{S}_2$ and order 1152, or type $\mathrm{GL}_2(3).2$ and order 96. The group of type $\mathrm{GL}_2(3).2$ can be checked using MAGMA (file Chap3calc) to be non-maximal, and its extension by δ is also not maximal. So let H be of type $\mathrm{Sp}_2(3) \wr \mathrm{S}_2$. Order considerations imply that $K \notin \mathscr{C}_3 \cup \mathscr{C}_6$, and Classes $\mathscr{C}_4$, $\mathscr{C}_5$, $\mathscr{C}_7$ and $\mathscr{C}_8$ are empty, so H is maximal amongst the geometric subgroups of $\mathrm{Sp}_4(3)$.

We therefore assume for the remainder of the proof that $q > 3$ so that, in particular, H is insoluble. Recall from the beginning of the proof that $K \notin \mathscr{C}_1$.

Assume that $K \in \mathscr{C}_2$. If K preserves a decomposition into four subspaces, then K is soluble, a contradiction, so K preserves a decomposition into two subspaces, and $\Omega \neq \mathrm{SL}_4(q)$. One of H or K preserves a decomposition into non-degenerate subspaces, and the other into totally singular subspaces. Lagrange's theorem applied to H^∞ and K^∞ eliminates both possibilities when $\Omega = \mathrm{SU}_4(q)$, so let $\Omega = \mathrm{Sp}_4(q)$. If H is of type $\mathrm{Sp}_2(q) \wr 2$ then $|H^\infty| > |K^\infty|$, so assume that H is of type $\mathrm{GL}_2(q).2$. Let the subspaces preserved by K be W_1 and W_2. By Lagrange's theorem, H is not contained in $L \wr 2$ with L a parabolic subgroup of $\mathrm{Sp}_2(q)$, so for at least one $i \in \{1, 2\}$ the restriction of the stabiliser in H of W_i to W_i must act irreducibly on W_i. But the irreducible H^∞-submodules are all isomorphic to V_1, and hence are totally singular, a contradiction.

If $K \in \mathscr{C}_3$ then $\Omega = \mathrm{SL}_4^\pm(q)$. First assume that $H^\infty \cong \mathrm{SL}_2(q) \times \mathrm{SL}_2(q)$. If K is of type $\mathrm{GL}_2(q^2).2$ or $\mathrm{Sp}_2(q^2).2$, then $\mathrm{SL}_2(q)^2 \leqslant \mathrm{SL}_2(q^2)$, which contradicts Lagrange's theorem, so K is of type $\mathrm{GU}_2(q).2$. However, then $\mathrm{SL}_2(q)^2 \leqslant \mathrm{SL}_2(q)$, a contradiction. Thus H is of type $\mathrm{GL}_2(q).2$, and so q is odd. If K is of type $\mathrm{Sp}_2(q^2).2$, then $|K| = 2q^2(q^4 - 1)$, whereas $|H| = 2q(q - 1)(q^2 - 1)$, a contradiction since $q > 3$. If K is of type $\mathrm{GU}_2(q).2$ then K^∞ is $\mathrm{SL}_2(q)$ acting irreducibly, but H^∞ is $\mathrm{SL}_2(q)$ acting reducibly, a contradiction.

Class $\mathscr{C}_4$ is empty when $n = 4$. By Lemma 2.3.10, if $K \in \mathscr{C}_5$, then the group $\Omega = \mathrm{SU}_4(q)$ and $H \cong \mathrm{SL}_2(q^2).(q - 1).2$. If K is of type $\mathrm{GO}_4^\pm(q)$, then q is odd and K^∞ is $\Omega_4^+(q) \cong \mathrm{SL}_2(q) \circ \mathrm{SL}_2(q)$ or $\Omega_4^-(q) \cong \mathrm{L}_2(q^2)$, a contradiction. A standard argument using traces shows that K is not of type $\mathrm{GU}_4(q_0)$, so $K = \mathrm{Sp}_4(q).(q + 1, 2)$. Since $(q^4 - 1) \mid |H^\infty|$, the group H^∞ is not contained in a $\mathscr{C}_1$-subgroup of $\mathrm{Sp}_4(q)$. Since $H^\infty \cong \mathrm{SL}_2(q^2)$ is reducible in its action on $\mathbb{F}_{q^2}^4$, this implies that H^∞ is not absolutely irreducible in its action on the natural module for $\mathrm{Sp}_4(q)$, so H^∞ is not an $\mathscr{S}$-subgroup of $\mathrm{Sp}_4(q)$. Examining composition factors of H shows that if $H^\infty \leqslant L^\infty$ for a geometric subgroup L of $\mathrm{Sp}_4(q)$ then L is contained in a member of $\mathscr{C}_3$, but $|H|/(q + 1, 2)$ is larger than the orders of the $\mathscr{C}_3$-subgroups of $\mathrm{Sp}_4(q)$, a contradiction.

If $K \in \mathscr{C}_6$ then q is an odd prime, and the only non-abelian composition factor of K is A_5 or A_6, a contradiction unless $|L_2(q)|$ divides $|A_6|$. Thus the group $\Omega = SL_4(5)$. If H is of type $GL_2(5) \wr S_2$ or $Sp_2(5) \wr S_2$ then H contains two composition factors isomorphic to A_5, but $|A_5|^2 > |A_6|$. If H is of type $GL_2(5).2$ then $|H| = 960$, and hence has index 2 in K. However, K is perfect, a contradiction.

Class $\mathscr{C}_7$ is empty because $n = 4$, and $K \notin \mathscr{C}_8$ by Lemma 2.3.12. $\qquad\square$

Recall Definition 2.2.5 of the $\mathscr{C}_3$-subgroups. Let H be a $\mathscr{C}_3$-subgroup in dimension 4. Then one of the following holds: $\Omega = SL_4(q)$ and H is of type $GL_2(q^2)$; $\Omega = Sp_4(q)$ and H is of type $Sp_2(q^2)$ or type $GU_2(q)$, with q odd in the latter case. Note in particular that Class $\mathscr{C}_3$ is void in $SU_4(q)$.

Proposition 3.3.4 *Let $n = 4$ and let H be a $\mathscr{C}_3$-subgroup of Ω. Then H is maximal amongst the geometric subgroups of Ω if and only if H is not of type $GU_2(3)$ in $Sp_4(3)$. If H is not maximal in Ω then H does not extend to a novel maximal subgroup.*

Proof We consider the non-maximal example, where $H \cong GU_2(3).2 \leqslant Sp_4(3)$. The group of type $GU_2(3)$ in $CSp_4(3)$ normalises an extraspecial subgroup of shape 2^{1+4}, and so is properly contained in a $\mathscr{C}_6$-subgroup K. There are unique classes of groups H and groups K in $Sp_4(3)$, and K contains a unique class of groups isomorphic to H, so H does not extend to a novelty.

For the remainder of this proof, we therefore assume that if H is of type $GU_2(q)$ then $q \geqslant 5$. Assume, by way of contradiction, that $H \leqslant K < \Omega$, where K is maximal amongst the geometric subgroups of Ω and is not of the same type as H.

If $K \in \mathscr{C}_1$, then by Lemma 2.3.14 H is of type $GU_2(q)$ in $Sp_4(q)$, so that $|H| = 2q(q+1)(q^2-1)$ and q is odd. Consulting Tables 2.2 and 2.3, there are two types of $\mathscr{C}_1$-subgroup in $Sp_4(q)$, and both have p'-order $(q-1)(q^2-1)$. Therefore, $K \notin \mathscr{C}_1$.

By Lemma 2.3.16, if $K \in \mathscr{C}_2$ then H is of type $GU_2(3)$.

It follows from Lemma 2.3.17 that if $K \in \mathscr{C}_3$, then H is of type $GU_2(q)$ in $Sp_4(q)$, so that K is of type $Sp_2(q^2)$. The order of K is $2q^2(q^4-1)$, and the order of H is $2q(q+1)(q^2-1)$, contradicting Lagrange's theorem.

Class $\mathscr{C}_4$ is empty. It is immediate from Lemma 2.3.19 that $K \notin \mathscr{C}_5$.

Suppose next that $K \in \mathscr{C}_6$. Then q is odd, and if $\Omega = SL_4(q)$ then K is of type $2^{2+4}.Sp_4(2)$, and if $\Omega = Sp_4(q)$ then K is of type $2^{1+4}_-.\Omega_4^-(2)$. In Case **L** the only non-abelian composition factor of K is A_6, which is smaller than $L_2(q^2)$ for $q > 3$. If $q = 3$ then $\mathscr{C}_6$ is void in Case **L**. In Case **S**, the only non-abelian composition factor of K is A_5, which is smaller than $L_2(q^2)$ for q odd, and smaller than $L_2(q)$ for $q > 5$. We have already considered $GU_2(3)$, so without loss of generality $H \cong \mathrm{GU}_2(5).2$. Then $|H|$ is divisible by 9, whilst $|K|$ is not, a contradiction.

Class $\mathscr{C}_7$ is empty, and it follows from Lemma 2.3.20 that $K \notin \mathscr{C}_8$. $\qquad\square$

Recall Definition 2.2.11 of the $\mathscr{C}_5$-subgroups.

Proposition 3.3.5 *Let $n = 4$ and let H be a $\mathscr{C}_5$-subgroup of Ω. If $\Omega = \mathrm{SU}_4(3)$ and H is of type $\Omega_4^+(3)$, then H is not maximal and does not extend to a novel maximal subgroup. Otherwise, H is maximal amongst the geometric subgroups of Ω.*

Proof In Cases **L** and **S** this is immediate from Proposition 2.3.29, and it is also immediate if H is of type $\mathrm{GU}_n(q_0)$ or $\mathrm{Sp}_n(q)$ in Case **U**. Thus, without loss of generality, $\Omega = \mathrm{SU}_4(q)$, the group H is of type $\mathrm{GO}_4^{\pm}(q)$, and q is odd.

Assume first that $q = 3$ and H is of shape $\mathrm{SO}_4^+(3).[4]$, which can be written as $(4 \circ Q_8 \circ Q_8).(3^2.[4])$, of order $2304 = 2^8.3^2$. Here, H has a characteristic symplectic-type subgroup $S = O_2(H)$ of order 2^{2+4}, so H is properly contained in a member K of $\mathscr{C}_6$. There are two classes of $\mathscr{C}_6$-subgroups in $\mathrm{SU}_4(3)$, and also two classes of groups of type $\mathrm{GO}_4^+(3)$. In each case the classes are interchanged by the diagonal automorphism δ, and normalised by $\langle \phi, \delta^2 \rangle \cong 2^2$. For extensions G of Ω, the group K_G is defined as the normaliser of S, so the fact that S is characteristic in H shows that H does not extend to a novelty. We assume for the rest of the proof that H is not of type $\mathrm{GO}_4^+(3)$.

Assume, by way of contradiction, that $H \leqslant K < \Omega$, where K is maximal amongst the geometric subgroups of Ω and is not of the same type as H. Then H^{∞} is isomorphic to one of $\Omega_4^+(q) \cong 2.\mathrm{L}_2(q)^2$ ($q > 3$ odd) or $\Omega_4^-(q) \cong \mathrm{L}_2(q^2)$ (q odd).

The group H^{∞} is absolutely irreducible, by Proposition 1.12.2, so the group $K \notin \mathscr{C}_1 \cup \mathscr{C}_2 \cup \mathscr{C}_3$. Classes $\mathscr{C}_4$, $\mathscr{C}_7$ and $\mathscr{C}_8$ are empty. It follows from Lemma 2.3.28 that $K \notin \mathscr{C}_5$.

Suppose finally that $K \in \mathscr{C}_6$, so that $p = q \equiv 3 \pmod 4$. Then by Table 2.9, the group K' is isomorphic to $(4 \circ 2^{1+4}).\mathrm{A}_6$. For odd $q > 3$ the groups $2.\mathrm{L}_2(q)^2$ and $\mathrm{L}_2(q^2)$ cannot be embedded into a group whose only non-abelian composition factor is A_6. This leaves only H of type $\mathrm{GO}_4^-(3)$, for which $H \cong 4 \times \mathrm{A}_6{}^{\cdot}2_3$. However, a MAGMA calculation (file `Chap3calc`) shows that the only subgroups of K of order $2|\mathrm{A}_6|$ are of shape $\mathrm{SL}_2(9)$, whereas H contains $\mathrm{A}_6{}^{\cdot}2 \cong \mathrm{M}_{10}$. $\qquad\square$

Recall Definition 2.2.13 of the $\mathscr{C}_6$-subgroups.

Proposition 3.3.6 *Let $n = 4$ and let H be a $\mathscr{C}_6$-subgroup of Ω. Then H is maximal amongst the geometric subgroups of Ω.*

Proof This is immediate from Proposition 2.3.31. $\qquad\square$

Recall Definition 2.2.17 of the $\mathscr{C}_8$-subgroups, and Definition 1.13.2 of $z_{q,n}$.

Proposition 3.3.7 *Let $n = 4$ and let H be a $\mathscr{C}_8$-subgroup of Ω. Then H is maximal amongst the geometric subgroups of Ω.*

Proof If $\mathscr{C}_8 \neq \varnothing$ then $\Omega = \mathrm{SL}_4(q)$ or $\mathrm{Sp}_4(q)$, and we assume that $q > 2$ in Case **S** as $\mathrm{Sp}_4(2)$ is not quasisimple. Suppose, by way of contradiction, that $H \leqslant K < \Omega$ where K is maximal amongst the geometric subgroups of Ω and is not of the same type as H.

We deal first with $H \cong \mathrm{SO}_4^+(3).2 \cong 2^{1+2+2}.3^2.2^2 < \mathrm{SL}_4(3)$, since this is the only occasion when H is soluble. The group $H' \cong \Omega_4^+(3)$ is absolutely irreducible by Proposition 1.12.2, so $K \notin \mathscr{C}_1 \cup \mathscr{C}_3$, and nor does K stabilise a decomposition into two subspaces in $\mathscr{C}_2$. We note that $\mathscr{C}_4 \cup \mathscr{C}_5 \cup \mathscr{C}_6 \cup \mathscr{C}_7 = \varnothing$. The $\mathscr{C}_2$-subgroup is not maximal, by Proposition 2.3.6. This leaves only $\mathscr{C}_8$, for which we note that $|\mathrm{SO}_4^-(3).2|$ is not divisible by $|H|$, whilst Lemma 1.12.4 shows that $\mathrm{Sp}_4(3) = K'$ does not contain H'.

After dealing with this exception, either $\Omega = \mathrm{SL}_4(q)$ and H^∞ is one of $\mathrm{Sp}_4(q)$, $\mathrm{SU}_4(q^{1/2})$ or $\Omega_4^\pm(q)$, or $\Omega = \mathrm{Sp}_4(2^e)$, and H^∞ is $\Omega_4^\pm(q)$. By Proposition 1.12.2 the group H^∞ is absolutely irreducible, so $K \notin \mathscr{C}_1 \cup \mathscr{C}_2 \cup \mathscr{C}_3$. Class $\mathscr{C}_4 = \varnothing$, and $K \notin \mathscr{C}_5$ as, by Proposition 1.12.7, H^∞ contains elements of all traces in $\mathbb{F}_q$, contradicting Lemma 2.2.12.

Suppose that $K \in \mathscr{C}_6$. Since $\mathscr{C}_6 = \varnothing$ for q even, $\Omega = \mathrm{SL}_4(q)$, and so $q \equiv 1 \bmod 4$ and q is prime. The only non-abelian composition factor of K is A_6, but unless $q = 5$ and $H^\infty = \Omega_4^+(5)$ the group H^∞ has larger non-abelian composition factors than A_6, a contradiction. In this exceptional case, we note that $\mathrm{L}_2(5)^2$ is larger than A_6.

Finally, suppose that $K \in \mathscr{C}_8$. If H is of type $\mathrm{GU}_4(q^{1/2})$ then $|H|$ is divisible by some $z_{q,3}$, contradicting Lagrange's theorem. If H is of type $\mathrm{Sp}_4(q)$ then $|H|$ is divisible by some $z_{q,4}$, so K is of type $\mathrm{GO}_4^-(q)$, contradicting Lemma 1.12.5.

Suppose next that H is of type $\mathrm{GO}_4^+(q)$. If K is of type $\mathrm{GU}_4(q^{1/2})$ then $|H|$ is divisible by a higher power of $z_{q^{1/2},4}$ than $|K|$. If K is of type $\Omega_4^-(q)$ then $|H|$ does not divide $|K|$. The group K is not of type $\mathrm{Sp}_4(q)$ by Lemma 1.12.4.

If H is of type $\mathrm{GO}_4^-(q)$ then $|H|$ is divisible by $z_{q,4}$, so K is of type $\mathrm{Sp}_4(q)$, again contradicting Lemma 1.12.4. $\qquad\square$

3.4 Dimension 5

By Definition 1.6.20, when $n = 5$ we find Cases **L** and **U** only. Recall Definition 2.2.1 of the $\mathscr{C}_1$-subgroups, and Definition 2.3.5 of *standard reducible behaviour*.

Proposition 3.4.1   *Let $n = 5$. Then Ω has standard reducible behaviour.*

Proof   This is immediate from Propositions 2.3.1, 2.3.2 and 2.3.4.   □

Recall Definition 2.2.3 of the $\mathscr{C}_2$-subgroups.

Proposition 3.4.2   *Let $n = 5$ and let H be a $\mathscr{C}_2$-subgroup of Ω. Then H is maximal amongst the geometric subgroups of Ω if and only if one of the following holds: $\Omega = \mathrm{SL}_5(q)$ and $q \geqslant 5$; $\Omega = \mathrm{SU}_5(q)$. If H is not maximal then H does not extend to a novel maximal subgroup.*

Proof   This follows immediately from Proposition 2.3.13.   □

Recall Definitions 2.2.5, 2.2.11, 2.2.13 and 2.2.17.

Proposition 3.4.3   *Let $n = 5$ and let H be a $\mathscr{C}_3$-, $\mathscr{C}_5$-, $\mathscr{C}_6$- or $\mathscr{C}_8$-subgroup of Ω. Then H is maximal amongst the geometric subgroups of Ω.*

Proof   This follows immediately from Propositions 2.3.21, 2.3.29, 2.3.31 and 2.3.32.   □

Classes $\mathscr{C}_4$ and $\mathscr{C}_7$ are empty.

3.5   Dimension 6

By Definition 1.6.20, when $n = 6$ we find Cases **L**, **U** and **S**. Recall Definition 2.2.1 of the $\mathscr{C}_1$-subgroups, and Definition 2.3.5 of *standard reducible behaviour*.

Proposition 3.5.1   *Let $n = 6$. Then Ω has standard reducible behaviour.*

Proof   This is immediate from Propositions 2.3.1, 2.3.2 and 2.3.4.   □

Recall Definition 2.2.3 of the $\mathscr{C}_2$-subgroups. If H is a $\mathscr{C}_2$-subgroup and $n = 6$, then one of the following holds: $\Omega = \mathrm{SL}_6(q)$ and H is of type $\mathrm{GL}_1(q) \wr S_6$, $\mathrm{GL}_2(q) \wr S_3$ or $\mathrm{GL}_3(q) \wr S_2$; $\Omega = \mathrm{SU}_6(q)$ and H is of type $\mathrm{GU}_1(q) \wr S_6$, $\mathrm{GU}_2(q) \wr S_3$, $\mathrm{GU}_3(q) \wr S_2$ or $\mathrm{GL}_3(q^2).2$; $\Omega = \mathrm{Sp}_6(q)$ and H is of type $\mathrm{Sp}_2(q) \wr S_3$ or $\mathrm{GL}_3(q).2$, with q odd in the latter case.

Proposition 3.5.2   *Let $n = 6$ and let H be a $\mathscr{C}_2$-subgroup of Ω, preserving a decomposition into six subspaces. Then H is maximal amongst the geometric subgroups of Ω if and only if $\Omega = \mathrm{SU}_6(q)$ or $q \geqslant 5$. If H is not maximal then H does not extend to a novel maximal subgroup.*

Proof The exceptions when $q \leqslant 4$ in Case **L** are shown in Proposition 2.3.6, so assume that $q \geqslant 5$ in Case **L**

Suppose, by way of contradiction, that $H \leqslant K < \Omega$, where K is maximal amongst the geometric subgroups of Ω and is not of the same type as H.

It is immediate from Lemma 2.3.7 (iv),(v) that $K \notin \mathscr{C}_1 \cup \mathscr{C}_3$, and from Lemma 2.3.8 (i) that $K \notin \mathscr{C}_2$.

Assume that $K \in \mathscr{C}_4$. By Lemma 2.2.4 (ii) the subgroup L of H is perfect, so $L \leqslant K^\infty$. In Case **L**, the order of K^∞ is $q^4(q^2 - 1)^2(q^3 - 1)$, whilst $|L|$ is divisible by $(q-1)^5$. Since $q \geqslant 5$ this contradicts Lagrange's theorem. In Case **U**, the order of K^∞ is $q^4(q^2 - 1)^2(q^3 + 1)$, whilst $|L|$ is divisible by $(q+1)^5$. This contradicts Lagrange's theorem.

It follows from Lemma 2.3.10 (i) that $K \notin \mathscr{C}_5$. Classes $\mathscr{C}_6$ and $\mathscr{C}_7$ are empty. It follows from Lemma 2.3.12 (ii) that $K \notin \mathscr{C}_8$. □

Proposition 3.5.3 *Let $n = 6$ and let H be a $\mathscr{C}_2$-subgroup of Ω, preserving a decomposition into three subspaces. Then H is maximal amongst the geometric subgroups of Ω if and only if $q \neq 2$. If $q = 2$ then H does not extend to a novel maximal subgroup.*

Proof The result for $q = 2$ is immediate from Proposition 2.3.6. We first assume that $q = 3$. By Lemma 2.3.7 (iv), the group H is irreducible. In Cases **L**, **U** and **S**, a short MAGMA calculation (file `Chap3calc`) shows that the only other subgroups of Ω that are maximal amongst the geometric groups and have order divisible by $|H|$ are reducible, a contradiction.

Suppose, by way of contradiction, that $H \leqslant K < \Omega$, where K is maximal amongst the geometric subgroups of Ω and is not of the same type as H. We now assume that $q \geqslant 4$, therefore $H^\infty \cong \mathrm{SL}_2(q)^3$. By Lemma 2.3.7 (iv), H is irreducible, so $K \notin \mathscr{C}_1$.

We show next that $K \notin \mathscr{C}_2$. Suppose first that K is of type $\mathrm{GL}_1^\pm(q) \wr \mathrm{S}_6$. The only non-abelian composition factor of K is A_6, however $|\mathrm{L}_2(q)^3|$ is greater than $|\mathrm{A}_6|$, a contradiction. Next, suppose that K is of type $\mathrm{GL}_3^\pm(q) \wr \mathrm{S}_2$, so that $K^\infty \cong \mathrm{SL}_3^\pm(q)^2$. This contradicts Lagrange's theorem, as $q \geqslant 4$. Finally, suppose that K is of type $\mathrm{GL}_3(q^u).2$, so that $\Omega = \mathrm{SU}_6(q)$ or $\mathrm{Sp}_6(q)$ and $K^\infty \cong \mathrm{SL}_3(q^u)$. This contradicts Lagrange's theorem as $q \geqslant 4$.

Suppose next that $K \in \mathscr{C}_3$. The order of H^∞ does not divide $|\mathrm{SL}_2(q^3)|$, so K is semilinear of degree 2. Therefore K has a non-absolutely irreducible subgroup of index 2. However, the largest non-absolutely irreducible subgroup of H has index 3, a contradiction.

We show next that $K \notin \mathscr{C}_4$. Suppose otherwise, then $K^\infty \cong \mathrm{SL}_2(q) \times \mathrm{SL}_3(q)$ in Case **L**, $K^\infty \cong \mathrm{SL}_2(q) \times \mathrm{SU}_3(q)$ in Case **U**, and $K^\infty \cong \mathrm{Sp}_2(q) \times \Omega_3(q)$ in Case **S**. Since $H \leqslant K$, there are homomorphisms $\phi_1 : H^\infty \to \mathrm{SL}_2(q)$, and $\phi_2 : H^\infty \to \mathrm{SL}_3^\pm(q)$ or $\Omega_3(q)$, and $\ker(\phi_1) \cap \ker(\phi_2) \leqslant \mathbb{F}_{q^u}^\times$. It is clear that

$\ker(\phi_1)$ contains at least two copies of $\mathrm{SL}_2(q)$, thus at least two commuting copies of $\mathrm{SL}_2(q)$ do not lie in $\ker(\phi_2)$, which contradicts Lagrange's theorem.

The fact that $K \not\leqslant \mathscr{C}_5$ follows from Lemma 2.3.10 (ii). Classes $\mathscr{C}_6$ and $\mathscr{C}_7$ are empty. By Lemma 2.3.12 (i), $K \not\leqslant \mathscr{C}_8$. □

Proposition 3.5.4 *Let $n = 6$ and let H be a $\mathscr{C}_2$-subgroup of Ω, preserving a decomposition into two subspaces. Then H is maximal amongst the geometric subgroups of Ω.*

Proof Suppose, by way of contradiction, that $H \leqslant K < \Omega$, where the group K is maximal amongst the geometric subgroups of Ω and is not of the same type of H.

We first deal with $\mathrm{SU}_6(2)$ as a special case. If H is of type $\mathrm{GU}_3(2) \wr S_2$ then 3^7 divides $|H|$, and does not divide the order of any other geometric group. If H is of type $\mathrm{GL}_3(4).2$ then the only other members of $\mathscr{C}_i$ for $1 \leqslant i \leqslant 8$ with order a multiple of $|H|$ are the $\mathscr{C}_1$-subgroup $2^9 {:} \mathrm{SL}_3(4)$ and the $\mathscr{C}_5$-subgroup of type $\mathrm{Sp}_6(2)$. The group H is irreducible by Lemma 2.3.7 (iii), and is not contained in a subfield group by Lemma 2.3.10 (iii).

So assume in the rest of the proof that $\Omega \neq \mathrm{SU}_6(2)$. Lemmas 2.3.7 (iii) and 2.3.8 (ii) show that $K \not\leqslant \mathscr{C}_1 \cup \mathscr{C}_2$.

Next we establish that $K \not\leqslant \mathscr{C}_3$. Otherwise, if H is of type $\mathrm{GL}_3^{\pm}(q) \wr S_2$ then $\Omega_1 \times \Omega_2$ is isomorphic to a subgroup of $\mathrm{SL}_{6/r}(q^r)$ for $r \in \{2, 3\}$, which violates Lagrange's theorem. Therefore H is of type $\mathrm{GL}_3(q^2).2$ or $\mathrm{GL}_3(q).2$. In Case **U**, $H^\infty \cong \mathrm{SL}_3(q^2)$, which is not a subgroup of $K^\infty = \mathrm{SU}_2(q^3)$, by Theorem 1.11.5. In Case **S**, $H^\infty \cong \mathrm{SL}_3(q)$, and K^∞ is isomorphic to $\mathrm{SL}_2(q^3)$ or $\mathrm{SU}_3(q)$, neither of which contains H^∞.

It follows from Lemmas 2.3.9, 2.3.10 (ii),(iii) and 2.3.12 (i) that $K \not\leqslant \mathscr{C}_4, \mathscr{C}_5$, and $\mathscr{C}_8$, respectively. Classes $\mathscr{C}_6$ and $\mathscr{C}_7$ are empty. □

Recall Definition 2.2.5 of the $\mathscr{C}_3$-subgroups. If $n = 6$ and H is a $\mathscr{C}_3$-subgroup of Ω, then one of the following holds: $\Omega = \mathrm{SL}_6(q)$ and H is of type $\mathrm{GL}_2(q^3)$ or $\mathrm{GL}_3(q^2)$; $\Omega = \mathrm{SU}_6(q)$ and H is of type $\mathrm{GU}_2(q^3)$; $\Omega = \mathrm{Sp}_6(q)$ and H is of type $\mathrm{Sp}_2(q^3)$ or $\mathrm{GU}_3(q)$, with q odd in the latter case.

Proposition 3.5.5 *Let $n = 6$, let H be a $\mathscr{C}_3$-subgroup of Ω, let G be almost simple with socle $\bar{\Omega}$, and let H_G be the corresponding $\mathscr{C}_3$-subgroup of G. Then H is maximal amongst the geometric subgroups of Ω if and only if $\Omega \neq \mathrm{SU}_6(2)$. If H is not maximal then H_G is maximal amongst the geometric subgroups of G if and only if G is not contained in a conjugate of $\mathrm{U}_6(2).\langle\phi\rangle$.*

Proof Suppose that $H \leqslant K < \Omega$ where K is maximal amongst the geometric subgroups of Ω and is not of the same type as H. We will prove that this implies that $\Omega = \mathrm{SU}_6(2)$, and deduce that there is a unique type for K in this case.

It is immediate from Lemmas 2.3.14, 2.3.16 and 2.3.17 that $K \notin \mathscr{C}_1, \mathscr{C}_2$, and $\mathscr{C}_3$, respectively. If $K \in \mathscr{C}_4$, then it follows from Lemma 2.3.18 and the possible types in Case **U** that $H^\infty \cong \mathrm{SL}_2(q^3)$. In Cases **L** and **U**, the group $K^\infty \cong \mathrm{SL}_2(q)^\infty \times \mathrm{SL}_3^\pm(q)$, by Table 2.7. Since $\mathrm{L}_2(q^3)$ is bigger than $\mathrm{L}_2(q)$, the image of H^∞ in $\mathrm{SL}_2(q)^\infty$ is trivial. Therefore $H^\infty \leqslant 1 \otimes \mathrm{SL}_3^\pm(q)$, and so H^∞ is reducible, contradicting Lemma 2.2.6. A similar but easier argument yields a contradiction in Case **S**. By Lemma 2.3.19, if $K \in \mathscr{C}_5$ then $\Omega = \mathrm{SU}_6(2)$, which we consider in the next paragraph. Classes $\mathscr{C}_6$ and $\mathscr{C}_7$ are empty. We proved in Lemma 2.3.20 that $K \notin \mathscr{C}_8$.

So let $\Omega = \mathrm{SU}_6(2)$, so that $H \cong 3 \times \mathrm{L}_2(8){:}3$. Then H is equal to the $\mathscr{C}_3$-subgroup of the $\mathscr{C}_5$-subgroup of type $\mathrm{Sp}_6(2)$, and hence is not maximal. Note that H is not properly contained in any other type of geometric subgroup, since $\mathrm{SU}_6(2)$ has only one type of $\mathscr{C}_5$-subgroup. There are three classes of groups of type $\mathrm{Sp}_6(2)$ in $\mathrm{U}_6(2)$, on which $\mathrm{Out}\,\mathrm{U}_6(2) = \langle \delta, \phi \rangle$ acts as S_3. Since, modulo scalars, H and K can be written over $\mathbb{F}_2$, they are both centralised modulo scalars by ϕ, so if $G \leqslant \mathrm{SU}_6(q).\langle \phi \rangle$ then H_G is not maximal. However, there is a unique $\mathrm{SU}_6(2)$-conjugacy class of groups of type $\mathrm{GU}_2(q^3)$, so if $G \nleqslant \mathrm{SU}_6(q).\langle \phi \rangle$ then H_G is maximal amongst the geometric subgroups of G. $\qquad\square$

Recall Definition 2.2.9 of the $\mathscr{C}_4$-subgroups. If H is a $\mathscr{C}_4$-subgroup in dimension 6 then one of the following holds: $\Omega = \mathrm{SL}_6(q)$ and H is of type $\mathrm{GL}_2(q) \otimes \mathrm{GL}_3(q)$; $\Omega = \mathrm{SU}_6(q)$ and H is of type $\mathrm{GU}_2(q) \otimes \mathrm{GU}_3(q)$; or $\Omega = \mathrm{Sp}_6(q)$ with q odd, and H is of type $\mathrm{Sp}_2(q) \otimes \mathrm{GO}_3(q)$.

Proposition 3.5.6 *Let $n = 6$ and let H be a $\mathscr{C}_4$-subgroup of Ω. Then H is maximal amongst the geometric subgroups of Ω if and only if $\Omega = \mathrm{SL}_6^\pm(q)$ and $q > 2$, or $\Omega = \mathrm{Sp}_6(q)$ and $q > 3$. If H is not maximal then H does not extend to a novel maximal subgroup.*

Proof The listed exceptions are all considered in Proposition 2.3.22, where it is shown that their behaviour is as stated, so assume for the rest of this proof that $\Omega \neq \mathrm{SL}_6^\pm(2)$ or $\mathrm{Sp}_6(3)$.

Suppose, by way of contradiction, that $H \leqslant K < \Omega$, where K is maximal amongst the geometric subgroups of Ω and is not of the same type as H. It is immediate from Lemma 2.3.23 that $K \notin \mathscr{C}_1 \cup \mathscr{C}_3$.

Suppose that $K \in \mathscr{C}_2$, preserving a decomposition into t subspaces of dimension $m = 6/t$. It follows immediately from Lemma 2.3.24 that $t = 6$ or $q = 3$, hence $\Omega = \mathrm{SL}_6^\pm(q)$. If $t = 6$ then the only non-abelian composition factor of K is A_6. However $|\mathrm{L}_3^\pm(q)| > |\mathrm{A}_6|$ since $q > 2$. Thus $q = 3$ and $2 \leqslant t \leqslant 3$. If $t = 2$ then K' is reducible. However, $\mathrm{SL}_2(3)'$ and $\mathrm{SL}_3^\pm(3)'$ are irreducible, so H' is irreducible by Lemma 2.2.10. Therefore $t = 3$. The group K^∞ is reducible, fixing three subspaces of dimension 2. The groups $\mathrm{SL}_3(3)$ and $\mathrm{SU}_3(3)$

have no representations in defining characteristic in dimension less than 3 by Theorem 1.11.5, a contradiction.

There is a unique type of tensor decomposition, so $K \not\in \mathscr{C}_4$. It follows from Lemma 2.3.25 that $K \not\in \mathscr{C}_5$. Classes $\mathscr{C}_6$ and $\mathscr{C}_7$ are empty. It follows from Lemma 2.3.26 that $K \not\in \mathscr{C}_8$, so we are done. $\square$

Recall Definition 2.2.11 of the $\mathscr{C}_5$-subgroups.

Proposition 3.5.7   *Let $n = 6$ and let H be a $\mathscr{C}_5$-subgroup of Ω. Then H is maximal amongst the geometric subgroups of Ω.*

Proof   This follows immediately from Proposition 2.3.29. $\square$

Classes $\mathscr{C}_6$ and $\mathscr{C}_7$ are empty. Recall Definition 2.2.17 of the $\mathscr{C}_8$-subgroups.

Proposition 3.5.8   *Let $n = 6$ and let H be a $\mathscr{C}_8$-subgroup of Ω. Then H is maximal amongst the geometric subgroups of Ω.*

Proof   This follows immediately from Proposition 2.3.32. $\square$

3.6 Dimension 7

By Definition 1.6.20, when $n = 7$ we find Cases **L**, **U** and **O**°. Recall Definition 2.2.1 of the $\mathscr{C}_1$-subgroups, and Definition 2.3.5 of *standard reducible behaviour*.

Proposition 3.6.1   *Let $n = 7$. Then Ω has standard reducible behaviour.*

Proof   This is immediate from Propositions 2.3.1, 2.3.2 and 2.3.4. $\square$

Recall Definition 2.2.3 of the $\mathscr{C}_2$-subgroups.

Proposition 3.6.2   *Let $n = 7$ and let H be a $\mathscr{C}_2$-subgroup of Ω. Then H is maximal amongst the geometric subgroups of Ω if and only if one of the following holds: $\Omega = \mathrm{SL}_7(q)$ and $q \geqslant 5$; $\Omega = \mathrm{SU}_7(q)$; $\Omega = \Omega_7(q)$. If H is not maximal in Ω then H does not extend to a novel maximal subgroup.*

Proof   This follows immediately from Proposition 2.3.13. $\square$

Recall Definitions 2.2.5, 2.2.11, 2.2.13 and 2.2.17.

Proposition 3.6.3   *Let $n = 7$ and let H be a $\mathscr{C}_3$-, $\mathscr{C}_5$-, $\mathscr{C}_6$- or $\mathscr{C}_8$-subgroup of Ω. Then H is maximal amongst the geometric subgroups of Ω.*

Proof   This follows immediately from Propositions 2.3.21, 2.3.29, 2.3.31 and 2.3.32. $\square$

Classes $\mathscr{C}_4$ and $\mathscr{C}_7$ are empty.

3.7 Dimension 8

By Definition 1.6.20, when $n = 8$ we find Cases **L**, **U**, **S** and **O**$^-$. The maximal subgroups of all almost simple groups with socle $\Omega_8^+(q)$ are classified in [62].

Recall Definition 2.2.1 of the $\mathscr{C}_1$-subgroups, and Definition 2.3.5 of *standard reducible behaviour*.

Proposition 3.7.1 *Let $n = 8$. Then Ω has standard reducible behaviour.*

Proof This is immediate from Propositions 2.3.1, 2.3.2, 2.3.3 and 2.3.4, unless $\Omega = \Omega_8^-(q)$ and H is the stabiliser of a non-degenerate k-space V_1 and its orthogonal complement V_2 for $2 \leqslant k \leqslant 4$. We consider these three exceptions. By Definitions 2.2.3, 2.2.9, 2.2.13, 2.2.15 and 2.2.17, Classes $\mathscr{C}_2$, $\mathscr{C}_4$, $\mathscr{C}_6$, $\mathscr{C}_7$ and $\mathscr{C}_8$ are all empty.

Assume, in the first instance, that $H \leqslant K < \Omega_8^-(q)$, where K is maximal amongst the geometric subgroups of $\Omega_8^-(q)$ and is not of the same type as H. We shall deduce that $q \leqslant 3$, and find just one possibility for K for each q.

If $K \in \mathscr{C}_1$ then H preserves a proper non-zero subspace of V other than V_1 and V_2 and so, by Lemma 1.8.11, H cannot act irreducibly on both V_1 and V_2. Now $\dim(V_2) \geqslant 4$ so H acts irreducibly on V_2 by Proposition 1.12.2. The matrices for H given in [46] show that H acts on V_1 as $\mathrm{GO}(V_1)$, so (again by Proposition 1.12.2) H acts irreducibly on V_1 unless $k = 2$, the form restricted to V_1 is of plus type, and $q \leqslant 3$.

For $q = 2$, notice that $\mathrm{GO}_2^+(2) \cong 2$ is reducible, and stabilises a non-singular 1-space. Thus $\mathrm{PGO}_8^-(2)_H$ is contained in a subgroup of $\mathrm{PGO}_8^-(2)$ of type $\mathrm{Sp}_6(2)$ (but no other $\mathscr{C}_1$-group). Since $\mathrm{Aut}\,\Omega_8^-(2) = \mathrm{PGO}_8^-(2)$, no extension of H is maximal.

For $q = 3$, notice that $\mathrm{GO}_2^+(3) \cong 2^2$ is completely reducible, and stabilises an orthogonal pair of non-degenerate 1-spaces. Thus $\mathrm{PGO}_8^-(3)_H$ is contained in two distinct subgroups K of $\mathrm{PGO}_8^-(3) = \Omega_8^-(3).\langle\varphi\rangle$ of type $\mathrm{GO}_1(3) \perp \mathrm{GO}_7(3)$ (but no other $\mathscr{C}_1$-groups). Now, δ can be chosen to interchange the two non-degenerate 1-spaces, whilst acting trivially on the 6-space stabilised by H, so $H_G \leqslant H_K$ if and only if $G \leqslant \mathrm{PGO}_8^-(3)$.

Assume now that $K \in \mathscr{C}_3$. Then K is of type $\mathrm{GO}_4^-(q^2) \cong \mathrm{L}_2(q^4)$, so unless $k = 4$ then considering $\Omega(V_2)$ we get a contradiction to Theorem 1.11.5. Noting that $\mathrm{L}_2(q^4)$ does not contain $(2, q-1).\mathrm{L}_2(q)^2 \times \mathrm{L}_2(q^2)$ completes this case.

Finally assume that $K \in \mathscr{C}_5$, so that $q \geqslant 8$ by Definition 2.2.11. Then H^∞ contains $\Omega(V_2)$ as a subgroup, acting naturally on $V_2 < V$, where $\dim V_2 \geqslant 4$. Thus by Proposition 1.12.7 and Lemma 2.2.12, $K \notin \mathscr{C}_5$.

Thus if H is not of type $\mathrm{GO}_2^+(q) \perp \mathrm{GO}_6^-(q)$, we have shown that H is maximal, and in these exceptional cases we have found a unique member of $\mathscr{C}_i$ for $1 \leqslant i \leqslant 8$ that contains H. Thus the maximality of H is as claimed. $\square$

Recall Definition 2.2.3 of the $\mathscr{C}_2$-subgroups. If $n = 8$ and H is a $\mathscr{C}_2$-subgroup, then one of the following holds:

(i) $\Omega = \mathrm{SL}_8(q)$ and H is of type $\mathrm{GL}_1(q) \wr \mathrm{S}_8, \mathrm{GL}_2(q) \wr \mathrm{S}_4$, or $\mathrm{GL}_4(q) \wr \mathrm{S}_2$;
(ii) $\Omega = \mathrm{SU}_8(q)$ and H is of type $\mathrm{GU}_1(q) \wr \mathrm{S}_8, \mathrm{GU}_2(q) \wr \mathrm{S}_4, \mathrm{GU}_4(q) \wr \mathrm{S}_2$, or $\mathrm{GL}_4(q^2).2$;
(iii) $\Omega = \mathrm{Sp}_8(q)$ and H is of type $\mathrm{Sp}_2(q) \wr \mathrm{S}_4, \mathrm{Sp}_4(q) \wr \mathrm{S}_2$, or $\mathrm{GL}_4(q).2$, with q odd in the latter case.

In particular, there are no $\mathscr{C}_2$-subgroups in $\Omega_8^-(q)$.

Proposition 3.7.2 *Let $n = 8$ and let H be a $\mathscr{C}_2$-subgroup of Ω, preserving a decomposition into eight subspaces. Then H is maximal amongst the geometric subgroups of Ω if and only if either $\Omega = \mathrm{SL}_8(q)$ and $q \geqslant 5$, or $\Omega = \mathrm{SU}_8(q)$. If H is not maximal then H does not extend to a novel maximal subgroup.*

Proof The exceptions in Case **L** are considered in Proposition 2.3.6, so we will assume that $q \geqslant 5$ in Case **L**. Assume, by way of contradiction, that $H \leqslant K < \Omega$, where K is maximal amongst the geometric subgroups of Ω and is not of the same type as H.

It is immediate from Lemma 2.3.7 (iv),(v) that $K \notin \mathscr{C}_1 \cup \mathscr{C}_3$. It follows from Lemma 2.3.8 (i) that $K \notin \mathscr{C}_2$. Suppose that $K \in \mathscr{C}_4$, and note that H^∞ has a subgroup isomorphic to $(q-1)^7$ in Case **L** and $(q+1)^7$ in Case **U**. We check that $(q\pm1)^7$ does not divide $|K^\infty| = q^7(q^2-1)^2(q^3\pm1)(q^4-1)/(q+1,2)$, unless $\Omega = \mathrm{SL}_8(q)$ with $q \leqslant 3$, which we have excluded. We proved in Lemma 2.3.10 (i) that $K \notin \mathscr{C}_5$, and in Lemma 2.3.11 that $K \notin \mathscr{C}_6$. Class $\mathscr{C}_7$ is empty. It follows from Lemma 2.3.12 (ii) that $K \notin \mathscr{C}_8$. $\square$

Proposition 3.7.3 *Let $n = 8$ and let H be a $\mathscr{C}_2$-subgroup of Ω, preserving a decomposition into four subspaces. Then H is maximal amongst the geometric subgroups of Ω if and only if $q > 2$. If $q = 2$ then H does not extend to a novel maximal subgroup.*

Proof The statement for $q = 2$ is immediate from Proposition 2.3.6, so we assume that $q > 2$. Assume, by way of contradiction, that $H \leqslant K < \Omega$, where K is maximal amongst the geometric subgroups of Ω and is not of the same type as H. Note that if $q > 3$ then $H^\infty \cong \mathrm{SL}_2(q)^4$.

It is immediate from Lemma 2.3.7 (iv) that $K \notin \mathscr{C}_1$. Suppose that $K \in \mathscr{C}_2$, preserving a decomposition into t subspaces, where $t \in \{2, 8\}$. The derived group of H is irreducible by Lemma 2.3.7 (vi), whereas if $t = 2$ then K' is reducible, a contradiction. So $t = 8$, $\Omega = \mathrm{SL}_8^\pm(q)$ and the only non-abelian composition factor of K is A_8. If $q = 3$ then $|H|$ does not divide $|K|$. If $q \geqslant 4$ then $|\mathrm{L}_2(q)|^4$ does not divide $|\mathrm{A}_8|$, a contradiction, so $K \notin \mathscr{C}_2$.

It is immediate from Lemma 2.3.7 (vi) that $K \notin \mathscr{C}_3$, so suppose that $K \in \mathscr{C}_4$.

If $q = 3$ then $|H|$ does not divide $|K|$. If $\Omega = \mathrm{SL}_8^{\pm}(q)$ with $q > 3$ then the group $K^\infty \cong \mathrm{SL}_2(q) \circ \mathrm{SL}_4^{\pm}(q)$, contradicting Lagrange's theorem. In Case **S**, q^4 divides $|H|$ but not $|K|$.

We proved in Lemma 2.3.10 (ii) that $K \not\in \mathscr{C}_5$ unless $\Omega = \mathrm{SU}_8(3)$, and we check that in this case $|H|$ does not divide the order of any $\mathscr{C}_5$-subgroup. Assume next that $K \in \mathscr{C}_6$, so that q is an odd prime. Consulting Table 2.9 we see that H is bigger than K for all q. If $K \in \mathscr{C}_7$ then by Definition 2.2.15 $\Omega = \mathrm{Sp}_8(q)$ with $q > 3$, and K^∞ is smaller than H^∞, a contradiction. If $K \in \mathscr{C}_8$ then, by Definition 2.2.17 and Lemma 2.3.12 (i), $\Omega = \mathrm{SL}_8(3)$. It is straightforward to check that if H is of type $\mathrm{GL}_2(3) \wr S_4$ then $|H|$ does not divide the order of any $\mathscr{C}_8$-subgroup. □

Proposition 3.7.4 *Let $n = 8$ and let H be a $\mathscr{C}_2$-subgroup of Ω, preserving a decomposition into two subspaces. Then H is maximal amongst the geometric subgroups of Ω.*

Proof Assume, by way of contradiction, that $H \leqslant K < \Omega$, where K is maximal amongst the geometric subgroups of Ω and is not of the same type as H.

It is immediate from Lemma 2.3.7 (iii) that if $K \in \mathscr{C}_1$ then $\Omega = \mathrm{Sp}_8(2)$ and H is of type $\mathrm{Sp}_4(2) \wr S_2$. But, as the wreath product of an irreducible group with a transitive permutation group, H is irreducible in this case also. It follows from Lemma 2.3.8 (ii) that $K \not\in \mathscr{C}_2$.

Assume that $K \in \mathscr{C}_3$, so that by Definition 2.2.5 the group $\Omega = \mathrm{SL}_8(q)$ or $\mathrm{Sp}_8(q)$, and by Lemma 2.3.7 (vii) the decomposition is into non-degenerate subspaces in Case **S**. In Case **L**, $|H^\infty| = q^{12}(q^2 - 1)^2(q^3 - 1)^2(q^4 - 1)^2$, whilst $|K^\infty| = q^{12}(q^4 - 1)(q^6 - 1)(q^8 - 1)$, which contradicts Lagrange's theorem. In Case **S**, H is of type $\mathrm{Sp}_4(q) \wr S_2$, then $|H^\infty| = q^8(q^2-1)^2(q^4-1)^2$. If K is of type $\mathrm{Sp}_4(q^2).2$ then $|K^\infty| = q^8(q^4 - 1)(q^8 - 1)$, whilst if K is of type $\mathrm{GU}_4(q)$ then $|K^\infty| = q^6(q^2 - 1)(q^3 + 1)(q^4 - 1)$. Each choice of K^∞ contradicts Lagrange's theorem.

Assume next that $K \in \mathscr{C}_4$, and recall the possibilities for K from Definition 2.2.9. For each choice of H and K we find that the non-abelian composition factors of H are larger than the smaller non-abelian composition factor of K, and that the product of their orders (or their order if there is only one) does not divide the order of the larger composition factor of K.

It is immediate from Lemma 2.3.10 (ii),(iii) that $K \not\in \mathscr{C}_5$, so assume next that $K \in \mathscr{C}_6$, so that q is odd. The only non-abelian composition factor of K is $S_6(2)$ or $O_6^-(2) \cong S_4(3)$. For all odd q these groups are smaller than $L_4(q)$, $U_4(q)$, and $S_4(q)^2$, a contradiction.

If $K \in \mathscr{C}_7$ then $\Omega = \mathrm{Sp}_8(q)$ and q is odd, by Definition 2.2.15. The largest composition factor of K is $L_2(q)$, a contradiction. We proved in Lemma 2.3.12 (i) that $K \not\in \mathscr{C}_8$. □

Recall Definition 2.2.5 of the $\mathscr{C}_3$-subgroups. Let $n = 8$, and let H be a $\mathscr{C}_3$-subgroup. Then one of the following holds: $\Omega = \mathrm{SL}_8(q)$ and H is of type $\mathrm{GL}_4(q^2)$; $\Omega = \mathrm{Sp}_8(q)$ and H is of type $\mathrm{Sp}_4(q^2)$ or $\mathrm{GU}_4(q)$, with q odd in the latter case; $\Omega = \Omega_8^-(q)$ and H is of type $\mathrm{GO}_4^-(q^2)$. Note in particular that Class $\mathscr{C}_3$ is empty in $\mathrm{SU}_8(q)$.

Proposition 3.7.5 *Let $n = 8$ and let H be a $\mathscr{C}_3$-subgroup of Ω. Then H is maximal amongst the geometric subgroups of Ω.*

Proof Suppose, by way of contradiction, that $H \leqslant K < \Omega$, where K is maximal amongst the geometric subgroups of Ω and is not of the same type as H. We proved in Lemma 2.3.14 that $K \notin \mathscr{C}_1$. Since H preserves a degree 2 field extension, it follows from Lemma 2.3.16 that $K \notin \mathscr{C}_2$. We proved that $K \notin \mathscr{C}_3$ in Lemma 2.3.17. If $K \in \mathscr{C}_4$ then $\Omega = \Omega_8^-(q)$ by Lemma 2.3.18, but Class $\mathscr{C}_4$ is empty in $\Omega_8^-(q)$, by Definition 2.2.9. We proved in Lemma 2.3.19 that $K \notin \mathscr{C}_5$.

If $K \in \mathscr{C}_6$ then by Definition 2.2.13 q is odd and $\Omega \neq \Omega_8^-(q)$. It is easy to check that H has a non-abelian composition factor that is larger than the unique non-abelian composition factor of K.

If $K \in \mathscr{C}_7$ then $\Omega = \mathrm{Sp}_8(q)$ and $q \geqslant 5$, by Definition 2.2.15. All non-abelian composition factors of K are isomorphic to $\mathrm{L}_2(q)$, but H has a composition factor $\mathrm{S}_4(q^2)$ or $\mathrm{U}_4(q)$, a contradiction. By Lemma 2.3.20, $K \notin \mathscr{C}_8$. $\square$

Recall Definition 2.2.9 of the $\mathscr{C}_4$-subgroups. If H is a $\mathscr{C}_4$-subgroup in dimension 8 then one of the following holds: $\Omega = \mathrm{SL}_8(q)$ and H is of type $\mathrm{GL}_2(q) \otimes \mathrm{GL}_4(q)$; $\Omega = \mathrm{SU}_8(q)$ and H is of type $\mathrm{GU}_2(q) \otimes \mathrm{GU}_4(q)$; or $\Omega = \mathrm{Sp}_8(q)$ with q odd, and H is of type $\mathrm{Sp}_2(q) \otimes \mathrm{GO}_4^+(q)$ or $\mathrm{Sp}_2(q) \otimes \mathrm{GO}_4^-(q)$.

We start with a lemma that shows that one of the types in Case **S** is non-maximal for all q.

Lemma 3.7.6 *Let H be a $\mathscr{C}_4$-subgroup of $\mathrm{Sp}_8(q)$, of type $\mathrm{Sp}_2(q) \otimes \mathrm{GO}_4^+(q)$. Then H is not maximal, and does not extend to a novel maximal subgroup.*

Proof By Definition 2.2.9, q is odd and $H \cong (\mathrm{Sp}_2(q) \circ \mathrm{GO}_4^+(q)).2$. The corresponding $\mathscr{C}_4$-subgroup of $\mathrm{P\Gamma Sp}_8(q)$, which we denote $H_{\mathrm{P\Gamma Sp}_8(q)}$, is of shape $(\mathrm{PCSp}_2(q) \times \mathrm{PCGO}_4^+(q)).\langle\phi\rangle$.

If $q = 3$ then H is properly contained in a $\mathscr{C}_6$-subgroup L: the group $\mathrm{Sp}_2(3) \circ \mathrm{GO}_4^+(3)$ has a characteristic normal subgroup 2^{1+6}, which is equal to the extraspecial subgroup of L. The group $\mathrm{Sp}_8(3)$ contains a single class of groups of each type, and the extension of H by the unique non-trivial outer automorphism, δ, is contained in the extension of L by δ.

Assume, therefore, that $q > 3$. By Lemma 1.12.3 the group $\mathrm{CGO}_4^+(q)$ is tensor induced, so $H_{\mathrm{PCSp}_8(q)}$ is tensor induced. Each tensor factor of $H_{\mathrm{PCSp}_8(q)}$ is a 2-space on which H^∞ acts as $\mathrm{Sp}_2(q)$, so the tensor factors respect the symplectic form, and hence $H_{\mathrm{PCSp}_8(q)}$ is (properly) contained in the $\mathscr{C}_7$-subgroup $K_{\mathrm{PCSp}_8(q)}$. The automorphism ϕ preserves Kronecker product decompositions of matrices, and hence preserves the tensor factors of both H and K, so $H_{\mathrm{PC\Gamma Sp}_8(q)} < K_{\mathrm{PC\Gamma Sp}_8(q)}$. $\qquad\square$

Proposition 3.7.7   *Let $n = 8$ and let H be a $\mathscr{C}_4$-subgroup of Ω. If $q = 2$ or H is of type $\mathrm{Sp}_2(q) \otimes \mathrm{GO}_4^+(q)$ then H is not maximal, and does not extend to a novel maximal subgroup. Otherwise, H is maximal amongst the geometric subgroups of Ω.*

Proof   The claim for $q = 2$ follows from Proposition 2.3.22, and the claim for type $\mathrm{Sp}_2(q) \otimes \mathrm{GO}_4^+(q)$ has just been proved in Lemma 3.7.6. We therefore assume that $q \geqslant 3$. Assume, by way of contradiction, that $H \leqslant K < \Omega$, where K is maximal amongst the geometric subgroups of Ω and is not of the same type as H.

If $q = 3$ then the 2-dimensional factor of H is soluble, whilst the other factor is insoluble. If $q > 3$ then $H^\infty \cong \mathrm{SL}_2(q) \circ \mathrm{SL}_4^\pm(q)$ or $H^\infty \cong \mathrm{SL}_2(q) \times \mathrm{L}_2(q^2)$.

It is immediate from Lemma 2.3.23 that $K \notin \mathscr{C}_1 \cup \mathscr{C}_3$. We assume next that $K \in \mathscr{C}_2$, preserving a decomposition into t subspaces, where $t \in \{2, 4, 8\}$. It follows from Lemma 2.3.24 that either $t = 8$ or $q = 3$. If $t = 8$, then $\Omega = \mathrm{SL}_8^\pm(q)$. The only non-abelian composition factor of K is A_8, whilst $\mathrm{L}_4(q)$ and $\mathrm{U}_4(q)$ are bigger than A_8 for all $q > 2$. If $t = 4$ (so that $q = 3$), then K is soluble, contradicting the insolubility of H. If $t = 2$ (so that $q = 3$) then K' is reducible. The group $\mathrm{SL}_2(3)'$ is irreducible, and hence H' is irreducible, a contradiction.

By Lemma 3.7.6, Ω contains at most one type of maximal $\mathscr{C}_4$-subgroup. It is immediate from Lemma 2.3.25 that $K \notin \mathscr{C}_5$.

If $K \in \mathscr{C}_6$ and $\Omega = \mathrm{SL}_8^\pm(q)$, then the only non-abelian composition factor of K is $\mathrm{Sp}_6(2)$. However, $\mathrm{L}_4(q)$ and $\mathrm{U}_4(q)$ are bigger than $\mathrm{S}_6(2)$ for all odd q, a contradiction. In Case **S**, $K \cong 2^{1+6}.\Omega_6^-(2) \cong 2^{1+6}.\mathrm{S}_4(3)$. If $q = 3$ then $H \cong \mathrm{SL}_2(3) \times \mathrm{L}_2(9)$, so if $H \leqslant K$, then $\mathrm{S}_4(3)$ contains a subgroup H_1 such that $\mathrm{L}_2(9) \cong T \leqslant H_1$ and $|H_1|$ is divisible by $3|\mathrm{L}_2(9)|$. The direct factor T of H is centralised by the element of order 3 in the factor $\mathrm{SL}_2(3)$, and since 3 is not a divisor of $3^i - 1$ for any i we deduce that T is reducible in $\mathrm{Sp}_4(3)$. If T stabilises a totally singular subspace, then T must be contained in a parabolic subgroup of $\mathrm{Sp}_4(3)$, but the parabolic subgroups of $\mathrm{Sp}_4(3)$ are soluble, a contradiction. If T stabilises a non-degenerate subspace W, then T also stabilises $W^\perp$, so T is contained in a $\mathscr{C}_2$-group of type $\mathrm{Sp}_2(3) \wr \mathrm{S}_2$. However, such groups are again soluble, a contradiction. If $q > 3$ then $|\mathrm{S}_4(3)|$ is not divisible by $|\mathrm{L}_2(q^2)|$.

If Class $\mathscr{C}_7$ is nonempty then $\Omega = \mathrm{Sp}_8(q)$. The group H contains a subgroup $\mathrm{L}_2(q^2)$, whereas all non-abelian composition factors of K are isomorphic to $\mathrm{L}_2(q)$, a contradiction. It follows from Lemma 2.3.26 that $K \notin \mathscr{C}_8$. $\qquad\square$

Recall Definition 2.2.11 of the $\mathscr{C}_5$-subgroups.

Proposition 3.7.8   *Let $n = 8$ and let H be a $\mathscr{C}_5$-subgroup of Ω. Then H is maximal amongst the geometric subgroups of Ω.*

Proof   This follows immediately from Proposition 2.3.29. $\square$

Recall Definition 2.2.13 of the $\mathscr{C}_6$-subgroups.

Proposition 3.7.9   *Let $n = 8$ and let H be a $\mathscr{C}_6$-subgroup of Ω. Then H is maximal amongst the geometric subgroups of Ω.*

Proof   This is immediate from Proposition 2.3.31. $\square$

Recall Definition 2.2.15 of the $\mathscr{C}_7$-subgroups.

Proposition 3.7.10   *Let $n = 8$ and let H be a $\mathscr{C}_7$-subgroup of Ω. Then H is maximal amongst the geometric subgroups of Ω.*

Proof   In dimension 8, Class $\mathscr{C}_7 = \varnothing$ in Cases **L**, **U** and $\mathbf{O}^-$. Thus $\Omega = \mathrm{Sp}_8(q)$, and $q > 3$ is odd by Definition 2.2.15.

Suppose, by way of contradiction, that $H \leqslant K < \mathrm{Sp}_8(q)$, where K is maximal amongst the geometric subgroups of $\mathrm{Sp}_8(q)$ and is not of the same type as H. Here $H^\infty \cong \mathrm{SL}_2(q) \circ \mathrm{SL}_2(q) \circ \mathrm{SL}_2(q) \cong (2, q - 1).\mathrm{L}_2(q)^3$, and H^∞ is absolutely irreducible by Lemma 2.2.16, so $K \notin \mathscr{C}_1 \cup \mathscr{C}_3$.

If $K \in \mathscr{C}_2$ then K preserves a decomposition into at most four subspaces, so K^∞ is reducible, a contradiction. If $K \in \mathscr{C}_4$ then K is of type $\mathrm{Sp}_2(q) \times \mathrm{GO}_4^-(q)$, by Lemma 3.7.6. Therefore $K^\infty \cong \mathrm{SL}_2(q) \times \mathrm{L}_2(q^2)$, which contradicts Lagrange's theorem since $q > 3$. It is immediate from Lemma 2.3.25 that $K \notin \mathscr{C}_5$. If $K \in \mathscr{C}_6$ then the only non-abelian composition factor of K is $\Omega_6^-(2) \cong S_4(3)$, contradicting the fact that $|\mathrm{L}_2(q)|^3 > |S_4(3)|$ for all $q > 3$. There is a unique type of $\mathscr{C}_7$-subgroup, so $K \notin \mathscr{C}_7$. Class $\mathscr{C}_8$ is empty. $\square$

Recall Definition 2.2.17 of the $\mathscr{C}_8$-subgroups.

Proposition 3.7.11   *Let $n = 8$ and let H be a $\mathscr{C}_8$-subgroup of Ω. Then H is maximal amongst the geometric subgroups of Ω.*

Proof   This follows immediately from Proposition 2.3.32. $\square$

3.8 Dimension 9

By Definition 1.6.20, when $n = 9$ we find Cases **L**, **U** and $\mathbf{O}^\circ$. Recall Definition 2.2.1 of the $\mathscr{C}_1$-subgroups, and Definition 2.3.5 of *standard reducible behaviour*.

Proposition 3.8.1 *Let $n = 9$. Then Ω has standard reducible behaviour.*

Proof This is immediate from Propositions 2.3.1, 2.3.2 and 2.3.4. □

Recall Definition 2.2.3 of the $\mathscr{C}_2$-subgroups.

Proposition 3.8.2 *Let $n = 9$ and let H be a $\mathscr{C}_2$-subgroup of Ω, preserving a decomposition into t spaces of dimension m. Then H is maximal amongst the geometric subgroups of Ω if and only if one of the following holds: $\Omega = \mathrm{SL}_9(q)$ and $q \geqslant 5$ or $m > 1$; $\Omega = \mathrm{SU}_9(q)$; $\Omega = \Omega_9(q)$ and $(m, q) \neq (3, 3)$. If H is not maximal then H does not extend to a novel maximal subgroup.*

Proof The non-maximal groups are as in Proposition 2.3.6, so assume that, if $m = 1$ and $\Omega = \mathrm{SL}_9(q)$ then $q \geqslant 5$, and that if $\Omega = \Omega_9(q)$ then $(m, q) \neq (3, 3)$. Suppose, by way of contradiction, that $H \leqslant K < \Omega$, where K is maximal among the geometric subgroups of Ω and is not of the same type as H.

We first consider some arguments which apply to both $m = 1$ and $m = 3$. It is immediate from Lemma 2.3.7 (iv) that H is irreducible, so $K \notin \mathscr{C}_1$. Class $\mathscr{C}_4$ is empty as $n = 9$. It follows from Lemma 2.3.10 (i), (ii) that, if $K \in \mathscr{C}_5$, then $\Omega = \mathrm{SU}_9(2)$ and $m = 3$, but by Definition 2.2.11 there are no $\mathscr{C}_5$-subgroups of $\mathrm{SU}_9(2)$. It follows from Lemma 2.3.12 that $K \notin \mathscr{C}_8$.

The remaining arguments depend on m, so assume first that $m = 1$. Then Lemma 2.3.8 (i) shows that $K \notin \mathscr{C}_2$, and Lemma 2.3.7 (v) that $K \notin \mathscr{C}_3$. It follows from Lemma 2.3.11 that $K \notin \mathscr{C}_6$. If $K \in \mathscr{C}_7$ then K has two non-abelian simple composition factors, both isomorphic to $\mathrm{L}_3(q)$, $\mathrm{U}_3(q)$, or $\mathrm{L}_2(q)$, whereas $\mathrm{A}_9 \leqslant H$. By Proposition 1.11.6, the group A_9 has no faithful projective representations in dimension less than 7, a contradiction.

Assume from now on that $m = 3$. Then H contains $T := \Omega_1 \times \Omega_2 \times \Omega_3$, where $\Omega_i \cong \mathrm{SL}_3(q)$ in Case **L**, $\Omega_i \cong \mathrm{SU}_3(q)$ in Case **U**, and $\Omega_i \cong \mathrm{L}_2(q)$ in Case **O**. In particular, T is perfect if and only if $\Omega_i \neq \mathrm{SU}_3(2)$. If $K \in \mathscr{C}_2$ then the only non-abelian composition factor of K is A_9. If $\Omega = \mathrm{SL}_9(q)$, then $q \geqslant 5$ by assumption, so $P(\mathrm{L}_3(q)) > 12$ by Theorem 1.11.2, a contradiction. If $\Omega = \mathrm{SU}_9(2)$, then $|H|$ does not divide $|K|$, whilst if $q > 2$ then $P(\mathrm{U}_3(q)) > 12$ by Theorem 1.11.2. If $\Omega = \Omega_9(q)$ then $q > 3$ is prime, and hence $|T|$ does not divide $|\mathrm{A}_9|$, a contradiction.

Suppose next that $K \in \mathscr{C}_3$. If $\Omega = \mathrm{SU}_9(2)$ then $|H|$ does not divide $|K|$, so assume that $\Omega \neq \mathrm{SU}_9(2)$. Then K^∞ is one of $\mathrm{SL}_3(q^3)$, $\mathrm{SU}_3(q^3)$ or $\mathrm{L}_2(q^3)$. In Cases **L** and **U**, $|K^\infty| = q^9(q^6 - 1)(q^9 \pm 1)$, whereas $|H^\infty| = q^3(q^2 - 1)^3(q^3 \pm 1)^3$. However $(q^3 \pm 1)^3$ does not divide $|K^\infty|$ for all q. In Case **O**, if $|H^\infty|$ divides $|K^\infty|$ then $q^2 - 1$ divides 3, contradicting the fact that q is odd.

If $K \in \mathscr{C}_6$ then $q = p$, with $q \equiv 1 \pmod 3$ in Case **L** and $q \equiv 2 \pmod 3$ in Case **U**. The only non-abelian composition factor of K is $\mathrm{S}_4(3)$. For Case **L** we note that $|\mathrm{L}_3(q)| > |\mathrm{S}_4(3)|$ for all $q \geqslant 7$, a contradiction. In Case **U**, it is straightforward to check that if H is of type $\mathrm{GU}_3(2) \wr S_3$ and $K \in \mathscr{C}_6$, then $|H|$ does not divide $|K|$. Since $|\mathrm{U}_3(q)| > |S_4(3)|$ for all $q \geqslant 5$, we have a contradiction. Finally, if $K \in \mathscr{C}_7$, then K is smaller than H. $\qquad \square$

Recall Definition 2.2.5 of the $\mathscr{C}_3$-subgroups.

Proposition 3.8.3 *Let* $n = 9$ *and let* H *be a* $\mathscr{C}_3$*-subgroup of* Ω. *Then* H *is maximal amongst the geometric subgroups of* Ω.

Proof Suppose, by way of contradiction, that $H \leqslant K < \Omega$, where K is maximal amongst the geometric subgroups of Ω and is not of the same type as H. If $\Omega = \mathrm{SL}_9^{\pm}(q)$ then $H^\infty \cong \mathrm{SL}_3^{\pm}(q^3)$. If $\Omega = \Omega_9(q)$ then $H^\infty \cong \mathrm{L}_2(q^3)$. Let X denote the non-abelian composition factor of H.

It is immediate from Lemmas 2.3.14 and 2.3.16 (ii) that $K \not\in \mathscr{C}_1, \mathscr{C}_2$, respectively. There is a unique type of $\mathscr{C}_3$-subgroup in each case, so $K \not\in \mathscr{C}_3$. Class $\mathscr{C}_4$ is empty. It follows from Lemma 2.3.19 that $K \not\in \mathscr{C}_5$. If $K \in \mathscr{C}_6$ then $\Omega = \mathrm{SL}_9^{\pm}(q)$ by Definition 2.2.13. The only non-abelian composition factor of K is $S_4(3)$, however $|S_4(3)| < |X|$ for all q, a contradiction. If $K \in \mathscr{C}_7$, then by Table 2.10 K has two non-abelian composition factors, both isomorphic to $\mathrm{L}_3^{\pm}(q)$ or $\mathrm{L}_2(q)$. Each non-abelian composition factor of K is smaller than X, a contradiction. It is immediate from Lemma 2.3.20 that $K \not\in \mathscr{C}_8$. $\qquad \square$

Recall Definition 2.2.11 of the $\mathscr{C}_5$-subgroups.

Proposition 3.8.4 *Let* $n = 9$ *and let* H *be a* $\mathscr{C}_5$*-subgroup of* Ω. *Then* H *is maximal amongst the geometric subgroups of* Ω.

Proof This follows immediately from Proposition 2.3.29. $\qquad \square$

Recall Definition 2.2.13 of the $\mathscr{C}_6$-subgroups.

Proposition 3.8.5 *Let* $n = 9$ *and let* H *be a* $\mathscr{C}_6$*-subgroup of* Ω. *Then* H *is maximal amongst the geometric subgroups of* Ω.

Proof This is immediate from Proposition 2.3.31. $\qquad \square$

Recall Definition 2.2.15 of the $\mathscr{C}_7$-subgroups.

Proposition 3.8.6 *Let* $n = 9$ *and let* H *be a* $\mathscr{C}_7$*-subgroup of* Ω. *Then* H *is maximal amongst the geometric subgroups of* Ω.

Proof If $\Omega = \mathrm{SL}_9(q)$ then $H^\infty \cong \mathrm{SL}_3(q) \circ \mathrm{SL}_3(q)$, if $\Omega = \mathrm{SU}_9(q)$ then $q \geqslant 3$ and $H^\infty \cong \mathrm{SU}_3(q) \circ \mathrm{SU}_3(q)$, and if $\Omega = \Omega_9(q)$ then $q \geqslant 5$ and $H^\infty \cong \mathrm{L}_2(q) \times \mathrm{L}_2(q)$. Suppose, by way of contradiction, that $H \leqslant K < \Omega$, where K is maximal amongst the geometric subgroups of Ω and is not of the same type as H.

It follows immediately from Lemma 2.2.16 that H^∞ is absolutely irreducible, so $K \not\in \mathscr{C}_1 \cup \mathscr{C}_3$.

Assume next that $K \in \mathscr{C}_2$. If K preserves a decomposition into three subspaces, then K^{∞} is reducible, contradicting the irreducibility of H^{∞}. If K preserves a decomposition into nine spaces, then the only non-abelian composition factor of K is A_9. In Case **L**, the group K is not maximal amongst the geometric subgroups of $\mathrm{SL}_9(q)$ when $q \leqslant 4$, by Proposition 2.3.6 (i),(ii),(iii), so without loss of generality $q \geqslant 5$ and hence $|\mathrm{L}_3(q)| > |A_9|$, a contradiction. In Case **U** we are assuming that $q \geqslant 3$, so $|\mathrm{U}_3(q)^2| > |A_9|$, which is also a contradiction. In Case **O°**, we are assuming that $q \geqslant 5$, so $|\Omega_3(q)^2|$ does not divide $|A_9|$.

Class $\mathscr{C}_4$ is empty, so $K \notin \mathscr{C}_4$. It is immediate from Lemma 2.3.25 that $K \notin \mathscr{C}_5$. There is a unique family of $\mathscr{C}_7$-subgroups, so $K \notin \mathscr{C}_7$.

Suppose next that $K \in \mathscr{C}_6$, so that $\Omega = \mathrm{SL}_9^{\pm}(q)$. The only non-abelian composition factor of K is $S_4(3)$, however $|S_4(3)| < |\mathrm{L}_3(q)|^2$ for all q, and $|S_4(3)| < |\mathrm{U}_3(q)|^2$ for all $q \geqslant 3$, a contradiction.

Finally, suppose that $K \in \mathscr{C}_8$, so $\Omega = \mathrm{SL}_9(q)$ and K consists of similarities of a non-degenerate unitary or orthogonal form f. Since the first central factor, Ω_1, of H^{∞} acts as $\mathrm{SL}_3(q)$ on $W := V_1 \otimes v$, for any fixed non-zero $v \in V_2$, the space W is totally singular under f. Therefore as an Ω_1-module $V/W^{\perp}$ is isomorphic to W^* or $W^{*\sigma}$. Since $\dim W > 2$ these are not isomorphic to W, contradicting the homogeneity of the action of Ω_1. □

Recall Definition 2.2.17 of the $\mathscr{C}_8$-subgroups.

Proposition 3.8.7 *Let $n = 9$ and let H be a $\mathscr{C}_8$-subgroup of Ω. Then H is maximal amongst the geometric subgroups of Ω.*

Proof This follows immediately from Proposition 2.3.32. □

3.9 Dimension 10

By Definition 1.6.20, when $n = 10$ we find Cases **L**, **U**, **S** and **O**$^{\pm}$. Recall Definition 2.2.1 of the $\mathscr{C}_1$-subgroups, Definition 2.3.5 of *standard reducible behaviour*, and Definition 1.13.2 of $z_{q,n}$.

Proposition 3.9.1 *Let $n = 10$. Then Ω has standard reducible behaviour.*

Proof This is immediate from Propositions 2.3.1, 2.3.2, 2.3.3 and 2.3.4, unless $\Omega = \Omega_{10}^{\pm}(q)$ and H is the stabiliser of a non-degenerate 4-space V_1 and a non-degenerate 6-space V_2. Thus H contains $\Omega_1 \times \Omega_2$, where Ω_1 is either $\Omega_4^+(q)$ or $\Omega_4^-(q)$, and Ω_2 is either $\Omega_6^+(q) = [\frac{(q-1,4)}{(q-1,2)}].\mathrm{L}_4(q)$ or $\Omega_6^-(q) = [\frac{(q+1,4)}{(q+1,2)}].\mathrm{U}_4(q)$.

Suppose, by way of contradiction, that $H \leqslant K < \Omega_{10}^{\pm}(q)$, where K is maximal amongst the geometric subgroups of $\Omega_{10}^{\pm}(q)$ and is not of the same type as H. Note that Classes $\mathscr{C}_4$, $\mathscr{C}_6$, $\mathscr{C}_7$, and $\mathscr{C}_8$ are empty.

If $K \in \mathscr{C}_1$, then H must stabilise a proper non-zero subspace of V other than V_1 and V_2, contradicting Proposition 1.12.2.

Suppose that $K \in \mathscr{C}_2$, stabilising a decomposition $\mathcal{D}_1$ into t_1 subspaces. By Proposition 1.10.2 and Theorem 1.11.2, if $q > 2$ then $P(\mathrm{L}_4(q)) > 10$ (and by Definition 2.2.3 there is no K with $t_1 = 10$ when $q = 2$), and $P(\mathrm{U}_4(q)) > 10$. Thus Ω_2 is a subgroup of $K_{(\mathcal{D}_1)}$, and hence $t_1 = 2$. We complete this case by noting that $\Omega_5(q) \cong \mathrm{S}_4(q) < \Omega_2$, so the subspaces are totally singular, and that the subgroup of $\Omega_{10}^+(q)$ of type $\mathrm{GL}_5(q).2$ stabilises a totally singular 5-space, whilst H stabilises only V_1 and V_2.

Suppose now that $K \in \mathscr{C}_3$. The fact that $R_p(\Omega_2) = 4$ implies that K preserves a field extension of degree 2, so either $K^\infty \cong \mathrm{SU}_5(q) \leqslant \Omega_{10}^-(q)$ or $K^\infty \cong \Omega_5(q^2) \leqslant \Omega_{10}^\pm(q)$. Considering Theorem 1.6.22, some $z_{q,3}$ or $z_{q,6}$ divides $|H^\infty|$ but not $|\Omega_5(q^2)|$. The group $\mathrm{SU}_5(q)$ does not have a section $\mathrm{L}_4(q)$, and whilst it does contain $\mathrm{SU}_4(q)$, it does not contain $\Omega_4^+(q) \times [\frac{(q+1,4)}{(q+1,2)}]{\cdot}\mathrm{U}_4(q)$.

The final possibility is that $K \in \mathscr{C}_5$, but since $\Omega_6^\pm(q)$ is acting naturally on V_2 this contradicts Proposition 1.12.7 and Lemma 2.2.12.   □

Recall Definition 2.2.3 of the $\mathscr{C}_2$-subgroups. Let $n = 10$, and let H be a $\mathscr{C}_2$-subgroup. Then one of the following holds:

(i) $\Omega = \mathrm{SL}_{10}(q)$ and H is of type $\mathrm{GL}_1(q) \wr \mathrm{S}_{10}$, $\mathrm{GL}_2(q) \wr \mathrm{S}_5$ or $\mathrm{GL}_5(q) \wr \mathrm{S}_2$;

(ii) $\Omega = \mathrm{SU}_{10}(q)$ and H is of type $\mathrm{GU}_1(q) \wr \mathrm{S}_{10}$, $\mathrm{GU}_2(q) \wr \mathrm{S}_5$, $\mathrm{GU}_5(q) \wr \mathrm{S}_2$, or $\mathrm{GL}_5(q^2).2$;

(iii) $\Omega = \mathrm{Sp}_{10}(q)$ and H is of type $\mathrm{Sp}_2(q) \wr \mathrm{S}_5$ or $\mathrm{GL}_5(q).2$, with q odd in the latter case;

(iv) $\Omega = \Omega_{10}^+(q)$ and H is of type $\mathrm{GO}_1^+(p) \wr \mathrm{S}_{10}$, $\mathrm{GO}_2^+(q) \wr \mathrm{S}_5$, $\mathrm{GO}_5(q) \wr \mathrm{S}_2$, $\mathrm{GO}_5(q)^2$ or $\mathrm{GL}_5(q).2$;

(v) $\Omega = \Omega_{10}^-(q)$ and H is of type $\mathrm{GO}_1(p) \wr \mathrm{S}_{10}$, $\mathrm{GO}_2^-(q) \wr \mathrm{S}_5$, $\mathrm{GO}_5(q) \wr \mathrm{S}_2$ or $\mathrm{GO}_5(q)^2$.

Proposition 3.9.2 *Let $n = 10$ and let H be a $\mathscr{C}_2$-subgroup of Ω, preserving a decomposition into ten subspaces. Then H is maximal amongst the geometric subgroups of Ω if and only if one of the following holds: $\Omega = \mathrm{SL}_{10}(q)$ and $q \geqslant 5$; $\Omega = \mathrm{SU}_{10}(q)$; $\Omega = \Omega_{10}^\pm(p)$. If H is not maximal then H does not extend to a novel maximal subgroup.*

Proof The claims for $q \leqslant 4$ in Case **L** follow from Proposition 2.3.6, so assume that $q \geqslant 5$ in Case **L**. Note from Definition 2.2.3 that a decomposition into ten subspaces does not define a $\mathscr{C}_2$-subgroup of $\mathrm{Sp}_{10}(q)$, and only defines a $\mathscr{C}_2$-subgroup of $\Omega_{10}^\pm(q)$ when $q = p$ is prime.

Suppose, by way of contradiction, that $H \leqslant K < \Omega$, where K is maximal amongst the geometric subgroups of Ω and is not of the same type as H.

We proved in Lemma 2.3.7 (iv),(v) that the subgroup L of H^∞ is absolutely irreducible, so $K \notin \mathscr{C}_1 \cup \mathscr{C}_3$.

Suppose next that $K \in \mathscr{C}_2$, preserving a decomposition into t subspaces. We proved in Lemma 2.3.8 (i) that $t \neq 2$, so $t = 5$. The non-abelian composition factors of K lie in the set $\{A_5, L_2(q)\}$. However, H contains a subgroup A_{10}. Now, $|A_{10}| > |A_5|$ and $R(A_{10}) = 8$ by Proposition 1.11.6, a contradiction.

If $K \in \mathscr{C}_4$, then $\Omega = \mathrm{SL}_{10}^\pm(q)$, and the highest rank composition factor of K is $L_5(q)$ or $U_5(q)$, contradicting the fact that $R(A_{10}) = 8$.

We proved in Lemma 2.3.10 (i) that $K \notin \mathscr{C}_5$. Classes $\mathscr{C}_6$ and $\mathscr{C}_7$ are empty. If $K \in \mathscr{C}_8$ then $\Omega = \mathrm{SL}_{10}(q)$, and Lemma 2.3.12 (ii) gives a contradiction. $\square$

Proposition 3.9.3 *Let $n = 10$ and let H be a $\mathscr{C}_2$-subgroup of Ω, preserving a decomposition into five subspaces, let G be almost simple with socle $\overline{\Omega}$, and let H_G be the corresponding $\mathscr{C}_2$-subgroup of G. Then H is maximal amongst the geometric subgroups of Ω if and only if one of the following holds: $\Omega = \mathrm{SL}_{10}^\pm(q)$ or $\mathrm{Sp}_{10}(q)$ and $q > 2$; $\Omega = \Omega_{10}^+(q)$ and $q > 5$; $\Omega = \Omega_{10}^-(q)$ and $q \neq 3$. If H is not maximal, then H_G is maximal amongst the geometric subgroups of G if and only if either $\Omega = \Omega_{10}^+(5)$ and $G \not\leqslant \mathrm{PGO}_{10}^+(5)$ or $\Omega = \Omega_{10}^-(3)$ and $G \not\leqslant \mathrm{PGO}_{10}^-(3)$.*

Proof Apart from $\Omega \in \{\Omega_{10}^+(5), \Omega_{10}^-(3)\}$ the non-maximal exceptions follow from Proposition 2.3.6. We therefore assume that $q > 2$ in Cases **L**, **U** and **S** and that $q > 4$ in Case **O**$^+$. Suppose that $H \leqslant K < \Omega$, where K is maximal amongst the geometric subgroups of Ω and is not of the same type as H: we shall deduce that $\Omega = \Omega_{10}^+(5)$ or $\Omega_{10}^-(3)$, with a list of possibilities for K.

It follows from Lemma 2.3.7 (iv),(v) that H contains an absolutely irreducible subgroup and is not semilinear, so $K \notin \mathscr{C}_1 \cup \mathscr{C}_3$.

Suppose that $K \in \mathscr{C}_2$, preserving a decomposition into t subspaces. From Lemma 2.3.8 (i) we see that $t = 10$, so $\Omega \neq \mathrm{Sp}_{10}(q)$. The only non-abelian composition factor of K is A_{10}. In Case **L**, $q \geqslant 5$ by Proposition 2.3.6 and the assumption that K is maximal, so $|H| > |K|$, a contradiction. In Case **U**, if $q > 9$ then $P(\mathrm{SU}_2(q)) > 10$ by Theorem 1.11.2, whilst if $3 \leqslant q \leqslant 9$ then $|H|$ does not divide $|K|$, a contradiction.

In Case **O**$^\varepsilon$, by Definition 2.2.3, since $t = 10$ we may assume that $q = p > 2$, with $\varepsilon = +$ if and only if $q \equiv 1 \bmod 4$. The group K is a subgroup of $2^9.S_{10}$, so if $\Omega \notin \{\Omega_{10}^+(5), \Omega_{10}^-(3)\}$ then $|H|$ does not divide $|K|$. The group $\mathrm{GO}_2^+(5)$ is imprimitive, preserving a decomposition into two non-degenerate subspaces. Therefore $\mathrm{GO}_2^+(5) \wr S_5$ is properly contained in $\mathrm{GO}_1(5) \wr S_{10}$, and the normaliser of H is not maximal in $\mathrm{GO}_{10}^+(5)$. However, there are two conjugacy classes in $\Omega_{10}^+(5)$ of groups of type $2^9.S_{10}$, and these are interchanged by δ, whilst δ stabilises each of the standard blocks for H. Thus, if G is almost simple with socle $\mathrm{O}_{10}^+(5)$ then $H_G \leqslant K_G$ if and only if $G \leqslant \mathrm{PGO}_{10}^+(5)$. Similarly, $\mathrm{GO}_2^-(3)$ is imprimitive, and an identical argument shows that if G is almost simple with socle $O_{10}^-(3)$, and K_G denotes the subgroup of G of type $GO_1(3) \wr S_{10}$, then $H_G \leqslant K_G$ if and only if $G \leqslant PGO_{10}^-(3)$. This completes the arguments for $\mathscr{C}_2$.

Suppose that $K \in \mathscr{C}_4$, so that $\Omega \neq \Omega_{10}^\pm(q)$ by Definition 2.2.9. Then the group $K^\infty \cong SL_2(q)^\infty \times SL_5^\pm(q)$ in Cases **L** and **U**, and $K^\infty \cong SL_2(q)^\infty \times \Omega_5(q)$ in Case **S**. If $H \leqslant K$ then $SL_2(q)^5.A_5 \leqslant K^\infty$. This implies that $|SL_2(q)^4|$ divides $|SL_5^\pm(q)|$ or $|Sp_4(q)|$, a contradiction since $q > 2$.

We proved in Lemma 2.3.10 (i) that $K \notin \mathscr{C}_5$. Classes $\mathscr{C}_6$ and $\mathscr{C}_7$ are empty. It follows from Lemma 2.3.12 (i) that if $K \in \mathscr{C}_8$ then $\Omega = SL_{10}(3)$. Then K is of type $Sp_{10}(3)$ or $GO_{10}^\pm(3)$. Denote the bilinear form for which K is a group of isometries by f. Let H_1 denote the subgroup of H that acts as $SL_2(3)$ on V_1, and centralises $V_2 \oplus \cdots \oplus V_5$. Then H_1 acts irreducibly on V_1 and centralises a complement to V_1, so V_1 is non-degenerate under f. However, H contains elements that multiply $f|_{V_1}$ and $f|_{V_2}$ by -1, whilst centralising $V_3 \oplus V_4 \oplus V_5$. These elements are not similarities of f, a contradiction.

Thus if $\Omega \neq \Omega_{10}^+(5)$ or $\Omega_{10}^-(3)$ then H is maximal in Ω. If $\Omega = \Omega_{10}^+(5)$ or $\Omega_{10}^-(3)$, then we have actually shown that only a single member of $\mathscr{C}_i$ for $1 \leqslant i \leqslant 8$ contains H. Thus when G is an almost simple group with socle $\bar{\Omega}$ to which this containment does not extend, H_G is maximal amongst the geometric subgroups of G. $\square$

Recall Definition 1.13.2 of $z_{q,n}$.

Proposition 3.9.4   *Let $n = 10$, let H be a $\mathscr{C}_2$-subgroup of Ω, preserving a decomposition into two subspaces, let G be almost simple with socle $\bar{\Omega}$, and let H_G be the corresponding $\mathscr{C}_2$-subgroup of G. Then H is maximal amongst the geometric subgroups of Ω if and only if H is not of type $GL_5(q).2$ in Case $\mathbf{O}^+$. If H is not maximal then H_G is maximal amongst the geometric subgroups of G if and only if $G \nleqslant O_{10}^+(q).\langle \phi, \delta \rangle$.*

Proof   Suppose that $H \leqslant K < \Omega$, where K is maximal amongst the geometric subgroups of Ω and is not of the same type as H.

By Lemma 2.3.7 (iii), if $K \in \mathscr{C}_1$ then H is of type $GL_5(q).2$ in Case $\mathbf{O}^+$. Let V_1 and V_2 be the totally singular subspaces preserved by H, so that without loss of generality K is a parabolic subgroup P_5, stabilising V_1, and since H is reducible (see, for example, the generators for H constructed in [46]) we find that $H < K$. When q is odd, by choosing V_1 and V_2 to be the subspaces spanned by $\{e_1, e_2, e_3, e_4, e_5\}$ and $\{f_1, f_2, f_3, f_4, f_5\}$ using our standard basis of V, as defined in Section 1.5, we see directly that the elements $\delta = \mathrm{diag}(\omega, \ldots, \omega, 1, \ldots, 1)$ and ϕ defined in Section 1.7 normalise both H and K, and so this containment extends to $O_{10}^+(q).\langle \delta, \phi \rangle$. However, there are two Ω-classes of groups of type K, which are interchanged by outer automorphisms not lying in $\langle \phi, \delta \rangle$, so if $G \nleqslant O_n^+(q).\langle \phi, \delta \rangle$ then $H_G \nleqslant K_G$.

It is immediate from Lemma 2.3.8 (ii) that $K \not\in \mathscr{C}_2$, so assume that $K \in \mathscr{C}_3$. By Lemma 2.3.7 (vii), either $\Omega = \mathrm{SL}_{10}(q)$ or V_1 and V_2 are non-degenerate, so H has composition factors $\mathrm{L}_5(q)$, $\mathrm{U}_5(q)$ or $\Omega_5(q)$, and $\Omega \neq \mathrm{Sp}_{10}(q)$. If the group $K^\infty \cong \mathrm{SL}_2(q^5)$, then the non-abelian composition factors of H are larger than those of K. So $\Omega \neq \mathrm{SU}_{10}(q)$. In Case **L**, $K^\infty \cong \mathrm{SL}_5(q^2)$ and $H^\infty \cong \mathrm{SL}_5(q)^2$, so $(q^5 - 1)^2$ divides $|H^\infty|$, whereas $|K^\infty|_{p'} = (q^4 - 1)(q^6 - 1)(q^8 - 1)(q^{10} - 1)$. A higher power of each $z_{q,5}$ divides $|H^\infty|$ than $|K^\infty|$, a contradiction. Finally, in Case **O**$^\pm$, the group K is of type $\mathrm{GO}_5(q^2)$ or type $\mathrm{SU}_5(q)$, whilst H is of type $\mathrm{GO}_5(q) \wr \mathrm{S}_2$ or $\mathrm{GO}_5(q)^2$. By Theorem 1.11.2 and Lemma 1.11.8, H^∞ has no faithful representations in dimension less than 8, a contradiction.

By Lemma 2.3.9, $K \not\in \mathscr{C}_4$. By Lemma 2.3.10 (ii),(iii), $K \not\in \mathscr{C}_5$. Classes $\mathscr{C}_6$ and $\mathscr{C}_7$ are empty. By Lemma 2.3.12 (i), the group $K \not\in \mathscr{C}_8$.

Thus if H is not of type $\mathrm{GL}_5(q).2$ in Case **O**$^+$ then H is maximal in Ω. If H is of type $\mathrm{GL}_5(q).2$ in $\Omega_{10}^+(q)$, then we have actually shown that the only other member of Class $\mathscr{C}_i$ for $1 \leqslant i \leqslant 8$ to contain H is P_5, so if G is an almost simple group with socle $\bar{\Omega}$ to which this containment does not extend, then H_G is maximal amongst the geometric subgroups of G. $\qquad\square$

Recall Definition 2.2.5 of the $\mathscr{C}_3$-subgroups. In Case **L** these are of type $\mathrm{GL}_2(q^5)$ or $\mathrm{GL}_5(q^2)$. In Case **U** these are of type $\mathrm{GU}_2(q^5)$. In Case **S** these are of type $\mathrm{Sp}_2(q^5)$ or $\mathrm{GU}_5(q)$, with q odd in the latter case. In Case **O**$^+$ these are of type $\mathrm{GO}_5(q^2)$. In Case **O**$^-$ these are of type $\mathrm{GO}_5(q^2)$ or $\mathrm{GU}_5(q)$.

Proposition 3.9.5 *Let $n = 10$ and let H be a $\mathscr{C}_3$-subgroup of Ω. Then H is maximal amongst the geometric subgroups of Ω.*

Proof Suppose, by way of contradiction, that $H \leqslant K < \Omega$, where K is maximal amongst the geometric subgroups of Ω and is not of the same type as H.

It is immediate from Lemma 2.3.14 that $K \not\in \mathscr{C}_1$, and from Lemma 2.3.16 (i) that if $K \in \mathscr{C}_2$ then H preserves a degree 5 field extension, so $H^\infty \cong \mathrm{SL}_2(q^5)$. Suppose that $K \in \mathscr{C}_2$ preserves a decomposition $\mathcal{D}$. Now, $P(\mathrm{L}_2(q^5)) > 12$ for all q by Theorem 1.11.2, so $H^\infty \leqslant K_{(\mathcal{D})}$. However, H^∞ is irreducible by Lemma 2.2.6, so $K \not\in \mathscr{C}_2$.

By Lemma 2.3.17, $K \not\in \mathscr{C}_3$, so suppose next that $K \in \mathscr{C}_4$. Then by Definition 2.2.9, $\Omega \neq \Omega_{10}^\pm(q)$. By Lemma 2.3.18, the group $H^\infty \cong \mathrm{SL}_2(q^5)$. In Cases **L** and **U**, $K^\infty \cong \mathrm{SL}_2(q)^\infty \times \mathrm{SL}_5^\pm(q)$, and in Case **S**, $K^\infty \cong \mathrm{SL}_2(q)^\infty \times \mathrm{O}_5(q)$. Then $|H^\infty|$ is divisible by $z_{q,10}$, by Proposition 1.13.4. If $\Omega = \mathrm{SL}_{10}(q)$ or $\mathrm{Sp}_{10}(q)$ then $|K^\infty|$ is not divisible by $z_{q,10}$, whilst if $\Omega = \mathrm{SU}_{10}(q)$ we conclude that H^∞ has trivial projection into the direct factor $\mathrm{SL}_2(q)$ of K^∞. But this implies that H^∞ is reducible, a contradiction.

It follows from Lemma 2.3.19 that $K \notin \mathcal{C}_5$. Classes $\mathcal{C}_6$ and $\mathcal{C}_7$ are empty. It follows from Lemma 2.3.20 that $K \notin \mathcal{C}_8$. $\square$

Recall Definition 2.2.9 of the $\mathcal{C}_4$-subgroups.

Proposition 3.9.6 *Let $n = 10$ and let H be a $\mathcal{C}_4$-subgroup of Ω. Then H is maximal amongst the geometric subgroups of Ω if and only if $q > 2$. If $q = 2$ then H does not extend to a novel maximal subgroup.*

Proof By Definition 2.2.9, $\Omega = \mathrm{SL}_{10}^{\pm}(q)$ or $\mathrm{Sp}_{10}(q)$, with $\Omega_1 \times \Omega_2$ contained in the $\mathcal{C}_4$-subgroup H, where $\Omega_2 \cong \mathrm{SL}_5^{\pm}(q)$ or $\mathrm{SO}_5(q)$. The claim for $q = 2$ is immediate from Proposition 2.3.22, so assume that $q \geqslant 3$.

Suppose, by way of contradiction, that $H \leqslant K < \Omega$, where K is maximal amongst the geometric subgroups of Ω and is not of the same type as H. It is immediate from Lemma 2.3.23 that $K \notin \mathcal{C}_1 \cup \mathcal{C}_3$.

Suppose next that $K \in \mathcal{C}_2$, preserving a decomposition into t subspaces. If $t = 10$ then $\Omega = \mathrm{SL}_{10}^{\pm}(q)$, and the only non-abelian composition factor of K is A_{10}. However, both $\mathrm{L}_5(q)$ and $\mathrm{U}_5(q)$ are larger than A_{10} for all q. Similarly, if $t = 5$ then the non-abelian composition factors of K are A_5 and $\mathrm{L}_2(q)$ (if $q \neq 3$). The simple group $\overline{\Omega_2}$ is larger than both $\mathrm{L}_2(q)$ and A_5, a contradiction. Thus $t = 2$. The derived group of H is irreducible for all q (note that $\mathrm{SL}_2(3)'$ is irreducible) whereas the derived group of K is reducible, a contradiction.

There is a unique type of $\mathcal{C}_4$-subgroup, so $K \notin \mathcal{C}_4$. By Lemmas 2.3.25 and 2.3.26, $K \notin \mathcal{C}_5 \cup \mathcal{C}_8$. Classes $\mathcal{C}_6$ and $\mathcal{C}_7$ are empty. $\square$

Recall Definition 2.2.11 of the $\mathcal{C}_5$-subgroups.

Proposition 3.9.7 *Let $n = 10$ and let H be a $\mathcal{C}_5$-subgroup of Ω. Then H is maximal amongst the geometric subgroups of Ω.*

Proof This follows immediately from Proposition 2.3.29. $\square$

Classes $\mathcal{C}_6$ and $\mathcal{C}_7$ are empty. Recall Definition 2.2.17 of the $\mathcal{C}_8$-subgroups.

Proposition 3.9.8 *Let $n = 10$ and let H be a $\mathcal{C}_8$-subgroup of Ω. Then H is maximal amongst the geometric subgroups of Ω.*

Proof This follows immediately from Proposition 2.3.32. $\square$

3.10 Dimension 11

By Definition 1.6.20, when $n = 11$ we find Cases **L**, **U** and **O**$^\circ$. Recall Definition 2.2.1 of the $\mathcal{C}_1$-subgroups, and Definition 2.3.5 of *standard reducible behaviour*.

Proposition 3.10.1 *Let $n = 11$. Then Ω has standard reducible behaviour.*

Proof This is immediate from Propositions 2.3.1, 2.3.2 and 2.3.4. □

Recall Definition 2.2.3 of the $\mathscr{C}_2$-subgroups.

Proposition 3.10.2 *Let $n = 11$ and let H be a $\mathscr{C}_2$-subgroup of Ω. Then H is maximal amongst the geometric subgroups of Ω if and only if one of the following holds: $\Omega = \mathrm{SL}_{11}(q)$ and $q \geqslant 5$; $\Omega = \mathrm{SU}_{11}(q)$, or $\Omega = \Omega_{11}(q)$. If H is not maximal in Ω then H does not extend to a novel maximal subgroup.*

Proof This follows immediately from Proposition 2.3.13. □

Recall Definitions 2.2.5, 2.2.11, 2.2.13 and 2.2.17.

Proposition 3.10.3 *Let $n = 11$ and let H be a $\mathscr{C}_3$-, $\mathscr{C}_5$-, $\mathscr{C}_6$- or $\mathscr{C}_8$-subgroup of Ω. Then H is maximal amongst the geometric subgroups of Ω.*

Proof This follows immediately from Propositions 2.3.21, 2.3.29, 2.3.31 and 2.3.32. □

Classes $\mathscr{C}_4$ and $\mathscr{C}_7$ are empty.

3.11 Dimension 12

By Definition 1.6.20, when $n = 12$ we find Cases **L**, **U S** and **O**$^\pm$. Recall Definition 2.2.1 of the $\mathscr{C}_1$-subgroups, and Definition 2.3.5 of *standard reducible behaviour*.

Proposition 3.11.1 *Let $n = 12$. Then Ω has standard reducible behaviour.*

Proof This is immediate from Propositions 2.3.1, 2.3.2, 2.3.3 and 2.3.4, unless $\Omega = \Omega_{12}^-(q)$ and H is the stabiliser of a pair of orthogonal non-degenerate non-isometric 6-spaces V_1 and V_2. In this case H contains

$$\Omega_6^+(q) \times \Omega_6^-(q) \cong [(q-1,4)/(q-1,2)].\mathrm{L}_4(q) \times [(q+1,4)/(q+1,2)].\mathrm{U}_4(q).$$

Assume, by way of contradiction, that $H \leqslant K < \Omega_{12}^-(q)$, where K is maximal amongst the geometric subgroups of $\Omega_{12}^-(q)$ and is not of the same type as H. Note that $\mathscr{C}_6 \cup \mathscr{C}_7 \cup \mathscr{C}_8 = \varnothing$.

If $K \in \mathscr{C}_1$ then $\Omega_6^+(q) \times \Omega_6^-(q)$ fixes some non-zero proper subspace of V other than V_1 and V_2, contradicting Lemma 1.8.11 and Proposition 1.12.2. If $K \in \mathscr{C}_2$ then, since $P(\mathrm{U}_4(q)) > 12$ by Theorem 1.11.2, we deduce that $\Omega_6^-(q)$ lies in the kernel of the action of K on blocks. By Theorem 1.11.5 $R_p(\mathrm{U}_4(q)) = 4$, so the blocks have dimension 4 or 6, and hence by Definition 2.2.3 have dimension 4. Thus $\Omega_6^-(q) \leqslant \Omega_4^-(q)$, a contradiction. If $K \in \mathscr{C}_3$ then as for $\mathscr{C}_2$ we deduce that the only possibility is $\Omega_6^+(q) \times \Omega_6^-(q) \leqslant \Omega_6^-(q^2)$, contradicting Lagrange's theorem. If $K \in \mathscr{C}_4$ then the largest non-abelian composition factor of K is $\Omega_4^-(q)$, which is too small. The result for $K \in \mathscr{C}_5$ follows from noting that $\Omega_6^+(q) \leqslant H$ and applying Proposition 1.12.7. $\qquad\square$

Recall Definition 2.2.3 of the $\mathscr{C}_2$-subgroups. If $n = 12$ and H is a $\mathscr{C}_2$-subgroup of Ω, then one of the following holds:

(i) $\Omega = \mathrm{SL}_{12}(q)$ and H is of type $\mathrm{GL}_1(q) \wr \mathrm{S}_{12}$, $\mathrm{GL}_2(q) \wr \mathrm{S}_6$, $\mathrm{GL}_3(q) \wr \mathrm{S}_4$, $\mathrm{GL}_4(q) \wr \mathrm{S}_3$ or $\mathrm{GL}_6(q) \wr \mathrm{S}_2$;

(ii) $\Omega = \mathrm{SU}_{12}(q)$ and H is of type $\mathrm{GU}_1(q) \wr \mathrm{S}_{12}$, $\mathrm{GU}_2(q) \wr \mathrm{S}_6$, $\mathrm{GU}_3(q) \wr \mathrm{S}_4$, $\mathrm{GU}_4(q) \wr \mathrm{S}_3$, $\mathrm{GU}_6(q) \wr \mathrm{S}_2$ or $\mathrm{GL}_6(q^2).2$;

(iii) $\Omega = \mathrm{Sp}_{12}(q)$ and H is of type $\mathrm{Sp}_2(q)\wr\mathrm{S}_6$, $\mathrm{Sp}_4(q)\wr\mathrm{S}_3$, $\mathrm{Sp}_6(q)\wr\mathrm{S}_2$ or $\mathrm{GL}_6(q).2$, with q odd in the latter case;

(iv) $\Omega = \Omega_{12}^+(q)$ and H is of type $\mathrm{GO}_1(p) \wr \mathrm{S}_{12}$, $\mathrm{GO}_2^{\pm}(q) \wr \mathrm{S}_6$, $\mathrm{GO}_3(q) \wr \mathrm{S}_4$, $\mathrm{GO}_4^+(q) \wr \mathrm{S}_3$, $\mathrm{GO}_6^{\pm}(q) \wr \mathrm{S}_2$ or $\mathrm{GL}_6(q).2$;

(v) $\Omega = \Omega_{12}^-(q)$ and H is of type $\mathrm{GO}_4^-(q) \wr \mathrm{S}_3$.

In each type we denote the decomposition preserved by H by

$$\mathcal{D} : V = V_1 \oplus \cdots \oplus V_t.$$

Proposition 3.11.2 *Let $n = 12$ and let H be a $\mathscr{C}_2$-subgroup of Ω, preserving a decomposition into twelve subspaces. Then H is maximal amongst the geometric subgroups of Ω if and only if either $\Omega = \mathrm{SL}_{12}(q)$ with $q \geqslant 5$; or $\Omega = \mathrm{SU}_{12}(q)$; or $\Omega = \Omega_{12}^+(p)$, with p prime. If H is not maximal then H does not extend to a novel maximal subgroup.*

Proof The non-maximal examples when $q \leqslant 4$ in Case **L** follow from Proposition 2.3.6, so we assume that $q \geqslant 5$ in Case **L**. By Definition 2.2.3 the decomposition into twelve subspaces only defines a $\mathscr{C}_2$-subgroup of $\Omega_{12}^\varepsilon(q)$ when $\varepsilon = +$ and $q = p$ is prime.

Suppose, by way of contradiction, that $H \leqslant K < \Omega$, where K is maximal amongst the geometric subgroups of Ω and is not of the same type as H. It is immediate from Lemma 2.3.7 (iv),(v) that $K \notin \mathscr{C}_1 \cup \mathscr{C}_3$, and from Lemma 2.3.8 (i) that, if $K \in \mathscr{C}_2$, then K preserves a decomposition into six subspaces. By Proposition 1.11.6, $R(\mathrm{A}_{12}) = 10$, so $K \notin \mathscr{C}_2 \cup \mathscr{C}_4$. It follows from Lemma 2.3.10 (i) that $K \notin \mathscr{C}_5$. Classes $\mathscr{C}_6$ and $\mathscr{C}_7$ are empty, and it follows from Lemma 2.3.12 (ii) that $K \notin \mathscr{C}_8$. $\qquad\square$

Proposition 3.11.3 *Let $n = 12$ and let H be a $\mathscr{C}_2$-subgroup of Ω, preserving a decomposition into six subspaces, let G be almost simple with socle $\bar{\Omega}$, and let H_G be the corresponding $\mathscr{C}_2$-subgroup of G. Then H is maximal amongst the geometric subgroups of Ω if and only if the following all hold: if H is not of* type $GO_2^-(q) \wr S_6$ then $q > 2$; if H is of type $GO_2^+(q) \wr S_6$ then $q \geqslant 7$; and if H is of type $GO_2^-(q) \wr S_6$ then $q \neq 3$.

If H is not maximal in Ω then H_G is maximal amongst the geometric subgroups of G if and only if either H is of type $GO_2^+(5) \wr S_6$ and $G \not\leqslant PGO_{12}^+(5)$ or H is of type $GO_2^-(3) \wr S_6$ and $G \not\leqslant PGO_{12}^+(3)$.

Proof The non-maximal exceptions when $q = 2$, along with the claims about groups of type $GO_2^+(q) \wr S_6$ for $q \leqslant 4$, are from Proposition 2.3.6, so assume that $q > 2$ in Cases **L**, **U** and **S**, and if H is of type $GO_2^+(q) \wr S_6$ then $q \geqslant 5$.

Suppose, in the first instance, that $H \leqslant K < \Omega$, where K is maximal amongst the geometric subgroups of Ω and is not of the same type as H. We will deduce that $\Omega = \Omega_{12}^+(q)$ with $q \in \{3, 5\}$, and that in these cases there is only a single choice for the type of K.

It is immediate from Lemma 2.3.7 (iv),(v) that $K \notin \mathscr{C}_1 \cup \mathscr{C}_3$, and from Lemma 2.3.8 (i) that if $K \in \mathscr{C}_2$ then either K preserves a decomposition into twelve subspaces, or H and K are of types $GO_2^+(q) \wr S_6$ and $GO_2^-(q) \wr S_6$. Considering the orders of $\Omega_2^+(q)$ and $\Omega_2^-(q)$, we see that this second possibility requires $q \leqslant 3$ and H to be of type $GO_2^+(q) \wr S_6$, which we have already considered.

Assume therefore that $K \in \mathscr{C}_2$ preserves a decomposition into twelve subspaces, so that the only non-abelian composition factor of K is A_{12}. Note that $q \geqslant 5$ in Case **L**, by Proposition 2.3.6, and that in $SU_{12}(3)$ the order of H does not divide the order of K. If $q \geqslant 4$ then $|L_2(q)^6| > A_{12}$, a contradiction. Thus $\Omega = \Omega_{12}^+(q)$ with q prime, and $K \leqslant 2^{11}.S_{12}$. If r is an odd prime then the highest power of r to divide $|S_{12}|$ is at most r^5. If H is of type $GO_2^+(q) \wr S_6$ and r divides $q - 1$, or if H is of type $GO_2^-(q) \wr S_6$ and r divides $q + 1$, then H contains an elementary abelian r-group of order r^6. Therefore, if H is of type $GO_2^+(q) \wr S_6$ then there exists an i such that $q - 1 = 2^i$, and if H is of type $GO_2^-(q) \wr S_6$ then there exists an i such that $q + 1 = 2^i$. Now, the 2-part of the order of K is at most 2^{21}, and the 2-part of the order of H is $2^{6(i-1)} \cdot 2^{10} \cdot 2^4$, so $i \leqslant 2$. If H is of type $GO_2^+(q) \wr S_6$ then $q = 3$ (which we have excluded) or $q = 5$. The standard copy of $GO_2^+(5)$ stabilises a decomposition $V = \langle e_1 + f_1 \rangle \oplus \langle e_1 - f_1 \rangle$ into non-degenerate subspaces. Thus $H_{GO_{12}^+(5)} < K_{GO_{12}^+(5)}$. However, there are two $GO_{12}^+(5)$-classes of groups of type $2^{11}.S_{12}$, interchanged by δ, so if $G \not\leqslant PGO_{12}^+(5)$ then $H_G \not\leqslant K_G$. If H is of type $GO_2^-(q) \wr S_6$ then $q = 3$. The group $GO_2^-(3)$ also stabilises the decomposition $V = \langle e_1 + f_1 \rangle \oplus \langle e_1 - f_1 \rangle$, and exactly the same arguments show that $H_G \leqslant K_G$ if and only if $G \leqslant PGO_{12}^+(3)$.

Assume next that $K \in \mathscr{C}_4$. The possibilities for K are listed in Table 2.7. Consider first Cases **L**, **U** and **S**. The order of H^∞ is divisible by $(q^2 - 1)^6$. In Case **L** the order of K^∞ is a divisor of $q^{16}(q^2-1)^2(q^3-1)(q^4-1)(q^5-1)(q^6-1)$ or $q^9(q^2-1)^2(q^3-1)^2(q^4-1)$, which contradicts Lagrange's theorem for all q. In

Case $\mathbf{U}$ the order of K^∞ is a divisor of $q^{16}(q^2-1)^2(q^3+1)(q^4-1)(q^5+1)(q^6-1)$ or $q^9(q^2-1)^2(q^3+1)^2(q^4-1)$, which contradicts Lagrange's theorem since $q > 2$. In Case $\mathbf{S}$ the order of K^∞ is a divisor of $q^7(q^2-1)^2(q^3\pm1)(q^4-1)$, contradicting Lagrange's theorem. So $\Omega = \Omega_{12}^+(q)$, and K is isomorphic to either $\mathrm{PGL}_2(q) \times (\mathrm{SL}_2(q) \circ \mathrm{SL}_2(q))$ or $(\mathrm{Sp}_2(q) \circ \mathrm{Sp}_6(q)).2$, with the outer involution extending $\mathrm{L}_2(q)$ to $\mathrm{PGL}_2(q)$. Let K preserve a decomposition $W_1 \otimes W_2$. The group A_6 is a subgroup of H, whilst S_6 is a quotient. Now, the group $\mathrm{L}_2(q)$ can contain A_6, but consulting Tables 8.1 and 8.2 we see that $\mathrm{PGL}_2(q)$ does not contain $\mathrm{S}_6 \cong \mathrm{P\Sigma L}_2(9)$. Recall the definition of L from just before Lemma 2.2.4. Since L is the normal closure of A_6, if the action of A_6 on W_1 is trivial, then the action of L on W_1 is reducible, and hence the action of L on V is reducible, a contradiction.

It is immediate from Lemma 2.3.10 (i) that $K \notin \mathscr{C}_5$. Classes $\mathscr{C}_6$ and $\mathscr{C}_7$ are empty. It is immediate from Lemma 2.3.12 (i) that if $K \in \mathscr{C}_8$ then $\Omega = \mathrm{SL}_{12}(3)$, but then $|H|$ does not divide $|K|$, a contradiction.

In our arguments for $\Omega_{12}^+(q)$ with $q \in \{3,5\}$, we have in fact shown that the non-maximal groups H are contained in a unique member of $\mathscr{C}_i$ for $1 \leqslant i \leqslant 8$. Our maximality claims for extensions of these groups therefore follow. $\qquad\square$

Proposition 3.11.4 *Let $n = 12$ and let H be a $\mathscr{C}_2$-subgroup of Ω, preserving a decomposition into four subspaces. Then H is maximal amongst the geometric subgroups of Ω if and only if $\Omega \neq \Omega_{12}^+(3)$. If H is not maximal then H does not extend to a novel maximal subgroup.*

Proof The non-maximality of H when $\Omega = \Omega_{12}^+(3)$ follows immediately from Proposition 2.3.6, so assume that $q \geqslant 5$ in Case $\mathbf{O}^+$.

Suppose, by way of contradiction, that $H \leqslant K < \Omega$, where K is maximal amongst the geometric subgroups of Ω and is not of the same type as H. It is immediate from Lemma 2.3.7 (vi) that $K \notin \mathscr{C}_1 \cup \mathscr{C}_3$ and that H' is irreducible.

First consider $\mathrm{U}_{12}(2)$ as a special case. Order considerations show that the group $K \notin \mathscr{C}_2 \cup \mathscr{C}_4 \cup \mathscr{C}_5$. Classes $\mathscr{C}_6$, $\mathscr{C}_7$ and $\mathscr{C}_8$ are empty. We therefore assume for the rest of the proof that $\Omega \neq \mathrm{SU}_{12}(2)$, so that, in particular, H is insoluble.

Assume that $K \in \mathscr{C}_2$. If K preserves a decomposition into two subspaces then K' is reducible, a contradiction. If K preserves a decomposition into three subspaces then K contains a reducible subgroup of index 3. However, the largest reducible subgroup of H has index 4, a contradiction, so K must preserve a decomposition into six or twelve subspaces, and the non-abelian composition factors of K lie in the set $\{\mathrm{L}_2(q), \mathrm{A}_6, \mathrm{A}_{12}\}$. In Case $\mathbf{L}$, $H^\infty = \mathrm{SL}_3(q)^4$, and $|\mathrm{L}_3(q)^4| > |\mathrm{A}_{12}|$, a contradiction. For Case $\mathbf{U}$, $|\mathrm{U}_3(q)^4| > |\mathrm{A}_{12}|$ (since $q > 2$). Thus $\Omega = \Omega_{12}^+(q)$, and the only non-abelian composition factor of K is A_6 or A_{12}. Since we are assuming that $q \geqslant 5$, $|\mathrm{L}_2(q)^4|$ does not divide $|\mathrm{A}_6|$ or $|\mathrm{A}_{12}|$, a contradiction.

Assume next that $K \in \mathscr{C}_4$, so that $K = K_1 \circ K_2$. The possibilities for K can be found in Table 2.7. By Lagrange's theorem, K preserves a tensor product decomposition into a 2-space and a 6-space. Let K_2 be the 6-dimensional factor, and suppose some direct factor C of H^∞ is contained in K_2. Then, since K_2 acts homogeneously, the action of C must have at least two non-trivial constituents, which is false. Thus every direct factor of H^∞ projects non-trivially on K_1, contradicting the fact that H contains $\mathrm{SL}_3^\pm(q)^4$ or $\mathrm{L}_2(q)^4$.

It is immediate from Lemma 2.3.10 (ii) that $K \notin \mathscr{C}_5$. Classes $\mathscr{C}_6$ and $\mathscr{C}_7$ are empty. It is immediate from Lemma 2.3.12 (i) that $K \notin \mathscr{C}_8$. $\qquad\square$

Proposition 3.11.5 *Let $n = 12$ and let H be a $\mathscr{C}_2$-subgroup of Ω, preserving a decomposition into three subspaces. Then H is maximal amongst the geometric subgroups of Ω if and only if $\Omega \neq \Omega_{12}^+(2)$. If H is not maximal then H does not extend to a novel maximal subgroup.*

Proof The non-maximal group is considered in Proposition 2.3.6, so assume that $\Omega \neq \Omega_{12}^+(2)$. Suppose, by way of contradiction, that $H \leqslant K < \Omega$, where K is maximal amongst the geometric subgroups of Ω and is not of the same type as H. It is immediate from Lemma 2.3.7 (iv) that $K \notin \mathscr{C}_1$.

Suppose that $K \in \mathscr{C}_2$, preserving a decomposition into t_1 subspaces. In Case $\mathbf{O}^-$ there is a unique type of $\mathscr{C}_2$-subgroup when $n = 12$, a contradiction. The group H does not contain an index 2 reducible subgroup, so $t_1 \neq 2$. If $t_1 = 4$ and $\Omega \neq \Omega_{12}^+(q)$, then the largest composition factor of K is $\mathrm{L}_3(q)$ or $\mathrm{U}_3(q)$ which is smaller than the largest composition factor of H. If $t_1 = 6$ and $\Omega \neq \Omega_{12}^+(q)$, then the largest composition factor of K is either A_6 or one of $\mathrm{L}_2(q)$, $\mathrm{U}_2(q)$ or $\mathrm{Sp}_2(q)$. These are all smaller than the largest composition factor of H. Thus $\Omega \neq \mathrm{Sp}_{12}(q)$, and if $\Omega = \mathrm{SL}_{12}^\pm(q)$ then $t = 12$. But then the only non-abelian composition factor of K is A_{12}, which is smaller than $\mathrm{L}_4(q)^3$ and $\mathrm{U}_4(q)^3$, a contradiction. It follows that $\Omega = \Omega_{12}^+(q)$. If $q = 3$, then $|H|$ does not divide the order of any other type of $\mathscr{C}_2$-subgroup. So we may assume that $q > 3$, and hence that H is insoluble. If $t_1 = 4$ then K contains four non-abelian composition factors, all isomorphic to $\mathrm{L}_2(q)$, whereas H contains six copies of $\mathrm{L}_2(q)$, a contradiction. If $t_1 \in \{6, 12\}$ then the only non-abelian composition factor of K is A_6 or A_{12}, which is smaller than $\mathrm{L}_2(q)^6$.

Suppose next that $K \in \mathscr{C}_3$, and recall the definition of L from just before Lemma 2.2.4. The group H' acts as A_3 on $\mathcal{D}$, so in particular $L \leqslant H'$. The group L is absolutely irreducible by Lemma 2.3.7 (iv), but K' is not absolutely irreducible, a contradiction.

It follows from Lemma 2.3.9 that if $K \in \mathscr{C}_4$ then $\Omega = \Omega_{12}^+(3)$. By Proposition 2.3.22, the $\mathscr{C}_4$-subgroups of type $\mathrm{GO}_3(3) \otimes \mathrm{GO}_4^+(3)$ are not maximal in $\Omega_{12}^+(3)$, so K is of type $\mathrm{Sp}_2(3) \otimes \mathrm{Sp}_6(3)$. Then $|H|$ does not divide $|K|$, a contradiction.

It is immediate from Lemma 2.3.10 (ii) that $K \notin \mathscr{C}_5$. Classes $\mathscr{C}_6$ and $\mathscr{C}_7$ are empty. It is immediate from Lemma 2.3.12 (i) that $K \notin \mathscr{C}_8$. $\qquad\square$

Proposition 3.11.6 *Let $n = 12$ and let H be a $\mathscr{C}_2$-subgroup of Ω, preserving a decomposition into two subspaces. Then H is maximal amongst the geometric subgroups of Ω.*

Proof Suppose, by way of contradiction, that $H \leqslant K < \Omega$, where K is maximal amongst the geometric subgroups of Ω and is not of the same type as H. It follows from Lemmas 2.3.7 (iii) and 2.3.8 (ii) that $K \notin \mathscr{C}_1$ or $\mathscr{C}_2$, respectively.

Suppose that $K \in \mathscr{C}_3$. By Lemma 2.3.7 (vii), the decomposition is into two non-degenerate subspaces, or $\Omega = \mathrm{SL}_{12}(q)$. In Case **L**, up to isomorphism $K^\infty \in \{\mathrm{SL}_6(q^2), \mathrm{SL}_4(q^3)\}$. The order of H^∞ is divisible by a higher power of a prime $z_{q,5}$ than $|\mathrm{SL}_6(q^2)|$, and H^∞ is larger than $\mathrm{SL}_4(q^3)$. In Case **U**, the group $K^\infty \cong \mathrm{SU}_4(q^3)$, and H^∞ is larger than K^∞. In Case **S**, the group $H^\infty \cong \mathrm{Sp}_6(q)^2$ whilst $K^\infty \in \{\mathrm{Sp}_4(q^3), \mathrm{Sp}_6(q^2), \mathrm{SU}_6(q)\}$. Thus $|H|$ does not divide $|K|$, a contradiction. In Case **O**$^+$, $K^\infty \in \{\Omega_4^+(q^3), \Omega_6^+(q^2), \mathrm{SU}_6(q)\}$. The order of $H^\infty \cong \Omega_6^+(q)^2$ is divisible by a higher power of a prime $z_{q,3}$ than $|\Omega_6^+(q^2)|$ or $|\mathrm{SU}_6(q)|$, and by a prime $z_{q,4}$, which does not divide $|\Omega_4^+(q^3)|$.

It follows from Lemmas 2.3.9 and 2.3.10 (ii),(iii) that $K \notin \mathscr{C}_4 \cup \mathscr{C}_5$. Classes $\mathscr{C}_6$ and $\mathscr{C}_7$ are empty, and it follows from Lemma 2.3.12 (i) that $K \notin \mathscr{C}_8$. $\quad\square$

Recall Definition 2.2.5 of the $\mathscr{C}_3$-subgroups. If $n = 12$ and $H < \Omega$ is a $\mathscr{C}_3$-subgroup, then one of the following holds:

(i) $\Omega = \mathrm{SL}_{12}(q)$ and H is of type $\mathrm{GL}_6(q^2)$ or $\mathrm{GL}_4(q^3)$;
(ii) $\Omega = \mathrm{SU}_{12}(q)$ and H is of type $\mathrm{GU}_4(q^3)$;
(iii) $\Omega = \mathrm{Sp}_{12}(q)$ and H is of type $\mathrm{Sp}_6(q^2)$ or $\mathrm{Sp}_4(q^3)$ or $\mathrm{GU}_6(q)$, with q odd in the latter case;
(iv) $\Omega = \Omega_{12}^+(q)$ and H is of type $\mathrm{GO}_6^+(q^2)$ or $\mathrm{GO}_4^+(q^3)$ or $\mathrm{GU}_6(q)$;
(v) $\Omega = \Omega_{12}^-(q)$ and H is of type $\mathrm{GO}_6^-(q^2)$ or $\mathrm{GO}_4^-(q^3)$.

Proposition 3.11.7 *Let $n = 12$ and let H be a $\mathscr{C}_3$-subgroup of Ω. Then H is maximal amongst the geometric subgroups of Ω.*

Proof Suppose, by way of contradiction, that $H \leqslant K < \Omega$, where K is maximal among the geometric subgroups of Ω and is not of the same type as H.

We proved in Lemma 2.3.14 that $K \notin \mathscr{C}_1$. Since H normalises a field extension of degree 2 or 3, it follows from Lemma 2.3.16 that if $K \in \mathscr{C}_2$ then H is of type $\mathrm{GO}_4^+(q^3)$, so that $H^\infty \cong \mathrm{SL}_2(q^3) \circ \mathrm{SL}_2(q^3)$. If $|\mathrm{L}_2(q^3)|$ divides $|A_d|$, for $d \in \{2, 3, 4, 6, 12\}$, then $q = 2$ and $d = 12$. However by Table 2.4 there is no such $\mathscr{C}_2$-subgroup when $q = 2$. Therefore in all cases $H^\infty \leqslant K_\mathcal{D}$, the pointwise stabiliser of the set of blocks. This implies that H^∞ is reducible, contradicting Lemma 2.2.6. It is immediate from Lemma 2.3.17 that $K \notin \mathscr{C}_3$.

Suppose next that $K \in \mathscr{C}_4$. We note first that by Lemma 2.3.18, and the possibilities for H in Case **U**, if $\Omega \neq \Omega_{12}^{\pm}(q)$ then H preserves a degree 3 field extension. By Table 2.7, for all Ω the group K is of one of the following types: Cases **L** and **U**, type $\mathrm{GL}_{n_1}^{\pm}(q) \otimes \mathrm{GL}_{n_2}^{\pm}(q)$, $(n_1, n_2) \in \{(2,6), (3,4)\}$; Case **S**, type $\mathrm{Sp}_{n_1}(q) \otimes \mathrm{GO}_{n_2}^{\varepsilon}(q)$, $(n_1, n_2) \in \{(2,6), (4,3)\}$; Case **O**$^+$, type $\mathrm{Sp}_2(q) \otimes \mathrm{Sp}_6(q)$ or type $\mathrm{GO}_3(q) \otimes \mathrm{GO}_4^+(q)$; Case **O**$^-$, type $\mathrm{GO}_3(q) \otimes \mathrm{GO}_4^-(q)$.

If $\Omega \neq \Omega_{12}^+(q)$, then $|H^\infty|$ is divisible by a prime $z_{q,12}$, by Proposition 1.13.4, whilst $z_{q,12}$ does not divide $|K^\infty|$. In Case **O**$^+$, if H is of type $\mathrm{GU}_6(q)$ then $|H^\infty|$ is divisible by a prime $z_{q,10}$ which does not divide $|K^\infty|$. If H is of type $\mathrm{GO}_6^+(q^2)$ then $|H^\infty|$ is divisible by a prime $z_{q,8}$, whilst $|K|$ is not. If H is of type $\mathrm{GO}_4^+(q^3)$ and $q \neq 2$, then $|H^\infty|$ is divisible by a higher power of a prime $z_{q,6}$ than $|K|$. In $\Omega_{12}^+(2)$ the only $\mathscr{C}_4$-subgroup is not maximal, by Proposition 2.3.22.

We proved in Lemma 2.3.19 that $K \notin \mathscr{C}_5$. Classes $\mathscr{C}_6$ and $\mathscr{C}_7$, are empty and it follows from Lemma 2.3.20 that $K \notin \mathscr{C}_8$. $\qquad\square$

Recall Definition 2.2.9 of the $\mathscr{C}_4$-subgroups, and in particular that in types $\mathrm{GL}_{n_1}^{\pm}(q) \otimes \mathrm{GL}_{n_2}^{\pm}(q)$ and $\mathrm{Sp}_{n_1}(q) \otimes \mathrm{Sp}_{n_2}(q)$ we assume that $n_1 < n_2$. For each type, the $\mathscr{C}_4$-subgroup H of Ω contains a subgroup $\Omega_1 \times \Omega_2$, acting on a tensor decomposition $V = V_1 \otimes V_2$ with factors of dimensions n_1 and n_2. In detail, H is of one of the following types:

(i) Cases **L** and **U**: $\mathrm{GL}_2^{\pm}(q) \otimes \mathrm{GL}_6^{\pm}(q)$, $\mathrm{GL}_3^{\pm}(q) \otimes \mathrm{GL}_4^{\pm}(q)$;
(ii) Case **S**: $\mathrm{Sp}_2(q) \otimes \mathrm{GO}_6^{\pm}(q)$, $\mathrm{Sp}_4(q) \otimes \mathrm{GO}_3(q)$;
(iii) Case **O**$^+$: $\mathrm{Sp}_2(q) \otimes \mathrm{Sp}_6(q)$, $\mathrm{GO}_3(q) \otimes \mathrm{GO}_4^+(q)$;
(iv) Case **O**$^-$: $\mathrm{GO}_3(q) \otimes \mathrm{GO}_4^-(q)$.

Note that q is odd in Case **S** and for the groups of type $\mathrm{GO}_3(q) \otimes \mathrm{GO}_4^{\pm}(q)$.

Proposition 3.11.8 *Let $n = 12$, let H be a $\mathscr{C}_4$-subgroup of Ω, let G be almost simple with socle $\bar{\Omega}$, and let H_G be the corresponding $\mathscr{C}_4$-subgroup of G. Then H is maximal amongst the geometric subgroups of Ω if and only if one of the following holds:*

(i) *$\Omega = \mathrm{SL}_{12}(q)$ or $\mathrm{SU}_{12}(q)$, and if $n_1 = 2$ then $q > 2$;*
(ii) *$\Omega = \mathrm{Sp}_{12}(q)$, and if $n_1 = 4$ then $q \neq 3$;*
(iii) *$\Omega = \Omega_{12}^+(q)$, H is not of type $\mathrm{GO}_3(q) \otimes \mathrm{GO}_4^+(q)$, and $q > 2$;*
(iv) *$\Omega = \Omega_{12}^-(q)$ and $q \neq 3$.*

If H is of type $\mathrm{GO}_3(q) \otimes \mathrm{GO}_4^+(q)$ and $q \neq 3$, then H_G is maximal amongst the geometric subgroups of G if and only if $G \nleq \Omega_{12}^+(q).\langle \delta, \delta', \phi \rangle$. Otherwise, if H is not maximal then H does not extend to a novel maximal subgroup.

Proof The groups of type $\mathrm{GL}_2^{\pm}(2) \otimes \mathrm{GL}_6^{\pm}(2)$, $\mathrm{Sp}_4(3) \otimes \mathrm{GO}_3(3)$, $\mathrm{Sp}_2(2) \otimes \mathrm{Sp}_6(2)$

and $GO_3(3) \otimes GO_4^\pm(3)$ are shown in Proposition 2.3.22 to be non-maximal, so assume that H is not one of these groups.

Suppose that $H \leqslant K < \Omega$, where K is maximal amongst the geometric subgroups of Ω and is not of the same type as H. We will show that H is of type $GO_3(q) \otimes GO_4^+(q)$ and that there is only one choice of $K \in \mathscr{C}_i$ for $1 \leqslant i \leqslant 8$.

It is immediate from Lemma 2.3.23 that $K \notin \mathscr{C}_1 \cup \mathscr{C}_3$, so suppose that $K \in \mathscr{C}_2$, preserving an imprimitive decomposition $\mathcal{D}$ into t subspaces. If $t = 12$, then $\Omega = SL_{12}^\pm(q)$ or $\Omega_{12}^+(q)$, and K has a unique non-abelian composition factor, namely A_{12}. Unless H is of type $GL_3^\pm(2) \otimes GL_4^\pm(2)$ or $GO_3(q) \otimes GO_4^+(q)$, the group $\overline{\Omega}_2$ is simple and $|\overline{\Omega}_2| \nmid |A_{12}|$, a contradiction. If $t = 12$ and $\Omega = SL_{12}(2)$ then K is not maximal by Proposition 2.3.6 (i), contrary to assumption. Theorem 1.11.2 states that $P(U_4(2)) > 12$. If H is of type $GO_3(q) \otimes GO_4^+(q)$ then by assumption $q > 3$, and $|A_{12}|$ is not divisible by $|L_2(q)|^3$.

If $t = 6$ then $\Omega \neq \Omega_{12}^-(q)$, and the non-abelian composition factors of K are each isomorphic to A_6 or $L_2(q)$. These are smaller than $\overline{\Omega}_2$ in Cases $\mathbf{L}$ and $\mathbf{U}$, type $Sp_2(q) \otimes GO_6^\pm(q)$ in Case $\mathbf{S}$, and type $Sp_2(q) \times Sp_6(q)$ in Case $\mathbf{O}^+$. The composition factors of K are also smaller than $S_4(q)$, eliminating type $Sp_4(q) \otimes GO_3(q)$. For type $GO_3(q) \otimes GO_4^+(q)$, we note that $|A_6|$ is smaller than $|L_2(q)|^2$, since $q > 3$.

If $t \in \{2, 3, 4\}$ then by Lemma 2.3.24 and our assumptions on H, the group H is of type $GL_2^\pm(3) \otimes GL_6^\pm(3)$, type $GL_3^\pm(2) \otimes GL_4^\pm(2)$, type $GL_3^\pm(3) \otimes GL_4^\pm(3)$, type $Sp_2(3) \otimes GO_6^\pm(3)$ or of type $Sp_2(3) \otimes Sp_6(3)$. In each type the derived group of H is absolutely irreducible, so $t \neq 2$. The second derived group of $SU_3(2)$ is absolutely irreducible, so if $\Omega = SU_{12}(2)$ then $t = 4$. But then K is soluble, and H is insoluble. Thus H is not of type $GU_3(2) \otimes GU_4(2)$. Therefore, if $n_1 \neq 2$ then H^∞ is absolutely irreducible, whereas K^∞ is reducible, so assume that $n_1 = 2$. In Cases $\mathbf{L}$, $\mathbf{U}$ and $\mathbf{O}^+$, the group Ω_2 has no non-trivial representations in defining characteristic in dimension at most 4, a contradiction. In Case $\mathbf{S}$ the group Ω_2 is isomorphic to $SL_4(3)$ or $SU_4(3)$, both of which are larger than $Sp_4(3)$, a contradiction. Thus $K \notin \mathscr{C}_2$.

Next assume that $K \in \mathscr{C}_4$, stabilising a tensor decomposition into spaces of dimensions d_1 and d_2. There is only a single family of tensor decompositions in $\Omega_{12}^-(q)$. Suppose first that $n_2 = 6$. For $\Omega = SL_{12}^\pm(q)$, the dimensions $R_p(SL_6^\pm(q))$ are greater than 4 by Theorem 1.11.5, a contradiction. In Case $\mathbf{S}$, the group $\overline{\Omega}_2$ is $O_6^+(q) \cong L_4(q)$ or $O_6^-(q) \cong U_4(q)$. The largest composition factor of K is either the other one of $L_4(q)$, $U_4(q)$, which cannot contain $\overline{\Omega}_2$ by Lagrange's theorem, or is $S_4(q)$, which is smaller than $\overline{\Omega}_2$. In Case $\mathbf{O}^+$, H contains $Sp_6(q)$, whilst the non-abelian composition factors of K are all isomorphic to $L_2(q)$, a contradiction. So $n_2 \neq 6$.

Assume next that H is of type $GL_3^\pm(q) \otimes GL_4^\pm(q)$. Then $d_1 = 2$ and $d_2 = 6$.

The groups Ω_i have a non-trivial representation in $\mathrm{SL}_2(q)$ if and only if Ω_i is soluble so $H^\infty \leqslant 1 \otimes \mathrm{SL}_6^{\pm}(q)$. If $\Omega_1 \neq \mathrm{SU}_3(2)$, these groups are reducible, whereas H^∞ is irreducible, a contradiction. If $\Omega_1 = \mathrm{SU}_3(2)$ then K is not maximal, contrary to assumption. Thus $K \notin \mathscr{C}_4$ in Cases **L** and **U**.

Assume next that H is of type $\mathrm{Sp}_4(q) \otimes \mathrm{GO}_3(q)$, so that $H^\infty \cong \mathrm{Sp}_4(q) \times \mathrm{L}_2(q)$ and $q \geqslant 5$. Here, $K^\infty = \Omega_3 \circ \Omega_4$, where $\Omega_3 = \mathrm{SL}_2(q)$ and $\Omega_4 = \Omega_6^{\pm}(q)$. Since $q \geqslant 5$ is odd, $\mathrm{L}_2(q)$ is not a subgroup of $\mathrm{SL}_2(q)$, and so $\mathrm{L}_2(q)$ must embed into $\Omega_6^{\pm}(q)$, and hence $\mathrm{L}_2(q) \times \mathrm{Sp}_4(q)$ or $\mathrm{L}_2(q) \times \mathrm{S}_4(q)$ is a subgroup of $\Omega_6^{\pm}(q)$, contradicting Lagrange's theorem.

Assume finally that H is of type $\mathrm{GO}_3(q) \otimes \mathrm{GO}_4^+(q)$, so that $q \geqslant 5$, and $H^\infty \cong \mathrm{L}_2(q) \times 2^{\cdot}\mathrm{L}_2(q)^2$, with $K^\infty \cong 2.(\mathrm{L}_2(q) \times \mathrm{S}_6(q))$. It is shown in [66, Proposition 6.3.4] that H is properly contained in two classes of groups of type K, as either of the factors of $\Omega_4^+(q)$ can be combined with $\Omega_3(q)$ to produce a $\mathscr{C}_4$-subgroup of $\mathrm{S}_6(q)$. There are two conjugacy classes in $\Omega_{12}^+(q)$ of groups of the same type as K, with stabiliser $S := \langle \delta, \delta', \phi \rangle$, and a single class of groups of the same type as H. By Lemma 1.12.3, automorphisms that lie in S are in the kernel of the action on the two tensor factors of $\Omega_4^+(q)$, so $H.S \leqslant K.S$. However, automorphisms of $\Omega_{12}^+(q)$ that do not lie in S extend $\Omega_4^+(q)$ to a tensor induced group, so that H is not contained in a group preserving tensor factors of dimensions 2 and 6. This concludes the arguments for $\mathscr{C}_4$.

Finally, $K \notin \mathscr{C}_5$ by Lemma 2.3.25, Classes $\mathscr{C}_6$ and $\mathscr{C}_7$ are empty, and $K \notin \mathscr{C}_8$ by Lemma 2.3.26. So the result follows for H not of type $\mathrm{GO}_3(q) \otimes \mathrm{GO}_4^+(q)$, and for this type we note that in fact we have shown that there is a unique other member K of $\mathscr{C}_i$, for $1 \leqslant i \leqslant 8$, that contains H (without requiring the assumption that K is maximal), so the claims follow regarding the extensions of H that are maximal. $\square$

Recall Definition 2.2.11 of the $\mathscr{C}_5$-subgroups.

Proposition 3.11.9 *Let $n = 12$ and let H be a $\mathscr{C}_5$-subgroup of Ω. Then H is maximal amongst the geometric subgroups of Ω.*

Proof This follows immediately from Proposition 2.3.29. $\square$

Classes $\mathscr{C}_6$ and $\mathscr{C}_7$ are empty. Recall Definition 2.2.17 of the $\mathscr{C}_8$-subgroups.

Proposition 3.11.10 *Let $n = 12$ and let H be a $\mathscr{C}_8$-subgroup of Ω. Then H is maximal amongst the geometric subgroups of Ω.*

Proof This follows immediately from Proposition 2.3.32. $\square$

4

Groups in Class $\mathscr{S}$: cross characteristic

4.1 Preamble

4.1.1 General strategy for determining the candidate $\mathscr{S}$-maximals

We now move on to the determination of the candidates for the almost simple groups in the Aschbacher Class $\mathscr{S}$ (see Definition 2.1.3) that can arise as maximal subgroups of almost simple extensions of simple classical groups of dimension at most 12. As we shall explain in detail in Subsection 4.1.2, these candidates are divided into classes $\mathscr{S}_1$ (cross characteristic) and $\mathscr{S}_2$ (defining characteristic). We shall call subgroups that are maximal among the $\mathscr{S}_1$- and $\mathscr{S}_2$-type subgroups $\mathscr{S}_1$-maximal and $\mathscr{S}_2$-maximal, respectively, and in this and the following chapter we shall determine the $\mathscr{S}_1$-maximal and $\mathscr{S}_2$-maximal subgroups, respectively. (This is not strictly true, because we shall save ourselves some effort by excluding from detailed consideration certain $\mathscr{S}_2$-maximal subgroups that are clearly contained in geometric subgroups of type $\mathscr{C}_4$ or $\mathscr{C}_7$.)

The descriptions of the $\mathscr{S}_1$-maximal subgroups, and their principal properties, are summarised in the final section of this chapter, Section 4.9. So the reader who simply wishes to know the $\mathscr{S}_1$-maximals in the extensions of some specific classical group of dimension up to 12 should look there first.

We start by summarising our methods of finding the candidate $\mathscr{S}$-maximals; more details will follow later. Recall that a group $G = Z^{\cdot}S$ is *quasisimple* if G is perfect, S is non-abelian simple and Z is central, and that by Lemma 1.3.4 for such a G, the group $\operatorname{Aut} G$ can naturally be regarded as a subgroup of $\operatorname{Aut} S$. In particular, the group H^{∞} in the definition (Definition 2.1.3) of a class $\mathscr{S}$ subgroup is quasisimple with cyclic centre Z. To find both the $\mathscr{S}_1$- and the $\mathscr{S}_2$-maximals, the first step is to determine all the quasisimple groups that possess a faithful absolutely irreducible representation of degree at most 12 in prime characteristic p. (It is an elementary result in representation theory that any group with a faithful irreducible representation has a cyclic centre.) The lists in [42] (cross characteristic) and [84] (defining characteristic), together with the ATLAS [12] and the Modular ATLAS [57], are our principal sources of information. The representations of such groups are classified up to representation equivalence. As a side effect, in the cross characteristic case, we obtain a classification of the characteristic 0 representations of such groups as well.

Once we have constructed a list of all such representations of quasisimple groups G, we determine the minimal fields $\mathbb{F}_q$ over which they can be realised. By Proposition 1.8.13, this is just the field generated by the character values.

We then determine the types of the forms A for which the images of the representations are groups of isometries when they are realised over their minimal fields, which includes finding the signs of symmetric bilinear and quadratic forms when appropriate. It is a consequence of Definition 2.1.3 of Class $\mathscr{S}$ and Lemma 1.8.8 that we thereby identify (up to conjugacy) the specific quasisimple classical group Ω in which an extension of $G\rho$ might be an $\mathscr{S}$-maximal subgroup, so that almost simple extensions of $G\rho/Z(G\rho)$ might be $\mathscr{S}$-maximal subgroups of almost simple extensions of $\Omega/Z(\Omega)$. Our definitions of the outer automorphisms of Ω were given in Subsection 1.7.1.

We also determine the action of Out G on the representations ρ, and thereby determine their quasi-equivalence classes (Definition 1.8.4). By Lemmas 1.8.6 and 1.8.10, this determines the conjugacy between, and the normalisers of, the images of the representations in the conformal classical group $C := \mathrm{N}_{\mathrm{GL}_n(q)}(\Omega)$.

So we can restrict our attention to a set of representatives of the quasi-equivalence classes of representations. For each such representative ρ, our first aim is to determine which of the elements of C that normalise $G\rho$ lie in $\Omega\mathbb{F}_q^\times$, and thereby find the normaliser in Ω of $G\rho$. This normaliser then becomes a candidate for an $\mathscr{S}$-maximal subgroup of Ω. To determine which elements of $\mathrm{N}_C(G\rho)$ lie in Ω, we need to calculate the determinants of the normalising elements, and their action on the form A. In the orthogonal cases, we may also need to compute their spinor norms or quasideterminants. It is generally possible to perform such determinant, action, and spinor norm/quasideterminant calculations either by using the information in [12, 57], or by direct computation or, for spinor norm or quasideterminant calculations, using Definition 1.6.10 or Proposition 1.6.11.

Finally, we compute the actions of any graph and field automorphisms of Ω on the representations, and thereby determine the stabiliser of the conjugacy class of $G\rho$ in the full automorphism group of Ω. Let β be a field or graph automorphism of Ω. If β has the same action on a representation ρ as some $\alpha \in \mathrm{Aut}\, G$ then by definition ρ^β and $^\alpha\rho$ are equivalent, and by Lemma 1.8.10 this equivalence is effected by some matrix $g \in C$. So the action of β on $G\rho$, followed by conjugation c_g by g, normalises and induces α on $G\rho$. Therefore the extension of $G/Z(G)$ by α occurs in the almost simple extension of $\bar{\Omega}$ by βc_g. So we need to identify c_g as an element of $\mathrm{Out}\,\bar{\Omega}$, which again involves computing its determinant, its action on the form A, and possibly its spinor norm or quasideterminant.

We remind the reader that, as we demonstrated in [6], for the unitary and orthogonal groups in even dimensions, the definition of the field automorphism ϕ or φ of Ω can depend on the specific form A for which Ω is a group of isometries, and so it is important that the results of our calculations are presented with respect to our chosen standard forms, which were listed in Table 1.1.

We had to carry out some of the computations described above by computer, and, for the most part, we used MAGMA for this purpose. It is straightforward, using standard MAGMA functionality, to do this for specific representations in a given characteristic p. However, many of the cross characteristic representations involved arise as reductions mod p of a characteristic 0 representation, and these occur for all but finitely many p. To perform the calculations generically for almost all primes p, we needed in some cases to construct the representations explicitly over a number field, and to deduce the behaviour mod p from the results of calculations in the characteristic 0 representation. This was necessary, for example, for calculating the signs of the symmetric bilinear or quadratic forms for which $G\rho$ is a group of isometries, for spinor norm calculations, and for identifying and studying the elements c_g described above. We remind the reader that the files of MAGMA calculations that we refer to are available on the webpage http://www.cambridge.org/9780521138604.

As remarked above, for representations over finite fields, the minimal field over which a representation ρ can be represented is generated by the character values of ρ, and these calculations can be carried out most effectively if the same is true for the corresponding characteristic 0 representation. This is the case if and only if the *Schur index* [19, Section 41, Page 292] of the representation is 1. Fortunately, this was the case for all of those representations that we actually needed to construct. (The smallest-dimensional example that we know of a quasisimple group with indicator $+$ or $\circ$ and Schur index not 1 is the 336-dimensional representation of J_2. It is an easy consequence of the Brauer–Speiser Theorem [20, (74.27)] that representations with indicator $-$ have Schur index 2, but it turned out that there was no need for us to construct any of these representations.)

4.1.2 Classes $\mathscr{S}_1$ and $\mathscr{S}_2$

The groups in Class $\mathscr{S}$ are divided into two subclasses, $\mathscr{S}_1$ and $\mathscr{S}_2$. To define these subclasses, we need a precise definition of the characteristic of a group of

Table 4.1 *Groups of Lie type in more than one characteristic, or which are alternating groups, plus possible cases of confusion*

Group	Characteristics for which it is group of Lie type
$A_5 \cong L_2(4) \cong L_2(5)$	2 and 5
$L_3(2) \cong L_2(7)$	2 and 7
$A_6 \cong L_2(9) \cong S_4(2)'$	3 (but not 2)
$A_8 \cong L_4(2)$	2
$U_4(2) \cong S_4(3)$	2 and 3
$L_2(8) \cong R(3)' = {}^2G_2(3)'$	2 (but not 3)
$U_3(3) \cong G_2(2)'$	3 (but not 2)
${}^2F_4(2)'$	none (not even 2)

Lie type, which we now present. This material is standard, and reader who is unfamiliar with it could consult [91], for example.

Most (by any reasonable measure) finite simple groups are groups of Lie type, and this includes all the classical groups. These have symbols ${}^tX_n(q)$ where $t \in \{1, 2, 3\}$ (and is related to the symmetries of the associated Dynkin diagram), $n \geqslant 1$ is an integer, $X \in \{A, B, C, D, E, F, G\}$ (these symbols are derived from the standard notation for the simple complex Lie algebras), and $q = p^e > 1$ is a power of the prime p, with various restrictions on the allowed combinations of t, X, n and q. The groups with $t = 1$ are the *Chevalley groups*, and we write $X_n(q)$ instead of ${}^1X_n(q)$ in this case. For these groups we have $X_n \in \{A_n, B_n, C_n, D_n, E_6, E_7, E_8, F_4, G_2\}$ where there are no further restrictions on n and q other than those given above, except that the case D_1 does not occur.

Let

$${}^tX_n \in \{A_n, B_n, C_n, D_n, E_6, E_7, E_8, F_4, G_2, {}^2A_n, {}^2D_n, {}^3D_4, {}^2E_6, {}^2B_2, {}^2F_4, {}^2G_2\}.$$

Then we consider ${}^tX_n(q)$ to be a group of Lie type in characteristic p if ${}^tX_n(q)$ is simple. Thus the sporadic groups and the groups A_7, A_m for $m \geqslant 9$, and ${}^2F_4(2)'$ are not groups of Lie type in any characteristic. In Table 4.1 we present all groups of Lie type that have more than one characteristic, and also all alternating groups that are groups of Lie type.

The remaining simple groups are groups of Lie type in precisely one characteristic. If H is a simple group of Lie type in characteristic p, then a quasisimple group $G = Z \cdot H$ is also considered to be a group of Lie type in characteristic p provided that $p \nmid |Z|$.

Definition 4.1.1 Let G be a subgroup in Class $\mathscr{S}$ of a classical group C in characteristic p. Then G lies in *Class $\mathscr{S}_2$* of C if G^∞ is isomorphic to a group of Lie type in characteristic p, and G lies in *Class $\mathscr{S}_1$* otherwise. Class $\mathscr{S}_1$ is the *cross characteristic* case, and Class $\mathscr{S}_2$ is the *defining characteristic* case. We say that a subgroup G of a classical group C is $\mathscr{S}_i$-*maximal* (with $i \in \{1,2\}$) if G is maximal amongst the $\mathscr{S}_i$-subgroups of C.

Cross characteristic representations are often (but not always) p-modular reductions of characteristic 0 representations. Moreover, groups arising in this class are usually defined over relatively low degree extensions of $\mathbb{F}_p$. For a given dimension the set of orders of the cross characteristic candidates is bounded above; in dimension up to 12 we shall see in Section 4.3 that the largest quasisimple cross characteristic candidate is 6ʻSuz, with order 2 690 072 985 600. In contrast, the candidates in defining characteristic have unbounded order, and require arbitrarily large extensions of $\mathbb{F}_p$ in order to write their representations.

The remainder of this chapter is devoted to classifying the $\mathscr{S}_1$-maximal subgroups of the classical groups in dimension up to 12: we shall handle the defining characteristic cases in Chapter 5. The chapter is structured as follows. We start, in Section 4.2, by enumerating the algebraic irrationalities that occur in the (Brauer) characters of the representations that will arise, and establishing a few of their elementary properties. In Section 4.3 we use [42, 12, 57] to produce a lengthy table containing a complete list of the required representations of quasisimple $\mathscr{S}_1$-candidates in dimensions up to 12. In Section 4.4, we describe how to calculate the normaliser of the quasisimple $\mathscr{S}_1$-subgroups in both the Ω-group and the conformal group, and the number of Ω-classes: details of these calculations for the candidates in dimension up to six are given in Section 4.5. Then in Section 4.6 we describe how to calculate the action on the Ω-classes of field and graph automorphisms, and in Section 4.7 we carry out these calculations in detail in dimension up to 6. Recall Definition 2.1.4 of a containment between two subgroups of a linear group. In Section 4.8 we analyse containments between the $\mathscr{S}_1$-subgroups, ultimately determining all $\mathscr{S}_1$-maximal subgroups in dimension up to 6. Having established and illustrated the techniques for performing all of the necessary calculations, we carry them out in slightly less detail for candidates in dimensions 7–12 in Section 4.9. Finally, in Section 4.10 we present a complete summary of the results of the chapter, with references back to where they are proved.

In addition to our principal source [42], there is a huge volume of literature on low-dimensional representations of quasisimple groups in cross characteristic, which we is too extensive to be adequately summarised here. The lower bounds on the degrees of representations of Chevalley groups established in [75], which were improved in [101, 47], were of particular significance. Other such papers of specific relevance to our work include [7, 34, 35, 36, 37, 38, 43, 72, 109], and the last of these is a useful survey paper. The methods described in Section 4.4 that we use to compute stabilisers of $\mathscr{S}_1$-subgroups in $\mathrm{GL}_n(q)$ in Case

L, are similar to those used in [73], where representations of quasisimple groups in coprime characteristic in dimensions 13 to 27 are considered, and $GL_n(q)$-conjugacy classes are determined.

4.2 Irrationalities

Our notation for algebraic irrationalities follows that in the ATLAS [12]. All such irrationalities are sums of roots of unity. Below we shall only define those irrationalities that are needed in this book; the ATLAS defines many more.

We use i to denote a fixed square root of -1. We define z_n (a particular primitive complex nth root of 1) as:

$$z_n := \exp(2\pi i/n) = \cos(2\pi/n) + i.\sin(2\pi/n).$$

Notice the identities $z_1 = 1$, $z_2 = -1$, and $z_4 = i$; we sometimes denote z_3 as ω.

For $n > 1$ odd, we then use z_n to define the number b_n:

$$b_n := \sum_{r=1}^{(n-1)/2} z_n^{r^2} = \frac{1}{2} \sum_{r=1}^{n-1} z_n^{r^2}.$$

Notice that $z_3 = b_3$. Gauss considered sums of this form, and proved that if $n \equiv 1 \pmod 4$ then $b_n = (-1 + \sqrt{n})/2$, whilst if $n \equiv 3 \pmod 4$ then $b_n = (-1 + i\sqrt{n})/2$. Thus, if $n \equiv 1 \pmod 4$ then b_n has minimal polynomial $X^2 + X - \frac{1}{4}(n-1)$ over $\mathbb{Q}$; whilst if $n \equiv 3 \pmod 4$ then b_n has minimal polynomial $X^2 + X + \frac{1}{4}(n+1)$ over $\mathbb{Q}$.

Next we use z_4 and z_8 to fix a square root of 2, namely $r_2 := (1 + z_4)/z_8$. For $n > 2$, we may then use b_n, r_2 and positive integer square roots to fix square roots $r_n = \sqrt{n}$ and $i_n = \sqrt{-n} = i.r_n$. Then $r_n := 1 + 2b_n$ if $n \equiv 1 \pmod 4$, and $i_n := 1 + 2b_n$ if $n \equiv 3 \pmod 4$.

Another useful irrationality is c_n for $n \equiv 1 \pmod 6$ and n prime, which is defined as $c_n := \frac{1}{3} \sum_{r=1}^{n-1} z_n^{r^3}$. We also define $y_n := z_n + z_n^{-1} = 2\cos(2\pi/n)$. It is a useful exercise for the reader to verify that $y_1 = 2$, $y_2 = -2$, $y_3 = -1$, $y_4 = 0$, $y_5 = b_5$, $y_6 = 1$, $y_7 = c_7$, $y_8 = r_2$ and $y_{12} = r_3$.

If θ is a quadratic irrationality, its non-trivial algebraic conjugate is easy to write down: if θ is a b_n irrationality this conjugate (denoted b_n^* if b_n is real and irrational, or b_n^{**} if b_n is not real) is $-1 - \theta$, while if θ is an i_n or r_n irrationality (including i) this conjugate is $-\theta$.

We next discuss p-modular reduction, namely the interpretation of these irrationalities as elements of finite fields. Fix a primitive multiplicative element $\omega_{p,e}$ of $\mathbb{F}_{p^e}^\times$. For all reasonably small fields we can choose $\omega_{p,e}$ to be a root of the Conway polynomial for that field; see Subsection 1.4.1. When n divides $p^e - 1$

but not $p^i - 1$ for $i < e$, we define the element z_n in characteristic p to be the smallest power of $\omega_{p,e}$ that has multiplicative order n. We define i to be z_4, and then follow the opening paragraphs of this section in defining b_n, r_n, i_n, c_n and y_n in terms of i and z_n. Note that z_n, b_n, c_n and y_n are undefined when $p \mid n$, and r_n and i_n are undefined when $p \mid n$ and when $p = 2$.

For n odd, b_n is equal to $\frac{1}{2}(-1 + \sqrt{\varepsilon n})$, where the sign $\varepsilon \in \{1, -1\}$ is such that $\varepsilon n \equiv 1 \pmod 4$. Now, $\varepsilon n = \prod_{i=1}^{r} \varepsilon_i q_i$, where the q_i are prime natural numbers and $\varepsilon_i q_i \equiv 1 \pmod 4$ for all i. Recall the Legendre symbol, and the basic facts regarding its behaviour, established in Section 1.13. For p odd, and not dividing n, the field element corresponding to b_n lies in $\mathbb{F}_p$ if and only if εn is a (non-zero) square modulo p. This occurs if and only if

$$\left(\frac{\varepsilon n}{p}\right) = \prod_{i=1}^{r} \left(\frac{\varepsilon_i q_i}{p}\right) = \prod_{i=1}^{r} \left(\frac{p}{q_i}\right) = 1,$$

by Proposition 1.13.7 (iii). For $p = 2$, we consider whether the minimal polynomial of b_n (whichever of $X^2 + X + \frac{\pm n + 1}{4}$ is in $\mathbb{Z}[X]$) factors into linear factors when reduced modulo 2. This is the case (and hence b_n is in $\mathbb{F}_2$) if and only if $n \equiv \pm 1 \pmod 8$, by Proposition 1.13.7 (ii) .

As an example, $b_{15} \in \mathbb{F}_p$ if and only if $p \neq 3, 5$ and $\left(\frac{-15}{p}\right) = \left(\frac{p}{3}\right)\left(\frac{p}{5}\right) = 1$. Now $\left(\frac{p}{3}\right) = \left(\frac{p}{5}\right) = 1$ if and only if $p \equiv 1 \pmod 3$ and $p \equiv 1, 4 \pmod 5$, that is $p \equiv 1, 4 \pmod{15}$. And $\left(\frac{p}{3}\right) = \left(\frac{p}{5}\right) = -1$ if and only if $p \equiv 2 \pmod 3$ and $p \equiv 2, 3 \pmod 5$, that is $p \equiv 2, 8 \pmod{15}$. So we see that b_{15} is in $\mathbb{F}_p$ if and only if $p \equiv 1, 2, 4, 8 \pmod{15}$.

We write $\mathbb{F}_q(y_n)$ to mean the smallest extension of $\mathbb{F}_q$ containing y_n, and $y_n \in \mathbb{F}_q$ will mean that $\mathbb{F}_q(y_n) = \mathbb{F}_q$.

Lemma 4.2.1 *If the prime p does not divide n and q is a power of p, then $y_n \in \mathbb{F}_q$ if and only if $q \equiv \pm 1 \pmod n$.*

Proof If $y_n = z_n + z_n^{-1} \in \mathbb{F}_q$, then z_n satisfies a quadratic equation over $\mathbb{F}_q$ and hence $z_n \in \mathbb{F}_{q^2}$. So $z_n + z_n^{-1} = z_n^q + z_n^{-q}$ and so, multiplying by z_n^q,

$$z_n^{2q} - z_n^{q+1} - z_n^{q-1} + 1 = (z_n^{q+1} - 1)(z_n^{q-1} - 1) = 0$$

and hence $q \equiv \pm 1 \pmod n$.

Conversely, if $q \equiv 1 \pmod n$ then $z_n \in \mathbb{F}_q$, so $y_n \in \mathbb{F}_q$, whereas if $q \equiv -1 \pmod n$ then $z_n \in \mathbb{F}_{q^2} \setminus \mathbb{F}_q$, and $y_n = z_n + z_n^{-1} = z_n + z_n^q \in \mathbb{F}_q$. □

In Tables 4.2 and 4.3 we present information about the algebraic irrationalities that we shall need; i.e. those that arise in character values of the cross characteristic candidates in Table 4.4. In column "Irrat" we give the name of the irrationality θ, and in Column "Real" we indicate whether θ is a real or complex number. Column "Deg" (only present in Table 4.3) gives the degree of the extension field $\mathbb{Q}(\theta)$: note that in Table 4.2 this number is always 2. Column "Cyc" gives the minimal n such that θ is a sum of n-th roots of unity. The minimal polynomial of θ over $\mathbb{Q}$ follows next, and finally in Column "p-modular reductions" we give congruences on primes p such that θ and all of its conjugates lie in $\mathbb{F}_{p^\alpha}$ for some α (with α minimised over all conjugates). For the irrationalities given in Tables 4.2 and 4.3, the p-modular reductions of the irrationality and each of its conjugates generate the same field.

To calculate the p-modular reductions in Table 4.2 we have made repeated use of Theorem 1.13.8. In Table 4.3 we include the irrationalities c_{13} and c_{19}, which we shall only need in characteristics 5 and 11, respectively, so we only give information about their behaviour in these specific characteristics. Lemma 4.2.1 can be used to determine the entries in Column "p-modular reductions" of Table 4.3 for the irrationalities y_n.

Table 4.2: Irrationality tables: quadratic irrationalities

Irrat	Real	Cyc	Min poly	p-modular reductions
z_3	no	3	X^2+X+1	Deg 1: $p \equiv 0, 1$ (3)
				Deg 2: $p \equiv 2$ (3)
b_5	yes	5	X^2+X-1	Deg 1: $p \equiv 0, 1, 4$ (5)
				Deg 2: $p \equiv 2, 3$ (5)
b_7	no	7	X^2+X+2	Deg 1: $p \equiv 0, 1, 2, 4$ (7)
				Deg 2: $p \equiv 3, 5, 6$ (7)
b_{11}	no	11	X^2+X+3	Deg 1: $p \equiv 0, 1, 3, 4, 5, 9$ (11)
				Deg 2: $p \equiv 2, 6, 7, 8, 10$ (11)
b_{13}	yes	13	X^2+X-3	Deg 1: $p \equiv 0, 1, 3, 4, 9, 10, 12$ (13)
				Deg 2: $p \equiv 2, 5, 6, 7, 8, 11$ (13)
b_{15}	no	15	X^2+X+4	Deg 1: $p \equiv 3, 5, 1, 2, 4, 8$ (15)
				Deg 2: $p \equiv 7, 11, 13, 14$ (15)
b_{17}	yes	17	X^2+X-4	Deg 1: $p \equiv 0, 1, 2, 4, 8, 9, 13, 15, 16$ (17)
				Deg 2: $p \equiv 3, 5, 6, 7, 10, 11, 12, 14$ (17)
b_{19}	no	19	X^2+X+5	Deg 1: $p \equiv 0, 1, 4, 5, 6, 7, 9, 11, 16, 17$ (19)
				Deg 2: $p \equiv 2, 3, 8, 10, 12, 13, 14,$
				$15, 18$ (19)
b_{21}	yes	21	X^2+X-5	Deg 1: $p \equiv 3, 7, 1, 4, 5, 16, 17, 20$ (21)
				Deg 2: $p \equiv 2, 8, 10, 11, 13, 19$ (21)

Table 4.2: Irrationality tables: quadratic irrationalities

Irrat	Real	Cyc	Min poly	p-modular reductions
b_{23}	no	23	X^2+X+6	Deg 1: $p \equiv 0,1,2,3,4,6,8,9,12,13,$
				$16,18\ (23)$
				Deg 2: $p \equiv 5,7,10,11,14,15,17,19,$
				$20,21,22\ (23)$
i	no	4	X^2+1	Deg 1: $p \equiv 2,1\ (4)$
				Deg 2: $p \equiv 3\ (4)$
i_2	no	8	X^2+2	Deg 1: $p \equiv 2,1,3\ (8)$
				Deg 2: $p \equiv 5,7\ (8)$
i_5	no	20	X^2+5	Deg 1: $p \equiv 2,5,1,3,7,9\ (20)$
				Deg 2: $p \equiv 11,13,17,19\ (20)$
r_2	yes	8	X^2-2	Deg 1: $p \equiv 2,1,7\ (8)$
				Deg 2: $p \equiv 3,5\ (8)$
r_3	yes	12	X^2-3	Deg 1: $p \equiv 2,3,1,11\ (12)$
				Deg 2: $p \equiv 5,7\ (12)$
r_6	yes	24	X^2-6	Deg 1: $p \equiv 2,3,1,5,19,23\ (24)$
				Deg 2: $p \equiv 7,11,13,17\ (24)$

Table 4.3: Irrationality tables: non-quadratic irrationalities

Irrat	Real	Deg	Cyc	Min poly	p-modular reductions
c_{13}	yes	3	13	X^3+X^2-4X+1	$c_{13} \in \mathbb{F}_5$
c_{19}	yes	3	19	X^3+X^2-6X-7	$c_{19} \in \mathbb{F}_{11}$
y_7	yes	3	7	X^3+X^2-2X-1	Deg 1: $p \equiv 0,1,6\ (7)$
					Deg 3: $p \equiv 2,3,4,5\ (7)$
y_9	yes	3	9	X^3-3X+1	Deg 1: $p \equiv 3,1,8\ (9)$
					Deg 3: $p \equiv 2,4,5,7\ (9)$

4.3 Cross characteristic candidates

In this section we calculate and tabulate the information that we shall require concerning the cross characteristic representations of quasisimple groups in dimensions up to 12. Representations in defining characteristics will be considered in Chapter 5, and are deliberately omitted from this chapter. Our main source for the information in this section is Theorem 4.3.1, the main result of [42].

Theorem 4.3.1 *Let G be a quasisimple finite group, and let V be an absolutely irreducible faithful FG-module of dimenion $d \leqslant 250$. If G is a group of Lie type, assume that the characteristic of F is not the defining characteristic of G. Then the values of (G, d), together with the Frobenius-Schur indicator of the representation and its character field, are contained in [42, Tables 2 and 3].*

We have also used the information in the ATLAS [12] and the Modular ATLAS [57]. We remind the reader that the modular character tables in [57] contain the irreducible *Brauer characters* (see, for example, [19, Page 588] or [56, Chapter 15]) of the groups G in characteristics dividing the group order. For $g \in G$ and an absolutely irreducible representation ρ in characteristic p, $\text{tr}(g\rho)$ is a sum $\sum \overline{w}_i$ of roots of unity, and the Brauer character $\chi(g)$ is equal to $\sum w_i$, where w_i is a complex root of unity that maps onto $\overline{w}_i$, as described earlier in Section 4.2. For primes p not dividing $|G|$, the Brauer character is equal to an ordinary character, and can be found in [12].

Table 4.4 contains the information that we require on the absolutely irreducible cross characteristic representations of quasisimple groups in dimensions up to 12. We need to explain our convention for when we include two representations ρ_1 and ρ_2 of a group G on the same row of Table 4.4.

In Section 1.8 we saw how automorphisms of G and F define actions on the representations of G over the field F.

In the case of complex representations, we recall the standard definition from representation theory that ρ_1 and ρ_2 are said to be *algebraically conjugate* if ρ_1^σ is equivalent to ρ_2 for some automorphism σ of $\mathbb{C}$.

We extend this concept to characteristic p representations in the following somewhat arbitrary and non-standard fashion. If ρ_1 and ρ_2 are absolutely irreducible and can both be defined as reductions modulo p of the absolutely irreducible complex representations ρ_1' and ρ_2', then ρ_1 and ρ_2 are algebraically conjugate if and only if ρ_1' and ρ_2' are. (This turns out to be independent of how exactly we define reduction modulo p, which depends on the choice of a maximal ideal above the ideal (p) in a suitable ring.)

If ρ_1 and ρ_2 are absolutely irreducible representations over a finite field F of characteristic p, and they do not arise as reductions modulo p of absolutely irreducible complex representations, then we shall say that they are algebraically conjugate if ρ_1^σ is equivalent to ρ_2 for some automorphism σ of F.

Definition 4.3.2 Two representations ρ_1 and ρ_2 of G are said to be *weakly equivalent* if one can be obtained from the other by application of group automorphisms, algebraic conjugacy, and duality of modules.

Hence quasi-equivalent representations (see Definition 1.8.4) are weakly equivalent. Our results are listed in Table 4.4, each row of which describes a class of weakly equivalent faithful absolutely irreducible representations.

How to read Table 4.4. The column "Group" gives the name of the quasisimple group G in ATLAS [12] notation.

The column "PmDivs" lists the prime divisors of $|G|$. If G is a group of Lie type then the defining characteristics of G are in bold.

Column "Dim" states the dimension (or degree) of the representations.

The column "Ind" refers to the Frobenius–Schur indicator (see, for example, [20, Page 725]). This is equal to ○ when the image of the representation preserves a unitary form or only an identically zero form, − when the image of the representation preserves a symplectic form but no quadratic form, and + when it preserves a quadratic form. We shall describe in Subsection 4.4.1 how to determine whether a representation with indicator ○ preserves a unitary form.

The column "#ρ" counts the number of equivalence classes of representations (or characters) that are represented by that line of the table.

Column "Stab" defines the stabiliser in Out G of one of the representations described in that line, by specifying its structure or generators.

Column "Charc" gives the characteristics over which the representation occurs, where by 0 we mean all primes that do not divide $|G|$. The bracketed [2,] in this entry for a 10-dimensional representation of $L_2(11)$ indicates that this representation is equivalent to the one in the line below in characteristic 2.

To determine the minimal field size for these representations on reduction modulo p, we require the *character ring* (i.e. the ring generated by the character values) of the representation, as an extension of $\mathbb{Z}$. Column "Ch Ring" lists algebraic irrationalities that generate the character ring over $\mathbb{Z}$. See Table 4.2 for the values of the irrationalities that occur, including their minimal polynomials.

The table is ordered first by degree, then by Frobenius–Schur indicator, then by order of the simple group, and then by order of the quasisimple group.

How Table 4.4 was calculated. Rather than giving all details of the calculations, which would be lengthy and repetitive, we describe how to carry them out and give a few examples.

Columns "Group", "PmDivs", "Deg" and "Ind" are in [42]. Column "#ρ", which is the number of inequivalent representations in the general linear group that are described by that row of the table. Several non-weakly-equivalent representations of a group can be listed in a single line in [42]. This does not in fact occur in dimensions up to 12 but, for example, if we were dealing with dimension 14 the two rational degree 14 representations of A_7 occupy one line of [42] but would appear twice in our list.

As an example of how to calculate the "$\#\rho$" value, we consider the 6-dimensional representation of $6\text{'}A_7$, which occurs in characteristics 0, 5 and 7. First, we note that in [12] and [57] there are four characters of degree 6 for $6\text{'}A_7$. These are all algebraically conjugate, with character field $K(z_3, r_2)$, where K is $\mathbb{Q}$, $\mathbb{F}_5$ or $\mathbb{F}_7$, so all four representations are in one row of our table. These representations are swapped in pairs by the outer automorphism α of $6\text{'}A_7$, since α acts in the same way as the automorphism $(z_3, r_2) \mapsto (z_3^{**}, -r_2)$ of $K(z_3, r_2)$ on these representations.

Similarly, for Column "Stab" of our $6\text{'}A_7$ example, [12] and [57] show that $6\text{'}S_7$ has no 6-dimensional representations in characteristic 0. Since $6\text{'}A_7$ has only one non-trivial outer automorphism, the entry for "Stab" is 1.

As a second example, consider the 5-dimensional representations of A_6, which occur in characteristics 0 and 5. There are two such representations in [12] and [57]. They are interchanged by the $.2_2$ and $.2_3$ automorphisms of A_6 and are stabilised by the $.2_1$ automorphism. Thus the two representations form a single row of our table, with "Stab" entry 2_1.

For almost all of the entries in Table 4.4, the character tables in [12] and [57] can be used to calculate the number of representations as we have just explained. The exceptions are:

(i) A_{13} in dimension 11 with $p = 13$;
(ii) A_{13} in dimension 12 with $p = 2, 3, 5, 7$ or 11;
(iii) A_{14} in dimension 12 with $p = 2$ or 7;
(iv) $2\text{'}Suz$ in dimension 12 with $p = 3$;
(v) $3\text{'}Suz$ in dimension 12 with $p = 2$;
(vi) $6\text{'}Suz$ in dimension 12 with $p = 5, 7, 11$ or 13.

None of these Brauer character tables are in [57], but they are all available in the GAP library of (Brauer) character tables, and can be accessed from within GAP. For example, for $3\text{'}Suz$ in dimension 12 with $p = 2$, the GAP commands

```
C:=CharacterTable("3.Suzmod2");
CharacterDegrees(C);
```

reveal that there are exactly two Brauer characters of this degree with $p = 2$.

The tables in [42] give generators for the character *field* of the representation, whereas we require the character ring. This can be read straightforwardly from the character values in [12] and [57].

In conclusion, we have illustrated how we have proved the following theorem.

Theorem 4.3.3 *Let S be an $\mathscr{S}_1$-subgroup of a classical group C in dimension at most 12. Then S^∞ is contained in Table 4.4.*

Table 4.4: Cross characteristic candidates

Group	PmDivs	Deg	Ind	#ρ	Stab	Charc	Ch Ring
$2^{\cdot}A_5$	$\mathbf{2,3,5}$	2	$-$	2	1	$0,3$	b_5
$L_3(2)$	$2,3,\mathbf{7}$	3	$\circ$	2	1	$0,3$	b_7
$3^{\cdot}A_6$	$2,\mathbf{3},5$	3	$\circ$	4	1	$0,2$	z_3,b_5
$3^{\cdot}A_6$	$2,\mathbf{3},5$	3	$\circ$	2	2_3	5	z_3
$3^{\cdot}A_7$	$2,3,5,7$	3	$\circ$	2	1	5	z_3,b_7
A_5	$\mathbf{2,3,5}$	3	$+$	2	1	$0,3$	b_5
$2^{\cdot}L_3(2)$	$\mathbf{2,3,7}$	4	$\circ$	2	1	$0,3$	b_7
A_7	$2,3,5,7$	4	$\circ$	2	1	2	b_7
$2^{\cdot}A_7$	$2,3,5,7$	4	$\circ$	2	1	$0,3,5$	b_7
$4_2{\cdot}L_3(4)$	$\mathbf{2,3,5,7}$	4	$\circ$	2	2_2	3	i,b_7
$2^{\cdot}U_4(2)$	$\mathbf{2,3,5}$	4	$\circ$	2	1	$0,5$	z_3
$2^{\cdot}A_5$	$\mathbf{2,3,5}$	4	$-$	1	2	0	—
A_6	$2,\mathbf{3},5$	4	$-$	2	2_1	2	—
$2^{\cdot}A_6$	$2,\mathbf{3},5$	4	$-$	2	2_1	$0,5$	—
$2^{\cdot}A_7$	$2,3,5,7$	4	$-$	1	2	7	—
A_5	$\mathbf{2,3,5}$	4	$+$	1	2	$0,3$	—
$L_2(11)$	$2,3,5,\mathbf{11}$	5	$\circ$	2	1	$0,2,3,5$	b_{11}
M_{11}	$2,3,5,11$	5	$\circ$	2	1	3	i_2,b_{11}
$U_4(2)$	$\mathbf{2,3,5}$	5	$\circ$	2	1	$0,5$	z_3
A_5	$\mathbf{2,3,5}$	5	$+$	1	2	0	—
A_6	$2,\mathbf{3},5$	5	$+$	2	2_1	$0,5$	—
A_7	$2,3,5,7$	5	$+$	1	2	7	—
$3^{\cdot}A_6$	$2,\mathbf{3},5$	6	$\circ$	2	2_3	0	z_3
						5	$2z_3$
$6^{\cdot}A_6$	$2,\mathbf{3},5$	6	$\circ$	4	1	0	z_3,r_2
						5	$2z_3,r_2$

Table 4.4: Cross characteristic candidates

Group	PmDivs	Deg	Ind	$\#\rho$	Stab	Charc	Ch Ring
$2 \dot{} L_2(11)$	$2, 3, 5, \mathbf{11}$	6	○	2	1	$0, 3, 5$	b_{11}
$3 \dot{} A_7$	$2, 3, 5, 7$	6	○	2	1	$0, 2, 5, 7$	z_3
$6 \dot{} A_7$	$2, 3, 5, 7$	6	○	4	1	$0, 5, 7$	z_3, r_2
$6 \dot{} L_3(4)$	$\mathbf{2}, 3, 5, 7$	6	○	2	2_1	$0, 5, 7$	z_3
$2 \dot{} M_{12}$	$2, 3, 5, 11$	6	○	2	1	3	i_2, i_5, b_{11}
$3 \dot{} M_{22}$	$2, 3, 5, 7, 11$	6	○	2	1	2	z_3, b_{11}
$3_1 \dot{} U_4(3)$	$2, \mathbf{3}, 5, 7$	6	○	2	2_2	2	z_3
$6_1 \dot{} U_4(3)$	$2, \mathbf{3}, 5, 7$	6	○	2	2_2	$0, 5, 7$	z_3
$2 \dot{} A_5$	$\mathbf{2, 3, 5}$	6	−	1	2	$0, 3$	—
$2 \dot{} L_3(2)$	$\mathbf{2, 3, 7}$	6	−	2	2	$0, 3$	r_2
$L_2(13)$	$2, 3, 7, \mathbf{13}$	6	−	2	1	2	b_{13}
$2 \dot{} L_2(13)$	$2, 3, 7, \mathbf{13}$	6	−	2	1	$0, 3, 7$	b_{13}
$2 \dot{} A_7$	$2, 3, 5, 7$	6	−	2	1	3	r_2
$U_3(3)$	$2, \mathbf{3}, 7$	6	−	1	2	$0, 2, 7$	—
J_2	$2, 3, 5, 7$	6	−	2	1	2	b_5
$2 \dot{} J_2$	$2, 3, 5, 7$	6	−	2	1	$0, 3, 7$	b_5
$2 \dot{} J_2$	$2, 3, 5, 7$	6	−	1	2	5	—
$L_3(2)$	$\mathbf{2, 3, 7}$	6	+	1	2	$0, 3$	—
A_7	$2, 3, 5, 7$	6	+	1	2	$0, 2, 3, 5$	—
$2 \dot{} L_3(4)$	$\mathbf{2}, 3, 5, 7$	6	+	1	2^2	3	—
$U_4(2)$	$\mathbf{2, 3}, 5$	6	+	1	2	$0, 5$	—
$U_3(3)$	$2, \mathbf{3}, 7$	7	○	2	1	$0, 7$	i
$L_3(2)$	$\mathbf{2}, 3, \mathbf{7}$	7	+	1	2	$0, 3$	—
$L_2(8)$	$2, 3, 7$	7	+	1	3	$0, 3, 7$	—
$L_2(8)$	$2, 3, 7$	7	+	3	1	$0, 7$	y_9
$L_2(13)$	$2, 3, 7, \mathbf{13}$	7	+	2	1	$0, 3, 7$	b_{13}
$U_3(3)$	$2, \mathbf{3}, 7$	7	+	1	2	$0, 7$	—
A_8	$\mathbf{2}, 3, 5, 7$	7	+	1	2	$0, 3, 5, 7$	—
J_1	$2, 3, 5,$ $7, 11, 19$	7	+	1	1	11	b_5, c_{19}

Table 4.4: Cross characteristic candidates

Group	PmDivs	Deg	Ind	#ρ	Stab	Charc	Ch Ring
A_9	$2, 3, 5, 7$	7	$+$	1	2	3	—
$S_6(2)$	$\mathbf{2}, 3, 5, 7$	7	$+$	1	1	$0, 3, 5, 7$	—
$4_1\!\cdot\!L_3(4)$	$\mathbf{2}, 3, 5, 7$	8	$\circ$	4	2_3	$0, 3, 7$	i, b_5
$4_1\!\cdot\!L_3(4)$	$\mathbf{2}, 3, 5, 7$	8	$\circ$	2	2_3	5	i
$2\!\cdot\!L_3(2)$	$\mathbf{2}, 3, \mathbf{7}$	8	$-$	1	2	0	—
$2\!\cdot\!A_6$	$2, \mathbf{3}, 5$	8	$-$	2	2_2	0	b_5
$L_2(17)$	$2, 3, \mathbf{17}$	8	$-$	2	1	2	b_{17}
$2\!\cdot\!L_2(17)$	$2, 3, \mathbf{17}$	8	$-$	2	1	$0, 3$	b_{17}
A_{10}	$2, 3, 5, 7$	8	$-$	1	2	2	—
$L_3(2)$	$\mathbf{2}, 3, \mathbf{7}$	8	$+$	1	2	0	—
A_6	$2, \mathbf{3}, 5$	8	$+$	2	2_2	$0, 2$	b_5
A_6	$2, \mathbf{3}, 5$	8	$+$	1	2^2	5	—
$L_2(8)$	$\mathbf{2}, 3, 7$	8	$+$	1	3	$0, 7$	—
A_7	$2, 3, 5, 7$	8	$+$	1	2	5	—
$2\!\cdot\!A_8$	$\mathbf{2}, 3, 5, 7$	8	$+$	1	2	$0, 3, 5, 7$	—
$2\!\cdot\!Sz(8)$	$\mathbf{2}, 5, 7, 13$	8	$+$	1	1	5	c_{13}
A_9	$2, 3, 5, 7$	8	$+$	1	2	$0, 2, 5, 7$	—
A_9	$2, 3, 5, 7$	8	$+$	2	1	2	—
$2\!\cdot\!A_9$	$2, 3, 5, 7$	8	$+$	2	1	$0, 5, 7$	—
$2\!\cdot\!A_9$	$2, 3, 5, 7$	8	$+$	1	2	3	—
$2\!\cdot\!S_6(2)$	$\mathbf{2}, 3, 5, 7$	8	$+$	1	1	$0, 3, 5, 7$	—
A_{10}	$2, 3, 5, 7$	8	$+$	1	2	5	—
$2\!\cdot\!A_{10}$	$2, 3, 5, 7$	8	$+$	2	1	5	r_6, b_{21}
$2\!\cdot\!O_8^+(2)$	$\mathbf{2}, 3, 5, 7$	8	$+$	1	2	$0, 3, 5, 7$	—
$3\!\cdot\!A_6$	$2, \mathbf{3}, 5$	9	$\circ$	2	2_3	$0, 2$	z_3
$3\!\cdot\!A_7$	$2, 3, 5, 7$	9	$\circ$	2	1	7	z_3
$L_2(19)$	$2, 3, 5, \mathbf{19}$	9	$\circ$	2	1	$0, 2, 3, 5$	b_{19}
$3\!\cdot\!J_3$	$2, 3, 5, 17, 19$	9	$\circ$	2	1	2	z_3, b_{17}, b_{19}
A_6	$2, \mathbf{3}, 5$	9	$+$	1	2^2	0	—
$L_2(8)$	$\mathbf{2}, 3, 7$	9	$+$	3	1	$0, 3$	y_7

Table 4.4: Cross characteristic candidates

Group	PmDivs	Deg	Ind	#ρ	Stab	Charc	Ch Ring
$L_2(17)$	2, 3, **17**	9	+	2	1	0, 3	b_{17}
M_{11}	2, 3, 5, 11	9	+	1	1	11	—
A_{10}	2, 3, 5, 7	9	+	1	2	0, 3, 7	—
A_{11}	2, 3, 5, 7, 11	9	+	1	2	11	—
A_7	2, 3, 5, 7	10	○	2	1	0, 3, 5	b_7
$2\dot{}L_2(19)$	2, 3, 5, **19**	10	○	2	1	0, 3, 5	b_{19}
M_{11}	2, 3, 5, 11	10	○	2	1	0, 3, 5, 11	i_2
$2\dot{}L_3(4)$	**2**, 3, 5, 7	10	○	2	2_2	0, 3, 5	b_7
$U_4(2)$	**2, 3**, 5	10	○	2	1	0, 5	z_3
$2\dot{}M_{12}$	2, 3, 5, 11	10	○	2	2	0, 3, 5, 11	i_2
M_{22}	2, 3, 5, 7, 11	10	○	2	2	2	b_7
$2\dot{}M_{22}$	2, 3, 5, 7, 11	10	○	2	2	0, 3, 5, 11	b_7
$2\dot{}A_6$	2, **3**, 5	10	−	2	2_2	0, 5	r_2
$2\dot{}L_2(11)$	2, 3, 5, **11**	10	−	1	2	0, 3, 5	—
$2\dot{}L_2(11)$	2, 3, 5, **11**	10	−	2	2	0, 5	r_3
$U_5(2)$	**2**, 3, 5, 11	10	−	1	2	0, 3, 5, 11	—
A_6	2, **3**, 5	10	+	1	2^2	0, 5	—
$L_2(11)$	2, 3, 5, **11**	10	+	1	2	0, [2,] 3, 5	—
$L_2(11)$	2, 3, 5, **11**	10	+	1	2	0, 2, 5	—
A_7	2, 3, 5, 7	10	+	1	2	7	—
M_{11}	2, 3, 5, 11	10	+	1	1	0, 2, 3, 5	—
$2\dot{}L_3(4)$	**2**, 3, 5, 7	10	+	1	2^2	7	—
M_{12}	2, 3, 5, 11	10	+	1	2	2	—
M_{12}	2, 3, 5, 11	10	+	2	1	3	—
$2\dot{}M_{22}$	2, 3, 5, 7, 11	10	+	1	2	7	—
A_{11}	2, 3, 5, 7, 11	10	+	1	2	0, 2, 3, 5, 7	—
A_{12}	2, 3, 5, 7, 11	10	+	1	2	2, 3	—
$L_2(23)$	2, 3, 11, **23**	11	○	2	1	0, 2, 3, 11	b_{23}
M_{23}	2, 3, 5, 7, 11, 23	11	○	2	1	2	b_7, b_{15}, b_{23}
$U_5(2)$	**2**, 3, 5, 11	11	○	2	1	0, 5, 11	z_3

Table 4.4: Cross characteristic candidates

Group	PmDivs	Deg	Ind	#ρ	Stab	Charc	Ch Ring
M_{24}	$2,3,5,7,11,23$	11	$\circ$	2	1	2	b_7, b_{15}, b_{23}
$L_2(11)$	$2,3,5,\mathbf{11}$	11	$+$	1	2	$0,5$	—
$L_3(3)$	$2,\mathbf{3},13$	11	$+$	1	2	13	—
M_{11}	$2,3,5,11$	11	$+$	1	1	$0,5,11$	—
M_{12}	$2,3,5,11$	11	$+$	2	1	$0,5,11$	—
A_{12}	$2,3,5,7,11$	11	$+$	1	2	$0,5,7,11$	—
A_{13}	$2,3,5,7,11,13$	11	$+$	1	2	13	—
$6{\cdot}A_6$	$2,\mathbf{3},5$	12	$\circ$	4	1	0	z_3, b_5
$6{\cdot}A_7$	$2,3,5,7$	12	$\circ$	2	1	5	z_3, b_7
$2{\cdot}L_2(23)$	$2,3,11,\mathbf{23}$	12	$\circ$	2	1	$0,3,11$	b_{23}
$12_2{\cdot}L_3(4)$	$\mathbf{2},3,5,7$	12	$\circ$	4	1	7	i, z_3, b_5
$3{\cdot}$Suz	$2,3,5,7,11,13$	12	$\circ$	2	1	2	z_3
$6{\cdot}$Suz	$2,3,5,7,11,13$	12	$\circ$	2	1	$0,5,7,11,13$	z_3
$2{\cdot}L_2(11)$	$2,3,5,\mathbf{11}$	12	$-$	2	2	$0,3$	b_5
$2{\cdot}L_2(13)$	$2,3,7,\mathbf{13}$	12	$-$	3	2	$0,3$	y_7
$L_2(25)$	$2,3,\mathbf{5},13$	12	$-$	2	2_2	2	—
$2{\cdot}L_2(25)$	$2,3,\mathbf{5},13$	12	$-$	2	2_2	$0,3,13$	—
$U_3(4)$	$\mathbf{2},3,5,13$	12	$-$	1	4	$0,3,5,13$	—
$S_4(5)$	$2,3,\mathbf{5},13$	12	$-$	2	1	2	b_5
$2{\cdot}S_4(5)$	$2,3,\mathbf{5},13$	12	$-$	2	1	$0,3,13$	b_5
$2{\cdot}G_2(4)$	$\mathbf{2},3,5,7,13$	12	$-$	1	2	$0,3,5,7,13$	—
A_{14}	$2,3,5,7,11,13$	12	$-$	1	2	2	—
$2{\cdot}$Suz	$2,3,5,7,11,13$	12	$-$	1	2	3	—
$L_2(11)$	$2,3,5,\mathbf{11}$	12	$+$	2	2	$0,2,3$	b_5
$L_2(13)$	$2,3,7,\mathbf{13}$	12	$+$	3	2	$0,2,3$	y_7
$L_2(13)$	$2,3,7,\mathbf{13}$	12	$+$	1	2	7	—
$L_3(3)$	$2,\mathbf{3},13$	12	$+$	1	2	$0,2$	—
$2{\cdot}M_{12}$	$2,3,5,11$	12	$+$	1	2	$0,5,11$	—
A_{13}	$2,3,5,7,11,13$	12	$+$	1	2	$0,2,3,5,7,11$	—
A_{14}	$2,3,5,7,11,13$	12	$+$	1	2	7	—

4.4 The type of the form and the stabilisers in Ω and C

4.4.1 Unitary forms

In this subsection we present some general methods to determine whether a representation over a finite field with indicator ○ preserves a unitary form, or only the identically zero form. Recall that an indicator of type $-$ indicates that the group preserves a symplectic form, and an indicator of type $+$ indicates that the group preserves a quadratic form: we will discuss in Subsection 4.9.3 how to determine the sign of the quadratic form in even dimension.

Since the definition of a unitary form requires a field automorphism of order 2, a representation can only preserve a unitary form when the field size is a square. The indicator ○ examples in Table 4.4 all have the property that they involve only quadratic irrationalities in their character rings, so in fact in each of these candidates the field size is p or p^2 for some prime p.

Lemma 4.4.1 *For a given absolutely irreducible representation of a group G over $\mathbb{F}_{q^2}$ with indicator ○, the image of G under the representation consists of isometries of a unitary form if and only if the action of the field automorphism $\sigma : x \mapsto x^q$ on the Brauer character is the same as complex conjugation.*

Proof Suppose that (the image of) G consists of isometries of a unitary form with matrix A. Then $gAg^{\sigma\mathsf{T}} = A$ for all $g \in G$. Rearranging, we see that $g^{-\mathsf{T}\sigma} = A^{-1}gA$ so, since $|\sigma| = 2$, the dual of the representation is equivalent to its image under σ. The effect of duality on the Brauer character is the same as complex conjugation [56, Lemma 15.3], so the result follows. □

From this we can immediately deduce the following result, which suffices to resolve this question in all examples up to dimension 12.

Corollary 4.4.2 *Suppose that the character ring of an absolutely irreducible representation with indicator ○ of a group G over $\mathbb{F}_{q^2}$ is generated over $\mathbb{Z}$ by the quadratic irrationalities $a_1, \ldots, a_r$, and let $\bar{a}_i$ denote a p-modular reduction of a_i to $\mathbb{F}_{q^2}$. Then the image of G under the representation consists of isometries of a unitary form if and only if $a_i \in \mathbb{R} \iff \bar{a}_i \in \mathbb{F}_q$ for $1 \leqslant i \leqslant r$.*

Thus, for example, for the listed representation of $6\!\cdot\!A_7$, the field size is p only when z_3 and r_2 both lie in $\mathbb{F}_p$, which is the case when $p \equiv 1$ or $7 \pmod{24}$. The group $6\!\cdot\!A_7$ preserves a unitary form over $\mathbb{F}_{p^2}$ when $r_2 \in \mathbb{F}_p$ but $z_3 \notin \mathbb{F}_p$; that is, when $p \equiv 17$ or $23 \pmod{24}$. When $p \equiv 5, 11, 13$ or $19 \pmod{24}$, the field size is p^2 and no form is preserved.

In Section 4.5 we shall use this lemma in dimensions up to 6 to determine which of the representations in Table 4.4 are unitary, whilst in Subsection 4.9.1 we shall apply it in dimensions 7 to 12.

4.4.2 Stabilisers in Ω and C, and quasishape

Theorem 4.3.3 is sufficient to determine the quasisimple subgroups G of the quasisimple classical groups Ω in dimensions up t 12. However, we wish to find the almost simple subgroups of the (projective) simple classical groups and of their almost simple extensions, so more work is needed.

Let Ω be a quasisimple classical group, let G be the image of a representation ρ of a quasisimple group, and assume that G is an $\mathscr{S}_1$- subgroup of Ω. Let C be the corresponding conformal group. In this subsection, we present some methods that can be used to determine the number of conjugacy classes of images of ρ in Ω and C, and of determining which of their stabilising outer automorphisms can be realised within Ω. These methods will be used in Section 4.5 to determine such stabilisers in dimensions up to 6, and in Section 4.9 in dimensions 7 to 12. Determining the effects of the remaining automorphisms is more complicated, and will be discussed in Section 4.6.

Lemma 4.4.3   *Let $G = S\rho_1$ be a quasisimple group for some faithful absolutely irreducible representation ρ_1 over a finite field, let $A = \mathrm{Out}\, G$, let $\{\rho_1, \ldots, \rho_r\}$ be a set of representatives of the equivalence classes of representations that are weakly equivalent to ρ_1, and let C be the corresponding conformal group of the smallest classical group Ω that contains G. Then*

  (i) *The orbits of A on $\{\rho_1, \ldots, \rho_r\}$ are in natural bijection with the conjugacy classes into which C partitions $\{S\rho_1, S\rho_2, \ldots, S\rho_r\}$.*

  (ii) *Each C-class of subgroups splits into $|C : \mathrm{N}_C(G)\Omega|$ classes in Ω.*

  (iii) *The outer automorphisms of G that are induced by elements of $\mathrm{N}_C(G)$ are precisely those that stabilise ρ_1.*

Proof   Let X be the general linear group naturally containing Ω. By Lemmas 1.8.6, the images of ρ_i and ρ_j are conjugate by an element $g \in X$ if and only if they are equivalent under an automorphism $\alpha \in A$ and, if $i = j$, then we may choose $g \in \mathrm{N}_X(S\rho_i)$ such that g induces α. Furthermore, by Lemmas 1.8.9 and 1.8.10, we may choose $g \in C$. Hence (i) and (iii) are true, and (ii) follows by the Orbit–Stabiliser Theorem. $\square$

The action of outer automorphisms of G on the representations can be found in [12, 57]. In many cases there is a single class of subgroups of C, but there are some exceptions. For example, the four representations of $6{\cdot}A_7$ in dimension 6 give rise to two classes of subgroups of $\mathrm{GL}_6^{\pm}(q)$.

We also need to determine $\mathrm{N}_\Omega(G)$. So, for $g \in \mathrm{N}_C(G)$, we need to decide whether some scalar multiple of g lies in Ω. This will depend in general on the effect of g on the invariant form (if non-zero), the determinant of g and, in the orthogonal cases, on the spinor norm or quasideterminant of any scalar multiple of g that is an isometry and has determinant 1. In dimension at most 12, the determinant of a suitable g and information on whether it is an isometry can be calculated from the character tables in [12, 57]: these calculations will be presented in Sections 4.5 and 4.9. Up to multiplication by scalars, we can always choose g to be an isometry in Case **U**, since $\mathrm{CGU}_n(q)$ is generated by $\mathrm{GU}_n(q)$ and scalar matrices.

Finally, we introduce some new notation, which we can use to denote an extension G of a quasisimple group, without having to specify precisely which scalars lie in G. Let $M.S$ be a quasisimple group with centre M and S simple, and let $S.A$ be almost simple. We say that G has *quasishape* $[\![M.S.A]\!]$, and write $G \approx [\![M.S.A]\!]$, if G has a normal subgroup H isomorphic to $M.S$ such that $G/\mathrm{C}_G(H) \cong S.A$. We use the notation $[\![M.S.A]\!]$ even if there is no group of shape $M.S.A$, and write $[\![M.S{:}A]\!]$ if the extension $S{:}A$ is split, even if any group $M.S.A$ is $M.S\dot{}A$ or if no such group exists. A notation such as $[\![6\dot{}\mathrm{PGL}_2(9)]\!]$ is the same as $[\![6\dot{}\mathrm{A}_6{:}2_2]\!]$. Note that we intend to use the notation $G \approx [\![M.S.A]\!]$ when G is an $\mathscr{S}$-subgroup of a classical group, so that $\mathrm{C}_G(H)$ should consist of the scalar matrices in G, but since G may include field and/or graph automorphisms, $\mathrm{C}_G(H)$ is not necessarily central in G. While the notions of quasishape and isoclinism are similar, they are not the same. For example, there are groups $\mathrm{A}_6.2^2$ and $\mathrm{A}_6.\mathrm{D}_8$ having quasishape $[\![\mathrm{Aut}\,\mathrm{A}_6]\!] = [\![1.\mathrm{A}_6.2^2]\!]$, but $\mathrm{A}_6.2^2$ and $\mathrm{A}_6.\mathrm{D}_8$ are not isoclinic. (The group $\mathrm{A}_6.\mathrm{D}_8$ arises when one attempts to extend the irreducible 10-dimensional representation of A_6 to $\mathrm{Aut}\,\mathrm{A}_6$.)

Definition 4.4.4 For subgroups $G \leqslant H \leqslant \mathrm{GL}_n(q)$, we say that G is *scalar-normalising* in H if $\mathrm{N}_H(G) \leqslant GZ$, where Z is the group of scalar matrices of $\mathrm{GL}_n(q)$.

4.5 Dimension up to 6: quasisimple and conformal groups

In this section, we carry out the calculations that we introduced in Section 4.4 in detail for dimensions up to 6. We deal with larger dimensions in less detail in Section 4.9. We remind the reader that, although we are including tables for the orthogonal groups in dimensions less than 7 in Chapter 8, we are not presenting details of the calculations in these cases, on the grounds that they are (projectively) isomorphic to other classical groups. This means that, in our calculations in this section, we are ignoring representations with indicator $+$.

4.5.1 Dimension 2

Since $\mathrm{SL}_2(q) = \mathrm{Sp}_2(q) \cong \mathrm{SU}_2(q)$ and $\Omega_2^{\pm}(q)$ is soluble, the only possible indicator for a 2-dimensional representation is $-$, so we consider these groups as subgroups of $\mathrm{Sp}_2(q)$. Furthermore, the only $\mathscr{S}_1$-candidate is $2\dot{\,}\mathrm{A}_5$.

Proposition 4.5.1   (i) *If $p \equiv \pm 1 \pmod 5$, then there are exactly two conjugacy classes of $\mathscr{S}_1$-subgroups of $\mathrm{Sp}_2(p)$ isomorphic to $2\dot{\,}\mathrm{A}_5$.*
  (ii) *If $p \equiv \pm 2 \pmod 5$ and $p \neq 2$, then there are exactly two conjugacy classes of $\mathscr{S}_1$-subgroups of $\mathrm{Sp}_2(p^2)$ isomorphic to $2\dot{\,}\mathrm{A}_5$.*
    In both cases, these subgroups are scalar-normalising in $\mathrm{CSp}_2(q) = \mathrm{GL}_2(q)$, and the two classes are fused by the diagonal automorphism of $\mathrm{Sp}_2(q)$. There are no other classes of $\mathscr{S}_1$-subgroups of $\mathrm{Sp}_2(q)$.

Proof   By Theorem 4.3.3, the only $\mathscr{S}_1$-candidate is $G = 2\dot{\,}\mathrm{A}_5$ in characteristics not equal to 2 or 5. Note that $\mathrm{Out}\, G = 2$.

The character ring of the relevant representations is the p-modular reduction of $\mathbb{Z}[\mathrm{b}_5]$. By Table 4.2 the quadratic irrationality b_5 lies in $\mathbb{F}_p$ if and only if p is a square modulo 5, so $G < \mathrm{Sp}_2(p)$ in that case ($p \equiv \pm 1 \pmod 5$) and $G < \mathrm{Sp}_2(p^2)$ otherwise ($p \equiv \pm 2 \pmod 5$). Thus these are the only possibilities for $\mathscr{S}_1$-subgroups of $\mathrm{Sp}_2(q)$.

There are two such representations of G, fused by the outer automorphism of G, and their stabilisers are trivial. So by Lemma 4.4.3, there is a single class of such groups $G < \mathrm{CSp}_2(q) = \mathrm{GL}_2(q)$. Since the outer automorphism of G is not induced by an element of $C := \mathrm{GL}_2(q)$, G is scalar-normalising. Let $Z = \mathrm{Z}(C)$. Then $\mathrm{N}_C(G) = GZ$, so $\mathrm{N}_C(G)\Omega = \Omega Z$. Since $|C : \Omega Z| = 2$, by Lemma 4.4.3 the single class in C splits into two Ω-classes, which are fused by an element of $\mathrm{GL}_2(q) \setminus \Omega Z$. $\qquad\square$

4.5.2 Dimension 3

There are no representations of indicator $-$, since the dimension is odd, and we are not considering subgroups of the orthogonal groups in dimension less than 7, so we consider only those representations with indicator $\circ$.

By Theorem 4.3.3, the quasisimple groups to consider are:

(i) $\mathrm{L}_3(2)$ in characteristic not 2 or 7;
(ii) $3\dot{\,}\mathrm{A}_6$ in characteristic not 3;
(iii) $3\dot{\,}\mathrm{A}_7$ in characteristic 5.

Proposition 4.5.2   (i) *If $p \equiv 1, 2, 4 \pmod 7$ and $p \neq 2$, then there are exactly $d := (p - 1, 3)$ conjugacy classes of $\mathscr{S}_1$-subgroups of $\mathrm{SL}_3(p)$ isomorphic to $\mathrm{L}_3(2)$. The subgroups are scalar-normalising in $\mathrm{GL}_3(p)$, and if $d = 3$ then the classes are fused by the diagonal automorphisms of $\mathrm{SL}_3(p)$.*

(ii) *If $p \equiv 3, 5, 6 \pmod 7$, then there are exactly $d := (p+1, 3)$ conjugacy classes of $\mathscr{S}_1$-subgroups of $\mathrm{SU}_3(p)$ isomorphic to $\mathrm{L}_3(2)$. The subgroups are scalar-normalising in $\mathrm{CGU}_3(p)$, and if $d = 3$ then the classes are fused by the diagonal automorphisms of $\mathrm{SU}_3(p)$.*

   For all other q, there are no $\mathscr{S}_1$-subgroups $\mathrm{L}_3(2)$ of $\mathrm{SL}_3(q)$ or $\mathrm{SU}_3(q)$.

Proof Let $G = \mathrm{L}_3(2)$. Then $|\operatorname{Out} G| = 2$. In characteristics other than 2 or 7, there are two 3-dimensional representations of G in Table 4.4, with character ring the p-modular reduction of $\mathbb{Z}[b_7]$. By Table 4.2, the complex quadratic irrationality b_7 lies in $\mathbb{F}_p$ if and only if $p \equiv 1, 2, 4 \pmod 7$. Corollary 4.4.2 gives $G < \mathrm{SL}_3(p)$ in that case, and $G < \mathrm{SU}_3(p)$ when $p \equiv 3, 5, 6 \pmod 7$. Thus there are no other values of q for which $\mathrm{SL}_3(q)$ or $\mathrm{SU}_3(q)$ has an $\mathscr{S}_1$-subgroup isomorphic to G.

We see from Table 4.4 that the representations have trivial stabiliser. Therefore the subgroups are scalar-normalising, and the two representations are interchanged by the outer automorphism of G. Hence, by Lemma 4.4.3 there is a single class of subgroups $G < C = \mathrm{GL}_3(q)$ or $\mathrm{CGU}_3(q)$. Let $Z = \mathrm{Z}(C)$. Since $|C : \Omega Z| = |C : \mathrm{N}_C(G)\Omega| = (q-1, 3)$ in Case **L** and $(q+1, 3)$ in Case **U**, the claim about the number of conjugacy classes in Ω follows.. $\qquad\square$

Proposition 4.5.3 (i) *If $p \equiv 1, 4 \pmod{15}$, then there are exactly three conjugacy classes of $\mathscr{S}_1$-subgroups of $\mathrm{SL}_3(p)$ isomorphic to $3{\cdot}\mathrm{A}_6$. The subgroups are scalar-normalising in $\mathrm{GL}_3(p)$, and the classes are fused by the diagonal automorphisms of $\mathrm{SL}_3(p)$.*

(ii) *If $p \equiv 11, 14 \pmod{15}$, then there are exactly three conjugacy classes of $\mathscr{S}_1$-subgroups of $\mathrm{SU}_3(p)$ isomorphic to $3{\cdot}\mathrm{A}_6$. The subgroups are scalar-normalising in $\mathrm{CGU}_3(p)$, and the classes are fused by the diagonal automorphisms of $\mathrm{SU}_3(p)$.*

(iii) *If $p \equiv 2, 7, 8, 13 \pmod{15}$, then there are exactly three conjugacy classes of $\mathscr{S}_1$-subgroups of $\mathrm{SL}_3(p^2)$ isomorphic to $3{\cdot}\mathrm{A}_6$. The subgroups are scalar-normalising in $\mathrm{GL}_3(p^2)$, and the classes are fused by the diagonal automorphisms of $\mathrm{SL}_3(p^2)$.*

(iv) *There are exactly three conjugacy classes of $\mathscr{S}_1$-subgroups of $\mathrm{SU}_3(5)$ isomorphic to $3{\cdot}\mathrm{A}_6.2_3$. The subgroups are scalar-normalising in $\mathrm{CGU}_3(5)$, and the classes are fused by the diagonal automorphisms of $\mathrm{SU}_3(5)$.*

   For all other q, there are no $\mathscr{S}_1$-subgroups $3{\cdot}\mathrm{A}_6$ of $\mathrm{SL}_3(q)$ or $\mathrm{SU}_3(q)$.

Proof Let $G = 3{\cdot}\mathrm{A}_6$. Then $\operatorname{Out} G \cong 2^2$.

By Theorem 4.3.3, in characteristics other than 3 and 5 the relevant representations of G have character ring the p-modular reduction of $\mathbb{Z}[z_3, b_5]$. By Table 4.2, the real quadratic irrationality b_5 lies in $\mathbb{F}_p$ if and only if $p \equiv \pm 1 \pmod 5$. If so, then Corollary 4.4.2 gives $G < \mathrm{SL}_3(p)$ when $z_3 \in \mathbb{F}_p$, namely when $p \equiv 1 \pmod 3$, and $G < \mathrm{SU}_3(p)$ when $p \equiv -1 \pmod 3$. If $p \equiv \pm 2 \pmod 5$ then $G < \mathrm{SL}_3(p^2)$ by Corollary 4.4.2. Thus there are no other powers q of p for which $\mathrm{SL}_3(q)$ or $\mathrm{SU}_3(q)$ has an $\mathscr{S}_1$-subgroup isomorphic to G.

By Theorem 4.3.3 there are four representations with trivial stabiliser. So the representations form a single orbit under $\mathrm{Out}\,G$, and hence, by Lemma 4.4.3, there is a single class of subgroups $G < C$. Since the stabiliser of the representations is trivial, G is scalar-normalising in C, so $\mathrm{N}_C(G)\Omega = \Omega\mathrm{Z}(C)$. Since $|C : \Omega\mathrm{Z}(C)| = (q-1,3)$ in Case **L** and $(q+1,3)$ in Case **U**, the result follows from Lemma 4.4.3.

In characteristic 5, the character ring is the 5-modular reduction of $\mathbb{Z}[z_3]$, so Corollary 4.4.2 gives $G < \mathrm{SU}_3(5)$, and there are no other powers q of 5 for which $\mathrm{SL}_3(q)$ or $\mathrm{SU}_3(q)$ has an $\mathscr{S}_1$-subgroup isomorphic to G. By Theorem 4.3.3, there are two representations, with stabiliser generated by the 2_3 automorphism of G. Hence the representations are interchanged by the 2_1 and 2_2 automorphisms. Now $|\mathrm{PGU}_3(5) : \mathrm{U}_3(5)| = |\mathrm{Z}(\mathrm{SU}_3(5))| = 3$, so $3{\cdot}\mathrm{A}_6.2_3 < \mathrm{SU}_3(5)$. (This can also be seen directly from the entry for $\mathrm{U}_3(5)$ in [12].) $\qquad\square$

Proposition 4.5.4 *There are exactly three conjugacy classes of $\mathscr{S}_1$-subgroups of $\mathrm{SU}_3(5)$ isomorphic to $3{\cdot}\mathrm{A}_7$. These groups are scalar-normalising in $\mathrm{CGU}_3(5)$ and the classes are fused by the diagonal automorphisms of $\mathrm{SU}_3(5)$.*

Proof Let $G = 3{\cdot}\mathrm{A}_7$. Then $|\mathrm{Out}\,G| = 2$. By Theorem 4.3.3, there are two relevant representations of $3{\cdot}\mathrm{A}_7$. These representations have character ring the 5-modular reduction of $\mathbb{Z}[z_3, b_7]$, so $G < \mathrm{SU}_3(5)$ by Table 4.2 and Corollary 4.4.2, and there are no other q for which $\mathrm{SL}_3(q)$ or $\mathrm{SU}_3(q)$ has an $\mathscr{S}_1$-subgroup isomorphic to G. The two representations have trivial stabiliser, so are interchanged by the outer automorphism of G. Thus, by Lemma 4.4.3 there is one conjugacy class of such subgroups of $\mathrm{CGU}_3(5)$, and G is scalar-normalising. Since $|\mathrm{CGU}_3(5) : \mathrm{Z}(\mathrm{CGU}_3(5))\mathrm{SU}_3(5)| = 3$, there are three conjugacy classes of such subgroups of $\mathrm{SU}_3(5)$. $\qquad\square$

4.5.3 Dimension 4

In dimension 4 we do not consider the representations with indicator $+$, since we are not considering the orthogonal groups in dimension less than 7.

Indicator $\circ$. By Theorem 4.3.3, the quasisimple groups to consider are:

  (i) $2{\cdot}\mathrm{L}_3(2)$ in characteristics not 2 or 7;
 (ii) A_7 in characteristic 2;
(iii) $2{\cdot}\mathrm{A}_7$ in characteristics not 2 or 7;

(iv) $4_2{}^{\cdot}L_3(4)$ in characteristic 3;

(v) $2^{\cdot}U_4(2)$ in characteristics greater than 3.

We first deal with $2^{\cdot}L_3(2)$, A_7 and $2^{\cdot}A_7$, since their behaviour is similar.

Proposition 4.5.5 (i) *If $p \equiv 1, 2, 4$ (mod 7), then $SL_4(p)$ has exactly $d :=$ $(p-1,4)$ conjugacy classes of $\mathscr{S}_1$-subgroups isomorphic to $2^{\cdot}L_3(2)$ (when $p \neq 2$) and exactly d such classes isomorphic to $2^{\cdot}A_7$ (or A_7 when $p = 2$). The subgroups are scalar-normalising in $GL_4(p)$, and if $d > 1$ then the classes are fused by the diagonal automorphisms of $SL_4(p)$.*

(ii) *If $p \equiv 3, 5, 6$ (mod 7), then $SU_4(p)$ has exactly $d := (p+1,4)$ conjugacy classes of $\mathscr{S}_1$-subgroups isomorphic to $2^{\cdot}L_3(2)$ and exactly d such classes isomorphic to $2^{\cdot}A_7$. The subgroups are scalar-normalising in $CGU_4(p)$, and if $d > 1$ then the classes are fused by the diagonal automorphisms of $SU_4(p)$.*

For all other q, there are no $\mathscr{S}_1$-subgroups of $SL_4(q)$ or $SU_4(q)$ isomorphic to $2^{\cdot}L_3(2)$, $2^{\cdot}A_7$ or A_7.

Proof Let G be one of $2^{\cdot}L_3(2)$, A_7 (with $p = 2$) or $2^{\cdot}A_7$. Then $|\operatorname{Out} G| = 2$. By Theorem 4.3.3, the relevant representations of G have character ring the p-modular reduction of $\mathbb{Z}[b_7]$. By Table 4.2, b_7 is a complex quadratic irrationality so, by Corollary 4.4.2, $G < SL_4(p)$ when $p \equiv 1, 2, 4$ (mod 7) and $G < SU_4(p)$ when $p \equiv 3, 5, 6$ (mod 7). Thus there are no other values of q for which $SL_4(q)$ or $SU_4(q)$ has an $\mathscr{S}_1$-subgroup isomorphic to G.

There are two representations of G. Each has trivial stabiliser, so they are interchanged by a group automorphism. Now by Lemma 4.4.3 there is a single class of subgroups $G < C$ in each case, the group G is scalar-normalising, and the number of classes in Ω follows from noting that $d = |C : \Omega Z|$. □

Proposition 4.5.6 *There are exactly two conjugacy classes of $\mathscr{S}_1$-subgroups G of $SU_4(3)$ isomorphic to $4_2{}^{\cdot}L_3(4)$. The normaliser in $CGU_4(3)$ of G is generated by G, scalars, and an element with determinant -1 in $GU_4(3) \setminus SU_4(3)$. The two classes are fused by the diagonal automorphism δ of $SU_4(3)$.*

Proof Let $G = 4_2{}^{\cdot}L_3(4)$. Then $\operatorname{Out} G \cong 2^2$. By Theorem 4.3.3, the relevant representations of G have character ring the 3-modular reduction of $\mathbb{Z}[i, b_7]$. Then Table 4.2 and Corollary 4.4.2 give $G < SU_4(3)$. Thus there are no other values of q for which $SU_4(q)$ has an $\mathscr{S}_1$-subgroup isomorphic to G.

There are two such representations, with stabiliser generated by the 2_2 automorphism of G. Therefore the representations are interchanged by the 2_1 and 2_3 automorphisms, and by Lemma 4.4.3 there is a single class of such subgroups of $CGU_4(3)$.

Furthermore, from [57], we find that the character values on elements of $G.2_2$

outside of G all lie in the same field as those of G itself, so $G.2_2 < \mathrm{GU}_4(3)$, but for g in Class 2C of $G.2_2$, the character value on g is 2, so the eigenvalues of g are 1^3 and -1, and hence $\det g = -1$. Since there is no scalar element with determinant -1 in $\mathrm{GU}_4(3)$, there is no class $\mathscr{S}_1$-subgroup of $\mathrm{SU}_4(3)$ with quasishape $[\![G.2_2]\!]$. The fact that $\det g = -1$ implies that the 2_2 automorphism is induced by the diagonal automorphism δ^2 of $\mathrm{SU}_4(3)$.　　□

Proposition 4.5.7　　(i) *If $p \equiv 1$ (mod 6), then there are exactly $d := (p - 1, 4)$ conjugacy classes of $\mathscr{S}_1$-subgroups of $\mathrm{SL}_4(p)$ isomorphic to $2^{\cdot}\mathrm{U}_4(2)$. The subgroups are scalar-normalising in $\mathrm{GL}_4(p)$, and if $d > 1$ then the classes are fused by the diagonal automorphisms of $\mathrm{SL}_4(p)$.*
　(ii) *If $p \equiv 5$ (mod 6), then there are exactly $d := (p+1, 4)$ conjugacy classes of $\mathscr{S}_1$-subgroups of $\mathrm{SU}_4(p)$ isomorphic to $2^{\cdot}\mathrm{U}_4(2)$. The subgroups are scalar-normalising in $\mathrm{CGU}_4(p)$, and if $d > 1$ then the classes are fused by the diagonal automorphisms of $\mathrm{SU}_4(p)$.*
　For all other q, there are no $\mathscr{S}_1$-subgroups $2^{\cdot}\mathrm{U}_4(2)$ of $\mathrm{SL}_4(q)$ or $\mathrm{SU}_4(q)$.

Proof　Let $G = 2^{\cdot}\mathrm{U}_4(2) \cong 2^{\cdot}\mathrm{Sp}_4(3)$. Then $|\mathrm{Out}\,G| = 2$. By Theorem 4.3.3, we have $p > 3$ and the relevant representations of G have character ring the p-modular reduction of $\mathbb{Z}[z_3]$. Therefore Table 4.2 and Corollary 4.4.2 imply that $G < \mathrm{SL}_4(p)$ when $p \equiv 1$ (mod 6) and $G < \mathrm{SU}_4(p)$ when $p \equiv 5$ (mod 6), and there are no other values of q for which $\mathrm{SL}_4(q)$ or $\mathrm{SU}_4(q)$ has an $\mathscr{S}_1$-subgroup isomorphic to G.

There are two such representations, with trivial stabiliser. Hence the representations are interchanged by the outer automorphism of G, so by Lemma 4.4.3 there is a single class of such groups in C. The fact that the representations have trivial stabiliser implies that G is scalar-normalising, and so the unique class in C splits into $d = |C : \Omega Z|$ conjugacy classes in Ω.　　□

Indicator $-$. By Theorem 4.3.3, the quasisimple groups to consider are:

　(i) $2^{\cdot}\mathrm{A}_5$ in characteristics greater than 5;
　(ii) A_6 in characteristic 2;
　(iii) $2^{\cdot}\mathrm{A}_6$ in characteristics greater than 3;
　(iv) $2^{\cdot}\mathrm{A}_7$ in characteristic 7.

Recall that $\mathrm{Sp}_4(2)$ is not quasisimple, and is not deemed to be a group of Lie type. We therefore shall not determine its maximal subgroups here.

Lemma 4.5.8　*For $e > 1$, the group $\mathrm{Sp}_4(2^e)$ has no $\mathscr{S}_1$-subgroups.*

Proof　The only possibility is A_6. However, $\mathrm{Sp}_4(2) \cong \mathrm{S}_6 = \mathrm{A}_6{:}2_1$.　　□

　　We can therefore assume that $p > 2$.

Proposition 4.5.9   (i) *If $p \equiv \pm 1 \pmod{12}$, then $\mathrm{Sp}_4(p)$ has exactly two conjugacy classes of $\mathscr{S}_1$-subgroups isomorphic to $2\,{}^{\cdot}S_5^-$. The subgroups are scalar-normalising in $\mathrm{CSp}_4(p)$ and the classes are fused by the diagonal automorphism of $\mathrm{Sp}_4(p)$.*

(ii) *If $p \equiv \pm 5 \pmod{12}$ with $p > 5$, then the group $\mathrm{Sp}_4(p)$ has exactly one conjugacy class of $\mathscr{S}_1$-subgroups G isomorphic to $2\,{}^{\cdot}A_5$. The normalisers of these subgroups G in $\mathrm{CSp}_4(p)$ are generated by G, scalars, and an element of $\mathrm{CSp}_6(p)$ that is a similarity but not an isometry of the symplectic form. For all other q, there are no $\mathscr{S}_1$-subgroups $2\,{}^{\cdot}A_5$ of $\mathrm{Sp}_4(q)$.*

Proof   Let $G = 2\,{}^{\cdot}A_5$. Then $|\mathrm{Out}\,G| = 2$. By Theorem 4.3.3, the relevant representation of G has character ring $\mathbb{Z}$. Thus $G < \mathrm{Sp}_4(p)$ for all $p > 5$, and there are no other values of q for which $\mathrm{Sp}_4(q)$ has an $\mathscr{S}_1$-subgroup isomorphic to G. There is a single such representation, so there is a single class of such groups in $C = \mathrm{CSp}_4(q)$.

The stabiliser of the representation in $\mathrm{Out}\,G$ has order 2, but the representation of $G.2$ in [12] involves the irrationality i_3 and so this version of $G.2$ does not preserve the symplectic form. By multiplying elements outside of G by a scalar element of order 4, we obtain a representation of the isoclinic group $G.2^- = 2\,{}^{\cdot}S_5^-$, which consists of isometries of the form, and involves the irrationality r_3. By Table 4.2 the quadratic irrationality r_3 lies in $\mathbb{F}_p$ if and only if $p \equiv \pm 1 \pmod{12}$. So $G.2^- < \mathrm{Sp}_4(p)$ if and only if $p \equiv \pm 1 \pmod{12}$, and otherwise the class of G is stabilised by the outer automorphism δ of $\mathrm{Sp}_4(p)$. $\square$

Proposition 4.5.10   (i) *If $p \equiv \pm 1 \pmod{12}$ then $\mathrm{Sp}_4(p)$ has exactly two conjugacy classes of $\mathscr{S}_1$-subgroups isomorphic to $2\,{}^{\cdot}A_6.2_1$. The subgroups are scalar-normalising in $\mathrm{CSp}_4(p)$ and the classes are fused by the diagonal automorphism of $\mathrm{Sp}_4(p)$.*

(ii) *If $p \equiv \pm 5 \pmod{12}$ then $\mathrm{Sp}_4(p)$ has exactly one conjugacy class of $\mathscr{S}_1$-subgroups G isomorphic to $2\,{}^{\cdot}A_6$. The normaliser in $\mathrm{CSp}_4(p)$ of G is generated by G, scalars, and an element in $\mathrm{CSp}_4(p)$ that is a similarity but not an isometry of the symplectic form, and induces the 2_1 automorphism of G.*

For all other q, there are no $\mathscr{S}_1$-subgroups $2\,{}^{\cdot}A_6$ of $\mathrm{Sp}_4(q)$.

Proof   Let $G = 2\,{}^{\cdot}A_6$. Then $\mathrm{Out}\,G \cong 2^2$, and by Theorem 4.3.3 the character ring of the relevant representations of G is the p-modular reduction of $\mathbb{Z}$, for all $p > 3$. Thus $G < \mathrm{Sp}_4(p)$ for all $p > 3$, and there are no other values of q for which $\mathrm{Sp}_4(q)$ has an $\mathscr{S}_1$-subgroup isomorphic to G.

We also find from Table 4.4 that there are two such representations, with stabiliser generated by the 2_1 automorphism of G. Therefore the representations are interchanged by the 2_2 and 2_3 automorphisms of G, and by Lemma 4.4.3

there is a single class of such groups in $\mathrm{CSp}_4(p)$. We find from [12, 57], that the extension of one of these two representations to $G.2_1$ consists of isometries of the symplectic form, but the character values of this representation on elements of $G.2_1 \setminus G$ involve the irrationality r_3, which by Table 4.2 lies in $\mathbb{F}_p$ if and only if $p \equiv \pm 1 \pmod{12}$. Note that $G.2_1^+ \cong G.2_1^-$, so there is no reason to consider the isoclinic variant of $G.2_1$.

So, if $p \equiv \pm 1 \pmod{12}$ then $G.2_1 < \mathrm{Sp}_4(p)$. Since the remaining automorphisms of G interchange the representations, the group $G.2_1$ is scalar-normalising in $\mathrm{CSp}_4(p)$, so there are two classes of groups $G.2_1 < \mathrm{Sp}_4(p)$, fused by the diagonal automorphism δ.

Similarly $p \equiv \pm 5 \pmod{12}$ then $r_3 \in \mathbb{F}_{p^2} \setminus \mathbb{F}_p$, and so there exists an element $g \in \mathrm{Sp}_4(p^2) \setminus \mathrm{Sp}_4(p)$ that induces the 2_1 automorphism of G. By Lemmas 1.8.6 and 1.8.9 this automorphism is also induced by conjugation by an element $g' \in \mathrm{CSp}_4(p)$. So $g'g^{-1}$ is a scalar matrix, which cannot be equal to $\pm I_4$, and so g' is not an isometry of the symplectic form. It follows that the 2_1 automorphism of G is induced by the diagonal automorphism δ of $\mathrm{Sp}_4(p)$ in this case.   □

Proposition 4.5.11   *There is a unique conjugacy class of $\mathscr{S}_1$-subgroups of $\mathrm{Sp}_4(7)$ isomorphic to $2^{\cdot}\mathrm{A}_7$. The normaliser in $\mathrm{CSp}_4(7)$ of G is generated by G, scalars, and an element in $\mathrm{CSp}_4(7)$ that negates the symplectic form.*

Proof   Let $G = 2^{\cdot}\mathrm{A}_7$, so $\mathrm{Out}\, G \cong 2$. By Theorem 4.3.3, the group $G < \mathrm{Sp}_4(7)$ and there are no other values of q for which $\mathrm{Sp}_4(q)$ has an $\mathscr{S}_1$-subgroup isomorphic to G. There is a single representation (with stabiliser $\mathrm{Out}\, G$), so there is a single class in $\mathrm{CSp}_4(7)$. From [57], we find that neither $G.2^+$ nor $G.2^-$ is contained in $\mathrm{Sp}_4(7)$. Therefore, the outer automorphism of G is induced by the outer automorphism δ of $\mathrm{Sp}_4(7)$.   □

4.5.4 Dimension 5

Since the dimension is odd and less than 7, we consider only indicator $\circ$.

By Theorem 4.3.3, the quasisimple groups to consider are:

(i) $\mathrm{L}_2(11)$ in characteristics not equal to 11;

(ii) M_{11} in characteristic 3;

(iii) $\mathrm{U}_4(2)$ in characteristics greater than 3.

Proposition 4.5.12   (i) *If $p \equiv 1, 3, 4, 5, 9 \pmod{11}$, then there are exactly $d := (p-1, 5)$ conjugacy classes of $\mathscr{S}_1$-subgroups of $\mathrm{SL}_5(p)$ isomorphic to $\mathrm{L}_2(11)$. The subgroups are scalar-normalising in $\mathrm{GL}_5(p)$, and if $d > 1$ then the classes are fused by the diagonal automorphisms of $\mathrm{SL}_5(p)$.*

(ii) *If $p \equiv 2, 6, 7, 8, 10 \pmod{11}$, then there are exactly $d := (p + 1, 5)$ conjugacy classes of $\mathscr{S}_1$-subgroups of $\mathrm{SU}_5(p)$ isomorphic to $\mathrm{L}_2(11)$. The subgroups are scalar-normalising in $\mathrm{CGU}_5(p)$, and if $d > 1$ then the classes are fused by the diagonal automorphisms of $\mathrm{SU}_5(p)$.*

(iii) *If $p \equiv 1 \pmod{6}$, then there are exactly $d := (p - 1, 5)$ conjugacy classes of $\mathscr{S}_1$-subgroups of $\mathrm{SL}_5(p)$ isomorphic to $\mathrm{U}_4(2)$. The subgroups are scalar-normalising in $\mathrm{GL}_5(p)$, and if $d > 1$ then the classes are fused by the diagonal automorphisms of $\mathrm{SL}_5(p)$.*

(iv) *If $p \equiv 5 \pmod{6}$, then there are exactly $d := (p + 1, 5)$ conjugacy classes of $\mathscr{S}_1$-subgroups of $\mathrm{SU}_5(p)$ isomorphic to $\mathrm{U}_4(2)$. The subgroups are scalar-normalising in $\mathrm{CGU}_5(p)$, and if $d > 1$ then the classes are fused by the diagonal automorphisms of $\mathrm{SU}_5(p)$.*

(v) *There are exactly two conjugacy classes of subgroups of $\mathrm{SL}_5(3)$ isomorphic to M_{11}. The subgroups are scalar-normalising in $\mathrm{GL}_5(3)$, and the classes are not fused in $\mathrm{GL}_5(3)$.*

For all q, there are no other $\mathscr{S}_1$-subgroups of $\mathrm{SL}_5(q)$ or $\mathrm{SU}_5(q)$.

Proof First, recall that $|\operatorname{Out} \mathrm{L}_2(11)| = 2$, $|\operatorname{Out} \mathrm{U}_4(2)| = 2$ and $|\operatorname{Out} \mathrm{M}_{11}| = 1$.

We consider the irrationalities involved. The character ring of the relevant representations of $\mathrm{L}_2(11)$ is the p-modular reduction of $\mathbb{Z}[\mathrm{b}_{11}]$. By Table 4.2 and Corollary 4.4.2, we find that $\mathrm{L}_2(11) \leqslant \mathrm{SL}_5(p)$ if p is a square modulo 11, and $\mathrm{L}_2(11) \leqslant \mathrm{SU}_5(p)$ otherwise. The relevant representations of $\mathrm{U}_4(2)$ have character ring the p-modular reduction of $\mathbb{Z}[\mathrm{z}_3]$, and so $\mathrm{U}_4(2) \leqslant \mathrm{SL}_5(p)$ when $p \equiv 1 \pmod{6}$ and $\mathrm{U}_4(2) \leqslant \mathrm{SU}_5(p)$ when $p \equiv 5 \pmod{6}$. The relevant representations of M_{11} have character ring the 3-modular reduction of $\mathbb{Z}[\mathrm{i}_2, \mathrm{b}_{11}]$, which is $\mathbb{F}_3$. Thus for no other q are there $\mathscr{S}_1$-subgroups of $\mathrm{SL}_5(q)$ or $\mathrm{SU}_5(q)$.

In each of these cases there are two representations, with trivial stabiliser. Therefore, in all cases, by Lemma 4.4.3 the group G is scalar-normalising in the group $C \in \{\mathrm{GL}_5(p), \mathrm{CGU}_5(p)\}$. In the first two cases, the representations are interchanged by the outer automorphism of G, so there is a single class $G < C$ and $d = (p \pm 1, 5)$ in Ω. But $G = \mathrm{M}_{11}$ has no outer automorphisms, so there are two classes $G < C$ and, since $\mathrm{SL}_5(3)$ has no non-trivial diagonal outer automorphisms, there are two classes in $\mathrm{SL}_5(3)$ also. $\qquad\square$

4.5.5 Dimension 6

Indicator $\circ$. By Theorem 4.3.3, the quasisimple groups to consider are:

(i) $3\dot{}\,\mathrm{A}_6$ in characteristics greater than 3;

(ii) $6\dot{}\,\mathrm{A}_6$ in characteristics greater than 3;

(iii) $2\dot{}\,\mathrm{L}_2(11)$ in characteristics not equal to 2 or 11;

(iv) $3\dot{}\,\mathrm{A}_7$ in characteristics not equal to 3;

  (v) $6{\cdot}A_7$ in characteristics greater than 3;
 (vi) $6{\cdot}L_3(4)$ in characteristics greater than 3;
(vii) $2{\cdot}M_{12}$ in characteristic 3;
(viii) $3{\cdot}M_{22}$ in characteristic 2;
 (ix) $3_1{\cdot}U_4(3)$ in characteristic 2;
  (x) $6_1{\cdot}U_4(3)$ in characteristics greater than 3.

Proposition 4.5.13   (i) *If $p \equiv 1$ or $19 \pmod{24}$, then there are exactly six conjugacy classes of $\mathscr{S}_1$-subgroups of $\mathrm{SL}_6(p)$ isomorphic to $3{\cdot}A_6.2_3$. The subgroups are scalar-normalising in $\mathrm{GL}_6(p)$, and the classes are fused under the diagonal automorphisms of $\mathrm{SL}_6(p)$.*

 (ii) *If $p \equiv 7$ or $13 \pmod{24}$, then there are exactly three conjugacy classes of $\mathscr{S}_1$-subgroups G of $\mathrm{SL}_6(p)$ isomorphic to $3{\cdot}A_6$. The normaliser of G in $\mathrm{GL}_6(p)$ is generated by G, scalars, and an element of $\mathrm{GL}_6(p) \setminus \mathrm{SL}_6(p)$ that induces the diagonal automorphism of order 2 of $\mathrm{SL}_6(p)$ and the 2_3 automorphism of G. The classes are fused under the diagonal automorphisms of order 3.*

(iii) *If $p \equiv 5$ or $23 \pmod{24}$, then there are exactly six conjugacy classes of $\mathscr{S}_1$-subgroups of $\mathrm{SU}_6(p)$ isomorphic to $3{\cdot}A_6.2_3$. The subgroups are scalar-normalising in $\mathrm{CGU}_6(p)$, and the classes are fused under the diagonal automorphisms of $\mathrm{SU}_6(p)$.*

 (iv) *If $p \equiv 11$ or $17 \pmod{24}$, then there are exactly three conjugacy classes of $\mathscr{S}_1$-subgroups G of $\mathrm{SU}_6(p)$ isomorphic to $3{\cdot}A_6$. The normaliser of G in $\mathrm{CGU}_6(p)$ is generated by G, scalars, and an element of $\mathrm{GU}_6(p) \setminus \mathrm{SU}_6(p)$ that induces the diagonal automorphism of order 2 of $\mathrm{SU}_6(p)$ and the 2_3 automorphism of G. The classes are fused under the diagonal automorphisms of order 3.*

For all other q, there are no $\mathscr{S}_1$-subgroups $3{\cdot}A_6$ of $\mathrm{SL}_6(q)$ or $\mathrm{SU}_6(q)$.

Proof Let $G = 3{\cdot}A_6$. Then $\mathrm{Out}\,G \cong 2^2$. By Theorem 4.3.3, the relevant representations of G have character ring the p-modular reduction of $\mathbb{Z}[z_3]$. So, by Corollary 4.4.2, $G \leqslant \mathrm{SL}_6(p)$ when $p \equiv 1 \pmod 6$ and $G \leqslant \mathrm{SU}_6(p)$ when $p \equiv 5 \pmod 6$, and for no other q are there $\mathscr{S}_1$-subgroups of $\mathrm{SL}_6(q)$ or $\mathrm{SU}_6(q)$ isomorphic to G.

By Theorem 4.3.3 there are two such representations with stabiliser generated by the 2_3 automorphism of G, so these two representations are swapped by group automorphisms, and thus by Lemma 4.4.3 there is a single class of $3{\cdot}A_6$ in $\mathrm{GL}_6(p)$ or $\mathrm{CGU}_6(p)$. Furthermore, by Lemmas 1.8.6 and 1.8.9, the group $3{\cdot}A_6$ extends to a subgroup of quasishape $[\![3{\cdot}A_6{\cdot}2_3]\!] = [\![3{\cdot}A_6.2_3]\!]$ of $\mathrm{GL}_6^{\pm}(q)$.

The character table of the (unique) group with the structure $3{\cdot}A_6.2_3$ is listed in [12] and in the 6-dimensional representation of this group, an element of order 4 lying outside of $3{\cdot}A_6$ has trace 0. Since its square lies in the unique class of order 2 in $3 \cdot A_6$ and has trace 2, this element of order 4 must have eigenvalues $1, 1, -1, -1, i, -i$, and hence its determinant is 1. The set of the values of this character on the elements lying outside of $3 \cdot A_6$ is $\{0, \pm i_2\}$, up to multiplication by the scalars lying in $3 \cdot A_6$ (which are $\{1, z_3, z_3^{**}\}$), so the representation can certainly be realised in $\mathrm{SL}_6(p^2)$.

By Table 4.2, the irrationality i_2 lies in $\mathbb{F}_p$ if and only if $p \equiv 1$ or 3 (mod 8). So $3 \cdot A_6.2_3 < \mathrm{SL}_6(p)$ when $p \equiv 1$ or 19 (mod 24). Since no other automorphisms of $3 \cdot A_6$ are induced by elements of $\mathrm{GL}_6(p)$, this subgroup is scalar-normalising in $\mathrm{GL}_6(p)$, and hence by Lemma 4.4.3 there are $|\mathrm{GL}_6(p) : Z(\mathrm{GL}_6(p))\mathrm{SL}_6(p)| = 6$ classes of subgroups of this type in $\mathrm{SL}_6(p)$.

When $p \equiv 5$ (mod 6), Corollary 4.4.2 tells us that this representation of $3 \cdot A_6.2_3$ has image in $\mathrm{SU}_6(p)$ whenever i_2 lies outside of $\mathbb{F}_p$, which by Table 4.2 is the case when $p \equiv 5$ or 23 (mod 24). In this case, the group $3 \cdot A_6.2_3$ is scalar-normalising, and there are six classes of such subgroups of $\mathrm{SU}_6(p)$.

For other congruences of p modulo 24, the image H of this representation in $\mathrm{SL}_6(p^2)$ does not lie in $\mathrm{SL}_6(p)$ or $\mathrm{SU}_6(p)$, but we still have to decide whether some other group G with the same quasishape $[\![3 \cdot A_6.2_3]\!]$ as $3 \cdot A_6.2_3$ could lie in $\mathrm{SL}_6(p)$ or $\mathrm{SU}_6(p)$. If this is the case then, since there is a unique class of $\mathscr{S}_1$-subgroups $3 \cdot A_6$ in $\mathrm{GL}_6(p^2)$, we can assume that $3 \cdot A_6 < G \cap H$. Then, since $3 \cdot A_6$ is absolutely irreducible, it is centralised in $\mathrm{SL}_6(p^2)$ only by scalars, and so the elements of H which induce the 2_3 automorphism of $3 \cdot A_6$ can be obtained from those of G by multiplying by scalar matrices of $\mathrm{SL}_6(p^2)$ of determinant 1. But all such scalar matrices lie in $\mathrm{SL}_6(p)$ (when $p \equiv 1$ (mod 6)) or $\mathrm{SU}_6(p)$ (when $p \equiv 5$ (mod 6)), so this would imply that $H < \mathrm{SL}_6(p)$ or $\mathrm{SU}_6(p)$, which is not the case.

So when $p \equiv 7$ or 13 (mod 24), there is a group with quasishape $[\![3 \cdot A_6.2_3]\!]$ in $\mathrm{GL}_6(p)$ but not in $\mathrm{SL}_6(p)$. Similarly, when $p \equiv 11$ or 17 (mod 24), there is a group with quasishape $[\![3 \cdot A_6.2_3]\!]$ in $\mathrm{GU}_6(p)$ but not in $\mathrm{SU}_6(p)$. The quasisimple group $3 \cdot A_6$ is self-normalising modulo scalars in $\mathrm{SL}_6(p)$ or $\mathrm{SU}_6(p)$ in these cases, and there are three conjugacy classes of such subgroups. $\qquad\square$

In the above proof, it is not essential for us to specify the precise structure of the minimal groups G with quasishape $[\![3 \cdot A_6.2_3]\!]$, but let us do that for $\mathrm{SL}_6(p)$. For $p \equiv 7$ (mod 24), we multiply outer elements in the image of the complex representation of $3 \cdot A_6.2_3$ by the scalar $i I_6$ to get a group $(2 \times 3) \cdot A_6.2_3$ that can be embedded in $\mathrm{GL}_6(p)$. With this embedding, outer elements have determinant -1. For $p \equiv 13$ (mod 24), we multiply outer elements by the scalar $z_8 = \frac{1}{2}(r_2 + i_2)$ to get a group $(4 \times 3) \cdot A_6.2_3 \cong \frac{1}{2}(3 \cdot A_6.2_3 \times 8)$ that can be embedded in $\mathrm{GL}_6(p)$. With this embedding outer elements have determinant $\pm i$. After multiplying by a scalar in $\langle z_3, i \rangle$ (corresponding to central elements of $3 \cdot A_6 \times 4$), the trace of an outer element can be taken to lie in $\{0, \pm 1 \pm i\}$. In these cases the presence of the additional scalars of order 2 and 4 is necessary to embed a group of type $[\![3{\cdot}A_6.2_3]\!]$ in $\mathrm{GL}_6(p)$.

Proposition 4.5.14   (i) *If $p \equiv 1$ or 7 (mod 24) then $\mathrm{SL}_6(p)$ has exactly six conjugacy classes of $\mathscr{S}_1$-subgroups isomorphic to $6{\cdot}A_6$. The subgroups are scalar-normalising in $\mathrm{GL}_6(p)$, and the classes are fused under diagonal automorphisms of $\mathrm{SL}_6(p)$.*

   (ii) *If $p \equiv 17$ or 23 (mod 24) then $\mathrm{SU}_6(p)$ has exactly six conjugacy classes of $\mathscr{S}_1$-subgroups isomorphic to $6{\cdot}A_6$. The subgroups are scalar-normalising in $\mathrm{CGU}_6(p)$, and the classes are fused under diagonal automorphisms of $\mathrm{SU}_6(p)$.*

   (iii) *If $p \equiv 5, 11, 13, 19$ (mod 24) then $\mathrm{SL}_6(p^2)$ has exactly six conjugacy classes of $\mathscr{S}_1$-subgroups isomorphic to $6{\cdot}A_6$. Each subgroup is scalar-normalising in $\mathrm{GL}_6(p^2)$, and the classes are fused under diagonal automorphisms of $\mathrm{SL}_6(p^2)$.*

   For all other q, there are no $\mathscr{S}_1$-subgroups $6{\cdot}A_6$ of $\mathrm{SL}_6(q)$ or $\mathrm{SU}_6(q)$.

Proof   Let $G = 6{\cdot}A_6$. Then $\mathrm{Out}\,G \cong 2^2$. By Theorem 4.3.3, the relevant representations of G have character ring the p-modular reduction of $\mathbb{Z}[z_3, r_2]$. By Table 4.2, the irrationality $r_2 \in \mathbb{F}_p$ if and only if $p \equiv \pm1$ (mod 8). If so, then $G < \mathrm{SL}_6^\pm(p)$ when $p \equiv \pm1$ (mod 6), respectively. If $p \equiv \pm3$ (mod 8) then $r_2 \notin \mathbb{F}_p$, so by Corollary 4.4.2 the image of G does not preserve a unitary form, and hence G is an $\mathscr{S}_1$-subgroup of $\mathrm{SL}_6(p^2)$. Thus for no other q are there $\mathscr{S}_1$-subgroups of $\mathrm{SL}_6(q)$ or $\mathrm{SU}_6(q)$ isomorphic to G.

By Theorem 4.3.3, there are four such representations, with trivial stabiliser, which are therefore permuted transitively by $\mathrm{Out}\,G$. Thus by Lemma 4.4.3 there is a single, scalar-normalising, class of subgroups $G < \mathrm{GL}_6^\pm(q)$. $\square$

Proposition 4.5.15   (i) *If $p \equiv 1, 3, 4, 5, 9$ (mod 11), then there are exactly $d := (p - 1, 6)$ conjugacy classes of $\mathscr{S}_1$-subgroups of $\mathrm{SL}_6(p)$ isomorphic to $2{\cdot}L_2(11)$. The subgroups are scalar-normalising in $\mathrm{GL}_6(p)$, and if $d > 1$ then the classes are fused by the diagonal automorphisms of $\mathrm{SL}_6(p)$.*

   (ii) *If $p \equiv 2, 6, 7, 8, 10$ (mod 11) and $p > 2$, then there are exactly $d := (p+1, 6)$ conjugacy classes of $\mathscr{S}_1$-subgroups of $\mathrm{SU}_6(p)$ isomorphic to $2{\cdot}L_2(11)$. The subgroups are scalar-normalising in $\mathrm{CGU}_6(p)$, and if $d > 1$ then the classes are fused by the diagonal automorphisms of $\mathrm{SU}_6(p)$.*

   For all other q, there are no $\mathscr{S}_1$-subgroups $2{\cdot}L_2(11)$ of $\mathrm{SL}_6(q)$ or $\mathrm{SU}_6(q)$.

Proof   Let $G = 2{\cdot}L_2(11)$. Then $|\mathrm{Out}\,G| = 2$. By Theorem 4.3.3, the relevant representations of G have character ring the p-modular reduction of $\mathbb{Z}[b_{11}]$, with $p \neq 2, 11$. By Table 4.2 the quadratic irrationality b_{11} is complex and lies in $\mathbb{F}_p$ if and only if p is a square mod 11. So in that case $G < \mathrm{SL}_6(p)$ and otherwise, by Corollary 4.4.2, $G < \mathrm{SU}_6(p)$. Therefore for no other q are there $\mathscr{S}_1$-subgroups of $\mathrm{SL}_6(q)$ or $\mathrm{SU}_6(q)$ isomorphic to G. There are two such representations, with trivial stabiliser. Therefore by Lemma 4.4.3 the representations are interchanged by a group automorphism, so there is a single class of such groups in $C \in \{\mathrm{GL}_6(p), \mathrm{CGU}_6(p)\}$, and G is scalar-normalising. $\qquad\square$

Proposition 4.5.16   (i) *If $p \equiv 1 \pmod{6}$, then there are exactly six conjugacy classes of type $\mathscr{S}_1$-subgroups of $\mathrm{SL}_6(p)$ isomorphic to $3\,\dot{}\,\mathrm{A}_7$. The subgroups are scalar-normalising in $\mathrm{GL}_6(p)$, and the classes are fused under the diagonal automorphisms of $\mathrm{SL}_6(p)$.*

(ii) *If $p \equiv 2, 5 \pmod{6}$, then there are exactly six (or three when $p = 2$) conjugacy classes of type $\mathscr{S}_1$-subgroups of $\mathrm{SU}_6(p)$ isomorphic to $3\,\dot{}\,\mathrm{A}_7$. The subgroups are scalar-normalising in $\mathrm{CGU}_6(p)$, and the classes are fused under the diagonal automorphisms of $\mathrm{SU}_6(p)$.*

For all other q, there are no $\mathscr{S}_1$-subgroups $3\,\dot{}\,\mathrm{A}_7$ of $\mathrm{SL}_6(q)$ or $\mathrm{SU}_6(q)$.

Proof   Let $G = 3\,\dot{}\,\mathrm{A}_7$. Then $|\mathrm{Out}\,G| = 2$. By Theorem 4.3.3 the relevant representations of G have character ring the p-modular reduction of $\mathbb{Z}[z_3]$, with $p \neq 3$. Thus by Table 4.2 and Corollary 4.4.2, if $p \equiv 1 \pmod{6}$ then $G \leqslant \mathrm{SL}_6(p)$, if $p \equiv 2 \pmod{3}$ then $G \leqslant \mathrm{SU}_6(p)$, and for no other q are there $\mathscr{S}_1$-subgroups of $\mathrm{SL}_6(q)$ or $\mathrm{SU}_6(q)$ isomorphic to G.

There are two such representations, with trivial stabiliser, which are therefore scalar-normalising, interchanged by an outer automorphism of G. Thus by Lemma 4.4.3 there is a single class in $\mathrm{GL}_6(p)$ or $\mathrm{CGU}_6(p)$, and $d = (p \mp 1, 6)$ classes in Ω. $\qquad\square$

Proposition 4.5.17   (i) *If $p \equiv 1, 7 \pmod{24}$ then the group $\mathrm{SL}_6(p)$ has exactly twelve conjugacy classes of $\mathscr{S}_1$-subgroups G isomorphic to $6\,\dot{}\,\mathrm{A}_7$. The group G is scalar-normalising in $\mathrm{GL}_6(p)$, and the classes form two orbits under the diagonal automorphisms of $\mathrm{SL}_6(p)$.*

(ii) *If $p \equiv 17, 23 \pmod{24}$ then $\mathrm{SU}_6(p)$ has exactly twelve conjugacy classes of $\mathscr{S}_1$-subgroups G isomorphic to $6\,\dot{}\,\mathrm{A}_7$. The group G is scalar-normalising in $\mathrm{CGU}_6(p)$, and the classes form two orbits under the diagonal automorphisms of $\mathrm{SU}_6(p)$.*

(iii) *If $p \equiv 5, 11, 13, 19 \pmod{24}$ then $\mathrm{SL}_6(p^2)$ has exactly twelve conjugacy classes of $\mathscr{S}_1$-subgroups G isomorphic to $6\,\dot{}\,\mathrm{A}_7$. The group G is scalar-normalising in $\mathrm{GL}_6(p^2)$, and the classes form two orbits under the diagonal automorphisms of $\mathrm{SL}_6(p^2)$.*

For all other q, there are no $\mathscr{S}_1$-subgroups $6\,\dot{}\,\mathrm{A}_7$ of $\mathrm{SL}_6(q)$ or $\mathrm{SU}_6(q)$.

Proof   Let $G = 6\,\dot{}\,\mathrm{A}_7$. Then $|\mathrm{Out}\,G| = 2$. By Theorem 4.3.3, the relevant representations have character ring the p-modular reduction of $\mathbb{Z}[z_3, r_2]$, where $p > 3$. We determine the appropriate congruences on p and q by an identical calculation to the proof of Proposition 4.5.14.

There are four such representations with trivial stabiliser. These representations are therefore permuted in pairs by the outer automorphism of G. Thus G is scalar-normalising, and there are two classes of subgroups $G < \mathrm{GL}_6^{\pm}(q)$, splitting into $2(q \mp 1)$ classes in Ω.   □

Proposition 4.5.18   (i) If $p \equiv 1$ or $19 \pmod{24}$, then there are exactly six conjugacy classes of $\mathscr{S}_1$-subgroups of $\mathrm{SL}_6(p)$ isomorphic to $6{\cdot}\mathrm{L}_3(4).2_1^-$. The subgroups are scalar-normalising in $\mathrm{GL}_6(p)$, and the classes are fused under the diagonal automorphisms of $\mathrm{SL}_6(p)$.

(ii) If $p \equiv 7$ or $13 \pmod{24}$, then there are exactly three conjugacy classes of $\mathscr{S}_1$-subgroups of $\mathrm{SL}_6(p)$ isomorphic to $6{\cdot}\mathrm{L}_3(4)$. The normalisers of these subgroups G in $\mathrm{GL}_6(p)$ are generated by G, scalars, and an element of $\mathrm{GL}_6(p) \setminus \mathrm{SL}_6(p)$ that induces the diagonal automorphism of order 2 of $\mathrm{SL}_6(p)$ and the 2_1 automorphism of G. The classes are fused under the diagonal automorphisms of order 3 of $\mathrm{SL}_6(p)$.

(iii) If $p \equiv 5$ or $23 \pmod{24}$, then there are exactly six conjugacy classes of $\mathscr{S}_1$-subgroups of $\mathrm{SU}_6(p)$ isomorphic to $6{\cdot}\mathrm{L}_3(4).2_1^-$. The subgroups are scalar-normalising in $\mathrm{CGU}_6(p)$, and the classes are fused under the diagonal automorphisms of $\mathrm{SU}_6(p)$.

(iv) If $p \equiv 11, 17 \pmod{24}$, then there are exactly three conjugacy classes of $\mathscr{S}_1$-subgroups of $\mathrm{SU}_6(p)$ isomorphic to $6{\cdot}\mathrm{L}_3(4)$. The normalisers of these subgroups G in $\mathrm{CGU}_6(p)$ are generated by G, scalars, and an element of $\mathrm{GU}_6(p) \setminus \mathrm{SU}_6(p)$ that induces the diagonal automorphism of order 2 of $\mathrm{SU}_6(p)$ and the 2_1 automorphism of G. The classes are fused under the diagonal automorphisms of order 3 of $\mathrm{SU}_6(p)$.

For all other q, there are no $\mathscr{S}_1$-subgroups $6{\cdot}\mathrm{L}_3(4)$ of $\mathrm{SL}_6(q)$ or $\mathrm{SU}_6(q)$.

Proof Let $G = 6{\cdot}\mathrm{L}_3(4)$. Then $\mathrm{Out}\,G \cong 2^2 \leqslant 2 \times \mathrm{S}_3 \cong \mathrm{Out}\,\mathrm{L}_3(4)$. By Theorem 4.3.3 the relevant representations of G have character ring the p-modular reduction of $\mathbb{Z}[z_3]$, with $p > 3$, so $G \leqslant \mathrm{SL}_6(p)$ when $p \equiv 1 \pmod 6$ and $G \leqslant \mathrm{SU}_6(p)$ when $p \equiv 5 \pmod 6$, and for no other q do $\mathrm{SL}_6(q)$ or $\mathrm{SU}_6(q)$ have an $\mathscr{S}_1$-subgroup isomorphic to G. There are two representations, with stabiliser generated by the 2_1 automorphism.

Consulting [12] and [57], in the representation of $6{\cdot}\mathrm{L}_3(4){:}2_1^+$ the elements outside of $6{\cdot}\mathrm{L}_3(4)$ have determinant -1. However, in the representation of $6{\cdot}\mathrm{L}_3(4){\cdot}2_1^-$, these determinants are 1. The set of character values of the displayed representation on the elements outside of $6{\cdot}\mathrm{L}_3(4)$ is $\{0, r_2\}$ up to multiplication by elements of $\langle -z_3 \rangle$, and hence that of the isoclinic representation is $\{0, i_2\}$ (up to multiplication by elements of $\langle -z_3 \rangle$). This is analogous to a situtation which we encountered in the proof of Proposition 4.5.13. If $p \equiv 1$ or $19 \pmod{24}$ in Case **L** then $r_2 \in \mathbb{F}_p$, so $6{\cdot}\mathrm{L}_3(4).2_1^- < \mathrm{SL}_6(p)$. If $p \equiv 5$ or

23 (mod 24) in Case **U** then $r_2 \notin \mathbb{F}_p$, so $6\dot{}L_3(4).2_1^- < SU_6(p)$. In the other cases, $6\dot{}L_3(4)$ is self-normalising in Ω. □

Proposition 4.5.19 (i) *There are exactly two classes of $\mathscr{S}_1$-subgroups of* $SL_6(3)$ *isomorphic to* $2\dot{}M_{12}$. *These subgroups are scalar-normalising in* $GL_6(3)$ *and the classes are fused by the diagonal automorphism of* $SL_6(3)$.

(ii) *There are exactly three classes of $\mathscr{S}_1$-subgroups of* $SU_6(2)$ *isomorphic to* $3\dot{}M_{22}$. *These subgroups are scalar-normalising in* $CGU_6(2)$ *and the classes are fused by the diagonal automorphisms of* $SU_6(2)$.

Proof First, let $G = 2\dot{}M_{12}$. Then $|\,\text{Out}\,G| = 2$. By Theorem 4.3.3 the character ring of the representation is the 3-modular reduction of $\mathbb{Z}[i_2, i_5, b_{11}]$, which by Table 4.2 is just $\mathbb{F}_3$. Thus for no other q are there $\mathscr{S}_1$-subgroups of $SL_6(q)$ or $SU_6(q)$ isomorphic to G.

There are two such representations, with trivial stabiliser and therefore interchanged by a group automorphism. Thus by Lemma 4.4.3 there is a single class of such subgroups in $GL_6(3)$, the group G is scalar-normalising, and there are $2 = (3 - 1, 6)$ classes of such subgroups in $SL_6(3)$.

The argument for $G = 3\dot{}M_{22}$ is similar, except that in this case the character ring of the representation is the 2-modular reduction of $\mathbb{Z}[z_3, b_{11}]$, so Table 4.2 and Corollary 4.4.2 imply that $G \leqslant SU_6(2)$. □

Proposition 4.5.20 (i) *There are exactly three conjugacy classes of $\mathscr{S}_1$-subgroups of* $SU_6(2)$ *isomorphic to* $3_1\dot{}U_4(3).2_2$. *The groups are scalar-normalising in* $CGU_6(3)$, *and the classes are fused under the diagonal automorphisms of* $SU_6(3)$.

(ii) *If $p \equiv 1$ (mod 12) then there are exactly six conjugacy classes of $\mathscr{S}_1$-subgroups of* $SL_6(p)$ *isomorphic to* $6_1\dot{}U_4(3)\dot{}2_2^-$. *Each of these subgroups is scalar-normalising in* $GL_6(p)$, *and the classes are fused under the diagonal automorphisms of* $SL_6(p)$.

(iii) *If $p \equiv 7$ (mod 12) then there are exactly three conjugacy classes of $\mathscr{S}_1$-subgroups of* $SL_6(p)$ *isomorphic to* $6_1\dot{}U_4(3)$. *The normalisers of these subgroups G in* $GL_6(p)$ *are generated by G, scalars, and an element of* $GL_6(p) \setminus SL_6(p)$ *that induces the diagonal automorphism of order 2 of* $SL_6(p)$. *The classes are fused under the diagonal automorphisms of order 3 of* $SL_6(p)$.

(iv) *If $p \equiv 11$ (mod 12) then there are exactly six conjugacy classes of $\mathscr{S}_1$-subgroups of* $SU_6(p)$ *isomorphic to* $6_1\dot{}U_4(3)\dot{}2_2^-$. *Each of these subgroups is scalar-normalising in* $CGU_6(p)$, *and the classes are fused under the diagonal automorphisms of* $SU_6(p)$.

(v) *If $p \equiv 5$ (mod 12) then there are exactly three conjugacy classes of $\mathscr{S}_1$-subgroups of* $SU_6(p)$ *isomorphic to* $6_1\dot{}U_4(3)$. *The normalisers of these sub-*

groups G in $\mathrm{CGU}_6(p)$ are generated by G, scalars, and an element of $\mathrm{GU}_6(p) \setminus \mathrm{SU}_6(p)$ that induces the diagonal automorphism of order 2. The classes are fused under the diagonal automorphisms of order 3 of $\mathrm{SU}_6(p)$. For no other q are there $\mathscr{S}_1$-subgroups $6_1{}^{\cdot}\mathrm{U}_4(3)$ or $3_1{}^{\cdot}\mathrm{U}_4(3)$ of $\mathrm{SL}_6^{\pm}(q)$.

Proof   Let $G = 6_1{}^{\cdot}\mathrm{U}_4(3)$, or $3_1{}^{\cdot}\mathrm{U}_4(3)$ when $p = 2$. Then $\mathrm{Out}\,G \cong 2^2$, which is a proper subgroup of $\mathrm{D}_8 \cong \mathrm{Out}\,\mathrm{U}_4(3)$. By Theorem 4.3.3, the character ring of the two relevant representations of G is the p-modular reduction of $\mathbb{Z}[z_3]$, where $p \neq 3$. Therefore $G \leqslant \mathrm{SL}_6(p)$ when $p \equiv 1 \pmod 3$ and $G \leqslant \mathrm{SU}_6(p)$ when $p \equiv 2 \pmod 3$: for no other q are there $\mathscr{S}_1$-subgroups of $\mathrm{SL}_6(q)$ or $\mathrm{SU}_6(q)$ isomorphic to G.

These two representations are swapped by the 2_1 automorphism of $\mathrm{U}_4(3)$ and stabilised by the 2_2 automorphism. We find that the elements outside $6_1{}^{\cdot}\mathrm{U}_4(3)$ of the representation of $6_1{}^{\cdot}\mathrm{U}_4(3){:}2_2^+$ listed in [12, 57] have determinant -1 and take character values in the set $\{0, \pm 1, \pm 2, 4, i_3, \pm i_3 - 2\}$, up to multiplication by elements of $\langle -z_3 \rangle$. So the corresponding set of character values for the isoclinic group $6_1{}^{\cdot}\mathrm{U}_4(3){}^{\cdot}2_2^-$, of which all elements have determinant 1, is $\{0, \pm i, \pm 2i, 4i, -r_3, \pm r_3 - 2i\}$ (up to multiplication by elements of $\langle -z_3 \rangle$). So, for $p \neq 2$, the stated results follow as in Proposition 4.5.13. When $p = 2$, $\mathrm{SU}_6(2)$ has no diagonal automorphism of order 2 and the images of the elements of $3_1{}^{\cdot}\mathrm{U}_4(3).2_2$ all have determinant 1, so we get $3_1{}^{\cdot}\mathrm{U}_4(3).2_2 < \mathrm{SU}_6(2)$. $\square$

Indicator $-$. By Theorem 4.3.3, the quasisimple groups to consider are:

   (i) $2{}^{\cdot}\mathrm{A}_5$ in characteristics not equal to 2 or 5;
  (ii) $2{}^{\cdot}\mathrm{L}_3(2)$ in characteristics not equal to 2 or 7;
 (iii) $\mathrm{L}_2(13)$ in characteristic 2;
 (iv) $2{}^{\cdot}\mathrm{L}_2(13)$ in characteristics not equal to 2 or 13;
  (v) $2{}^{\cdot}\mathrm{A}_7$ in characteristic 3;
 (vi) $\mathrm{U}_3(3)$ in characteristics not equal to 3;
(vii) J_2 in characteristic 2;
(viii) $2{}^{\cdot}\mathrm{J}_2$ in characteristics not equal to 2.

Proposition 4.5.21   (i) *If $p \equiv \pm 1 \pmod 8$, then $\mathrm{Sp}_6(p)$ has exactly two conjugacy classes of $\mathscr{S}_1$-subgroups isomorphic to $2{}^{\cdot}\mathrm{S}_5^-$. The subgroups are scalar-normalising in $\mathrm{CSp}_6(p)$ and the classes are fused by the diagonal automorphism of $\mathrm{Sp}_6(p)$.*

(ii) *If $p \equiv \pm 3 \pmod 8$ and $p \neq 5$, then $\mathrm{Sp}_6(p)$ has a unique conjugacy class of $\mathscr{S}_1$-subgroups G isomorphic to $2{}^{\cdot}\mathrm{A}_5$. The normalisers of these subgroups G in $\mathrm{CSp}_6(p)$ are generated by G, scalars, and an element of $\mathrm{CSp}_6(p)$ that is a similarity but not an isometry of the symplectic form.*

For all other q, there are no $\mathscr{S}_1$-subgroups $2{}^{\cdot}\mathrm{A}_5$ of $\mathrm{Sp}_6(q)$.

Proof Let $G = 2\cdot A_5$. Then $|\operatorname{Out} G| = 2$. By Theorem 4.3.3, the relevant representations of G have character ring the p-modular reduction of $\mathbb{Z}$, with $p \neq 2, 5$. Therefore $G < \operatorname{Sp}_6(p)$ for all $p \neq 2, 5$, and for no other q are there $\mathscr{S}_1$-subgroups of $\operatorname{Sp}_6(q)$ isomorphic to G.

There is a single such representation, with stabiliser of order 2. The representation of $G.2^+$ in [12, 57] involves the irrationality i_2 and does not consist of isometries of the symplectic form. By multiplying elements outside of G by a scalar element of order 4, we obtain a representation of $G.2^- = 2\cdot S_5^-$, which consists of isometries and has character ring $\mathbb{Z}[r_2]$. The irrationality r_2 lies in $\mathbb{F}_p$ if and only if $p \equiv \pm 1 \pmod 8$. So $G.2^- < \operatorname{Sp}_6(p)$ if and only if $p \equiv \pm 1 \pmod 8$, and otherwise the class of G is stabilised by the outer automorphism δ of Ω. $\square$

Proposition 4.5.22 (i) *If $p \equiv \pm 1 \pmod{16}$, then $\operatorname{Sp}_6(p)$ has exactly four conjugacy classes of $\mathscr{S}_1$-subgroups isomorphic to $2\cdot L_3(2).2$. The subgroups are scalar-normalising in $\operatorname{CSp}_6(p)$, and the classes form two orbits of length 2 under the action of the diagonal automorphism of $\operatorname{Sp}_6(p)$.*

(ii) *If $p \equiv \pm 7 \pmod{16}$ and $p \neq 7$, then $\operatorname{Sp}_6(p)$ has exactly two conjugacy classes of $\mathscr{S}_1$-subgroups isomorphic to $2\cdot L_3(2)$. The normalisers of these subgroups G in $\operatorname{CSp}_6(p)$ are generated by G, scalars, and an element of $\operatorname{CSp}_6(p)$ that is a similarity but not an isometry of the symplectic form. The two conjugacy classes remain distinct in $\operatorname{CSp}_6(p)$.*

(iii) *If $p \equiv \pm 3 \pmod 8$, then $\operatorname{Sp}_6(p^2)$ has exactly two conjugacy classes of $\mathscr{S}_1$-subgroups isomorphic to $2\cdot L_3(2)$. The normalisers of these subgroups G in $\operatorname{CSp}_6(p^2)$ are generated by G, scalars, and an element of $\operatorname{CSp}_6(p^2)$ that is a similarity but not an isometry of the symplectic form. The two conjugacy classes remain distinct in $\operatorname{CSp}_6(p^2)$.*

For all other q, there are no $\mathscr{S}_1$-subgroups $2\cdot L_3(2)$ of $\operatorname{Sp}_6(q)$.

Proof Let $G = 2\cdot L_3(2)$. Then $|\operatorname{Out} G| = 2$. By Theorem 4.3.3, the character ring of the relevant representations is the p-modular reduction of $\mathbb{Z}[r_2]$, where $p \neq 2, 7$. Therefore, by Table 4.2, $G < \operatorname{Sp}_6(p)$ when $p \equiv \pm 1 \pmod 8$, $G < \operatorname{Sp}_6(p^2)$ when $p \equiv \pm 3 \pmod 8$, and for no other q are there $\mathscr{S}_1$-subgroups of $\operatorname{Sp}_6(q)$ isomorphic to G.

There are two (algebraically conjugate) representations, both with stabiliser of order 2, and so by Lemma 4.4.3 there are two classes of subgroups $G < \operatorname{CSp}_6(q)$. The representations of $G.2^+$ in [12, 57] consist of isometries and have character ring $\mathbb{Z}[r_2, y_{16}]$. By Lemma 4.2.1, $y_{16} \in \mathbb{F}_q$ if and only if $q \equiv \pm 1 \pmod{16}$. So $G.2 < \operatorname{Sp}_6(p)$ if and only if $p \equiv \pm 1 \pmod{16}$, and otherwise (the $\operatorname{Sp}_6(p)$-class of) G is stabilised by the outer automorphism δ of Ω. (Note that if $p \equiv \pm 3 \pmod 8$ then $p^2 \not\equiv \pm 1 \pmod{16}$.) $\square$

Proposition 4.5.23 (i) *If $p \equiv \pm 1, \pm 3, \pm 4 \pmod{13}$, then $\operatorname{Sp}_6(p)$ has ex-*

actly two conjugacy classes of $\mathscr{S}_1$-subgroups isomorphic to $2^{\cdot}L_2(13)$. The subgroups are scalar-normalising in $\mathrm{CSp}_6(p)$ and the classes are fused by the diagonal automorphism of $\mathrm{Sp}_6(p)$.

(ii) If $p \equiv \pm 2, \pm 5, \pm 6 \pmod{13}$, then $\mathrm{Sp}_6(p^2)$ has exactly $(p-1,2)$ conjugacy classes of $\mathscr{S}_1$-subgroups isomorphic to $2^{\cdot}L_2(13)$ (or $L_2(13)$ when $p = 2$). The subgroups are scalar-normalising in $\mathrm{CSp}_6(p^2)$ and when p is odd the classes are fused by the diagonal automorphism of $\mathrm{Sp}_6(p^2)$.

For all other q, there are no $\mathscr{S}_1$-subgroups $2^{\cdot}L_2(13)$ or $L_2(13)$ of $\mathrm{Sp}_6(q)$.

Proof Let $G = 2^{\cdot}L_2(13)$ or $L_2(13)$. Then $|\operatorname{Out}G| = 2$. For $G = L_2(13)$ in characteristic 2 or $2^{\cdot}L_2(13)$ in characteristics not equal to 2 or 13, there are two representations interchanged by the outer automorphism of G, and hence there is a single class $G < \mathrm{CSp}_6(q)$. They have character ring the p-modular reduction of $\mathbb{Z}[b_{13}]$, which lies in $\mathbb{F}_p$ if and only if p is a square modulo 13. So $G < \mathrm{Sp}_6(p)$ in those cases, while $G < \mathrm{Sp}_6(p^2)$ when p is a non-square modulo 13. For no other q are there $\mathscr{S}_1$-subgroups of $\mathrm{Sp}_6(q)$ isomorphic to G. $\square$

Proposition 4.5.24 *The group $\mathrm{Sp}_6(9)$ has exactly two conjugacy classes of $\mathscr{S}_1$-subgroups isomorphic to $2^{\cdot}A_7$. The subgroups are scalar-normalising in $\mathrm{CSp}_6(9)$ and the classes are fused by the diagonal automorphism of $\mathrm{Sp}_6(9)$.*

Proof Let $G = 2^{\cdot}A_7$. Then $|\operatorname{Out}G| = 2$. By Theorem 4.3.3, the relevant representations of G have character ring the 3-modular reduction of $\mathbb{Z}[r_2]$, which is $\mathbb{F}_9$ by Table 4.2. There are two such representations, interchanged by the outer automorphism of G, and hence there is a single class of scalar-normalising groups $G < \mathrm{CSp}_6(9)$, by Lemma 4.4.3. $\square$

Proposition 4.5.25 (i) *The group $\mathrm{Sp}_6(2)$ has a single conjugacy class of $\mathscr{S}_1$-subgroups isomorphic to $U_3(3).2$. These groups are scalar-normalising in $\mathrm{CSp}_6(2)$.*

(ii) *If $p \equiv \pm 1 \pmod{12}$, then $\mathrm{Sp}_6(p)$ has exactly two conjugacy classes of $\mathscr{S}_1$-subgroups isomorphic to $(2 \times U_3(3)).2$. Each of these subgroups is scalar-normalising in $\mathrm{CSp}_6(p)$, and the classes are fused by the diagonal automorphism of $\mathrm{Sp}_6(p)$.*

(iii) *If $p \equiv \pm 5 \pmod{12}$, then $\mathrm{Sp}_6(p)$ has a single conjugacy class of $\mathscr{S}_1$-subgroups isomorphic to $U_3(3)$. The normalisers of these subgroups G in $\mathrm{CSp}_6(p)$ are generated by G, scalars, and an element of $\mathrm{CSp}_6(p)$ that is a similarity but not an isometry of the symplectic form.*

For all other q, there are no $\mathscr{S}_1$-subgroups $U_3(3)$ of $\mathrm{Sp}_6(q)$.

Proof Let $G = U_3(3)$. Then $|\operatorname{Out}G| = 2$. By Theorem 4.3.3, the relevant representations of G have character ring the p-modular reduction of $\mathbb{Z}$, for all

$p \neq 3$, so $G \leqslant \mathrm{Sp}_6(p)$ for all p, and for no other q are there $\mathscr{S}_1$-subgroups of $\mathrm{Sp}_6(q)$ isomorphic to G.

There is a single representation, with stabiliser of order 2. In characteristic 2, we find from [57] that $G.2 < \mathrm{Sp}_6(2)$. Otherwise, the representations of $G.2$ in [12, 57] involve i_3, and contain similarities of the symplectic form that are not isometries. By multiplying elements outside of G by a scalar of order 4, we obtain a representation of a group with the structure $(2 \times \mathrm{U}_3(3)).2$, which consists of isometries and involves the irrationality r_3, which lies in $\mathbb{F}_p$ if and only if $p \equiv \pm 1 \pmod{12}$. So $(2 \times \mathrm{U}_3(3)).2 < \mathrm{Sp}_6(p)$ if and only if $p \equiv \pm 1 \pmod{12}$, and otherwise the class of G is stabilised by the outer automorphism δ of Ω. $\square$

Proposition 4.5.26   (i) *The group* $\mathrm{Sp}_6(5)$ *has exactly one class of* $\mathscr{S}_1$-*subgroups* G *isomorphic to* $2^{\cdot}\mathrm{J}_2$. *The normalisers of these subgroups* G *in* $\mathrm{CSp}_6(5)$ *are generated by* G, *scalars, and an element of* $\mathrm{CSp}_6(5)$ *that is a similarity but not an isometry of the symplectic form.*

(ii) *If* $p \equiv \pm 1 \pmod 5$, *then* $\mathrm{Sp}_6(p)$ *has exactly two conjugacy classes of* $\mathscr{S}_1$-*subgroups isomorphic to* $2^{\cdot}\mathrm{J}_2$. *The subgroups are scalar-normalising in* $\mathrm{CSp}_6(p)$ *and the classes are fused by the diagonal automorphism of* $\mathrm{Sp}_6(p)$.

(iii) *If* $p \equiv \pm 2 \pmod 5$, *then* $\mathrm{Sp}_6(p^2)$ *has exactly* $(p-1, 2)$ *conjugacy classes of* $\mathscr{S}_1$-*subgroups isomorphic to* $2^{\cdot}\mathrm{J}_2$ *(or* J_2 *when* $p = 2$). *The subgroups are scalar-normalising in* $\mathrm{CSp}_6(p^2)$ *and, when* p *is odd, the two classes are fused by the diagonal automorphism of* $\mathrm{Sp}_6(p^2)$.

For all other q, *there are no* $\mathscr{S}_1$-*subgroups* J_2 *or* $2^{\cdot}\mathrm{J}_2$ *of* $\mathrm{Sp}_6(q)$.

Proof Let $G = 2^{\cdot}\mathrm{J}_2$, or J_2 when $p = 2$. Then $|\mathrm{Out}\,G| = 2$. By Theorem 4.3.3, in characteristic $p \neq 5$, the character ring of the relevant representation of G is the p-modular reduction of $\mathbb{Z}[\mathrm{b}_5]$. The irrationality b_5 lies in $\mathbb{F}_p$ if and only if $p \equiv \pm 1 \pmod 5$. So $G < \mathrm{Sp}_6(p)$ in those cases, $G < \mathrm{Sp}_6(p^2)$ if $p \equiv \pm 2 \pmod 5$, and for no other power of p are there $\mathscr{S}_1$-subgroups of $\mathrm{Sp}_6(q)$ isomorphic to G. There are two representations interchanged by $\mathrm{Out}\,G$, and hence there is a single class of scalar-normalising groups $G < \mathrm{CSp}_6(q)$, by Lemma 4.4.3.

In characteristic 5, there is a single representation of $G = 2^{\cdot}\mathrm{J}_2$, stabilised by the outer automorphism of G. From [57] we find that $G < \mathrm{Sp}_6(5)$ but elements in $G.2 \setminus G$ involve irrationalities that lie outside of $\mathbb{F}_5$, so the class of G is stabilised by the outer automorphism of Ω. $\square$

4.6 Determining the effects of duality and field automorphisms

At this point, we have determined the conjugacy classes of $\mathscr{S}_1$-subgroups of the quasisimple and conformal classical groups in dimensions up to 6. It remains to consider the actions of the duality and field automorphisms, and in this section we will present some general theory for how to do so. The calculations will then be carried out in Section 4.7 for dimension up to 6 and in Section 4.9 for dimensions 7 to 12.

The ordinary and modular character tables of the quasisimple groups G that arise as $\mathscr{S}_1$-subgroups of a quasisimple classical group Ω enable us to determine which automorphisms of G can be realised by conjugation by an element of the general linear group in which Ω lies (that is, which automorphisms can be realised inside the conformal classical group C corresponding to Ω): see Section 1.8 and Subsection 4.4.2. The information in [12] also enables the experienced user to determine which outer automorphisms of G are induced by graph and field automorphisms of Ω.

Let $\beta = \gamma$ or ϕ be a duality or field automorphism of the general linear group containing Ω: note that γ acts as complex conjugation on Brauer character values, whilst ϕ replaces each eigenvalue by its p-th power. Recall from Section 1.7 that, with the exception of the case $\Omega = \Omega_n^-(q)$ for certain values of n and q, which we are not considering in this section, β normalises Ω. If β has the same action on a representation ρ as some $\alpha \in \operatorname{Aut} G$ then ρ^β and $^\alpha\rho$ are equivalent, and by Lemma 1.8.10 this equivalence is effected by some element g of the conformal classical group C containing Ω. So the action of β on $G\rho$, followed by conjugation c_g by g, normalises and induces α on $G\rho$. Therefore the extension of $G/Z(G)$ by α occurs in the almost simple extension of $\overline{\Omega}$ by βc_g, and we need to identify c_g as an element of $\operatorname{Out} \overline{\Omega}$.

The reader might want to recall our general notation for automorphisms of Ω from Section 1.6.3, together with our general discussion of outer automorphisms in Section 1.3 and Lemma 1.3.1.

4.6.1 Cases L and U

In Case **L**, the automorphisms to consider are γ, which we will take to be the inverse-transpose automorphism, and ϕ, the automorphism that replaces each matrix entry by its p-th power.

In Case **U**, the rest of the outer automorphism group is generated by the field automorphism ϕ; recall from just before Definition 1.6.17 that the action of ϕ, and hence the stabiliser of the class of a quasisimple $\mathscr{S}$-subgroup, can depend on the choice of the unitary form, and we are using I_n as that form. We write γ for the duality automorphism of $U_n(q)$, and σ for the q-th power map. In many cases of interest, q is prime so that for the unitary form with matrix I_n the automorphisms ϕ, σ and γ are all equal: in this instance, to achieve uniformity with Case **L**, we will normally write γ for the outer automorphism.

As discussed in the preamble to this section, for many of the representations

ρ in Cases **L** and **U** in which we are interested, there exists an automorphism $\alpha \in \operatorname{Aut} \Omega$ such that ${}^{\alpha}\rho = \rho \beta c_g$, where $\beta \in \{\phi, \gamma\}$ and c_g is conjugation by the element g of the corresponding conformal classical group. Therefore $\beta \delta^i$ will stabilise the Ω-class of $G\rho$ for some i with $0 \leqslant i < d$, where $d = |\delta| = (q - \varepsilon, n)$, (where $\varepsilon = 1$ and -1 in Cases **L** and **U**, respectively).

In the tables in Chapter 8 (with the exception of those for $\operatorname{Sp}_4(2^e)$ and $\Omega_8^+(q)$, which do not concern us here), each row represents a set of c conjugacy classes of subgroups of the classical group Ω, which together form an orbit of the action of $\operatorname{Out} \Omega$ on the set of its conjugacy classes of subgroups. The class stabiliser listed in the tables is the stabiliser in $\operatorname{Out} \Omega$ of one of these classes. The stabilisers of the other classes are the conjugates under $\operatorname{Out} \Omega$ of the specified stabiliser. So we are only interested in determining the stabiliser up to conjugacy in $\operatorname{Out} \Omega$, and we generally try to choose the class for which the stabiliser has the nicest set of generators.

Lemma 4.6.1 *Let $\rho : G \to \operatorname{SL}_n^\varepsilon(q) = \Omega$ be a representation, and let $d = (q - \varepsilon, n)$ be odd. If there exists $\alpha \in \operatorname{Aut} G$ such that ${}^{\alpha}\rho$ is equivalent to ρ^{γ}, then the stabiliser of the class of $G\rho$ in Ω contains a conjugate of γ in $\operatorname{Out} \Omega$.*

Proof Since $\delta^{\gamma} = \delta^{-1}$ and d is odd, all elements $\gamma \delta^i$ are conjugate in $\langle \gamma, \delta \rangle$. $\square$

We shall now describe how to determine the stabiliser in the case where d is even. We first present the theory involved and then discuss how to carry out the necessary calculations.

Theory for Case L.

Lemma 4.6.2 *Let $\rho : G \to \operatorname{SL}_n(q)$ be a representation with $d = (q - 1, n)$ even. Let $\beta \in \langle \phi, \gamma \rangle$, and assume that $\beta \delta^i$ and $\beta \delta^j$ are conjugate by a power of δ when $i - j$ is even. In particular, this assumption holds with $\beta = \gamma$.*

Assume that there exists $\alpha \in \operatorname{Out} G$ and $L \in \operatorname{GL}_n(q)$ such that $L^{-1}(x\rho)^{\beta} L = (x^{\alpha})\rho$ for all $x \in G$. Then the class of $G\rho$ in $\operatorname{SL}_n(q)$ is stabilised by a conjugate of β in $\operatorname{Out} \operatorname{SL}_n(q)$ if $\det L$ is a square in $\mathbb{F}_q^{\times}$, and by a conjugate of $\beta \delta$ otherwise.

Proof Note that $L^{-1}(G\rho)^{\beta} L = (G^{\alpha})\rho = G\rho$, so βc_L normalises $G\rho$, where c_L is conjugation by L. Thus the class of $G\rho$ is stabilised by $\beta \delta^i$, where δ^i is the image in $\operatorname{Out} \operatorname{SL}_n(q)$ of c_L. By assumption, $\beta \delta^i$ is conjugate in $\operatorname{Out} \Omega$ to β when i is even and to $\beta \delta$ when i is odd. The result follows from the fact that i is even if and only if $\det L$ is a square in $\mathbb{F}_q^{\times}$. The assumption holds with $\beta = \gamma$, because $\langle \delta, \gamma \rangle$ is dihedral of order $2d$ with d even. $\square$

Theory for Case U. This is complicated by the fact that we shall need to carry out calculations in the images of representations that preserve unitary forms B other than I_n. Some of the results proved here will also be used in Chapter 5 to solve corresponding problems in the defining characteristic candidates. Note that the maps $\gamma = -\mathsf{T}$, σ and ϕ preserve $\mathrm{GL}_n(q^2)$, $\mathrm{SL}_n(q^2)$, $\mathrm{GU}_n(q, I_n)$ and $\mathrm{SU}_n(q, I_n)$ and are automorphisms thereof, but they need not normalise $\mathrm{SU}_n(q, B)$ for other unitary forms B.

The following lemma is also proved in [6, Lemma 5].

Lemma 4.6.3   *If q is odd, n is even, and $d := (q+1, n)$, then the sets*

$$\{\, \phi\delta^{2i} \;:\; 0 \leqslant i \leqslant d/2 - 1 \,\} \text{ and } \{\, \phi\delta^{2i+1} \;:\; 0 \leqslant i \leqslant d/2 - 1 \,\}$$

are conjugacy classes in $\mathrm{Out}\,\mathrm{U}_n(q)$. Otherwise $\{\, \phi\delta^i \;:\; 0 \leqslant i \leqslant d - 1 \,\}$ is a single class.

Proof   Note that $(\phi\delta^i)^\delta = \delta^{-1}\phi\delta^{i+1} = \phi\phi^{-1}\delta^{-1}\phi\delta^{i+1} = \phi\delta^{i+1-p}$. Therefore the conjugacy class of $\phi\delta^i$ contains $S_i := \{\, \phi\delta^{i+j(1-p)} \;:\; 0 \leqslant j \leqslant d - 1 \,\}$. The number of such sets S_i is $(d, p-1) = (n, q+1, p-1)$, and the size of S_i is independent of i. Since $p-1$ divides $q-1$, the greatest common divisor of $p-1$ and $q+1$ is $(p-1, 2)$. Hence the number of such conjugacy classes is at most 2 if p is odd and n is even, and is 1 otherwise. If p is odd and n is even then, since $(\phi\delta^i)^\phi = \phi\delta^{ip}$, we see that S_i is invariant under conjugation by δ and ϕ, and hence that there are indeed two conjugacy classes, S_0 and S_1. $\square$

Lemma 4.6.4   *Let $\rho : G \to \mathrm{GL}_n(q^2)$ be an absolutely irreducible representation, with $d := (q+1, n)$ even, and $G\rho \leqslant \Omega = \mathrm{SU}_n(q, B) \cong \mathrm{SU}_n(q)$ for some non-degenerate unitary form B, and let β be one of the maps γ or ϕ. Assume that there exists $\alpha \in \mathrm{Aut}\,G$ such that ρ^β is equivalent to $^\alpha\rho$. Then:*

(i) *There exists $A \in \mathrm{GL}_n(q^2)$ such that $AA^{\sigma\mathsf{T}} = B$, and hence the conjugate $(G\rho)^A \leqslant \mathrm{SU}_n(q, I_n) = \mathrm{SU}_n(q)$.*

(ii) *There exists $L \in \mathrm{GL}_n(q^2)$ conjugating $(x\rho)^\beta$ to $(x^\alpha)\rho$ for all $x \in G$, and $LBL^{\sigma\mathsf{T}} = \lambda B^\beta$ for some $\lambda \in \mathbb{F}_q^\times$.*

(iii) *There exists $C \in \mathrm{GU}_n(q)$ conjugating $(A^{-1}(x\rho)A)^\beta$ to $A^{-1}(x^\alpha)\rho A$, for all $x \in G$.*

(iv) *For any such element C, the class of $(G\rho)^A$ in $\mathrm{SU}_n(q)$ is stabilised by a conjugate of β in $\mathrm{Out}\,\mathrm{SU}_n(q)$ when $\det C$ is a square in the cyclic group $X := \{\, \xi : \xi \in \mathbb{F}_{q^2}^\times \mid \xi\xi^\sigma = 1 \,\}$, and by a conjugate of $\beta\delta$ otherwise.*

Proof   As we saw in Section 1.5, all non-degenerate unitary forms of dimension n over $\mathbb{F}_{q^2}$ are isometric, and so Part (i) holds. Then $\Omega^A = \mathrm{SU}_n(q, I_n)$.

The fact that ρ^β is equivalent to $^\alpha\rho$ immediately implies the existence of an L inducing the equivalence, proving the first claim of Part (ii). Let $D = A^{-\beta}LA$, so that $L = A^\beta DA^{-1}$. Since $G\rho$ is absolutely irreducible, the matrices D and L are determined up to scalar multiplication.

The sets

$$\{(A^{-1}(x\rho)A)^\beta \mid x \in G\} = (A^{-1}(G\rho)A)^\beta = ((G\rho)^A)^\beta$$

and

$$\{A^{-1}(x^\alpha)\rho A \mid x \in G\} = A^{-1}(G\rho)A = (G\rho)^A$$

both form subgroups of $\Omega^A = \mathrm{SU}_n(q, I_n)$. Since $G\rho$ is absolutely irreducible and D conjugates $((G\rho)^A)^\beta$ to $(G\rho)^A$, it follows from Lemma 1.8.9 that the matrix $D \in \mathrm{CGU}_n(q)$. Therefore $DD^{\sigma\mathsf{T}} = \lambda I_n$ for some $\lambda \in \mathbb{F}_q^\times$.

Since β commutes with $\sigma\mathsf{T}$ and is an automorphism of $\mathrm{GL}_n(q^2)$,

$$LBL^{\sigma\mathsf{T}} = A^\beta DA^{-1}AA^{\sigma\mathsf{T}}A^{-\sigma\mathsf{T}}D^{\sigma\mathsf{T}}(A^\beta)^{\sigma\mathsf{T}}$$
$$= A^\beta DD^{\sigma\mathsf{T}}(A^{\sigma\mathsf{T}})^\beta = \lambda A^\beta (A^{\sigma\mathsf{T}})^\beta = \lambda B^\beta,$$

as required. This completes the proof of Part (ii).

Since $\lambda \in \mathbb{F}_q^\times$, there exists $\mu \in \mathbb{F}_{q^2}^\times$ such that $\mu\mu^\sigma = \mu^{1+q} = \lambda$. We can define $C := (1/\mu)D = (1/\mu)A^{-\beta}LA \in \mathrm{GU}_n(q)$, proving Part (iii).

The determinants of elements of $\mathrm{GU}_n(q)$ lie in $X \leqslant \mathbb{F}_{q^2}^\times$, so Part (iv) follows as in the proof of Lemma 4.6.2, using Lemma 4.6.3 for the case $\beta = \phi$. $\square$

Finally we record the following result, which will be useful when carrying out p-modular reduction of characteristic 0 representations.

Lemma 4.6.5 *With the same notation and assumptions as in Lemma 4.6.4, let L satisfy the conclusion of Lemma 4.6.4 (ii). Then $\det L$ is a square in $\mathbb{F}_{q^2}^\times$.*

Let $\kappa \in \mathbb{F}_{q^2}^\times$ with $\kappa^2 = \det L$. Define η to be equal to $(1/\lambda^{n/2})\kappa^{1+\sigma} \det B$ when $\beta = -\mathsf{T}$, and to $(1/\lambda^{n/2})\kappa^{1+\sigma}(\det B)^{(1-p)/2}$ when $\beta = \phi$. Then the class of $(G\rho)^A$ in $\mathrm{SU}_n(q)$ is stabilised by a conjugate of β in $\mathrm{Out}\,\mathrm{SU}_n(q)$ when $\eta = 1$, and by a conjugate of $\beta\delta$ when $\eta = -1$.

Proof Let the matrix C satisfy the conclusion of Lemma 4.6.4 (iii). Since $\det C \in X$ (with X as defined in Lemma 4.6.4 (iv)) and $(q^2 - 1)/(q+1) = q - 1$ is even, $\det C$ is a square in $\mathbb{F}_{q^2}^\times$.

By Lemma 4.6.4 (iv), the class of $(G\rho)^A$ is stabilised by a conjugate of β when $\vartheta := \sqrt{\det C} \in X$, and by a conjugate of $\beta\delta$ otherwise. But $\vartheta \in X$ if and only if the norm $\vartheta^{1+\sigma}$ of ϑ is 1; otherwise it is -1. (Note that ξ and $-\xi$ have the same norm for all $\xi \in \mathbb{F}_{q^2}^\times$, so the choice of ϑ is unimportant.) Since $L = \mu A^\beta C A^{-1}$, n is even, and $\vartheta \in \mathbb{F}_{q^2}^\times$, the element $\kappa = \sqrt{\det L} \in \mathbb{F}_{q^2}^\times$, and

$$\vartheta = \begin{cases} (1/\mu^{n/2})\kappa \det A & \text{if } \beta = -\mathsf{T}, \\ (1/\mu^{n/2})\kappa(\det A)^{(1-p)/2} & \text{if } \beta = \phi, \end{cases}$$

and so $\eta = \vartheta^{1+\sigma}$, which proves the lemma. $\square$

Representations in characteristic 0. We now consider the situation in which the representation ρ arises as the p-modular reduction of a characteristic 0 representation $\hat{\rho}$. As we remarked in Subsection 4.1.1, for dimensions up to 12, each representation $\hat{\rho}$ that arises in Cases **L** and **U** can in fact be realised over its character field F, for all ρ in which we are interested. Lemma 4.4.1 states that in Case **U** the map σ corresponds to the p-modular reduction of complex conjugation.

Note that the ring generated by the entries of $x\hat{\rho}$ for $x \in G$ will be a subring of $R[\frac{1}{p_1}, \ldots, \frac{1}{p_s}]$ for some (finite number of) primes $p_1, \ldots, p_s$, where R is the character ring of $\hat{\rho}$. We cannot reduce $\hat{\rho}$ modulo these primes, and we shall call these *exceptional primes*.

Let β be the duality map $-\mathsf{T}$ (defined now on matrices over F), and suppose that $\hat{\rho}^\beta$ is equivalent to $\,^\alpha\hat{\rho}$, for some $\alpha \in \mathrm{Out}\, G$.

In the linear case, let $\hat{L} \in \mathrm{GL}_n(F)$ satisfy $\hat{L}^{-1}(x\hat{\rho})^\beta \hat{L} = (x^\alpha)\hat{\rho}$ for all $x \in G$. In the situations in which we are interested, $\hat{L}$ reduces modulo p to the matrix L in Lemma 4.6.2, and $\det L$ is the reduction modulo p of $\det \hat{L}$. This will work, provided that $\det L \neq 0$ and both $\hat{L}$ and $\hat{L}^{-1}$ lie in a ring $R[\frac{1}{p_1}, \ldots, \frac{1}{p_s}]$ with $p \neq p_i$ for all i. Since $\hat{L}$ is only determined modulo a scalar, we can attempt to achieve this by multiplying it by a suitable scalar. If we do not succeed, then we shall again refer to p as an *exceptional prime*.

Now suppose that we are in the unitary case. Then the image of $\hat{\rho}$ consists of isometries of a positive definite σ-Hermitian form $\hat{B}$ over F, and we can take B to be the p-modular reduction of $\hat{B}$, provided that this is non-degenerate.

By the standard theory of σ-Hermitian matrices, all forms $\hat{B}$ are equivalent over $\mathbb{C}$, so there exists a complex matrix $\hat{A}$ with $\hat{A}\hat{A}^{\sigma\mathsf{T}} = \hat{B}$. But, as we shall see in Proposition 4.7.7, there may be no such $\hat{A}$ with entries in F.

For $\hat{B}$ to be reducible modulo p to the matrix of a non-degenerate unitary form we require both $\hat{B}$ and $\hat{B}^{-1}$ to lie in a ring $R[\frac{1}{p_1}, \ldots, \frac{1}{p_s}]$ with $p \neq p_i$ for all i. Since $\hat{B}$ is only determined modulo a scalar, we can attempt to achieve this by multiplying it by a suitable scalar. If we do not succeed, then we shall once again call p an *exceptional prime*.

The results that we need to calculate are for the unitary group that is the isometry group of I_n; that is, we need to work in the subgroup G^A inside Ω^A, where A is the p-modular reduction of $\hat{A}$. This is inconvenient in characteristic 0, mainly because of the problem just described, that we cannot necessarily choose $\hat{A}$ with entries in F.

We start by finding a matrix $\hat{L} \in \mathrm{GL}_n(F)$ satisfying $\hat{L}^{-1}(x\hat{\rho})^\beta \hat{L} = (x^\alpha)\hat{\rho}$ for all $x \in G$ and, by multiplying $\hat{L}$ by a suitable scalar, we attempt to choose $\hat{L}$ such that both $\hat{L}$ and $\hat{L}^{-1}$ lie in a ring $R[\frac{1}{p_1}, \ldots, \frac{1}{p_s}]$ with $p \neq p_i$ for all i. If we do not succeed, then we shall yet again refer to p as an *exceptional prime*.

The proposition below shows how, in the case $\beta = -\mathsf{T}$, we can extract the information we require without knowing $\hat{A}$ explicitly. (We know of no such method when $\beta = \phi$, but fortunately this turned out not to be required in the examples up to dimension 12.)

Proposition 4.6.6 *Suppose that G has an absolutely irreducible representation ρ with image in $\mathrm{SU}_n(q, B)$ that arises as the p-modular reduction of a characteristic 0 representation $\hat{\rho}$ over the character field $F \subset \mathbb{C}$ of $\hat{\rho}$, whose image preserves a form $\hat{B}$. Suppose that, for some $\alpha \in \mathrm{Aut}\, G$, there exists $\hat{L} \in \mathrm{GL}_n(F)$ that conjugates $(x\hat{\rho})^{\gamma}$ to $(x^{\alpha})\hat{\rho}$ for all $x \in G$. Assume also that $\hat{B}$, $\hat{B}^{-1}$, $\hat{L}$ and $\hat{L}^{-1}$ have entries in a ring $S := R[\frac{1}{p_1}, \ldots, \frac{1}{p_s}]$ with $p \neq p_i$ for all i.*

Suppose that $\det \hat{L}$ factorises in S as $\hat{\nu}^2\hat{\zeta}$, with $\hat{\zeta} \in \mathbb{R}$. Let ζ be the p-modular reduction of $\hat{\zeta}$, and define $\varepsilon \in \{1, -1\}$ by $\varepsilon = 1$ if $\sqrt{\zeta} \in \mathbb{F}_q^{\times}$ and $\varepsilon = -1$ otherwise.

Then, for a suitable $A \in \mathrm{GL}_n(q^2)$, the group $(G\rho)^A \leqslant \mathrm{SU}_n(q)$. The class of $(G\rho)^A$ in $\mathrm{SU}_n(q)$ is stabilised by a conjugate of γ in $\mathrm{Out}\,\mathrm{SU}_n(q)$ when $\varepsilon\,\mathrm{sgn}(\hat{\zeta}) = 1$, and by a conjugate of $\gamma\delta$ otherwise.

Proof As explained above, there is a complex matrix $\hat{A}$ with $\hat{A}\hat{A}^{\sigma\mathsf{T}} = \hat{B}$. As in the proof of Lemma 4.6.4, $\hat{D} := \hat{A}^{-\beta}\hat{L}\hat{A}$ conjugates $((G\hat{\rho})^{\hat{A}})^{\beta}$ to $(G\hat{\rho})^{\hat{A}}$, which are both absolutely irreducible subgroups of $\mathrm{SU}_n(\mathbb{C}, I_n)$ so, by Lemma 1.8.9, $\hat{D} \in \mathrm{CGU}_n(\mathbb{C}, I_n)$ and hence $\hat{D}\hat{D}^{\sigma\mathsf{T}} = \hat{\lambda}I_n$, where $\hat{\lambda}$ is real and positive. In fact $\hat{B} = \hat{A}\hat{A}^{\sigma\mathsf{T}}$ is positive definite, so $\det \hat{B}$ is also real and positive. Then $\hat{L}\hat{B}\hat{L}^{\sigma\mathsf{T}} = \hat{\lambda}\hat{B}^{\beta}$, so $(\det \hat{L})(\det \hat{L})^{\sigma}(\det \hat{B})^2 = \hat{\lambda}^n$.

We have assumed that we can factorise $\det \hat{L}$ in S as $\hat{\nu}^2\hat{\zeta}$ with $\hat{\zeta} \in \mathbb{R}$. Then $(\det \hat{L})^{\sigma} = \hat{\nu}^{2\sigma}\hat{\zeta}$, and $\hat{\nu}^2\hat{\nu}^{2\sigma}\hat{\zeta}^2(\det \hat{B})^2 = \hat{\lambda}^n$. Since $\hat{\nu}\hat{\nu}^{\sigma}$, $\det \hat{B}$ and $\hat{\lambda}$ are all real and positive,

$$\frac{\hat{\nu}\hat{\nu}^{\sigma}\hat{\zeta}\det \hat{B}}{\hat{\lambda}^{n/2}} = \mathrm{sgn}(\hat{\zeta}).$$

Now let L be the reduction modulo p of $\hat{L}$, and let ν and ζ be the reduction modulo p of $\hat{\nu}$ and $\hat{\zeta}$. Then $\hat{\zeta} \in \mathbb{R}$ implies that $\zeta \in \mathbb{F}_q^{\times}$, and we can choose $\kappa = \sqrt{\det L} = \nu\sqrt{\zeta} \in \mathbb{F}_{q^2}^{\times}$. Then $\kappa^{1+\sigma} = \varepsilon\nu\nu^{\sigma}\zeta$, where $\varepsilon \in \{1, -1\}$ and $\varepsilon = 1$ if and only if $\sqrt{\zeta} \in \mathbb{F}_q^{\times}$. The result now follows from Lemma 4.6.5. (But note that the matrix A is not necesarrily the p-modular reduction of $\hat{A}$.)   □

Computational considerations. We now discuss how we use the theory just described to carry out the necessary calculations in practice.

Representations are defined by specifying the images of the two *standard generators* of a quasisimple group G, which are generators for which it is computationally easy to find corresponding elements (i.e. images under an isomorphism) in any permutation or matrix representation of G; see [111] for details.

The representations over finite fields can generally be found easily using MeatAxe techniques; see, for example [44]. There are various sources from which representations are available over the integers and number fields. One such source is [111], and MAGMA also has its own library. These are available over the character field of the representation whenever possible, which is the case for all required representations in Cases **L** and **U** up to dimension 12. (As we remarked in the preface, all of the matrices required for our calculations are stored in data files on the webpages that accompany this book.)

For a representation ρ, it is straightforward to compute ρ^β when β is duality or a field automorphism. For automorphisms α of G, words for x^α and y^α in the standard generators x and y of G have been computed for all groups that we will need to consider [111], which makes the computation of (the matrices for) $^\alpha\rho$ straightforward.

Finding a matrix inducing a specific equivalence between representations can be done easily over finite fields, again using MeatAxe techniques [44]. This problem is more difficult in characteristic 0, but for the dimensions in question, it can be done routinely using the MAGMA command GHom.

For computations in specific characteristics, we can compute all of the matrices involved in order to apply Lemma 4.6.2 (Case **L**) or 4.6.4 (Case **U**) and the problem reduces to deciding whether $\det L$ is a square in $\mathbb{F}_q^\times$ or $\det C$ is a square in the subgroup X of $\mathbb{F}_{q^2}^\times$ defined in Lemma 4.6.4, respectively.

In cases in which we are using a characteristic 0 representation, if there are any exceptional primes dividing the denominators in the matrix entries of the generators, or the matrices defining module equivalences, or the form matrix (in Case **U**) then there are only finitely many, so it is straightforward to carry out any necessary calculations in these finitely many cases. However (as we shall see in our calculations, later) it transpired that for the representations of interest to us, there were no such primes.

Thus in the linear case we compute the matrix $\hat{g}$ effecting the relevant equivalence, and we typically need to decide whether the reduction of $\det \hat{g}$ is a square in $\mathbb{F}_q^\times$. In the unitary case, we can often use Lemma 4.6.6 to reduce the problem to deciding whether the reduction modulo p of $\hat{\zeta}$ (with $\hat{\zeta}$ as defined in Lemma 4.6.6) is a square in $\mathbb{F}_q^\times$. Such problems are not difficult in most examples although, as we shall see later, there are some challenging cases.

4.6.2 Case S

This case is easier than Cases **L** and **U**. As we shall see in Sections 4.7 and 4.9, the following proposition will be sufficient to enable us to carry out the required calculations.

Lemma 4.6.7 *All involutions in* $\mathrm{PC\Gamma Sp}_n(q)$ *lie in* $\mathrm{S}_n(q)\langle\phi\rangle \cup \mathrm{PCSp}_n(q)$.

Proof Let $g \in \mathrm{PC\Gamma Sp}_n(q)$ be an involution, and assume that g is the image of $A\sigma$, where $A \in \mathrm{CSp}_n(q)$ and $\sigma \in \mathrm{C\Gamma Sp}_n(q) \setminus \mathrm{CSp}_n(q)$ induces by conjugation a field automorphism of $\mathrm{S}_n(q)$. Furthermore, we may assume that $\mathrm{CSp}_n(q)$ is a standard copy, preserving the form $B = \mathrm{antidiag}(1, 1, \ldots, -1, -1)$. Clearly σ has order at most 2, while if $\sigma = 1$ then $g \in \mathrm{PCSp}_n(q)$. So we may assume that σ has order 2 (so that q is a square and σ induces the field automorphism $x \mapsto x^{\sqrt{q}}$). Now $AA^\sigma = (A\sigma)^2 = \lambda I_n$ for some $\lambda \in \mathbb{F}_q^\times$. Conjugating by A and σ respectively gives $A^\sigma A = \lambda I_n$ and $A^\sigma A = \lambda^\sigma I_n$, and so $\lambda = \lambda^\sigma \in \mathbb{F}_{\sqrt{q}}^\times$. Now $ABA^\mathsf{T} = \mu B$ for some $\mu \in \mathbb{F}_q^\times$, whence conjugating by σ gives $A^\sigma B A^{\sigma\mathsf{T}} = \mu^\sigma B$. Thus $AA^\sigma = \lambda I_n$ scales B by $\mu^{1+\sigma} = \lambda^2$. Since $\lambda \in \mathbb{F}_{\sqrt{q}}^\times$, the element μ is a square in $\mathbb{F}_q^\times$, and so $g \in \mathrm{S}_n(q)\langle\phi\rangle$. $\qquad\square$

4.7 Dimension up to 6: graph and field automorphisms

In this section we calculate the actions of duality and field automorphisms on the conjugacy classes of $\mathscr{S}_1$-subgroups in dimensions up to 6, using the general methods described in the previous section.

4.7.1 Cases L and U

We will consider $\mathrm{SL}_2(q) = \mathrm{Sp}_2(q)$ under Case **S**: see Subsection 4.7.2.

Dimension 3. The arguments in dimension 3 are reasonably straightforward, but we present them in full.

Theorem 4.7.1 *Let Ω be either $\mathrm{SL}_3(q)$ or $\mathrm{SU}_3(q)$, let G be an $\mathscr{S}_1$-subgroup of Ω, and let $d = (q - 1, 3)$ in Case **L** and $(q + 1, 3)$ in Case **U**. Then one of the following holds:*

(i) $G = d \times \mathrm{L}_3(2)$ *with $q = p$, and the stabiliser in $\mathrm{Out}\,\bar{\Omega}$ of at least one of the d classes of G in Ω is $\langle\gamma\rangle$.*

(ii) $G = 3^{\cdot}\mathrm{A}_6$, *with $q = p \equiv \pm 1 \pmod 5$, and the stabiliser in $\mathrm{Out}\,\bar{\Omega}$ of at least one of the three classes of G in Ω is $\langle\gamma\rangle$, which induces the 2_2 automorphism of G.*

(iii) $G = 3^{\cdot}\mathrm{A}_6$, *with $p \equiv \pm 2 \pmod 5$, and the stabiliser in $\mathrm{Out}\,\mathrm{L}_3(p^2)$ of at least one of the three classes of G in $\mathrm{SL}_3(p^2)$ is $\langle\gamma, \phi\rangle$, where γ induces the 2_2 automorphism of G, and ϕ induces the 2_1 automorphism of G when $p \equiv 2, 8 \pmod{15}$ and the 2_3 automorphism of G otherwise.*

(iv) $G = 3\,\dot{}A_6\,\dot{}2_3$, and the stabiliser in $\operatorname{Out} U_3(5)$ of at least one of the three classes of G in $\operatorname{SU}_3(5)$ is $\langle \gamma \rangle$.

(v) $G = 3\,\dot{}A_7$, and the stabiliser in $\operatorname{Out} U_3(5)$ of at least one of the three classes of G in $\operatorname{SU}_3(5)$ is $\langle \gamma \rangle$.

Proof By Theorem 4.3.3 the quasisimple groups that we must consider are $L_3(2)$, $3\,\dot{}A_6$ and $3\,\dot{}A_7$.

Let $G = L_3(2)$. Then $|\operatorname{Out} G| = 2$, and by Proposition 4.5.2 there are $d = (p \pm 1, 3)$ classes of G in $\Omega = \operatorname{SL}_3(p)$ or $\operatorname{SU}_3(p)$. The automorphism δ acts transitively on the classes and d is odd, so the result follows from Lemma 4.6.1.

Similarly, if $G = 3\,\dot{}A_7$ then by Proposition 4.5.4 there are three classes of G in $\operatorname{SU}_3(5)$ permuted transitively by δ. Therefore by Lemma 4.6.1 we may choose the class stabiliser of G to be $\langle \gamma \rangle$.

Finally, let $G = 3\,\dot{}A_6$ or $3\,\dot{}A_6.2_3$. The number of conjugacy classes of G in $\Omega = \operatorname{SL}_3(p), \operatorname{SU}_3(p), \operatorname{SL}_3(p^2)$ is described by Proposition 4.5.3. If q is prime then by Lemma 4.6.1 we may assume that the class stabiliser of G is $\langle \gamma \rangle$. If q is not prime then $\operatorname{Out} \overline{\Omega} = \langle \delta, \gamma, \phi \rangle$, of shape 3.2^2, so up to conjugacy the class stabiliser of G must be $\langle \gamma, \phi \rangle$. The outer automorphisms 2_1 and 2_2 of $3\,\dot{}A_6$ are equivalent modulo $3\,\dot{}A_6.2_3$, so if $G = 3\,\dot{}A_6.2_3$ we are done.

It remains to work out which automorphisms induce which actions when $G = 3\,\dot{}A_6$. The automorphism γ of Ω induces complex conjugation on the character of G. Consulting [12], we see that γ therefore normalises the two conjugacy classes of elements of order 5. Since both the 2_1 and the 2_3 automorphisms fuse these two classes, we deduce that γ induces the 2_2 automorphism. This leaves only the automorphism ϕ of $L_3(p^2)$, with $p \equiv \pm 2 \pmod 5$. The two central elements of order 3 in $3\,\dot{}A_6$ are conjugate under the 2_1 automorphism (since the Schur multiplier of S_6 has order 2), but not under the 2_3 automorphism. Since ϕ interchanges z_3 and z_3^{**} exactly when $p \equiv 2 \pmod 3$, the result follows.　　□

Let us pause to explain exactly what we are asserting about the outer automorphisms, and their actions on the conjugacy classes of $3\,\dot{}A_6$. We describe the case when $p \equiv 2, 8 \pmod{15}$; the other cases are similar. Here the outer automorphism group of $\operatorname{SL}_3(p^2)$ is isomorphic to $2 \times S_3$ and has presentation

$$\langle \delta, \phi, \gamma \mid \delta^3 = \phi^2 = \gamma^2 = 1, \ \delta^\phi = \delta^\gamma = \delta^{-1}, \ \phi^\gamma = \phi \rangle.$$

There are three classes of $3\,\dot{}A_6$, and δ acts on them as a 3-cycle. Since γ and ϕ both invert δ and $\gamma\phi$ centralises δ it follows that the elements $\gamma\delta^i$ and $\phi\delta^i$ for $i \in \{0, 1, 2\}$ act as involutions, whilst $\gamma\phi$ normalises all three classes. Thus one of the classes of $3\,\dot{}A_6$ is normalised by $\{1, \gamma, \phi, \gamma\phi\}$; one by $\{1, \gamma\delta, \phi\delta, \gamma\phi\}$; and one by $\{1, \gamma\delta^2, \phi\delta^2, \gamma\phi\}$. Table 8.4 contains just the first of these stabilisers (as we consider it to be the 'neatest' one).

Dimension 4. By Theorem 4.3.3, the quasisimple groups that we must consider in dimension 4, Cases **L** and **U**, are $2^{\cdot}L_3(2)$ with $p \neq 2, 7$; $2^{\cdot}A_7$ with $p \neq 2, 7$; $2^{\cdot}U_4(2)$ with $p \neq 2, 3$; along with $A_7 < L_4(2) \cong A_8$ and $4_2^{\cdot}L_3(4) < SU_4(3)$.

Dimension 4 in Case **U** is unusual in that all of the required characteristic 0 representations can be written over the character field in such a manner that the resulting matrices are isometries of the form with matrix I_4. This is not always possible to achieve, and even in cases where it is possible it may be hard to do, and result in matrices with extremely complicated entries. Moreover, in some cases we have been able to ensure that even the conjugating matrix $\hat{L}$ from Proposition 4.6.6 is the identity.

The reader can also find details of computer calculations to check Propositions 4.7.2, 4.7.3 and 4.7.4 in files 2u42d4calc, 2a7d4calc, sl27d4calc, but they do not involve exactly the same matrices as we present here. For expositional purposes, we analyse the groups in the order $2^{\cdot}U_4(2)$, $(2, q-1)^{\cdot}A_7$, $2^{\cdot}L_2(7)$, $4_2^{\cdot}L_3(4)$.

Proposition 4.7.2   (i) *If $p \equiv 1 \pmod 6$, then the stabiliser in* $\mathrm{Out}\, L_4(p)$
of at least one of the $(p-1, 4)$ classes of $2^{\cdot}U_4(2)$ in $SL_4(p)$ is $\langle \gamma \rangle$.

(ii) *If $p \equiv 5 \pmod 6$, then the stabiliser in* $\mathrm{Out}\, U_4(p)$ *of at least one of the $(p+1, 4)$ classes of $2^{\cdot}U_4(2)$ in $SU_4(p)$ is* $\langle \gamma \rangle$.

Proof Let $G = 2^{\cdot}U_4(2)$, and recall from Proposition 4.5.7 that $SL_4(p)$ has $(p-1, 4)$ classes of $\mathscr{S}_1$-subgroups isomorphic to G when $p \equiv 1 \pmod 6$ whereas $SU_4(p)$ has $(p+1, 4)$ such classes when $p \equiv 5 \pmod 6$.

We represent G using standard generators x and y, as defined in [111], where x is an involution. An outer automorphism α of $U_4(2)$ can be taken to map the image of (x, y) to the image of $(x^\alpha, y^\alpha) = (x, y^{-1})$, and one can check that this induces an automorphism of G. (Normally, the words in [111] need some adjustment by central elements for covers of groups.) We choose matrices

$$ x = \begin{bmatrix} 1 & 0 & 0 & 0 \\ 0 & 0 & 1 & 0 \\ 0 & 1 & 0 & 0 \\ 0 & 0 & 0 & -1 \end{bmatrix} \quad \text{and} \quad y = \frac{1}{\sqrt{-3}} \begin{bmatrix} -1 & 0 & 1 & \omega \\ 0 & -\omega & \bar{\omega} & -1 \\ 1 & \bar{\omega} & 1 & 0 \\ \omega & -1 & 0 & \bar{\omega} \end{bmatrix}, $$

where $\omega = z_3$ and $\bar{\omega} = \omega^2$ are the primitive cube roots of 1. We check that these matrices are isometries of the σ-Hermitian form I_4, so the γ-automorphism of $SL_4^{\pm}(p)$ is the inverse-transpose map. But x and y are symmetric matrices, and so $(x^\gamma, y^\gamma) = (x^{-T}, y^{-T}) = (x^{-1}, y^{-1}) = (x, y^{-1}) = (x^\alpha, y^\alpha)$. That is, the element γ (without adjustment by inner elements of $SL_4^{\pm}(p)$) normalises this representation of $2^{\cdot}U_4(2)$, and induces its outer automorphism. $\square$

We deal with the cases $2^{\cdot}L_2(7)$ and $2^{\cdot}A_7$ together over the next two proofs, as these embed in the same groups $\mathrm{SL}_4^{\pm}(p)$, and in such a way that we may conjugate the $2^{\cdot}A_7$ inside $\mathrm{GL}_4^{\varepsilon}(p)$ to get $2^{\cdot}L_2(7) < 2^{\cdot}A_7$.

Proposition 4.7.3   (i) *The stabiliser in* $\mathrm{Out}\, L_4(2)$ *of the single class of* A_7
   in $\mathrm{SL}_4(2)$ *is* $\langle\gamma\rangle$.

   (ii) *If* $p \equiv 1, 2, 4 \pmod 7$ *and* $p \neq 2$, *then the stabiliser in* $\mathrm{Out}\, L_4(p)$ *of at
   least one of the* $(p-1,4)$ *classes of* $2^{\cdot}A_7$ *in* $\mathrm{SL}_4(p)$ *is* $\langle\gamma\rangle$.

   (iii) *If* $p \equiv 3, 5, 6 \pmod 7$, *then the stabiliser in* $\mathrm{Out}\, U_4(p)$ *of at least one of
   the* $(p+1,4)$ *classes of* $2^{\cdot}A_7$ *in* $\mathrm{SU}_4(p)$ *is* $\langle\gamma\rangle$.

Proof These $\mathscr{S}_1$-subgroups of $\mathrm{SL}_4^{\pm}(p)$ were described in Proposition 4.5.5.

We shall use generators x, y, z of $2^{\cdot}A_7$ that correspond to elements of A_7 as follows: $x \sim (0,1)(2,5)$, $y \sim (1,2,4)(3,6,5)$ and $z \sim (2,5)(3,4)$. A presentation of A_7 on these generators is given below.

$$\langle\, x, y, z \mid x^2, y^3, (xy)^7, [x,y]^4, z^2, [x,z], (yz)^3, (y^{-1}xyz)^3, (xyz)^7 \,\rangle.$$

We then choose x, y, z inside $2^{\cdot}A_7$ so that xy, y and yz have orders 7, 3 and 3 respectively. Throughout this discussion we shall regard x, y, z as standard generators for $2^{\cdot}A_7$ (although they do not correspond to the generators defined in [111]). An outer automorphism of $2^{\cdot}A_7$ may be taken to be τ (corresponding to $(1,6)(2,5)(3,4)$) that maps (x,y,z) to $(x^{y^{-1}zyz}, y, z)$.

Let x, y and z correspond respectively to the matrices below over $\mathbb{Q}(\sqrt{-7})$,

$$\frac{-1}{\sqrt{-7}}\begin{bmatrix} -1 & 2 & b & 0 \\ 2 & 1 & 0 & b \\ c & 0 & 1 & -2 \\ 0 & c & -2 & -1 \end{bmatrix}, \quad \begin{bmatrix} 1 & 0 & 0 & 0 \\ 0 & 0 & 0 & 1 \\ 0 & 1 & 0 & 0 \\ 0 & 0 & 1 & 0 \end{bmatrix} \quad \text{and} \quad \begin{bmatrix} 0 & -1 & 0 & 0 \\ 1 & 0 & 0 & 0 \\ 0 & 0 & 0 & 1 \\ 0 & 0 & -1 & 0 \end{bmatrix},$$

where $b = \mathrm{b}_7$ and $c = \mathrm{b}_7^{**}$.

These matrices are isometries of the σ-Hermitian form with matrix I_4, and so the γ automorphism of $\mathrm{SL}_4^{\pm}(p)$ into which we shall embed $2^{\cdot}A_7$ can be taken to be the inverse-transpose map. One may check that $(x^\gamma, y^\gamma, z^\gamma) = (x^\tau, y^\tau, z^\tau)$, so $\langle\gamma\rangle$ normalises this particular copy of $2^{\cdot}A_7$.

The case $A_7 < \mathrm{SL}_4(2) \cong L_4(2) \cong A_8$ is easy to deal with. There is just one class of subgroups of A_7 in $L_4(2)$, and since $\mathrm{Out}\, L_4(2) = \langle\gamma\rangle$, we conclude that A_7 has class stabiliser $\langle\gamma\rangle$. $\square$

The calculations for $2^{\cdot}L_2(7)$ are more complicated, and we shall use some of the matrices defined in the previous proof.

Proposition 4.7.4   (i) *If* $p \equiv 1, 9, 15, 23, 25, 39 \pmod{56}$ *then the stabiliser* in $\operatorname{Out} L_4(p)$ *of at least one of the* $(p-1,4)$ *classes of* $2^{\cdot}L_2(7)$ *in* $\operatorname{SL}_4(p)$ *is* $\langle \gamma \rangle$.

(ii) *If* $p \equiv 11, 29, 37, 43, 51, 53 \pmod{56}$ *then the stabiliser in* $\operatorname{Out} L_4(p)$ *of at least one of the* $(p-1,4)$ *classes of* $2^{\cdot}L_2(7)$ *in* $\operatorname{SL}_4(p)$ *is* $\langle \gamma\delta \rangle$.

(iii) *If* $p \equiv 17, 31, 33, 41, 47, 55 \pmod{56}$ *then the stabiliser in* $\operatorname{Out} U_4(p)$ *of at least one of the* $(p+1,4)$ *classes of* $2^{\cdot}L_2(7)$ *in* $\operatorname{SU}_4(p)$ *is* $\langle \gamma \rangle$.

(iv) *If* $p \equiv 3, 5, 13, 19, 27, 45 \pmod{56}$ *then the stabiliser in* $\operatorname{Out} U_4(p)$ *of at least one of the* $(p+1,4)$ *classes of* $2^{\cdot}L_2(7)$ *in* $\operatorname{SU}_4(p)$ *is* $\langle \gamma\delta \rangle$.

Proof We say that x, y are standard generators for $2^{\cdot}L_2(7)$ if x, y and xy have orders 4, 3 and 7 respectively, as in [111]. If x and y are standard generators of $2^{\cdot}L_2(7)$ then an outer automorphism α can be taken to map (x, y) to (x^{-1}, y^{-1}).

Note that the generators x and y of $2^{\cdot}A_7$ from the previous proof are standard generators of a copy of $2^{\cdot}L_2(7)$ inside of $2^{\cdot}A_7$. There is also a second class of such subgroups $2^{\cdot}L_2(7)$, with generators $\langle x^\tau, y^\tau \rangle = \langle x^{y^{-1}zyz}, y \rangle$, where τ is the automorphism of $2^{\cdot}A_7$ from the previous proof. The automorphism γ of $\operatorname{SL}_4^{\pm}(p)$ can again be taken to be the inverse-transpose map.

Standard MeatAxe techniques find a matrix

$$L_1 = \frac{1}{\sqrt{-7}} \begin{bmatrix} 1 & b & b & b \\ b & b+2 & 0 & (-b-2)/2 \\ b & 0 & (-b-2)/2 & b+2 \\ b & (-b-2)/2 & b+2 & 0 \end{bmatrix}$$

that conjugates (x^γ, y^γ) to $(x^\alpha, y^\alpha) = (x^{-1}, y^{-1})$, where $b = b_7$. We also calculate that $L_2 := L_1^{-1} = \overline{L}_1$ conjugates the pair of images $(x^{\tau\gamma}, y^{\tau\gamma})$ to $(x^{\tau\alpha}, y^{\tau\alpha}) = ((x^\tau)^{-1}, (y^\tau)^{-1})$. It also happens that L_2 conjugates $\langle x, y \rangle$ to $\langle x^\tau, y^\tau \rangle$, since $(x^{\tau\tau}, y^{\tau\tau}) = (x, y)$. The matrices L_1 and L_2 are both isometries of the σ-Hermitian form with matrix I_4, and have determinants $(b/c)^3$ and $(c/b)^3$ respectively, where $c = b_7^{**}$.

First suppose we are in Case **L**: that is, $p \equiv 1, 2, 4 \pmod{7}$, with $p \neq 2$. Since $\det L_1 = b^3 c/c^4 = 2b^2/c^4$, we see that the reduction modulo p of $\det L_1$ is a square in $\mathbb{F}_p^\times$ if and only if 2 is a square in $\mathbb{F}_p^\times$; that is, if $p \equiv 1, 7 \pmod{8}$. It follows from Lemma 4.6.2 applied to $\langle x, y \rangle$ with $\beta = \gamma$ that one of the classes of $2^{\cdot}L_2(7)$ in $\operatorname{SL}_4(p)$ is stabilised by γ if $p \equiv 1, 7 \pmod{8}$ and by $\gamma\delta$ if $p \equiv 3, 5 \pmod{8}$.

Next suppose that we are in Case **U**: that is, $p \equiv 3, 5, 6 \pmod{7}$. Then we can apply Lemma 4.6.4 to $\langle x, y \rangle$ with $\beta = \gamma$, $A = B = I_4$ and $C = L = L_1$, and we have to determine whether, on reduction modulo p, $\det L_1 = (b/c)^3$ is a square in the group $X \cong C_{p+1}$ of norm 1 elements of $\mathbb{F}_{p^2}^\times$. This is the case if and only if the reduction of $b/c = b^2/2$ is a square in X. Observe that $b/c = (b/\sqrt{2})^2$

(with $\sqrt{2} \in \mathbb{F}_{p^2}^{\times}$), so that $b/\sqrt{2} \in \mathbb{F}_{p^2}^{\times}$ has norm $1 = 2/2$ if $\sqrt{2} \in \mathbb{F}_p^{\times}$, and norm $-1 = 2/(-2)$ if $\sqrt{2} \notin \mathbb{F}_p^{\times}$. Since $-b/\sqrt{2}$ (the other square root of b/c) has the same norm as $b/\sqrt{2}$, we conclude that b/c is a square in X if and only if 2 is a square in $\mathbb{F}_p^{\times}$; that is, if and only if $p \equiv 1, 7 \pmod 8$. So, by Lemma 4.6.4 (iv), one of the classes of $2{\cdot}L_3(7)$ in $U_4(p)$ is stabilised by γ if $p \equiv 1, 7 \pmod 8$ and by $\gamma\delta$ if $p \equiv 3, 5 \pmod 8$. $\qquad \square$

The following proposition can be proved by a straightforward computer calculation (file `4134d4calc`) using Lemma 4.6.4 (iv).

Proposition 4.7.5 *The stabiliser in* $\mathrm{Out}\, U_4(3)$ *of each of the two classes of* $4_2{\cdot}L_3(4)$ *in* $\mathrm{SU}_4(3)$ *is* $\langle \delta^2, \gamma\delta \rangle$, *where* δ^2 *induces the* 2_2 *automorphism of* $L_3(4)$, *and* $\gamma\delta$ *induces the* 2_1 *and* 2_3 *automorphisms in the two classes of subgroups.*

We briefly expand our description of the actions of the outer automorphisms of $\mathrm{SU}_4(3)$ on the conjugacy classes of $4_2{\cdot}L_3(4)$. Here $\mathrm{Out}\,\mathrm{SU}_4(3) \cong D_8$ and has presentation $\langle\, \delta, \gamma \mid \delta^4 = \gamma^2 = 1,\ \delta^\gamma = \delta^{-1} \,\rangle$. There are two classes of $4_2{\cdot}L_3(4)$, and δ acts on them as a transposition (so that δ^2 normalises both classes). The elements $\gamma\delta^i$ for $i \in \{0, 2\}$ act as involutions, whilst $\gamma\delta^i$ for $i \in \{1, 3\}$ normalises both classes. Thus both classes are normalised by $\{1, \delta^2, \gamma\delta, \gamma\delta^3\}$ and they are interchanged by $\{\delta, \delta^3, \gamma, \gamma\delta^2\}$: see Table 8.11.

Dimension 5. The situation in dimension 5 is straightforward, and we summarise it briefly.

Theorem 4.7.6 *Let* Ω *be either* $\mathrm{SL}_5(q)$ *or* $\mathrm{SU}_5(q)$, *let* G *be an* $\mathscr{S}_1$-*subgroup of* Ω, *and let* $d = (q-1, 5)$ *in Case* **L** *and* $(q+1, 5)$ *in Case* **U**. *Then,* $q = p$ *in all cases, and one of the following holds:*

(i) $G = d \times L_2(11)$, *and the stabiliser in* $\mathrm{Out}\,\bar{\Omega}$ *of at least one of the* d *classes of* G *in* Ω *is* $\langle \gamma \rangle$.

(ii) $G = d \times U_4(2)$, *and the stabiliser in* $\mathrm{Out}\,\bar{\Omega}$ *of at least one of the* d *classes of* G *in* Ω *is* $\langle \gamma \rangle$.

(iii) $G = M_{11}$, *and the stabiliser in* $\mathrm{Out}\,L_5(3)$ *of each of the two classes of* G *in* $\mathrm{SL}_5(3)$ *is trivial.*

Proof By Theorem 4.3.3, the quasisimple groups G to consider are $L_2(11)$, $U_4(2)$ and M_{11}. In the first two cases $|\mathrm{Out}\,G| = 2$, whilst $\mathrm{Out}\,M_{11}$ is trivial. By Proposition 4.5.12 in each case $q = p$, and for $L_2(11)$ and $U_4(2)$ the automorphism δ permutes the d Ω-classes transitively, whilst for M_{11} there are two classes in Ω and δ is trivial. For $L_2(11)$ and $U_4(2)$, the result therefore follows from Lemma 4.6.1, whilst for $G = M_{11}$ the automorphism γ must interchange the two classes. $\qquad \square$

Dimension 6. By Theorem 4.3.3, the quasisimple groups to consider are:

$$3\,\dot{}\,A_6, 6\,\dot{}\,A_6, 2\,\dot{}\,L_2(11), 3\,\dot{}\,A_7, 6\,\dot{}\,A_7, 6\,\dot{}\,L_3(4), 2\,\dot{}\,M_{12}, 3\,\dot{}\,M_{22}, 3_1\,\dot{}\,U_4(3), 6_1\,\dot{}\,U_4(3).$$

As in dimension four, for expositional purposes it is easier to consider $2\,\dot{}\,L_2(11)$ first, as it is a reasonably straightforward example and we can present it in full detail. We shall gradually become briefer as we work through the remaining groups.

Proposition 4.7.7 *Let $d = (p-1, 6)$ with $p \equiv 1, 3, 4, 5, 9 \pmod{11}$ in Case* **L**, *and $d = (p+1, 6)$ with $p \equiv 2, 6, 7, 8, 10 \pmod{11}$ in Case* **U**.

(i) *If $p \equiv \pm 1 \pmod 8$, then the stabiliser in $\mathrm{Out}\,L_6(p)$ of at least one of the d classes of $d \circ 2\,\dot{}\,L_2(11)$ in $\mathrm{SL}_6(p)$ is $\langle \gamma \rangle$.*

(ii) *If $p \equiv \pm 3 \pmod 8$, then the stabiliser in $\mathrm{Out}\,L_6(p)$ of at least one of the d classes of $d \circ 2\,\dot{}\,L_2(11)$ in $\mathrm{SL}_6(p)$ is $\langle \gamma\delta \rangle$.*

(iii) *If $p \equiv \pm 1 \pmod 8$, then the stabiliser in $\mathrm{Out}\,U_6(p)$ of at least one of the d classes of $d \circ 2\,\dot{}\,L_2(11)$ in $\mathrm{SU}_6(p)$ is $\langle \gamma \rangle$.*

(iv) *If $p \equiv \pm 3 \pmod 8$, then the stabiliser in $\mathrm{Out}\,U_6(p)$ of at least one of the d classes of $d \circ 2\,\dot{}\,L_2(11)$ in $\mathrm{SU}_6(p)$ is $\langle \gamma\delta \rangle$.*

Proof The reader should recall Proposition 4.5.15, where we previously considered $G = 2\,\dot{}\,L_2(11)$. There are two dual representations interchanged by the outer automorphism α of G.

By [111], standard generators for G are x of order 4 and y of order 3 such that xy has order 11. A straightforward computation shows that the map $\alpha :$ $(x, y) \to (x^{-1}, y^{-1})$ induces an outer automorphism of G.

The representation of G in $\mathrm{SL}_6(p)$ is the p-modular reduction of a representation over the ring $\mathbb{Z}[b_{11}]$ with $b = b_{11}$, for which the images of x and y are respectively:

$$
\begin{bmatrix}
0 & 1 & 0 & 0 & 0 & 0 \\
-1 & 0 & 0 & 0 & 0 & 0 \\
0 & 0 & 0 & 1 & 0 & 0 \\
0 & 0 & -1 & 0 & 0 & 0 \\
0 & 0 & 0 & 0 & 0 & 1 \\
0 & 0 & 0 & 0 & -1 & 0
\end{bmatrix}
\quad \text{and} \quad
\begin{bmatrix}
1 & 1 & 1 & -b & -b-1 & -1 \\
1 & 0 & 0 & 0 & 0 & 0 \\
0 & 0 & 0 & 0 & -1 & 0 \\
0 & 0 & 0 & 0 & 0 & -1 \\
-b & -1 & -b-1 & -2 & b-1 & b+2 \\
-1 & b+1 & b-1 & b+2 & 2 & -b
\end{bmatrix}.
$$

Since α and the duality map γ have the same effect on the representation, there exists a matrix $L \in \mathrm{GL}_6(\mathbb{C})$ that conjugates the images (x^γ, y^γ) to $(x^\alpha, y^\alpha) = (x^{-1}, y^{-1})$. The MeatAxe techniques mentioned in Section 4.6.1

yield

$$L := \begin{bmatrix} 2 & -2b-1 & -b+3 & -2b-3 & -b-4 & 2b-1 \\ -2b-1 & -2 & -2b-3 & b-3 & 2b-1 & b+4 \\ -b+3 & -2b-3 & -3b+2 & -2b-6 & -6 & 4b+2 \\ -2b-3 & b-3 & -2b-6 & 3b-2 & 4b+2 & 6 \\ -b-4 & 2b-1 & -6 & 4b+2 & 3b+5 & -2b+4 \\ 2b-1 & b+4 & 4b+2 & 6 & -2b+4 & -3b-5 \end{bmatrix}.$$

We first consider Case **L**. The determinant of L is 2, which is a square modulo p if and only if $p \equiv \pm1 \pmod 8$, by Table 4.2. Hence by Lemma 4.6.2, up to conjugacy in $\mathrm{Out}\,\mathrm{SL}_6(p)$, we may take the stabiliser of the conjugacy class of G to be $\langle\gamma\rangle$ when $p \equiv \pm1 \pmod 8$ and $\langle\gamma\delta\rangle$ when $p \equiv \pm3 \pmod 8$.

Now we consider Case **U**. To calculate a σ-Hermitian form B preserved by G, we compute a suitable matrix that conjugates (x, y) to $(\bar{x}^{-\mathsf{T}}, \bar{y}^{-\mathsf{T}})$, where $\bar{}$ denotes complex conjugation. We can choose

$$B = \begin{bmatrix} 11 & -\sqrt{-11} & \sqrt{-11} & \sqrt{-11} & \sqrt{-11} & -\sqrt{-11} \\ \sqrt{-11} & 11 & -\sqrt{-11} & \sqrt{-11} & \sqrt{-11} & \sqrt{-11} \\ -\sqrt{-11} & \sqrt{-11} & 11 & \sqrt{-11} & \sqrt{-11} & -\sqrt{-11} \\ -\sqrt{-11} & -\sqrt{-11} & -\sqrt{-11} & 11 & \sqrt{-11} & \sqrt{-11} \\ -\sqrt{-11} & -\sqrt{-11} & -\sqrt{-11} & -\sqrt{-11} & 11 & \sqrt{-11} \\ \sqrt{-11} & -\sqrt{-11} & \sqrt{-11} & -\sqrt{-11} & -\sqrt{-11} & 11 \end{bmatrix}.$$

We apply Proposition 4.6.6 with the matrices $\hat{L}$ and $\hat{B}$ in the proposition replaced by L and B. Since $\det L = 2$, in the notation of Proposition 4.6.6, we can choose $\hat{\nu} = 1$, $\hat{\zeta} = 2$, and we conclude that some $\mathrm{GL}_6(p^2)$-conjugate of G in $\mathrm{SU}_6(p)$ is stabilised by $\langle\gamma\rangle$ when 2 is a square in $\mathbb{F}_p^\times$: that is, when $p \equiv \pm1 \pmod 8$. Otherwise, when $p \equiv \pm3 \pmod 8$, some $\mathrm{GL}_6(p^2)$-conjugate of G in $\mathrm{SU}_6(p)$ is stabilised by $\langle\gamma\delta\rangle$.

There are no non-unit denominators in the entries in B, but we must also check that there is no exceptional behaviour for divisors of $\det B = 2^5 \times 11^3$. Since $p \neq 2, 11$, there is no exceptional behaviour for small primes.

Details of computer calculations for this example (but not using the identical matrices as in the proof given here) can be found in file `sl211d6calc`. $\qquad\square$

As an aside, we note that in the previous example the equation $\xi\bar{\xi} = 2^5 11^2$ is not soluble in $\mathbb{Q}(\sqrt{-11})$, so B is not equivalent over $\mathbb{Q}(\sqrt{-11})$ to I_6.

The proofs of the remaining results in this section have a similar structure, and will be given with less detail.

Proposition 4.7.8　　(i) *If $p \equiv 1 \pmod{24}$, then the stabiliser in $\mathrm{Out}\, L_6(p)$ of at least one of the six classes of $2 \times 3^{\cdot}A_6.2_3$ in $\mathrm{SL}_6(p)$ is $\langle \gamma \rangle$.*

(ii) *If $p \equiv 19 \pmod{24}$, then the stabiliser in $\mathrm{Out}\, L_6(p)$ of at least one of the six classes of $2 \times 3^{\cdot}A_6.2_3$ in $\mathrm{SL}_6(p)$ is $\langle \gamma\delta \rangle$.*

(iii) *If $p \equiv 7, 13 \pmod{24}$, then the stabiliser in $\mathrm{Out}\, L_6(p)$ of at least one of the three classes of $2 \times 3^{\cdot}A_6$ in $\mathrm{SL}_6(p)$ is $\langle \delta^3, \gamma \rangle$, where δ^3 induces the 2_3 automorphism of A_6. The 2_2 and 2_1 automorphisms of A_6 are induced by γ when $p \equiv 7, 13 \pmod{24}$, respectively.*

(iv) *If $p \equiv 23 \pmod{24}$, then the stabiliser in $\mathrm{Out}\, U_6(p)$ of at least one of the six classes of $2 \times 3^{\cdot}A_6.2_3$ in $\mathrm{SU}_6(p)$ is $\langle \gamma \rangle$.*

(v) *If $p \equiv 5 \pmod{24}$, then the stabiliser in $\mathrm{Out}\, U_6(p)$ of at least one of the six classes of $2 \times 3^{\cdot}A_6.2_3$ in $\mathrm{SU}_6(p)$ is $\langle \gamma\delta \rangle$.*

(vi) *If $p \equiv 11, 17 \pmod{24}$, then the stabiliser in $\mathrm{Out}\, U_6(p)$ of at least one of the three classes of $2 \times 3^{\cdot}A_6$ in $\mathrm{SU}_6(p)$ is $\langle \delta^3, \gamma \rangle$, where δ^3 induces the 2_3 automorphism of A_6. The 2_1 and 2_2 automorphisms of A_6 are induced by γ when $p \equiv 11, 17 \pmod{24}$, respectively.*

Proof　The reader should recall Proposition 4.5.13, where we previously considered $G = 3^{\cdot}A_6$. There are two dual 6-dimensional representations, which are stabilised by the 2_3 automorphism and interchanged by the 2_1 and 2_3 automorphisms of $3^{\cdot}A_6$; so at least one conjugacy class representative of these groups is normalised by either the duality automorphism γ of $\mathrm{SL}_6^\varepsilon(p)$ or by $\gamma\delta$.

We wrote G as a group of classically unitary matrices (i.e. preserving the form with matrix I_6) with entries in $\mathbb{Z}[\omega]$, where $\omega = z_3$. We found matrices L_1 and L_2, with $\det L_1 = -\omega = -1.(\bar{\omega}^2)^2$ and $\det L_2 = -216\omega = 2.(6\bar{\omega}i_3)^2$, such that the map $\gamma = -\mathsf{T}$ applied to $\mathrm{GL}_6(\mathbb{Q}(\omega))$ followed by conjugation by L_1 or by L_2 normalises G and induces, respectively, the 2_1 or the 2_2 automorphism of G (modulo inner automorphisms). The denominators of the entries in $L_i^{\pm 1}$ are divisible only by the primes 2 and 3, so there are no exceptional primes.

For the linear case, by Lemma 4.6.2, these outer automorphisms are effected by γ, $\gamma\delta^2$ or $\gamma\delta^4$ modulo inner automorphisms if the p-modular reduction of $\det L_i$ is a square in $\mathbb{F}_p^\times$, and otherwise by $\gamma\delta$, $\gamma\delta^3$ or $\gamma\delta^5$. Note that the reduction of $\det L_i$ is in $\{-1, 2\}$ modulo squares of $\mathbb{F}_p^\times$. By Table 4.2, -1 is a square modulo p if $p \equiv 1 \pmod 4$, and 2 is a square modulo p if $p \equiv \pm 1 \pmod 8$. Parts (i), (ii) and (iii) now follow from Proposition 4.5.13 (i) and (ii). (In fact the calculations involving either of L_1 or L_2 alone would suffice for this conclusion!)

In the unitary case, since q is prime, Proposition 4.6.6 tells us that outer automorphisms are induced by γ when the square-free part z of $\det L$ reduces to a square in $\mathbb{F}_p^\times$ and z is positive, or when z is a non-square in $\mathbb{F}_p^\times$ and z is negative. Otherwise outer automorphisms are induced by $\gamma\delta$. Thus, up to conjugacy, the automorphisms are induced by γ when 2 is a square or when -1

is a non-square, respectively. Parts (iv), (v) and (vi) now follow from Proposition 4.5.13 (iii) and (iv).

Details of the computer calculations for this example can be found in file `3a6d6calc`.	□

For clarity, let us slightly expand our description of the actions of the outer automorphisms on the conjugacy classes of $2 \times 3^{\cdot}A_6$. We concentrate on the case when $p \equiv 7, 13 \pmod{24}$, as the other cases are similar but easier. The outer automorphism group of $\mathrm{SL}_6(p)$ is isomorphic to D_{12} and has presentation

$$\langle \delta, \phi, \gamma \mid \delta^6 = \gamma^2 = 1, \delta^\gamma = \delta^{-1} \rangle.$$

There are three classes of $2 \times 3^{\cdot}A_6$, and δ acts (non-faithfully) as a 3-cycle on them, with δ^3 normalising all three classes. The elements $\gamma\delta^i$ for $i \in \{0, \dots, 5\}$ all act as involutions. Thus one of the classes of $2 \times 3^{\cdot}A_6$ is normalised by $\{1, \delta^3, \gamma, \gamma\delta^3\}$; one by $\{1, \delta^3, \gamma\delta, \gamma\delta^4\}$; and one by $\{1, \delta^3, \gamma\delta^2, \gamma\delta^5\}$, but Table 8.25 contains just the first of these stabilisers.

Proposition 4.7.9	(i) *If $p \equiv 1, 31 \pmod{48}$ then the stabiliser in $\mathrm{Out}\,L_6(p)$ of at least one of the six classes of $6^{\cdot}A_6$ in $\mathrm{SL}_6(p)$ is $\langle\gamma\rangle$, and γ induces the 2_2 automorphism of A_6.*

(ii) *If $p \equiv 7$ or $25 \pmod{48}$ then the stabiliser in $\mathrm{Out}\,L_6(p)$ of at least one of the six classes of $6^{\cdot}A_6$ in $\mathrm{SL}_6(p)$ is $\langle\gamma\delta\rangle$, and $\gamma\delta$ induces the 2_2 automorphism of A_6.*

(iii) *If $p \equiv 5$ or $11 \pmod{24}$ then the stabiliser in $\mathrm{Out}\,L_6(p^2)$ of at least one of the six classes of $6^{\cdot}A_6$ in $\mathrm{SL}_6(p^2)$ is $\langle\gamma\delta^3, \phi\rangle$, where $\gamma\delta^3$ and ϕ induce the 2_2 and the 2_1 automorphisms of A_6, respectively.*

(iv) *If $p \equiv 13$ or $19 \pmod{24}$ then the stabiliser in $\mathrm{Out}\,L_6(p^2)$ of at least one of the six classes of $6^{\cdot}A_6$ in $\mathrm{SL}_6(p^2)$ is $\langle\gamma\delta^3, \phi\gamma\rangle$, where $\gamma\delta^3$ and $\phi\gamma$ induce the 2_2 and the 2_1 automorphisms of A_6.*

(v) *If $p \equiv 17$ or $47 \pmod{48}$ then the stabiliser in $\mathrm{Out}\,U_6(p)$ of at least one of the six classes of $6^{\cdot}A_6$ in $\mathrm{SU}_6(p)$ is $\langle\gamma\rangle$, and γ induces the 2_2 automorphism of A_6.*

(vi) *If $p \equiv 23$ or $41 \pmod{48}$ then the stabiliser in $\mathrm{Out}\,U_6(p)$ of at least one of the six classes of $6^{\cdot}A_6$ in $\mathrm{SU}_6(p)$ is $\langle\gamma\delta\rangle$, and $\gamma\delta$ induces the 2_2 automorphism of A_6.*

Proof	The reader should recall Proposition 4.5.14, where we previously considered $G = 6^{\cdot}A_6$. We saw there that G is potentially a maximal subgroup of Ω if $p \equiv 1$ or $7 \pmod{24}$ with $\Omega = \mathrm{SL}_6(p)$, $p \equiv 5, 11, 13$ or $19 \pmod{24}$ with $\Omega = \mathrm{SL}_6(p^2)$, and $p \equiv 17$ or $23 \pmod{24}$ with $\Omega = \mathrm{SU}_6(p)$.

There are four representations with character ring $\mathbb{Z}[z_3, r_2]$, permuted transitively by $\mathrm{Out}\,G$. Duality maps (z_3, r_2) to (z_3^{**}, r_2). From the character tables in [12] (or [57] for $p = 5$), it can be seen firstly that the 2_2 automorphism of G inverts the central 3-element (this is indicated by the incomplete square for $3.G.2_2$ in the character table map) and secondly that the 2_2 automorphism normalises the two conjugacy classes of elements of order 8 in G (because there are elements of order 16 in $G.2_2$). Since the character values of these 6-dimensional representations are $\pm z_3$ and $\pm r_2$ on the central 3-element and the elements of order 8, respectively, it follows that duality has the same action on the representations as the 2_2 automorphism.

We wrote G as a subgroup of $GL_6(R)$ with $R = \mathbb{Z}[z_3, r_2, 1/2]$. We found a matrix B for a unitary form for which G is a group of isometries, with entries of $B^{\pm 1}$ in $R[1/3]$. We found a matrix L, with entries of $L^{\pm 1}$ in R, of determinant $(2 + r_2)\nu^2$ with $\nu \in R$, such that $\gamma = -\mathsf{T}$ followed by conjugation by L normalises and induces the 2_2 automorphism of G (modulo inner automorphisms). So neither B nor L result in exceptional primes.

Thus, in Case **L**, to determine whether a conjugate of γ or a conjugate of $\gamma\delta$ induces the 2_3 automorphism, by Lemma 4.6.2 we need to determine when $2 + r_2$ reduces to a square in $\mathbb{F}_q^\times$. Recall from Section 4.2 that $r_2 = y_8$. Then $2 + r_2 = 2 + y_8 = 2 + z_8 + z_8^{-1} = 2 + z_{16}^2 + z_{16}^{-2} = y_{16}^2$. We can calculate $\mathbb{F}_p(y_{16})$ for all primes p by using Lemma 4.2.1. We get $\mathbb{F}_p(y_{16}) = \mathbb{F}_p$ if $p = 2$ or $p \equiv \pm 1 \pmod{16}$, $\mathbb{F}_p(y_{16}) = \mathbb{F}_{p^2}$ if $p \equiv \pm 7 \pmod{16}$, and $\mathbb{F}_p(y_{16}) = \mathbb{F}_{p^4}$ if $p \equiv \pm 3$ or $\pm 5 \pmod{16}$.

If $p \equiv 5, 11, 13$ or $19 \pmod{24}$ then $p \equiv \pm 3$ or $\pm 5 \pmod{16}$, so $\det L$ is a non-square in $\mathbb{F}_{p^2}^\times$, and a conjugate of $\gamma\delta$ induces the 2_2 automorphism. If $p \equiv 7$ or $25 \pmod{48}$ then $p \equiv \pm 7 \pmod{16}$, so $\det L$ is a non-square in $\mathbb{F}_p^\times$, and once again a conjugate of $\gamma\delta$ induces the 2_2 automorphism. Finally, if $p \equiv 1$ or $31 \pmod{48}$ then $\det L$ is a square in $\mathbb{F}_p^\times$, so that a conjugate of γ induces the 2_2 automorphism. This completes the analysis of the action of the duality automorphism on the conjugacy classes of $6{\cdot}A_6$ in Ω in the linear cases.

We still have to consider the action of the field automorphism ϕ of Ω when $\Omega = SL_6(p^2)$; that is, when $p \equiv 5, 11, 13, 19 \pmod{24}$. First let $p \equiv 5, 11 \pmod{24}$. Then $\delta^\phi = \delta^{-1}$, so we can apply Lemma 4.6.2 with $\beta = \gamma$. From character values we calculate that ϕ and $\gamma\phi$ induce the 2_1 and 2_3 automorphisms of G, modulo diagonal and inner automorphisms. In the 2_1 case, we found a conjugating matrix M, with entries of $M^{\pm 1}$ in R, with determinant $-\nu^2$ for some $\nu \in R$, a square in $\mathbb{F}_{p^2}^\times$ so, by Lemma 4.6.2, a conjugate of ϕ induces the 2_1 automorphism. Hence the class of $6{\cdot}A_6$ is stabilised by a conjugate of $\gamma\delta$ and by a conjugate of ϕ in $\langle \delta, \gamma, \phi \rangle$. Since its stabiliser in $\langle \delta, \gamma, \phi \rangle$ has order 4 and $|\langle \gamma\delta, \phi \rangle| = |\langle \gamma\delta^5, \phi \rangle| = 12$, it follows that this stabiliser must be a conjugate of $\langle \gamma\delta^3, \phi \rangle$.

The other linear case is $p \equiv 13$ or $19 \pmod{24}$. Then $\delta^{\phi\gamma} = \delta^{-1}$, so we can apply Lemma 4.6.2 with $\beta = \phi\gamma$. From character values we deduce that $\gamma\phi$

and ϕ induce the 2_1 and 2_3 automorphisms of G modulo diagonal and inner automorphisms. In the 2_1 case, we found a conjugating matrix M, with entries of $M^{\pm 1}$ in R, with determinant 1, so the 2_1 automorphism is induced by a conjugate of $\phi\gamma$. The stabiliser of the class of $6\dot{}A_6$ in $\langle \delta, \gamma, \phi \rangle$ is therefore a conjugate of $\langle \gamma\delta^3, \phi\gamma \rangle$.

Turning now to the Case **U**, for which $p \equiv 17$ or $23 \pmod{24}$, we recall that $\det L = (2 + r_2)\nu^2$ with $\nu \in R$. So, in the notation of Proposition 4.6.6, we can take $z = 2 + r_2$ which is positive. As we saw above, $\sqrt{z} \in \mathbb{F}_p$ if and only if $p \equiv \pm 1 \pmod{16}$ so, by Proposition 4.6.6, some $\mathrm{GL}_6(p^2)$-conjugate of G is stabilised by γ when $p \equiv \pm 1 \pmod{16}$ and by $\gamma\delta$ when $p \equiv \pm 7 \pmod{16}$, and the 2_2 automorphism of $6\dot{}A_6$ is induced.

Details of the computer calculations for this example can be found in file `6a6d6calc`. □

Once again, let us expand our description of the actions of the outer automorphisms on the conjugacy classes of $6\dot{}A_6$. We concentrate on the case when $p \equiv 5, 11 \pmod{24}$. Here the outer automorphism group of $\mathrm{SL}_6(p^2)$ is isomorphic to $D_{12} \times 2$ and has presentation

$$\langle \delta, \phi, \gamma \mid \delta^6 = \phi^2 = \gamma^2 = 1, \; \delta^\phi = \delta^\gamma = \delta^{-1}, \; \phi^\gamma = \phi \rangle.$$

There are six classes of $6\dot{}A_6$, and δ acts as a 6-cycle on them. The elements $\gamma\delta^i$ for $i \in \{0, 2, 4\}$ and $\phi\delta^i$ for $i \in \{1, 3, 5\}$ have cycle type 2^3 on the six classes of $6\dot{}A_6$, whereas $\gamma\delta^i$ ($i \in \{1, 3, 5\}$) and $\phi\delta^i$ ($i \in \{0, 2, 4\}$) act with cycle shape $2^2.1^2$. Lastly $\gamma\phi\delta^3$ normalises all six classes of $6\dot{}A_6$, while the other $\gamma\phi\delta^i$ normalise none. Thus two of the classes of $6\dot{}A_6$ are normalised by $\{1, \gamma\delta^3, \phi, \gamma\phi\delta^3\}$; two by $\{1, \gamma\delta, \phi\delta^4, \gamma\phi\delta^3\}$; and two by $\{1, \gamma\delta^{-1}, \phi\delta^{-4}, \gamma\phi\delta^3\}$. Table 8.27 contains just the first of these stabilisers (as we consider it to be the 'neatest' one).

Proposition 4.7.10   (i) *If $p \equiv 1 \pmod{12}$ then the stabiliser in $\mathrm{Out}\, L_6(p)$ of at least one of the six classes of $3\dot{}A_7$ in $\mathrm{SL}_6(p)$ is $\langle \gamma \rangle$.*

(ii) *If $p \equiv 7 \pmod{12}$ then the stabiliser in $\mathrm{Out}\, L_6(p)$ of at least one of the six classes of $3\dot{}A_7$ in $\mathrm{SL}_6(p)$ is $\langle \gamma\delta \rangle$.*

(iii) *If $p \equiv 11 \pmod{12}$ then the stabiliser in $\mathrm{Out}\, U_6(p)$ of at least one of the six classes of $3\dot{}A_7$ in $\mathrm{SU}_6(p)$ is $\langle \gamma \rangle$.*

(iv) *If $p \equiv 5 \pmod{12}$ then the stabiliser in $\mathrm{Out}\, U_6(p)$ of at least one of the six classes of $3\dot{}A_7$ in $\mathrm{SU}_6(p)$ is $\langle \gamma\delta \rangle$.*

(v) *If $p = 2$ then the stabiliser in $\mathrm{Out}\, U_6(p)$ of at least one of the three classes of $3\dot{}A_7$ in $\mathrm{SU}_6(p)$ is $\langle \gamma \rangle$.*

Proof The arguments here are similar to those for $3\dot{}A_6$ in Proposition 4.7.8, and will be sketched only briefly. The reader should recall Proposition 4.5.16, where we previously considered $G = 3{\cdot}A_7$. There are two dual representations interchanged by the outer automorphism of G.

When $p = 2$ there are only three classes of subgroups in $SU_6(2)$, and by Lemma 4.6.1 one of these must be stabilised by γ. So assume that $p > 2$.

We wrote G as a subgroup of $GL_6(R)$ with $R = \mathbb{Z}[\omega]$, where $\omega = z_3$. We found a matrix B for a unitary form for which G is a group of isometries with entries of $B^{\pm 1}$ in R. We found a matrix L, with entries of $L^{\pm 1}$ in R, of determinant -1 such that $\gamma = -\mathsf{T}$ followed by conjugation by L normalises and induces the outer automorphism of G (modulo inner automorphisms). So neither B nor L result in exceptional primes.

For the linear case, by Lemma 4.6.2, up to inner automorphisms and conjugacy in $\mathrm{Out}\,SL_6(p)$ these outer automorphisms will be effected by γ if -1 is a square in $\mathbb{F}_p^{\times}$ (that is, when $p \equiv 1 \pmod 4$), and otherwise one will require $\gamma\delta$. In the unitary case, Proposition 4.6.6 tells us that outer automorphisms are induced by a conjugate of γ when -1 is a non-square in $\mathbb{F}_p^{\times}$, and by a conjugate of $\gamma\delta$ otherwise.

Details of the computer calculations for this example can be found in file `3a7d6calc`. □

Proposition 4.7.11   (i) *If $p \equiv 1, 7, 17, 23 \pmod{24}$, then the stabilisers in* $\mathrm{Out}\,\overline{\Omega}$ *of the twelve classes of* $6{\cdot}A_7$ *in* $\Omega = SL_6(p)$ *or* $SU_6(p)$ *are trivial.*

(ii) *When $p \equiv 5, 11 \pmod{24}$, the stabiliser in $\mathrm{Out}\,L_6(p^2)$ of at least one of the twelve classes of $6{\cdot}A_7$ in $SL_6(p^2)$ is $\langle\phi\rangle$.*

(iii) *When $p \equiv 13, 19 \pmod{24}$, the stabiliser in $\mathrm{Out}\,L_6(p^2)$ of at least one of the twelve classes of $6{\cdot}A_7$ in $SL_6(p^2)$ is $\langle\phi\gamma\rangle$.*

Proof   Since the group $6{\cdot}A_7$ has a single conjugacy class of subgroups isomorphic to $6{\cdot}A_6$, an element of $\mathrm{Out}\,\overline{\Omega}$ stabilising the class of $6{\cdot}A_7$ also stabilises the class of $6{\cdot}A_6$. Therefore, for each of the congruences on p, the stabiliser of a class of subgroups of $6{\cdot}A_7$ is a subgroup of the corresponding stabiliser of $6{\cdot}A_6$, as given in Proposition 4.7.9. Furthermore, there are twice as many classes of $6{\cdot}A_7$ in $\overline{\Omega}$ as there are of $6{\cdot}A_6$, so the stabiliser must be of index 2. Thus the result for $p \equiv 1, 7, 17, 23 \pmod{24}$ is clear. Since $6{\cdot}S_7$ contains $6{\cdot}S_6$ but no other extension of $6{\cdot}A_6$, the stabiliser of the class of $6{\cdot}A_7$ in $SL_6(p^2)$ must be the automorphism that induces the 2_1 automorphism of $6{\cdot}A_6$, so the result holds for the remaining congruences on p. □

Proposition 4.7.12   (i) *If $p \equiv 1, 19 \pmod{24}$, then the stabiliser in* $\mathrm{Out}\,L_6(p)$ *of at least one of the six classes of* $6{\cdot}L_3(4).2_1^-$ *in* $SL_6(p)$ *is* $\langle\gamma\rangle$, *extending it to* $6{\cdot}L_3(4).2^2$.

(ii) *If $p \equiv 7, 13 \pmod{24}$, then the stabiliser in $\mathrm{Out}\,L_6(p)$ of at least one of* the three classes of $6\text{'}L_3(4)$ in $\mathrm{SL}_6(p)$ is $\langle \delta^3, \gamma \rangle$, where δ^3 and γ induce the 2_1 and 2_2 automorphisms of $L_3(4)$, respectively.

(iii) When $p \equiv 5, 23 \pmod{24}$, the stabiliser in $\mathrm{Out}\,U_6(p)$ of at least one of the six classes of $6\text{'}L_3(4).2_1^-$ in $\mathrm{SU}_6(p)$ is $\langle \gamma \rangle$, extending it to $6\text{'}L_3(4).2^2$.

(iv) When $p \equiv 11, 17 \pmod{24}$, the stabiliser in $\mathrm{Out}\,U_6(p)$ of at least one of the three classes of $6\text{'}L_3(4)$ in $\mathrm{SU}_6(p)$ is $\langle \delta^3, \gamma \rangle$, where δ^3 and γ induce the 2_1 and 2_2 automorphisms of $L_3(4)$, respectively.

Proof The reader should recall Proposition 4.5.18, where we previously considered $G = 6\text{'}L_3(4)$. There are two dual representations, stabilised by the 2_1 automorphism and interchanged by the 2_2 and 2_3 automorphisms of G.

We wrote G as a subgroup of $\mathrm{GL}_6(R)$ with $R = \mathbb{Z}[\omega]$, where $\omega = z_3$. We found a matrix B for a unitary form for which G is a group of isometries with entries of $B^{\pm 1}$ in $R[1/2]$. We found matrices L_1 and L_2, with entries of $L_i^{\pm 1}$ in $R[1/2]$, and $\det L_1 = 1$, $\det L_2 = -8 = -2.2^2$, such that $\gamma = -\mathsf{T}$ followed by conjugation by L_1 or by L_2 normalises G and induces, respectively, the 2_2 or the 2_3 automorphism of G (modulo inner automorphisms). So neither B nor L_i result in exceptional primes. Note that the 2_2 and 2_3 outer automorphisms of G are equivalent modulo $G.2_1$.

First consider Case **L**. If $p \equiv 1, 19 \pmod{24}$, then $\det L_1$ and $\det L_2$ are both squares, so up to conjugacy the stabiliser of the class of $6\text{'}L_3(4).2_1$ is $\langle \gamma \rangle$. If $p \equiv 7, 13 \pmod{24}$ then $\det L_1$ is square but $\det L_2$ is non-square, so up to conjugacy the 2_2 automorphism is induced by γ, and the 2_3 by $\delta^3\gamma$.

Now consider Case **U**. If $p \equiv 5, 23 \pmod{24}$, then by Proposition 4.6.6 the stabiliser of the class of $6\text{'}L_3(4).2_1$ can be assumed to be $\langle \gamma \rangle$. Conversely, if $p \equiv 11, 17 \pmod{24}$, then -2 reduces to a square in $\mathbb{F}_p^\times$, so by Proposition 4.6.6 we deduce that γ induces the 2_2 automorphism of G and $\gamma\delta^3$ the 2_3.

Details of the computer calculations for this example can be found in file `6134d6calc`.   □

Proposition 4.7.13   *Each of the two classes of $\mathscr{S}_1$-subgroups of $\mathrm{SL}_6(3)$ isomorphic to $2\text{'}M_{12}$ has stabiliser $\langle \gamma\delta \rangle$ in $\mathrm{Out}\,L_6(3)$.*

Proof We considered this group in Proposition 4.5.19. From the character table, we find that the subgroup $2\text{'}M_{12} < \mathrm{SL}_6(3)$ is extendible by an automorphism of type $\gamma\delta^i$ for some i, and $2\text{'}M_{12}$ restricts absolutely irreducibly to a subgroup $2\text{'}L_2(11)$ with representation as in Proposition 4.7.7. Moreover $2\text{'}L_2(11)$ extends to a subgroup $2\text{'}L_2(11).2$ of $2\text{'}M_{12}.2$, and when $p = 3$, up to conjugacy the stabiliser of the class of $2\text{'}L_2(11)$ in $\mathrm{SU}_6(p)$ is $\langle \gamma\delta \rangle$. Since $2\text{'}M_{12}$ has a unique class of subgroups isomorphic to $2\text{'}L_2(11)$, it follows that (the class of) $2\text{'}M_{12} < \mathrm{SL}_6(3)$ is also stabilised by a conjugate of $\gamma\delta$. This result can also be checked directly by a computer calculation (file `2m12d6f3calc`).   □

Proposition 4.7.14 *The stabiliser in* $\mathrm{Out}\,\mathrm{U}_6(2)$ *of at least one of the three classes of* $3\dot{\,}\mathrm{M}_{22}$ *in* $\mathrm{SU}_6(2)$ *is* $\langle\gamma\rangle$.

Proof We considered $3\dot{\,}\mathrm{M}_{22}$ in Proposition 4.5.19. The duality automorphism of $\mathrm{SU}_6(2)$ has the same effect on $3\dot{\,}\mathrm{M}_{22}$ as the outer automorphism. Since $|\delta| = 3$ is odd, by Lemma 4.6.1 we can take the class stabiliser to be $\langle\gamma\rangle$. □

Proposition 4.7.15 *The stabiliser in* $\mathrm{Out}\,\mathrm{U}_6(2)$ *of at least one of the three classes of* $3_1\dot{\,}\mathrm{U}_4(3).2_2$ *in* $\mathrm{SU}_6(2)$ *is* $\langle\gamma\rangle$.

Proof We considered this example in Proposition 4.5.20. The automorphism γ of $\mathrm{SU}_6(2)$ has the same effect on the character of $3_1\dot{\,}\mathrm{U}_4(3).2_2$ as the 2_1 automorphism. Since $|\delta| = 3$, we can take the class stabiliser to be $\langle\gamma\rangle$, by Lemma 4.6.1. □

Proposition 4.7.16   (i) *If* $p \equiv 1 \pmod{12}$, *then the stabiliser in* $\mathrm{Out}\,\mathrm{L}_6(p)$ *of at least one of the six classes of* $6_1\dot{\,}\mathrm{U}_4(3).2_2^-$ *in* $\mathrm{SL}_6(p)$ *is* $\langle\gamma\rangle$, *extending it to* $6_1\dot{\,}\mathrm{U}_4(3).2_{122}^2$.

(ii) *If* $p \equiv 7 \pmod{12}$, *then the stabiliser in* $\mathrm{Out}\,\mathrm{L}_6(p)$ *of at least one of the three classes of* $6_1\dot{\,}\mathrm{U}_4(3)$ *in* $\mathrm{SL}_6(p)$ *is* $\langle\delta^3, \gamma\rangle$, *where* δ^3 *and* γ *induce the* 2_2 *and* 2_1 *automorphisms of* $\mathrm{U}_4(3)$, *respectively.*

(iii) *If* $p \equiv 11 \pmod{12}$, *then the stabiliser in* $\mathrm{Out}\,\mathrm{U}_6(p)$ *of at least one of the six classes of* $6_1\dot{\,}\mathrm{U}_4(3).2_2^-$ *in* $\mathrm{SU}_6(p)$ *is* $\langle\gamma\rangle$, *extending it to* $6_1\dot{\,}\mathrm{U}_4(3).2_{122}^2$.

(iv) *If* $p \equiv 5 \pmod{12}$, *then the stabiliser in* $\mathrm{Out}\,\mathrm{U}_6(p)$ *of at least one of the three classes of* $6_1\dot{\,}\mathrm{U}_4(3)$ *in* $\mathrm{SU}_6(p)$ *is* $\langle\delta^3, \gamma\rangle$, *where* δ^3 *and* γ *induce the* 2_2 *and* 2_1 *automorphisms of* $\mathrm{U}_4(3)$, *respectively.*

Proof The reader should recall Proposition 4.5.20, where we previously considered $G = 6_1\dot{\,}\mathrm{U}_3(4)$. There are two dual representations, stabilised by the 2_2 automorphism and interchanged by the 2_1 and $2_2' = 2_12_2$ automorphisms of G. (Note that $2_2'$ is conjugate to 2_2 in $\mathrm{Out}\,\mathrm{U}_4(3)$ but not in $\mathrm{Out}\,G$.)

We wrote G as a subgroup of $\mathrm{GL}_6(R)$ with $R = \mathbb{Z}[\omega]$, where $\omega = z_3$. We found a matrix B for a unitary form for which G is a group of isometries, with entries of $B^{\pm 1}$ in $R[1/2]$. We found matrices L_1 and L_2, with entries of $L_i^{\pm 1}$ in $R[1/2]$, and $\det L_1 = 1$, $\det L_2 = -1$, such that $\gamma = -\mathsf{T}$ followed by conjugation by L_1 or by L_2 normalises G and induces, respectively, the 2_1 or the $2_2'$ automorphism of G (modulo inner automorphisms). So neither B nor L_i result in exceptional primes.

First consider Case **L**. If $p \equiv 1 \pmod{12}$ then L_1 has square determinant, so by Lemma 4.6.2 we may assume that the stabiliser of the class of $6_1\dot{\,}\mathrm{U}_4(3).2_2$ in $\mathrm{Out}\,\mathrm{SL}_6(p)$ is $\langle\gamma\rangle$, which extends the group to $6_1\dot{\,}\mathrm{U}_4(3).2_{122}^2$. If $p \equiv 7 \pmod{12}$ then 1 is square and -1 is non-square, so γ induces the 2_1 automorphism of G and $\gamma\delta^3$ induces the $2_2'$ automorphism.

Now consider Case **U**. If $p \equiv 5 \pmod{12}$ then 1 and -1 are both square in $\mathbb{F}_p^\times$, so by Proposition 4.6.6 γ induces the 2_1 automorphism of G, and $\gamma\delta^3$ induces the $2_2'$ automorphism. Finally, if $p \equiv 11 \pmod{12}$ then the stabiliser of the class of $6_1{}^{\cdot}U_4(3).2_1$ is $\langle\gamma\rangle$, extending the group to $6_1{}^{\cdot}U_4(3).2_{122}^2$.

Details of the computer calculations for this example can be found in file `6u43d6calc`.   □

4.7.2 Case S

In dimension 4 when $q > 2$ is even, the outer automorphism group of $\mathrm{Sp}_n(q)$ contains a graph automorphism. However by Lemma 4.5.8, there are no $\mathscr{S}_1$-subgroups of $\mathrm{Sp}_4(q)$ when q is even, so we can ignore that case. Recall also that $\mathrm{Sp}_4(2)$ is excluded, as it is not quasisimple. Therefore we need only calculate the action of the field automorphism ϕ, and can clearly restrict to those cases where ϕ acts non-trivially.

Theorem 4.7.17 *Let G be an $\mathscr{S}_1$-subgroup of $\mathrm{Sp}_n(q)$ with $n \leqslant 6$, and assume that q is a proper power. Then one of the following holds:*

  (i) $G = 2{}^{\cdot}\mathrm{A}_5$ *in dimension 2, and the class stabiliser of G in $\mathrm{Out}\,\overline{\Omega}$ is $\langle\phi\rangle$;*
 (ii) $G = 2{}^{\cdot}\mathrm{L}_3(2)$ *in dimension 6, and the class stabiliser of G in $\mathrm{Out}\,\overline{\Omega}$ is $\langle\delta\rangle$;*
(iii) $G = (2{}^{\cdot})\mathrm{L}_2(13)$, $2{}^{\cdot}\mathrm{A}_7$ *or* $(2{}^{\cdot})\mathrm{J}_2$ *in dimension 6, and the class stabiliser of G in $\mathrm{Out}\,\overline{\Omega}$ is $\langle\phi\rangle$.*

Proof By Theorem 4.3.3, the possibilities for G are considered in Propositions 4.5.1, 4.5.9, 4.5.10, 4.5.11 and 4.5.21–4.5.25.

Consulting these results, we see that the only examples in which q is a proper power (so that ϕ is non-trivial) are $2{}^{\cdot}\mathrm{A}_5$ in dimension 2, and $2{}^{\cdot}\mathrm{L}_3(2)$, $(2{}^{\cdot})\mathrm{L}_2(13)$, $2{}^{\cdot}\mathrm{A}_7$, $(2{}^{\cdot})\mathrm{J}_2$ in dimension 6.

Consider first $G = 2{}^{\cdot}\mathrm{L}_3(2) \leqslant \mathrm{Sp}_6(p^2)$. The character ring of the representation is the p-modular reduction of $\mathbb{Z}[r_2]$, and by Proposition 4.5.22 we require $p \equiv \pm3, \pm5 \pmod{16}$. There are two algebraically conjugate representations. Both of these are stabilised by the automorphism δ of $\mathrm{Sp}_6(q)$ by Proposition 4.5.22, and consulting [12, 57], we see that their characters are interchanged by ϕ. The two conjugacy classes of $\mathscr{S}_1$-subgroups isomorphic to G are therefore swapped by ϕ and by $\phi\delta$.

Next consider $G = \mathrm{L}_2(13)$ or J_2, with $G \leqslant \mathrm{Sp}_6(4)$. By Theorem 4.3.3 there are two representations, interchanged by the outer automorphism α of G. Consulting [57], we see that the two representations are interchanged by ϕ, so $\phi = \alpha c_g$ for some $g \in \mathrm{CSp}_6(4)$. Since there are no diagonal automorphisms, it follows that ϕ stabilises the conjugacy class of G.

For $G = 2{}^{\cdot}\mathrm{A}_5$ in dimension 2, or $2{}^{\cdot}\mathrm{L}_2(13)$, $2{}^{\cdot}\mathrm{A}_7$, $2{}^{\cdot}\mathrm{J}_2$ in dimension 6, there are two representations that are swapped by both the outer automorphism α of G and by ϕ. For each of these groups, we have $G = Z.S$ with S simple, and we find from [12] that there is a unique almost simple group with the structure $S.2$, and this extension is split; that is, there are involutions in $S.2 \setminus S$.

In each of these three cases, there are two Ω-classes of subgroups that are swapped by δ, so we must determine whether it is ϕ or $\phi\delta$ that stabilises the Ω-class and induces an outer automorphism of G. But we know from Lemma 4.6.7 that all elements of order 2 in $\mathrm{PC\Gamma Sp}_n(q)$ lie in $\mathrm{P\Sigma Sp}_n(q) \cup \mathrm{PCSp}_n(q)$, and it follows that the class stabiliser is $\langle \phi \rangle$. $\qquad\square$

4.8 Dimension up to 6: containments

Recall Definition 2.1.4 of a containment between two subgroups of a classical group. In this section we determine all containments between the candidate maximal $\mathscr{S}_1$-subgroups in dimensions up to 6. We recall that the stabilisers of the classes of candidates in the conformal group normalising Ω were determined in Section 4.5, and their full stabilisers in $\mathrm{Out}\,\overline{\Omega}$ in Section 4.7. The reader may need to refer back to these sections to find the structure of the candidates and the extensions of $\overline{\Omega}$ in which they lie.

To avoid specifying precise structures of extensions of quasisimple groups, we shall often make statements about containments projectively, but we shall include in brackets the centres of the quasisimple groups involved in the representations.

To verify that there is a containment $H_1 < H_2$ between two $\mathscr{S}_1$-subgroups of Ω, we of course need not only to check that H_1 is a subgroup of H_2 as an abstract group, but also that the restriction of the relevant irreducible representation of H_2^∞ to H_1^∞ is the relevant irreducible representation of H_1^∞. In most cases, this is straightforward using [12, 57], either because H_1^∞ has no non-trivial irreducible representations of lower degree, or by looking at the character values. We shall supply details only when this might not be clear.

Our overall conclusions about which groups are maximal amongst the $\mathscr{S}_1$-subgroups of Ω and its extensions are presented in Section 4.10, where we also present results for dimensions greater than 6 and at most 12.

Proposition 4.8.1 *Let* $\mathrm{SL}_2(q) \leqslant G \leqslant \mathrm{\Gamma L}_2(q)$, *and let* H *be an* $\mathscr{S}_1$-subgroup *of* G *such that* $H\mathrm{SL}_2(q) = G$ *and* $H = \mathrm{N}_G(H)$. *Then* H *is* $\mathscr{S}_1$-maximal *in* G.

Proof By Theorem 4.3.3 there is only one $\mathscr{S}_1$-candidate, and hence no possible containments. $\qquad\square$

Proposition 4.8.2 *Let* $\overline{\Omega} \leqslant G \leqslant \mathrm{Aut}\,\overline{\Omega}$, *with* $\overline{\Omega} = \mathrm{L}_3(q)$ *or* $\mathrm{U}_3(q)$. *Let* H *and*

K be distinct $\mathscr{S}_1$-subgroups of G such that $H\bar{\Omega} = K\bar{\Omega} = G$, with $H = N_G(H)$ and $K = N_G(K)$. There is a containment $H < K$ if and only if $H = L_3(2)$, $K = (3^{\cdot})A_7$ and $G = U_3(5)$.

Proof By Theorem 4.3.3, the quasisimple examples are $L_3(2)$ (characteristic not 2 or 7), $3^{\cdot}A_6$ (characteristic not 3), and $3^{\cdot}A_7$ (characteristic 5 only). The stabilisers of these groups in Out $\bar{\Omega}$ have been calculated in Propositions 4.5.2, 4.5.3 and 4.5.4 and Theorem 4.7.1, and the reader should recall these results.

The group $L_3(2)$ is not a subgroup of $(3^{\cdot})A_6$, but is a subgroup of $3^{\cdot}A_7$ in $SU_3(5)$. Consulting [12], the group $L_3(2)$ has no non-trivial representations in dimension less than 3, so $L_3(2)$ is irreducible in $3^{\cdot}A_7$. Similarly, since $3^{\cdot}A_6$ has no non-trivial representations in dimension less than 3, there is a containment $3^{\cdot}A_6 < 3^{\cdot}A_7$ in $SU_3(5)$, by Propositions 4.5.3 and 4.5.4.

We saw in Proposition 4.5.3 that $3^{\cdot}A_6$ extends to $3^{\cdot}A_6.2_3$ within $SU_3(5)$, whereas $A_6.2_3$ is not contained in $A_7.2 = S_7$, so $3^{\cdot}A_6.2_3$ and its extensions are in fact $\mathscr{S}_1$-maximal.

We saw in Proposition 4.5.2 that $L_3(2)$ is scalar-normalising in $CGU_3(5)$, so the containment $L_3(2) < 3^{\cdot}A_7$ prevents $3 \times L_3(2)$ from being $\mathscr{S}_1$-maximal in $SU_3(5)$. But $PGL_2(7) = L_3(2).2$ is not a subgroup of $A_7.2 = S_7$, so (projectively) $L_3(2).2$ is $\mathscr{S}_1$-maximal in $U_3(5)\langle\gamma\rangle$. □

Proposition 4.8.3 *Let* $\bar{\Omega} \leqslant G \leqslant \operatorname{Aut} \bar{\Omega}$, *with* $\bar{\Omega} = L_4(q)$ *or* $U_4(q)$. *Let* H *and* K *be distinct* $\mathscr{S}_1$-*subgroups of* G *such that* $H\bar{\Omega} = K\bar{\Omega} = G$, *with* $H = N_G(H)$ *and* $K = N_G(K)$. *There is a containment* $H < K$ *if and only if* $H^\infty = (2^{\cdot})L_2(7)$ *and either* $K^\infty = (2^{\cdot})A_7$ *and* $G = \bar{\Omega}$, *or* $K^\infty = (4_2{}^{\cdot})L_3(4)$ *and* $G \leqslant U_4(3)\langle\gamma\delta\rangle$.

Proof By Theorem 4.3.3 the quasisimple examples are $2^{\cdot}L_3(2) \cong 2^{\cdot}L_2(7)$ (characteristic not 2 or 7), A_7 (characteristic 2 only), $2^{\cdot}A_7$ (characteristic not 2 or 7), $4_2{}^{\cdot}L_3(4)$ (characteristic 3 only), and $2^{\cdot}U_4(2)$ (characteristic not 2 or 3). By Lagrange's theorem there are no containments involving $U_4(2)$.

The smallest degree of a non-trivial representation of $2^{\cdot}L_3(2)$ in characteristic not equal to 2 is 4 [12, 57], so there is a containment $2^{\cdot}L_3(2) < 2^{\cdot}A_7$. However, there is no containment $L_3(2).2 \not\leqslant A_7.2$. The stabilisers of these groups in Out $\bar{\Omega}$ are described in Propositions 4.7.4 and 4.7.3, and in each case have order 2. From these results we deduce that $(2^{\cdot})L_3(2)$ is not $\mathscr{S}_1$-maximal in $L_4(p)$ ($p \equiv 1, 2, 4 \pmod 7$) or in $U_4(p)$ ($p \equiv 3, 5, 6 \pmod 7$), but $(2^{\cdot})L_3(2).2$ is $\mathscr{S}_1$-maximal in $L_4^{\pm}(p)\langle\gamma\rangle$ or $L_4^{\pm}(p)\langle\gamma\delta\rangle$, depending on its stabiliser.

Also, in characteristic 3, there are containments $2^{\cdot}L_3(2) < 4_2{}^{\cdot}L_3(4)$ and $2^{\cdot}L_3(2).2 < 4_2{}^{\cdot}L_3(4).2_1$, so $(2^{\cdot})L_3(2).2$ is not $\mathscr{S}_1$-maximal in $U_4(3)\langle\gamma\delta\rangle$. However, $L_3(4)$ has no subgroups of index less than 21, so does not contain A_7. □

It will follow from the results proved in Chapter 6 that the groups of shape $2^{\cdot}L_2(7).2$ are indeed maximal subgroups of the corresponding $SL_4^{\pm}(p)$, for all

$p \neq 2, 3, 7$. The question of whether they are type 1 or type 2 novelties with respect to A_7, as defined in Subsection 1.3.1, is not relevant for our main objective of finding the maximal subgroups of the almost simple classical groups, but the following result may nevertheless be of some interest.

Proposition 4.8.4 *For infinitely many primes p, there is an almost simple extension of $L_4(p)$ that contains $L_2(7).2$ as a type 1 novel maximal subgroup with respect to A_7. Additionally, for infinitely many primes p, there is an almost simple extension of $L_4(p)$ that contains $L_2(7).2$ as a type 2 novel maximal subgroup with respect to A_7. The same is true for almost simple extensions of $U_4(p)$.*

Proof Let $p \equiv 1, 2, 4 \pmod 7$: so $\Omega = SL_4(p)$ in Propositions 4.7.3 and 4.7.4. By Dirichlet's Theorem, there are infinitely many such primes p with $p \equiv 3$ or $5 \pmod 8$. For such primes, $L_2(7).2 < L_4(p)\langle \gamma\delta \rangle$ by Proposition 4.7.4. But, by Proposition 4.7.3, the stabilisers in $\mathrm{Out}\, L_4(p)$ of the class of A_7 are conjugates of $\langle \gamma \rangle$ and, since $\langle \gamma\delta \rangle$ is not conjugate to $\langle \gamma \rangle$, the group A_7 is self-normalising in $L_4(p).\langle \gamma\delta \rangle$. Hence $L_2(7).2$ is a type 1 novelty with respect to A_7.

There are also infinitely many primes p satisfying $p \equiv 1, 2, 4 \pmod 7$ and $p \equiv 7 \pmod 8$ and, for such primes, Proposition 4.7.4 gives $L_2(7).2 < L_4(p).\langle \gamma \rangle$. In this case $\mathrm{Out}\, L_4(p) = \langle \gamma, \delta \rangle \cong 2^2$ is abelian, so the only conjugate of $\langle \gamma \rangle$ in $\mathrm{Out}\, L_4(p)$ is $\langle \gamma \rangle$ itself. Hence $L_2(7) < A_7 < L_4(p)$, where both of these subgroups $L_2(7)$ and A_7 are strictly contained in their normalisers in $L_4(p)\langle \gamma \rangle$. So $L_2(7).2$ is a type 2 novelty with respect to A_7.

Similarly, if $p \equiv 2, 3, 6 \pmod 7$ then we get type 1 novelties when $p \equiv 3$ or $5 \pmod 8$ and of type 2 when $p \equiv 1 \pmod 8$ inside some almost simple extension of $U_4(p)$. □

We remark that the situation is more complicated when $p \equiv 1 \pmod 8$ in Case **L** and when $p \equiv 7 \pmod 8$ in Case **U** because, in those cases, we have to consider the possibility that, in the chain of subgroups $L_2(7) < A_7 < L_4^{\pm}(p)$, $L_2(7)$ is normalised by $\langle \gamma \rangle$, but A_7 is normalised by its conjugate subgroup $\langle \gamma\delta^2 \rangle$ of $\mathrm{Out}\, L_4^{\pm}(p)$. It turns out that both type 1 and type 2 novel maximal subgroups occur for infinitely many primes, but the complete analysis is more complicated, and will be omitted.

Proposition 4.8.5 *Let $\bar{\Omega} \leqslant G \leqslant \mathrm{Aut}\, \bar{\Omega}$, with $\bar{\Omega} = S_4(q)$. Let H and K be distinct $\mathscr{S}_1$-subgroups of G such that $H\bar{\Omega} = K\bar{\Omega} = G$, with $H = N_G(H)$ and $K = N_G(K)$. There is a containment $H < K$ if and only if one of the following holds:*

(i) $H^\infty = (2^{\cdot})A_5$, $K^\infty = (2^{\cdot})A_6$ *and* $q = p > 5$;
(ii) $H^\infty = (2^{\cdot})A_5$ *or* $(2^{\cdot})A_6$, *and* $K^\infty = (2^{\cdot})A_7$, *with* $q = 7$.

Proof By Lemma 4.5.8 the field size q is odd. By Theorem 4.3.3, the quasisimple groups to consider are $2^{\cdot}A_5$ (characteristic not 2, 3 or 5), $2^{\cdot}A_6$ (characteristic not 2 or 3), and $2^{\cdot}A_7$ (characteristic 7 only).

We find from [12] that there are 2 classes of $2^{\cdot}A_5$ in $2^{\cdot}A_6$, and the restriction of the 4-dimensional representation of $2^{\cdot}A_6$ to one of them is irreducible, so there is a containment $2^{\cdot}A_5 < 2^{\cdot}A_6$. By Propositions 4.5.9 and 4.5.10, these groups extend under the outer automorphism of $\bar{\Omega}$ to $(2^{\cdot})A_5.2$ and $(2^{\cdot})A_6.2_1$ Since $A_5.2 \cong S_5$ and $A_6.2_1 \cong S_6$, we have $A_5.2 < A_6.2_1$. So $(2^{\cdot})A_5$ and its extension are never $\mathscr{S}_1$-maximal.

By [12], the smallest degree of a non-trivial representation of $2^{\cdot}A_6$ in characteristic not equal to 2 is 4, so there is also a containment $2^{\cdot}A_6 < 2^{\cdot}A_7$ in characteristic 7. By Propositions 4.5.10 and 4.5.11 and the fact that $S_6 \cong A_6.2_1 < S_7$, this containment extends under $\langle\delta\rangle$ to $(2^{\cdot})A_6.2_1 < (2^{\cdot})A_7.2$. Hence $(2^{\cdot})A_6$ and $(2^{\cdot})A_6.2_1$ are not $\mathscr{S}_1$-maximal in $S_4(7)$ or $PCSp_4(7)$. $\square$

Proposition 4.8.6 *Let $\bar{\Omega} \leqslant G \leqslant \operatorname{Aut}\bar{\Omega}$, with $\bar{\Omega} = L_5(q)$ or $U_5(q)$. Let H and K be distinct $\mathscr{S}_1$-subgroups of G such that $H\bar{\Omega} = K\bar{\Omega} = G$, with $H = \mathrm{N}_G(H)$ and $K = \mathrm{N}_G(K)$. There is a containment $H < K$ if and only if $H = L_2(11)$, $K = M_{11}$ and $G = L_5(3)$.*

Proof By Theorem 4.3.3, the quasisimple groups are $L_2(11)$ ($p \neq 11$), M_{11} ($p = 3$ only) and $U_4(2)$ ($p \neq 2, 3$).

Lagrange's theorem rules out all possibilities, except for the containment $L_2(11) < M_{11}$ in $SL_5(3)$: note that by [57] the group $L_2(11)$ has no non-trivial representations in dimension less than 5 in characteristic 3. Thus $L_2(11)$ is not $\mathscr{S}_1$-maximal in this case. However, by Theorem 4.7.6 the stabiliser of (the class of) M_{11} is trivial and the stabiliser of (the class of) $L_2(11)$ is $\langle\gamma\rangle$, so $L_2(11).2$ is $\mathscr{S}_1$-maximal in $L_5(3)\langle\gamma\rangle$. $\square$

Dimension 6, Indicator $\circ$. This is the most complicated case. By Theorem 4.3.3 the quasisimple groups to consider are $3^{\cdot}A_6$ (characteristic not 2 or 3), $6^{\cdot}A_6$ (characteristic not 2 or 3), $2^{\cdot}L_2(11)$ (characteristic not 2 or 11), $3^{\cdot}A_7$ (characteristic not 3), $6^{\cdot}A_7$ (characteristic not 2 or 3), $6^{\cdot}L_3(4)$ (characteristic not 2 or 3), $2^{\cdot}M_{12}$ (characteristic 3 only), $3^{\cdot}M_{22}$ (characteristic 2 only), $3_1^{\cdot}U_4(3)$ (characteristic 2 only) and $6_1^{\cdot}U_4(3)$ (characteristic not 2 or 3).

Because some of the containments are considerably easier to eliminate than others, we will deal with these groups in a different order from their occurrence in Theorem 4.3.3.

We first consider all containments where one of the groups is $2^{\cdot}L_2(11)$, $2^{\cdot}M_{12}$ or $3^{\cdot}M_{22}$.

Proposition 4.8.7 *Let $\bar{\Omega} \leqslant G \leqslant \operatorname{Aut}\bar{\Omega}$, with $\bar{\Omega} = L_6(q)$ or $U_6(q)$. Let H and*

K be distinct $\mathscr{S}_1$-subgroups of G such that $H\bar{\Omega} = K\bar{\Omega} = G$, with $H = N_G(H)$ and $K = N_G(K)$. Assume in addition that at least one of H or K is isomorphic to $(2^{\cdot})L_2(11)$, $(2^{\cdot})M_{12}$ or $(3^{\cdot})M_{22}$. There is a containment $H < K$ if and only if either $H^{\infty} = (2^{\cdot})L_2(11)$, $K^{\infty} = (2^{\cdot})M_{12}$ and $\bar{\Omega} = L_6(3)$; or $H = (3^{\cdot})A_7$, $K = (3^{\cdot})M_{22}$ and $G = U_6(2)$.

Proof The behaviour of these three groups has previously been analysed in Propositions 4.5.15, 4.5.19, 4.7.7, 4.7.13 and 4.7.14, and the reader should remind themself of these results. In particular, $2^{\cdot}M_{12}$ only occurs as a subgroup of $SL_6(3)$ and $3^{\cdot}M_{22}$ only occurs as a subgroup of $SU_6(2)$.

The smallest faithful representation of $2^{\cdot}L_2(11)$ in characteristic 3 has degree 6 [57], so there is a containment $2^{\cdot}L_2(11) < 2^{\cdot}M_{12}$, extending to a containment $2^{\cdot}L_2(11).2 < 2^{\cdot}M_{12}.2$. This prevents extensions of $(2^{\cdot})L_2(11)$ from being $\mathscr{S}_1$-maximal when $\bar{\Omega} = L_6(3)$.

Similarly, the smallest faithful representation of $3^{\cdot}A_7$ in characteristic 2 has degree 6 [57], so there is a containment $3^{\cdot}A_7 < 3^{\cdot}M_{22}$. However, this does not extend, even as an abstract inclusion, to $3^{\cdot}A_7.2 < 3^{\cdot}M_{22}.2$. Using Lagrange's theorem and Theorem 4.3.3, we see that there are no other possible containments involving the quasisimple candidates $2^{\cdot}L_2(11)$, $2^{\cdot}M_{12}$ and $3^{\cdot}M_{22}$. $\square$

Next we deal with all containments involving $6^{\cdot}A_6$ and $6^{\cdot}A_7$.

Proposition 4.8.8 *Let $\bar{\Omega} \leqslant G \leqslant \operatorname{Aut}\bar{\Omega}$, with $\bar{\Omega} = L_6(q)$ or $U_6(q)$. Let H and K be distinct $\mathscr{S}_1$-subgroups of G such that $H\bar{\Omega} = K\bar{\Omega} = G$, with $H = N_G(H)$ and $K = N_G(K)$. Assume in addition that at least one of H or K is isomorphic to $(6^{\cdot})A_6$ or $(6^{\cdot})A_7$. There is a containment $H < K$ if and only if $H^{\infty} = (6^{\cdot})A_6$, $K^{\infty} = (6^{\cdot})A_7$ and one of the following holds:*

(i) $G = \bar{\Omega}$;
(ii) $G = L_6(p^2)\langle\phi\rangle$ with $p \equiv 5, 11 \pmod{24}$;
(iii) $G = L_6(p^2)\langle\phi\gamma\rangle$ with $p \equiv 13, 19 \pmod{24}$.

Proof The stabilisers of $6^{\cdot}A_6$ and $6^{\cdot}A_7$ are described in Propositions 4.7.9 and 4.7.11. By [12, 57], the minimal degree of a faithful representation of $6^{\cdot}A_6$ is 6, so there are containments $6^{\cdot}A_6 < 6^{\cdot}A_7$ and $(6^{\cdot})A_6.2_1 < (6^{\cdot})S_7$, so extensions of $(6^{\cdot})A_6$ contained in $(6^{\cdot})A_6.2_1$ are not maximal among $\mathscr{S}_1$-subgroups. But extensions involving the 2_2 or 2_3 automorphisms of A_6 are not contained in extensions of $6^{\cdot}A_7$.

The potential containments of $6^{\cdot}A_6$ or $6^{\cdot}A_7$ in $6^{\cdot}L_3(4)$ or $6_1^{\cdot}U_4(3)$ can be ruled out by using [12] to show that involutions in A_6 and A_7 have inverse images of order 4 in $6^{\cdot}A_6$ and $6^{\cdot}A_7$, whereas those in $L_3(4)$ and $U_4(3)$ have inverse images of order 2 in $6^{\cdot}L_3(4)$ and $6_1^{\cdot}U_4(3)$. It is clear that $3^{\cdot}A_6$ and $3^{\cdot}A_7$

are not subgroups of $6\dot{}A_6$ or $6\dot{}A_7$, and all other containments are eliminated by Lagrange's theorem.   □

Proposition 4.8.9   *No extension of the $\mathscr{S}_1$-subgroup $(3\dot{})A_7$ is $\mathscr{S}_1$-maximal in any almost simple extension of $L_6(p)$ or $U_6(p)$.*

Proof   When $p = 2$, the minimal degree of a faithful representation of $3\dot{}A_7$ is 6 by [57], so there is a containment $3\dot{}A_7 < 3_1\dot{}U_4(3){:}2_2$. By Propositions 4.7.10 and 4.7.15 and the abstract containment $A_7.2 < U_4(3){:}2^2$, this containment extends to $3\dot{}A_7.2 < 3_1\dot{}U_4(3){:}2^2$, so assume for the rest of the proof that $p > 2$.

From Proposition 4.7.3, we find that the group $U_4(3)$ has four classes of scalar-normalising $\mathscr{S}_1$-subgroups isomorphic to A_7. These are permuted in pairs by the 2_1 $(= \delta^2)$ automorphism of $U_4(3)$ and two of them are normalised by the 2_2 $(= \gamma)$ automorphism, while the other two are normalised by the $2_2'(= 2_1 2_2)$ automorphism. (All four classes are conjugate under the outer automorphism of order 4, but this does not act on $6_1\dot{}U_4(3)$.) Computer calculations (file `containmentsd6`) using a 6-dimensional representation of $6_1\dot{}U_4(3)$ over a finite field, reveal that the classes normalised by $2_2'$ lift to $2 \times 3\dot{}A_7$, where the subgroup $3\dot{}A_7$ acts absolutely irreducibly in the 6-dimensional representation.

Consulting Propositions 4.7.10 and 4.7.16, we now conclude that whenever $(3\dot{})A_7.2$ is a subgroup of either $L_6^\varepsilon(p)\langle\gamma\rangle$ or $L_6^\varepsilon(p)\langle\gamma\delta^3\rangle$, there are containments $(3\dot{})A_7.2 < (6_1\dot{})U_4(3).2_2' < L_6^\varepsilon(p)\langle\gamma\rangle$ or $L_6^\varepsilon(p)\langle\gamma\delta^3\rangle$, and the result follows.   □

It remains to deal with containments between extensions of $3\dot{}A_6$, $6\dot{}L_3(4)$, and $6_1\dot{}U_4(3)$ (or $3_1\dot{}U_4(3)$ when $p = 2$), in extensions of $SL_6(p)$ and $SU_6(p)$.

Proposition 4.8.10   *Let $\bar{\Omega} \leqslant G \leqslant \mathrm{Aut}\,\bar{\Omega}$, with $\bar{\Omega} = L_6(q)$ or $U_6(q)$. Let H and K be distinct $\mathscr{S}_1$-subgroups of G such that $H\bar{\Omega} = K\bar{\Omega} = G$, with $H = N_G(H)$ and $K = N_G(K)$. If $H < K$ then $H^\infty \neq (6_1\dot{})U_4(3)$ and $H^\infty \neq (3_1)\dot{}U_4(3)$.*

Proof   This follows from Lagrange's theorem.   □

In the course of the following two proofs, we shall make a number of assertions about the structure and irreducibility of various subgroups of the groups concerned. These assertions are straightforward to verify in MAGMA, working either in permutation representations of the almost simple groups involved or, in cases involving assertions of irreducibility of subgroups, in the images of 6-dimensional representations of the quasisimple groups over finite fields or number fields.

Proposition 4.8.11   *Let $\bar{\Omega} \leqslant G \leqslant \mathrm{Aut}\,\bar{\Omega}$, with $\bar{\Omega} = L_6(q)$ or $U_6(q)$. Let H and K be distinct $\mathscr{S}_1$-subgroups of G such that $H\bar{\Omega} = K\bar{\Omega} = G$, with $H = N_G(H)$ and $K = N_G(K)$. Assume that $H^\infty = (6\dot{})L_3(4)$. There is a* containment $H < K$ if and only if $K^\infty = (6_1{}^\cdot)U_4(3)$, $\bar\Omega \leqslant G \leqslant \bar\Omega\langle\gamma\rangle$, and $p \equiv \pm 7, \pm 11 \pmod{24}$.

Proof By Theorem 4.3.3 and Lagrange's theorem, of the $\mathscr{S}_1$-subgroups under consideration, only $U_4(3)$ can contain $L_3(4)$. The stabiliser of $L_3(4)$ in Out $\bar\Omega$ is calculated in Proposition 4.7.12, and of $U_4(3)$ in Proposition 4.7.16. Recall in particular that Out $\bar\Omega$ induces non-trivial automorphisms 2_1, 2_2 and $2_3 = 2_1 2_2$ of $L_3(4)$ and $2_1, 2_2$ and $2'_2 = 2_1 2_2$ of $U_4(3)$, and that for $L_3(4)$ we require characteristic not 2 or 3.

From Proposition 4.7.5, we find that $U_4(3)$ has two classes of $\mathscr{S}_1$-subgroups isomorphic to $L_3(4)$, which are self-normalising. As discussed just after Proposition 4.7.5, both classes are normalised by the 2_1 $(= \delta^2)$ and 2_3 $(= \gamma\delta)$ automorphism of $U_4(3)$ but fused by the 2_2 $(= \gamma)$ automorphism. Since $6{}^\cdot L_3(4)$ is the only quasisimple cover of $L_3(4)$ with a faithful representation of degree at most 6 (in characteristic 0 or $p \geqslant 5$) [12, 57], the irreducible 6-dimensional representation of $6_1{}^\cdot U_4(3)$ must restrict to the irreducible 6-dimensional representation of $6{}^\cdot L_3(4)$.

By Proposition 4.7.5, there are containments $L_3(4){:}2_2 < U_4(3){:}2_1$ and from Propositions 4.7.12 and 4.7.16 we find that whenever $(6{}^\cdot)L_3(4){:}2_2$ is a subgroup of $L_6^\varepsilon(p)\langle\gamma\rangle$, there is a containment $(6{}^\cdot)L_3(4){:}2_2 < (6_1{}^\cdot)U_4(3){:}2_1 < L_6^\varepsilon(p)\langle\gamma\rangle$. So $L_3(4)$ and $L_3(4){:}2_2$ are never $\mathscr{S}_1$-maximal in almost simple extensions of $L_6^\varepsilon(p)$. Since $L_3(4)$ can only occur as an $\mathscr{S}_1$-subgroup of $U_4(3)$, it follows from Proposition 4.7.5 that $L_3(4){:}2_1$, $L_3(4){:}2_3$ and $L_3(4){:}2^2$ are not contained in $U_4(3){:}2^2_{122}$, so these subgroups are $\mathscr{S}_1$-maximal in the appropriate almost simple extension of $L_6^\varepsilon(p)$ whenever they arise. $\qquad\square$

We finish with the most complicated case, which is determining the $\mathscr{S}_1$-maximality of extensions of $3{}^\cdot A_6$.

Proposition 4.8.12 *Let $\bar\Omega \leqslant G \leqslant \operatorname{Aut}\bar\Omega$, with $\bar\Omega = L_6(q)$ or $U_6(q)$. Let H and K be distinct $\mathscr{S}_1$-subgroups of G such that $H\bar\Omega = K\bar\Omega = G$, with $H = N_G(H)$ and $K = N_G(K)$. Assume that $H^\infty = (3{}^\cdot)A_6$. There is a containment $H < K$ for some such K if and only if one of the following holds:*

(i) $G = \bar\Omega$;

(ii) $p = q \equiv \pm 7, \pm 11 \pmod{24}$ *and* $G = \bar\Omega\langle\delta^3\rangle$;

(iii) $p = q \equiv \pm 11 \pmod{24}$ *and* $G = \bar\Omega\langle\gamma\rangle$;

(iv) $p = q \equiv \pm 7 \pmod{24}$ *and* $G = \bar\Omega\langle\gamma\delta^3\rangle$.

Proof The behaviour of $3{}^\cdot A_6$ has previously been analysed in Propositions 4.5.13 and 4.7.8, and the reader should recall these results. Using [12], and considering character values on involutions, we see that there is a containment $3{}^\cdot A_6 < 3{}^\cdot A_7$. Using Proposition 4.7.10 we see that $(3{}^\cdot)A_6.2_1 \cong (3{}^\cdot)S_6 < (3{}^\cdot)S_7$

in all occurrences of either of these subgroups in almost simple extensions of $L_6(p)$ and $U_6(p)$. So $(3^{\cdot})A_6$ and $(3^{\cdot})A_6.2_1$ are never $\mathscr{S}_1$-maximal in such extensions. However the extension $(3^{\cdot})A_6.2_2 \cong (3^{\cdot})\mathrm{PGL}_2(9)$ is not contained in any almost simple extension of $(3^{\cdot})A_7$.

Since A_6 can only occur as an $\mathscr{S}_1$-subgroup of $L_3(4)$, it follows from Theorem 4.7.1 that $L_3(4)$ has three classes of self-normalising subgroups isomorphic to A_6. It is explained in the discussion following the proof of Theorem 4.7.1 that these classes are all normalised by the 2_1 $(= \gamma\phi)$ automorphism of $L_3(4)$, which induces the 2_3 automorphism of A_6, whereas just one of them is normalised by the 2_2 $(= \phi)$ automorphism of $L_3(4)$, which induces the 2_1 automorphism of A_6. A computer calculation (file `containmentsd6`) in a 6-dimensional representation of $6^{\cdot}L_3(4)$ over a suitable finite field reveals that all three have inverse image $2 \times 3^{\cdot}A_6$ in $6^{\cdot}L_3(4)$, but that one of them, namely the one normalised by the 2_2 automorphism of $L_3(4)$, has reducible inverse image, whereas the other two have absolutely irreducible inverse images. Noting the behaviour of extensions in Proposition 4.7.12, with the 2_1 automorphism of $L_3(4)$ corresponding to the 2_3 automorphism of A_6, we find that $(3^{\cdot})A_6.2_3 \cong (3^{\cdot})M_{10} < (6^{\cdot})L_3(4).2_1$ in all occurrences of either of these subgroups in almost simple extensions of $L_6(p)$ and $U_6(p)$. So $(3^{\cdot})A_6.2_3$ is never $\mathscr{S}_1$-maximal in such extensions. However, once again, the extension $(3^{\cdot})A_6.2_2 \cong (3^{\cdot})\mathrm{PGL}_2(9)$ is not contained in any almost simple extension of $(6^{\cdot})L_3(4)$.

Finally, computer calculations (file `containmentsd6`) show that $U_4(3)$ has 11 classes of subgroups isomorphic to A_6. Six of these classes have as inverse image in $6_1^{\cdot}U_4(3)$ a group with the structure $2 \times 3^{\cdot}A_6$ that is absolutely irreducible in the 6-dimensional representation of $6_1^{\cdot}U_4(3)$. Of these 6 classes, four have images that are self-normalising in $\mathrm{Aut}(6_1^{\cdot}U_4(3))$, and two have normalisers of image isomorphic to S_6. So, using Proposition 4.7.16, we find no further instances of extensions of $3^{\cdot}A_6$ being not $\mathscr{S}_1$-maximal.

So the groups H with $H < K$ for some K, as in the proposition statement, are precisely those isomorphic to A_6, $A_6{:}2_1$ or $A_6{:}2_3$ which, according to Proposition 4.7.8, correspond to those listed in its conclusion. $\square$

We remind the reader that our overall conclusions on which groups are maximal amongst the $\mathscr{S}_1$-subgroups of Ω and its extensions are presented in Section 4.10.

Dimension 6, Indicator $-$. By Theorem 4.3.3, the quasisimple groups to consider are $2^{\cdot}A_5$ (characteristic not 2 or 5); $2^{\cdot}L_3(2)$ (characteristic not 2 or 7); $L_2(13)$ (characteristic 2 only); $2^{\cdot}L_2(13)$ (characteristic not 2 or 13); $2^{\cdot}A_7$ (characteristic 3 only); $U_3(3)$ (characteristic not 3); J_2 (characteristic 2 only); and $2^{\cdot}J_2$ (characteristic not 2).

Proposition 4.8.13 *Let* $\bar{\Omega} \leqslant G \leqslant \mathrm{Aut}\,\bar{\Omega}$, *with* $\bar{\Omega} = \mathrm{S}_6(q)$. *Let H and K be distinct* $\mathscr{S}_1$*-subgroups of G such that* $H\bar{\Omega} = K\bar{\Omega} = G$, *with* $H = \mathrm{N}_G(H)$ *and* $K = \mathrm{N}_G(K)$. *Assume that* $H^\infty = (2\dot{})\mathrm{L}_3(2)$. *There is a containment* $H < K$ *if and only if* $K = (2\dot{})\mathrm{A}_7$ *and* $G = \mathrm{S}_6(9)$.

Proof We consider the candidate groups K in turn. Lagrange's theorem eliminates $(2\dot{})\mathrm{A}_5$ and $(2\dot{})\mathrm{L}_2(13)$, and computer calculations (file containmentsd6) show that $2\dot{}\mathrm{L}_3(2) \not< \mathrm{U}_3(3)$ or $2\dot{}\mathrm{J}_2$. Note that this calculation makes use of the maximal subgroups of J_2, which are determined in [24].

The behaviour of H in $\mathrm{PCSp}_6(q)$ is described in Proposition 4.5.22, of $(2\dot{})\mathrm{A}_7$ in $\mathrm{PCSp}_6(q)$ in Proposition 4.5.24, and of both in $\mathrm{Aut}\,\bar{\Omega}$ in Theorem 4.7.17. We see that $2\dot{}\mathrm{A}_7$ only arises as a subgroup of $\mathrm{Sp}_6(9)$. From [57] we find that the restriction of this representation of $2\dot{}\mathrm{A}_7$ to $2\dot{}\mathrm{L}_3(2)$ is absolutely irreducible, so we have the containment described in the proposition but, since $\mathrm{L}_3(2).2$ is not a subgroup of $\mathrm{A}_7.2$, it does not extend to $(2\dot{})\mathrm{L}_3(2).2$. $\square$

Proposition 4.8.14 *Let* $\bar{\Omega} \leqslant G \leqslant \mathrm{Aut}\,\bar{\Omega}$, *with* $\bar{\Omega} = \mathrm{S}_6(q)$. *Let H and K be distinct* $\mathscr{S}_1$*-subgroups of G such that* $H\bar{\Omega} = K\bar{\Omega} = G$, *with* $H = \mathrm{N}_G(H)$ *and* $K = \mathrm{N}_G(K)$. *Assume that* $H^\infty = \mathrm{U}_3(3)$. *There is a containment* $H < K$ *if and only if* $K^\backsim = 2\dot{}\mathrm{J}_2$ *and either* $p \equiv \pm19, \pm29 \pmod{60}$ *and* $G = \bar{\Omega}$; *or* $q = 5$.

Proof Lagrange's theorem eliminates all possibilities for K except $(2\dot{})\mathrm{J}_2$. The behaviour of H is described in Proposition 4.5.25, and of $2\dot{}\mathrm{J}_2$ in Proposition 4.5.26: note that $2\dot{}\mathrm{J}_2$ is scalar-normalising in $\mathrm{CSp}_6(p)$ except when $p = 5$.

We find in [12, 57] that in characteristic not 3 the smallest degree of a nontrivial representation of $\mathrm{U}_3(3)$ is 6, and that J_2 has a unique class of subgroups $\mathrm{U}_3(3)$ [24]. Since the Schur multiplier of $\mathrm{U}_3(3)$ is trivial, it follows that there is a containment $\mathrm{U}_3(3) < 2\dot{}\mathrm{J}_2$. When $p \equiv \pm2 \pmod 5$, the group $\mathrm{U}_3(3)$ is a subgroup of $\mathrm{S}_6(p)$ but $(2\dot{})\mathrm{J}_2 \not\leqslant \mathrm{S}_6(p)$, so this containment is only relevant when $p \equiv \pm1 \pmod 5$ or $p = 5$. Furthermore, when $p = 2$ or $p \equiv \pm1 \pmod{12}$, $\mathrm{U}_3(3).2 < \mathrm{S}_6(p)$, but $\mathrm{U}_3(3).2$ is not a subgroup of $2\dot{}\mathrm{J}_2$ ([12] or [24]).

If $p \equiv \pm19, \pm29 \pmod{60}$ or $p = 5$, then $\mathrm{U}_3(3)$ is not $\mathscr{S}_1$-maximal in $\mathrm{S}_6(p)$. However, $\mathrm{U}_3(3).2$ is $\mathscr{S}_1$-maximal in $\mathrm{PCSp}_6(p)$ when $p \equiv \pm19, \pm29 \pmod{60}$. There is a containment $\mathrm{U}_3(3).2 < (2\dot{})\mathrm{J}_2.2 < \mathrm{PCSp}_6(5)$. $\square$

Proposition 4.8.15 *Let* $\bar{\Omega} \leqslant G \leqslant \mathrm{Aut}\,\bar{\Omega}$, *with* $\bar{\Omega} = \mathrm{S}_6(q)$. *Let H and K be distinct* $\mathscr{S}_1$*-subgroups of G such that* $H\bar{\Omega} = K\bar{\Omega} = G$, *with* $H = \mathrm{N}_G(H)$ *and* $K = \mathrm{N}_G(K)$. *If* $H < K$ *then H and K are described in Proposition 4.8.13 or 4.8.14.*

Proof By Theorem 4.3.3 the remaining possibilities for the group H are $(2\dot{})\mathrm{A}_5$, $(2\dot{})\mathrm{L}_2(13)$, $(2\dot{})\mathrm{A}_7$ and $(2\dot{})\mathrm{J}_2$.

If H is $(2\dot{})\mathrm{A}_5$ then by Lagrange's theorem K is $(2\dot{})\mathrm{A}_7$ (characteristic 3 only) or $(2^{\cdot})J_2$ (all characteristics). By Propositions 4.5.21 and 4.5.24 the group $2^{\cdot}A_5$ is a subgroup of $Sp_6(3)$, but $2^{\cdot}A_7$ is only a subgroup of $Sp_6(9)$. By considering the restriction of the 6-dimensional representation of $2^{\cdot}J_2$ to the 5-elements (see [12]), we see that $2^{\cdot}A_5$ does not occur as an irreducible subgroup of $2^{\cdot}J_2$.

By Lagrange's theorem, H is not $(2^{\cdot})L_2(13)$ or $(2^{\cdot})J_2$, and if H is $(2^{\cdot})A_7$ then K could only be $(2^{\cdot})J_2$. However, consulting [12] (or [24]), we see that $(2^{\cdot})J_2$ does not have $(2^{\cdot})A_7$ as a subgroup. □

4.9 Dimensions greater than 6

Now we determine all $\mathscr{S}_1$-maximal subgroups of the classical groups in dimensions up to 12 and greater than 6. In Section 4.4 we described how to calculate the normaliser of the quasisimple $\mathscr{S}_1$-subgroup in both the Ω-group and the conformal group, and the number of Ω-classes: details of these calculations in dimension up to six were given in Section 4.5. Then in Section 4.6 we described how to calculate the action on the Ω-classes of field and graph automorphisms, and in Section 4.7 we carried out these calculations in detail in dimension up to 6. Finally, in Section 4.8 we analysed containments between the $\mathscr{S}_1$-subgroups, ultimately determining all $\mathscr{S}_1$-maximal subgroups in dimensions up to 6. Now that we have acquainted ourselves with the techniques necessary to solve the various aspects of this analysis, it becomes more efficient to carry them out together for the candidates in each case and each dimension.

Information on candidates is from Theorem 4.3.3, [12] and [57]. Information on containments can usually be determined relatively easily from the lists of maximal subgroups and the character tables in [12, 57].

We saw in the discussion immediately before Lemma 4.6.1 that we are only interested in describing the class stabilisers modulo conjugacy in Out Ω. From now on, we shall simply describe the class stabilisers of the groups G as specific subgroups of Out Ω. This really means that the stabiliser of the class of G is a conjugate of the specified subgroup in Out Ω or, equivalently, that the Ω-class of some Out Ω-conjugate of G is stabilised by the specified subgroup.

4.9.1 Cases L and U

In this subsection we determine the $\mathscr{S}_1$-maximal subgroups of the linear and unitary groups in dimensions 7 to 12: recall that justifications of our method, and detailed versions of similar calculations, can be found in Sections 4.4 to 4.8. The reader may wish to recall our notation developed in Section 1.7 for the outer automorphism groups of the linear and unitary groups. Recall that representations with Frobenius-Schur indicator ∘ always occur in dual pairs.

Dimension 7. We now determine the $\mathscr{S}_1$-maximal subgroups of $\mathrm{SL}_7^{\pm}(q)$. By Theorem 4.3.3, the only quasisimple candidate is $\mathrm{U}_3(3)$ with $p \neq 2, 3$.

Proposition 4.9.1 *Let $\Omega = \mathrm{SL}_7^{\pm}(q)$, and let $G = \mathrm{U}_3(3)$ be an $\mathscr{S}_1$-subgroup of $\bar{\Omega}$. Then $\mathrm{N}_{\bar{\Omega}}(G) = G$, with class stabiliser $\langle \gamma \rangle$. If $p \equiv 1 \pmod 4$ then $\Omega = \mathrm{SL}_7(p)$. If $p \equiv 3 \pmod 4$ then $p > 3$ and $\Omega = \mathrm{SU}_7(p)$. The group G is $\mathscr{S}_1$-maximal, there is a single $\mathrm{Aut}\,\bar{\Omega}$-class of such groups G, and for no other q are there $\mathscr{S}_1$-subgroups of $\mathrm{L}_7^{\pm}(q)$ isomorphic to G.*

Proof The congruences on q follow from the character ring given in Theorem 4.3.3. Also by Theorem 4.3.3, there are two dual representations, interchanged by the unique outer automorphism of G, so the other claims follow. □

Dimension 8. We now determine the $\mathscr{S}_1$-maximal subgroups of $\mathrm{SL}_8^{\pm}(q)$. By Theorem 4.3.3, the only quasisimple candidate is $G = 4_1 \dot{} \mathrm{L}_3(4)$ with $p \neq 2$, but it is a complicated one!

Proposition 4.9.2 *Let $\Omega = \mathrm{SL}_8^{\pm}(q)$, let $G = (4_1\dot{})\mathrm{L}_3(4)$ be an $\mathscr{S}_1$-subgroup of $\bar{\Omega}$, and let $S = \mathrm{N}_{\bar{\Omega}}(G)$. If $p \equiv 1, 5, 9 \pmod{20}$ then the group $\Omega = \mathrm{SL}_8(p)$, if $p \equiv 11, 19 \pmod{20}$ then $\Omega = \mathrm{SU}_8(p)$, and if $p = +2 \pmod 5$ then $p \neq 2$ and $\Omega = \mathrm{SL}_8(p^2)$.*

If $\Omega = \mathrm{SL}_8(q)$ and $q \equiv 1 \pmod{16}$ then $S = (4_1\dot{})\mathrm{L}_3(4).2_3$. If $\Omega = \mathrm{SU}_8(q)$ and $q \equiv 15 \pmod{16}$ then $S = (4_1\dot{})\mathrm{L}_3(4).2_3$. Otherwise $S = G$.

If $q = 5$ then the class stabiliser is $\langle \delta^2, \gamma \rangle$, with δ^2 inducing the 2_3 automorphism of G and γ inducing the 2_1 automorphism on one of the two $\bar{\Omega}$-classes and the 2_2 automorphism on the other.

If $S = (4_1\dot{})\mathrm{L}_3(4).2_3$ and $q = p$ then the class stabiliser is trivial. Otherwise, if $q = p \neq 5$ then the class stabiliser is $\langle \delta^{d/2} \rangle$, inducing 2_3 on G.

If $q = p^2$ and $p \equiv 1 \pmod 8$ then the class stabiliser is $\langle \phi \gamma \rangle$. If $q = p^2$ and $p \equiv 5 \pmod 8$ then the class stabiliser is $\langle \phi \gamma, \delta^4 \rangle$, with δ^4 inducing the 2_3 automorphism and $\phi \gamma$ the 2_1 or the 2_2. If $q = p^2$ and $p \equiv 7 \pmod 8$ then the class stabiliser is $\langle \phi \rangle$. If $q = p^2$ and $p \equiv 3 \pmod 8$, then the class stabiliser is $\langle \phi, \delta^4 \rangle$, with δ^4 inducing the 2_3 automorphism and ϕ the 2_1 or the 2_2.

The group S is $\mathscr{S}_1$-maximal, there is a single $\mathrm{Aut}\,\bar{\Omega}$-class of such groups G, and for no other q are there $\mathscr{S}_1$-subgroups $\mathrm{L}_3(4)$ of $\mathrm{SL}_8^{\pm}(q)$.

Proof By Theorem 4.3.3, there are four representations except in characteristic 5, when there are just two, so let us first consider characteristic 5. Then, by Theorem 4.3.3, the group $G < \mathrm{L}_8(5)$, and there are two dual representations. The 2_3 automorphism stabilises the representations, and they are interchanged by the 2_1 and 2_2 automorphisms, so there is a single $\mathrm{Aut}\,\bar{\Omega}$-class of groups G. From [57] we find that the character values on elements of $4_1\dot{}\mathrm{L}_3(4).2_3 \setminus 4_1\dot{}\mathrm{L}_3(4)$ all lie in the same field as those of $4_1\dot{}\mathrm{L}_3(4)$, so $G.2_3 < \mathrm{PGL}_8(5)$, but the character value on g in Class 2D is 2, so its eigenvalues are $1^5, -1^3$ and $\det g = -1$. Since there is no primitive 16th root of unity in $\mathbb{F}_5$, there is no isoclinic extension of G contained in $\mathrm{SL}_8(5)$, so there are two $\bar{\Omega}$-classes with $\mathrm{N}_{\bar{\Omega}}(G) = G$, both stabilised by δ^2. A MAGMA calculation (file `4134d8calc`) shows that $\gamma\delta^2$ induces the 2_1 (or the 2_2) automorphism of G, so the class stabiliser is $\langle \delta^2, \gamma \rangle$.

For all other p, the congruences on q for Ω follow from the character ring given in Theorem 4.3.3. There are two dual pairs of representations, all of which are stabilised by the 2_3 automorphism of G, whereas the 2_1 and 2_2 automorphisms interchange the representations in pairs. As in the case $p = 5$, we find from [12, 57] that $4_1{\cdot}\mathrm{L}_3(4).2_3 < \mathrm{GL}_8^{\pm}(q)$ (with $q = p$ or p^2), and that elements outside of $4_1{\cdot}\mathrm{L}_3(4)$ have determinant -1. If $q \equiv 1 \pmod{16}$ or $15 \pmod{16}$ respectively in the linear and unitary cases, then we can use scalar elements of order 16 to construct an isoclinic subgroup $4_1{\cdot}\mathrm{L}_3(4).2_3^-$ of $\mathrm{SL}_8^{\pm}(q)$. Otherwise $\mathrm{SL}_8^{\pm}(q)$ contains no subgroup of quasishape $[\![4_1{\cdot}\mathrm{L}_3(4).2_3]\!]$ and the class of subgroups is stabilised by $\delta^{d/2}$.

The pairs interchanged by the 2_1 and 2_2 automorphisms are not dual pairs, so if $q = p$ then the class stabiliser contains no other elements.

It remains to analyse the effect of ϕ when $G < \mathrm{L}_8(p^2)$. First suppose that $p \equiv 1 \pmod 4$. Then $\delta^{\phi\gamma} = \delta^{-1}$ when $p \equiv 1 \pmod 8$, whilst $\delta^{\phi\gamma} = \delta^3$ when $p \equiv 5 \pmod 8$, and in either case we can apply Lemma 4.6.2 with $\beta = \phi\gamma$. Computer calculations (file `4134d8calc`) using a representation over the field $\mathbb{Q}(\mathrm{i}, b_5)$ show that $\phi\gamma$ composed with a diagonal automorphism with square determinant normalises G and induces the 2_1 automorphism. So, by Lemma 4.6.2, we may assume that the class stabiliser contains $\phi\gamma$. As we saw earlier, when $p \equiv 5 \pmod 8$ it also contains δ^4.

If $p \equiv 3 \pmod 4$, then $\delta^{\phi} = \delta^{-1}$ when $p \equiv 7 \pmod 8$, whilst $\delta^{\phi} = \delta^3$ when $p \equiv 3 \pmod 8$, and we can apply Lemma 4.6.2 with $\beta = \phi$. The computer calculations show that ϕ composed with a diagonal automorphism with square determinant normalises G and induces the 2_1 automorphism. So, by Lemma 4.6.2, we may assume that the class stabiliser contains ϕ and, when $p \equiv 3 \pmod 8$, it also contains δ^4. $\qquad\square$

Dimension 9. We now determine the $\mathscr{S}_1$-maximal subgroups of $\mathrm{SL}_9^{\pm}(q)$. By Theorem 4.3.3, the quasisimple candidates are $3{\cdot}\mathrm{A}_6$ ($p \neq 3, 5$); $3{\cdot}\mathrm{A}_7$ ($p = 7$); $\mathrm{L}_2(19)$ ($p \neq 19$); and $3{\cdot}\mathrm{J}_3$ ($p = 2$). We consider them in reverse order.

Proposition 4.9.3 *Let $\Omega = \mathrm{SL}_9^{\pm}(q)$, and let $G = (3{\cdot})\mathrm{J}_3$ be an $\mathscr{S}_1$-subgroup of $\bar{\Omega}$. Then $\Omega = \mathrm{SU}_9(2)$ and $\mathrm{N}_{\bar{\Omega}}(G) = G$, with class stabiliser $\langle \gamma \rangle$. The group G is $\mathscr{S}_1$-maximal and there is a single $\mathrm{Aut}\,\bar{\Omega}$-class of such groups G.*

Proof Theorem 4.3.3 shows that $G < \mathrm{U}_9(2)$, and there are two (dual) representations, interchanged by the unique outer automorphism of G. $\qquad\square$

Proposition 4.9.4 *Let* $\Omega = \mathrm{SL}_9^{\pm}(q)$, *and let* $G = \mathrm{L}_2(19)$ *be an* $\mathscr{S}_1$*-subgroup of* $\overline{\Omega}$. *Then* $\mathrm{N}_{\overline{\Omega}}(G) = G$, *with class stabiliser* $\langle \gamma \rangle$. *If* $p \equiv 1, 4, 5, 6, 7, 9, 11, 16, 17$ (mod 19) *then* $\Omega = \mathrm{SL}_9(p)$. *Otherwise, if* $p \not\equiv 0$ (mod 19) *then* $\Omega = \mathrm{SU}_9(p)$. *If* $p \neq 2$ *then* G *is* $\mathscr{S}_1$*-maximal, but if* $p = 2$ *then* G *is not* $\mathscr{S}_1$*-maximal but* $\mathrm{N}_{\overline{\Omega}\langle\gamma\rangle}(G)$ *is* $\mathscr{S}_1$*-maximal. There is a single* $\mathrm{Aut}\,\overline{\Omega}$*-class of such groups* G, *and for no other* q *are there* $\mathscr{S}_1$*-subgroups* $\mathrm{L}_2(19)$ *of* $\mathrm{L}_9^{\pm}(q)$.

Proof The congruences on q follow from Theorem 4.3.3. There are two dual representations, interchanged by the outer automorphism of G, so the stabiliser and conjugacy claims follow. We note that by [12] or [25], as abstract groups $\mathrm{L}_2(19) < 3^{\cdot}\mathrm{J}_3$, but $\mathrm{L}_2(19).2 \nleq 3^{\cdot}\mathrm{J}_3.2$. By [57], the group $\mathrm{L}_2(19)$ has no faithful representations of degree less than 9 in characteristic 2, so this is a containment of $\mathscr{S}_1$-subgroups. □

Proposition 4.9.5 *Let* $\Omega = \mathrm{SL}_9^{\pm}(q)$, *and let* $G = (3^{\cdot})\mathrm{A}_7$ *be an* $\mathscr{S}_1$*-subgroup of* $\overline{\Omega}$. *Then* $\Omega = \mathrm{SL}_9(7)$, *and* $\mathrm{N}_{\overline{\Omega}}(G) = G$, *with class stabiliser* $\langle \gamma \rangle$. *The group* G *is* $\mathscr{S}_1$*-maximal and there is a single* $\mathrm{Aut}\,\overline{\Omega}$*-class of such groups* G.

Proof Theorem 4.3.3 shows that $G < \mathrm{L}_9(7)$, and there are two dual representations, interchanged by the outer automorphism of G. □

Proposition 4.9.6 *Let* $\Omega = \mathrm{SL}_9^{\pm}(q)$, *and let* $G = (3^{\cdot})\mathrm{A}_6$ *be an* $\mathscr{S}_1$*-subgroup of* $\overline{\Omega}$. *Then* $S = \mathrm{N}_{\overline{\Omega}}(G) = G.2_3$, *with class stabiliser* $\langle \gamma \rangle$. *If* $p \equiv 1$ (mod 3) *then* $\Omega = \mathrm{SL}_9(p)$. *If* $p \equiv 2$ (mod 3) *then* $p \neq 5$ *and* $\Omega = \mathrm{SU}_9(p)$. *The group* S *is* $\mathscr{S}_1$*-maximal, there is a single* $\mathrm{Aut}\,\overline{\Omega}$*-class of such groups* G, *and for no other* q *are there* $\mathscr{S}_1$*-subgroups of* $\mathrm{L}_9^{\pm}(q)$ *isomorphic to* G.

Proof From Theorem 4.3.3 we get the congruences on q, and that there are two dual representations, stabilised by the 2_3 automorphism and interchanged by the 2_1 and 2_2 automorphisms of G. By [12, 57], the character values of the representation of $3^{\cdot}\mathrm{A}_6.2_3$ all lie in the character ring of $(3^{\cdot})\mathrm{A}_6$, so $3^{\cdot}\mathrm{A}_6.2_3$ is a subgroup of $\mathrm{GL}_9^{\pm}(p)$. Considering g in Class 4C, we see that g has trace 1, and squares to an element of trace 1. So the element g has determinant 1 and S is as given. Also, γ induces the 2_1 and 2_2 automorphisms. Since $\mathrm{A}_6.2_3 \nleq \mathrm{S}_7$ (see [12]), the containment of G in $(3^{\cdot})\mathrm{A}_7$ when $p = 7$ is not relevant. Similarly, J_3 has a subgroup $(3 \times \mathrm{A}_6){:}2_2$ (see [12, 25]), but none isomorphic to $\mathrm{A}_6.2_3$. □

Dimension 10. We now determine the $\mathscr{S}_1$-maximal subgroups of $\mathrm{SL}_{10}^{\pm}(q)$. By Theorem 4.3.3, the quasisimple candidates are A_7 ($p \neq 2, 7$); $2^{\cdot}\mathrm{L}_2(19)$ ($p \neq 2, 19$); M_{11} ($p \neq 2$); $2^{\cdot}\mathrm{L}_3(4)$ ($p \neq 2, 7$); $\mathrm{U}_4(2)$ ($p \neq 2, 3$); $2^{\cdot}\mathrm{M}_{12}$ ($p \neq 2$); M_{22} ($p = 2$); and $2^{\cdot}\mathrm{M}_{22}$ ($p \neq 2, 7$). We consider them in reverse order.

Proposition 4.9.7 *Let* $\Omega = \mathrm{SL}_{10}^{\pm}(q)$, *let* $G = ((2, q - 1)^{\cdot})\mathrm{M}_{22}$ *be an* $\mathscr{S}_1$*-subgroup of* $\overline{\Omega}$, *and let* $S = \mathrm{N}_{\overline{\Omega}}(G)$. *If* $p \equiv 1, 2, 4$ (mod 7) *then* $\Omega = \mathrm{SL}_{10}(p)$, *and if $p \equiv 3, 5, 6$ (mod 7) then $\Omega = \mathrm{SU}_{10}(p)$. If $p \equiv 3$ (mod 4) in Case* **L**, *or $p \equiv 1$ (mod 4) in Case* **U**, *then $S = G$, with class stabiliser $\langle \delta^5 \rangle$; otherwise, $S = G.2$, with trivial class stabiliser. The group S is $\mathcal{S}_1$-maximal, there is a single $\mathrm{Aut}\,\overline{\Omega}$-class of such groups G, and for no other q are there $\mathcal{S}_1$-subgroups of $\mathrm{L}_{10}^{\pm}(q)$ isomorphic to G.*

Proof From Theorem 4.3.3 we get the congruences on p for Ω, and that there are two dual representations, both stabilised by the unique outer automorphism of G. Thus, there are two C-classes of groups G and one $\mathrm{Aut}\,\overline{\Omega}$-class. If $p = 2$, then the claims follow easily from [57], so suppose that $p > 2$.

From [12, 57] we find that that the character values of these representations on the group $2\!\cdot\!\mathrm{M}_{22}.2$ lie in the same field as the character values on $2\!\cdot\!\mathrm{M}_{22}$ itself, so $2\!\cdot\!\mathrm{M}_{22}.2 < \mathrm{GL}_{10}^{\pm}(p)$. Furthermore, the character value on an element g in class 2B is 4, so its eigenvalues are $1^7, -1^3$ and hence $\det g = -1$.

When $p \equiv 1$ (mod 4) in Case **L**, or $p \equiv 3$ (mod 4) in Case **U**, there is a scalar element of $\mathrm{GL}_{10}^{\pm}(p)$ of order 4 and determinant -1, which can be used to construct an isoclinic subgroup $2\!\cdot\!\mathrm{M}_{22}.2^-$ of $\mathrm{SL}_{10}^{\pm}(p)$. Otherwise, $2\!\cdot\!\mathrm{M}_{22}$ is scalar-normalising in $\mathrm{SL}_{10}^{\pm}(p)$. $\square$

Proposition 4.9.8 *Let $\Omega = \mathrm{SL}_{10}^{\pm}(q)$, let $G = (2\!\cdot\!)\mathrm{M}_{12}$ be an $\mathcal{S}_1$-subgroup of $\overline{\Omega}$, and let $S = \mathrm{N}_{\overline{\Omega}}(G)$. If $p \equiv 1, 3$ (mod 8) then the group $\Omega = \mathrm{SL}_{10}(p)$, and if $p \equiv 5, 7$ (mod 8) then $\Omega = \mathrm{SU}_{10}(p)$. If $p \equiv \pm 1$ (mod 8) then $S = G.2$, with trivial class stabiliser. If $p \equiv \pm 3$ (mod 8) then $S = G$, with class stabiliser $\langle \delta^5 \rangle$. The group S is $\mathcal{S}_1$-maximal, there is a single $\mathrm{Aut}\,\overline{\Omega}$-class of such groups G, and for no other q are there $\mathcal{S}_1$-subgroups of $\mathrm{L}_{10}^{\pm}(q)$ isomorphic to G.*

Proof From Theorem 4.3.3 we get the congruences on p for Ω, and that there are two dual representations, both stabilised by the unique outer automorphism of G. Thus, there are two C-classes of groups G and one $\mathrm{Aut}\,\overline{\Omega}$-class. From [12, 57] we find that the character values of these representations of $2\!\cdot\!\mathrm{M}_{12}.2$ lie in the same field as the character values on $2\!\cdot\!\mathrm{M}_{12}$ itself, so $2\!\cdot\!\mathrm{M}_{12}.2 < \mathrm{GL}_{10}^{\pm}(p)$. Furthermore, the character value on $g \in 2\!\cdot\!\mathrm{M}_{12}.2 \setminus 2\!\cdot\!\mathrm{M}_{12}$ in class 2C is 0, so its eigenvalues are $1^5, -1^5$ and hence $\det g = -1$.

When $p \equiv 1$ (mod 8) in Case **L**, or $p \equiv 7$ (mod 8) in Case **U**, there is a scalar element of $\mathrm{GL}_{10}^{\pm}(p)$ of order 4 and determinant -1, which can be used to construct an isoclinic subgroup $2\!\cdot\!\mathrm{M}_{12}.2^-$ of $\mathrm{SL}_{10}^{\pm}(p)$. Otherwise, no such subgroup exists.

The group $(2\!\cdot\!)\mathrm{M}_{12}$ is not a subgroup of $(2\!\cdot\!)\mathrm{M}_{22}$, by [12]. $\square$

Proposition 4.9.9 *Let $\Omega = \mathrm{SL}_{10}^{\pm}(q)$, and let $G = \mathrm{U}_4(2)$ be an $\mathcal{S}_1$-subgroup of $\overline{\Omega}$. Then $\mathrm{N}_{\overline{\Omega}}(G) = G$, with class stabiliser $\langle \gamma \rangle$. If $p \equiv 1$ (mod 3) then the group $\Omega = \mathrm{SL}_{10}(p)$, and if $p \equiv 2$ (mod 3) then $p \neq 2$ and $\Omega = \mathrm{SU}_{10}(p)$. The* group G is $\mathscr{S}_1$-maximal, there is a single $\operatorname{Aut}\overline{\Omega}$-class of such groups G, and for no other q are there $\mathscr{S}_1$-subgroups of $\mathrm{L}_{10}^{\pm}(q)$ isomorphic to G.

Proof From Theorem 4.3.3 we get the congruences on p, and that there are two dual representations, interchanged by the unique outer automorphism of G. Computer calculations (file `u42d10calc`) over the field $\mathbb{Q}(z_3)$ show that γ induces the outer automorphism of G composed with a diagonal automorphism with determinant a square. $\qquad\square$

Proposition 4.9.10 *Let* $\Omega = \mathrm{SL}_{10}^{\pm}(q)$, *let* $G = (2^{\cdot})\mathrm{L}_3(4)$ *be an* $\mathscr{S}_1$-subgroup *of* $\overline{\Omega}$, *and let* $S = \mathrm{N}_{\overline{\Omega}}(G)$. *If* $p \equiv 1, 2, 4 \pmod 7$ *then* $p \neq 2$ *and* $\Omega = \mathrm{SL}_{10}(p)$, *and if* $p \equiv 3, 5, 6 \pmod 7$ *then* $\Omega = \mathrm{SU}_{10}(p)$.

If $p \equiv 1 \pmod 8$ *in Case* **L**, *or* $p \equiv 7 \pmod 8$ *in Case* **U**, *then* $S = G.2_2$, *with class stabiliser* $\langle\gamma\rangle$. *If* $p \equiv 5 \pmod 8$ *in Case* **L**, *or* $p \equiv 3 \pmod 8$ *in Case* **U**, *then* $S = G.2_2$, *with class stabiliser* $\langle\gamma\delta\rangle$.

Otherwise, $S = G$ *with class stabiliser* $\langle\gamma, \delta^5\rangle$. *The automorphism* δ^5 *induces the* 2_2 *automorphism of* G. *The automorphism* γ *induces the* 2_1 *automorphism of* G *if* $p \equiv \pm 3 \pmod 8$, *and the* $2_1 2_2$ *automorphism of* G *if* $p \equiv \pm 1 \pmod 8$.

If $K \leqslant \overline{\Omega}\langle\delta^5\rangle$ *then* $\mathrm{N}_{\overline{\Omega}K}(G)$ *is not* $\mathscr{S}_1$-maximal, *but otherwise* $\mathrm{N}_{\overline{\Omega}K}(G)$ *is* $\mathscr{S}_1$-maximal. *There is a single* $\operatorname{Aut}\overline{\Omega}$-class *of such groups* G, *and for no other* q *are there* $\mathscr{S}_1$-subgroups *of* $\mathrm{L}_{10}^{\pm}(q)$ *isomorphic to* G.

Proof From Theorem 4.3.3, we get the congruences on p for Ω, and that there are two dual representations, interchanged by the 2_1 and 2_3 automorphisms of G and stabilised by the 2_2 automorphism of G.

From [12, 57] we find that the character value of an element g in class 2C is 4, so its eigenvalues are $1^7, -1^3$ and hence $\det g = -1$. When $p \equiv 1 \pmod 4$ in Case **L**, or $p \equiv 3 \pmod 4$ in Case **U**, there is a scalar element of $\mathrm{GL}_{10}^{\pm}(p)$ of order 4 and determinant -1, which can be used to construct an isoclinic subgroup $2^{\cdot}\mathrm{L}_3(4).2_2^-$ of $\mathrm{SL}_{10}^{\pm}(p)$. Otherwise, $\mathrm{L}_3(4)$ is self-normalising in $\overline{\Omega}$, and the class stabiliser contains δ^5, inducing the 2_2 automorphism of G.

Computer calculations (file `2134d10calc`) over $\mathbb{Q}(b_7)$ show that γ induces the $2_1 2_2$ automorphism of G composed with a diagonal automorphism with determinant twice a square, which is a square in $\mathbb{F}_p^{\times}$ if and only if $p \equiv \pm 1 \pmod 8$.

By [12], the group $2^{\cdot}\mathrm{L}_3(4).2_2$ is a subgroup of $2^{\cdot}\mathrm{M}_{22}.2$. By character values in [12, 57], this is a containment of $\mathscr{S}_1$-subgroups. Since $\operatorname{Out}\mathrm{M}_{22} = 2$, there is no containment of other extensions of G. $\qquad\square$

Proposition 4.9.11 *No extension of* M_{11} *is* $\mathscr{S}_1$-maximal *in any extension of* $\mathrm{L}_{10}(q)$ *or* $\mathrm{U}_{10}(q)$.

Proof By [12] there is an abstract containment $\mathrm{M}_{11} < 2^{\cdot}\mathrm{M}_{12}$, and character values on elements of order 4 show that this is a containment of $\mathscr{S}_1$-subgroups in all characteristics [12, 57]. Since Out M_{11} is trivial, we are done. □

Proposition 4.9.12 *Let $\Omega = \mathrm{SL}_{10}^{\pm}(q)$, and let $G = (2\cdot)L_2(19)$ be an $\mathscr{S}_1$-subgroup of $\bar{\Omega}$. Then $N_{\bar{\Omega}}(G) = G$. If $p \equiv 1, 4, 5, 6, 7, 9, 11, 16, 17 \pmod{19}$ then $\Omega = \mathrm{SL}_{10}(p)$, and if $p \equiv 2, 3, 8, 10, 12, 13, 14, 15, 18 \pmod{19}$ then $p \neq 2$ and $\Omega = \mathrm{SU}_{10}(p)$. The class stabiliser is $\langle \gamma \rangle$ for $p \equiv \pm 1 \pmod 8$, and $\langle \gamma\delta \rangle$ for $p \equiv \pm 3 \pmod 8$. The group G is $\mathscr{S}_1$-maximal, there is a single Aut $\bar{\Omega}$-class of such groups G, and for no other q are there $\mathscr{S}_1$-subgroups $L_2(19)$ of $L_{10}^{\pm}(q)$.*

Proof From Theorem 4.3.3 we get the congruences on p for Ω, and that there are two dual representations, interchanged by the unique outer automorphism of G. Computer calculations (file sl219d10calc) over $\mathbb{Q}(b_{19})$ show that γ induces the outer automorphism of G composed with a diagonal automorphism with determinant twice a square, which is a square in $\mathbb{F}_p^{\times}$ if and only if $p \equiv 1, 7 \pmod 8$. □

Proposition 4.9.13 *Let $\Omega = \mathrm{SL}_{10}^{\pm}(q)$, and let $G = A_7$ be an $\mathscr{S}_1$-subgroup of $\bar{\Omega}$. Then $N_{\bar{\Omega}}(G) = G$, with class stabiliser $\langle \gamma \rangle$. If $p \equiv 1, 2, 4 \pmod 7$ then $p \neq 2$ and $\Omega = \mathrm{SL}_{10}(p)$, and if $p \equiv 3, 5, 6 \pmod 7$ then $\Omega = \mathrm{SU}_{10}(p)$. The group G is not $\mathscr{S}_1$-maximal, but $N_{\bar{\Omega}\langle\gamma\rangle}(G)$ is $\mathscr{S}_1$-maximal. There is a single Aut $\bar{\Omega}$-class of such groups G, and for no other q are there $\mathscr{S}_1$-subgroups A_7 of $L_{10}^{\pm}(q)$.*

Proof From Theorem 4.3.3 we get the congruences on q, and that there are two dual representations, interchanged by the outer automorphism of G. Computer calculations (file a7d10calc) over $\mathbb{Q}(b_7)$ show that γ induces the outer automorphism of G composed with a diagonal automorphism with determinant a square. Note by [12, 57], there is a containment of $\mathscr{S}_1$-subgroups $A_7 < 2\cdot M_{22}$ in all characteristics, but $A_7.2 \not\leqslant 2\cdot M_{22}.2$. Also, A_7 is not a subgroup of $L_3(4)$. □

Dimension 11. We now determine the $\mathscr{S}_1$-maximal subgroups of $\mathrm{SL}_{11}^{\pm}(q)$. By Theorem 4.3.3, the quasisimple candidates are $L_2(23)$ ($p \neq 23$); M_{23} ($p = 2$); $U_5(2)$ ($p \neq 2, 3$); and M_{24} ($p = 2$). We consider them in reverse order.

Proposition 4.9.14 *Let $\Omega = \mathrm{SL}_{11}^{\pm}(q)$, and let $G = M_{24}$ be an $\mathscr{S}_1$-subgroup of $\bar{\Omega}$. Then $\Omega = \mathrm{SL}_{11}(2)$, with $N_{\bar{\Omega}}(G) = G$ and trivial class stabiliser. The group G is $\mathscr{S}_1$-maximal and there is a single Aut $\bar{\Omega}$-class of such groups G.*

Proof We see from Theorem 4.3.3 that $\Omega = \mathrm{SL}_{11}(2)$, and there are two dual representations. Since Out G is trivial, there are two C-classes of groups G and one Aut $\bar{\Omega}$-class. □

Proposition 4.9.15 *Let $\Omega = \mathrm{SL}_{11}^{\pm}(q)$, and let $G = U_5(2)$ be an $\mathscr{S}_1$-subgroup of $\bar{\Omega}$. Then $N_{\bar{\Omega}}(G) = G$, with class stabiliser $\langle \gamma \rangle$. If $p \equiv 1 \pmod 3$ then the* group $\Omega = \mathrm{SL}_{11}(p)$, and if $p \equiv 2 \pmod 3$ then $p \neq 2$ and $\Omega = \mathrm{SU}_{11}(p)$. The group G is $\mathscr{S}_1$-maximal, there is a single $\mathrm{Aut}\,\bar{\Omega}$-class of such groups G, and for no other q are there $\mathscr{S}_1$-subgroups of $\mathrm{L}_{11}^{\pm}(q)$ isomorphic to G.

Proof From Theorem 4.3.3, we get the congruences on p, and that there are two dual representations, interchanged by the outer automorphism of G. $\quad\square$

Proposition 4.9.16 *No extension of* M_{23} *is* $\mathscr{S}_1$-maximal in any extension of $\mathrm{L}_{11}(q)$ or $\mathrm{U}_{11}(q)$.

Proof The group M_{23} only arises in characteristic 2. There is a containment $\mathrm{M}_{23} < \mathrm{M}_{24}$, which is a containment of $\mathscr{S}_1$-subgroups by [57]. Since $\mathrm{Out}\,\mathrm{M}_{23}$ is trivial, no extension is $\mathscr{S}_1$-maximal. $\quad\square$

Proposition 4.9.17 *Let* $\Omega = \mathrm{SL}_{11}^{\pm}(q)$, and let $G = \mathrm{L}_2(23)$ be an $\mathscr{S}_1$-subgroup of $\bar{\Omega}$. Then $\mathrm{N}_{\bar{\Omega}}(G) = G$, with class stabiliser $\langle\gamma\rangle$. If $p \equiv 1,2,3,4,6,8,9,12,$ $13,16,18 \pmod{23}$ then $\Omega = \mathrm{SL}_{11}(p)$, and if $p \equiv 5,7,10,11,\ 14,15,17,19,20,$ $21,22 \pmod{23}$ then $\Omega = \mathrm{SU}_{11}(p)$. If $p \neq 2$ then G is $\mathscr{S}_1$-maximal. If $p = 2$ then G is not $\mathscr{S}_1$-maximal in $\bar{\Omega}$ but $\mathrm{N}_{\bar{\Omega}\langle\gamma\rangle}(G)$ is $\mathscr{S}_1$-maximal. There is a single $\mathrm{Aut}\,\bar{\Omega}$-class of such groups G, and for no other q are there $\mathscr{S}_1$-subgroups of $\mathrm{L}_{11}^{\pm}(q)$ isomorphic to G.

Proof From Theorem 4.3.3, we get the congruences on p, and that there are two dual representations, interchanged by the unique outer automorphism of G. By [12], the group M_{24} has a subgroup $\mathrm{L}_2(23)$. This is a containment of $\mathscr{S}_1$-subgroups by [57], but does not extend to $\mathrm{L}_2(23).2$. $\quad\square$

Dimension 12. We now determine the $\mathscr{S}_1$-maximal subgroups of $\mathrm{SL}_{12}^{\pm}(q)$. By Theorem 4.3.3, the quasisimple candidates are $6^{\cdot}\mathrm{A}_6$ $(p \neq 2,3,5)$; $6^{\cdot}\mathrm{A}_7$ $(p = 5)$; $2^{\cdot}\mathrm{L}_2(23)$ $(p \neq 2,23)$; $12_2{}^{\cdot}\mathrm{L}_3(4)$ $(p = 7)$; $3^{\cdot}\mathrm{Suz}$ $(p = 2)$; and $6^{\cdot}\mathrm{Suz}$ $(p \neq 2,3)$. We consider them in reverse order.

Proposition 4.9.18 *Let* $\Omega = \mathrm{SL}_{12}^{\pm}(q)$, and let $G = (6^{\cdot})\mathrm{Suz}$ *(or* $(3^{\cdot})\mathrm{Suz}$ *when* $p = 2$*) be an* $\mathscr{S}_1$-subgroup of $\bar{\Omega}$. Then $\mathrm{N}_{\bar{\Omega}}(G) = G$, with class stabiliser $\langle\gamma\rangle$. If $p \equiv 1 \pmod 3$ then $\Omega = \mathrm{SL}_{12}(p)$, and if $p \equiv 2 \pmod 3$ then $\Omega = \mathrm{SU}_{12}(p)$. The group G is $\mathscr{S}_1$-maximal, there is a single $\mathrm{Aut}\,\bar{\Omega}$-class of such groups G, and for no other q are there $\mathscr{S}_1$-subgroups of $\mathrm{L}_{12}^{\pm}(q)$ isomorphic to G.

Proof From Theorem 4.3.3, we get the congruences on p for Ω, and that there are two dual representations interchanged by the outer automorphism of G. A computer calculation (file **6suzd12calc**) in $\mathrm{SL}_{12}(K)$ with $K := \mathbb{Q}(z_3)$ shows that γ induces the outer automorphism of $6^{\cdot}\mathrm{Suz}$ composed with a diagonal automorphism with determinant a square. $\quad\square$

Proposition 4.9.19 *Let $\Omega = \mathrm{SL}_{12}^{\pm}(q)$, and let $G = (12_2{}^{\boldsymbol{\cdot}})\mathrm{L}_3(4)$ be an $\mathscr{S}_1$-subgroup of $\bar{\Omega}$. Then $\Omega = \mathrm{SL}_{12}(49)$ and $\mathrm{N}_{\bar{\Omega}}(G) = G$. The class stabiliser is $\langle \phi, \gamma \rangle$, and ϕ induces the 2_1 automorphism of G, whilst γ induces the 2_3. The group G is $\mathscr{S}_1$-maximal, and there is a single $\mathrm{Aut}\,\bar{\Omega}$-class of such groups G.*

Proof By Theorem 4.3.3, the group $G < \mathrm{L}_{12}(49)$. By [57], there are four representations, permuted transitively by $\mathrm{Out}\,G$. A computer calculation (file 12134d12calc) in $\mathrm{SL}_{12}(49)$ establishes the remaining claims.   □

Proposition 4.9.20 *Let $\Omega = \mathrm{SL}_{12}^{\pm}(q)$, and let $G = (2{}^{\boldsymbol{\cdot}})\mathrm{L}_2(23)$ be an $\mathscr{S}_1$-subgroup of $\bar{\Omega}$. Then $\mathrm{N}_{\bar{\Omega}}(G) = G$. If p is a non-zero square modulo 23 then $p \neq 2$ and $\Omega = \mathrm{SL}_{12}(p)$, and if p is a non-square modulo 23 then $\Omega = \mathrm{SU}_{12}(p)$. The class stabiliser is $\langle \gamma \rangle$ if $p \equiv \pm 1$ (mod 8), and $\langle \gamma \delta \rangle$ if $p \equiv \pm 3$ (mod 8). The group G is $\mathscr{S}_1$-maximal, there is a single $\mathrm{Aut}\,\bar{\Omega}$-class of such groups G, and for no other q are there $\mathscr{S}_1$-subgroups $\mathrm{L}_2(23)$ of $\mathrm{L}_{12}^{\pm}(q)$.*

Proof From Theorem 4.3.3 we get the congruences on p for Ω, and that there are two dual representations, interchanged by the unique outer automorphism of G. Computer calculations (file sl223d12calc) in $\mathrm{SL}_{12}(K)$ with $K := \mathbb{Q}(\mathrm{b}_{23})$ show that γ induces the outer automorphism of G composed with a diagonal automorphism with determinant a square times 2, which is a square in $\mathbb{F}_p^{\times}$ if and only if $p \equiv \pm 1$ (mod 8).   □

Proposition 4.9.21 *Let $\Omega = \mathrm{SL}_{12}^{\pm}(q)$, and let $G = (6{}^{\boldsymbol{\cdot}})\mathrm{A}_7$ be an $\mathscr{S}_1$-subgroup of $\bar{\Omega}$. Then $\Omega = \mathrm{SU}_{12}(5)$, and $\mathrm{N}_{\bar{\Omega}}(G) = G$, with class stabiliser $\langle \gamma \rangle$. The group G is $\mathscr{S}_1$-maximal, and there is a single $\mathrm{Aut}\,\bar{\Omega}$-class of such groups G.*

Proof It follows from Theorem 4.3.3 that $\Omega = \mathrm{SU}_{12}(5)$, and that there are two dual representations interchanged by the outer automorphism of G. A computer calculation (file 6a7d12f5calc) in $\mathrm{SU}_{12}(5)$ shows that γ induces the outer automorphism. We see by considering character values on elements of order 7 [12, 57] that the restriction of the 12-dimensional representation of $6{}^{\boldsymbol{\cdot}}\mathrm{Suz}$ to $6{}^{\boldsymbol{\cdot}}\mathrm{A}_7$ in characteristic 5 is reducible, so there is no containment here.   □

We finish with $6{}^{\boldsymbol{\cdot}}\mathrm{A}_6$, which is one of the most complicated examples in the book! As we shall see in Chapter 6, it is also difficult to determine exactly which extensions of G are maximal in the corresponding extensions of $\mathrm{SL}_{12}^{\pm}(q)$.

Proposition 4.9.22 *Let $\Omega = \mathrm{SL}_{12}^{\pm}(q)$, and let $G = (6{}^{\boldsymbol{\cdot}})\mathrm{A}_6$ be an $\mathscr{S}_1$-subgroup of $\bar{\Omega}$. Then $\mathrm{N}_{\bar{\Omega}}(G) = G$. If $p \equiv 1, 4$ (mod 15) then $\Omega = \mathrm{SL}_{12}(p)$, if $p \equiv 11$ or 14 (mod 15) then $\Omega = \mathrm{SU}_{12}(p)$, and if $p \equiv \pm 2$ (mod 5) then $p > 3$ and $\Omega = \mathrm{SL}_{12}(p^2)$.*

If $p \equiv \pm 1, \pm 9$ (mod 40) then the class stabiliser is $\langle \gamma \rangle$, inducing the 2_2 automorphism of G. If $p \equiv \pm 11, \pm 19$ (mod 40) then the class stabiliser is $\langle \gamma \delta \rangle$, *inducing the 2_2 automorphism of G. Otherwise, the prime $p \equiv \pm 2$ (mod 5), and if $p \equiv \pm 5$ (mod 12) then the class stabiliser is $\langle \gamma, \phi \rangle$, whilst if $p \equiv \pm 1$ (mod 12) then the class stabiliser is $\langle \gamma, \phi\delta^6 \rangle$. In these two cases γ induces the 2_2 automorphism of G, whilst ϕ and $\phi\delta^6$ induce the 2_1 automorphism if $p \equiv 2$ (mod 3) and the 2_3 automorphism if $p \equiv 1$ (mod 3).*

If $q = 49$ and $K \leqslant \langle \phi \rangle$, then $N_{\overline{\Omega}K}(G)$ is not $\mathscr{S}_1$-maximal. Otherwise, $N_{\overline{\Omega}K}(G)$ is $\mathscr{S}_1$-maximal for all subgroups K of the class stabiliser of G. There is a single $\mathrm{Aut}\,\overline{\Omega}$-class of such groups G, and for no other q are there $\mathscr{S}_1$-subgroups of $\mathrm{L}_{12}^{\pm}(q)$ isomorphic to G.

Proof The congruences on p for Ω follow from the character ring given in Theorem 4.3.3. By [12], there are four representations permuted transitively by the outer automorphisms of G, and γ permutes them in the same way as the 2_2 automorphism of G.

Computer calculations (file `6a6d12calc`) in $\mathrm{SL}_{12}(K)$ with $K := \mathbb{Q}(z_3, b_5)$ show that the 2_2 automorphism of $6{\cdot}A_6$ is induced by γ composed with a diagonal automorphism of $\mathrm{SL}_{12}(K)$ with determinant a square times 2. This determinant maps onto a square in $\mathbb{F}_p^{\times}$ if and only if $p \equiv \pm 1$ (mod 8) and so, if $p \equiv \pm 1$ (mod 5), then the class stabiliser is as given.

Suppose then that $p \equiv \pm 2$ (mod 5). So $6{\cdot}A_6 < \mathrm{SL}_{12}(p^2)$ and, since 2 is a square in $\mathbb{F}_{p^2}^{\times}$ for all p, we may assume that the class stabiliser contains γ. We find from [12] that ϕ induces the 2_1 automorphism of G when $p \equiv 2$ (mod 3) and the 2_3 automorphism when $p \equiv 1$ (mod 3). In either case the class stabiliser must be conjugate to either $\langle \phi, \gamma \rangle$ or $\langle \phi\delta^6, \gamma \rangle$. We observe also that ϕ and $\phi\gamma$ respectively centralise δ^3 when $p \equiv 1$ (mod 4) and $p \equiv 3$ (mod 4), and to decide which is the correct class stabiliser it suffices to decide whether the class stabiliser contains ϕ or $\phi\delta^6$ when $p \equiv 1$ (mod 4), and whether the class stabiliser contains $\phi\gamma$ or $\phi\gamma\delta^6$ when $p \equiv 3$ (mod 4). This involves determining whether various elements in K map onto a fourth power in $\mathbb{F}_{p^2}^{\times}$.

Let α_1 and α_2 be the automorphisms of K mapping $z_3 \mapsto z_3$, $r_5 \mapsto -r_5$, and $z_3 \mapsto z_3^{-1}$, $r_5 \mapsto -r_5$, respectively. We use α_1 and α_2 also to denote the induced automorphisms of $\mathrm{SL}_{12}(K)$, and we use γ to denote the duality automorphism of $\mathrm{SL}_{12}(K)$. So γ on $\mathrm{SL}_{12}(K)$ induces γ on $\mathrm{SL}_{12}(p^2)$, whereas α_1 and α_2 on $\mathrm{SL}_{12}(K)$ induce ϕ on $\mathrm{SL}_{12}(p^2)$ when $p \equiv 1, 2$ (mod 3), respectively. We shall need to consider the four possible values of p (mod 12) separately.

We find by computer calculation (file `6a6d12calc`) that:

(i) 2_1 is induced by α_2 composed with a diagonal automorphism of $\mathrm{SL}_{12}(K)$ of determinant v^2, where $-v$ is a square in K.

(ii) 2_1 is induced by $\alpha_1\gamma$ composed with a diagonal automorphism of determinant a fourth power in K.

It follows from (i) that ϕ is in the class stabiliser when $p \equiv 5 \pmod{12}$, and hence the full class stabiliser is $\langle \gamma, \phi \rangle$ in this case. Similarly, it follows from (ii) that the class stabiliser is $\langle \gamma, \phi \rangle$ when $p \equiv 7 \pmod{12}$.

So it remains to consider the cases $p \equiv \pm 1 \pmod{12}$. When $p \equiv 1 \pmod{12}$, ϕ is induced by $(\alpha_1 \gamma)\gamma$ and when $p \equiv -1 \pmod{12}$ $\phi\gamma$ is induced by $\alpha_2\gamma$, so in both cases we need to re-examine the automorphism of $6{\cdot}A_6 < \mathrm{SL}_{12}(K)$ induced by γ. We observed in the second paragraph of this proof that 2_2 on $6{\cdot}A_6$ is induced by γ composed with a diagonal automorphism of which the determinant has the form $2w^2$ for a certain $w \in K$. So it follows from (ii) (when $p \equiv 1 \pmod{12}$) and from (i) (when $p \equiv -1 \pmod{12}$) that the class stabiliser is $\langle \gamma, \phi \rangle$ if $\sqrt{2}w$ is a square in $\mathbb{F}_{p^2}^\times$, and $\langle \gamma, \phi\delta^6 \rangle$ otherwise. Hence we need to decide for which primes p the algebraic number $\sqrt{2}w$ reduces to a square in $\mathbb{F}_{p^2}^\times$. (This does not depend on our choice of r_2, z_3 and r_5 in $\mathbb{F}_{p^2}^\times$.)

The element w is messy, but with a little experimenting, we found an element $z \in K$ with minimal polynomial $x^4 + 10x^3 + 35x^2 + 50x + 100$ such that $2zw$ is a square in K. Since 2 is a square in $\mathbb{F}_{p^2}^\times$, it is sufficient to decide when $\sqrt{2}z$ is a square in $\mathbb{F}_{p^2}^\times$. In fact $z = (-\sqrt{-15} + 2\sqrt{5} - 5)/2$.

In general, $y \in \mathbb{F}_{p^2}^\times$ is a square in $\mathbb{F}_{p^2}^\times$ if and only if $y\bar{y}$ is a square in $\mathbb{F}_p^\times$, where $\bar{y} = y^p$. Thus if $p \equiv \pm 1 \pmod{8}$ then $\sqrt{2} \in \mathbb{F}_p^\times$, and so $\sqrt{2}$ is a square in $\mathbb{F}_{p^2}^\times$. If $p \equiv \pm 3 \pmod{8}$ then $t := \sqrt{2} \in \mathbb{F}_{p^2}^\times \setminus \mathbb{F}_p^\times$, so that $\bar{t} = -\sqrt{2}$, and $t\bar{t} = -2$, which is a square when $p \equiv 3 \pmod{8}$, but not when $p \equiv 5 \pmod{8}$. So $\sqrt{2}$ is a square in $\mathbb{F}_{p^2}^\times$ except when $p \equiv 5 \pmod{8}$. The images of z under $\langle \alpha_1, \alpha_2 \rangle$ are the p-modular reductions of the four numbers:

$$\tfrac{1}{2}(\pm\sqrt{5}\sqrt{-3} \pm 2\sqrt{5} - 5).$$

If $\sqrt{5}, \sqrt{-3} \in \mathbb{F}_p^\times$, then all the values of z are in $\mathbb{F}_p^\times$, and so z is a square in $\mathbb{F}_{p^2}^\times$. If $\sqrt{5} \in \mathbb{F}_p^\times$ but $\sqrt{-3} \notin \mathbb{F}_p^\times$, then $z\bar{z} = 10(\frac{1\pm\sqrt{5}}{2})^2$, which is a square in $\mathbb{F}_p^\times$ if and only if 2 is a square in $\mathbb{F}_p^\times$. If $\sqrt{5} \notin \mathbb{F}_p$ but $\sqrt{-3} \in \mathbb{F}_p^\times$, so that $p \equiv 1 \pmod{6}$, then $z\bar{z} = 10\frac{1\pm\sqrt{-3}}{2}$ is a square in $\mathbb{F}_p^\times$ if precisely one of 2 and $\frac{1\pm\sqrt{-3}}{2}$ is a square in $\mathbb{F}_p^\times$. Since $(\frac{1\pm\sqrt{-3}}{2})^3 = -1$, the latter is a square if and only if $p \equiv 1 \pmod{4}$, so that $z\bar{z}$ is a square in $\mathbb{F}_p^\times$ if and only if $p \equiv 5, 7 \pmod{8}$. The final case is when $\sqrt{5}, \sqrt{-3} \notin \mathbb{F}_p$, from which it follows that $\sqrt{5}\sqrt{-3} \in \mathbb{F}_p^\times$. We must determine when $z\bar{z} = \frac{5}{2}(-1 \pm \sqrt{5}\sqrt{-3}) = -(\frac{-5\pm\sqrt{5}\sqrt{-3}}{2})^2$ is a square in $\mathbb{F}_p^\times$, which is the case if and only if -1 is a square in $\mathbb{F}_p^\times$, that is when $p \equiv 1 \pmod{4}$.

Putting all this information together, we find that w is a square in $\mathbb{F}_{p^2}^\times$ if and only if $\sqrt{2}z$ is a square in $\mathbb{F}_{p^2}^\times$, which is the case if and only if

$$p \equiv 1, 7, 17, 19, 29, 31, 41, 49, 71, 79, 89, 91, 101, 103, 113, 119 \pmod{120}.$$

Since none of these values satisfy $p \equiv 2, 3 \pmod 5$ and $p \equiv 1, 11 \pmod{12}$, we conclude that the class stabiliser is $\langle \gamma, \phi\delta^6 \rangle$ in these cases.

From the character tables in [12] we find that the character values of elements of order 5 in the 12-dimensional representations of $6\!\cdot\!\mathrm{Suz}$ are 2 and -3, whereas those on the 12-dimensional representation of $6\!\cdot\!\mathrm{A}_6$ involve the irrationality b_5, so there is no containment of $\mathscr{S}_1$-subgroups $6\!\cdot\!\mathrm{A}_6 < 6\!\cdot\!\mathrm{Suz}$.

By Proposition 4.5.3 (iii), there is an abstract containment of A_6 in $\mathrm{L}_3(4)$, which by Theorem 4.3.3 is only relevant when $q = 49$. Using computer calculations (file `12134d12calc`), we find that the inverse images in $12_2\!\cdot\!\mathrm{L}_3(4)$ of the three classes of subgroups of $\mathrm{L}_3(4)$ isomorphic to A_6 are $2 \times 6\!\cdot\!\mathrm{A}_6$ twice and $4 \times 3\!\cdot\!\mathrm{A}_6$ once. From Theorem 4.7.1, all three of these classes are normalised by the 2_1 automorphism of $\mathrm{L}_3(4)$ (that is, the $\gamma\phi$ automorphism), which induces 2_3 on A_6, and one of them is normalised and the other two interchanged by the 2_2 ($= \phi$) and 2_3 ($= \gamma$) automorphisms of $\mathrm{L}_3(4)$, which induce the 2_1 and 2_3 automorphisms of A_6, respectively. So clearly it is the two classes of $2 \times 6\!\cdot\!\mathrm{A}_6$ that are interchanged by the 2_2 automorphism. Hence $6\!\cdot\!\mathrm{A}_6.2_3 < 12_2\!\cdot\!\mathrm{L}_3(4).2_1$, but $6\!\cdot\!\mathrm{A}_6.2_2 \not< 12_2\!\cdot\!\mathrm{L}_3(4).2_3$ and $6\!\cdot\!\mathrm{A}_6.2_1 \not< 12_2\!\cdot\!\mathrm{L}_3(4).2_2$. $\qquad\square$

4.9.2 Case S

In this subsection we calculate the $\mathscr{S}_1$-maximal subgroups of $\mathrm{Sp}_n(q)$ for $n = 8, 10, 12$: recall that justifications of our method, and detailed versions of similar calculations, can be found in Sections 4.4 to 4.8.

In Case **S** in dimension not 4, the only outer automorphisms are the diagonal automorphism δ, of order $(q - 1, 2)$, and the field automorphism ϕ, of order e, where $q = p^e$.

Dimension 8. We now determine the $\mathscr{S}_1$-maximal subgroups of $\mathrm{Sp}_8(q)$. By Theorem 4.3.3, the quasisimple candidates are $2\!\cdot\!\mathrm{L}_3(2)$ ($p \neq 2, 3, 7$); $2\!\cdot\!\mathrm{A}_6$ ($p \neq 2, 3, 5$); $\mathrm{L}_2(17)$ ($p = 2$); $2\!\cdot\!\mathrm{L}_2(17)$ ($p \neq 2, 17$); and A_{10} ($p = 2$). We consider them in reverse order.

Proposition 4.9.23 *Let* $\Omega = \mathrm{Sp}_8(q)$, *and let* $G = \mathrm{A}_{10}$ *be an* $\mathscr{S}_1$-subgroup *of* $\overline{\Omega}$. *Then* $q = 2$ *and* $\mathrm{N}_{\overline{\Omega}}(G) = G.2$, *with trivial class stabiliser. The group* $\mathrm{N}_{\overline{\Omega}}(G)$ *is* $\mathscr{S}_1$-maximal, *and there is a single* $\mathrm{Aut}\,\overline{\Omega}$-*class of such groups* G.

Proof The claims are immediate from Theorem 4.3.3 and [57]. $\qquad\square$

Proposition 4.9.24 *Let* $\Omega = \mathrm{Sp}_8(q)$, *and let* $G = ((2, q - 1)\!\cdot\!)\mathrm{L}_2(17)$ *be an* $\mathscr{S}_1$-subgroup *of* $\overline{\Omega}$. *Then* $\mathrm{N}_{\overline{\Omega}}(G) = G$. *If* $p \equiv \pm 1, \pm 2, \pm 4, \pm 8 \pmod{17}$ *then* $q = p$, *with trivial class stabiliser. If* $p \equiv \pm 3, \pm 5, \pm 6, \pm 7 \pmod{17}$ *then* $q = p^2$, *with class stabiliser* $\langle \phi \rangle$. *The group* G *is* $\mathscr{S}_1$-maximal, *there is a single* $\mathrm{Aut}\,\overline{\Omega}$-*class of such groups* G, *and for no other* q *are there* $\mathscr{S}_1$-subgroups G *of* $\mathrm{S}_8(q)$.

Proof The congruences on p for q follow from Theorem 4.3.3, which also states that there are two representations, interchanged by the outer automorphism of G. Hence, the class stabiliser is trivial when $q = p$. If $q = p^2$ then by [12] the two representations are also interchanged by ϕ. Since $L_2(17).2 \backslash L_2(17)$ contains involutions, Lemma 4.6.7 implies that the class stabiliser is $\langle \phi \rangle$. □

Proposition 4.9.25 *Let $\Omega = Sp_8(q)$, let $G = (2\dot{\,})A_6$ be an $\mathscr{S}_1$-subgroup of $\bar{\Omega}$, and let $S = N_{\bar{\Omega}}(G)$. If $p \equiv \pm 1 \pmod{20}$ then $q = p$, with $S = G.2_2$ and trivial class stabiliser. If $p \equiv \pm 9 \pmod{20}$ then $q = p$, with $S = G$ and class stabiliser $\langle \delta \rangle$. If $p \equiv \pm 2 \pmod 5$ then $q = p^2 \neq 4, 9$, with $S = G$ and class stabiliser $\langle \delta, \phi \rangle$. If δ stabilises the class of G then δ induces the 2_2 automorphism of G, whilst if $q = p^2$ then ϕ induces the 2_1 automorphism of G. The group S is $\mathscr{S}_1$-maximal, there is a single $Aut\,\bar{\Omega}$-class of such groups G, and for no other q are there $\mathscr{S}_1$-subgroups of $S_8(q)$ isomorphic to G.*

Proof The congruences on p for q follow from Theorem 4.3.3, which also states that there are two representations, stabilised by the 2_2 automorphism and interchanged by the 2_1 and 2_3 automorphisms of G. Thus there is a unique $Aut\,\bar{\Omega}$-class, and if δ stabilises the class then δ induces the 2_2 automorphism.

From [12] we find that elements of $2\dot{\,}A_6.2_2 \backslash 2\dot{\,}A_6$ are isometries, but their character values involve y_{20} which lies in $\mathbb{F}_q^\times$ if and only if $q \equiv \pm 1 \pmod{20}$, by Lemma 4.2.1. If $q = p^2$ then by [12] the two representations are interchanged by ϕ, and the class stabiliser is $\langle \delta, \phi \rangle$. Since $A_6.2_1 \backslash A_6$ contains involutions, ϕ induces the 2_1 automorphism of G. By Theorem 4.3.3, the group A_{10} only arises in characteristic 2, so S is $\mathscr{S}_1$-maximal. □

Proposition 4.9.26 *Let $\Omega = Sp_8(q)$, let $G = (2\dot{\,})L_3(2)$ be an $\mathscr{S}_1$-subgroup of $\bar{\Omega}$, and let $S = N_{\bar{\Omega}}(G)$. Then $q = p$ and $p \neq 2, 3, 7$. If $p \equiv \pm 1 \pmod{12}$ then $S = G.2$, with trivial class stabiliser. If $p \equiv \pm 5 \pmod{12}$ then $S = G$, with class stabiliser $\langle \delta \rangle$. The group S is $\mathscr{S}_1$-maximal, there is a single $Aut\,\bar{\Omega}$-class of such groups G, and for no other q are there $\mathscr{S}_1$-subgroups G of $S_8(q)$.*

Proof By Theorem 4.3.3, $q = p \neq 2, 3, 7$, and there is a unique representation, so a single $Aut\,\bar{\Omega}$-class. From [12] we find that elements of $2\dot{\,}L_3(2).2 \backslash 2\dot{\,}L_3(2)$ are isometries of the form, but their character values involve r_3, which lies in $\mathbb{F}_p^\times$ if and only if $p \equiv \pm 1 \pmod{12}$. By Theorem 4.3.3, the group A_{10} only arises in characteristic 2, so S is $\mathscr{S}_1$-maximal. □

Dimension 10. We now determine the $\mathscr{S}_1$-maximal subgroups of $Sp_{10}(q)$. By Theorem 4.3.3, the quasisimple candidates are $2\dot{\,}A_6$ ($p \neq 2, 3$); two representations of $2\dot{\,}L_2(11)$, with $p \neq 2, 11$ in the first and $p \neq 2, 3, 11$ in the second; and $U_5(2)$ ($p \neq 2$). Note that the two representations of $2\dot{\,}L_2(11)$ are distinct in the sense that neither is an algebraic conjugate or dual of the other, and neither can be obtained from the other by applying a group automorphism. To distinguish between them, we shall refer to them as $2^{\cdot}L_2(11)_1$ and $2^{\cdot}L_2(11)_2$.

Proposition 4.9.27 *Let $\Omega = \mathrm{Sp}_{10}(q)$, let $G = \mathrm{U}_5(2)$ be an $\mathscr{S}_1$-subgroup of $\bar{\Omega}$, and let $S = \mathrm{N}_{\bar{\Omega}}(G)$. Then $q = p$. If $p \equiv \pm 1 \pmod 8$ then $S = G.2$, with trivial class stabiliser. If $p \equiv \pm 3 \pmod 8$ then $S = G$, with class stabiliser $\langle \delta \rangle$. The group S is $\mathscr{S}_1$-maximal, there is a single $\mathrm{Aut}\,\bar{\Omega}$-class of such groups G, and for no other q are there $\mathscr{S}_1$-subgroups of $\mathrm{S}_{10}(q)$ isomorphic to G.*

Proof Theorem 4.3.3 implies that $q = p \neq 2$, and there is a single such representation of G. From [12, 57] we find elements of $G.2 \setminus G$ negate the fixed form. Hence, by multiplying the outer elements by a scalar element of order 4, which also negates the form, we obtain a representation of a group H with structure $(2 \times G)^{\cdot}2$ with $H/[H, H]$ cyclic of order 4, which consists of isometries. We calculate from the character tables of $G.2$ in [12, 57] that the character values of this representation on elements of $H \setminus G$ involve r_2, which lies in $\mathbb{F}_p^\times$ if and only if $p \equiv \pm 1 \pmod 8$. □

Proposition 4.9.28 *Let $\Omega = \mathrm{Sp}_{10}(q)$, let $G = (2^{\cdot})L_2(11)_1$ be an $\mathscr{S}_1$-subgroup of $\bar{\Omega}$, and let $S = \mathrm{N}_{\bar{\Omega}}(G)$. Then $q = p$. If $p \equiv \pm 1 \pmod 8$ then $S = G.2$, with trivial class stabiliser. If $p \equiv \pm 3 \pmod 8$ then $p \neq 11$, with $S = G$ and class stabiliser $\langle \delta \rangle$. The group S is $\mathscr{S}_1$-maximal, there is a single $\mathrm{Aut}\,\bar{\Omega}$-class of such groups G, and the only other $\mathscr{S}_1$-subgroups of $\mathrm{S}_{10}(q)$ isomorphic to G are as given in Proposition 4.9.29.*

Proof The congruences on p for q follow from Theorem 4.3.3, which also states that there is a unique representation of this type. From [12, 57] we find that, in this representation of $2^{\cdot}L_2(11).2$, elements of $2^{\cdot}L_2(11).2 \setminus 2^{\cdot}L_2(11)$ are isometries, but their character values involve r_2, which lies in $\mathbb{F}_p^\times$ if and only if $p \equiv \pm 1 \pmod 8$. Since the image of $2^{\cdot}L_2(11)$ in $\mathrm{Sp}_{10}(q)$ has centre of order 2, but the image of the simple group $\mathrm{U}_5(2)$ has trivial centre, there can be no containment of $2^{\cdot}L_2(11)$ in $\mathrm{U}_5(2)$. □

Proposition 4.9.29 *Let $\Omega = \mathrm{Sp}_{10}(q)$, let $G = (2^{\cdot})L_2(11)_2$ be an $\mathscr{S}_1$-subgroup of $\bar{\Omega}$, and let $S = \mathrm{N}_{\bar{\Omega}}(G)$. If $p \equiv \pm 1 \pmod{24}$ then $q = p$, with $S = G.2$, trivial class stabiliser and exactly two $\mathrm{Aut}\,\bar{\Omega}$-classes of groups G. If $p \equiv \pm 11 \pmod{24}$ then $p \neq 11$ and $q = p$, with $S = G$, class stabiliser $\langle \delta \rangle$ and exactly two $\mathrm{Aut}\,\bar{\Omega}$-classes. If $p \equiv \pm 5 \pmod{12}$ then $q = p^2$, with $S = G.2$, trivial class stabiliser and a single $\mathrm{Aut}\,\bar{\Omega}$-class of groups G. The group S is $\mathscr{S}_1$-maximal, and the only other $\mathscr{S}_1$-subgroups of $\mathrm{S}_{10}(q)$ isomorphic to G are those given in Proposition 4.9.28, above.*

Proof The congruences on p for q follow from Theorem 4.3.3, which also states that there are two representations of this type, both stabilised by the outer automorphism of G. From [12, 57] we find that, in this representation of $2\dot{}L_2(11).2$, elements of $2\dot{}L_2(11).2 \setminus 2\dot{}L_2(11)$ are all isometries, but their character values involve the irrationalities r_2 and y_{24}, which both lie in $\mathbb{F}_p^\times$ if and only if $p \equiv \pm 1 \pmod{24}$.

If $p \equiv \pm 5 \pmod{12}$, then $p^2 \equiv 1 \pmod{24}$, so $G.2 < S_{10}(p^2)$. The two representations are interchanged by ϕ, so the class stabiliser is trivial. As in the previous proposition, there can be no containment of $2\dot{}L_2(11)$ in $U_5(2)$. □

Proposition 4.9.30 *Let $\Omega = \mathrm{Sp}_{10}(q)$, let $G = (2\dot{})A_6$ be an $\mathscr{S}_1$-subgroup of $\bar\Omega$, and let $S = N_{\bar\Omega}(G)$. If $p \equiv \pm 1 \pmod{16}$ then $q = p$, with $S = G.2_2$ and trivial class stabiliser. If $p \equiv \pm 7 \pmod{16}$ then $q = p$, with $S = G$ and class stabiliser $\langle\delta\rangle$, inducing the 2_2 automorphism of G. If $p \equiv \pm 3 \pmod{8}$ then $q = p^2$, with $p \neq 3$, $S = G$ and class stabiliser $\langle\delta, \phi\rangle$; here δ induces the 2_2 automorphism of G and ϕ the 2_1. The group S is $\mathscr{S}_1$-maximal, there is a single $\mathrm{Aut}\,\bar\Omega$-class of groups G, and for no other q are there $\mathscr{S}_1$-subgroups A_6 of $S_{10}(q)$.*

Proof The congruences on p for q follow from Theorem 4.3.3, which also states that there are two representations, stabilised by the 2_2 automorphism and interchanged by the 2_1 and 2_3 automorphisms of G, so there is a unique $\mathrm{Aut}\,\bar\Omega$-class.

From [12, 57] we find that elements of $2\dot{}A_6.2_2 \setminus 2\dot{}A_6$ are isometries, but their character values involve the irrationality y_{16} which, by Lemma 4.2.1, lies in $\mathbb{F}_q^\times$ if and only if $q \equiv \pm 1 \pmod{16}$. If $q = p^2$ then the two representations are interchanged by ϕ [12, 57]. Since $A_6.2_1 \setminus A_6$ contains involutions but $A_6.2_3 \setminus A_6$ does not, ϕ induces the 2_1 automorphism of G, by Lemma 4.6.7.

Since the image of $2\dot{}A_6$ in $\mathrm{Sp}_{10}(q)$ has centre of order 2, but the image of $U_5(2)$ has trivial centre, there can be no containment of $2\dot{}A_6$ in $U_5(2)$. □

Dimension 12. We now determine the $\mathscr{S}_1$-maximal subgroups of $\mathrm{Sp}_{12}(q)$. By Theorem 4.3.3, the quasisimple candidates are $2\dot{}L_2(11)$ $(p \neq 2, 5, 11)$; $2\dot{}L_2(13)$ $(p \neq 2, 7, 13)$; $L_2(25)$ $(p = 2)$; $2\dot{}L_2(25)$ $(p \neq 2, 5)$; $U_3(4)$ $(p \neq 2)$; $S_4(5)$ $(p = 2)$; $2\dot{}S_4(5)$ $(p \neq 2, 5)$; $2\dot{}G_2(4)$ $(p \neq 2)$; A_{14} $(p = 2)$; and $2\dot{}\mathrm{Suz}$ $(p = 3)$. We consider them in reverse order.

Proposition 4.9.31 *Let $\Omega = \mathrm{Sp}_{12}(q)$ and let $G = (2\dot{})\mathrm{Suz}$ be an $\mathscr{S}_1$-subgroup of $\bar\Omega$. Then $q = 3$ and $N_{\bar\Omega}(G) = G$, with class stabiliser $\langle\delta\rangle$. The group G is $\mathscr{S}_1$-maximal, and there is a unique $\mathrm{Aut}\,\bar\Omega$-class of such groups G.*

Proof By Theorem 4.3.3, there is a single representation and $q = 3$. This group is not in [57], but the representation of $2\dot{}\mathrm{Suz}.2 < \mathrm{SL}_{12}(3)$ is available in [111], and we can verify in MAGMA, for example, that $2\dot{}\mathrm{Suz}$ consists of isometries of a symplectic form, but $2\dot{}\mathrm{Suz}.2$ does not. □

Proposition 4.9.32 *Let* $\Omega = \mathrm{Sp}_{12}(q)$, *let* $G = \mathrm{A}_{14}$ *be an* $\mathscr{S}_1$-*subgroup of* $\overline{\Omega}$, *and let* $S = \mathrm{N}_{\overline{\Omega}}(G)$. *Then* $q = 2$ *and* $S = G.2$, *with trivial class stabiliser. The group* S *is* $\mathscr{S}_1$-*maximal, and there is a unique* $\mathrm{Aut}\,\overline{\Omega}$-*class of such groups* G

Proof This follows from Theorem 4.3.3: there is a single representation, which is over $\mathbb{F}_2$, so $S = G.2 < \mathrm{Sp}_{12}(2)$ with trivial class stabiliser. $\square$

Proposition 4.9.33 *Let* $\Omega = \mathrm{Sp}_{12}(q)$, *let* $G = (2^{\cdot})\mathrm{G}_2(4)$ *be an* $\mathscr{S}_1$-*subgroup of* $\overline{\Omega}$, *and let* $S = \mathrm{N}_{\overline{\Omega}}(G)$. *Then* $q = p$. *If* $p \equiv \pm 1 \pmod 8$ *then* $S = G.2$, *with trivial class stabiliser. If* $p \equiv \pm 3 \pmod 8$ *then* $S = G$, *with class stabiliser* $\langle \delta \rangle$. *The group* S *is* $\mathscr{S}_1$-*maximal if* $p \neq 3$; *otherwise* $(p = 3)$ *no extension of* G *is* $\mathscr{S}_1$-*maximal. There is a unique* $\mathrm{Aut}\,\overline{\Omega}$-*class of such groups* G, *and for no other* q *are there* $\mathscr{S}_1$-*subgroups of* $\mathrm{S}_{12}(q)$ *isomorphic to* G.

Proof By Theorem 4.3.3, there is a single representation and $q = p \neq 2$. From [12, 57], we find that the representation of $2^{\cdot}\mathrm{G}_2(4).2$ consists of isometries, but the character values on elements of $2^{\cdot}\mathrm{G}_2(4).2 \setminus 2^{\cdot}\mathrm{G}_2(4)$ involve r_2. By [12], the group $2^{\cdot}\mathrm{Suz}.2$ has a subgroup isomorphic to $2^{\cdot}\mathrm{G}_2(4).2$, and this is a genuine containment of $\mathscr{S}_1$-subgroups, because 12 is the minimal degree of a non-trivial representation of $2^{\cdot}\mathrm{G}_2(4)$ in odd characteristic. $\square$

Proposition 4.9.34 *Let* $\Omega = \mathrm{Sp}_{12}(q)$ *and let* $G = ((q-1,2)^{\cdot})\mathrm{S}_4(5)$ *be an* $\mathscr{S}_1$-*subgroup of* $\overline{\Omega}$. *Then* $\mathrm{N}_{\overline{\Omega}}(G) = G$. *If* $p \equiv \pm 1 \pmod 5$ *then* $q = p$, *with trivial class stabiliser. If* $p \equiv \pm 2 \pmod 5$ *then* $q = p^2$, *with class stabiliser* $\langle \phi \rangle$. *The group* G *is* $\mathscr{S}_1$-*maximal, there is a unique* $\mathrm{Aut}\,\overline{\Omega}$-*class of such groups* G, *and for no other* q *are there* $\mathscr{S}_1$-*subgroups of* $\mathrm{S}_{12}(q)$ *isomorphic to* G.

Proof The congruences on p for q, and the fact that there are two representations, interchanged by the outer automorphism of $\mathrm{S}_4(5)$, follow from Theorem 4.3.3. If $q = p^2$ then by [12, 57] the representations are interchanged by ϕ. From [12], the coset $\mathrm{S}_4(5).2 \setminus \mathrm{S}_4(5)$ contains involutions, so the class stabiliser is $\langle \phi \rangle$ by Lemma 4.6.7. $\square$

Proposition 4.9.35 *No extension of* $\mathrm{U}_3(4)$ *is* $\mathscr{S}_1$-*maximal in any extension of* $\mathrm{Sp}_{12}(q)$.

Proof The group $2^{\cdot}\mathrm{G}_2(4).2$ has a subgroup isomorphic to $\mathrm{U}_3(4).4$ by [14] (see our Table 8.30). We see in [12, 57] that the minimal degree of a non-trivial representation of $\mathrm{U}_3(4)$ in odd characteristic is 12, so this is a containment of $\mathscr{S}_1$-subgroups. $\square$

Proposition 4.9.36 *Let* $\Omega = \mathrm{Sp}_{12}(q)$, *let* $G = ((2,q-1)^{\cdot})\mathrm{L}_2(25)$ *be an* $\mathscr{S}_1$-*subgroup of* $\overline{\Omega}$, *and let* $S = \mathrm{N}_{\overline{\Omega}}(G)$. *Then* $q = p$. *If* $p \equiv \pm 1 \pmod 5$ *or* $p = 2$ *then* $S = G.2_2$, *with trivial class stabiliser. If* $p \equiv \pm 2 \pmod 5$ *and* $p \neq 2$ *then* $S = G$, *with class stabiliser* $\langle \delta \rangle$, *inducing the* 2_2 *automorphism of* G. *The group*

S is $\mathscr{S}_1$-*maximal if* $p \equiv \pm 2$ (mod 5) *and* $p \neq 3$. *Otherwise, no extension of* G *is* $\mathscr{S}_1$-*maximal. There is a unique* $\mathrm{Aut}\,\bar{\Omega}$-*class of such groups* G, *and for no other* q *are there* $\mathscr{S}_1$-*subgroups of* $\mathrm{S}_{12}(q)$ *isomorphic to* G.

Proof Theorem 4.3.3 implies the congruences on p for q, and that there are two representations, both stabilised by the 2_2 outer automorphism of G, and interchanged by the 2_1 and 2_3. When $p = 2$ we find that matrices lying in $\mathrm{L}_2(25).2_2 \setminus \mathrm{L}_2(25)$ are isometries of trace zero. When p is odd, from [12, 57] we find that, in the extension of one of these two representations to $2\dot{}\mathrm{L}_2(25).2_2$, elements of $2\dot{}\mathrm{L}_2(25).2_2 \setminus 2\dot{}\mathrm{L}_2(25)$ are isometries, but the character involves r_5, which lies in $\mathbb{F}_p$ if and only if $p \equiv \pm 1$ (mod 5). (For $p \neq 2$, there is also a representation of an isoclinic group $2\dot{}\mathrm{L}_2(25).2_2^-$ with the same property, which is derived from the other representation of $2\dot{}\mathrm{L}_2(25).2_2$.)

By [12], the group $2\dot{}\mathrm{Suz}.2$ contains a subgroup isomorphic to $2\dot{}\mathrm{L}_2(25).2$. By Theorem 4.3.3, this is relevant only when $p = 3$, in which case this is a genuine containment of $\mathscr{S}_1$-subgroups, because by [12, 57] the minimal degree of a non-trivial representation of $2\dot{}\mathrm{L}_2(25)$ in characteristic not 5 is 12.

By Table 2.6, the group $2\dot{}\mathrm{Sp}_4(5)$ has a subgroup isomorphic to $2\dot{}\mathrm{L}_2(25).2$. If $p \equiv \pm 2$ (mod 5), then Theorem 4.3.3 implies that $2\dot{}\mathrm{Sp}_4(5) \not\leqslant \mathrm{Sp}_{12}(p)$, but for other p this is a containment of $\mathscr{S}_1$-subgroups, as in the previous paragraph.

Of the remaining possibilities for containments compatible with Lagrange's theorem, we note that all subgroups of A_{14} have faithful permutation actions on at most 14 points, and that $2\dot{}\mathrm{G}_2(4)$ has no elements of order 24. $\qquad\square$

Proposition 4.9.37 *Let* $\Omega = \mathrm{Sp}_{12}(q)$, *let* $G = (2\dot{})\mathrm{L}_2(13)$ *be an* $\mathscr{S}_1$-*subgroup of* $\bar{\Omega}$, *and let* $S = \mathrm{N}_{\bar{\Omega}}(G)$. *If* $p \equiv \pm 1$ (mod 28) *then* $q = p$, *with* $S = G.2$, *trivial class stabiliser and exactly three* $\mathrm{Aut}\,\bar{\Omega}$-*classes of groups* G. *If the prime* $p \equiv \pm 13$ (mod 28) *then* $q = p \neq 13$, *with* $S = G$, *class stabiliser* $\langle \delta \rangle$ *and exactly three* $\mathrm{Aut}\,\bar{\Omega}$-*classes of groups* G. *If* $p \equiv \pm 2, \pm 3$ (mod 7) *then* $q = p^3 \neq 2^3$, *with* $S = G$, *class stabiliser* $\langle \delta \rangle$ *and a single* $\mathrm{Aut}\,\bar{\Omega}$-*class of groups* G. *The group* S *is* $\mathscr{S}_1$-*maximal, and for no other* q *are there* $\mathscr{S}_1$-*subgroups* G *of* $\mathrm{S}_{12}(q)$.

Proof Theorem 4.3.3 implies the congruences on p for q, and that there are three representations, all stabilised by the unique outer automorphism of G. Thus, if $q = p$ then there are three $\mathrm{Aut}\,\bar{\Omega}$-classes. From [12, 57] we find that elements of $2\dot{}\mathrm{L}_2(13).2 \setminus 2\dot{}\mathrm{L}_2(13)$ are isometries, but the character involves y_{28} which, by Lemma 4.2.1, lies in $\mathbb{F}_q$ if and only if $q \equiv \pm 1$ (mod 28). If $q = p^3$ then the three representations are permuted by ϕ, by [12, 57].

Note that, although $2\dot{}\mathrm{G}_2(4)$ contains $\mathrm{SL}_2(13)$ as a subgroup, we find from the character values of elements of order 7 in [12] that this subgroup acts reducibly, with two components of degree 6, in the 12-dimensional irreducible representation of $2\cdot G_2(4)$. In characteristic 3, we find from [12] that the only containment of $SL_2(13)$ in $2\cdot Suz$ is via $SL_2(13) < 2\cdot G_2(4) < 2\cdot Suz$.   □

Proposition 4.9.38 *Let* $\Omega = Sp_{12}(q)$, *let* $G = (2\cdot)L_2(11)$ *be an* $\mathscr{S}_1$-*subgroup of* $\overline{\Omega}$, *and let* $S = N_{\overline{\Omega}}(G)$. *If* $p \equiv \pm 1 \pmod{20}$ *then* $q = p$, *with* $S = G.2$, *trivial class stabiliser and exactly two* $Aut\,\overline{\Omega}$-*classes. If* $p \equiv \pm 9 \pmod{20}$ *then* $q = p \neq 11$, *with* $S = G$, *class stabiliser* $\langle \delta \rangle$, *and exactly two* $Aut\,\overline{\Omega}$-*classes. If* $p \equiv \pm 2 \pmod 5$ *then* $q = p^2 \neq 4$, *with* $S = G$, *class stabiliser* $\langle \delta \rangle$, *and exactly one* $Aut\,\overline{\Omega}$-*class. The group* S *is* $\mathscr{S}_1$-*maximal and for no other* q *are there* $\mathscr{S}_1$-*subgroups of* $S_{12}(q)$ *isomorphic to* G.

Proof Theorem 4.3.3 implies the congruences on p for q, and that there are two representations, both stabilised by the outer automorphism of G. Thus, if $q = p$ then there are two $Aut\,\overline{\Omega}$-classes. From [12, 57] we find that elements of $2\cdot L_2(11).2 \setminus 2\cdot L_2(11)$ are isometries, but the character involves y_{20} which, by Lemma 4.2.1, lies in $\mathbb{F}_q$ if and only if $q \equiv \pm 1 \pmod{20}$. If $p \not\equiv \pm 1 \pmod 5$, then $p^2 \equiv \pm 1 \pmod{20}$, so the class stabiliser is again $\langle \delta \rangle$, as by [12, 57] the two representations are interchanged by ϕ. In characteristic 3, the group $2\cdot L_2(11) \leqslant Sp_{12}(9)$, whilst $2\cdot Suz \leqslant Sp_{12}(3)$, so there is no containment here.   □

4.9.3 Cases O^ε

In this section we determine the $\mathscr{S}_1$-maximal subgroups of the orthogonal groups in dimension 7 to 12: recall that justifications of our methods, and detailed versions of similar calculations, can be found in Sections 4.4 to 4.8.

Since the simple orthogonal groups in dimension less than 7 are isomorphic to other classical groups, we shall not analyse these cases. We have, however, included them in the tables in Chapter 8. Recall also that we exclude $\Omega_8^+(q)$ from our considerations, since the maximal subgroups of all almost simple extensions of $O_8^+(q)$ are determined in [62]: see Table 8.50. The reader may wish to recall our notation, developed in Section 1.7, for the outer automorphism groups of the orthogonal groups. In particular:

(i) δ is defined only for odd q. It denotes a diagonal automorphism induced by an element of $CGO_n^\pm(q) \setminus GO_n^\pm(q)$ when n is even, or an element of $SO_n(q) \setminus \Omega_n(q)$ when n is odd. As an element of $Out\,\overline{\Omega}$, it has order 4 when $n \equiv 2 \pmod 4$ and the form has square discriminant. Otherwise it has order 2.

(ii) δ' is the diagonal automorphism induced by an element of $SO_n^\pm(q) \setminus \Omega_n^\pm(q)$ when n is even, q is odd, and the form has square discriminant. Otherwise it is undefined.

(iii) γ is defined only for even n. It is the graph automorphism induced by an element of $\mathrm{GO}_n^\pm(q) \setminus \mathrm{SO}_n^\pm(q)$ when q is odd, or of $\mathrm{SO}_n^\pm(q) \setminus \Omega_n^\pm(q)$ when q is even.

(iv) The field automorphism ϕ of $\overline{\Omega}$ is induced by raising matrix entries to the p-th power for forms of $\circ$−type and of +-type, and for forms of −-type when p is odd and the discriminant of the form is a square in $\mathbb{F}_q$. For forms of −-type with p even or with p odd and non-square discriminant, this operation is not a semi-similarity, and is composed with conjugation by an element of $\mathrm{GL}_n(q)$ to give an outer automorphism φ of $\overline{\Omega}$.

When n is even, the sign of the form for which $\mathrm{Im}\, G$ is a group of isometries will generally depend on the prime p. When p is odd, we resolve this question by computing the form preserved in the image of the representation over a suitable number field, and determining for which primes p its determinant is square on reduction modulo p. When $p = 2$, a straightforward calculation over a finite field suffices, using Definition 1.5.40. For spinor norm and quasideterminant calculations, we use the method presented in Proposition 1.6.11. In some examples, we first compute the matrices A, F, B and calculate $\det(BAFB^\mathsf{T})$ over a suitable number field, and then determine whether this determinant reduces to a square in $\mathbb{F}_q^\times$.

Several examples that arise are groups for which the corresponding module is a deleted permutation module (that is, the quotient of a permutation module by the all-1 vector), so let us analyse that case.

Lemma 4.9.39 *Let $G \leqslant \mathrm{GL}_n(q)$ be the action matrices of a deleted permutation module, with q odd and $n + 1 \not\equiv 0 \pmod{p}$. Then G consists of isometries of a symmetric bilinear form. If n is even, then the discriminant of the form is square if and only if $n + 1 \pmod{p}$ is a square in $\mathbb{F}_q^\times$. All elements of G corresponding to even permutations have spinor norm 1. If n is odd and $g \in G$ corresponds to an odd permutation, then the spinor norm of $-g$ is 1 if and only if $(n + 1)/2 \pmod{p}$ is a square in $\mathbb{F}_q^\times$.*

Proof Let $(e_i \mid 1 \leqslant i \leqslant n + 1)$ be the basis of the permutation module, and without loss of generality take $(e_i - e_{n+1} \mid 1 \leqslant i \leqslant n)$ as the basis of the deleted permutation module. It can then be verified that G consists of isometries of a symmetric bilinear form with matrix $F = (f_{ij})$ where, for $1 \leqslant i, j \leqslant n$, $f_{ii} = 2$ and $f_{ij} = 1$ for $i \neq j$. It is easy to calculate that $\det F = n + 1 \pmod{p}$.

Since G is isomorphic to a subgroup of S_{n+1} within $\mathrm{GO}_n(q, F)$, of which A_{n+1} is the unique subgroup of index 2, all elements in G corresponding to even permutations must have spinor norm 1.

Matrices $g \in G$ corresponding to odd permutations have determinant -1, but if n is odd then $-g$ is also an isometry, and $-g$ has determinant 1. Since the matrices of all odd permutations have the same spinor norm, we may assume that g is the $n \times n$ permutation matrix for $(1,2)$. Then $A := I_n + g$ has rank $n-1$, and we can take the matrix B of Proposition 1.6.11 to be defined by removing the first row from the identity matrix. Let $M = (m_{ij}) = BAFB^\mathsf{T}$. Then $m_{11} = 3$, $m_{ii} = 4$ for $1 < i \leqslant n-1$, and $m_{ij} = 2$ for $i \neq j$ with $1 \leqslant i, j \leqslant n-1$. It is not hard to show that $\det M = 2^{n-2}(n+1) \pmod p$. $\square$

The following result, which will be useful for determining the class stabiliser in some cases, is similar to Lemma 4.6.7, so we just sketch its proof.

Lemma 4.9.40 *All involutions in* $\mathrm{P\Gamma O}_n^\circ(q)$ *lie in* $\mathrm{O}_n^\circ(q)\langle\phi\rangle \cup \mathrm{PCSO}_n^\circ(q)$.

Proof We can work in $\mathrm{SO}_n^\circ(q)\langle\phi\rangle \cong \mathrm{P\Gamma O}_n^\circ(q)$. Let $g \in \mathrm{P\Gamma O}_n^\circ(q)$ be an involution, and assume that g is the image of $A\sigma$, where $A \in \mathrm{SO}_n^\circ(q)$ and σ is a power of ϕ. As in Lemma 4.6.7, we may assume that σ has order 2, and so q is a square, and $AA^\sigma = (A\sigma)^2 = I_n$. Now the spinor norm of A is 1 if and only if $\mu := \prod_{i=1}^k \beta(v_i, v_i)$, as defined in Definition 1.6.10, is a square in $\mathbb{F}_q^\times$. Similarly, the spinor norm of A^σ is 1 if and only if μ^σ is a square in $\mathbb{F}_q^\times$. Since $AA^\sigma = I_n$, which has spinor norm 1 in $\mathrm{SO}_n^\circ(\sqrt{q})$, the element $\mu^{1+\sigma}$ is a square in $\mathbb{F}_{\sqrt{q}}^\times$. Hence μ is a square in $\mathbb{F}_q^\times$, and so $A \in \Omega_n^\circ(q)$ and $g \in \mathrm{O}_n^\circ(q)\langle\phi\rangle$. $\square$

Dimension 7. We now determine the $\mathscr{S}_1$-maximal subgroups of $\Omega_7(q)$. By Theorem 4.3.3, the quasisimple candidates are $\mathrm{L}_3(2)$ $(p \neq 2, 7)$; $\mathrm{L}_2(8)$ (twice, with $p \neq 2$ for the first and $p \neq 2, 3$ for the second); $\mathrm{L}_2(13)$ $(p \neq 2, 13)$; $\mathrm{U}_3(3)$ $(p \neq 2, 3)$; A_8 $(p \neq 2)$; J_1 $(p = 11)$; A_9 $(p = 3)$; and $\mathrm{S}_6(2)$ $(p \neq 2)$. Note that since $\Omega_7(q) = \mathrm{O}_7(q)$, all candidates are simple. Recall that we assume throughout that q is odd in Case $\mathbf{O}^\circ$.

Note that the two representations of $\mathrm{L}_2(8)$ are distinct in the sense that neither is an algebraic conjugate or dual of the other and neither can be obtained from the other by applying a group automorphism. To distinguish between them, we shall refer to them as $\mathrm{L}_2(8)_1$ and $\mathrm{L}_2(8)_2$, where $\mathrm{L}_2(8)_2$ is the representation with character values involving y_9.

We shall first show that four of these possibilities can be eliminated, and then consider the remaining $\mathscr{S}_1$-candidates in reverse order.

Proposition 4.9.41 *No extension of any of* $\mathrm{L}_3(2)$, $\mathrm{L}_2(8)_1$, $\mathrm{U}_3(3)$ *or* A_8 *is $\mathscr{S}_1$-maximal in any extension of* $\Omega_7(q)$.

Proof In characteristics other than 3 it follows from Theorem 4.7.1 that $\mathrm{L}_3(2) < \mathrm{U}_3(3)$ and $\mathrm{L}_3(2).2 < \mathrm{U}_3(3).2$. From [12] we see that this corresponds to a containment of $\mathscr{S}_1$-subgroups. Over $\mathbb{F}_3$ a straightforward computer calculation (file `containmentsd7`) shows that $\mathrm{L}_3(2).2 < \mathrm{S}_6(2)$ as $\mathscr{S}_1$-subgroups. It follows from Proposition 4.5.25 that $\mathrm{U}_3(3).2 < \mathrm{S}_6(2)$, and from [12, 57] that this is a containment of $\mathscr{S}_1$-subgroups.

The group $\operatorname{Aut} L_2(8) = L_2(8).3 < S_6(2)$, by Theorem 3.5.5. By [12, 57], this can only be representation $L_2(8)_1$, and the smallest non-trivial irreducible representation of $L_2(8)$ with $p \neq 2$ has dimension 7.

Now, $\operatorname{Aut} A_8 \cong SO_6^+(2) < S_6(2)$, by Theorem 3.5.8. By [12, 57], the smallest non-trivial irreducible representation of A_8 with $p \neq 2$ has dimension 7. $\square$

Proposition 4.9.42 *Let $\Omega = \Omega_7(q) = \bar{\Omega}$, and let $G = S_6(2)$ be an $\mathscr{S}_1$-subgroup of Ω. Then $q = p \neq 2$, with $N_\Omega(G) = G$ and trivial class stabiliser. The group G is $\mathscr{S}_1$-maximal, there is a unique $\operatorname{Aut} \bar{\Omega}$-class of groups G, and for no other q are there $\mathscr{S}_1$-subgroups of $\Omega_7(q)$ isomorphic to G.*

Proof This follows straightforwardly from Theorem 4.3.3, since $\operatorname{Out} G = 1$ and there is a single representation. $\square$

Proposition 4.9.43 *Let $\Omega = \Omega_7(q) = \bar{\Omega}$, and let $G = A_9$ be an $\mathscr{S}_1$-subgroup of Ω. Then $q = 3$, with $N_\Omega(G) = G.2$ and trivial class stabiliser. The group $N_\Omega(G)$ is $\mathscr{S}_1$-maximal, and there is a unique $\operatorname{Aut} \bar{\Omega}$-class of groups G.*

Proof By Theorem 4.3.3, there is a unique representation and $q = 3$. Most claims are easy computer calculations (file `a9d7f3calc`). For maximality, note that A_9 would have index 4 in $S_6(2)$, violating Theorem 1.11.2. $\square$

Proposition 4.9.44 *Let $\Omega = \Omega_7(q) = \bar{\Omega}$, and let $G = J_1$ be an $\mathscr{S}_1$-subgroup of Ω. Then $q = 11$, with $N_\Omega(G) = G$ and trivial class stabiliser. The group G is $\mathscr{S}_1$-maximal, and there is a unique $\operatorname{Aut} \bar{\Omega}$-class of groups G.*

Proof This follows straightforwardly from Theorem 4.3.3, since $\operatorname{Out} G = 1$ and there is a single such representation. $\square$

Proposition 4.9.45 *Let $\Omega = \Omega_7(q) = \bar{\Omega}$, and let $G = L_2(13)$ be an $\mathscr{S}_1$-subgroup of Ω. Then $N_\Omega(G) = G$. If $p \equiv \pm 1, \pm 3, \pm 4 \pmod{13}$ then $q = p$, with trivial class stabiliser. If $p \equiv \pm 2, \pm 5, \pm 6 \pmod{13}$ then $q = p^2 \neq 4$, with class stabiliser $\langle \phi \rangle$. The group G is $\mathscr{S}_1$-maximal, there is a unique $\operatorname{Aut} \bar{\Omega}$-class of groups G, and for no other q are there $\mathscr{S}_1$-subgroups of $\Omega_7(q)$ isomorphic to G.*

Proof From Theorem 4.3.3, we get $q = p$ and that there are two representations, swapped by the unique outer automorphism of G. If $q = p^2$ then the representations are also swapped by ϕ [12, 57]. Since $L_2(13).2 \setminus L_2(13)$ contains involutions, Lemma 4.9.40 implies that the class stabiliser is $\langle \phi \rangle$. $\square$

Proposition 4.9.46 *Let $\Omega = \Omega_7(q) = \bar{\Omega}$, and let $G = L_2(8)_2$ be an $\mathscr{S}_1$-subgroup of Ω. Then $N_\Omega(G) = G$. If $p \equiv \pm 1 \pmod 9$ then $q = p$, with trivial class stabiliser. If $p \equiv \pm 2, \pm 4 \pmod 9$ then $q = p^3 \neq 8$, with class stabiliser $\langle \phi \rangle$. The group G is $\mathscr{S}_1$-maximal, there is a unique $\operatorname{Aut} \bar{\Omega}$-class of groups G, and the only other $\mathscr{S}_1$-subgroups of $\Omega_7(q)$ isomorphic to G are equivalent to the image of the representation $L_2(8)_1$.*

Proof From Theorem 4.3.3, we get the congruences on p for q, and that there are three representations, permuted by the unique outer automorphism of G. When $G < \Omega_7(p^3)$, they are also permuted by ϕ [12, 57]. The group $L_2(8)$ is a subgroup of $S_6(2)$, but character values on elements of order 9 in [12, 57] show that the restriction of this representation of $S_6(2)$ is to $L_2(8)_1$. $\qquad\square$

Dimension 8. We now determine the $\mathscr{S}_1$-maximal subgroups of $\Omega_8^-(q)$. By Theorem 4.3.3, the quasisimple candidates are $L_3(2)$ $(p \neq 2, 3, 7)$; A_6 $(p \neq 3)$; $L_2(8)$ $(p \neq 2, 3)$; A_7 $(p = 5)$; $2{\cdot}A_8$ $(p \neq 2)$; $2{\cdot}Sz(8)$ $(p = 5)$; A_9 $(p \neq 3$, with two weak equivalence classes of representations when $p = 2)$; $2{\cdot}A_9$ $(p \neq 2)$; $2{\cdot}S_6(2)$ $(p \neq 2)$; A_{10} $(p = 5)$; $2{\cdot}A_{10}$ $(p = 5)$; and $2{\cdot}\Omega_8^+(2)$ $(p \neq 2)$.

Recall that the maximal subgroups of $\Omega_8^+(q)$ are classified in detail in [62], so we shall not be considering that case: see Table 8.50.

Proposition 4.9.47 *Let $\Omega = \Omega_8^\pm(q)$, and let G be an $\mathscr{S}_1$-subgroup of $\bar{\Omega}$. If $G \neq L_3(2)$ then $\Omega = \Omega_8^+(q)$.*

Proof Computer calculations (file o8+calc) over appropriate fields show that the orthogonal form preserved has square discriminant. $\qquad\square$

Proposition 4.9.48 *Let $\Omega = \Omega_8^\pm(q)$, and let $G = L_3(2)$ be an $\mathscr{S}_1$-subgroup of $\bar{\Omega}$. If $p \equiv \pm1, \pm4, \pm5 \pmod{21}$ then $\Omega = \Omega_8^+(p)$, and if $p \equiv \pm2, \pm8, \pm10 \pmod{21}$ then $p \neq 2$ and $\Omega = \Omega_8^-(p)$. The class stabiliser is $\langle \gamma \rangle$, and $N_{\bar{\Omega}}(G) = G$. There is a unique $\mathrm{Aut}\,\bar{\Omega}$-class of groups G, and for no other q are there $\mathscr{S}_1$-subgroups of $O_8^\pm(q)$ isomorphic to G. If $\Omega = \Omega_8^-(q)$ then G is $\mathscr{S}_1$-maximal.*

Proof From Theorem 4.3.3, we get $q = p$ and that there is a unique representation, so a unique $\mathrm{Aut}\,\bar{\Omega}$-class. Also, there is a unique $\mathscr{S}_1$-candidate for $\Omega_8^-(q)$, so any such groups are $\mathscr{S}_1$-maximal. A computer calculation (file 132d8calc) shows that the discriminant of the form on which G acts via isometries is 21 times a square. By [12], in the representation of $G.2$, elements of $G.2 \setminus G$ are isometries, and have determinant -1 and entries in $\mathbb{Q}$. So they lie in $GO_8^\pm(p) \setminus SO_8^\pm(p)$. $\qquad\square$

Dimension 9. We now determine the $\mathscr{S}_1$-maximal subgroups of $\Omega_9(q)$. By Theorem 4.3.3, the quasisimple candidates are A_6 $(p \neq 2, 3, 5)$; $L_2(8)$ $(p \neq 2, 7)$; $L_2(17)$ $(p \neq 2, 17)$; M_{11} $(p = 11)$; A_{10} $(p \neq 2, 5)$; and A_{11} $(p = 11)$. Note that since $\Omega_9(q) = O_9(q)$, all candidates are simple. Recall that we assume throughout that q is odd in Case $\mathbf{O}^\circ$. We first eliminate two of these groups, and then consider the rest in reverse order.

Proposition 4.9.49 *No extension of A_6 or M_{11} is $\mathscr{S}_1$-maximal in any extension of $\Omega_9(q)$.*

Proof As abstract groups, $A_6.2_3 < A_{10}$ and $A_6.2^2 < A_{10}.2$, by [12]. By Theorem 4.3.3, there are rational 10-dimensional irreducible representations of A_6 in all characteristics other than $2, 3$ and 5, and of A_{10} in all characteristics other than 2 and 5. By [12], the representations of S_{10} and $\mathrm{Aut}\,A_6$ preserve an orthogonal form, and considering the character values we see that this is a containment of $\mathscr{S}_1$-subgroups. By Lemma 4.9.39 this containment extends to all extensions of $\Omega_9(q)$.

We note that $M_{11} < A_{11}$: since $\mathrm{Out}\,M_{11} = 1$ and by [57] the smallest degree of a non-trivial representation of M_{11} in characteristic 11 is 9, this is a containment of $\mathscr{S}_1$-subgroups. □

Proposition 4.9.50 *Let $\Omega = \Omega_9(q)$, and let $G = A_{11}$ be an $\mathscr{S}_1$-subgroup of Ω. Then $q = 11$, $S = \mathrm{N}_\Omega(G) = G.2$ with trivial class stabiliser, S is $\mathscr{S}_1$-maximal and there is a unique $\mathrm{Aut}\,\Omega$-class of groups G.*

Proof By Theorem 4.3.3 there is a unique representation, and the claims follow from [57] and a straightforward computer calculation (file `a11d9calc`) using Proposition 1.6.11. □

Proposition 4.9.51 *Let $\Omega = \Omega_9(q)$, let $G = A_{10}$ be an $\mathscr{S}_1$-subgroup of Ω, and let $S = \mathrm{N}_\Omega(G)$. Then $q = p \neq 2, 5$. If $p \equiv \pm 1 \pmod 5$ then $S = G.2$, with trivial class stabiliser. If $p \equiv \pm 2 \pmod 5$ then $S = G$, with class stabiliser $\langle \delta \rangle$. If $q \neq 11$ then S is $\mathscr{S}_1$-maximal, otherwise it is not $\mathscr{S}_1$-maximal. There is a unique $\mathrm{Aut}\,\Omega$-class of groups G, and for no other q are there $\mathscr{S}_1$-subgroups of $\Omega_9(q)$ isomorphic to G.*

Proof By Theorem 4.3.3 there is a unique representation, and from [12, 57], we find that $G.2 < \mathrm{SO}_9(p)$ for all $p \neq 2, 5$. The representation arises from a deleted permutation module, so Lemma 4.9.39 implies that $S = G.2$ if and only if $p \equiv \pm 1 \pmod 5$. For $q = 11$ we note that $S < S_{11}$ (see Proposition 4.9.50), and by [12] this is a containment of $\mathscr{S}_1$-subgroups. □

Proposition 4.9.52 *Let $\Omega = \Omega_9(q)$, and let $G = L_2(17)$ be an $\mathscr{S}_1$-subgroup of Ω. Then $\mathrm{N}_\Omega(G) = G$. If $p \equiv \pm 1, \pm 2, \pm 4, \pm 8 \pmod{17}$ then $q = p \neq 2$, with trivial class stabiliser. If $p \equiv \pm 3, \pm 5, \pm 6, \pm 7 \pmod{17}$ then $q = p^2$, with class stabiliser $\langle \phi \rangle$. The group G is $\mathscr{S}_1$-maximal, there is a unique $\mathrm{Aut}\,\Omega$-class of groups G, and for no other q are there $\mathscr{S}_1$-subgroups G of $\Omega_9(q)$.*

Proof From Theorem 4.3.3, we get the congruences on p for q and that there are two representations, interchanged by the outer automorphism of G. Thus $\mathrm{N}_\Omega(G) = G$ and there is a unique $\mathrm{Aut}\,\Omega$-class. If $q = p^2$ then the representations are also permuted by ϕ [12, 57]. Since $L_2(17).2 \setminus L_2(17)$ contains involutions, Lemma 4.9.40 implies that the class stabiliser is $\langle \phi \rangle$. □

Proposition 4.9.53 *Let $\Omega = \Omega_9(q)$, and let $G = L_2(8)$ be an $\mathscr{S}_1$-subgroup of Ω. Then $N_\Omega(G) = G$. If $p \equiv \pm 1 \pmod 7$ then $q = p$, with trivial class stabiliser. If $p \equiv \pm 2, \pm 3 \pmod 7$ then $q = p^3 \neq 8$, with class stabiliser $\langle \phi \rangle$. The group G is $\mathscr{S}_1$-maximal, there is a unique $\operatorname{Aut}\Omega$-class of groups G, and for no other q are there $\mathscr{S}_1$-subgroups of $\Omega_9(q)$ isomorphic to G.*

Proof By Theorem 4.3.3, there are three representations, permuted by the outer automorphism of G, and the field is as given. By [12, 57], if $q = p^3$ then they are also permuted by ϕ. Whilst $L_2(8)$ is a subgroup of A_{10} and A_{11}, character values on elements of order 7 show that this is not a containment of $\mathscr{S}_1$-subgroups. $\qquad\square$

Dimension 10. We now determine the $\mathscr{S}_1$-maximal subgroups of $\Omega_{10}^\pm(q)$. By Theorem 4.3.3, the quasisimple candidates are A_6 ($p \neq 2, 3$); $L_2(11)$ ($p \neq 11$, two representations when $p > 3$); A_7 ($p = 7$); M_{11} ($p \neq 11$); $2^{\boldsymbol{\cdot}}L_3(4)$ ($p = 7$); M_{12} ($p = 2, 3$); $2^{\boldsymbol{\cdot}}M_{22}$ ($p = 7$); A_{11} ($p \neq 11$); and A_{12} ($p = 2, 3$). We consider them in reverse order.

When $p > 3$, the two representations of $L_2(11)$ are distinct, in the sense that neither is an algebraic conjugate or dual of the other, and neither can be obtained from the other by applying a group automorphism. To distinguish between them we shall refer to them as $L_2(11)_1$ and $L_2(11)_2$. The second of these is the deleted permutation module arising from its degree 11 permutation representation $L_2(11)$. The reductions of the two modules modulo 2 are equivalent, and it is more convenient to consider this module as $L_2(11)_2$ when $p = 2$, so we shall do that. When $p = 3$, the reduction of $L_2(11)_1$ remains absolutely irreducible, but that of $L_2(11)_2$ is the sum of two 5-dimensional modules.

Proposition 4.9.54 *Let $\Omega = \Omega_{10}^\pm(q)$, and let $G = A_{12}$ be an $\mathscr{S}_1$-subgroup of $\overline{\Omega}$. Then $\Omega = \Omega_{10}^-(2)$ or $\Omega_{10}^+(3)$, with $N_{\overline{\Omega}}(G) = G$ and class stabiliser $\langle \gamma \rangle$. The group G is $\mathscr{S}_1$-maximal and there is a single $\operatorname{Aut}\overline{\Omega}$-class of groups G.*

Proof From Theorem 4.3.3 we get $p = 2$ or 3, and that there is a unique representation in each case. The other claims follow from [57] and routine computer calculations (file `a12d10f2and3calc`). $\qquad\square$

Proposition 4.9.55 *Let $\Omega = \Omega_{10}^\pm(q)$, and let A_{11} be an $\mathscr{S}_1$-subgroup of $\overline{\Omega}$. Then $N_{\overline{\Omega}}(G) = G$, with class stabiliser $\langle \gamma \rangle$. If $p \equiv 1, 3, 4, 5, 9 \pmod{11}$ then $\Omega = \Omega_{10}^+(p)$, and if $p \equiv 2, 6, 7, 8, 10 \pmod{11}$ then $\Omega = \Omega_{10}^-(p)$. If $p \neq 2, 3$ then G is $\mathscr{S}_1$-maximal, and if $p = 2, 3$ then no extension of G is $\mathscr{S}_1$-maximal. There is a single $\operatorname{Aut}\overline{\Omega}$-class of groups G, and for no other q are there $\mathscr{S}_1$-subgroups of $O_{10}^\pm(q)$ isomorphic to G.*

Proof From Theorem 4.3.3, we get $q = p$ and that there is a unique representation. It follows from [12, 57] that if p is odd then elements of $G.2 \setminus G$ stabilise the form, have determinant -1, and have character values in $\mathbb{Q}$, so are induced (up to conjugacy) by γ. This is a deleted permutation module, and we get the congruences on p from Lemma 4.9.39. For $p = 2$ we check the class stabiliser and form by direct calculation (file a11d10calc).

The smallest dimension of a non-trivial representation of A_{11} in characteristic 2 or 3 is 10, by [57], so by Proposition 4.9.54 there are containments $A_{11} < A_{12}$ and $A_{11}.2 < A_{12}.2$ when $p = 2$ or 3. $\qquad\square$

Proposition 4.9.56 *Let $\Omega = \Omega_{10}^{\pm}(q)$ and let $G = (2^{\cdot})M_{22}$ be an $\mathscr{S}_1$-subgroup of $\bar{\Omega}$. Then $\Omega = \Omega_{10}^{-}(7)$, with $N_{\bar{\Omega}}(G) = G$ and class stabiliser $\langle\gamma\rangle$. The group G is $\mathscr{S}_1$-maximal, and there is a single $\mathrm{Aut}\,\bar{\Omega}$-class of groups G.*

Proof By [57], Theorem 4.3.3, and a MAGMA calculation (file 2m22d10f7calc), there is a unique representation, $N_{\bar{\Omega}}(G) = G < \Omega_{10}^{-}(7)$, and elements in $G.2\setminus G$ are isometries, have determinant -1, and have character values in $\mathbb{Q}$, so are induced by γ. There are no containments, as M_{22} has no permutation representation on fewer than 22 points. $\qquad\square$

Proposition 4.9.57 *Let $\Omega = \Omega_{10}^{\pm}(q)$, and let $G = M_{12}$ be an $\mathscr{S}_1$-subgroup of $\bar{\Omega}$. Then $\Omega = \Omega_{10}^{-}(2)$ or $\Omega_{10}^{+}(3)$, with $N_{\bar{\Omega}}(G) = G$. If $p = 2$ then the class stabiliser is $\langle\gamma\rangle$, and G is not $\mathscr{S}_1$-maximal, but $G.2$ is $\mathscr{S}_1$-maximal. If $p = 3$ then no extension of G is $\mathscr{S}_1$-maximal. In both cases there is a single $\mathrm{Aut}\,\bar{\Omega}$-class of groups G.*

Proof It is immediate from Theorem 4.3.3 that $q = 2$ or 3, and that if $q = 2$ then there is a unique representation, whilst if $q = 3$ there are two, interchanged by the outer automorphism of G.

Using [57] and routine computer calculations (file m12d10f2and3calc) we find that $G < \Omega_{10}^{-}(2)$, with class stabiliser $\langle\gamma\rangle$. The smallest degree of a non-trivial representation of M_{12} in characteristic 2 is 10 [57], so there is a containment of $\mathscr{S}_1$-subgroups $M_{12} < A_{12}$, but this does not extend to $M_{12}.2$.

Similarly, $G < A_{12} < \Omega_{10}^{+}(3)$ and G has trivial class stabiliser, so no extension of G is $\mathscr{S}_1$-maximal in characteristic 3. $\qquad\square$

Proposition 4.9.58 *Let $\Omega = \Omega_{10}^{\pm}(q)$ and let $G = (2^{\cdot})L_3(4)$ be an $\mathscr{S}_1$-subgroup of $\bar{\Omega}$. Then $\Omega = \Omega_{10}^{-}(7)$, with $N_{\bar{\Omega}}(G) = G$ and class stabiliser $\langle\delta', \gamma\rangle$. Here δ' induces the 2_3 automorphism of G, whilst γ induces the 2_1 automorphism on one Ω-class and the 2_2 automorphism on the other. The group G is not $\mathscr{S}_1$-maximal, and for $K \leqslant \langle\delta', \gamma\rangle$ the group $N_{\bar{\Omega}K}(G)$ is $\mathscr{S}_1$-maximal if and only if $N_{\bar{\Omega}K}(G) \not\leqslant (2^{\cdot})L_3(4).2_2$. There is a single $\mathrm{Aut}\,\bar{\Omega}$-class of groups G.*

Proof From Theorem 4.3.3 and [57], we find that there is a unique representation, $q = 7$, and elements in $G.2_2 \setminus G$ are isometries, have determinant -1, and have character values in $\mathbb{Q}$, whereas elements in $G.2_1 \setminus G$ are isometries, have determinant -1 and have character values that involve the irrationality r_2, which lies in $\mathbb{F}_7$. Thus elements of $G.2_3 \setminus G$ (note that $2_2 = 2_1 2_2$) have determinant 1, and we must determine their spinor norm.

A computer calculation (file `2134d10calc`) using Proposition 1.6.11 shows that $G < \Omega_{10}^-(7)$, $G.2_3 < \mathrm{SO}_{10}^-(7)$ but $G.2_3$ is not contained in $\Omega_{10}^-(7)$. So the class stabiliser is $\langle \delta', \gamma \rangle$, and there are two classes, interchanged by δ. Since $\gamma^\delta = \gamma \delta'$, it follows that γ induces the 2_1 and 2_2 automorphisms of G in the two classes.

The group $\mathrm{L}_3(4)$ is not a subgroup of A_{12}, which also rules out containments in A_{11} and M_{12}. A further computer calculation (file `containmentsd10`) shows that there are containments $2{\cdot}\mathrm{L}_3(4) < 2{\cdot}\mathrm{M}_{22}$ and $2{\cdot}\mathrm{L}_3(4).2_2 < 2{\cdot}\mathrm{M}_{22}.2$. □

Proposition 4.9.59 *No extension of* M_{11} *is* $\mathscr{S}_1$-*maximal in any extension of* $\mathrm{O}_{10}^\pm(q)$.

Proof The group M_{11} is a subgroup of A_{11}, and it follows from [12, 57] that this is an inclusion of $\mathscr{S}_1$-subgroups. Since $\mathrm{Out}\,\mathrm{M}_{11} = 1$, we are done. □

Proposition 4.9.60 *Let* $\Omega = \Omega_{10}^\pm(q)$, *and let* $G = \mathrm{A}_7$ *be an* $\mathscr{S}_1$-*subgroup of* $\overline{\Omega}$. *Then* $\Omega = \Omega_{10}^-(7)$, *with* $\mathrm{N}_{\overline{\Omega}}(G) = G$, *and class stabiliser* $\langle \delta' \rangle$. *The group* G *is not* $\mathscr{S}_1$-*maximal, but* $\mathrm{N}_{\mathrm{PSO}_{10}^-(7)}(G)$ *is* $\mathscr{S}_1$-*maximal. There is a single* $\mathrm{Aut}\,\overline{\Omega}$-*class of groups* G.

Proof By Theorem 4.3.3 and [57], we find that there is a unique representation, with $G < \Omega_{10}^\pm(7)$ and $G.2 < \mathrm{SO}_{10}^\pm(7)$. A computer calculation (file `a7d10calc`) shows that $G < \Omega_{10}^-(7)$ and $G.2$ is not contained in $\Omega_{10}^-(7)$, so the class stabiliser is $\langle \delta' \rangle$.

Neither A_{12} not M_{12} arise in characteristic 7. The 10-dimensional module for A_{11} is the deleted permutation module and, by Theorem 1.11.2, the only subgroups of A_{11} isomorphic to A_7 have four fixed points, so the restriction of this module to A_7 is reducible. By [12], A_7 is a subgroup of $2{\cdot}\mathrm{M}_{22}$ (see Proposition 4.9.56), and by looking at character values on elements of order 3 [57], we see that this is a containment of $\mathscr{S}_1$-subgroups. However, $\mathrm{M}_{22}.2$ does not contain S_7. There is no subgroup of $2{\cdot}\mathrm{L}_3(4)$ isomorphic to A_7 by Theorem 4.3.3. □

Proposition 4.9.61 *Let* $\Omega = \Omega_{10}^\pm(q)$, *and let* $G = \mathrm{L}_2(11)_1$ *be an* $\mathscr{S}_1$-*subgroup of* $\overline{\Omega}$ *with* $p > 2$. *Then* $\mathrm{N}_{\overline{\Omega}}(G) = G$, *with class stabiliser* $\langle \gamma \rangle$. *If* $p \equiv 1, 3, 4, 5, 9$ (mod 11) *then* $\Omega = \Omega_{10}^+(p)$. *If* $p \equiv 2, 6, 7, 8, 10$ (mod 11) *then* $\Omega = \Omega_{10}^-(p)$. *The group* G *is* $\mathscr{S}_1$-*maximal if and only if* $q \neq 3$ *or* 7, *and if* $q = 3$ *or* 7 *then no extension of* G *is* $\mathscr{S}_1$-*maximal in any extension of* $\overline{\Omega}$. *There is a single* $\mathrm{Aut}\,\overline{\Omega}$-*class of groups* G, *and the only other* $\mathscr{S}_1$-*subgroups of* $\mathrm{O}_{10}^\pm(q)$ *isomorphic to* G *are those in Proposition 4.9.62, below.*

Proof   From Theorem 4.3.3, we get $q = p$ for all $p \neq 2, 11$ (for $p = 2$, the module is equivalent to $L_2(11)_2$ and will be considered as such), and that there is a unique representation. Computer calculations (file `1211ad10calc`) in $SL_{10}(\mathbb{Q})$ show that G preserves a form with determinant 11 times a square. From [12, 57] we find that $G.2$ consists of isometries, and that elements of $G.2 \setminus G$ have determinant -1 and character values in $\mathbb{Q}$, so are induced by a conjugate of γ.

A straightforward computer calculation (file `containmentsd10`) shows that if $q = 3$ then there are containments of $\mathscr{S}_1$-subgroups $L_2(11)_1 < A_{12}$ and $L_2(11)_1.2 < A_{12}.2$. Alternatively, we can deduce this from the fact that the module in question is the 10-dimensional constituent of the permutation module arising from the embedding $L_2(11)_1.2 < A_{12}.2 \cong S_{12}$. So we assume in the rest of this proof that $q > 3$, and do not consider containments in M_{12}.

The group A_{11} contains two classes of transitive subgroups $L_2(11)$ that are conjugate in S_{11} and, since the 10-dimensional module for A_{11} is the deleted permutation module, its restriction to $L_2(11)$ is also the deleted permutation module arising from its degree 11 permutation representation, which is $L_2(11)_2$. So there is no containment $L_2(11)_1 < A_{11}$.

Suppose now that $q = p = 7$. We find from [12] that M_{22} has a unique class of subgroups $L_2(11)$ which extends to $L_2(11).2 < M_{22}.2$. A computer calculation (file `containmentsd10`) shows that the subgroup $2.M_{22}$ of $\Omega_{10}^-(7)$ contains $(2\times)L_2(11)$ as a subgroup acting irreducibly as $L_2(11)_1$. So $L_2(11)_1$ and $L_2(11)_1.2$ are not $\mathscr{S}_1$-maximal when $q = 7$. $\square$

Proposition 4.9.62   *Let $\Omega = \Omega_{10}^{\pm}(q)$, let $G = L_2(11)_2$ be an $\mathscr{S}_1$-subgroup of $\bar{\Omega}$. Then $N_{\bar{\Omega}}(G) = G$ and $p \neq 3$. If $p \equiv 1, 3, 4, 5, 9 \pmod{11}$ then $\Omega = \Omega_{10}^+(p)$, and if $p \equiv 2, 6, 7, 8, 10 \pmod{11}$ then $\Omega = \Omega_{10}^-(p)$. If $p \equiv \pm 1, 2 \pmod{12}$ then the class stabiliser is $\langle \gamma \rangle$, and if $p \equiv \pm 5 \pmod{12}$ then the class stabiliser is $\langle \gamma\delta \rangle$. The group G is never $\mathscr{S}_1$-maximal, and $G.2$ is not maximal when $p = 2$, but if $p \neq 2$ then $G.2$ is $\mathscr{S}_1$-maximal in the corresponding extension of $\bar{\Omega}$. There is a single $\operatorname{Aut}\bar{\Omega}$-class of groups G, and the only other $\mathscr{S}_1$-subgroups of $O_{10}^{\pm}(q)$ isomorphic to G are those in Proposition 4.9.61.*

Proof   From Theorem 4.3.3, we get $q = p \neq 3, 11$ and that there is a unique representation. Computer calculations (file `1211bd10calc`) show that G consists of isometries of a bilinear form F with determinant 11 times a square, so the sign of the form is as given. (The case $p = 2$ is done seperately.) These calculations also show that there exists $A \in GL_{10}(\mathbb{Z})$ that normalises and induces the outer automorphism of G, has determinant -3^5, and transforms F to $3F$.

When $p = 2$, the matrix A has quasideterminant -1, so the class stabiliser is $\langle \gamma \rangle$. If $p \equiv \pm 1 \pmod{12}$, then 3 is a square in $\mathbb{F}_p^{\times}$ and so by multiplying A by the scalar matrix $3^{-1/2}I_{10} \in GL_{10}(p)$, we find a matrix in $GO_{10}^{\pm}(p) \setminus SO_{10}^{\pm}(p)$ that normalises and induces the outer automorphism of G. So this matrix represents an outer automorphism that is conjugate in $\mathrm{Out}\,\overline{\Omega}$ to γ.

If $p \equiv \pm 5 \pmod{12}$ then 3 is a non-square in $\mathbb{F}_p^\times$. Let D be the matrix that induces the outer automorphism δ of Ω, as defined in Subsection 1.7.1. We saw there that, irrespective of the sign of the bilinear form F, the determinant of D is w^5 and D transforms F to wF, where w is a primitive element of $\mathbb{F}_p$. Since 3 is a non-square, $3 = w^{2k+1}$ in $\mathbb{F}_p$ for some k. So $w^k D A^{-1}$ fixes F and has determinant -1. It follows that the element of $\mathrm{Out}\,\overline{\Omega}$ induced by A on reduction modulo p is conjugate in $\mathrm{Out}\,\overline{\Omega}$ to $\delta\gamma$.

For all $p \neq 3, 11$ there is a containment of $\mathscr{S}_1$-subgroups $\mathrm{L}_2(11)_2 < \mathrm{A}_{11}$ (see Proposition 4.9.55 and [12, 57]), but S_{11} does not have a subgroup $\mathrm{L}_2(11).2$. A computer calculation (file `containmentsd10`) shows that there is a containment $\mathrm{L}_2(11).2 < \mathrm{M}_{12}.2 < \mathrm{SO}_{10}^-(2)$, so assume from now that $p > 3$. As we observed in the proof of Proposition 4.9.61, the group M_{22} has a unique class of subgroups $\mathrm{L}_2(11)$ and, in characteristic 7, the restriction of the 10-dimensional representation of M_{22} to $\mathrm{L}_2(11)$ is to $\mathrm{L}_2(11)_1$, and so there is no containment of $\mathrm{L}_2(11)_2$ in $2\cdot\mathrm{M}_{22}$. $\square$

Proposition 4.9.63 *Let* $\Omega = \Omega_{10}^\pm(q)$, *let* $G = \mathrm{A}_6$ *be an* $\mathscr{S}_1$-*subgroup of* $\overline{\Omega}$, *and let* $S = \mathrm{N}_{\overline{\Omega}}(G)$. *If* $p \equiv 1 \pmod 4$ *then* $\Omega = \Omega_{10}^+(p)$. *If* $p \equiv 3 \pmod 4$ *then* $p \neq 3$ *and* $\Omega = \Omega_{10}^-(p)$. *If* $p \equiv \pm 1 \pmod{24}$ *then* $S = G.2_1$, *with class stabiliser* $\langle\gamma\rangle$. *If* $p \equiv \pm 5 \pmod{24}$ *then* $S = G$, *with class stabiliser* $\langle\gamma\delta, \delta'\rangle$, *where* δ' *induces the* 2_1 *automorphism, and* $\gamma\delta$ *induces the* 2_2. *If* $p \equiv \pm 7 \pmod{24}$ *then* $S = G$, *with class stabiliser* $\langle\gamma, \delta'\rangle$, *where* δ' *induces the* 2_1 *automorphism and* γ *the* 2_2. *If* $p \equiv \pm 11 \pmod{24}$ *then* $S = G.2_1$, *with class stabiliser* $\langle\delta\gamma\rangle$.

The group S *is* $\mathscr{S}_1$-*maximal if and only if* $q \neq 7$. *If* $q = 7$ *then, for any subgroup* K *of* $\mathrm{Out}\,\overline{\Omega}$, *the group* $\mathrm{N}_{\overline{\Omega}K}(G)$ *is* $\mathscr{S}_1$-*maximal if and only if* $\mathrm{N}_{\overline{\Omega}K}(G) \not\leq (2\times)\mathrm{A}_6.2_1$ *and* $\mathrm{N}_{\overline{\Omega}K}(G) \not\leq (2\times)\mathrm{A}_6.2_3$. *There is a single* $\mathrm{Aut}\,\overline{\Omega}$-*class of groups* G, *and for no other* q *are there* $\mathscr{S}_1$-*subgroups* G *of* $\mathrm{O}_{10}^\pm(q)$.

Proof It follows from Theorem 4.3.3 that $q = p \neq 2, 3$, and there is a unique representation.

Computer calculations (file `a6d10calc`) in $\mathrm{SL}_{10}(\mathbb{Z})$ show that G preserves a form F with square determinant. From the computer calculation or [12, 57], we find $G.2_1 < \mathrm{SO}_{10}^\pm(p)$. By a computer calculation (file `a6d10calc`) using Proposition 1.6.11, the spinor norm of an element of $G.2_1 \setminus G$ is 1 if and only if 3 is a square modulo p, which is the case if and only if $p \equiv \pm 1 \pmod{12}$. So this spinor norm is -1 when $p \equiv \pm 5 \pmod{12}$, and hence δ' induces the 2_1 automorphism of G in this case.

The computer calculations show also that there is a matrix $A \in \mathrm{GL}_{10}(\mathbb{Z})$ that normalises and induces the outer automorphism 2_2 of G, has determinant -2^5, and transforms F to $2F$.

If $p \equiv \pm 1 \pmod 8$, then 2 is a square in $\mathbb{F}_p^\times$ and so by multiplying A by the scalar matrix $2^{-1/2}\mathrm{I}_{10} \in \mathrm{GL}_{10}(p)$, we find a matrix in $\mathrm{GO}_{10}^\pm(p) \setminus \mathrm{SO}_{10}^\pm(p)$ that normalises and induces the 2_2 outer automorphism of G. So this matrix represents an outer automorphism that is conjugate in $\mathrm{Out}\,\bar\Omega$ to γ.

If $p \equiv \pm 3 \pmod 8$, then 2 is a non-square in $\mathbb{F}_p^\times$. By comparing the determinant and action on F of the matrix A with those of the matrix that induces the outer automorphism δ of Ω (as defined in Subsection 1.7.1), we can deduce that the outer automorphism of $\bar\Omega$ induced by A on reduction modulo p is conjugate in $\mathrm{Out}\,\bar\Omega$ to $\delta\gamma$.

We note that A_6 is a subgroup of A_{11}, but by looking at character values on elements of order 2 [12, 57], we see that this is not a containment of $\mathscr{S}_1$-subgroups.

Suppose now that $p = 7$. We note that (by [12]) $\mathrm{A}_6 < 2{\cdot}\mathrm{M}_{22}$, $\mathrm{A}_6 < \mathrm{A}_7$ and $\mathrm{A}_6.2_1 < \mathrm{A}_7.2$. A computer calculation (file `containmentsd10`) shows that these are containments of $\mathscr{S}_1$-subgroups. However, the same calculation also shows that, as $\mathscr{S}_1$-subgroups, $\mathrm{A}_6.2_2$ and $\mathrm{A}_6.2_3$ are not contained in $2{\cdot}\mathrm{M}_{22}.2$. There are also three classes of subgroups A_6 of $2{\cdot}\mathrm{L}_3(4)$, which are fused under the automorphism of order 3 of $(2{\cdot})\mathrm{L}_3(4)$. The three classes of elements of order 4 in $\mathrm{L}_3(4)$ are also fused under this automorphism and, by considering character values on these elements in [57], we see that the restriction of the 10-dimensional representation of $2{\cdot}\mathrm{L}_3(4)$ to two of the A_6 classes is irreducible, but it is reducible on the third such class. From Theorem 4.7.1 or [12], we find that the 2_1 automorphism $\gamma\phi$ of $2{\cdot}\mathrm{L}_3(4)$ normalises all three A_6 classes and induces the 2_3 automorphism of A_6, whereas the 2_2 and 2_3 automorphisms normalise just one of these classes. So the two classes for which the restriction of the representation of $2{\cdot}\mathrm{L}_3(4)$ is irreducible are normalised by the 2_1 automorphism only, and we have containments of $\mathscr{S}_1$-subgroups $\mathrm{A}_6.2_3 < 2{\cdot}\mathrm{L}_3(4).2_1$, but no such containment of $\mathrm{A}_6.2_1$ or $\mathrm{A}_6.2_2$. This completes the proof. $\square$

Dimension 11. We now determine the $\mathscr{S}_1$-maximal subgroups of $\Omega_{11}(q)$. By Theorem 4.3.3, the quasisimple candidates are $\mathrm{L}_2(11)$ $(p \neq 2, 3, 11)$; $\mathrm{L}_3(3)$ $(p = 13)$; M_{11} $(p \neq 2, 3)$; M_{12} $(p \neq 2, 3)$; A_{12} $(p \neq 2, 3)$; and A_{13} $(p = 13)$. Note that since $\Omega_{11}(q) = \mathrm{O}_{11}(q)$, all candidates are simple. Recall that we assume throughout that q is odd in Case $\mathbf{O}^\circ$. We first eliminate three of these groups, then consider the remaining possibilities in reverse order.

Proposition 4.9.64   *No extension of any of* $\mathrm{L}_2(11)$, M_{11} *or* M_{12} *is* $\mathscr{S}_1$-*maximal in any extension of* $\Omega_{11}(q)$.

Proof   The 11-dimensional irreducible representations of $\mathrm{L}_2(11)$, M_{11} and M_{12} are all deleted permutation modules arising from their degree 12 permutation representations, so their images are all contained in that of A_{12}, and similarly $L_2(11).2 < S_{12}$ (as $\mathscr{S}_1$-subgroups).

The group M_{11} is a subgroup of A_{12}, which is a containment of $\mathscr{S}_1$-subgroups for all fields of characteristic not 2 or 3, by character values [12, 57]. Since $\text{Out}\,M_{11} = 1$, we are done.

There is an abstract containment of $M_{12} < A_{12}$. The smallest degree of a non-trivial representation of M_{12} is 11, so this is a genuine containment of M_{12} in A_{12} for all fields of characteristic not 2 or 3. By Theorem 4.3.3, the class stabiliser of M_{12} is trivial, so no extension of M_{12} arises. □

Proposition 4.9.65 *Let $\Omega = \Omega_{11}(q)$, and let $G = A_{13}$ be an $\mathscr{S}_1$-subgroup of Ω. Then $q = 13$ and $N_\Omega(G) = G$, with class stabiliser $\langle \delta \rangle$. The group G is $\mathscr{S}_1$-maximal and there is a unique $\text{Aut}\,\Omega$-class of groups G.*

Proof By Theorem 4.3.3 there is a unique representation. A MAGMA calculation (file `a13d12calc`) shows that $G.2 \leqslant SO_{11}(13)$, with $G.2 \nleqslant \Omega_{11}(13)$. □

Proposition 4.9.66 *Let $\Omega = \Omega_{11}(q)$, let $G = A_{12}$ be an $\mathscr{S}_1$-subgroup of Ω, and let $S = N_\Omega(G)$. Then $q = p$. If $p \equiv \pm 1, \pm 5 \pmod{24}$ then $S = G.2$, with trivial class stabiliser. If $p \equiv \pm 7, \pm 11 \pmod{24}$ then $S = G$, with class stabiliser $\langle \delta \rangle$. If $q \neq 13$ then S is $\mathscr{S}_1$-maximal, otherwise all extensions of S are not $\mathscr{S}_1$-maximal. There is a unique $\text{Aut}\,\Omega$-class of groups G, and for no other q are there $\mathscr{S}_1$-subgroups of $\Omega_{11}(q)$ isomorphic to G.*

Proof By Theorem 4.3.3, there is a unique representation and $q = p$. From [57] we see that $G.2 < SO_{11}(p)$ for all p. This is a deleted permutation module so, by Lemma 4.9.39, $G.2 < \Omega_{11}(p)$ when 6 is a square modulo p, whereas $G.2$ is not contained in $\Omega_{11}(p)$ otherwise. The group S_{12} is a subgroup of S_{13} so, by [12] and Proposition 4.9.65, in characteristic 13 there are containments of $\mathscr{S}_1$-subgroups A_{12} in A_{13} and S_{12} in S_{13}. □

Proposition 4.9.67 *Let $\Omega = \Omega_{11}(q)$, and let $G = L_3(3)$ be an $\mathscr{S}_1$-subgroup of Ω. Then $q = 13$ and $S = N_\Omega(G) = G.2$, with trivial class stabiliser. The group G is not $\mathscr{S}_1$-maximal, but S is $\mathscr{S}_1$-maximal, and there is a unique $\text{Aut}\,\Omega$-class of groups G.*

Proof There is a unique representation. A straightforward calculation (file `1211d12calc`) shows that $S = G.2 < \Omega_{11}(13)$, so the class stabiliser is trivial. From [12, 57] we find $G < A_{13}$ as $\mathscr{S}_1$-subgroups but $G.2 \nleqslant A_{13}$. No extension of A_{12} is maximal when $q = 13$. □

Dimension 12. We now determine the $\mathscr{S}_1$-maximal subgroups of $\Omega_{12}^\pm(q)$. By Theorem 4.3.3, the quasisimple candidates are $L_2(11)$ $(p \neq 5, 11)$; $L_2(13)$ $(p \neq$

13); $L_3(3)$ ($p \neq 3, 13$); $2\cdot M_{12}$ ($p \neq 2, 3$); A_{13} ($p \neq 13$); and A_{14} ($p = 7$). We consider these groups in reverse order.

Proposition 4.9.68   *Let $\Omega = \Omega_{12}^{\pm}(q)$, and let $G = A_{14}$ be an $\mathscr{S}_1$-subgroup of $\bar{\Omega}$. Then $\Omega = \Omega_{12}^{-}(7)$, with $N_{\bar{\Omega}}(G) = G$, and class stabiliser $\langle \gamma \rangle$. There is a single Aut $\bar{\Omega}$-class of groups G, and G is $\mathscr{S}_1$-maximal.*

Proof   By Theorem 4.3.3, there is a unique representation and $q = 7$, so it is straightforward to verify these claims in MAGMA.   □

Proposition 4.9.69   *Let $\Omega = \Omega_{12}^{\pm}(q)$, and let $G = A_{13}$ be an $\mathscr{S}_1$-subgroup of $\bar{\Omega}$. Then $N_{\bar{\Omega}}(G) = G$, with class stabiliser $\langle \gamma \rangle$. If $p \equiv \pm 1, \pm 3, \pm 4 \pmod{13}$ then $\Omega = \Omega_{12}^{+}(p)$. If $p \equiv \pm 2, \pm 5, \pm 6 \pmod{13}$ then $\Omega = \Omega_{12}^{-}(p)$. If $q \neq 7$ then G is $\mathscr{S}_1$-maximal in $\bar{\Omega}$, otherwise ($q = 7$) no extension of G is $\mathscr{S}_1$-maximal in any extension of $\bar{\Omega}$. There is a single Aut $\bar{\Omega}$-class of groups G and for no other q are there $\mathscr{S}_1$-subgroups of $O_{12}^{\pm}(q)$ isomorphic to G.*

Proof   This is just the deleted permutation representation, so $q = p$ and there is a unique representation. Computer calculations (file a13d12calc) in $SL_{12}(\mathbb{Q})$ show that the form preserved has determinant 13: for $p = 2$ we carry out an additional calculation over $\mathbb{F}_2$.

For $p > 11$, we see from [12] that $G.2$ consists of isometries, and elements of $G.2 \setminus G$ have determinant -1. So the outer automorphism of G is induced by a conjugate of γ. Computer calculations (file a13d12calc) show that the same is true for $p = 2, 3, 5, 7, 11$.

In characteristic 7 a straightforward calculation shows that there is an $\mathscr{S}_1$-containment $A_{13}.2 < A_{14}.2$, but for all other p the group G is $\mathscr{S}_1$-maximal.   □

Proposition 4.9.70   *Let $\Omega = \Omega_{12}^{\pm}(q)$, let $G = (2\cdot)M_{12}$ be an $\mathscr{S}_1$-subgroup of $\bar{\Omega}$, and let $S = N_{\bar{\Omega}}(G)$. Then $\Omega = \Omega_{12}^{+}(p)$. If $p \equiv \pm 1 \pmod{24}$ then the group $S = G.2$, with trivial class stabiliser. If $p \equiv \pm 11 \pmod{24}$ then $S = G$, with class stabiliser $\langle \delta' \rangle$. If $p \equiv \pm 5 \pmod{12}$ then $S = G$, with class stabiliser $\langle \delta \rangle$. There is a single Aut $\bar{\Omega}$-class of groups G, the group G is $\mathscr{S}_1$-maximal, and for no other q are there $\mathscr{S}_1$-subgroups of $O_{12}^{\pm}(q)$ isomorphic to G.*

Proof   By Theorem 4.3.3, there is a unique representation and $q = p$. A computer calculation (file 2m12d12calc) show that the form preserved has determinant 1.

From [12, 57], we find that $G.2$ consists of determinant 1 isometries, but their character values involve the irrationality r_3, which lies in $\mathbb{F}_p$ if and only if $p \equiv \pm 1 \pmod{12}$. So if $p \equiv \pm 5 \pmod{12}$ then the class stabiliser is $\langle \delta \rangle$.

If $p \equiv \pm 1 \pmod{12}$, then $2\cdot M_{12}.2 < SO_{12}^{+}(p)$ and we have to determine the spinor norm of elements of $2\cdot M_{12}.2 \setminus 2\cdot M_{12}$. Using a computer calculation (file 2m12d12calc) over $\mathbb{Q}(r_3)$ and Proposition 1.6.11 we find that the spinor norm of an element of $2^{\cdot}M_{12}.2 \setminus 2^{\cdot}M_{12}$ is 1 if and only if $3(2 - r_3)$ is a square in $\mathbb{F}_p^{\times}$. Now $2 - r_3$ has a square root in the cyclotomic field $\mathbb{Q}(z_{24})$ (one such root is $z_{24}{}^5 - z_{24}{}^7$), so whether $3(2 - r_3)$ is a square in $\mathbb{F}_p^{\times}$ depends on the value of p modulo 24. We find that it is a square, in which case $2^{\cdot}M_{12}.2 < \Omega_{12}^{+}(p)$, when $p \equiv \pm 1 \pmod{24}$, and a non-square when $p \equiv \pm 11 \pmod{24}$.

Maximality is clear, since $2^{\cdot}M_{12}$ is not a subgroup of A_{14}, by [12]. □

Proposition 4.9.71 *Let* $\Omega = \Omega_{12}^{\pm}(q)$, *let* $G = L_3(3)$ *be an* $\mathscr{S}_1$-*subgroup of* $\overline{\Omega}$, *and let* $S = N_{\overline{\Omega}}(G)$. *Then* $p \neq 3, 13$. *If* $p \equiv \pm 1, \pm 3, \pm 4 \pmod{13}$ *then* $\Omega = \Omega_{12}^{+}(p)$. *If* $p \equiv \pm 2, \pm 5, \pm 6 \pmod{13}$ *then* $\Omega = \Omega_{12}^{-}(p)$. *If* $p \equiv \pm 5 \pmod{12}$ *then* $S = G$, *with class stabiliser* $\langle \delta \rangle$. *If* $p \equiv 1, 2, 11 \pmod{12}$ *with* $\Omega = \Omega_{12}^{-}(p)$, *then* $S = G.2$, *with trivial class stabiliser. If* $p \equiv \pm 1 \pmod{12}$ *with* $\Omega = \Omega_{12}^{+}(p)$, *and* $x^4 - 10x^2 + 13$ *has four linear factors modulo* p, *then* $S = G.2$, *with trivial class stabiliser. If* $p \equiv \pm 1 \pmod{12}$ *with* $\Omega = \Omega_{12}^{+}(p)$, *and* $x^4 - 10x^2 + 13$ *has no linear factors modulo* p, *then* $S = G$ *with class stabiliser* $\langle \delta' \rangle$. *There is a single* Aut $\overline{\Omega}$-*class of groups* G. *The group* G *is not* $\mathscr{S}_1$-*maximal in* $\overline{\Omega}$, *however* $G.2$ *is* $\mathscr{S}_1$-*maximal wherever it occurs. For no other* q *are there* $\mathscr{S}_1$-*subgroups of* $O_{12}^{\pm}(q)$ *isomorphic to* G.

Proof By Theorem 4.3.3, there is a unique representation and $q = p$. Computer calculations (file `133d12calc`) using a representation of G in $GL_{12}(\mathbb{Z})$ show that the determinant of the form F for which G is a group of isometries is 13. For $p = 2$, we require an independent computation. They show also that there is a matrix $A \in GL_{12}(\mathbb{Z})$ that normalises and induces the outer automorphism of G, has determinant 3^6, and transforms F to $3F$.

Suppose first that $p \equiv \pm 5 \pmod{12}$. Then 3 is a non-square in $\mathbb{F}_p^{\times}$. By comparing the determinant and action on F of the matrix A with those of the matrix that induces the outer automorphism δ of $\Omega_{12}^{\pm}(p)$, as defined in Subsection 1.7.1, we can deduce that the outer automorphism of G induced by A on reduction modulo p is conjugate in Out $\overline{\Omega}$ to δ.

Suppose then that $p \equiv 1, 2, 11 \pmod{12}$. Then 3 is a square in $\mathbb{F}_p^{\times}$ and so by multiplying A by the scalar matrix $I_{12}/\sqrt{3} \in GL_{12}(p)$, we can find a matrix in $SO_{12}^{\pm}(p)$ that normalises and induces the outer automorphism of G. If $G < \Omega_{12}^{-}(p)$ then $SO_{12}^{-}(p) = \Omega_{12}^{-}(p) \times Z(SO_{12}^{-}(p))$, so $G.2 < \Omega_{12}^{-}(p)$. If $G < \Omega_{12}^{+}(p)$ then $G.2 < SO_{12}^{+}(p)$, and we need to determine the spinor norm or quasideterminant of elements of $G.2 \setminus G$. When p is odd, using a computer calculation (file `133d12calc`)) over $\mathbb{Q}(r_3)$ and Proposition 1.6.11 we find that this spinor norm is 1 if and only if $5 - 2r_3$ is a square in $\mathbb{F}_p^{\times}$. This is the case if and only if the minimal polynomial $x^4 - 10x^2 + 13$ of $\sqrt{5 - 2r_3}$ over $\mathbb{Q}$ has four linear factors when reduced modulo p. There appears to be no alternative method of characterising such primes. We do a separate computer calculation of the quasideterminant for the case $p = 2$.

By [12], $L_3(3) < A_{13}$ and by [12, 57] this is an $\mathscr{S}_1$-containment. However, $L_3(3).2 \nleq S_{13}, S_{14}$. □

Proposition 4.9.72 *Let* $\Omega = \Omega_{12}^{\pm}(q)$, *and let* $G = L_2(13)$ *be an* $\mathscr{S}_1$-*subgroup of* $\bar{\Omega}$. *Then* $N_{\bar{\Omega}}(G) = G$ *and the class stabiliser is* $\langle \gamma \rangle$. *If* $q = 7$ *then* $\Omega = \Omega_{12}^{-}(7)$. *If* $p \equiv \pm 1 \pmod 7$ *then* $q = p$, *and there are exactly three* $\text{Aut}\,\bar{\Omega}$-*classes of groups* G. *If* $p \equiv \pm 2, \pm 3 \pmod 7$ *then* $q = p^3$, *and there is exactly one* $\text{Aut}\,\bar{\Omega}$-*class of groups* G. *If* $p \equiv \pm 1, \pm 3, \pm 4 \pmod{13}$ *then* $\Omega = \Omega_{12}^{+}(q)$. *If* $p \equiv \pm 2, \pm 5, \pm 6 \pmod{13}$ *then* $\Omega = \Omega_{12}^{-}(q)$. *If* $q \neq 7$ *then* G *is* $\mathscr{S}_1$-*maximal, and if* $q = 7$ *then no extension of* G *is* $\mathscr{S}_1$-*maximal in any extension of* $\bar{\Omega}$. *For no other* q *are there* $\mathscr{S}_1$-*subgroups of* $O_{12}^{\pm}(q)$ *isomorphic to* G.

Proof When $p = 7$ there is a unique representation, by Theorem 4.3.3, which extends to $G.2$ by [57]. However, a MAGMA calculation (file 1213d12calc) shows that $\Omega = \Omega_{12}^{-}(7)$ and there are $\mathscr{S}_1$-containments $G < A_{14}$ and $G.2 < S_{14}$.

By Theorem 4.3.3, $q = p$ when $p \equiv \pm 1 \pmod 7$ and $q = p^3$ when $p \equiv \pm 2, \pm 3 \pmod 7$. There are three representations, all stabilised by the outer automorphism of G, so three $\text{Aut}\,\bar{\Omega}$-classes when $q = p$.

Computer calculations (file 1213d12calc) in $SL_{12}(K)$ with $K := \mathbb{Q}(y_7)$ show that the form preserved has determinant d, where $13d$ is a square in K. So the sign of this form over $\mathbb{F}_q$ depends on whether 13 is a square in $\mathbb{F}_q^{\times}$, and hence on whether p is a square modulo 13. For $p = 2$ we require an additional calculation over $\mathbb{F}_8$.

From [12, 57], we find that the representation of $G.2$ consists of isometries, and elements of $G.2 \setminus G$ have determinant -1 so, when p is odd, the outer automorphism of G is induced by a conjugate of γ. A computer calculation (file 1213d12calc) shows that the same is true when $p = 2$. The three representations are cycled by ϕ when $q = p^3$ [12, 57].

The group $L_2(13)$ is not a subgroup of A_{13}. □

Proposition 4.9.73 *Let* $\Omega = \Omega_{12}^{\pm}(q)$, *and let* $G = L_2(11)$ *be an* $\mathscr{S}_1$-*subgroup of* $\bar{\Omega}$. *Then* $N_{\bar{\Omega}}(G) = G$, *with class stabiliser* $\langle \gamma \rangle$. *If* $p \equiv \pm 1, \pm 16, \pm 19, \pm 24, \pm 26 \pmod{55}$ *then* $\Omega = \Omega_{12}^{+}(p)$. *If* $p \equiv \pm 4, \pm 6, \pm 9, \pm 14, \pm 21 \pmod{55}$ *then* $\Omega = \Omega_{12}^{-}(p)$. *If* $p \equiv \pm 2 \pmod 5$ *then* $\Omega = \Omega_{12}^{-}(p^2)$. *If* $q = p$ *then there are exactly two* $\text{Aut}\,\bar{\Omega}$-*classes of groups* G, *but if* $q = p^2$ *then there is only one. For no other* q *are there* $\mathscr{S}_1$-*subgroups* $L_2(11)$ *of* $O_{12}^{\pm}(q)$, *and* G *is* $\mathscr{S}_1$-*maximal.*

Proof By Theorem 4.3.3, $q = p$ when $p \equiv \pm 1 \pmod 5$ and $q = p^2$ when $p \equiv \pm 2 \pmod 5$. There are two representations, both stabilised by the outer automorphism of G, so if $q = p$ then there are two $\text{Aut}\,\bar{\Omega}$-classes.

Computer calculations (file 1211d12calc) in $SL_{12}(K)$ with $K := \mathbb{Q}(b_5)$ show that the form β for which G is a group of isometries has determinant d, where $11(2 + b_5)d$ is a square in K. Now $2 + b_5$ has a square root in the cyclotomic field $\mathbb{Q}(z_{20})$ (one such root is $z_{20}{}^7 - z_{20}{}^5 + z_{20}{}^3 - 2z_{20}$), so whether $2 + b_5$ is a square in $\mathbb{F}_q^\times$ depends on the value of q modulo 20. So the sign of β over $\mathbb{F}_q$, which is 1 and -1 when d is respectively a square and a non-square in $\mathbb{F}_q^\times$, depends on whether $11(2 + b_5)$ is a square in $\mathbb{F}_q^\times$, which is a function of $q \pmod{220}$, and yields the stated congruences (which turn out not to depend on $q \pmod 4$). Note that $p = 2$ requires an independent computation over $\mathbb{F}_4$.

From [12, 57], we find that elements of $G.2$ are isometries, and elements of $G.2 \setminus G$ have determinant -1 so, when p is odd, the outer automorphism of G is induced by a conjugate of γ. A computer calculation (file `1211d12calc`) shows that this is also the case when $p = 2$. The two representations are interchanged by ϕ when $p \equiv \pm 2 \pmod 5$ [12, 57], so there is a unique $\operatorname{Aut}\overline{\Omega}$-class.

The group $L_2(11)$ is a subgroup of A_{13}, but character values [12] on elements of order 5 show that this is not a containment of $\mathscr{S}_1$-subgroups for $p > 3$. Conversely, if $p \leqslant 3$, then $L_2(11)$ is a subgroup of $\Omega_{12}^-(p^2)$ but not of $\Omega_{12}^\pm(p)$, unlike A_{13}, so there is no containment here either. For the same reason, there is no containment in A_{14} with $p = 7$. A computer calculation (file `containmentsd12`) shows that the maximal subgroup $L_2(11)$ of M_{12} has inverse image $2{\cdot}L_2(11)$ in $2{\cdot}M_{12}$, The only other maximal subgroup of M_{12} containing $L_2(11)$ is M_{11}, which has inverse image $2 \times M_{11}$ in $2{\cdot}M_{12}$. Hence $L_2(11)$ arises as a subgroup of $2{\cdot}M_{12}$ only via $L_2(11) < M_{11} < 2{\cdot}M_{12}$, but by Theorem 4.3.3 the group M_{11} does not have an irreducible representation of degree 12, so this does not correspond to a containment of $\mathscr{S}_1$-subgroups. $\qquad\square$

4.10 Summary of the $\mathscr{S}_1$-maximal subgroups

In this section we state theorems that summarise the results proved earlier in the chapter, and refer back to those results for proofs. Our aim is for this section to be the most convenient starting point for the reader who requires lists of the $\mathscr{S}_1$-maximals of an almost simple extension of a simple classical group of dimension up to 12.

These theorems provide similar information to the tables in Chapter 8. We now present an explanation of how to interpret the information given in the lists following the theorem statements. We warn the reader that the groups denoted by 'G' in this section are not the same as in the rest of this chapter.

Convention 4.10.1   (i) *The group Ω is a quasisimple classical group, $Z = Z(\Omega)$, $\overline{\Omega} = \Omega/Z$, and G is a group with $\overline{\Omega} \leqslant G \leqslant \operatorname{Aut}\overline{\Omega}$.*

(ii) *The structure of a proper subgroup S of Ω with $Z < S$ is specified, and $\overline{S} := S/Z$. This subgroup represents a single conjugacy class of subgroups of Ω under the action of $\operatorname{Aut}\overline{\Omega}$.*

(iii) *The values of $q = p^e$ for which this list item may represent $\mathscr{S}_1$-maximal subgroups of G are specified. Different values of q may correspond to different cases for Ω.*

(iv) *The stabiliser of the conjugacy class of S in Ω under the action of $\mathrm{Out}\,\bar{\Omega}$ is specified as a subgroup of $\mathrm{Out}\,\bar{\Omega}$. See Section 1.7 for a definition of the generators of $\mathrm{Out}\,\bar{\Omega}$.*

(v) *For the specified values of q, the list item represents $\mathscr{S}_1$-maximal subgroups of G only if $G/\bar{\Omega}$ is a subgroup of the class stabiliser. The default assumption is that this is the case if and only if $G/\bar{\Omega}$ is a subgroup of the class stabiliser. In cases where this is not true (that is, for the so-called novel maximal subgroups), the subgroups of the class stabiliser for which it is true are specified.*

(vi) *If the list item does represent $\mathscr{S}_1$-maximal subgroups of G, then one such subgroup is $\mathrm{N}_G(\bar{S})$. Representatives of the G-classes of subgroups represented by this item are obtained by conjugating $\mathrm{N}_G(\bar{S})$ by coset representatives of $\mathrm{N}_T(G)$ in $\mathrm{N}_{\mathrm{Aut}\,\bar{\Omega}}(G)$, where T is the inverse image in $\mathrm{Aut}\,\bar{\Omega}$ of the class stabiliser.*

4.10.1 Cases L and U

Recall that we treat $\mathrm{SL}_2(q)$ as $\mathrm{Sp}_2(q)$.

Theorem 4.10.2 *Let G and Ω be as in Convention 4.10.1, with $\Omega = \mathrm{SL}_3(q)$ or $\mathrm{SU}_3(q)$. Then representatives of the conjugacy classes of $\mathscr{S}_1$-maximal subgroups of G are described in the list below, using Convention 4.10.1.*

Proof See Theorem 4.3.3 for the list of candidates. See Proposition 4.5.2 for Item 1, Proposition 4.5.3 for Item 2, Proposition 4.5.4 for Item 3, and Theorem 4.7.1 and Proposition 4.8.2 for all three items.   □

1. $S = \mathrm{L}_3(2) \times Z$ with $p \neq 2, 7$. If $p \equiv 1, 2, 4 \pmod 7$ then $S < \mathrm{SL}_3(p)$, and if $p \equiv 3, 5, 6 \pmod 7$ then $S < \mathrm{SU}_3(p)$. The class stabiliser is $\langle \gamma \rangle$ in all cases. The group $\bar{S}$ is not $\mathscr{S}_1$-maximal in $\mathrm{U}_3(5)$ but $\bar{S}.2$ is $\mathscr{S}_1$-maximal in $\mathrm{U}_3(5).2$.

2. $S = 3 \cdot \mathrm{A}_6$ with $p \neq 3, 5$, or $S = 3 \cdot \mathrm{A}_6.2_3$ with $p = 5$. If $p \equiv 1, 4 \pmod{15}$ then $S < \mathrm{SL}_3(p)$, if $p \equiv \pm 2 \pmod 5$ then $S < \mathrm{SL}_3(p^2)$, and if $p = 5$ or $p \equiv 11, 14 \pmod{15}$ then $S < \mathrm{SU}_3(p)$. The class stabiliser is $\langle \gamma \rangle$ when $q = p$ and $\langle \gamma, \phi \rangle$ when $q = p^2$. If $\bar{S} = \mathrm{A}_6$ then the automorphism γ induces the 2_2 automorphism of A_6, and ϕ induces the 2_1 automorphism of A_6 when $p \equiv 2, 8 \pmod{15}$ and the 2_3 automorphism of A_6 otherwise.

3. $S = 3 \cdot \mathrm{A}_7 < \mathrm{SU}_3(5)$, with class stabiliser $\langle \gamma \rangle$.

Theorem 4.10.3 *Let G and Ω be as in Convention 4.10.1, with $\Omega = \mathrm{SL}_4(q)$*

or $\mathrm{SU}_4(q)$. *Then representatives of the conjugacy classes of $\mathscr{S}_1$-maximal subgroups of G are described in the list below, using Convention 4.10.1.*

Proof See Theorem 4.3.3 for the list of candidates. See Propositions 4.5.5 and 4.7.4 for Item 1, Propositions 4.5.5 and 4.7.3 for Item 2, Propositions 4.5.6 and 4.7.5 for Item 3, Propositions 4.5.7 and 4.7.2 for Item 4, and Proposition 4.8.3 for all items. □

1. $S = 2^{\cdot}\mathrm{L}_3(2)Z$ with $p \neq 2, 3, 7$. If $p \equiv 1, 2, 4 \pmod 7$ then $S < \mathrm{SL}_4(p)$, and if $p \equiv 3, 5, 6 \pmod 7$ then $S < \mathrm{SU}_4(p)$. The class stabiliser is $\langle \gamma \rangle$ when $p \equiv \pm 1 \pmod 8$ and $\langle \gamma \delta \rangle$ when $p \equiv \pm 3 \pmod 8$. The group S is not $\mathscr{S}_1$-maximal in Ω, but $\overline{S}.2$ is $\mathscr{S}_1$-maximal in $\overline{\Omega}.2$.

2. $S = \mathrm{A}_7$ with $p = 2$ or $S = 2^{\cdot}\mathrm{A}_7 Z$ with $p \neq 2, 7$. If $p \equiv 1, 2, 4 \pmod 7$ then $S < \mathrm{SL}_4(p)$, and if $p \equiv 3, 5, 6 \pmod 7$ then $S < \mathrm{SU}_4(p)$. The class stabiliser is $\langle \gamma \rangle$.

3. $S = 4_2^{\cdot}\mathrm{L}_3(4) < \mathrm{SU}_4(3)$. The class stabiliser is $\langle \delta^2, \gamma \delta \rangle$. The automorphism δ^2 induces the 2_2 automorphism of $\mathrm{L}_3(4)$, and $\gamma \delta$ induces the 2_1 automorphism.

4. $S = 2^{\cdot}\mathrm{U}_4(2)Z$ with $p \neq 2, 3$. If $p \equiv 1 \pmod 6$ then $S < \mathrm{SL}_4(p)$, and if $p \equiv 5 \pmod 6$ then $S < \mathrm{SU}_4(p)$. The class stabiliser is $\langle \gamma \rangle$.

Theorem 4.10.4 *Let G and Ω be as in Convention 4.10.1, with $\Omega = \mathrm{SL}_5(q)$ or $\mathrm{SU}_5(q)$. Then representatives of the conjugacy classes of $\mathscr{S}_1$-maximal subgroups of G are described in the list below, using Convention 4.10.1.*

Proof See Theorem 4.3.3 for the list of candidates. See Propositions 4.5.12 and 4.8.6, and Theorem 4.7.6. □

1. $S = \mathrm{L}_2(11) \times Z$ with $p \neq 11$. If $p \equiv 1, 3, 4, 5, 9 \pmod{11}$ then $S < \mathrm{SL}_5(p)$, and if $p \equiv 2, 6, 7, 8, 10 \pmod{11}$ then $S < \mathrm{SU}_5(p)$. The class stabiliser is $\langle \gamma \rangle$. The group S is not $\mathscr{S}_1$-maximal in $\mathrm{L}_5(3)$, but $S.2$ is $\mathscr{S}_1$-maximal in $\mathrm{L}_5(3).2$.

2. $S = \mathrm{M}_{11} < \mathrm{SL}_5(3)$, with trivial class stabiliser.

3. $S = \mathrm{U}_4(2) \times Z$ with $p \neq 2, 3$. If $p \equiv 1 \pmod 6$ then $S < \mathrm{SL}_5(p)$, and if $p \equiv 5 \pmod 6$ then $S < \mathrm{SU}_5(p)$. The class stabiliser is $\langle \gamma \rangle$.

Theorem 4.10.5 *Let G and Ω be as in Convention 4.10.1, with $\Omega = \mathrm{SL}_6(q)$ or $\mathrm{SU}_6(q)$. Then representatives of the conjugacy classes of $\mathscr{S}_1$-maximal subgroups of G are described in the list below, using Convention 4.10.1.*

Proof See Theorem 4.3.3 for the list of candidates. See Propositions 4.5.13 and 4.7.8 for Item 1, Propositions 4.5.14 and 4.7.9 for Item 2, Propositions 4.5.15 and 4.7.7 for Item 3, Propositions 4.5.17 and 4.7.11 for Item 4, Propositions 4.5.18 and 4.7.12 for Item 5, Propositions 4.5.19, 4.7.13 and 4.7.14 for Items 6 and 7, Propositions 4.5.20, 4.7.15 and 4.7.16 for Item 8, and Propositions 4.8.7 – 4.8.12, for the analysis of containments between candidates. □

1. $S = 3\dot{} A_6 Z$ with $p \neq 3$ ($p \equiv 5, 7 \pmod 8$ in Case **L**, or $p \equiv 1, 3 \pmod 8$ in Case **U**), or $S = 3\dot{} A_6 . 2_3 Z$ with $p \neq 3$ ($p \equiv 1, 3 \pmod 8$ in Case **L**, or $p \equiv 5, 7 \pmod 8$ in Case **U**). If $p \equiv 1 \pmod 6$ then $S < \mathrm{SL}_6(p)$, and if $p \equiv 5 \pmod 6$ then $S < \mathrm{SU}_6(p)$. The class stabiliser is $\langle \delta^3, \gamma \rangle$ when $\bar{S} = A_6$ and either $\langle \gamma \rangle$ ($p \equiv \pm 1 \pmod 8$) or $\langle \gamma \delta \rangle$ ($p \equiv \pm 3 \pmod 8$) when $\bar{S} = A_6 . 2_3$. If the class stabiliser is $\langle \delta^3, \gamma \rangle$ then δ^3 induces the 2_3 automorphism of A_6, and γ induces the 2_2 automorphism of A_6 when $p \equiv 7, 17 \pmod{24}$, and the 2_1 automorphism otherwise. The groups A_6, $A_6 . 2_1$ and $A_6 . 2_3$ are not $\mathscr{S}_1$-maximal, but all other extensions of A_6 are $\mathscr{S}_1$-maximal in G.

2. $S = 6\dot{} A_6$ with $p \neq 2, 3$. If $p \equiv 1, 7 \pmod{24}$ then $S < \mathrm{SL}_6(p)$, and if $p \equiv 17, 23 \pmod{24}$ then $S < \mathrm{SU}_6(p)$. In both of these cases, the class stabiliser is $\langle \gamma \rangle$ for $p \equiv \pm 1 \pmod{16}$ and $\langle \gamma \delta \rangle$ for $p \equiv \pm 7 \pmod{16}$. If $p \equiv \pm 5, \pm 11 \pmod{24}$ then $S < \mathrm{SL}_6(p^2)$, with class stabiliser $\langle \gamma \delta^3, \phi \rangle$ when $p \equiv 5, 11 \pmod{24}$ and $\langle \gamma \delta^3, \phi \gamma \rangle$ when $p \equiv 13, 19 \pmod{24}$. If the class stabiliser has order 2, then it induces the 2_2 automorphism of A_6. If $p \equiv 5, 11 \pmod{24}$ then $\gamma \delta^3$ and ϕ induce the 2_2 and the 2_1 automorphisms of A_6, respectively. If $p \equiv 13, 19 \pmod{24}$ then $\gamma \delta^3$ and $\gamma \phi$ induce the 2_2 and the 2_1 automorphisms of A_6, respectively. The groups A_6 and $A_6 . 2_1$ are not $\mathscr{S}_1$-maximal, but all other extensions of A_6 are $\mathscr{S}_1$-maximal in G.

3. $S = 2\dot{} L_2(11) Z$ with $p \neq 2, 3, 11$. If $p \equiv 1, 3, 4, 5, 9 \pmod{11}$ then $S < \mathrm{SL}_6(p)$, and if $p \equiv 2, 6, 7, 8, 10 \pmod{11}$ then $S < \mathrm{SU}_6(p)$. The class stabiliser is $\langle \gamma \rangle$ when $p \equiv \pm 1 \pmod 8$, and $\langle \gamma \delta \rangle$ when $p \equiv \pm 3 \pmod 8$.

4. $S = 6\dot{} A_7$ with $p \neq 2, 3$. If $p \equiv 1, 7 \pmod{24}$ then $S < \mathrm{SL}_6(p)$, and if $p \equiv 17, 23 \pmod{24}$ then $S < \mathrm{SU}_6(p)$. In both of these cases, the class stabiliser is trivial. If $p \equiv \pm 5, \pm 11 \pmod{24}$ then $S < \mathrm{SL}_6(p^2)$, with class stabiliser $\langle \phi \rangle$ when $p \equiv 5, 11 \pmod{24}$ and $\langle \phi \gamma \rangle$ when $p \equiv 13, 19 \pmod{24}$.

5. $S = 6\dot{} L_3(4)$ with $p \neq 3$ ($p \equiv 5, 7 \pmod 8$ in Case **L**, or $p \equiv 1, 3 \pmod 8$ in Case **U**), or $S = 6\dot{} L_3(4) . 2_1$ with $p \neq 3$ ($p \equiv 1, 3 \pmod 8$ in Case **L**, or $p \equiv 5, 7 \pmod 8$ in Case **U**). If $p \equiv 1 \pmod 6$ then $S < \mathrm{SL}_6(p)$, and if $p \equiv 5 \pmod 6$ then $S < \mathrm{SU}_6(p)$. The class stabiliser is $\langle \delta^3, \gamma \rangle$ when $\bar{S} = L_3(4)$, and $\langle \gamma \rangle$ when $\bar{S} = L_3(4) . 2_1$. If the class stabiliser is $\langle \delta^3, \gamma \rangle$ then δ^3 and γ induce the 2_1 and 2_2 automorphisms of $L_3(4)$, respectively. If the class stabiliser is $\langle \gamma \rangle$ then γ extends $\bar{S}$ to $L_3(4) . 2^2$. The groups $L_3(4)$ and $L_3(4) . 2_2$ are not $\mathscr{S}_1$-maximal, but all other extensions of $L_3(4)$ are $\mathscr{S}_1$-maximal in G.

6. $S = 2\dot{} M_{12} < \mathrm{SL}_6(3)$, with class stabiliser $\langle \gamma \delta \rangle$.

7. $S = 3\dot{} M_{22} < \mathrm{SU}_6(2)$, with class stabiliser $\langle \gamma \rangle$.

8. $S = 3_1\dot{} U_4(3) . 2_2$ with $p = 2$, $S = 6_1\dot{} U_4(3)$ with $p \neq 3$ ($p \equiv 3 \pmod 4$ in Case **L**, or $p \equiv 1 \pmod 4$ in Case **U**), or $S = 6_1\dot{} U_4(3) . 2_2$ with $p \neq 3$ ($p \equiv 1 \pmod 4$ in Case **L**, or $p \equiv 3 \pmod 4$ in Case **U**). If $p \equiv 1 \pmod 6$

then $S < \mathrm{SL}_6(p)$, and if $p = 2$ or $p \equiv 5 \pmod 6$ then $S < \mathrm{SU}_6(p)$. The class stabiliser is $\langle \delta^3, \gamma \rangle$ when $\bar{S} = \mathrm{U}_4(3)$, and $\langle \gamma \rangle$ when $\bar{S} = \mathrm{U}_4(3).2_2$. If the class stabiliser is $\langle \delta^3, \gamma \rangle$ then δ^3 and γ induce the 2_2 and 2_1 automorphisms of $\mathrm{U}_4(3)$, respectively. If $\bar{S} = \mathrm{U}_4(3).2_2$, then γ extends $\bar{S}$ to $\mathrm{U}_4(3).2_{122}^2$.

Theorem 4.10.6 *Let G and Ω be as in Convention 4.10.1, with $\Omega = \mathrm{SL}_7(q)$ or $\mathrm{SU}_7(q)$. Then representatives of the conjugacy classes of $\mathscr{S}_1$-maximal subgroups of G are described in the list below, using Convention 4.10.1.*

Proof See Theorem 4.3.3 and Proposition 4.9.1. ∎

1. $S = \mathrm{U}_3(3) \times Z$ with $p \neq 2, 3$. If $p \equiv 1 \pmod 4$ then $S < \mathrm{SL}_7(p)$, and if $p \equiv 3 \pmod 4$ then $S < \mathrm{SU}_7(p)$. The class stabiliser is $\langle \gamma \rangle$.

Theorem 4.10.7 *Let G and Ω be as in Convention 4.10.1, with $\Omega = \mathrm{SL}_8(q)$ or $\mathrm{SU}_8(q)$. Then representatives of the conjugacy classes of $\mathscr{S}_1$-maximal subgroups of G are described in the list below, using Convention 4.10.1.*

Proof See Theorem 4.3.3 and Proposition 4.9.2. ∎

1. $S = 4_1{}^{.}\mathrm{L}_3(4).2_3 Z$ for $p \equiv \pm 1, \pm 7, \pm 17, \pm 23, \pm 31, \pm 33 \pmod{80}$, or $S = 4_1{}^{.}\mathrm{L}_3(4)Z$ for all other values of $p \neq 2$. If $p \equiv 1, 9 \pmod{20}$ or $p = 5$ then $S < \mathrm{SL}_8(p)$, if $p \equiv 11, 19 \pmod{20}$ then $S < \mathrm{SU}_8(p)$, and if $p \equiv \pm 2 \pmod 5$ then $S < \mathrm{SL}_8(p^2)$.

The class stabiliser O_S is as follows, and if a non-trivial power of δ is in O_S, then this induces the 2_3 automorphism:

   (i) If $p = 5$ then $O_S = \langle \delta^2, \gamma \rangle$, with γ inducing the 2_1 automorphism.
   (ii) If $\bar{S} = \mathrm{L}_3(4) < \mathrm{L}_8(p)$, then $O_S = \langle \delta^{(q-1,8)/2} \rangle$.
   (iii) If $\bar{S} = \mathrm{L}_3(4) < \mathrm{U}_8(p)$, then $O_S = \langle \delta^{(q+1,8)/2} \rangle$.
   (iv) If $\bar{S} = \mathrm{L}_3(4) < \mathrm{L}_8(p^2)$, then $O_S = \langle \delta^4, \phi \rangle$ when $p \equiv 3 \pmod 4$, and $O_S = \langle \delta^4, \phi\gamma \rangle$ when $p \equiv 1 \pmod 4$. Here ϕ or $\phi\gamma$ induces the 2_1 automorphism.
   (v) If $\bar{S} = \mathrm{L}_3(4).2_3 < \mathrm{L}_8(p)$, then $O_S = 1$.
   (vi) If $\bar{S} = \mathrm{L}_3(4).2_3 < \mathrm{U}_8(p)$, then $O_S = 1$.
   (vii) If $\bar{S} = \mathrm{L}_3(4).2_3 < \mathrm{L}_8(p^2)$, then $O_S = \langle \phi \rangle$ for $p \equiv 7 \pmod 8$, and $O_S = \langle \phi\gamma \rangle$ for $p \equiv 1 \pmod 8$.

Theorem 4.10.8 *Let G and Ω be as in Convention 4.10.1, with $\Omega = \mathrm{SL}_9(q)$ or $\mathrm{SU}_9(q)$. Then representatives of the conjugacy classes of $\mathscr{S}_1$-maximal subgroups of G are described in the list below, using Convention 4.10.1.*

Proof See Theorem 4.3.3 for the list of candidates. See Propositions 4.9.3 to 4.9.6 for the other claims. ∎

1. $S = 3{}^{.}\mathrm{A}_6.2_3 Z$ with $p \neq 3, 5$. If $p \equiv 1 \pmod 3$ then $S < \mathrm{SL}_9(p)$, and if $p \equiv 2 \pmod 3$ then $S < \mathrm{SU}_9(p)$. The class stabiliser is $\langle \gamma \rangle$.

2. $S = 3 \cdot A_7 < SL_9(7)$, with class stabiliser $\langle \gamma \rangle$.

3. $S = L_2(19) \times Z$ with $p \neq 19$. If $p \equiv 1, 4, 5, 6, 7, 9, 11, 16, 17 \pmod{19}$ then $S < SL_9(p)$, and if $p \equiv 2, 3, 8, 10, 12, 13, 14, 15, 18 \pmod{19}$ then $S < SU_9(p)$. The class stabiliser is $\langle \gamma \rangle$. If $p = 2$ then S is not $\mathscr{S}_1$-maximal, but $N_{U_9(2)\langle \gamma \rangle}(\bar{S})$ is $\mathscr{S}_1$-maximal.

4. $S = 3 \cdot J_3 < SU_9(2)$, with class stabiliser $\langle \gamma \rangle$.

Theorem 4.10.9 *Let G and Ω be as in Convention 4.10.1, with $\Omega = SL_{10}(q)$ or $SU_{10}(q)$. Then representatives of the conjugacy classes of $\mathscr{S}_1$-maximal subgroups of G are described in the list below, using Convention 4.10.1.*

Proof See Theorem 4.3.3 for the list of candidates. See Propositions 4.9.7 to 4.9.13 for the other claims. □

1. $S = A_7 \times Z$ with $p \neq 2, 7$. If $p \equiv 1, 2, 4 \pmod{7}$ then $S < SL_{10}(p)$, and if $p \equiv 3, 5, 6 \pmod{7}$ then $S < SU_{10}(p)$. The class stabiliser is $\langle \gamma \rangle$. The group S is not $\mathscr{S}_1$-maximal, but $\bar{S}.2$ is $\mathscr{S}_1$-maximal in $\bar{\Omega}\langle \gamma \rangle$.

2. $S = 2 \cdot L_2(19)Z$ with $p \neq 2, 19$. If $p \equiv 1, 4, 5, 6, 7, 9, 11, 16, 17 \pmod{19}$ then $S < SL_{10}(p)$, and if $p \equiv 2, 3, 8, 10, 12, 13, 14, 15, 18 \pmod{19}$ then $S < SU_{10}(p)$. The class stabiliser is $\langle \gamma \rangle$ when $p \equiv \pm 1 \pmod{8}$, and $\langle \gamma \delta \rangle$ when $p \equiv \pm 3 \pmod{8}$.

3. $S = 2 \cdot L_3(4).2_2 Z$ with $p \neq 7$ ($p \equiv 1 \pmod{4}$ in Case **L**, or $p \equiv 3 \pmod{4}$ in Case **U**), or $S = 2 \cdot L_3(4)Z$ with $p \neq 7$ ($p \equiv 3 \pmod{4}$ in Case **L**, or $p \equiv 1 \pmod{4}$ in Case **U**). If $p \equiv 1, 2, 4 \pmod{7}$ then $S < SL_{10}(p)$, and if $p \equiv 3, 5, 6 \pmod{7}$ then $S < SU_{10}(p)$. If $\bar{S} = L_3(4).2_2$, then the class stabiliser is $\langle \gamma \rangle$ when $p \equiv \pm 1 \pmod{8}$ and $\langle \gamma \delta \rangle$ when $p \equiv \pm 3 \pmod{8}$. If $\bar{S} = L_3(4)$, then the class stabiliser is $\langle \delta^5, \gamma \rangle$, with δ^5 inducing the 2_2 automorphism and γ inducing the 2_1 automorphism when $p \equiv \pm 3 \pmod{8}$ and the $2_1 2_2$ automorphism otherwise. The groups $L_3(4)$ and $L_3(4).2_2$ are not $\mathscr{S}_1$-maximal, but all other extensions of $\bar{S}$ are $\mathscr{S}_1$-maximal in the corresponding extensions of $\bar{\Omega}$.

4. $S = U_4(2)Z$ with $p \neq 2, 3$. If $p \equiv 1 \pmod{3}$ then $S < SL_{10}(p)$, and if $p \equiv 2 \pmod{3}$ then $S < SU_{10}(p)$. The class stabiliser is $\langle \gamma \rangle$.

5. $S = 2 \cdot M_{12}.2Z$ with $p \equiv \pm 1 \pmod{8}$, or $S = 2 \cdot M_{12}Z$ with $p \equiv \pm 3 \pmod{8}$. If $p \equiv 1, 3 \pmod{8}$ then $S < SL_{10}(p)$, and if $p \equiv 5, 7 \pmod{8}$ then $S < SU_{10}(p)$. The class stabiliser is trivial when $\bar{S} = M_{12}.2$, and is $\langle \delta^5 \rangle$ when $\bar{S} = M_{12}$.

6. $S = M_{22}.2$ with $p = 2$, $S = 2 \cdot M_{22}Z$ with $p \neq 7$ ($p \equiv 3 \pmod{4}$ in Case **L**, or $p \equiv 1 \pmod{4}$ in Case **U**), or $S = 2 \cdot M_{22}.2Z$ with $p \neq 7$ ($p \equiv 1 \pmod{4}$ in Case **L**, or $p \equiv 3 \pmod{4}$ in Case **U**). If $p \equiv 1, 2, 4 \pmod{7}$ then $S < SL_{10}(p)$, and if $p \equiv 3, 5, 6 \pmod{7}$ then $S < SU_{10}(p)$. The class stabiliser is trivial when $\bar{S} = M_{22}.2$, and is $\langle \delta^5 \rangle$ when $\bar{S} = M_{22}$.

Theorem 4.10.10 *Let G and Ω be as in Convention 4.10.1, with $\Omega = \mathrm{SL}_{11}(q)$ or $\mathrm{SU}_{11}(q)$. Then representatives of the conjugacy classes of $\mathscr{S}_1$-maximal subgroups of G are described in the list below, using Convention 4.10.1.*

Proof See Theorem 4.3.3 for the list of candidates. See Propositions 4.9.14 to 4.9.17 for the other claims. □

1. $S = \mathrm{L}_2(23) \times Z$ with $p \neq 23$. If $p \equiv 1, 2, 3, 4, 6, 8, 9, 12, 13, 16, 18 \pmod{23}$ then $S < \mathrm{SL}_{11}(p)$, and if $p \equiv 5, 7, 10, 11, 14, 15, 17, 19, 20, 21, 22 \pmod{23}$ then $S < \mathrm{SU}_{11}(p)$. The class stabiliser is $\langle \gamma \rangle$. If $p = 2$ then S is not $\mathscr{S}_1$-maximal, but $\bar{S}.2$ is $\mathscr{S}_1$-maximal in $\mathrm{L}_5(2)\langle \gamma \rangle$.

2. $S = \mathrm{U}_5(2) \times Z$ with $p \neq 2, 3$. If $p \equiv 1 \pmod 3$ then $S < \mathrm{SL}_{11}(p)$, and if $p \equiv 2 \pmod 3$ then $S < \mathrm{SU}_{11}(p)$. The class stabiliser is $\langle \gamma \rangle$.

3. $S = \mathrm{M}_{24} < \mathrm{SL}_{11}(2)$, with trivial class stabiliser.

Theorem 4.10.11 *Let G and Ω be as in Convention 4.10.1, with $\Omega = \mathrm{SL}_{12}(q)$ or $\mathrm{SU}_{12}(q)$. Then representatives of the conjugacy classes of $\mathscr{S}_1$-maximal subgroups of G are described in the list below, using Convention 4.10.1.*

Proof See Theorem 4.3.3 for the list of candidates. See Propositions 4.9.18 to 4.9.22 for the other claims. □

1. $S = 6 \dot{} \mathrm{A}_6 Z$ with $p \neq 2, 3, 5$. If $p \equiv 1, 4 \pmod{15}$ then $S < \mathrm{SL}_{12}(p)$, if $p \equiv 11, 14 \pmod{15}$ then $S < \mathrm{SU}_{12}(p)$, and if $p \equiv \pm 2 \pmod 5$ then $G < \mathrm{SL}_{12}(p^2)$. If $q = p$ then the class stabiliser is $\langle \gamma \rangle$ when $p \equiv \pm 1 \pmod 8$, and $\langle \gamma \delta \rangle$ when $p \equiv \pm 3 \pmod 8$, and the class stabiliser induces the 2_2 automorphism of A_6. If $q = p^2$ then the class stabiliser is $\langle \gamma, \phi \rangle$ when $p \equiv \pm 5 \pmod{12}$ and $\langle \gamma, \phi \delta^6 \rangle$ when $p \equiv \pm 1 \pmod{12}$. Here γ induces the 2_2 automorphism of A_6, and ϕ or $\phi \delta^6$ induces the 2_1 automorphism when $p \equiv 5, 11 \pmod{12}$, and the 2_3 automorphism otherwise. If $p = 7$ then $\bar{S}$ and $\bar{S}.2_3$ are not $\mathscr{S}_1$-maximal, but in all other cases, $\overline{S}$ and its extensions are $\mathscr{S}_1$-maximal in the corresponding extensions of $\overline{\Omega}$.

2. $S = 6 \dot{} \mathrm{A}_7 < \mathrm{SU}_{12}(5)$, with class stabiliser $\langle \gamma \rangle$.

3. $S = 2 \dot{} \mathrm{L}_2(23) Z$ with $p \neq 2, 23$. If $p \equiv 1, 2, 3, 4, 6, 8, 9, 12, 13, 16, 18 \pmod{23}$ then $S < \mathrm{SL}_{12}(p)$, and if $p \equiv 5, 7, 10, 11, 14, 15, 17, 19, 20, 21, 22 \pmod{23}$ then $S < \mathrm{SU}_{12}(p)$. The class stabiliser is $\langle \gamma \rangle$ when $p \equiv \pm 1 \pmod 8$ and $\langle \gamma \delta \rangle$ when $p \equiv \pm 3 \pmod 8$.

4. $S = 12_2 \dot{} \mathrm{L}_3(4) < \mathrm{SL}_{12}(49)$, with class stabiliser $\langle \gamma, \phi \rangle$, where ϕ induces the 2_1 automorphism of $\mathrm{L}_3(4)$ and γ the 2_3.

5. $S = 3 \dot{} \mathrm{Suz}$ with $p = 2$ or $S = 6 \dot{} \mathrm{Suz} Z$ with $p \neq 2, 3$. If $p \equiv 1 \pmod 3$ then $S < \mathrm{SL}_{12}(p)$, and if $p \equiv 2 \pmod 3$ then $S < \mathrm{SU}_{12}(p)$. The class stabiliser is $\langle \gamma \rangle$.

4.10.2 Case S

In this subsection, we list the $\mathcal{S}_1$-maximal subgroups of the symplectic groups in dimensions 2 to 12.

Theorem 4.10.12 *Let G and Ω be as in Convention 4.10.1, with $\Omega = \mathrm{Sp}_2(q)$. Then representatives of the conjugacy classes of $\mathcal{S}_1$-maximal subgroups of G are described in the list below, using Convention 4.10.1.*

Proof See Theorem 4.3.3 for the list of candidates. See Proposition 4.5.1 and Theorem 4.7.17 for the other claims. □

1. $S = 2^{\cdot}A_5$ with $p \neq 2,5$. If $p \equiv \pm 1 \pmod 5$ then $S < \mathrm{Sp}_2(p)$, and if $p \equiv \pm 2 \pmod 5$ then $S < \mathrm{Sp}_2(p^2)$. The class stabiliser is trivial in the first case and $\langle \phi \rangle$ in the second.

Theorem 4.10.13 *Let G and Ω be as in Convention 4.10.1, with $\Omega = \mathrm{Sp}_4(q)$. Then representatives of the conjugacy classes of $\mathcal{S}_1$-maximal subgroups of G are described in the list below, using Convention 4.10.1.*

Proof See Theorem 4.3.3 for the list of candidates. See Proposition 4.5.10 for Item 1, Proposition 4.5.11 for Item 2, and Proposition 4.8.5 for containment information. □

1. $S = 2^{\cdot}A_6.2_1$ with $p \equiv \pm 1 \pmod{12}$, or $S = 2^{\cdot}A_6$ with $p \equiv \pm 5 \pmod{12}$ and $p \neq 7$. The group $S < \mathrm{Sp}_4(p)$ for all such p. The class stabiliser is trivial for $\bar{S} = A_6.2_1$, and $\langle \delta \rangle$ for $\bar{S} = A_6$, and δ induces the 2_1 automorphism of A_6.
2. $S = 2^{\cdot}A_7 < \mathrm{Sp}_4(7)$, with class stabiliser $\langle \delta \rangle$.

Theorem 4.10.14 *Let G and Ω be as in Convention 4.10.1, with $\Omega = \mathrm{Sp}_6(q)$. Then representatives of the conjugacy classes of $\mathcal{S}_1$-maximal subgroups of G are described in the list below, using Convention 4.10.1.*

Proof See Theorem 4.3.3 for the list of candidates. See Proposition 4.5.21 for Item 1, Proposition 4.5.22 for Items 2 and 3, Proposition 4.5.23 for Item 4, Proposition 4.5.24 for Item 5, Proposition 4.5.25 for Item 6, Proposition 4.5.26 for Item 7, Theorem 4.7.17 for the class stabiliser when q is not prime, and Propositions 4.8.13 to 4.8.15 for an analysis of the containments between the candidates. □

1. $S = 2^{\cdot}A_5.2$ with $p \equiv \pm 1 \pmod 8$, or $S = 2^{\cdot}A_5$ with $p \equiv \pm 3 \pmod 8$ and $p \neq 5$. The group $S < \mathrm{Sp}_6(p)$ for all such p. The class stabiliser is trivial for $\bar{S} = A_5.2$, and $\langle \delta \rangle$ for $\bar{S} = A_5$.

2. $S = 2\dot{}L_3(2).2$ with $p \equiv \pm 1 \pmod{16}$, or $S = 2\dot{}L_3(2)$ with $p \equiv \pm 3, \pm 5, \pm 7$ (mod 16) and $p \neq 7$. If $p \equiv \pm 1 \pmod 8$ then $S < \mathrm{Sp}_6(p)$, and if $p \equiv \pm 3 \pmod 8$ then $S < \mathrm{Sp}_6(p^2)$. The class stabiliser is trivial for $\bar{S} = L_3(2).2$ and $\langle \delta \rangle$ for $\bar{S} = L_3(2)$. If $p = 3$ then S is not $\mathscr{S}_1$-maximal in $\mathrm{Sp}_6(9)$, but $\bar{S}.2$ is $\mathscr{S}_1$-maximal in $\mathrm{S}_6(9)\langle \delta \rangle$.

3. $S = 2\dot{}L_3(2).2$ with $p \equiv \pm 1 \pmod{16}$ or $S = 2\dot{}L_3(2)$ with $p \equiv \pm 7 \pmod{16}$ and $p \neq 7$. The group $S < \mathrm{Sp}_6(p)$ for all such p. The class stabiliser is trivial for $\bar{S} = L_3(2).2$ and $\langle \delta \rangle$ for $\bar{S} = L_3(2)$.

4. $S = 2\dot{}L_2(13)$ with $p \neq 2, 13$, or $S = L_2(13)$ with $p = 2$. If $p \equiv \pm 1, \pm 3, \pm 4$ (mod 13) then $S < \mathrm{Sp}_6(p)$, and if $p \equiv \pm 2, \pm 5, \pm 6 \pmod{13}$ then $S < \mathrm{Sp}_6(p^2)$. The class stabiliser is trivial in the first case and $\langle \phi \rangle$ in the second.

5. $S = 2\dot{}A_7 < \mathrm{Sp}_6(9)$, with class stabiliser $\langle \phi \rangle$.

6. $S = (2 \times U_3(3)).2$ with $p \equiv \pm 1 \pmod{12}$, or $S = 2 \times U_3(3)$ with $p \equiv \pm 5 \pmod{12}$ and $p \neq 5$, or $S = U_3(3) : 2$ for $p = 2$. The group $S < \mathrm{Sp}_6(p)$ for all such p. The class stabiliser is trivial when $p \equiv \pm 1 \pmod{12}$ or $p = 2$, and $\langle \delta \rangle$ when $p \equiv \pm 5 \pmod{12}$. If $p \equiv \pm 19, \pm 29 \pmod{60}$ then S is not $\mathscr{S}_1$-maximal, but $\bar{S}.2$ is $\mathscr{S}_1$-maximal in $\mathrm{S}_6(p).2$.

7. $S = J_2$ with $p = 2$, or $S = 2\dot{}J_2$ with $p \neq 2$. If $p \equiv \pm 1 \pmod 5$ or $p = 5$ then $S < \mathrm{Sp}_6(p)$, and if $p \equiv \pm 2 \pmod 5$ then $S < \mathrm{Sp}_6(p^2)$. The class stabiliser is trivial when $p \equiv \pm 1 \pmod 5$, equal to $\langle \delta \rangle$ when $p = 5$, and $\langle \phi \rangle$ when $p \equiv \pm 2 \pmod 5$.

Theorem 4.10.15 *Let G and Ω be as in Convention 4.10.1, with $\Omega = \mathrm{Sp}_8(q)$. Then representatives of the conjugacy classes of $\mathscr{S}_1$-maximal subgroups of G are described in the list below, using Convention 4.10.1.*

Proof See Theorem 4.3.3 for the list of candidates. See Propositions 4.9.26 to 4.9.23 for the other claims. □

1. $S = 2\dot{}L_3(2).2$ for $p \equiv \pm 1 \pmod{12}$, or $S = 2\dot{}L_3(2)$ for $p \equiv \pm 5 \pmod{12}$ and $p \neq 7$. The group $S < \mathrm{Sp}_8(p)$ for all such p. The class stabiliser is trivial when $\bar{S} = L_3(2).2$, and $\langle \delta \rangle$ when $\bar{S} = L_3(2)$.

2. $S = 2\dot{}A_6.2_2$ with $p \equiv \pm 1 \pmod{20}$, or $S = 2\dot{}A_6$ for all other $p \neq 2, 3, 5$. If $p \equiv \pm 1 \pmod 5$ then $S < \mathrm{Sp}_8(p)$, and if $p \equiv \pm 2 \pmod 5$ then $S < \mathrm{Sp}_8(p^2)$. The class stabiliser is trivial when $\bar{S} = A_6.2_2$, equal to $\langle \delta \rangle$ when $\bar{S} = A_6 < \mathrm{S}_8(p)$, and $\langle \delta, \phi \rangle$ when $q = p^2$. If δ stabilises S then δ induces the 2_2 automorphism of A_6, and similarly ϕ induces the 2_1 automorphism.

3. $S = L_2(17)$ with $p = 2$, or $S = 2\dot{}L_2(17)$ with $p \neq 2, 17$. If $p \equiv \pm 1, \pm 2, \pm 4, \pm 8 \pmod{17}$ then $S < \mathrm{Sp}_8(p)$, and if $p \equiv \pm 3, \pm 5, \pm 6, \pm 7 \pmod{17}$ then $S < \mathrm{Sp}_8(p^2)$. The class stabiliser is trivial when $q = p$, and $\langle \phi \rangle$ when $q = p^2$.

4. $S = A_{10}.2 < Sp_8(2)$, with trivial class stabiliser.

Theorem 4.10.16   *Let G and Ω be as in Convention 4.10.1, with $\Omega = Sp_{10}(q)$. Then representatives of the conjugacy classes of $\mathscr{S}_1$-maximal subgroups of G are described in the list below, using Convention 4.10.1.*

Proof   See Theorem 4.3.3 for the list of candidates. See Propositions 4.9.30 to 4.9.27 for the other claims.   □

1. $S = 2^{\cdot}A_6.2_2$ with $p \equiv \pm1 \pmod{16}$, or $S = 2^{\cdot}A_6$ for all other $p \neq 2, 3$. If $p \equiv \pm1 \pmod 8$ then $S < Sp_{10}(p)$, and if $p \equiv \pm3 \pmod 8$ then $S < Sp_{10}(p^2)$. The class stabiliser is trivial when $\bar{S} = A_6.2_2$, equal to $\langle \delta \rangle$ when $\bar{S} = A_6 < S_{10}(p)$, and $\langle \delta, \phi \rangle$ when $q = p^2$. If δ stabilises S then δ induces the 2_2 automorphism of A_6, and similarly ϕ induces the 2_1 automorphism.

2. $S = 2^{\cdot}L_2(11).2$ with $p \equiv \pm1 \pmod 8$ or $S = 2^{\cdot}L_2(11)$ with $p \equiv \pm3 \pmod 8$ and $p \neq 11$. The group $S < Sp_{10}(p)$ for all $p \neq 2, 11$. The class stabiliser is trivial when $\bar{S} = L_2(11).2$, and $\langle \delta \rangle$ when $\bar{S} = L_2(11)$.

3. $S = 2^{\cdot}L_2(11).2$ with $p \equiv \pm1 \pmod{24}$ or $p \equiv \pm5 \pmod{12}$, or $S = 2^{\cdot}L_2(11)$ with $p \equiv \pm11 \pmod{24}$ and $p \neq 11$. If $p \equiv \pm1 \pmod{12}$ then $S < Sp_{10}(p)$, and if $p \equiv \pm5 \pmod{12}$ then $S < Sp_{10}(p^2)$. The class stabiliser is trivial when $\bar{S} = L_2(11).2$, and $\langle \delta \rangle$ when $\bar{S} = L_2(11)$.

4. $S = 2^{\cdot}L_2(11).2$ with $p \equiv \pm1 \pmod{24}$, or $S = 2^{\cdot}L_2(11)$ with $p \neq 11$ and $p \equiv \pm11 \pmod{24}$. The group $S < Sp_{10}(p)$ for all such p. The class stabiliser is trivial when $\bar{S} = L_2(11).2$, and $\langle \delta \rangle$ when $\bar{S} = L_2(11)$.

5. $S = (2 \times U_5(2)).2$ with $p \equiv \pm1 \pmod 8$, or $S = 2 \times U_5(2)$ with $p \equiv \pm3 \pmod 8$. Here $S < Sp_{10}(p)$ for all $p \neq 2$. The class stabiliser is trivial when $\bar{S} = U_5(2).2$, and $\langle \delta \rangle$ when $\bar{S} = U_5(2)$.

Theorem 4.10.17   *Let G and Ω be as in Convention 4.10.1, with $\Omega = Sp_{12}(q)$. Then representatives of the conjugacy classes of $\mathscr{S}_1$-maximal subgroups of G are described in the list below, using Convention 4.10.1.*

Proof   See Theorem 4.3.3 for the list of candidates. See Propositions 4.9.31 to 4.9.38 for the other claims.   □

1. $S = 2^{\cdot}L_2(11).2$ with $p \equiv \pm1 \pmod{20}$, or $S = 2^{\cdot}L_2(11)$ for all other $p \neq 2, 5, 11$. If $p \equiv \pm1 \pmod 5$ then $S < Sp_{12}(p)$, and if $p \equiv \pm2 \pmod 5$ then $S < Sp_{12}(p^2)$. The class stabiliser is trivial when $\bar{S} = L_2(11).2$, and $\langle \delta \rangle$ when $\bar{S} = L_2(11)$.

2. $S = 2^{\cdot}L_2(11).2$ with $p \equiv \pm1 \pmod{20}$, or $S = 2^{\cdot}L_2(11)$ with $p \neq 11$ and $p \equiv \pm9 \pmod{20}$. The group $S < Sp_{12}(p)$ for all such p. The class stabiliser is trivial when $\bar{S} = L_2(11).2$, and $\langle \delta \rangle$ when $\bar{S} = L_2(11)$.

3. $S = 2\dot{}L_2(13).2$ with $p \equiv \pm 1 \pmod{28}$, or $S = 2\dot{}L_2(13)$ for all other $p \neq 2, 7, 13$. If $p \equiv \pm 1 \pmod 7$ then $S < \mathrm{Sp}_{12}(p)$, and if $p \equiv \pm 2, \pm 3 \pmod 7$ then $S < \mathrm{Sp}_{12}(p^3)$. The class stabiliser is trivial when $\bar{S} = L_2(13).2$, and $\langle \delta \rangle$ when $\bar{S} = L_2(13)$.

4. $S = 2\dot{}L_2(13).2$ with $p \equiv \pm 1 \pmod{28}$, or $S = 2\dot{}L_2(13)$ with $p \neq 13$ and $p \equiv \pm 13 \pmod{28}$. The group $S < \mathrm{Sp}_{12}(p)$ for all such p. The class stabiliser is trivial when $\bar{S} = L_2(13).2$, and $\langle \delta \rangle$ when $\bar{S} = L_2(13)$.

5. The same description as Item 4.

6. $S = L_2(25).2_2 < \mathrm{Sp}_{12}(2)$, or $S = 2\dot{}L_2(25)$ with $p \equiv \pm 2 \pmod 5$ and $p \neq 2, 3$. The group $S < \mathrm{Sp}_{12}(p)$ for all such p. The class stabiliser is trivial when $p = 2$, and $\langle \delta \rangle$ otherwise, in which case δ induces the 2_2 automorphism of $L_2(25)$.

7. $S = S_4(5)$ with $p = 2$, or $S = 2\dot{}S_4(5)$ with $p \neq 2, 5$. If $p \equiv \pm 1 \pmod 5$ then $S < \mathrm{Sp}_{12}(p)$, and if $p \equiv \pm 2 \pmod 5$ then $S < \mathrm{Sp}_{12}(p^2)$. The class stabiliser is trivial when $q = p$, and $\langle \phi \rangle$ when $q = p^2$.

8. $S = 2\dot{}G_2(4).2$ with $p \equiv \pm 1 \pmod 8$, or $S = 2\dot{}G_2(4)$ with $p \equiv \pm 3 \pmod 8$ and $p \neq 3$. The group $S < \mathrm{Sp}_{12}(p)$ for all $p \neq 2, 3$. The class stabiliser is trivial when $S = 2\dot{}G_2(4).2$, and $\langle \delta \rangle$ when $S = 2\dot{}G_2(4)$.

9. $S = A_{14}.2 < \mathrm{Sp}_{12}(2)$, with trivial class stabiliser.

10. $S = 2\dot{}\mathrm{Suz} < \mathrm{Sp}_{12}(3)$, with class stabiliser $\langle \delta \rangle$.

4.10.3 Cases O^ε

In this section we list the $\mathscr{S}_1$-maximal subgroups of the orthogonal groups in dimensions 7 to 12.

Theorem 4.10.18 *Let G and Ω be as in Convention 4.10.1, with $\Omega = \Omega_7(q)$. Then representatives of the conjugacy classes of $\mathscr{S}_1$-maximal subgroups of G are described in the list below, using Convention 4.10.1.*

Proof See Theorem 4.3.3 for the list of candidates. See Propositions 4.9.41 to 4.9.46 for the other claims. $\qquad \square$

1. $S = L_2(8)_2$ with $p \neq 2, 3$. If $p \equiv \pm 1 \pmod 9$ then $S < \Omega_7(p)$, and if $p \equiv \pm 2, \pm 4 \pmod 9$ then $S < \Omega_7(p^3)$. The class stabiliser is trivial when $q = p$, and $\langle \phi \rangle$ when $q = p^3$.

2. $S = L_2(13)$ with $p \neq 2, 13$. If $p \equiv \pm 1, \pm 3, \pm 4 \pmod{13}$ then $S < \Omega_7(p)$, and if $p \equiv \pm 2, \pm 5, \pm 6 \pmod{13}$ then $S < \Omega_7(p^2)$. The class stabiliser is trivial when $q = p$ and $\langle \phi \rangle$ when $q = p^2$.

3. $S = J_1 < \Omega_7(11)$, with trivial class stabiliser.

4. $S = A_9.2 < \Omega_7(3)$, with trivial class stabiliser.

5. $S = S_6(2)$ with $p \neq 2$. For all such p, the group $S < \Omega_7(p)$, with trivial class stabiliser.

Theorem 4.10.19 *Let G and Ω be as in Convention 4.10.1, with $\Omega = \Omega_8^-(q)$. Then representatives of the conjugacy classes of $\mathscr{S}_1$-maximal subgroups of G are described in the list below, using Convention 4.10.1.*

Proof See Theorem 4.3.3 for the list of candidates. See Propositions 4.9.47 and 4.9.48 for the other claims. □

1. $S = L_3(2)$, with $p \equiv \pm2, \pm8, \pm10 \pmod{21}$ and $p \neq 2$. For all such p the group $S < \Omega_8^-(p)$, with class stabiliser $\langle \gamma \rangle$.

Theorem 4.10.20 *Let G and Ω be as in Convention 4.10.1, with $\Omega = \Omega_9(q)$. Then representatives of the conjugacy classes of $\mathscr{S}_1$-maximal subgroups of G are described in the list below, using Convention 4.10.1.*

Proof See Theorem 4.3.3 for the list of candidates. See Propositions 4.9.49 to 4.9.53 for the other claims. □

1. $S = L_2(8)$ with $p \neq 2, 7$. If $p \equiv \pm1 \pmod{7}$ then $S < \Omega_9(p)$, and if $p \equiv \pm2, \pm3 \pmod{7}$ then $S < \Omega_9(p^3)$. The class stabiliser is trivial when $q = p$, and $\langle \phi \rangle$ when $q = p^3$.

2. $S = L_2(17)$ with $p \neq 2, 17$. If $p \equiv \pm1, \pm2, \pm4, \pm8 \pmod{17}$ then $S < \Omega_9(p)$, and if $p \equiv \pm3, \pm5, \pm6, \pm7 \pmod{17}$ then $S < \Omega_9(p^2)$. The class stabiliser is trivial when $q = p$, and $\langle \phi \rangle$ when $q = p^2$.

3. $S = A_{10}.2$ with $p \equiv \pm1 \pmod{5}$, or $S = A_{10}$ with $p \equiv \pm2 \pmod{5}$. If $p \neq 2, 11$ then $S < \Omega_9(p)$. The class stabiliser is trivial when $S = A_{10}.2$, and $\langle \delta \rangle$ when $S = A_{10}$.

4. $S = A_{11}.2 < \Omega_9(11)$, with trivial class stabiliser.

Theorem 4.10.21 *Let G and Ω be as in Convention 4.10.1, with $\Omega = \Omega_{10}^+(q)$ or $\Omega_{10}^-(q)$. Then representatives of the conjugacy classes of $\mathscr{S}_1$-maximal subgroups of G are described in the list below, using Convention 4.10.1.*

Proof See Theorem 4.3.3 for the list of candidates. See Propositions 4.9.54 to 4.9.63 for the other claims. □

1. $S = A_6.2_1 \times Z$ with $p \equiv \pm1 \pmod{12}$ or $S = A_6 \times Z$ with $p \equiv \pm5 \pmod{12}$. If $p \equiv 1 \pmod{4}$ then $S < \Omega_{10}^+(p)$, and if $p \equiv 3 \pmod{4}$ then $S < \Omega_{10}^-(p)$. Let X be the class stabiliser. Then $X = \langle \delta\gamma \rangle$ if $p \equiv \pm11 \pmod{24}$, $X = \langle \gamma, \delta' \rangle$ if $p \equiv \pm7 \pmod{24}$, and $X = \langle \delta\gamma, \delta' \rangle$ if $p \equiv \pm5 \pmod{24}$. If $|X| = 4$ then δ' induces the 2_1 automorphism of A_6, and γ or $\delta\gamma$ induces the 2_2 automorphism.

If $p \neq 7$, then $\bar{S}$ and all of its extensions are $\mathscr{S}_1$-maximal in the corresponding extensions of $\bar{\Omega}$. If $p = 7$ then $\bar{S}$, $\bar{S}.2_1$ and $\bar{S}.2_3$ are not $\mathscr{S}_1$-maximal, but all other extensions are.

2. $S = L_2(11) \times Z$ with $p \neq 2,3,7,11$. If $p \equiv 1,3,4,5,9 \pmod{11}$ then $S < \Omega_{10}^+(p)$, and if $p \equiv 2,6,7,8,10 \pmod{11}$ then $S < \Omega_{10}^-(p)$. The class stabiliser is $\langle \gamma \rangle$.

3. $S = L_2(11) \times Z$ with $p \neq 2,3,11$. If $p \equiv 1,3,4,5,9 \pmod{11}$ then $S < \Omega_{10}^+(p)$ and if $p \equiv 2,6,7,8,10 \pmod{11}$ then $S < \Omega_{10}^-(p)$. The class stabiliser is $\langle \gamma \rangle$ when $p \equiv \pm 1 \pmod{12}$ and $\langle \delta\gamma \rangle$ when $p \equiv \pm 5 \pmod{12}$. The group S is not $\mathscr{S}_1$-maximal, but $\bar{S}.2$ is $\mathscr{S}_1$-maximal in the relevant extension $\bar{\Omega}.2$.

4. $S = 2 \times A_7 < \Omega_{10}^-(7)$, with class stabiliser $\langle \delta' \rangle$. The group S is not $\mathscr{S}_1$-maximal, but $S.2$ is $\mathscr{S}_1$-maximal in $\Omega\langle \delta' \rangle = SO_{10}^-(7)$.

5. $S = 2^{\cdot}L_3(4) < \Omega_{10}^-(7)$, with class stabiliser $\langle \delta', \gamma \rangle$. Here, δ' induces the 2_3 automorphism, and γ the 2_1 on one class and the 2_2 on the other. The groups $\bar{S}$ and $\bar{S}.2_2$ are not $\mathscr{S}_1$-maximal, but all other extensions of $\bar{S}$ are $\mathscr{S}_1$-maximal in the corresponding extensions of $\bar{\Omega}$.

6. $S = M_{12} < \Omega_{10}^-(2)$, with class stabiliser $\langle \gamma \rangle$. The group S is not $\mathscr{S}_1$-maximal, but $S.2$ is $\mathscr{S}_1$-maximal in $\Omega.2 = SO_{10}^-(2)$.

7. $S = 2^{\cdot}M_{22} < \Omega_{10}^-(7)$, with class stabiliser $\langle \gamma \rangle$.

8. $S = A_{11} \times Z$ with $p \neq 2,3,11$. If $p \equiv 1,3,4,5,9 \pmod{11}$ then $S < \Omega_{10}^+(p)$, and if $p \equiv 2,6,7,8,10 \pmod{11}$ then $S < \Omega_{10}^-(p)$. The class stabiliser is $\langle \gamma \rangle$.

9. $S = A_{12} \times Z$ with $p = 2$ or 3. Here $S < \Omega_{10}^-(2)$, and $S < \Omega_{10}^+(3)$. The class stabiliser is $\langle \gamma \rangle$.

Theorem 4.10.22 *Let G and Ω be as in Convention 4.10.1, with $\Omega = \Omega_{11}(q)$. Then representatives of the conjugacy classes of $\mathscr{S}_1$-maximal subgroups of G are described in the list below, using Convention 4.10.1.*

Proof See Theorem 4.3.3 for the list of candidates. See Propositions 4.9.64 to 4.9.67 for the other claims. $\qquad\square$

1. $S = L_3(3).2 < \Omega_{11}(13)$, with trivial class stabiliser.

2. $S = A_{12}.2$ with $p \equiv \pm 1, \pm 5 \pmod{24}$, or $S = A_{12}$ with $p \neq 13$ and $p \equiv \pm 7, \pm 11 \pmod{24}$. The group $S < \Omega_{11}(p)$ for all such p. The class stabiliser is trivial when $S = A_{12}.2$, and $\langle \delta \rangle$ when $S = A_{12}$.

3. $S = A_{13} < \Omega_{11}(13)$, with class stabiliser $\langle \delta \rangle$.

Theorem 4.10.23 *Let G and Ω be as in Convention 4.10.1, with $\Omega = \Omega_{12}^+(q)$ or $\Omega_{12}^-(q)$. Then representatives of the conjugacy classes of $\mathscr{S}_1$-maximal subgroups of G are described in the list below, using Convention 4.10.1.*

Proof See Theorem 4.3.3 for the list of candidates. See Propositions 4.9.68 to 4.9.73 for the other claims. $\square$

1. $S = \mathrm{L}_2(11) \times Z$ with $p \neq 5, 11$. If $p \equiv \pm1, \pm16, \pm19, \pm24, \pm26 \pmod{55}$ then $S < \Omega_{12}^+(p)$, if $p \equiv \pm4, \pm6, \pm9, \pm14, \pm21 \pmod{55}$ then $S < \Omega_{12}^-(p)$, and if $p \equiv \pm2 \pmod 5$ then $S < \Omega_{12}^-(p^2)$. The class stabiliser is $\langle \gamma \rangle$.

2. $S = \mathrm{L}_2(11) \times Z$ with $p \neq 5, 11$. If $p \equiv \pm1, \pm16, \pm19, \pm24, \pm26 \pmod{55}$ then $S < \Omega_{12}^+(p)$, and if $p \equiv \pm4, \pm6, \pm9, \pm14, \pm21 \pmod{55}$ then $S < \Omega_{12}^-(p)$. The class stabiliser is $\langle \gamma \rangle$.

3. $S = \mathrm{L}_2(13) \times Z$ with $p \neq 7, 13$. If $p \equiv \pm1 \pmod 7$ then $q = p$, and if $p \equiv \pm2, \pm3 \pmod 7$ then $q = p^3$. If $p \equiv \pm1, \pm3, \pm4 \pmod{13}$ then $S < \Omega_{12}^+(q)$, and if $p \equiv \pm2, \pm5, \pm6 \pmod{13}$ then $S < \Omega_{12}^-(q)$. The class stabiliser is $\langle \gamma \rangle$.

4. $S = \mathrm{L}_2(13) \times Z$ with $p \equiv \pm1 \pmod 7$. If $p \equiv \pm1, \pm3, \pm4 \pmod{13}$ then $S < \Omega_{12}^+(p)$, and if $p \equiv \pm2, \pm5, \pm6 \pmod{13}$ then $S < \Omega_{12}^-(p)$. The class stabiliser is $\langle \gamma \rangle$.

5. The same description as Item 4.

6. $S = \mathrm{L}_3(3).2 \times Z$ or $\mathrm{L}_3(3) \times Z$, with $p \neq 3, 13$. If $p \equiv \pm1, \pm3, \pm4 \pmod{13}$ then $S < \Omega_{12}^+(p)$, and if $p \equiv \pm2, \pm5, \pm6 \pmod{13}$ then $S < \Omega_{12}^-(p)$. If $S < \Omega_{12}^+(p)$, then $\bar{S} = \mathrm{L}_3(3).2$ if and only if $p \equiv \pm1 \pmod{12}$ and $x^4 - 10x^2 + 13$ has four linear factors when reduced modulo p. If $S < \Omega_{12}^-(p)$, then $\bar{S} = \mathrm{L}_3(3).2$ if and only if $p \equiv \pm1 \pmod{12}$ or $p = 2$. Otherwise $\bar{S} = \mathrm{L}_3(3)$. When $\bar{S} = \mathrm{L}_3(3).2$, the class stabiliser is trivial. When $\bar{S} = \mathrm{L}_3(3)$, the class stabiliser is $\langle \delta \rangle$ when $p \equiv \pm5 \pmod{12}$ and $\langle \delta' \rangle$ when $p \equiv \pm1 \pmod{12}$. If $\bar{S} = \mathrm{L}_3(3)$ then $\bar{S}$ is not $\mathscr{S}_1$-maximal, but $\bar{S}.2$ is $\mathscr{S}_1$-maximal in the relevant extension $\bar{\Omega}.2$.

7. $S = 2\,\dot{}\,\mathrm{M}_{12}.2$ with $p \equiv \pm1 \pmod{24}$, or $S = 2\,\dot{}\,\mathrm{M}_{12}$ with $p \equiv \pm5, \pm7, \pm11 \pmod{24}$. Here, $S < \Omega_{12}^+(p)$ for all such p. The class stabiliser is trivial when $p \equiv \pm1 \pmod{24}$, equal to $\langle \delta \rangle$ when $p \equiv \pm5 \pmod{12}$, and $\langle \delta' \rangle$ when $p \equiv \pm11 \pmod{24}$.

8. $S = \mathrm{A}_{13} \times Z$ with $p \neq 7, 13$. If $p \equiv \pm1, \pm3, \pm4 \pmod{13}$ then $S < \Omega_{12}^+(p)$, and if $p \equiv \pm2, \pm5, \pm6 \pmod{13}$ then $S < \Omega_{12}^-(p)$. The class stabiliser is $\langle \gamma \rangle$.

9. $S = \mathrm{A}_{14} < \Omega_{12}^-(7)$, with class stabiliser $\langle \gamma \rangle$.

5

Groups in Class $\mathscr{S}$: defining characteristic

The theory of representations of finite simple groups of Lie type in defining characteristic is somewhat advanced. The representations arise from those of the associated algebraic groups, and so some familiarity with the theory of algebraic groups is necessary in order to understand it. For an introduction to this theory see, for example, the survey article by Humphreys [51]. The enthusiastic reader may wish to consult Jantzen [58] for a more detailed exposition. Humphrey's classic book [50] provide a general exposition of the theory of algebraic groups and their representations, whilst Malle and Testerman's book [91] gives an excellent introduction to the general theory, subgroup structure, and representation theory of the finite and algebraic groups of Lie type, including a fuller discussion of all of the introductory material in this chapter.

In many respects, the study of the $\mathscr{S}_2$-candidates is easier than that of the $\mathscr{S}_1$-candidates, simply because there are far fewer of them: we just need to know about the representations in dimensions up to 12, and to be able to determine some of their properties, such as forms preserved and their behaviour under the actions of group and field automorphisms. Fortunately it is possible to extract this information starting from a superficial familiarity with the main results of the theory, principally the Steinberg Tensor Product Theorems. These theorems, together with the tables in [84], suffice to determine the representations. Most of these modules arise as symmetric or anti-symmetric powers or as easily-defined constituents of tensor products, and in those cases for which we need to carry out detailed calculations, we can use these descriptions to write down explicit matrices for group generators and forms preserved. We remind the reader that the files of MAGMA calculations that we refer to are available on the webpage http://www.cambridge.org/9780521138604.

This chapter is organised in a similar way to the previous chapter. However, rather than examine the candidates in order of case and dimension, it is more convenient to study all defining characteristic representations of each of the types of groups of Lie type (such as $\mathrm{Sp}_n(q)$) together. In Section 5.1 we summarise the results from the general theory of representations of groups of Lie type in defining characteristic that we shall need to classify the $\mathscr{S}_2$-subgroups, and in Definition 5.1.15 we define a class $\mathscr{S}_2^*$, which roughly consists of those $\mathscr{S}_2$-subgroups which are not *obviously* contained in a member of $\mathscr{C}_4 \cup \mathscr{C}_7$. In Section 5.2, we present a thorough study of symmetric and anti-symmetric powers of modules, since they will occur frequently in the remainder of the chapter. The theory of representations of $\mathrm{SL}_2(q)$ in defining characteristic is more elementary than the general case, so in Section 5.3 we determine the $\mathscr{S}_2^*$-subgroups that arise from representations of extensions of $\mathrm{SL}_2(q)$, as well as finding their class stabilisers in the relevant full automorphism group of the classical group. After this, in Sections 5.4 to 5.7 we analyse each of the remaining types of groups of Lie type in turn, determining each possible intersection with Ω, and the corresponding class stabiliser in the conformal group containing Ω. We thereby compute the same information as we did in Section 4.4 for the $\mathscr{S}_1$-candidates. All of this information is summarised in Section 5.8. The remaining sections of the chapter then follow the same course as in Chapter 4: in Section 5.9 we find the class stabilisers under graph and field automorphisms, then in Section 5.10 we analyse containments between all of the $\mathscr{S}_2^*$-candidates (including the groups $\mathrm{SL}_2(q)$), and finally in Section 5.11 we present the complete list of $\mathscr{S}_2^*$-maximals.

5.1 General theory of $\mathscr{S}_2$-subgroups

5.1.1 The Steinberg Theorems

We start by summarising the main results that we shall need from the theory of algebraic groups and their defining characteristic representations. Let $q = p^e$ be a power of a prime p, and let $\sigma : \lambda \mapsto \lambda^q$ and $\phi : \lambda \mapsto \lambda^p$ be field automorphisms, applicable to any field of characteristic p. Let ${}^t X_\ell(q)$ denote a group of simple Lie type and Lie rank ℓ, with ${}^t \widehat{X}_\ell(q)$ being the covering group of ${}^t X_\ell(q)$ by the p'-part of its Schur multiplier (so ${}^t \widehat{X}_\ell(q)$ is the *simply connected* version of ${}^t X_\ell(q)$). The groups ${}^t \widehat{X}_\ell(q)$ can all be obtained as the centraliser of a suitable automorphism of the algebraic group $\widehat{X}_\ell(\overline{\mathbb{F}_p})$, which we regard as being a matrix group in some standard form.

If $t = 1$, then this is simply the field automorphism $\sigma = \phi^e$, which is applied to the entries of the elements of $\widehat{X}_\ell(\overline{\mathbb{F}_p})$. If ${}^t X_\ell$ is one of ${}^2 A_\ell$, ${}^2 D_\ell$, ${}^3 D_4$ and ${}^2 E_6$ then $\widehat{X}_\ell(\overline{\mathbb{F}_p})$ has a graph automorphism γ of order t, and ${}^t \widehat{X}_\ell(q)$ is the centraliser therein of the automorphism $\gamma^{-1}\sigma$. In the case of $\widehat{A}_\ell(\overline{\mathbb{F}_p}) = \mathrm{SL}_{\ell+1}(\overline{\mathbb{F}_p})$, we take the automorphism γ to be the inverse-transpose map (note that there are other possibilities, which can yield non-isomorphic groups in the definitions that follow). The centraliser of $\gamma^{-1}\sigma = \gamma\sigma$ is then the group ${}^2\widehat{A}_\ell(q) = \mathrm{SU}_{\ell+1}(q)$, where the invariant σ-Hermitian form has matrix $I_{\ell+1}$.

If ${}^t X_\ell$ is one of ${}^2B_2$, ${}^2G_2$ and ${}^2F_4$ then $\widehat{X}_\ell(\overline{\mathbb{F}_p})$ has a graph automorphism γ such that $\gamma^2 = \phi$, but only in characteristics $p = 2, 3$ and 2 respectively. The group ${}^t\widehat{X}_\ell(p^e) = {}^t X_\ell(p^e)$ is then the centraliser in $\widehat{X}_\ell(\overline{\mathbb{F}_p})$ of γ^e, but only if e is odd and p is the relevant characteristic (2, 3 or 2 respectively). The automorphism γ of ${}^2B_2(\overline{\mathbb{F}_2}) = \mathrm{Sp}_4(\overline{\mathbb{F}_2})$ is described more explicitly in Section 7.2.

The following result about automorphism groups of finite groups of Lie type is standard. See, for example, [32, Theorem 2.5.1] or [12, Section 3.3]. We discussed diagonal, field and graph automorphisms of the classical groups in Section 1.7.

Proposition 5.1.1 *Any automorphism of a finite group ${}^t X_\ell(p^e)$ of Lie type can be written as a product $idfg$, where i, d, f, g are respectively inner, diagonal, field, and graph automorphisms.*

By a result of Chevalley [50, p190], the finite-dimensional irreducible algebraic $\overline{\mathbb{F}_p}$-representations of $\widehat{X}_\ell(\overline{\mathbb{F}_p})$ are indexed by elements $\lambda \in \mathbb{N}^\ell$ (with $0 \in \mathbb{N}$), where the associated irreducible modules $M(\lambda)$ and $M(\mu)$ are isomorphic if and only if $\lambda = \mu$. We call λ the *highest weight* of $M(\lambda)$. (We shall say more about the definition of weights in Subsection 5.1.2 below.) The trivial module of $\widehat{X}_\ell(\overline{\mathbb{F}_p})$ is $M(0)$.

A weight $\lambda = (a_1, \ldots, a_\ell)$ is *m-restricted* if $0 \leqslant a_i \leqslant m - 1$ for all i. Two results of Steinberg relate the representations of $\widehat{X}_\ell(\overline{\mathbb{F}_p})$ to the irreducible $\overline{\mathbb{F}_p}$-representations of ${}^t\widehat{X}_\ell(q)$.

Theorem 5.1.2 (Steinberg's Tensor Product Theorem [104]) *Let $\lambda_0, \ldots, \lambda_r$ be p-restricted weights associated with $\widehat{X}_\ell(\overline{\mathbb{F}_p})$. Then, as $\widehat{X}_\ell(\overline{\mathbb{F}_p})$-modules,*

$$M(\lambda_0 + p\lambda_1 + \cdots + p^r\lambda_r) \cong M(\lambda_0) \otimes {}^\phi M(\lambda_1) \otimes \cdots \otimes {}^{\phi^r} M(\lambda_r).$$

Note that, according to the notation introduced in Section 1.8, ${}^\phi M(\lambda_1)$ is defined by the application of the group automorphism ϕ of $\widehat{X}_\ell(\overline{\mathbb{F}_p})$ to $M(\lambda_1)$, and that ${}^\phi M(\lambda_1) = M(p\lambda_1)$. We shall see below in Proposition 5.1.9 that ${}^\phi M(\lambda_1)$ is isomorphic to the module $M(\lambda_1)^\phi$ defined by applying the field automorphism to the image of the associated representation.

Theorem 5.1.3 (Steinberg [104]) *Let ${}^t X_\ell(q)$ be other than ${}^2B_2(q)$, ${}^2G_2(q)$ or ${}^2F_4(q)$. Then any irreducible $\overline{\mathbb{F}_p}$-module for ${}^t X_\ell(q)$ is isomorphic to the restriction of the $\widehat{X}_\ell(\overline{\mathbb{F}_p})$ module $M(\lambda)$ to ${}^t\widehat{X}_\ell(q)$ for some q-restricted weight λ. Moreover if λ and μ are q-restricted and not equal then the restrictions of $M(\lambda)$ and $M(\mu)$ to ${}^t\widehat{X}_\ell(q)$ are not isomorphic as $\overline{\mathbb{F}_p}{}^t\widehat{X}_\ell(q)$-modules.*

The statement of the above theorem is somewhat more complicated for the cases $^2\mathrm{B}_2(q)$, $^2\mathrm{G}_2(q)$ or $^2\mathrm{F}_4(q)$, but we shall not require this result, as a much simpler result that is similarly useful holds in these cases.

Let us ignore the cases $^t X_\ell(q) = {}^2\mathrm{B}_2(q)$, $^2\mathrm{G}_2(q)$ and $^2\mathrm{F}_4(q)$ for now. Then $^t\widehat{X}_\ell(q)$ has q^ℓ absolutely irreducible modules over $\overline{\mathbb{F}_p}$ (up to isomorphism) and it follows, from Corollary 1.8.14 and Proposition 5.1.9 (ii) below, that each of them can be written over $\mathbb{F}_{q^t}$. In light of the above two theorems (and also Theorem 5.1.5 below), we define a set $\mathscr{M}$ of p^ℓ absolutely irreducible modules for $^t\widehat{X}_\ell(q)$. The set $\mathscr{M}$ consists of the restrictions to $^t\widehat{X}_\ell(q)$ of the $M(\lambda)$, for all p-restricted weights λ, and is called the set of *p-restricted modules* or *p-restricted representations* for $^t\widehat{X}_\ell(q)$.

Our main reference for classifying the (small dimensional) modules in $\mathscr{M}$ is Lübeck [84]. The following remark is an explanation of how we use the tables.

Remark 5.1.4 The following information can be obtained from Lübeck [84], using the Dynkin diagram labelling conventions of **CHEVIE** as detailed in Appendix A.1 of [84].

  (i) The p-restricted modules for the groups $^t\widehat{X}_\ell(q)$ and $\widehat{X}_\ell(\overline{\mathbb{F}_p})$ of dimension up to at least 300 are those listed in the table for X_ℓ in [84] for which the entries in the vector in the λ-column of that table are all less than p, or in Lübeck [84, Remark 4.5] when $X_\ell = \mathrm{A}_1$. However, in the case of A_ℓ for $\ell > 1$, only one of each pair of dual p-restricted modules is listed in the table, and the dual module is obtained by reversing the vector in the λ-column.

 (ii) The graph automorphism of order 2 of $\mathrm{D}_\ell(q)$ interchanges the first two coordinates of the vectors in the λ-column. This corresponds to module duality when ℓ is odd, but when ℓ is even all modules of $\mathrm{D}_\ell(q)$ are self-dual.

(iii) The graph automorphisms of $\mathrm{D}_4(q)$ and $\mathrm{E}_6(q)$ permute the coordinates of the entry in the λ-column according to the numbering of the vertices in the Dynkin diagrams defined in [84, Appendix A.1].

The low-dimensional p-restricted modules can generally be expressed in terms of well-known modules for the group in question, like the natural module. Thus we shall not need to delve too deeply into the technicalities of module weights.

Let 1 denote the trivial module, and define $\mathscr{M}_i = {}^{\phi^i}\mathscr{M} = \{\,{}^{\phi^i}M : M \in \mathscr{M}\,\}$ for $i \in \mathbb{Z}$. Thus $\mathscr{M}_0 = \mathscr{M}$. It is clear from Theorems 5.1.2 and 5.1.3 that $\mathscr{M} \cap \mathscr{M}_i = \{1\}$ for $0 < i < e$, and that $^\sigma\mathscr{M} = \mathscr{M}$; thus the set $\mathscr{M}_i$ only depends on the congruence class of i modulo e.

A similar situation also pertains when $^t X_\ell(q) = {}^2\mathrm{B}_2(q)$, $^2\mathrm{G}_2(q)$ or $^2\mathrm{F}_4(q)$, except that there are $q^{\ell/2}$ irreducible modules (i.e. q, q and q^2 modules respectively), and $\mathscr{M}$ contains $p^{\ell/2}$ irreducible modules (i.e. 2, 3 and 4 modules respectively), each defined over $\mathbb{F}_q$. We shall define $\mathscr{M}$ in these cases in Section 5.7.

Theorem 5.1.5 (Steinberg's Twisted Tensor Product Theorem [104]**)** *Let* $^tX_\ell(q)$, $\mathscr{M}$ *and* $\mathscr{M}_i$ *be as above. Then any irreducible module for* $^t\widehat{X}_\ell(q)$ *over* $\overline{\mathbb{F}_p}$ *has the form:*

$$M_0 \otimes M_1 \otimes \cdots \otimes M_{e-1},$$

where $M_i \in \mathscr{M}_i$ *for all* i. *Furthermore, the* $|\mathscr{M}|^e = q^\ell$ *(or* $q^{\ell/2}$ *when* $^tX_\ell(q)$ *is* $^2B_2(q)$, $^2G_2(q)$ *or* $^2F_4(q)$*) possibilities are pairwise non-isomorphic.* □

The following lemma is an immediate consequence of the definitions of $^t\widehat{X}_\ell(q)$, and of the p-restricted modules for $\widehat{X}_\ell(q)$ as restrictions of the p-restricted $\overline{\mathbb{F}_p}$-modules for $\widehat{X}_\ell$.

Lemma 5.1.6 *There are natural embeddings of* $\widehat{X}_\ell(q)$ *in* $\widehat{X}_\ell(q^s)$ *for all* $s \geqslant 1$, *and of* $^t\widehat{X}_\ell(q)$ *in* $\widehat{X}_\ell(q^t)$, *and the restriction of any p-restricted module for* $\widehat{X}_\ell(q^s)$ *to* $\widehat{X}_\ell(q)$ *is a p-restricted module of the same weight.*

5.1.2 Weights

We remind the reader that the material in this section is standard, and a more thorough introduction to this material can be found in, for example, [91]. A *torus* of $\widehat{X}_\ell(\overline{\mathbb{F}_p})$ is a subgroup that is isomorphic as an algebraic group to a direct product of copies of $\overline{\mathbb{F}_p}^\times$, the multiplicative group of $\overline{\mathbb{F}_p}$. Every torus lies in a maximal torus, each maximal torus T is isomorphic to $(\overline{\mathbb{F}_p}^\times)^\ell$ (where ℓ is the Lie rank of $\widehat{X}_\ell(\overline{\mathbb{F}_p})$), and all maximal tori are conjugate in $\widehat{X}_\ell(\overline{\mathbb{F}_p})$ [50, p123]. A torus of $\widehat{X}_\ell(q)$ is the centraliser of σ in a torus for $\widehat{X}_\ell(\overline{\mathbb{F}_p})$.

Definition 5.1.7 A *weight* of T is an element of the the set of algebraic group homomorphisms from T to $\overline{\mathbb{F}_p}^\times$, which is the *character group* $X = X(T)$.

The character group X is isomorphic to (and identified with) the additive group $\mathbb{Z}^\ell$, since each element of X sends $(\alpha_1, \ldots, \alpha_l) \in T$ to $\alpha_1^{a_1}\alpha_2^{a_2}\cdots\alpha_l^{a_l}$ for some $a_i \in \mathbb{Z}$.

The characterisation of the algebraic irreducible $\widehat{X}_\ell(\overline{\mathbb{F}_p})$-modules M by their highest weights means that they can be identified by their restrictions to a maximal torus T, using the decomposition into eigenspaces $M = \oplus_{\mu \in X}M_\mu$, and $M_\mu = \{v \in M \mid vt = \mu(t)v \,\forall t \in T\}$. Note that M_μ is non-zero for only finitely many $\mu \in X$.

Definition 5.1.8 Let M be an algebraic irreducible $\widehat{X}_\ell(\overline{\mathbb{F}_p})$-module. The μ for which M_μ is non-zero are the *weights of M* (and of its corresponding representation).

Weights can be given an additive partial order, and it turns out that each algebraic irreducible $\overline{\mathbb{F}_p}$-module M has a *highest weight* μ, such that $\mu > \mu'$ for all other weights μ' of M. Some weights of M can occur with multiplicity > 1 but the highest weight cannot. As remarked earlier, by a result of Chevalley, these highest weights characterise the algebraic irreducible $\widehat{X}_\ell(\overline{\mathbb{F}_p})$-modules. It is clear from the above definition that if N is a constituent of a module M then the weights of N are a subset of the weights of M.

Proposition 5.1.9 (i) *The diagonal automorphisms of $^t\widehat{X}_\ell(q)$ stabilise each irreducible module of $^t\widehat{X}_\ell(q)$.*

(ii) *The field automorphism ϕ_X of the group $^t\widehat{X}_\ell(q)$ induced by $\lambda \mapsto \lambda^p$ has the same effect on its (isomorphism classes of) irreducible modules as the field automorphism $\phi_{\mathrm{GL}_n(q^t)}$ applied to the images of the associated representations.*

(iii) *When $^tX_\ell(q) \neq {}^2B_2(q)$, $^2G_2(q)$ or $^2F_4(q)$, duality of modules stabilises each $\mathscr{M}_i$ setwise. The graph automorphism of $A_\ell(q)$ has the same effect on its irreducible modules as duality of modules; that is, $^\gamma M = M^*$.*

(iv) *When $X_\ell(q) = A_\ell(q)$, $D_\ell(q)$, $D_4(q)$ or $E_6(q)$, each graph automorphism of $\widehat{X}_\ell(q)$ stabilises each $\mathscr{M}_i$ setwise.*

(v) *For twisted groups $G = {}^t\widehat{X}_\ell(q)$ different from $^2B_2(q)$, $^2G_2(q)$ and $^2F_4(q)$, the field automorphism σ_G of G induced by $\lambda \mapsto \lambda^q$ has the same effect on the irreducible modules as a graph automorphism of G of order t, and also stabilises each $\mathscr{M}_i$ setwise.*

Proof Part (i) follows from the fact (see [32, Theorem 2.5.1] or [105, Exercise, page 158]) that the diagonal automorphisms of $^t\widehat{X}_\ell(q)$ are induced by conjugation by elements of a maximal torus of $^t\widehat{X}_\ell(\overline{\mathbb{F}_p})$ which, since the torus is abelian, stabilise all representations of that torus. The remaining parts are stated in [66, Propositions 5.4.2 and 5.4.3]. $\square$

In view of Proposition 5.1.9, we shall usually write M^ϕ rather than $^\phi M$ in future, since the former is a little easier to read.

Corollary 5.1.10 Let $G = {}^t\widehat{X}_\ell(q)$ have a representation $\rho : G \to \Omega$, where Ω is a quasisimple classical group over $\mathbb{F}_{q^t}$ and dimension $n \leqslant 12$. Assume that ρ is absolutely irreducible and tensor indecomposable. Let C be the conformal group corresponding to Ω, and let $S = \operatorname{Im} \rho \leqslant \Omega$. Then S is C-conjugate to the image of a representation listed in [84].

Proof By Theorem 5.1.5, the representation ρ is equivalent to a representation ρ_1 with module $M_0 \otimes \cdots \otimes M_{e-1}$, where $M_i \in \mathscr{M}_i$ for all i. By Lemma 1.8.10 this implies that S is C-conjugate to $\operatorname{Im} \rho_1$. The assumption that S is tensor indecomposable implies that all but one of the M_i are trivial. By Proposition 5.1.9 (ii), for any p-restricted representation τ with module M, the module $\phi^i M$ has representation $\phi^i \tau$ and hence, by Lemmas 1.8.6 and 1.8.10, S is C-conjugate to Im μ for some p-restricted representation μ.

If $G \neq A_\ell(q)$ we are done, by Remark 5.1.4, but if $G = A_\ell(q)$ then μ might be the dual of one of the representations listed in [84]. However, by Proposition 5.1.9 (iii) duality of modules acts in the same way as the duality automorphism of G, so once again S is C-conjugate to the image of a representation listed in [84]. □

Lemma 5.1.11 *If M_1 and M_2 are (finite-dimensional algebraic) irreducible $\overline{\mathbb{F}_p}\widehat{X}_\ell(\overline{\mathbb{F}_p})$-modules, then the weights of $M_1 \otimes M_2$ are of the form $\mu_1 + \mu_2$ for weights μ_1 and μ_2 of M_1 and M_2. In particular, $M(\lambda + \mu)$ is a constituent of $M(\lambda) \otimes M(\mu)$ of multiplicity 1.*

Proof Let W_1 and W_2 be the sets of weights of M_1 and M_2. Then, as modules for a maximal torus, $M_1 = \bigoplus_{\mu_1 \in W_1} M_{\mu_1}$, $M_2 = \bigoplus_{\mu_2 \in W_2} M_{\mu_2}$, and so (since addition in $X \equiv \mathbb{Z}^\ell$ corresponds to multiplication of group homomorphisms), $M := M_1 \otimes M_2 = \bigoplus_{\mu_1 \in X_1, \mu_2 \in X_2} M_{\mu_1 + \mu_2}$. So $\lambda + \mu$ is a weight of $M(\lambda) \otimes M(\mu)$, and since $\lambda \geqslant \lambda'$ for all weights λ' of $M(\lambda)$ and $\mu \geqslant \mu'$ for all weights μ' of $M(\mu)$ we have $\lambda + \mu \geqslant \lambda' + \mu'$ for all weights $\lambda' + \mu'$ of $M(\lambda) \otimes M(\mu)$. Therefore $M(\lambda) \otimes M(\mu)$ must have a constituent with highest weight $\lambda + \mu$, that is the module $M(\lambda + \mu)$, and since $\lambda + \mu$ occurs just once as a weight of $M(\lambda) \otimes M(\mu)$, $M(\lambda + \mu)$ occurs just once as a constituent of $M(\lambda) \otimes M(\mu)$. □

Proposition 5.1.12 ([50, 3.1.6]) *For types B_ℓ, C_ℓ, D_ℓ (ℓ even), G_2, F_4, E_7 and E_8, all irreducible $\overline{\mathbb{F}_p}$-modules are self-dual.*

A consequence of the above discussion is that a module $M_0 \otimes \cdots \otimes M_{e-1}$ for a group of Lie type in defining characteristic is self-dual if and only if each M_i is self-dual. There is no such general result regarding tensor products of arbitrary modules, for example the Mathieu group M_{12} has a 176-dimensional self-dual representation in characteristic 0 that is a tensor product of a self-dual 11-dimensional representation and a non-self-dual 16-dimensional representation.

5.1.3 Minimal fields and fixed forms

We need to be able to determine the minimal field of the relevant representations of the irreducible $^t X_\ell(q)$-modules. This will be a subfield of $\mathbb{F}_{q^t} = \mathbb{F}_{p^{et}}$ (since they can all be written over that field). A similar result to the following is proved in [66, Proposition 5.4.6]. There is a corresponding but more complicated result for $^3 D_4(q)$, which we shall not need.

Theorem 5.1.13 *Suppose that either $t = 1$, or $^tX_\ell$ is one of $^2A_\ell$, $^2D_\ell$ or $^2E_6$. Let M be an irreducible module for $^tX_\ell(q)$ with $q = p^e$, and suppose that $M \cong \otimes_{i=0}^{e-1} M_i^{\phi^i}$ with $M_i \in \mathscr{M}$ (as in Theorem 5.1.5). Let $f \mid te$. Then M can be realised over $\mathbb{F}_{p^f}$ if and only if one of the following conditions hold:*

(i) *$t = 1$ and $M_i \cong M_j$ whenever $i \equiv j$ (mod f).*
(ii) *$t = 2$, and one of the following occurs:*
  (a) *$f \mid e$, $M_i \cong M_i^{\phi^e}$ for all i, and $M_i \cong M_j$ whenever $i \equiv j$ (mod f).*
  (b) *$f \nmid e$, $M_i \not\cong M_i^{\phi^e}$ for some i, $M_i \cong M_j$ whenever $i \equiv j$ (mod f), and $M_i \cong M_j^{\phi^e}$ whenever $i \equiv j$ (mod $f/2$) but $i \not\equiv j$ (mod f).*

(Observe that these conditions are satisfied vacuously when $f = te$.)

Proof By Corollary 1.8.14, the module can be written over the subfield $\mathbb{F}_{p^f}$ of $\mathbb{F}_{p^{te}}$ if and only if it is stabilised by the power of ϕ that centralises that subfield. In particular, M is stabilised by ϕ^{et}.

The trivial module belongs to $\mathscr{M}$ and is stabilised by ϕ. It follows immediately from Theorem 5.1.5 that, for any non-trivial module $M_i \in \mathscr{M}$, the modules $M_i, M_i^\phi, M_i^{\phi^2}, \ldots, M_i^{\phi^{e-1}}$ are pairwise non-isomorphic. Part (i) now follows immediately from Theorem 5.1.5 and Corollary 1.8.14.

So assume that $t = 2$. Then, by Proposition 5.1.9, for a non-trivial $M_i \in \mathscr{M}$, either $M_i \cong M_i^{\phi^e}$ (in which case M_i can be written over $\mathbb{F}_q$), or else M_i and $M_i^{\phi^e}$ are distinct modules in $\mathscr{M}$, which cannot be written over $\mathbb{F}_q$. So Part (ii) also follows from Theorem 5.1.5 and Corollary 1.8.14. $\qquad\square$

Recall Propositions 1.9.4 and 1.9.5, concerning the forms on tensor products of modules, and Definition 1.9.6 of an induced form.

Proposition 5.1.14 *Let S be the image of an irreducible representation of $^t\widehat{X}_\ell(q)$ over $\mathbb{F}_{q^t}$, with corresponding $\mathbb{F}_{q^t}$-module $M = M_0 \otimes M_1 \otimes \cdots \otimes M_{e-1}$, as in Theorem 5.1.5, with $k \geqslant 2$ of the M_i non-trivial. Suppose that S is an $\mathscr{S}_2$-subgroup of a quasisimple classical group Ω, and let G be almost simple with socle $\bar{\Omega}$. If $N_G(\bar{S})$ is a maximal subgroup of G then one of the following holds:*

(i) *Ω is defined over a proper subfield of $\mathbb{F}_{q^t}$.*
(ii) *Ω preserves an invariant classical form other than the induced form arising from the tensor factors M_i (if any).*
(iii) *(i) and (ii) do not occur, and some outer automorphism of S that is induced by G does not permute the M_i: this can only arise if $^tX_\ell(q) = {}^2B_2(2^e)$, $B_2(2^e)$, $^2G_2(3^e)$, $G_2(3^e)$, $^2F_4(2^e)$ or $F_4(2^e)$.*

Proof Let the non-trivial M_i be $M_{i_1}, \ldots, M_{i_k}$, so that $M \cong M_{i_1} \otimes \cdots \otimes M_{i_k}$. If the M_{i_j} are each equipped with a bilinear form, or are each equipped with a sesquilinear form, then let β be the induced form on M and let Ω_1 be the quasisimple group of isometries of M. Otherwise, let $\Omega_1 = \mathrm{SL}_n(q^t)$ and β be identically zero.

Let T be the image of the representation of ${}^t\widehat{X}_\ell(q)^k$ with module $M_{i_1} \otimes \cdots \otimes M_{i_k}$. Then T consists of isometries of β, so $S < T < \Omega_1$. Furthermore, if each outer automorphism α of $\bar{S}$ that can be realised in $\mathrm{Aut}\,\bar{\Omega}_1$ permutes the tensor factors M_{i_j}, then each almost simple extension $\bar{S}.Y$ of $\bar{S} = S/Z(S)$ that can be realised in $\mathrm{Aut}\,\bar{\Omega}_1$ is properly contained in the corresponding extension $\bar{S}^k.Y$ of $\bar{S}^k$.

Assume that all automorphisms of $\bar{S}$ induced by G permute the M_{i_j}. If $N_G(\bar{S})$ is maximal then $\Omega \neq \Omega_1$, by the previous paragraph. Thus Ω is another classical group over a subfield of $\mathbb{F}_{q^t}$, and so either Ω is defined over a smaller field than Ω_1, or Ω is a group of isometries of a classical form other than β.

Suppose now that ${}^t X_\ell(q) \neq {}^2\mathrm{B}_2(2^e)$, $\mathrm{B}_2(2^e)$, ${}^2\mathrm{G}_2(3^e)$, $\mathrm{G}_2(3^e)$, ${}^2\mathrm{F}_4(2^e)$ or $\mathrm{F}_4(2^e)$. We know from Proposition 5.1.1 that every outer automorphism α of ${}^t X_\ell(q)$ is a product of a diagonal, a field, and a graph automorphism. By Proposition 5.1.9, the diagonal and graph automorphisms fix each set $\mathscr{M}_i$, whereas the field automorphisms permute the set $\{\mathscr{M}_0, \ldots, \mathscr{M}_{e-1}\}$. It follows now from Theorem 5.1.5 that, if $M \cong {}^\alpha M$, then α must permute the tensor factors M_i of M, and so (iii) does not arise. $\qquad\square$

Note that the outer automorphisms of ${}^2\mathrm{B}_2(2^e)$, ${}^2\mathrm{G}_2(3^e)$ and ${}^2\mathrm{F}_4(2^e)$ do in fact permute the as-yet-undefined set $\{\mathscr{M}_0, \ldots, \mathscr{M}_{e-1}\}$, and so do not give exceptions to Part (iii) of the above Proposition. Note also that we have not assumed in the above proof that the modules M_i are tensor indecomposable.

The possible examples satisfying the first of these conditions in small dimensions can be gleaned from Theorem 5.1.13. As an example, let M be the natural module for $\mathrm{SL}_3(q^2)$. Then $M \otimes M^\sigma$ gives rise to an $\mathscr{S}_2$-maximal subgroup of $\mathrm{SL}_9(q)$, and $M \otimes M^{\sigma*}$ (where $*$ denotes duality) gives rise to an $\mathscr{S}_2$-maximal subgroup of $\mathrm{SU}_9(q)$.

In the situations described by Proposition 5.1.14, although S and its extensions may still be $\mathscr{S}_2$-maximal subgroups of Ω and its extensions, we shall simplify matters by not including them in our lists of candidates.

Definition 5.1.15 Let S be a quasisimple $\mathscr{S}_2$-subgroup of Ω, arising from a module $M \cong \otimes_{i=0}^{e-1} M_i^{\phi^i}$ for ${}^t\widehat{X}_\ell(q)$, as in Theorem 5.1.5, and suppose that exactly $k \geqslant 1$ of the modules M_i are non-trivial.

Let G be almost simple with socle $\bar{\Omega}$. Then $N_G(\bar{S})$ is an $\mathscr{S}_2^*$-subgroup of G if either $k = 1$, or Ω is defined over a smaller field than $\mathbb{F}_{q^t}$, or Ω is the isometry group of a form other than the induced form on M (if any), or there exists an automorphism of S that is induced by an element of $\mathrm{Aut}\,\bar{\Omega}$ that does not permute the non-trivial tensor factors.

An $\mathscr{S}_2^*$-subgroup of G is called $\mathscr{S}_2^*$-*maximal* if it is maximal among the $\mathscr{S}_2^*$-subgroups of G.

This chapter will classify the $\mathscr{S}_2^*$-maximal subgroups, rather than the $\mathscr{S}_2$-maximal subgroups, of almost simple extensions of $\overline{\Omega}$ for classical groups Ω of dimension at most 12.

5.2 Symmetric and anti-symmetric powers

Many of the modules for the groups of Lie type that we shall need to study arise as symmetric or anti-symmetric powers of the natural module, so we start by considering their elementary properties. The reader may wish to revisit Section 1.9, where we introduced tensor products.

Let V be an FG-module, where F is a field and G is a group. There are various possible definitions of the symmetric and anti-symmetric powers $S^m(V)$ and $\Lambda^m(V)$ of V. We prefer to define them as suitable quotients of the tensor power module $V^{\otimes m} = V \otimes \cdots \otimes V$ (with m factors). This is because it is easier to calculate G-actions in these versions. For some applications, such as computing invariant forms, defining $S^m(V)$ and $\Lambda^m(V)$ as submodules might be more convenient, but the reader should be aware that the submodule and quotient versions of $S^m(V)$ need not be isomorphic and, even when they are, their natural bases need not correspond. So we must be careful to use the submodule versions only in situations where they can be proved to be isomorphic to the quotient versions. In what follows, V has F-basis $(e_1, \ldots, e_n)$.

5.2.1 Symmetric and anti-symmetric squares

The module $V \otimes V$ has FG-submodules $A := \langle u \otimes v - v \otimes u : u, v \in V \rangle_F$, which has basis $(e_i \otimes e_j - e_j \otimes e_i : 1 \leqslant i < j \leqslant n)$, and $S := \langle v \otimes v : v \in V \rangle_F$, with basis $(e_i \otimes e_i : 1 \leqslant i \leqslant n) \cup \{ e_i \otimes e_j + e_j \otimes e_i : 1 \leqslant i < j \leqslant n \}$. Hence $\dim A = n(n-1)/2$ and $\dim S = n(n+1)/2$.

Definition 5.2.1 We define the *anti-symmetric square* of V to be

$$\Lambda^2(V) := (V \otimes V)/S.$$

We define the *symmetric square* of V to be

$$S^2(V) := (V \otimes V)/A.$$

We denote the image of $u \otimes v$ in $\Lambda^2(V)$ by $u \wedge v$, and the image of $u \otimes v$ in $S^2(V)$ by uv. We shall often write v^2 instead of vv.

It is straightforward to check that $v \wedge u = -(u \wedge v)$ and $vu = uv$ for all $u, v \in V$, and that $v \wedge v = 0$ for all $v \in V$ (even in characteristic 2).

Definition 5.2.2 Our *standard bases* for $\Lambda^2(V)$ and $S^2(V)$ are respectively $(e_i \wedge e_j : 1 \leqslant i < j \leqslant n)$ and $(e_i e_j : 1 \leqslant i \leqslant j \leqslant n)$, which we shall order lexicographically, giving

$$e_1 \wedge e_2, e_1 \wedge e_3, \ldots, e_1 \wedge e_n, e_2 \wedge e_3, \ldots, e_{n-1} \wedge e_n \text{ for } \Lambda^2(V)$$

and $\quad e_1^2, e_1 e_2, e_1 e_3, \ldots, e_1 e_n, e_2^2, e_2 e_3, \ldots, e_{n-1} e_n, e_n^2 \text{ for } S^2(V).$

First we show that $\Lambda^2(V)$ is isomorphic to a submodule of $V \otimes V$, and that if the characteristic of F is not 2 then the same is true for $S^2(V)$.

Lemma 5.2.3 *Define maps* $f : A \to \Lambda^2(V)$ *and* $g : S \to S^2(V)$ *to be the F-linear extensions (respectively) of*

$$\left. \begin{aligned} f &: u \otimes v - v \otimes u \mapsto u \wedge v \\ g &: \tfrac{1}{2}(u \otimes v + v \otimes u) \mapsto uv \end{aligned} \right\} \text{ for all } u, v \in V,$$

where g is defined only when $\operatorname{char} F \neq 2$. *Then both f and g induce FG-isomorphisms.*

Proof It is routine to check that these maps are FG-homomorphisms. They are clearly surjective and, since $\dim S = \dim S^2(V)$ and $\dim A = \dim \Lambda^2(V)$, they are isomorphisms. $\qquad\square$

In odd characteristic, S and A have trivial intersection, so as FG-modules $V \otimes V = S \oplus A$, and hence $V \otimes V \cong S^2(V) \oplus \Lambda^2(V)$. However, in characteristic 2, $A < S$, with $S/A \cong V^\phi$ where $\phi : x \mapsto x^2$ is a field endomorphism of F, and there are cases where $S \ncong S^2(V)$.

If G consists of isometries of a bilinear or sesquilinear form β on V, then G preserves an induced form $\beta^{\otimes 2} = \beta \otimes \beta$ on $V \otimes V$, as described in Definition 1.9.6, which can be restricted to the submodules A and S. Closely related to this are the forms on $\Lambda^2(V)$ and $S^2(V)$ given in the next result.

Proposition 5.2.4 *Suppose that G consists of isometries of a σ-Hermitian, alternating or symmetric form β on V. Then G preserves forms β^{2-} and β^{2+} on $\Lambda^2(V)$ and $S^2(V)$, respectively, such that*

$$\beta^{2-}(e_i \wedge e_j, e_k \wedge e_l) = \beta(e_i, e_k)\beta(e_j, e_l) - \beta(e_i, e_l)\beta(e_j, e_k)$$

and

$$\beta^{2+}(e_i e_j, e_k e_l) = \beta(e_i, e_k)\beta(e_j, e_l) + \beta(e_i, e_l)\beta(e_j, e_k).$$

Furthermore, if β is σ-Hermitian, then so are β^{2-} and β^{2+}, whereas if β is alternating or symmetric, then β^{2-} and β^{2+} are symmetric.

Proof We can extend the given definitions of $\beta^{2\pm}$ on the basis elements to sesquilinear or bilinear forms on $\Lambda^2(V)$ and $S^2(V)$, and verify that G acts via isometries. One sees easily that these forms have the specified types. □

There is no guarantee that the forms we obtain are non-degenerate, although they will be zero or non-degenerate on any of $\Lambda^2(V)$ or $S^2(V)$ that happen to be irreducible (otherwise $\Lambda^2(V)^\perp$ or $S^2(V)^\perp$ would be a submodule).

For future reference, we record the following standard result. It is proved in all characteristics other than 2 in [87, Theorem 2.7.4].

Proposition 5.2.5 *Let χ be a complex character or Brauer character in odd characteristic of the group G. Then, for $g \in G$, the values on g of the symmetric and anti-symmetric squares of the $\mathbb{C}G$-module corresponding to χ are respectively $(\chi(g)^2 + \chi(g^2))/2$ and $(\chi(g)^2 - \chi(g^2))/2$.*

5.2.2 Symmetric and anti-symmetric cubes and higher powers

For $\pi \in S_3$, we write $(v_1 \otimes v_2 \otimes v_3)\pi$ to denote $v_{1\pi} \otimes v_{2\pi} \otimes v_{3\pi}$. In this subsection we shall use $\varepsilon(\pi)$ to mean the sign of the permutation $\pi \in S_n$.

The FG-module $V^{\otimes 3}$ has a submodule A with basis

$$(e_i \otimes e_j \otimes e_k - (e_i \otimes e_j \otimes e_k)\pi : 1 \leqslant i < j < k \leqslant n, \pi \in \mathrm{Sym}(\{i,j,k\}) \setminus \{1\})$$
$$\cup\, (e_i \otimes e_i \otimes e_j - e_i \otimes e_j \otimes e_i : 1 \leqslant i, j \leqslant n, i \neq j)$$
$$\cup\, (e_i \otimes e_i \otimes e_j - e_j \otimes e_i \otimes e_i : 1 \leqslant i, j \leqslant n, i \neq j).$$

The reader may check that the dimension of A is $5n^3/6 - n^2/2 - n/3$, and that A contains a submodule

$$B = \Big\langle \sum_{\pi \in S_3} \varepsilon(\pi)(u \otimes v \otimes w)\pi : u, v, w \in V \Big\rangle_F,$$

of dimension $\binom{n}{3}$.

The module $V^{\otimes 3}$ also has a submodule S with basis

$$(e_i \otimes e_i \otimes e_i \ : \ 1 \leqslant i \leqslant n)$$
$$\cup\, (e_i \otimes e_i \otimes e_j, \ e_i \otimes e_j \otimes e_i, \ e_j \otimes e_i \otimes e_i \ : \ 1 \leqslant i, j \leqslant n, i \neq j)$$
$$\cup\, (e_i \otimes e_j \otimes e_k - \varepsilon(\pi)(e_i \otimes e_j \otimes e_k)\pi \ : \ 1 \leqslant i < j < k \leqslant n, \pi \in S_3 \setminus \{1\}).$$

The dimension of S is $5n^3/6 + n^2/2 - n/3$, and S contains submodules

$$T_1 = \langle u \otimes u \otimes u : u \in V \rangle_F \quad \text{and} \quad T_2 = \Big\langle \sum_{\pi \in S_3}(u \otimes v \otimes w)\pi : u, v, w \in V \Big\rangle_F.$$

The dimension of T_1 is $n + 2\binom{n}{2} + \binom{n}{3} = \binom{n+2}{3}$ except for the case $F = \mathbb{F}_2$ (for $n \geqslant 2$), and if char $F \neq 2$ or 3 then $T_2 = T_1$. If char $F = 2$ or 3 then $T_2 < T_1$ (for $n \geqslant 1$).

Definition 5.2.6 We define the *symmetric cube* of V to be

$$S^3(V) := V^{\otimes 3}/A.$$

We define the *anti-symmetric cube* of V to be

$$\Lambda^3(V) = V^{\otimes 3}/S.$$

We denote the image of $u \otimes v \otimes w$ in $S^3(V)$ by uvw, and in $\Lambda^3(V)$ by $u \wedge v \wedge w$.

Thus in $S^3(V)$ we identify $u \otimes v \otimes w$ with all of its images under permutations of the three vectors, whilst in $\Lambda^3(V)$, applying an odd permutation to the entries of $u \otimes v \otimes w$ also multiplies the tensor by -1. Notice that the dimension of $S^3(V)$ is $\binom{n+2}{3}$, whilst the dimension of $\Lambda^3(V)$ is $\binom{n}{3}$.

Definition 5.2.7 Our *standard bases* for $S^3(V)$ and $\Lambda^3(V)$ are respectively $(e_i e_j e_k : 1 \leqslant i \leqslant j \leqslant k \leqslant n)$ and $(e_i \wedge e_j \wedge e_k : 1 \leqslant i < j < k \leqslant n)$, in each case ordered lexicographically.

The proof of the next lemma is similar to that of Lemma 5.2.3, and is left as an exercise.

Lemma 5.2.8 *Define maps* $f : B \to \Lambda^3(V)$ *and* $g : T_2 \to S^3(V)$ *to be the F-linear extensions (respectively) of*

$$\left.\begin{array}{l} f : \sum_{\pi \in S_3} \varepsilon(\pi)(u \otimes v \otimes w)\pi \mapsto u \wedge v \wedge w \\[2mm] g : \frac{1}{6}\left(\sum_{\pi \in S_3}(u \otimes v \otimes w)\pi\right) \mapsto uvw \end{array}\right\} \quad \text{for all } u, v, w \in V,$$

where g *is defined only when* char $F \neq 2$ *or* 3. *Then both* f *and* g *induce FG-isomorphisms.*

If char $F \neq 2$ or 3 then one may show that $S \cong T_1 \oplus M^3(V) \oplus M^3(V)$, where $M^3(V)$ denotes a tensor cube of dimension $\frac{1}{3}(n^3 - n) = n(n+1)(n-1)/3$. Similarly, if char $F \neq 2$ or 3, then one may show that $A \cong B \oplus M^3(V) \oplus M^3(V)$, and that $A \cap T_1$ and $B \cap S$ are trivial. From this it follows that the module $V^{\otimes 3} \cong S^3(V) \oplus \Lambda^3(V) \oplus M^3(V) \oplus M^3(V)$.

We shall also need higher symmetric powers of the natural modules for $SL_2(q)$. For more on this see [28, Appendix B.2].

Definition 5.2.9 The *symmetric power* $S^k(V)$ is the quotient of $V^{\otimes k}$ by the submodule $\langle v_1 \otimes \cdots \otimes v_k - (v_1 \otimes \cdots \otimes v_k)\sigma : \sigma \in S_k \setminus \{1\}\rangle$. It has *standard basis* $(e_{i_1} e_{i_2} \cdots e_{i_k} : 1 \leqslant i_1 \leqslant \cdots \leqslant i_k \leqslant n)$, ordered lexicographically.

The anti-symmetric power $\Lambda^k(V)$ can be defined as the quotient of $V^{\otimes k}$ by the submodule generated by all $v_1 \otimes \cdots \otimes v_k$ for which $v_i = v_j$ for some $i \neq j$, but we shall not need to consider these further for $k \geqslant 3$. One reason for this is that it is straightforward to show that, for V of dimension n, $\Lambda^n(V)$

is the one-dimensional module with action of $g \in G$ given by multiplication by $\det g$. Hence, if $G \leqslant \mathrm{SL}_n(F)$, then $\Lambda^n(V)$ is the trivial module, and it can be verified, using the FG-bilinear map $\Lambda^k(V) \times \Lambda^{n-k}(V) \to \Lambda^n(V)$ defined by $((u_1 \wedge \cdots \wedge u_k), (v_1 \wedge \cdots \wedge v_{n-k})) \to u_1 \wedge \cdots \wedge u_k \wedge v_1 \wedge \cdots \wedge v_{n-k}$, that $\Lambda^k(V) \cong \Lambda^{n-k}(V)^*$ for $0 \leqslant k \leqslant n$, where $\Lambda^0(V) = F$ and $\Lambda^1(V) = V$.

5.3 The groups $\mathrm{SL}_2(q) = \mathrm{Sp}_2(q)$

In this section we study the $\mathscr{S}_2^*$-subgroups that are isomorphic to $\mathrm{SL}_2(q)$ and to $\mathrm{L}_2(q)$. We shall determine the quasisimple classical groups Ω in which they arise, as well as their normalisers in Ω, and their class stabilisers in $\mathrm{Out}\,\Omega$. From this we can deduce the number of conjugacy classes of such groups in Ω.

We are interested in the absolutely irreducible modules of $\mathrm{SL}_2(q) = \mathrm{Sp}_2(q)$ in characteristic p whenever $q \geqslant 4$, since the groups $\mathrm{SL}_2(2)$ and $\mathrm{SL}_2(3)$ are soluble. These were first classified by Brauer and Nesbitt [4], and we shall state this classification in Theorem 5.3.2 below.

Let F be a field. We define V to be the natural module of $\mathrm{SL}_2(F)$, and for $n \geqslant 1$ we define V_{n+1} to be the $(n+1)$-dimensional module $\mathrm{S}^n(V)$, as described in Definition 5.2.1. By an abuse of notation we shall also use V_{n+1} and $\mathrm{S}^n(V)$ to refer to the corresponding modules of $\mathrm{GL}_2(F)$. The module V_1 is the trivial module, for both $\mathrm{SL}_2(F)$ and $\mathrm{GL}_2(F)$. If $n > 1$, then V_n is a faithful module for $\mathrm{SL}_2(q)$ when n is even, and a faithful module for $\mathrm{L}_2(q)$ when n is odd.

Lemma 5.3.1 *Suppose that* $\mathrm{char}\,F = 0$ *or* $n + 1 \leqslant \mathrm{char}\,F$. *Then* V_{n+1} *is absolutely irreducible. Furthermore,* V_{n+1} *is self-dual for* $\mathrm{SL}_2(F)$.

Proof For prime characteristics, the absolute irreducibility claim is proved in [4], by a straightforward calculation that also works in characteristic 0. The self-duality is immediate from the fact that $\mathrm{SL}_2(F) = \mathrm{Sp}_2(F)$. $\qquad\square$

It can be shown by the methods of Section 5.2 that the quotient and submodule versions of $\mathrm{S}^n(V)$ are isomorphic for both $\mathrm{SL}_2(F)$ and $\mathrm{GL}_2(F)$. The modules V_{n+1} with $n + 1 \leqslant p$ are the p-restricted modules for $\mathrm{SL}_2(F)$.

For $n \in \mathbb{N} = \mathbb{N}_0$, we define the module $M(n)$ of $\mathrm{SL}_2(F)$ (and of $\mathrm{GL}_2(F)$) as follows. If $\mathrm{char}\,F = 0$, then $M(n) := V_{n+1}$. If $\mathrm{char}\,F = p > 0$ then there exist $s \geqslant 0$ and $a_0, \ldots, a_s \in \{0, \ldots, p-1\}$ such that $n = a_0 + a_1 p + a_2 p^2 + \cdots + a_s p^s$. Then

$$M(n) := V_{a_0+1} \otimes V_{a_1+1}^{\phi} \otimes \cdots \otimes V_{a_s+1}^{\phi^s},$$

which has dimension $(a_0 + 1)(a_1 + 1) \cdots (a_s + 1) \leqslant n + 1$. If a higher value of s is chosen than the smallest possible, then the only effect is that more trivial modules being involved in the tensor product for $M(n)$. The module $M(n)$ is a self-dual module for $\mathrm{SL}_2(F)$, since it is a tensor product of self-dual modules.

The following is a special case of Theorems 5.1.2, 5.1.3 and 5.1.5.

Theorem 5.3.2 (Brauer and Nesbitt [4]) *Let* $q = p^e$ *be a power of a prime* p. *Then each of the modules* $M(n)$ *with* $0 \leqslant n \leqslant q - 1$ *is an absolutely irreducible module for* $\mathrm{SL}_2(q)$, *and if* $0 \leqslant i < j \leqslant q - 1$ *then* $M(i)$ *is not isomorphic to* $M(j)$. *Conversely each absolutely irreducible module for* $\mathrm{SL}_2(q)$ *in characteristic* p *is isomorphic to* $M(n)$ *for some* n *with* $0 \leqslant n \leqslant q - 1$.

We can determine the minimal field of realisation of the modules $M(n)$ of $\mathrm{SL}_2(q)$ from the following immediate corollary of Theorem 5.1.13.

Corollary 5.3.3 *Let* $n = a_0 + a_1 p + a_2 p^2 + \cdots + a_{e-1} p^{e-1}$, *where the coefficients satisfy* $0 \leqslant a_i \leqslant p - 1$ *for all* i. *With* q, p, e *as above, the absolutely irreducible* $\mathrm{SL}_2(q)$-*module* $M(n) = M(a_0) \otimes M(a_1)^\phi \otimes \cdots \otimes M(a_{e-1})^{\phi^{e-1}}$ *has minimal field of realisation* $\mathbb{F}_{p^f}$ *if and only if*

(i) $f \mid e$;

(ii) f *is minimal such that* $a_i = a_j$ *whenever* $i \equiv j \pmod{f}$.

In view of Proposition 5.1.14, some of the work below is devoted to showing that the majority of the non-trivial tensor product modules of $\mathrm{SL}_2(q)$ give rise to embeddings of $L_2(q)$ in $\bar{\Omega}$ such that all almost simple extensions of $L_2(q)$ that arise embed into $\mathscr{C}_4$- or $\mathscr{C}_7$-subgroups, and hence are not $\mathscr{S}_2^*$-maximal in the simple classical group $\bar{\Omega}$. Another part of the work proves that the p-restricted modules V_n are not tensor decomposable, and so the only absolutely irreducible defining characteristic $\mathrm{SL}_2(q)$-modules that are tensor products are the ones that arise as non-trivial tensor products in Theorem 5.3.2.

Let F be a field. Then $\mathrm{SL}_2(F)$ is generated by matrices:

$$x(a) := \begin{bmatrix} 1 & 0 \\ a & 1 \end{bmatrix}, \quad y(b) := \begin{bmatrix} b & 0 \\ 0 & b^{-1} \end{bmatrix} \quad \text{and} \quad z := \begin{bmatrix} 0 & 1 \\ -1 & 0 \end{bmatrix},$$

where a ranges over the whole of F, and b ranges over $F^\times$. We extend to $\mathrm{GL}_2(F)$ by adding elements $w(c) := \mathrm{diag}(c, 1)$ where c ranges over $F^\times$. If $F = \mathbb{F}_q$ is finite, then we let λ be a primitive element of $F^\times$ and define $x := x(1)$, $y := y(\lambda)$ and $w := w(\lambda)$, so that $\langle x, y, z \rangle = \mathrm{SL}_2(q)$ and $\langle x, y, z, w \rangle = \mathrm{GL}_2(q)$.

If $(\varepsilon_0, \varepsilon_1)$ is a basis for V then $(e_0, \ldots, e_n)$ is a basis for V_{n+1}, where $e_i = \varepsilon_0^{n-i} \varepsilon_1^i$ for all i, as in Definition 5.2.2. For $m \in \mathbb{Z}$, we define the linear representation $\det^m$ of $\mathrm{SL}_2(F)$ and $\mathrm{GL}_2(F)$ by $g . \det^m = [(\det g)^m]$ for $g \in \mathrm{GL}_2(F)$: this is of course trivial for $\mathrm{SL}_2(F)$. The following is straightforward.

Lemma 5.3.4   *The actions of $x(a)$, $y(b)$, z and $w(c)$ on V_{n+1} are given by:*

$$e_i.x(a) = \sum_{j=0}^{i} a^{i-j} \binom{i}{j} e_j, \qquad e_i.z = (-1)^i e_{n-i},$$
$$e_i.y(b) = b^{n-2i} e_i, \qquad e_i.w(c) = c^{n-i} e_i,$$

where $0 \leqslant i \leqslant n$.

We first prove that if char $F \geqslant n$ then V_n is equipped with a non-degenerate bilinear form, of symplectic type when n is even and orthogonal type otherwise.

Proposition 5.3.5   (i) *The action $\mathrm{SL}_2(F)$ on V_{n+1} is via isometries of a bilinear form on V_{n+1} with matrix $B_{n+1} := \mathrm{antidiag}(a_0, \ldots, a_n)$, where*

$$a_i = (-1)^i i!(n-i)! = (-1)^i \frac{n!}{\binom{n}{i}}.$$

(ii) *The element $w(c)$ scales the form by a factor of c^n.*

(iii) *If char $F > n$, or char $F = 0$, then this form is non-degenerate and unique up to scalar multiplication.*

Proof   The non-degeneracy of $B := B_{n+1}$ when char $F > n$ is clear, and its uniqueness follows from Lemma 5.3.1. So we just need to verify the equations $XBX^{\mathsf{T}} = B$, $YBY^{\mathsf{T}} = B$, $ZBZ^{\mathsf{T}} = B$ and $WBW^{\mathsf{T}} = c^n B$, where X, Y, Z, W are the matrices of the actions of $x(a)$, $y(b)$, z and $w(c)$, respectively, on V_{n+1}. This is straightforward for Y, Z and W, by Lemma 5.3.4.

It is convenient to index the rows and columns of X and B by integers ranging from 0 to n. So $X_{ij} = a^{i-j} \binom{i}{j}$ by Lemma 5.3.4 and

$$(BX^{\mathsf{T}})_{ij} = (-1)^i\, i!\, (n-i)! \binom{j}{n-i} a^{i+j-n},$$

and hence

$$(XBX^{\mathsf{T}})_{ij} = \sum_{k=0}^{n} (-1)^k\, k!\, (n-k)! \binom{i}{k} \binom{j}{n-k} a^{i+j-n}$$

$$= a^{i+j-n}\, i!\, j! \sum_{k=n-j}^{i} \frac{(-1)^k}{(i-k)!\,(k-(n-j))!}.$$

Putting $l = k - (n-j)$ and $t = i+j-n$, the sum in this expression simplifies to $(-1)^{n-j} \sum_{l=0}^{t} \frac{(-1)^l}{(t-l)!\,l!}$, which (by considering the binomial expansion of $(1-1)^t$) is $(-1)^{n-j}$ when $t = 0$ and 0 otherwise. It now follows that $XBX^{\mathsf{T}} = B$, and so X acts via isometries as claimed. $\square$

In the following proposition, we analyse the behaviour of all of the p-restricted representations. Recall that the images of all p-restricted representations lie in $\mathscr{S}_2^*$: see Definition 5.1.15. Since there is more than one family of classical group involved, we write δ_Ω, for example, to denote the diagonal automorphism of the group Ω: we shall use this notation frequently during this chapter.

Proposition 5.3.6 (i) *If n is even and $p \geqslant n > 2$, then there is a single conjugacy class of self-normalising $\mathscr{S}_2$-subgroups of $\Omega = \mathrm{Sp}_n(q)$ isomorphic to $S = \mathrm{SL}_2(q)$. This is stabilised by δ_Ω and ϕ_Ω, which induce δ_S and ϕ_S, respectively.*

(ii) *If $n \equiv \pm 3 \pmod 8$ and $p \geqslant n$, then there is a single conjugacy class of self-normalising $\mathscr{S}_2$-subgroups of $\Omega = \Omega_n(q)$ isomorphic to $S = \mathrm{L}_2(q)$. This is stabilised by δ_Ω and ϕ_Ω which induce δ_S and ϕ_S, respectively.*

(iii) *If $n \equiv \pm 1 \pmod 8$ and $p \geqslant n > 1$, then there are exactly two conjugacy classes of self-normalising $\mathscr{S}_2$-subgroups of $\Omega = \Omega_n(q)$ isomorphic to $\mathrm{L}_2(q).2$. They are interchanged by δ_Ω and stabilised by ϕ_Ω, which induces the field automorphism of $\mathrm{L}_2(q)$.*

Proof Recall that $\mathrm{Out}\, \mathrm{L}_2(p^e)$ is of order $(2, p - 1)e$. We shall first determine the isomorphism type of the normaliser of the image of the representation associated with the module V_{n+1} of $\mathrm{SL}_2(q)$ or $\mathrm{L}_2(q)$ in $\Omega = \mathrm{Sp}_n(q)$ or $\Omega_n(q)$, and the action of any diagonal automorphisms of Ω, and hence the number of Ω-classes: recall that by Corollary 5.1.10 each p-restricted representation yields a single C-class of subgroups. Then we shall consider the action of any field automorphisms. Note that by Theorem 5.3.2 we are interested in the case $1 < n < \mathrm{char}\, F$, and so $p \neq 2$.

Suppose first that n is odd, and hence the dimension $n + 1$ of V_{n+1} is even. Then the invariant form is anti-symmetric by Proposition 5.3.5 and $-I_2$ induces $-I_{n+1}$, so we have an embedding of $\mathrm{SL}_2(F)$ into $\Omega = \mathrm{Sp}_{n+1}(F)$. If $\mathrm{char}\, F > 2$ is finite, then $w(\lambda)$ with λ a primitive multiplicative field element scales the form by a non-square, so the diagonal automorphism of $\mathrm{L}_2(q)$ is not induced by an element of $\overline{\Omega}$.

Next assume that $n \geqslant 2$ is even, and consider $W_{n+1} := V_{n+1} \otimes \det^{-n/2}$. For this module $w(c)$, which by Lemma 5.3.4 acts via $e_i.w(c) = c^{n/2-i}e_i$, is an isometry of B_{n+1} of determinant 1. The scalars of $\mathrm{GL}_2(F)$ all act trivially on W_{n+1}, so we get an embedding of $\mathrm{PGL}_2(F)$ into $\mathrm{SO}_{n+1}(F)$. We must now calculate spinor norms (recall that q is odd), as in Definition 1.6.10. Restricting to $\langle e_i, e_{n-i} \rangle$ (for $0 \leqslant i < n/2$), we see that $w(c)$ acts as:

$$\begin{bmatrix} c^{n/2-i} & 0 \\ 0 & c^{i-n/2} \end{bmatrix} = \begin{bmatrix} 0 & 1 \\ 1 & 0 \end{bmatrix} \begin{bmatrix} 0 & c^{n/2-i} \\ c^{i-n/2} & 0 \end{bmatrix},$$

a product of reflections in $(1, -1)$ and $(1, -c^{n/2-i})$, of norms $-a_i$ and $-a_i c^{n/2-i}$. Also $w(c)$ acts as the identity on $\langle e_{n/2} \rangle$. Therefore $w(c)$ has spinor norm (modulo squares) $\prod_{i=0}^{n/2-1} a_i^2 c^{n/2-i}$, which is $c^{n(n+2)/8}$ modulo squares. If c is a non-square then this is a square when $n \equiv 0, 6 \pmod 8$ and a non-square when $n \equiv 2, 4 \pmod 8$. So if $n \equiv 0, 6 \pmod 8$ we get an embedding of $\mathrm{PGL}_2(F)$ as an absolutely irreducible subgroup of $\Omega_{n+1}(F, B_{n+1})$ and if $n \equiv 2, 4 \pmod 8$ then $\mathrm{L}_2(F)$ is an absolutely irreducible subgroup of $\Omega_{n+1}(F, B_{n+1})$.

The above deals with the diagonal automorphisms of $\mathrm{SL}_2(F)$. We now let $F = \mathbb{F}_q$, and consider the effect of the field automorphism ϕ. In Section 1.7.1, we defined the action of ϕ on matrices over $\mathbb{F}_q$ to be the application of ϕ to all entries of the matrix. In all cases but $\mathbf{O}^-$, the field automorphism ϕ of the classical group Ω was defined simply by applying ϕ to the matrices in Ω. We therefore see from our definitions of the generators of the action $\mathrm{SL}_2(q)_{n+1}$ of $\mathrm{SL}_2(q)$ on V_{n+1} that the field automorphism ϕ of $\mathrm{SL}_2(q)$ induces the field automorphism ϕ on the entries of the matrices in $\mathrm{SL}_2(q)_{n+1}$ for all n. So, in the conjugate of $\Omega_{n+1}(q)$ (n even) or $\mathrm{Sp}_{n+1}(q)$ (n odd) that consists of isometries of B_{n+1} described above, the field automorphism ϕ normalises $\mathrm{SL}_2(q)_{n+1}$ and induces ϕ on $\mathrm{SL}_2(q)$. This form is not our standard form for $\mathrm{Sp}_{n+1}(q)$ and $\mathrm{SO}_{n+1}(q)$ that we defined in Subsection 1.5.7, but [6, Theorem 4, Proposition 11] states that the field automorphisms are independent of the form for these groups. So we conclude that the (class of the) image of $\mathrm{SL}_2(q)$ in $\mathrm{Sp}_{n+1}(q)$ or $\Omega_{n+1}(q)$ is stabilised by ϕ.

The above calculations show also that the image of the group automorphism ϕ of $\mathrm{SL}_2(q)$ on the (equivalence class of the) module V_{n+1} is the same as the image of the field automorphism ϕ on the module. Assuming that F is finite with char $F > n$, we know from Theorem 5.3.2 that V_{n+1} is absolutely irreducible and is not stabilised by any non-trivial field automorphisms of F. Therefore the field automorphisms of $\mathrm{SL}_2(q)$ are not induced within the conformal group in which $\mathrm{SL}_2(q)_{n+1}$ lies. It follows that the normaliser of $\mathrm{SL}_2(q)_{n+1}$ within the projective conformal group is $\mathrm{PGL}_2(q)$ in all cases. $\qquad\square$

Having dealt with the p-restricted modules, we now consider the non-trivial tensor product representations of $\mathrm{SL}_2(q)$ that arise in Theorem 5.3.2. As we saw in Proposition 5.1.14, there are three ways in which such candidates could belong to $\mathscr{S}_2^*$, the third of which does not occur for $\mathrm{SL}_2(q)$. We shall consider certain modules, and later prove that these are the only ones which arise.

Proposition 5.3.7 *If $q > 2$ then there is a single conjugacy class of self-normalising $\mathscr{S}_2$-subgroups of $\mathrm{Sp}_8(q)$ isomorphic to $\mathrm{SL}_2(q^3).3$. The class stabiliser is $\langle \delta, \phi \rangle$, where δ induces the diagonal automorphism of $\mathrm{SL}_2(q^3)$.*

Proof Let $q = p^e$, let $\sigma = \phi^e$ be the automorphism $x \mapsto x^q$ of $\mathbb{F}_{q^3}$, let ρ_1 be the natural representation of $\mathrm{SL}_2(q^3)$, and let $\rho := \rho_1 \otimes \rho_1^\sigma \otimes \rho_1^{\sigma^2}$, with associated module $M = V_2 \otimes V_2^\sigma \otimes V_2^{\sigma^2}$. By Proposition 1.9.4, the image of ρ over

$\mathbb{F}_{q^3}$ preserves a form of symplectic or orthogonal plus type when q is odd or even, respectively. We are tensoring three faithful representations of $\mathrm{SL}_2(q^3)$, so $\mathrm{Im}\,\rho \cong \mathrm{SL}_2(q^3)$. By Proposition 1.8.12, self-dual modules over $\mathbb{F}_{q^3}$ remain self-dual when written over $\mathbb{F}_q$. Since orthogonal and symplectic forms over $\mathbb{F}_q$ have the same type when regarded as being over $\mathbb{F}_{q^3}$, the forms preserved by these subgroups remain symplectic and orthogonal plus type when the representations are written over $\mathbb{F}_q$. Since we are not dealing with Case $\mathbf{O}^+$ in dimension 8, we will assume from now on that p is odd, so $\Omega = \mathrm{Sp}_8(q)$.

Let c be a non-square in $\mathbb{F}_q^\times$, and hence also in $\mathbb{F}_{q^3}^\times$. The element $w(c)$ of $\mathrm{GL}_2(q^3)$ is a similarity which multiplies the form in dimension 8 by a factor c^3, which is also a non-square, and so no scalar multiple of $w(c)$ lies in $\mathrm{S}_8(q)$. Therefore the diagonal outer automorphism of $\mathrm{SL}_2(q^3)$ is induced by a diagonal outer automorphism of $\mathrm{Sp}_8(q)$.

The class of the image of ρ is also stabilised by the field automorphism σ of $\mathrm{SL}_2(q^3)$ of order 3, since σ permutes the three copies of ρ_1. By Corollary 1.8.14 we deduce that σ is therefore induced by an element of $\mathrm{GL}_n(q)$, and then by Lemma 1.8.9 that σ is induced by an element of $\mathrm{CSp}_8(q)$. Since the index of $\mathrm{S}_8(q)$ in $\mathrm{PCSp}_8(q)$ is 2, the automorphism σ must be induced by conjugation by an element of $\mathrm{Sp}_8(q)$.

By Theorem 5.3.2, the automorphisms ϕ^k for $1 \leqslant k < e$ do not stabilise the module, and so these automorphisms of $\mathrm{SL}_2(q^3)$ are not induced by elements of $\mathrm{GL}_8(q)$. Hence $\mathrm{SL}_2(q^3).3$ is a self-normalising subgroup of $\mathrm{Sp}_8(q)$. Since there is a single class of such subgroups of $\mathrm{Sp}_8(q)$, the field automorphism ϕ of $\mathrm{Sp}_8(q)$ must lie in the class stabiliser. $\qquad\square$

Proposition 5.3.8 *For odd q, there is a single scalar-normalising conjugacy class of $\mathscr{S}_2$-subgroups of $\Omega_9(q)$ isomorphic to $\mathrm{L}_2(q^2).2$. The outer automorphism of $\mathrm{L}_2(q^2)$ of order 2 in this extension is the field automorphism when $q \equiv \pm 1 \pmod 8$, and the product of the field automorphism of order 2 and a diagonal automorphism when $q \equiv \pm 3 \pmod 8$. The class stabiliser is $\langle \delta, \phi \rangle$, where δ induces the diagonal automorphism of $\mathrm{SL}_2(q^2).2$.*

Proof Let ρ_1 be the 3-dimensional p-restricted representation of $\mathrm{SL}_2(q^2)$, and let $\rho := \rho_1 \otimes \rho_1^\sigma$, with module $M = V_3 \otimes V_3^\sigma$, where $q = p^e$ and $\sigma = \phi^e$.

The form preserved by $\mathrm{Im}\,\rho$ is orthogonal, by Proposition 1.9.4, so Ω is $\Omega_9(q)$. Recall that $\mathrm{Out}\,\Omega_9(p^e) \cong 2 \times e$, where the first direct factor extends $\Omega_9(q)$ to $\mathrm{SO}_9(q)$, and the second induces field automorphisms.

The representation ρ is a tensor product of faithful representations of $\mathrm{L}_2(q^2)$, so $\mathrm{Im}\,\rho \cong \mathrm{L}_2(q^2)$. In $\mathrm{Im}\,\rho$, the diagonal automorphism is induced by conjugation by elements of $\mathrm{CGO}_9(q)$ by Proposition 5.3.5. Since σ interchanges V_3 and V_3^σ, it stabilises ρ and hence by Corollary 1.8.14 and Lemma 1.8.9 also lies in

$CGO_9(q)$. Now, $CGO_9(q) \cong SO_9(q) \times C_{q-1}$ and so both of these automorphisms are induced by elements of $SO_9(q)$.

We have to decide which automorphisms are induced by elements of $\Omega_9(q)$. For this calculation, it is slightly more convenient to take the form preserved in V_3 to be antidiag$(1, 1, 1)$. The diagonal automorphism of the image of the action of $L_2(q^2)$ on V_3 is induced by the matrix diag$(\lambda, 1, \lambda^{-1})$, where λ is a primitive element of $\mathbb{F}_{q^2}^\times$. So, in the image of the action on $V_3 \otimes V_3^\sigma$, the form preserved is antidiag$(1, 1, 1, 1, 1, 1, 1, 1, 1)$ and the diagonal automorphism is induced by

$$g := \mathrm{diag}(\lambda^{q+1}, \lambda, \lambda^{1-q}, \lambda^q, 1, \lambda^{-q}, \lambda^{q-1}, \lambda^{-1}, \lambda^{-(q+1)}).$$

We need to determine the spinor norm of g when rewritten as an element of $\Omega_9(q)$. By decomposing g as a direct sum of four matrices, one of which is the identity, we reduce this to the following problems.

(i) Find the spinor norm of $g_1 := \mathrm{diag}(\lambda, \lambda^q, \lambda^{-q}, \lambda^{-1})$, with preserved form antidiag$(1, 1, 1, 1)$, when rewritten as an element of $\Omega_4^\pm(q)$.

(ii) Find the spinor norm of $g_2 := \mathrm{diag}(\lambda^{1-q}, \lambda^{q-1})$, with preserved form given by antidiag$(1, 1)$, when rewritten as an element of $\Omega_2^\pm(q)$.

(iii) Find the spinor norm of $g_3 := \mathrm{diag}(\lambda^{q+1}, \lambda^{-(1+q)})$ with preserved form antidiag$(1, 1)$ as an element of $\Omega_2^+(q)$.

As we observed in Lemma 1.12.3, $\Omega_4^+(q)$ is isomorphic to a central product of two copies of $SL_2(q)$, whereas $\Omega_4^-(q)$ is isomorphic to $L_2(q^2)$, and neither of these have elements of order $|g_1| = q^2 - 1$ so g_1 has spinor norm -1. In (ii) $|g_2| = q + 1$, and neither $\Omega_2^+(q)$ nor $\Omega_2^-(q)$ have elements of this order, so the spinor norm is -1, whilst in (iii), $|g_3| = q - 1$, and $\Omega_2^+(q)$ has no element of this order, so the spinor norm is again -1. So when we rewrite g over $\mathbb{F}_q$, its spinor norm is -1 and hence the diagonal automorphism of $SL_2(q^2)$ is induced by conjugation by an element of $SO_9(q) \setminus \Omega_9(q)$ for all (odd) q.

The field automorphism σ of $\mathrm{Im}\,\rho$ is induced by $-P \in SO_9(q^2)$, where P is the permutation matrix of $(2, 4)(3, 7)(6, 8) \in S_9$. Since P is already over $\mathbb{F}_q$, it follows easily from Defintion 1.6.10 that $-P$ has spinor norm 1 when $q \equiv \pm 1 \pmod 8$ and -1 when $q \equiv \pm 3 \pmod 8$. So, if $q \equiv \pm 1 \pmod 8$, then σ is induced by an element of $\Omega_9(q)$. If $q \equiv \pm 3 \pmod 8$, the product of σ with the diagonal automorphism of $SL_2(q^2)$ is induced by an element of $\Omega_9(q)$.

As in the previous examples, the automorphisms ϕ^k for $1 \leqslant k < e$ do not stabilise ρ, and so are not induced by elements of $GL_9(q)$. However, there is a single class of such subgroups in $\Omega_9(q)$, so all outer automorphisms of $\Omega_9(q)$ lie in the class stabiliser. □

The following theorem is our main description of the $\mathscr{S}_2$-subgroups that are isomorphic to extensions of $SL_2(q)$ or $L_2(q)$ in dimensions up to 12.

Theorem 5.3.9 *The only candidates for $\mathscr{S}_2^*$-maximal subgroups of the classical groups in dimensions up to 12, other than in extensions of $\mathrm{O}_8^+(q)$, that are isomorphic to extensions of $\mathrm{SL}_2(q)$ or $\mathrm{L}_2(q)$ are as described in Propositions 5.3.6, 5.3.7, and 5.3.8.*

Proof The p-restricted modules are described in Proposition 5.3.6, and all other modules that are expressed in Theorem 5.3.2 as a tensor product with only one non-trivial factor are algebraic conjugates of the p-restricted modules, and hence yield conjugate subgroups by Corollary 5.1.10. Therefore we only need to consider modules M with more than one non-trivial tensor factor.

For these modules M, at least one of the first two conditions specified in Proposition 5.1.14 must apply to M. If the first condition applies, then M can be written over a proper subfield so (changing notation), we consider representations of $\mathrm{SL}_2(q^s)$ with $s > 1$ that can be realised over $\mathbb{F}_q$. By Corollary 5.3.3, up to algebraic conjugacy M is therefore one of the following:

  (i) the $\mathrm{SL}_2(q^2)$-module $V_2 \otimes V_2^\sigma$;
 (ii) the $\mathrm{SL}_2(q^3)$-module $V_2 \otimes V_2^\sigma \otimes V_2^{\sigma^2}$;
(iii) the $\mathrm{SL}_2(q^2)$-module $V_3 \otimes V_3^\sigma$ for q odd.

The $\mathrm{SL}_2(q^2)$-module $V_2 \otimes V_2^\sigma$ is the natural module for $\Omega_4^-(q)$ and so does not correspond to an $\mathscr{S}_2$-subgroup. The other two cases are as considered in Propositions 5.3.7, and 5.3.8.

Thus (reverting to the original notation), we may assume that M is a module for $\mathrm{SL}_2(q)$ that cannot be written over a proper subfield of $\mathbb{F}_q$. For the second condition of Proposition 5.1.14 to apply, M would have to have a preserved form other than the induced symplectic or symmetric form. By Lemma 1.8.8 (iii), there is no σ-Hermitian form on M. The only other possibility is that M has a quadratic form when $p = 2$. By Proposition 1.9.4 (iii), this does indeed happen, but the quadratic form is of plus-type, so the representation associated with M is a subgroup of $\Omega_4^+(q)$ or $\Omega_8^+(q)$. $\qquad\qquad\square$

We finish with a result which, whilst it will not be used until Chapter 6, is most convenient to prove here.

Proposition 5.3.10 *Let $n \geqslant m \geqslant 1$. Then $V_{n+1} \otimes V_{m+1}$ has a proper submodule for $\mathrm{GL}_2(F)$ isomorphic to $V_{n-m+1} \otimes \det^m$.*

Proof The basis we take for this submodule is:

$$h_i := \sum_{j=0}^m (-1)^j \binom{m}{j} e_{i+j} \otimes f_{m-j}, \quad \text{for } 0 \leqslant i \leqslant n - m,$$

where $e_0, \ldots, e_n$ and $f_0, \ldots, f_m$ are our standard bases of V_{n+1} and V_{m+1}. It is quite straightforward to check the required actions of $y(b)$, z and $w(c)$ are as expected. Namely that:

$$h_i.y(b) = b^{n-m-2i}h_i, \quad h_i.z = (-1)^i h_{n-m-i} \quad \text{and} \quad h_i.w(c) = c^{n-i}h_i.$$

It is considerably harder to check that the action of $x(a)$ is as it should be. We first need a couple of identities involving binomial coefficients. Note that some of what we produce below involves binomial coefficients $\binom{N}{M}$ where N and M are integers with $N \geqslant 0$ and either $M < 0$ or $M > N$, and of course $\binom{N}{M} = 0$ for such N and M. The identities are:

(i) $\binom{m}{j}\binom{m-j}{\ell} = \binom{m}{\ell}\binom{m-\ell}{j}$ for all (integers) $m, j, \ell \geqslant 0$; and

(ii) $\sum_{j=0}^{N}(-1)^j\binom{N}{j}\binom{i+j}{k} = (-1)^N\binom{i}{k-N}$ for all $N, i, k \geqslant 0$.

For the first of these, we simply write out the formulae for both sides in terms of factorials; they both evaluate to $\frac{m!}{j!\ell!(m-j-\ell)!}$. The second seems to be best approached by induction on N, and is a special case of [33, Equation (5.24)]. Thus we can now calculate that:

$$\begin{aligned}
h_i.x(a) &= \sum_{j=0}^{m}(-1)^j\binom{m}{j}e_{i+j}x(a) \otimes f_{m-j}x(a) \\
&= \sum_{j=0}^{m}\sum_{\ell=0}^{m-j}\sum_{k=0}^{i+j}(-1)^j\binom{m}{j}a^{i+j-k}\binom{i+j}{k}a^{m-j-\ell}\binom{m-j}{\ell}e_k \otimes f_\ell \\
&= \sum_{\ell=0}^{m}\sum_{j=0}^{m-\ell}\sum_{k=0}^{i+j}(-1)^j a^{i-(k+\ell-m)}\binom{m}{\ell}\binom{m-\ell}{j}\binom{i+j}{k}e_k \otimes f_\ell,
\end{aligned}$$

where the last of these equalities uses the first of the above binomial identities and expresses the summation for the range $0 \leqslant j, \ell$, $j + \ell \leqslant m$ differently. We now focus attention to the inner two summations. Clearly $0 \leqslant k \leqslant i + m - \ell$, and if $k > i + j$ then $\binom{i+j}{k} = 0$, so we can extend the range for k in this manner without affecting the sum. We then swap the inner two sums to get:

$$\begin{aligned}
h_i.x(a) &= \sum_{\ell=0}^{m}\sum_{k=0}^{i+m-\ell}\sum_{j=0}^{m-\ell}(-1)^j a^{i-(k+\ell-m)}\binom{m}{\ell}\binom{m-\ell}{j}\binom{i+j}{k}e_k \otimes f_\ell \\
&= \sum_{\ell=0}^{m}\sum_{k=0}^{i+m-\ell}a^{i-(k+\ell-m)}\binom{m}{\ell}(-1)^{m-\ell}\binom{i}{k+\ell-m}e_k \otimes f_\ell \\
&= \sum_{j=0}^{m}\sum_{k=0}^{i+j}(-1)^j a^{i-(k-j)}\binom{m}{j}\binom{i}{k-j}e_k \otimes f_{m-j},
\end{aligned}$$

where the second of the above lines uses our second binomial identity, and the third line uses the substitution $j = m - \ell$ (and of course the identity $\binom{m}{\ell} = \binom{m}{m-\ell}$). But $\binom{i}{k-j} = 0$ if $k < j$, so the inner summation has range $j \leqslant k \leqslant i + j$. Letting $\ell = k - j$, we get:

$$\begin{aligned}
h_i.x(a) &= \sum_{j=0}^{m}\sum_{\ell=0}^{i}(-1)^j a^{i-\ell}\binom{m}{j}\binom{i}{\ell}e_{\ell+j} \otimes f_{m-j} \\
&= \sum_{\ell=0}^{i}a^{i-\ell}\binom{i}{\ell}\left(\sum_{j=0}^{m}(-1)^j\binom{m}{j}e_{\ell+j} \otimes f_{m-j}\right) \\
&= \sum_{\ell=0}^{i}a^{i-\ell}\binom{i}{\ell}h_\ell,
\end{aligned}$$

as required. $\square$

5.4 The groups $\mathrm{SL}_n(q)$ and $\mathrm{SU}_n(q)$ for $n \geqslant 3$

In this section we shall determine the $\mathscr{S}_2^*$-subgroups that are isomorphic to $\mathrm{SL}_n(q)$, $\mathrm{SU}_n(q)$ and their central quotients for $n \geqslant 3$. We shall find their normalisers in the quasisimple groups, and their class stabilisers in the corresponding conformal group, and hence the number of conjugacy classes in the quasisimple groups. The groups $\mathrm{SL}_n(q)$ and $\mathrm{SU}_n(q)$ correspond to the groups of Lie type $A_{n-1}(q)$ and $^2A_{n-1}(q)$. Let $G \cong \mathrm{SL}_n(q)$ or $\mathrm{SU}_n(q)$, and let V_n be a natural (n-dimensional) module for G, with V_n^* denoting the dual of V_n. Note that $V_n^* \cong V_n^\sigma$ for $\mathrm{SU}_n(q)$, but not for $\mathrm{SL}_n(q)$.

It is helpful to relate V_n, V_n^* and the constituents of some of their tensor products and powers to the modules listed by highest weight in the tables for Lie type A_{n-1} in [84].

Following [84], we choose $\lambda_{n-1} = (0, \ldots, 0, 1)$ and $\lambda_1 = (1, 0, \ldots, 0)$ to correspond to the natural module V_n and its dual V_n^*, respectively. It is shown in [77, Theorem 1.1] that $\Lambda^2(V_n)$ is (absolutely) irreducible and corresponds to $\lambda_{n-2} = (0, \ldots, 0, 1, 0)$ (although the numbering of nodes is in the reverse order there). From Lemma 5.1.11 we see that the highest weight of $V \otimes V$ is $2\lambda_{n-1} = (0, \ldots, 0, 2)$. From [84, Table 2 and §§A.6–A.15] we find that, for $p > 2$, the irreducible module with highest weight $2\lambda_{n-1}$ has dimension $n(n+1)/2$ and, since $V \otimes V$ has the submodule $\mathrm{S}^2(V_n)$ with this dimension (which is more than half that of $V \otimes V$), it follows that $\mathrm{S}^2(V_n)$ corresponds to $2\lambda_{n-1}$ when $p > 2$.

Similarly, $V^* \otimes V$ has a constituent W with highest weight equal to the sum $\lambda_1 + \lambda_{n-1} = (1, 0, \ldots 0, 1)$, which has dimension $n^2 - 2$ if $p \mid n$ and $n^2 - 1$ otherwise. This is the *adjoint module*, which we shall study in detail in Subsection 5.4.1. It can also be shown that, for $p > 3$, $\mathrm{S}^3(V_n)$ corresponds to the absolutely irreducible module with highest weight $(0, \ldots, 0, 3)$, but we shall only need to consider this module for $n = 3$.

Generically, the non-trivial p-restricted representations of $\mathrm{SL}_n^\pm(q)$ of degree at most $\mathrm{O}(n^2)$ are as in Table 5.1. Moreover these modules are all distinct if $n \geqslant 5$. By Lübeck [84, Table 2], Table 5.1 contains all non-trivial members of $\mathscr{M}$ of dimension at most $(n-1)^3/8$ when $n \geqslant 12$. Note that $(n-1)^3/8 \geqslant n^2$ for $n \geqslant 11$. Inspecting Appendices A.6–A.15 of Lübeck [84] allows us to complete the determination of members of $\mathscr{M}$ of dimension 12.

Proposition 5.4.1   *The p-restricted modules for $\mathrm{SL}_n(q)$ and $\mathrm{SU}_n(q)$, with $n \geqslant 3$, of dimension at most 12 are as given in Table 5.2, except that the modules $\mathrm{S}^2(V_n)$ and $\mathrm{S}^2(V_n^*)$ require $p > 2$, and the modules $\mathrm{S}^3(V_n)$ and $\mathrm{S}^3(V_n^*)$ require $p > 3$.*

We now determine which of these p-restricted modules give rise to $\mathscr{S}_2$-subgroups. Note that the modules V_n and V_n^* are the natural modules, and

Table 5.1 *Low-dimensional modules of* $\mathrm{SL}_n(q)$ *and* $\mathrm{SU}_n(q)$

module(s)	dimension	primes p
V_n and V_n^*	n	all
$\Lambda^2(V_n)$ and $\Lambda^2(V_n^*)$	$\frac{1}{2}n(n-1)$	all
$\mathrm{S}^2(V_n)$ and $\mathrm{S}^2(V_n^*)$	$\frac{1}{2}n(n+1)$	$p \neq 2$
W	$n^2 - 2$	$p \mid n$
W	$n^2 - 1$	$p \nmid n$

Table 5.2 *Modules of* $\mathrm{SL}_n(q)$ *and* $\mathrm{SU}_n(q)$ *of dimension at most* 12

dimension n	members of $\mathcal{M}$ of dimension at most 12
$n \geqslant 13$	1
$6 \leqslant n \leqslant 12$	$1, V_n, V_n^*$
$n = 5$	$1, V_5, V_5^*, \Lambda^2(V_5), \Lambda^2(V_5^*)$
$n = 4$	$1, V_4, V_4^*, \Lambda^2(V_4) \cong \Lambda^2(V_4^*), \mathrm{S}^2(V_4), \mathrm{S}^2(V_4^*)$
$n = 3$	$1, V_3, V_3^*, \mathrm{S}^2(V_3), \mathrm{S}^2(V_3^*), W, \mathrm{S}^3(V_3), \mathrm{S}^3(V_3^*)$

so do not give rise to $\mathscr{S}_2$-candidates. Recall that as discussed in the proof of Corollary 5.1.10, we need consider p-restricted modules only up to duality.

Proposition 5.4.2 *The module* $\Lambda^2(V_4)$ *does not yield any* $\mathscr{S}_2$-candidates.

Proof As we observed earlier, the highest weight of $\Lambda^2(V_4)$ is $(0,1,0)$, so $\Lambda^2(V_4)$ is self-dual. Therefore the image of the representation consists of isometries of a symmetric bilinear form, and in fact $\Lambda^2(V_4)$ for $\mathrm{SL}_4(q)$ is isomorphic to the natural module for $\Omega_6^+(q)$. In the case of $\mathrm{SU}_4(q)$, the image of the representation corresponding to $\Lambda^2(V_4)$ acts via isometries of both an σ-Hermitian and a symmetric bilinear form, by Proposition 5.2.4. It can therefore be written over $\mathbb{F}_q$, and then becomes the natural module for $\Omega_6^-(q)$ respectively. $\square$

Proposition 5.4.3 *Let* M *be one of* $\mathrm{S}^2(V_3), \mathrm{S}^3(V_3), \mathrm{S}^2(V_4)$ *or* $\Lambda^2(V_5)$, *with corresponding representation* $\rho_M : G \to \Omega$. *For* $G = \mathrm{SL}_n(q)$ *the image of* ρ_M *preserves no form, and for* $G = \mathrm{SU}_n(q)$ *it preserves only a unitary form. Each outer automorphism of* G *is induced in* $\mathrm{Aut}\,\bar{\Omega}$, *and the diagonal automorphisms of* G *are precisely the ones arising in the conformal group of* Ω.

Proof By looking at the weights of $\mathrm{S}^2(V_3), \mathrm{S}^3(V_3), \mathrm{S}^2(V_4)$ and $\Lambda^2(V_5)$ calculated above, we see that these modules are not self-dual, and so $\mathrm{Im}\,\rho_M$ does not consist of similarities of a symplectic or orthogonal form. Furthermore, in the case of $G = \mathrm{SL}_n(q)$, if $q = p^e$ is square let $\tau = \phi^{e/2}$. By Proposition 5.1.9 (ii), (iii), the automorphism τ sends p-restricted modules to non-p-restricted modules, whilst duality sends p-restricted modules to p-restricted modules. Thus in particular $M^\tau \neq M^\gamma$, and so $M^{\tau\gamma} \neq M$ and the module M cannot be equipped with a unitary form.

In the case when $G = \mathrm{SU}_n(q)$ and for the modules $\mathrm{S}^2(V_3)$, $\mathrm{S}^2(V_4)$ and $\Lambda^2(V_5)$, we can use Proposition 5.2.4 to conclude that $\mathrm{Im}\,\rho_M$ preserves a nondegenerate σ-Hermitian form. For the module $\mathrm{S}^3(V_3)$ we restrict the induced σ-Hermitian form on $V^{\otimes 3}$ to the submodule version of $\mathrm{S}^3(V_3)$ to give us the same conclusion: we proved in all relevant circumstances that the submodule and quotient versions of $\mathrm{S}^3(V_3)$ are isomorphic in Lemma 5.2.8.

The final claim follows from Proposition 5.1.9.   □

The adjoint module W for $\mathrm{SL}_3^{\pm}(q)$ has highest weight $\lambda = (1,1)$, and so is self-dual by Remark 5.1.4. This module will be studied in Subsection 5.4.1.

We shall use the following result to determine which diagonal automorphisms can be realised in the quasisimple classical group of the image of the representation.

Lemma 5.4.4 *Let* $g_\lambda = \mathrm{diag}(\lambda, 1, 1, \ldots, 1)$ *be a typical outer automorphism of diagonal type for* $\mathrm{SL}_n(q)$ *or* $\mathrm{SU}_n(q)$ *(so that* $\lambda\lambda^q = 1$ *for* $\mathrm{SU}_n(q)$*). Then the eigenvalues and determinant of* g_λ *in its action on* $\mathrm{S}^2(V_n)$ *(*$3 \leqslant n \leqslant 4$*),* $\mathrm{S}^3(V_3)$ *and* $\Lambda^2(V_5)$ *are as given in Table 5.3.*

Proof This is a straightforward computation using the bases of $\Lambda^2(V)$ and $\mathrm{S}^i(V)$ described in Definitions 5.2.2 and 5.2.7.   □

Table 5.3 *Determinants and eigenvalues of some low degree representations*

module M	dim M	det g_λ	eigenvalues of g_λ
$\mathrm{S}^2(V_3)$	6	λ^4	$\lambda^2, \lambda, \lambda, 1, 1, 1$
$\mathrm{S}^3(V_3)$	10	λ^{10}	$\lambda^3, \lambda^2, \lambda^2, \lambda, \lambda, \lambda, 1, 1, 1, 1$
$\mathrm{S}^2(V_4)$	10	λ^5	$\lambda^2, \lambda, \lambda, \lambda, 1, 1, 1, 1, 1, 1$
$\Lambda^2(V_5)$	10	λ^4	$\lambda, \lambda, \lambda, \lambda, 1, 1, 1, 1, 1, 1$

We now study each representation associated with the above lemma in turn, and start by considering $\mathrm{S}^2(V_3)$.

Proposition 5.4.5   (i) *For odd* q*, there are exactly two conjugacy classes of* $\mathscr{S}_2$-*subgroups* G *of* $\mathrm{SL}_6(q)$ *isomorphic to* $\mathrm{SL}_3(q)$*. Their normaliser in* $\mathrm{GL}_6(q)$ *is generated by* G*, scalars and* δ^2*.*

(ii) *For odd* q*, there are exactly two conjugacy classes of* $\mathscr{S}_2$-*subgroups* G *of* $\mathrm{SU}_6(q)$ *isomorphic to* $\mathrm{SU}_3(q)$*. Their normaliser in* $\mathrm{CGU}_6(q)$ *is generated by* G*, scalars and* δ^2*.*

Proof The scalar μI_3 induces $\mu^2 I_6$ on $S^2(V_3)$, and q is odd by Proposition 5.4.1, so $SL_3^{\pm}(q)$ acts faithfully on $S^2(V_3)$.

Let g_λ be as in Lemma 5.4.4. If there exists $\mu I_6 \in GL_6^{\pm}(q)$ such that $\det \mu g_\lambda = 1$, then λ must be a cube in the cyclic group of order $q \pm 1$. Since the diagonal automorphisms of G have order $(3, q \pm 1)$, no non-trivial diagonal automorphism of G is induced by an element of $SL_6^{\pm}(q)$. Therefore G is scalar-normalising in $SL_6^{\pm}(q)$, by Proposition 5.4.3.

Since $\det g_\lambda$ in the action on $S^2(V_3)$ is a square, the class stabiliser in $GL_6^{\pm}(q)$ is generated by elements with square determinant. There is a single class of $\mathscr{S}_2$-subgroups isomorphic to G in the conformal group by Proposition 5.4.1 and Corollary 5.1.10. Therefore, since q is odd, there are two classes in $SL_6^{\pm}(q)$. $\square$

We now consider $S^3(V_3)$.

Proposition 5.4.6   (i) *For $p \geqslant 5$, there are exactly $(q-1, 10)$ conjugacy classes of $\mathscr{S}_2$-subgroups G of $SL_{10}(q)$ isomorphic to $PGL_3(q)$, which are scalar-normalising in $GL_{10}(q)$.*

(ii) *For $p \geqslant 5$, there are exactly $(q+1, 10)$ conjugacy classes of $\mathscr{S}_2$-subgroups G of $SU_{10}(q)$ isomorphic to $PGU_3(q)$, which are scalar-normalising in $CGU_{10}(q)$.*

Proof The scalar μI_3 induces $\mu^3 I_{10}$ on $S^3(V_3)$, so the image of this representation is isomorphic to $L_3^{\pm}(q)$. Note that $p \geqslant 5$ by Proposition 5.4.1.

Let g_λ be as in Lemma 5.4.4, then in the action on $S^3(V_3)$ the determinant of g_λ is a 10-th power. The diagonal automorphisms of $L_3(q)$ have order $(3, q \pm 1)$, so all diagonal automorphisms of G are induced by elements of $SL_{10}^{\pm}(q)$. Thus by Proposition 5.4.1 and Corollary 5.1.10, there is a single class of such groups G in the conformal group, with trivial stabiliser. $\square$

We next consider the module $S^2(V_4)$. Note that, for odd q, the quotient $SL_4^{\pm}(q)/\langle -I_4 \rangle$ has the structure $\frac{(q \pm 1, 4)}{2} \cdot SL_4^{\pm}(q)$.

Proposition 5.4.7   (i) *For odd q, there are exactly $(q-1, 5)$ conjugacy classes of $\mathscr{S}_2$-subgroups G of $SL_{10}(q)$ isomorphic to $\frac{(q-1,4)}{2} \cdot SL_4(q) \cdot \frac{(q-1,4)}{2}$. Their normaliser in $GL_{10}(q)$ is generated by G, scalars and δ^5.*

(ii) *For odd q, there are exactly $(q+1, 5)$ conjugacy classes of $\mathscr{S}_2$-subgroups G of $SU_{10}(q)$ isomorphic to $\frac{(q+1,4)}{2} \cdot SU_4(q) \cdot \frac{(q+1,4)}{2}$. Their normaliser in $CGU_{10}(q)$ is generated by G, scalars and δ^5.*

Proof The scalar μI_4 induces $\mu^2 I_{10}$ on $S^2(V_4)$ so, since q is odd by Proposition 5.4.1, the image of this representation is isomorphic to $\frac{(q \pm 1, 4)}{2} \cdot SL_4^{\pm}(q)$.

Let g_λ be as in Lemma 5.4.4, then $\det g$ in the action on $S^2(V_4)$ is a fifth power, whilst scalars in $GL_{10}(q)$ have determinant a tenth power. Therefore, squares of diagonal automorphisms of $L_4^{\pm}(q)$ are induced by elements of $SL_{10}^{\pm}(q)$, and by Proposition 5.4.3 the class stabiliser in the projective conformal group is $\langle \delta^5 \rangle$. The number of conjugacy classes then follows from Proposition 5.4.1 and Corollary 5.1.10.   □

Finally, we consider the module $\Lambda^2(V_5)$.

Proposition 5.4.8   (i) *For all q, there are exactly $(q{-}1, 2)$ conjugacy classes of $\mathscr{S}_2$-subgroups G of $\mathrm{SL}_{10}(q)$ isomorphic to $\mathrm{SL}_5(q)$. Their normaliser in $\mathrm{GL}_{10}(q)$ is generated by G, scalars and δ^2.*

(ii) *For all q, there are exactly $(q-1, 2)$ conjugacy classes of $\mathscr{S}_2$-subgroups of $\mathrm{SU}_{10}(q)$ isomorphic to $\mathrm{SU}_5(q)$. Their normaliser in $\mathrm{CGU}_{10}(q)$ is generated by G, scalars and δ^2.*

Proof   The scalar $\mu \mathrm{I}_5$ induces $\mu^2 \mathrm{I}_{10}$ on $\Lambda^2(V_5)$, and hence the image of the representation is $\mathrm{SL}_5^{\pm}(q)$.

Let g_λ be as in Lemma 5.4.4, then $\det g$ in the action on $\Lambda^2(V_5)$ is a fourth power. Since scalars in $\mathrm{GL}_{10}(q)$ have determinant a tenth power, no non-trivial diagonal automorphisms of $\mathrm{L}_5^{\pm}(q)$ are induced by elements of $\mathrm{SL}_{10}^{\pm}(q)$. Hence by Proposition 5.4.3 the class stabiliser in the projective conformal group is $\langle \delta^2 \rangle$. The number of conjugacy classes then follows from Proposition 5.4.1 and Corollary 5.1.10.   □

5.4.1 The adjoint module

We deal now with the adjoint modules W for $\mathrm{SL}_n^{\pm}(q)$. Since the concept of duality is critical here, we encourage the reader to read the discussion of duality in Section 1.8.

Let ρ be a representation of a group G, with FG-module V with basis $(e_1, \ldots, e_n)$, and let ρ^* be the corresponding dual representation with module V^* and dual basis $(e_1^*, \ldots, e_n^*)$. Let π be a representation of G corresponding to an FG-module U with basis $(f_1, \ldots, f_m)$. Then, for $w \in V^*$, $u \in U$, where w and u are regarded as row vectors with respect to the bases e_i^* and f_i, we identify $w \otimes u$ with the $n \times m$ matrix $w^\mathsf{T} u$ (so $e_i^* \otimes f_j$ is identified with the elementary matrix $\mathrm{E}_{i,j}$) and thus identify $V^* \otimes U$ with $\mathrm{M}_{n \times m}(F)$. Since, by Proposition 1.8.3, $w(g\,\rho^*) = w(g^{-\mathsf{T}} \rho)$, the action of G on $\mathrm{M}_{n \times m}(F)$ is given by $M.g = (g\rho)^{-1} M (g\pi)$.

Recall that our choice of unitary form means that $g \in \mathrm{GL}_n(q^2)$ lies in $\mathrm{GU}_n(q)$ if and only if $g g^{\sigma \mathsf{T}} = 1$, where σ is the field automorphism $x \mapsto x^q$ applied to the matrix entries. The following definition introduces some notation. The assertions therein are straightforward to verify.

Definition 5.4.9   Let V be the natural module for $G := \mathrm{GL}_n(F)$, or for $G := \mathrm{GU}_n(q)$ with $F = \mathbb{F}_{q^2}$, and consider the module $V^* \otimes V$, represented as $M_{n \times n}(F)$ with the conjugation action of G. For $G = GL_n(F)$, let M be the FG-module $M_{n \times n}(F)$ and, for $G = GU_n(q)$, let M be the $\mathbb{F}_q G$-module $\{ A \in M_{n \times n}(\mathbb{F}_{q^2}) \mid A^\mathsf{T} = A^\sigma \}$. Let U be the submodule of M consisting of all matrices of trace 0, and U' be the submodule of M consisting of all scalar matrices. Define the *adjoint module* W by

$$W = U/(U \cap U').$$

(Note that the Lie algebra for $GU_n(q)$ consists of anti-Hermitian matrices, rather than the Hermitian ones we have chosen to use here. Technically, U is the adjoint module, but we are interested in irreducible modules, hence our abuse of language.)

We leave it to the reader to verify that we can extend M to a module for $\langle GL_n(F), -\mathsf{T} \rangle$ or $\langle GU_n(q), -\mathsf{T} \rangle$ by defining $A.(-\mathsf{T}) := A^\mathsf{T}$ for all $A \in M$. Notice that the action of scalar matrices is trivial.

Lemma 5.4.10 *Let U, U' and W be as in Definition 5.4.9 with $F = \mathbb{F}_q$ in the linear case; so $G = GL_n^\pm(q)$. Then $\dim_{\mathbb{F}_q} M = n^2$, $\dim_{\mathbb{F}_q} U = n^2 - 1$, $\dim_{\mathbb{F}_q} U' = 1$; and $\dim_{\mathbb{F}_q} W = n^2 - 2$ when $p \mid n$ and $n^2 - 1$ otherwise. The $\mathbb{F}_q GL_n^\pm(q)$-modules M, U and U' are self-dual, the module W is self-dual, and W is absolutely irreducible as an $\mathbb{F}_q SL_n^\pm(q)$-module.*

Proof Clearly $\dim M = n^2$ when $G = GL_n(q)$. When $G = GU_n(q)$, the entries of a matrix (α_{ij}) in M with $i < j$ are determined by those with $i > j$, and $\alpha_{ii} \in \mathbb{F}_q$ for all i. So again $\dim M = n^2$. In both cases, the module U has codimension 1 in M, whereas U' has dimension 1. Note that $U' < U$ when $p \mid n$ and $U' \cap U = 0$ otherwise, so the dimension of W is as given.

It follows from Proposition 1.8.3 that the map $A \to A^\mathsf{T}$ defines a G-isomorphism from M to its dual. Under this map it is clear that both U and U' are self-dual, so W is also self-dual.

We then see from [84, Theorem 5.1 and Appendices A.6–A.15] that, for $n \geqslant 3$, the only non-trivial irreducible p-restricted self-dual module for $SL_n^\pm(q)$ of dimension at most n^2 has the same dimension as W, so W must be absolutely irreducible as an $SL_n^\pm(q)$-module. $\square$

Define a quadratic form Q on M by

$$Q(A) = \sum_{1 \leqslant i < j \leqslant n} (\alpha_{ij} \alpha_{ji} - \alpha_{ii} \alpha_{jj}),$$

and let β be the associated bilinear form. By observing that $Q(A)$ is the negative of the coefficient of x^2 in the characteristic polynomial of A, we see that Q is invariant under G. In odd characteristic, $Q(A) = \frac{1}{2}[\mathrm{tr}(A^2) - (\mathrm{tr}\, A)^2]$ and $\beta(A, B) = \mathrm{tr}(AB) - (\mathrm{tr}\, A)(\mathrm{tr}\, B)$. Notice that $U' = U^\perp$ with respect to β.

All of the relevant assertions in the remainder of this section will be with respect to the forms Q and β. It remains to determine when W is non-degenerate and, if so, to find the sign of the form on W.

Let D be the set of diagonal matrices of M, and let E be the set of matrices of M with all diagonal entries 0. For $1 \leqslant i < j \leqslant n$, let $E(i,j)$ be the subspace $\langle E_{i,j}, E_{j,i} \rangle$ of E when $G = \mathrm{GL}_n(q)$, and the subspace $\langle E_{i,j} + E_{j,i}, \lambda E_{i,j} + \lambda^\sigma E_{j,i} \rangle$ of E for some fixed $\lambda \in \mathbb{F}_{q^2} \setminus \mathbb{F}_q$ when $G = \mathrm{GU}_n(q)$.

Lemma 5.4.11 (i) E *is the orthogonal direct sum of the* $E(i,j)$, *which are non-degenerate, and* $W \cong E \perp (D \cap U)/U'$.

(ii) *If* $G = \mathrm{GL}_n(q)$ *then* E *is a non-degenerate space of plus type.*

(iii) *If* $G = \mathrm{GU}_n(q)$ *then* E *is a non-degenerate space of type* $(-1)^{\binom{n}{2}}$.

(iv) β *is degenerate on* $D \cap U$ *if and only if* $p \mid n$.

(v) *If* $p \nmid n$ *and* n *is even then* W *is non-degenerate of odd dimension.*

(vi) *If* $p \mid n$ *and* n *is odd then* β *induces a non-degenerate symmetric bilinear form on* $U/U' = W$.

Proof (i) The statements about E and $E(i,j)$ are straightforward. The module M is the orthogonal direct sum of D and E with respect to β. Furthermore, we check that $E \leqslant U$, $E \cap U' = 0$ and $U' \leqslant D$, so that $U = E \oplus (D \cap U)$ and $W \cong E \oplus (D \cap U)/U'$.

(ii) Observe that if $i \neq j$ then $Q(E_{i,j}) = Q(E_{j,i}) = 0$ and $\beta(E_{i,j}, E_{j,i}) = 1$. Suppose first that $G = \mathrm{GL}_n(q)$. Then the restriction of β to $E(i,j)$ has matrix $\mathrm{antidiag}(1,1)$, so $E(i,j)$ is of plus-type.

(iii) Now suppose that $G = \mathrm{GU}_n(q)$. Each $E(i,j)$ is an orthogonal space of minus-type. To see this, note that the matrix of the quadratic form on $E(i,j)$ is $Q_2 := \begin{pmatrix} 1 & \lambda + \lambda^\sigma \\ 0 & \lambda\lambda^\sigma \end{pmatrix}$. In odd characteristic the determinant of the bilinear form matrix $Q_2 + Q_2^\mathsf{T}$ is $4\lambda\lambda^\sigma - (\lambda + \lambda^\sigma)^2 = -(\lambda - \lambda^\sigma)^2$, and $\lambda - \lambda^\sigma \notin \mathbb{F}_q$, so $(\lambda - \lambda^\sigma)^2$ is a non-square in $\mathbb{F}_q^\times$. In even characteristic, the only vectors on which Q_2 is 0 are multiples of $(\lambda, 1)$ and $(\lambda^\sigma, 1)$, none of which lie in $\mathbb{F}_q^2$, so the Witt index over $\mathbb{F}_q$ of the form defined by Q_2 is 0, and hence it is of minus type.

(iv) Now let $d_i = E_{i,i} - E_{n,n}$ for $1 \leqslant i \leqslant n - 1$. Then $D \cap U$ has $\mathbb{F}_q$-basis $(d_1, d_2, \ldots, d_{n-1})$, and the matrix of β with respect this basis is:

$$I_{n-1} + J_{n-1} = \begin{pmatrix} 2 & 1 & 1 & \cdots \\ 1 & 2 & 1 & \cdots \\ 1 & 1 & 2 & \cdots \\ \vdots & \vdots & \vdots & \ddots \end{pmatrix},$$

which has determinant n, while $Q(d_i) = 1$ for $1 \leqslant i \leqslant n-1$. The form β is thus degenerate on $D \cap U$ if and only if $p \mid n$, which is precisely when $U' \leqslant D \cap U$.
(v) If $p \nmid n$ and n is even then $D \cap U$ is non-degenerate of odd dimension, and hence so is W, which proves Part (v). If $p \mid n$ and p is odd then U' is β-orthogonal to U (in particular β is identically zero on U') and $Q(\lambda I_n) = -\binom{n}{2}\lambda^2$. Thus β gives rise to a well-defined form $\widehat{\beta}$ on $W = U/U'$. The matrix of $\widehat{\beta}$ with respect to the images of $d_1, d_2, \ldots, d_{n-2}$ is $I_{n-2} + J_{n-2}$ and thus has determinant $n-1$. So $\widehat{\beta}$ is non-degenerate. $\qquad\square$

If $p \nmid n$ and n is odd then $\dim W$ is even, and we have to calculate the sign of the restriction of Q to $U \cong W$. Note that the case $p \mid n$ with n even does not occur when $n \leqslant 12$.

Lemma 5.4.12 *Suppose that $q = p^e$ and $p \nmid n$, with n odd. For $G = \mathrm{GL}_n(q)$, the space W is of plus-type if and only if either:*

(i) p is odd and $(-1)^{(n-1)/2}n$ is a square in $\mathbb{F}_q^\times$; or
(ii) $p = 2$ and either e is even or $n \equiv \pm 1 \pmod 8$.

The type of the quadratic form for the adjoint module of $\mathrm{GU}_n(q)$ is $(-1)^{\binom{n}{2}}$ times the type of the form for $\mathrm{GL}_n(q)$.

Proof We have shown that the restriction of β to E has plus-type, when G is $\mathrm{GL}_n(q)$, and $(-1)^{\binom{n}{2}}$-type when G is $\mathrm{GU}_n(q)$. Thus the form-type of W is determined by that of the restriction of Q to $D \cap U$, and we saw above that the associated bilinear form β on $D \cap U$ has form matrix with determinant n. By Proposition 1.5.42 (ii), if p is odd then the form has plus-type if and only if either the discriminant is square in $\mathbb{F}_q^\times$ and $(n-1)(q-1)/4$ is even, or the discriminant is non-square in $\mathbb{F}_q^\times$ and $(n-1)(q-1)/4$ is odd. Since -1 is square in $\mathbb{F}_q^\times$ if and only if $(q-1)/2$ is even, this condition is equivalent to $(-1)^{(n-1)/2}n$ being a square in $\mathbb{F}_q^\times$.

So suppose that $p = 2$, and let A_{n-1} be the $(n-1) \times (n-1)$-matrix of the form $Q \mid_{D \cap U}$, as in Definition 1.5.17. Then A_{n-1} has entries 1 on and above the diagonal and 0 below the diagonal, and in particular has all entries in $\mathbb{F}_2$. Therefore, Proposition 1.5.42 (v) implies Q has plus type if e is even.

Thus we assume for the rest of the proof that e is odd. Define T_n to be a matrix with 1's on the diagonal, 1 in position (i,j) for $i > 4$ and $j \leqslant 4$, and 0 elsewhere. Then it can be checked that the quadratic form defined by $T_{n-1}A_{n-1}T_{n-1}^\mathsf{T}$ has matrix $A_4 \oplus A_{n-5}$ for $n > 4$. The form defined by A_2 has minus-type by Proposition 1.5.42 (iii), because e is odd and so $X^2 + X + 1$ is irreducible over $\mathbb{F}_q$ in that case. Define U to be the 4×4-matrix with 1's on the diagonal, 1 in position (i,j) for $i > 2$ and $j \leqslant 2$, and 0 elsewhere. Then we find that the form defined by UA_4U^T is the sum of 2-dimensional forms of plus- and minus-types when e is odd, so it also has minus-type. So given any odd n, the matrix A_{n-1} is equivalent to a direct sum of copies of A_4 and at most one copy of A_2, and the result follows. □

We now consider the action of $\mathrm{Out}\,\mathrm{SL}_n^{\pm}(q)$ on the adjoint module.

Lemma 5.4.13 *Let g be a generator for the group of diagonal automorphisms of $\mathrm{SL}_n^{\pm}(q)$, and let ρ be the adjoint representation of $\mathrm{SL}_n(q)$. Then $\rho(g)$ is an element of $\mathrm{SO}_k^{\varepsilon}(q)$, with $\rho(g) \in \Omega_k^{\varepsilon}(q)$ if and only if n is odd or q is even.*

Proof For $\mathrm{SL}_n(q)$, we take $g = \mathrm{diag}(\omega, 1, 1, \ldots, 1)$, where ω is a primitive element of $\mathbb{F}_q^{\times}$. Then g centralises D, and acts on the $n-1$ spaces $E(1,j)$ as $\mathrm{diag}(\omega^{-1}, \omega)$, but centralises $E(i,j)$ if $i,j \geqslant 2$. Therefore this action of g has determinant 1. If n is odd, then $n-1$ is even, so g defines an element of $\Omega_{\dim W}^{\varepsilon}(q)$. If n is even, then this is the case if and only if the element $\mathrm{diag}(\omega^{-1}, \omega)$ has spinor norm or quasideterminant 1 as an element of $\Omega_2^+(\mathbb{F}_q, \mathrm{antidiag}(1,1))$. By Proposition 1.6.11, the quasideterminant is 1 if q is even, but (using $\omega + \omega^{-1} - 2 = \omega(1 + \omega^{-1})^2$) the spinor norm is -1 if q is odd.

For $\mathrm{SU}_n(q)$, we take $g = \mathrm{diag}(\lambda, 1, 1, \ldots, 1)$ with $\lambda = \omega^{q-1}$, and we choose the same value of λ for our bases $\langle \mathrm{E}_{i,j} + \mathrm{E}_{j,i}, \lambda \mathrm{E}_{i,j} + \lambda^{\sigma} \mathrm{E}_{j,i} \rangle$ of the spaces $E(i,j)$, so in fact $\lambda^{\sigma} = \lambda^{-1}$. Again g centralises D, and the action of g on $E(i,j)$ has

matrix $A := \begin{pmatrix} \lambda + \lambda^{-1} & -1 \\ 1 & 0 \end{pmatrix}$, which again has determinant 1. If q is even,

then $\lambda + \lambda^{-1} \neq 0$, so $\mathrm{I}_2 - A$ has rank 2, and Proposition 1.6.11 (i) tells us that A has quasideterminant 1. If q is odd then $-\det(\mathrm{I}_2 - A) = \lambda + \lambda - 2 = (\mu - \mu^{-1})^2$, where $\mu = \omega^{(q-1)/2}$, and it is straightforward to check that $(\mu - \mu^{-1})^q = \mu - \mu^{-1}$ and hence $\mu - \mu^{-1} \in \mathbb{F}_q$, so $-\det(A - \mathrm{I}_2)$ is a square in $\mathbb{F}_q^{\times}$. We saw in the proof of Lemma 5.4.11 (iii) that the restriction of the form to $E(i,j)$ has matrix $Q_2 + Q_2^{\mathsf{T}}$, and that $-\det(Q_2 + Q_2^{\mathsf{T}})$ is a non-square in $\mathbb{F}_q^{\times}$, so Proposition 1.6.11 (ii) implies that A has spinor norm -1. We therefore get the same conclusion as we did for $\mathrm{SL}_n(q)$. □

Recall that the map $-\mathsf{T}$ acts in the same way as the map σ on $\mathrm{SU}_n(q)$.

Lemma 5.4.14 *Let $G = \mathrm{SL}_n^{\pm}(q)$, and let W be the adjoint representation of G, of dimension k. Then $-\mathsf{T}$ is induced by an element $t \in \mathrm{GO}_k^{\varepsilon}(q)$, and at least one of t or $-t$ lies in $\Omega_k^{\varepsilon}(q)$ if and only if $\binom{n}{2}$ is even or qk is odd. If q is even and $\binom{n}{2}$ is odd then $t \in \mathrm{SO}_k^{\varepsilon}(q) \setminus \Omega_n^{\varepsilon}(q)$, whilst if q is odd, k is even and $\binom{n}{2}$ is odd then $t \in \mathrm{GO}_k^{\varepsilon}(q) \setminus \mathrm{SO}_k^{\varepsilon}(q)$.*

Proof The duality automorphism $-\mathsf{T}$ of G maps X to X^{T} for $X \in M$, and therefore centralises D but swaps $\mathrm{E}_{i,j}$ with $\mathrm{E}_{j,i}$ for all i and j. So duality acts the same way on each of the subspaces $E(i,j)$. The matrix A of these actions for

$G = \mathrm{SL}_n(q)$ and $G = \mathrm{SU}_n(q)$ is respectively $\begin{pmatrix} 0 & 1 \\ 1 & 0 \end{pmatrix}$ and $\begin{pmatrix} 1 & 0 \\ \lambda + \lambda^{-1} & -1 \end{pmatrix}$,

of determinant -1 in both cases. If q is even, then A acts as as a transvection. Let t be the element of $\mathrm{GO}_k^\varepsilon(q)$ defined by the action of $-\mathsf{T}$.

If $\binom{n}{2}$ is even then t is a product of an even number of isometric reflections, so $t \in \Omega_k^\varepsilon(q)$. Suppose instead that $\binom{n}{2}$ is odd. If q is even then $\mathrm{I}_2 - A$ has rank 1, so it has quasideterminant -1 by Proposition 1.6.11 (i), and hence t does as well. Suppose therefore that q is odd. Then $\det t = -1$, so $t \in \mathrm{GO}_k^\varepsilon(q) \setminus \mathrm{SO}_k^\varepsilon(q)$. If k is odd, then using Proposition 1.6.11 (ii), we find that $\mathrm{I}_2 + A$ (which has rank 1) has spinor norm 1. Since $k - \dim E$ is even in this case, $-t \in \Omega_k^\varepsilon(q)$. $\square$

Proposition 5.4.15 *For $p = 3$, there are exactly two conjugacy classes of $\mathscr{S}_2$-subgroups G of $\Omega_7(q)$ isomorphic to $\mathrm{L}_3(q).2$, which are scalar-normalising in $\mathrm{CGO}_7(q)$.*

Proof It follows from Lemmas 5.4.10 and 5.4.11 that the adjoint representation of $\mathrm{SL}_3(q)$ with $q = 3^e$ is irreducible, 7-dimensional, and preserves an orthogonal form, and the image of the representation is $\mathrm{SL}_3(q) \cong \mathrm{L}_3(q)$.

There are no non-trivial diagonal automorphisms of G. By Proposition 5.1.9 the field automorphisms of G are not induced by an element of $\mathrm{CGO}_7(q)$. It follows from Lemma 5.4.14 that the normaliser of G in $\Omega_7(q)$ induces the duality automorphism of $\mathrm{L}_3(q)$. The number of conjugacy classes then follows from Proposition 5.4.1 and Corollary 5.1.10. $\square$

Proposition 5.4.16 (i) *If $q \equiv 1 \pmod 3$, there are exactly $(q - 1, 2)^2$ conjugacy classes of $\mathscr{S}_2$-subgroups G of $\Omega_8^+(q)$ isomorphic to $\mathrm{L}_3(q).3$. Their normaliser in $\mathrm{CGO}_8^+(q)$ is generated by G, scalars and γ.*

(ii) *If $q \equiv 2 \pmod 3$, there are exactly $(q - 1, 2)$ conjugacy classes of $\mathscr{S}_2$-subgroups G of $\Omega_8^-(q)$ isomorphic to $\mathrm{L}_3(q)$. Their normaliser in $\mathrm{CGO}_8^-(q)$ is generated by G, scalars and γ.*

Proof It follows from Lemma 5.4.12 that, since $n = 3$, the form preserved is non-degenerate and of plus-type when $q \equiv 1 \pmod 3$ and of minus-type when $q \equiv -1 \pmod 3$. This is the case for both even and odd q.

It is immediate from Lemma 5.4.13 that any diagonal automorphisms of $\mathrm{L}_3(q)$ are induced within $\Omega_8^\pm(q)$. It follows from Lemma 5.4.14 that the normaliser of G in $\Omega_8^\pm(q)$ does not induce $\gamma_{\mathrm{L}_3(q)}$, but that the γ automorphism of $\Omega_8^\pm(q)$ induces $\gamma_{\mathrm{L}_3(q)}$. By Proposition 5.1.9 the field automorphism of G is not induced by an element of $\mathrm{CGO}_7(q)$.

There is a single class of such subgroups in $\mathrm{CGO}_n^\pm(q)$ by Proposition 5.4.1 and Corollary 5.1.10, so the number of classes in $\Omega_8^\pm(q)$ is immediate. $\square$

Proposition 5.4.17 *For $p = 3$, there are exactly two conjugacy classes of*

$\mathscr{S}_2$-subgroups G of $\Omega_7(q)$ isomorphic to $\mathrm{U}_3(q).2$, which are scalar-normalising in $\mathrm{CGO}_7(q)$.

Proof It follows from Lemmas 5.4.10 and 5.4.11 that the adjoint representation of the group $\mathrm{SU}_3(q)$ with $q = 3^e$ is irreducible, 7-dimensional, and preserves an orthogonal form, and the image of the representation is $\mathrm{SU}_3(q) \cong \mathrm{U}_3(q)$.

There are no non-trivial diagonal automorphisms of G. It follows from Lemma 5.4.14 that the normaliser of G in $\Omega_7(q)$ induces the σ automorphism of G, and from Proposition 5.1.9 that no other field automorphisms of G are induced by an element of $\mathrm{CGO}_7(q)$. The number of conjugacy classes in $\Omega_7(q)$ then follows from Proposition 5.4.1 and Corollary 5.1.10. □

Proposition 5.4.18 (i) *If* $q \equiv 2 \pmod 3$, *there are exactly* $(q-1,2)^2$ *conjugacy classes of* $\mathscr{S}_2$-*subgroups* G *of* $\Omega_8^+(q)$ *isomorphic to* $\mathrm{U}_3(q).3$. *Their normaliser in* $\mathrm{CGO}_8^+(q)$ *is generated by* G, *scalars and* γ.

(ii) *If* $q \equiv 1 \pmod 3$, *there are exactly* $(q-1,2)$ *conjugacy classes of* $\mathscr{S}_2$-*subgroups* G *of* $\Omega_8^-(q)$ *isomorphic to* $\mathrm{U}_3(q)$. *Their normaliser in* $\mathrm{CGO}_8^-(q)$ *is generated by* G, *scalars and* γ.

Proof The proof of this is left as an exercise: copy Proposition 5.4.16. □

5.4.2 Irreducible tensor products

In this section, we consider the possibilities given in Proposition 5.1.14 to determine the candidate $\mathscr{S}_2^*$-maximals that are tensor products.

Lemma 5.4.19 *Let* $\rho : G = \mathrm{SL}_n^\pm(q) \to \Omega$ *with* $n \geqslant 3$ *be a representation with image an* $\mathscr{S}_2^*$-*maximal subgroup of* Ω *in dimension at most 12. Suppose that the corresponding module* M *decomposes as a non-trivial tensor product* $M_1 \otimes M_2$. *Then* $G = \mathrm{SL}_3(q)$ *and either* $\Omega = \mathrm{SL}_9(\sqrt{q})$, *with* M *quasi-equivalent to* $V_3 \otimes V_3^{\phi^{e/2}}$, *or* $\Omega = \mathrm{SU}_9(\sqrt{q})$, *with* M *quasi-equivalent to* $V_3 \otimes (V_3^*)^{\phi^{e/2}}$.

Proof The non-trivial modules for $\mathrm{SL}_n^\pm(q)$ have dimension at least n and $\mathrm{SL}_3^\pm(q)$ has no irreducible modules of dimension 4 so, since $n \geqslant 3$, the only way in which a tensor product M of modules with more than one non-trivial factor in Theorem 5.1.5 could give rise to an $\mathscr{S}_2$-subgroup is with $n = 3$ and a module of dimension 9. Hence $\dim(M_1) = \dim(M_2) = 3$. The p-restricted modules of dimension 3 are the natural module V and its dual V^*. Therefore, by replacing M by a quasi-equivalent module, we may assume that $M_1 = V$ and $M_2 = V^{\phi^i}$ or $(V^*)^{\phi^i}$ for some i with $1 \leqslant i \leqslant e$.

It was shown in Proposition 5.1.14 that the third possibility listed there cannot arise for $\mathrm{SL}_3(q)$ or $\mathrm{SU}_3(q)$, so we need only consider the first two.

We show first that there are no such examples for $SU_3(q)$. If Proposition 5.1.14 (i) is satisfied, then the module is defined over a proper subfield $\mathbb{F}_{p^f}$ of $\mathbb{F}_{p^{2e}}$, and we can apply Theorem 5.1.13 (ii). Since $M_1 = V$ is not isomorphic to $M_1^{\phi^e} = V^*$, Theorem 5.1.13 (ii) (b) applies. But in that case, since $f \mid 2e$ with $f < 2e$, but $f \nmid e$, the number of non-trivial tensor factors must be at least 3, which is not the case.

If, on the other hand, Proposition 5.1.14 (ii) is satisfied, then there must be a form on M other than the induced unitary form, and this new form would necessarily be bilinear. So M would be self-dual. But the dual of $M_1 = V$ is the p-restricted module V^*, which is a contradiction, because $M_2 \neq V^*$ by Theorem 5.1.5.

So we only need consider examples for $SL_3(q)$. If Proposition 5.1.14 (i) is satisfied, then we can apply Theorem 5.1.13 (i), and we find that the only possibility for M is $V \otimes V^{\phi^{e/2}}$ with e even, for which the minimal field of representation is $\mathbb{F}_{p^{e/2}} = \mathbb{F}_{\sqrt{q}}$. By Theorem 5.1.5, the image of this representation is not self-dual and preserves no unitary form, so it gives rise to an $\mathscr{S}_2^*$-subgroup of $SL_9(\sqrt{q})$ isomorphic to a central quotient of $SL_3(q)$.

Suppose, on the other hand, that Proposition 5.1.14 (ii) applies. None of the modules $V \otimes V^{\phi^i}$ or $V \otimes (V^*)^{\phi^i}$ with $1 \leqslant i < e$ are self-dual, so any form on them is not bilinear, and the only such module M that is isomorphic to $(M^*)^{\phi^{e/2}}$ (with e even) is $V \otimes (V^*)^{\phi^{e/2}}$. This module gives rise to an $\mathscr{S}_2^*$-subgroup of $SU_9(\sqrt{q})$ isomorphic to a central quotient of $SL_3(q)$. $\qquad\square$

Since e is even in both of these situations, let us change notation and replace q by q^2 (still with $q = p^e$). So there are $\mathscr{S}_2^*$-subgroups of $SL_9(q)$ and $SU_9(q)$ arising from the modules $M_L := V \otimes V^{\phi^e}$ and $M_U := V \otimes (V^*)^{\phi^e}$ of $SL_3(q^2)$.

Proposition 5.4.20   (i) *When $q \equiv 0 \pmod 3$, there is a unique class of $\mathscr{S}_2$-subgroups G of $SL_9(q)$ isomorphic to $L_3(q^2).2$, and these subgroups are scalar-normalising in $GL_9(q)$.*

  (ii) *When $q \equiv 1 \pmod 9$, there are exactly three classes of $\mathscr{S}_2$-subgroups G of $SL_9(q)$ isomorphic to $3{\cdot}L_3(q^2).2$. Their normaliser in $GL_9(q)$ is generated by G, scalars and δ^3.*

 (iii) *When $q \equiv 4$ or $7 \pmod 9$, there are exactly three classes of $\mathscr{S}_2$-subgroups G of $SL_9(q)$ isomorphic to $3{\cdot}L_3(q^2).6$, and these subgroups are scalar-normalising in $GL_9(q)$.*

  (iv) *When $q \equiv 2 \pmod 3$, there is a unique class of $\mathscr{S}_2$-subgroups G of $SL_9(q)$ isomorphic to $L_3(q^2).S_3$, and these are scalar-normalising in $GL_9(q)$.*

Proof   By considering what happens to the action of the scalars in $SL_3(q^2)$ on M_L we find that the image of $SL_3(q^2)$ is as given.

It follows from Theorem 5.1.5 that M_L is not stabilised by the automorphisms ϕ^k for $1 \leqslant k < e$ of $\mathrm{SL}_3(q^2)$, or its duality automorphism γ; however M_L is stabilised by $\sigma := \phi^e$.

We need to calculate what happens to the diagonal automorphisms of $\mathrm{L}_3(q^2)$. By considering the determinants of the images in $\mathrm{GL}_9(q)$ of elements of $\mathrm{GL}_3(q^2)$ inducing these automorphisms, we find that if $q \equiv 4, 7 \pmod 9$ then they are effected by conjugation by elements of $\mathrm{SL}_9(q)$, but if $q \equiv 1 \pmod 9$, then they are induced by diagonal automorphisms of order 3 of $\mathrm{SL}_9^{\pm}(q)$.

The field automorphism $\sigma : x \mapsto x^q$ in its action on the image of $\mathrm{SL}_3(q^2)$ is realised by an element of $\mathrm{GL}_9(q)$ that permutes coordinates. Since $\mathrm{L}_9(q)$ has no diagonal automorphisms of even order, σ must be effected by conjugation by an element of $\mathrm{SL}_9(q)$. The structure of the outer automorphism group of $\mathrm{L}_3(q^2)$ is described in Subsection 1.7.2.   □

Proposition 5.4.21   (i) *When $q \equiv 0 \pmod 3$ there is a unique class of $\mathscr{S}_2$-subgroups G of $\mathrm{SU}_9(q)$ isomorphic to $\mathrm{L}_3(q^2).2$, and these subgroups are scalar-normalising in $\mathrm{CGU}_9(q)$.*

(ii) *When $q \equiv 8 \pmod 9$ there are exactly three classes of $\mathscr{S}_2$-subgroups G of $\mathrm{SU}_9(q)$ isomorphic to $3^{\cdot}\mathrm{L}_3(q^2).2$. Their normaliser in $\mathrm{CGU}_9(q)$ is generated by G, scalars, and δ^3.*

(iii) *When $q \equiv 2, 5 \pmod 9$ there are exactly three classes of $\mathscr{S}_2$-subgroups G of $\mathrm{SU}_9(q)$ isomorphic to $3^{\cdot}\mathrm{L}_3(q^2).6$, which are scalar-normalising in the group $\mathrm{CGU}_9(q)$.*

(iv) *When $q \equiv 1 \pmod 3$ there is a unique class of $\mathscr{S}_2$-subgroups of $\mathrm{SU}_9(q)$ isomorphic to $\mathrm{L}_3(q^2).\mathrm{S}_3$, which are scalar-normalising in $\mathrm{CGU}_9(q)$.*

Proof The proof of this is similar to that of Proposition 5.4.20, so we only sketch it briefly. By considering the scalars of $\mathrm{SL}_3(q^2)$ in their action on M_U, we deduce that the image of $\mathrm{SL}_3(q^2)$ is as given.

As before, M_U is not stabilised by ϕ^k for $1 \leqslant k < e$, or by γ, but is stabilised by $\sigma\gamma$. By considering the determinants of the images in $\mathrm{GL}_9(q^2)$ of elements of $\mathrm{GL}_3(q^2)$ that induce diagonal automorphisms, we find that if $q \equiv 2, 5 \pmod 9$ then they are effected by elements of $\mathrm{SU}_9(q)$, but if $q \equiv 8 \pmod 9$ then they are induced by diagonal automorphisms $\mathrm{SU}_9(q)$.

As before, $\gamma\sigma$ is realised by an element of $\mathrm{GL}_9(q^2)$ that permutes coordinates. Since $\mathrm{U}_9(q)$ has no diagonal elements of even order, $\gamma\sigma$ is effected by an element of $\mathrm{SU}_9(q)$.   □

We can summarise the results in this section as follows.

Theorem 5.4.22 *The only candidates for maximal $\mathscr{S}_2^*$-subgroups of the classical groups in dimensions up to 12, other than in extensions of $\mathrm{O}_8^+(q)$ or $\Omega_8^+(q)$, with a composition factor $\mathrm{L}_n^{\pm}(q)$ for $n \geqslant 3$ are those described in Propositions 5.4.5, 5.4.6, 5.4.7, 5.4.8, 5.4.15–5.4.18, 5.4.20 and 5.4.21.*

5.5 The groups $\mathrm{Sp}_n(q)$

We assume here that $n \geqslant 4$ and that $(n, q) \neq (4, 2)$. We start by considering $G := \mathrm{Sp}_4(q)$ with q odd. The candidate members of $\mathscr{M}$ listed in [84, Appendix A.22] of dimension up to 12 have dimensions 4, 5 and 10, and also 12 in characteristic 5. In order to carry out calculations with these modules, we need to see how to construct them explicitly using tensor products. We may assume that the natural module V_4 for G is the module with highest weight $\lambda_1 = (1, 0)$.

Lemma 5.5.1 (i) *The 5-dimensional p-restricted module V_5 with highest weight $(0, 1)$ is a constituent of $\Lambda^2(V_4)$.*

(ii) *The 10-dimensional p-restricted module with highest weight $(2, 0)$ is isomorphic to $\mathrm{S}^2(V_4)$.*

(iii) *In characteristic 5, the 12-dimensional p-restricted module with highest weight $(1, 1)$ is a constituent of $V_4 \otimes V_5$.*

There are no further non-natural p-restricted representations of G in dimension at most 12.

Proof As noted, the final claim is from [84, Appendix A.22]. Part (i) is proved in [77, Theorem 1.1]. Using Lemma 5.1.11, we see that the 10-dimensional module with highest weight $(2, 0)$ must be the constituent $\mathrm{S}^2(V_4)$ of $V_4 \otimes V_4$, so (ii) is true. Part (iii) follows from (i) and Lemma 5.1.11. (In odd characteristics other than 5, the module with highest weight $(1, 1)$ has dimension 16.) $\square$

The module V_5 is the natural module for $\Omega_5(q)$, and so does not give rise to $\mathscr{S}_2$-candidates. So we first consider $\mathrm{S}^2(V_4)$ of dimension 10.

Proposition 5.5.2 (i) *When $q \equiv 1 \pmod 4$ there are exactly four classes of $\mathscr{S}_2$-subgroups G of $\Omega_{10}^+(q)$ isomorphic to $\mathrm{S}_4(q)$. The normaliser of these groups in $\mathrm{CGO}_{10}^+(q)$ is generated by G, scalars and δ'; and δ' induces δ_G.*

(ii) *When $q \equiv 3 \pmod 4$ there are exactly four classes of $\mathscr{S}_2$-subgroups G of $\Omega_{10}^-(q)$ isomorphic to $\mathrm{S}_4(q)$. Their normaliser in $\mathrm{CGO}_{10}^-(q)$ is generated by G, scalars and δ'; and δ' induces δ_G.*

Proof The matrix $-\mathrm{I}_4 \in \mathrm{Sp}_4(q)$ acts trivially on $V_{10} = \mathrm{S}^2(V_4)$, so the image of $\mathrm{Sp}_4(q)$ in this representation is $\mathrm{S}_4(q)$. Using our usual basis for $\mathrm{S}^2(V_4)$, we find that $\mathrm{PCSp}_4(q)$ consists of isometries of the orthogonal form F given by $F_1 = 2(\mathrm{E}_{1,10} + \mathrm{E}_{5,8}) + \mathrm{E}_{2,9} - \mathrm{E}_{3,7} - 1/2(\mathrm{E}_{4,4} + \mathrm{E}_{6,6})$ and $F = F_1 + F_1^\mathsf{T}$. Now F has determinant 16, a square, so $\mathrm{S}_4(q) < \Omega_{10}^+(q)$ or $\Omega_{10}^-(q)$ when $q \equiv 1, 3 \bmod 4$, respectively.

Let ω be a primitive element of $\mathbb{F}_q^\times$. The action of $\mathrm{diag}(\omega, \omega, 1, 1)$ on $\mathrm{S}^2(V_4)$ after multiplying by $\omega^{-1}\mathrm{I}_{10}$ is

$$\mathrm{diag}(\omega, \omega, 1, 1, \omega, 1, 1, \omega^{-1}, \omega^{-1}, \omega^{-1}) \in \mathrm{SO}_{10}^\pm(q).$$

Calculating the spinor norm of this, as described in Proposition 1.6.11, we find that $BAFB^{\mathsf{T}} = \mathrm{antidiag}(1 - \omega, 2(1 - \omega), 1 - \omega, 1 - \omega^{-1}, 2(1 - \omega^{-1}), 1 - \omega^{-1})$, which has determinant $-4(1 - w)^3(1 - w^{-1})^3 = w^{-1}(2(1 - w)^2(1 - w^{-1}))^2$, a non-square. So the images of elements of $\mathrm{CSp}_4(q)\backslash\mathrm{Sp}_4(q)$ lie in $\mathrm{SO}_{10}^{\pm}(q)\backslash\Omega_{10}^{\pm}(q)$, and hence the automorphism δ' of $\Omega_{10}^{\pm}(q)$ induces δ on $\mathrm{Sp}_4(q)$. The conjugacy classes then follow from Lemma 5.5.1 and Corollary 5.1.10. $\qquad\square$

Proposition 5.5.3 *When $p = 5$ there is a single class of $\mathscr{S}_2$-subgroups G of $\mathrm{Sp}_{12}(q)$ isomorphic to $\mathrm{Sp}_4(q)$. Their normaliser in $\mathrm{CSp}_{12}(q)$ is generated by G, scalars and δ, which induces δ_G.*

Proof For $q = 5$, we can show by computer calculation (file s45d12calc) that the module $V_4 \otimes V_5$ is uniserial with constituents V_4, V_{12}, V_4, where V_{12} is irreducible with dimension 12. Since we know from Lemma 5.5.1 that $V_4 \otimes V_5$ has a constituent with dimension 12 for all $q = 5^e$, by considering the restriction to $\mathbb{F}_5$ and using Lemma 5.1.6, we see that $V_4 \otimes V_5$ must be uniserial with constituents of the same degrees for all such q. Let W_1 and W_2 be the unique submodules of $V_4 \otimes V_5$ of dimensions 4 and 16. Since W_1 is irreducible, it is either non-degenerate or totally singular with respect to the induced symplectic form on $V_4 \otimes V_5$. If it were non-degenerate, then W_1 would have an orthogonal complement in $V_4 \otimes V_5$, which it does not, so W_1 is totally singular. Therefore the induced symplectic form of $V_4 \otimes V_5$ induces a non-degenerate symplectic form on V_{12}, and hence there is an embedding $\mathrm{Sp}_4(q) < \mathrm{Sp}_{12}(q)$.

The diagonal automorphism of $\mathrm{Sp}_4(q)$ multiplies the forms on V_4 and V_5 by a non-square and a square, respectively, and hence multiplies the form on V_{12} by a non-square. So it is induced by the diagonal automorphism of $\mathrm{Sp}_{12}(q)$. The number of conjugacy classes follows from Lemma 5.5.1 and Corollary 5.1.10. $\quad\square$

Theorem 5.5.4 *The only candidates for maximal $\mathscr{S}_2^*$-subgroups of the classical groups in dimensions up to 12, other than in extensions of $\mathrm{O}_8^+(q)$ or $\Omega_8^+(q)$, that have a composition factor $\mathrm{S}_n(q)$ for $n \geqslant 4$ are those described in Propositions 5.5.2 and 5.5.3.*

Proof If $n \geqslant 6$ then the lists of Lübeck [84] and Corollary 5.1.10 show that the only relevant representations of $\mathrm{Sp}_n(q)$ are quasi-equivalent to the natural representations of $\mathrm{Sp}_6(q)$, $\mathrm{Sp}_8(q)$, $\mathrm{Sp}_{10}(q)$ and $\mathrm{Sp}_{12}(q)$ or to the 8-dimensional spin representations of $\mathrm{Sp}_6(q)$ for q even. The image of the spin representation of $\mathrm{S}_6(q)$ for q even in dimension 8 lies in $\Omega_8^+(q)$ by [66, Proposition 5.4.9], so this need not concern us further. For $n \geqslant 4$ the lowest dimension of a non-trivial p-restricted module is 4, so there is no need to consider tensor products of more than one non-trivial factor in Theorem 5.1.5.

For $\mathrm{Sp}_4(q)$ with q even, the only non-zero modules of dimension up to 12 listed in [84] are two of dimension 4, which are the natural module and its image under the exceptional graph automorphism. (We shall study these modules in more detail in Section 7.2.) So there are no $\mathscr{S}_2^*$-maximal subgroups in this case.

Thus the only possibility is $\mathrm{Sp}_4(q)$ with q odd, which is considered in Lemma 5.5.1 and Propositions 5.5.2 and 5.5.3. $\qquad\square$

5.6 The groups $O_n^\varepsilon(q)$, $^3D_4(q)$, and their covers

For odd q and $n \geqslant 7$, the full covering group of $O_n^\varepsilon(q)$ is not $\Omega_n^\varepsilon(q)$, but is $\mathrm{Spin}_n^\varepsilon(q)$, which is twice as large. See, for example, [114, Section 3.9] for a brief introduction to the spin groups, which are defined using Clifford algebras. Fortunately, we shall not need to carry out any calculations with them!

Theorem 5.6.1 *There are no maximal $\mathscr{S}_2^*$-subgroups of the classical groups in dimensions up to 12, other than in extensions of $O_8^+(q)$ or $\Omega_8^+(q)$, that have composition factor $O_n^\varepsilon(q)$ or $^3D_4(q)$ for $n \geqslant 7$.*

Proof The only non-trivial p-restricted representations of dimension at most 12 for the groups $(2\dot{\ })O_7(q)$ have dimensions 7 and 8 by [84], so tensor products have dimension at least 49. The 7-dimensional modules are natural, and hence do not give rise to $\mathscr{S}_2$-subgroups. The 8-dimensional modules are spin representations, that have image in $\Omega_8^+(q) \cong 2\dot{\ }O_8^+(q)$ by [66, Proposition 5.4.9], and hence we need not consider them.

There are three non-trivial p-restricted representations associated with the groups $O_8^+(q)$, $O_8^-(q)$, $^3D_4(q)$ and their covers in dimension at most 12, and they all have dimension 8 by [84]. Therefore all non-trivial tensor products have dimension at least 64.

The candidate representations for $O_8^+(q)$ are the natural and $[\frac{1}{2}]$-spin representations, which all have dimension 8 and are equivalent under the triality automorphism. Since these representations are quasi-equivalent, and one of them is the natural representation of $\Omega_8^+(q)$, their images all lie in $\Omega_8^+(q)$ so they need not concern us further.

The candidate representations of $^3D_4(q)$ have minimal field $\mathbb{F}_{q^3}$, and preserve a quadratic form of $+$-type, since $^3D_4(q)$ is defined as the centraliser in $O_8^+(q^3)$ of a suitable outer automorphism of order 3: see the beginning of Section 5.1. Therefore they are subgroups of $\Omega_8^+(q)$.

For $O_8^-(q)$, the p-restricted representations are either natural or $\frac{1}{2}$-spin. The natural representation does not give rise to an $\mathscr{S}_2$-subgroup. The $\frac{1}{2}$-spin representations of $\mathrm{Spin}_8^-(q)$ require $\mathbb{F}_{q^2}$ to represent them, and can be obtained by restricting the $\frac{1}{2}$-spin representations of $\mathrm{Spin}_8^+(q^2)$. So a quadratic form of $+$-type is preserved here. See [66, Proposition 5.4.9].

The natural modules of $\Omega_n^\varepsilon(q)$ for $9 \leqslant n \leqslant 12$ are their only non-trivial

Table 5.4 *Degrees of p-restricted representations of* $G_2(\overline{\mathbb{F}_p})$

p	Degrees
2	$1, 6, 14, 64.$
3	$1, 7, 7, 27, 27, 49, 189, 189, 729.$
$\geqslant 5$	$1, 7, > 12 \ldots$

p-restricted modules of dimension as most 12, by [84]. Therefore no tensor products of dimension at most 12 occur. □

5.7 The remaining groups and their covers

In this section we consider the remaining groups of Lie type, namely $E_6(q)$, $E_7(q)$, $E_8(q)$, $F_4(q)$, $G_2(q)$, ${}^2B_2(2^{2\mu+1})$, ${}^2E_6(q)$, ${}^2F_4(2^{2\mu+1})$ and ${}^2G_2(3^{2\mu+1})$. As in previous sections, we will first analyse the $\mathscr{S}_2^*$-subgroups that arise, before proving that these are the only examples.

The groups $G_2(q)$ have trivial p'-multiplier, and the degrees of some of their p-restricted representations are given in Table 5.4, as taken from [84]. Recall that $G_2(2)$ is not simple and hence is not a group of Lie type in characteristic 2; see Table 4.1. The outer automorphism group of $G_2(p^e)$ has order $2e$ when $p = 3$, and e when $p \neq 3$.

Proposition 5.7.1 *For q even, there is a single class of $\mathscr{S}_2^*$-subgroups G of $\mathrm{Sp}_6(q)$ isomorphic to $G_2(q)$, which are scalar-normalising in $\mathrm{CSp}_6(q)$.*

Proof By Table 5.4 there is a single non-trivial p-restricted representation ρ of degree $\leqslant 6$, which is self-dual by Proposition 5.1.12. We must determine whether $\mathrm{Im}\,\rho$ preserves an orthogonal form.

By Lemma 5.1.6 the group $G_2(2^e)$ contains $G_2(2)$ as a subgroup. We see from [57] that the 6-dimensional 2-modular representation of $G_2(2)' \cong U_3(3)$ has indicator $-$, so the natural representation of $G_2(2^e)$ is symplectic but not orthogonal. Now, $S_6(2^e) \cong \mathrm{PCSp}_6(2^e)$, and the conjugacy claims follow from Corollary 5.1.10. □

Proposition 5.7.2 *For q odd, there are exactly two classes of $\mathscr{S}_2^*$-subgroups G of $\Omega_7(q)$ isomorphic to $G_2(q)$, interchanged by the diagonal automorphism. They are scalar-normalising in $\mathrm{CGO}_7(q)$.*

Proof By Table 5.4 there is a single non-trivial p-restricted representation ρ of degree $\leqslant 7$ if $p \geqslant 5$, and a pair if $p = 3$. The 7-dimensional representations

Table 5.5 *p-Restricted representations of Suzuki and small Ree groups*

group	module	field size	dimension	description
$\mathrm{Sz}(2^{2\mu+1})$	1	2	1	trivial
	V	$2^{2\mu+1}$	4	natural
$\mathrm{R}(3^{2\mu+1})$	1	3	1	trivial
	V	$3^{2\mu+1}$	7	natural
	W	$3^{2\mu+1}$	27	$\mathrm{S}^2(V) \cong W \oplus 1$

of $\mathrm{G}_2(q)$ are orthogonal, since they are self-dual by Proposition 5.1.12 and of odd dimension. From the λ-column in [84, Appendix A.49] we see that if $p = 3$ the two p-restricted modules are interchanged by the graph automorphism of $\mathrm{G}_2(q)$, and hence the corresponding subgroups are conjugate in $\mathrm{CGO}_7(q)$. The outer automorphism group of $\mathrm{G}_2(q)$ acts regularly on the e (or $2e$) images of the p-restricted modules, so G is scalar-normalising in $\mathrm{CGO}_7(q)$. $\qquad\square$

The Suzuki and small Ree groups, namely $\mathrm{Sz}(2^{2\mu+1}) = {}^2\mathrm{B}_2(2^{2\mu+1})$ and $\mathrm{R}(3^{2\mu+1}) = {}^2\mathrm{G}_2(3^{2\mu+1})$ with $\mu \in \mathbb{N}_{\geqslant 0}$, both occur in just a single characteristic, and all have trivial Schur multiplier apart from $\mathrm{Sz}(8)$, which has multiplier 2^2. The p-restricted representations of these groups are described in Table 5.5, taken from [84]. The outer automorphism groups of $\mathrm{Sz}(2^{2\mu+1})$ and $\mathrm{R}(3^{2\mu+1})$ are generated by field automorphisms, and a non-inner automorphism does not stabilise the natural module of the group in question.

Proposition 5.7.3 *For $q = 2^e$ with $e > 1$ odd, there is a single class of $\mathscr{S}_2^*$-subgroups G of $\mathrm{Sp}_4(q)$ isomorphic to $\mathrm{Sz}(q)$, which are scalar-normalising in $\mathrm{CSp}_4(q)$.*

Proof All modules for $\mathrm{Sz}(q)$ are self-dual by Proposition 5.1.12, so G must preserve a symplectic or orthogonal form. The non-abelian composition factors of $\mathrm{GO}_4^{\pm}(q)$ are $\mathrm{L}_2(q)$ or $\mathrm{L}_2(q^2)$, and hence all non-abelian composition factors of their subgroups are of the form $\mathrm{L}_2(r)$ where $r | q^2$. However, $\mathrm{Sz}(q)$ has no faithful 2-dimensional representations by Table 5.5, and so its 4-dimensional representation does not preserve an orthogonal form.

There is a unique $\mathrm{CSp}_4(q)$-conjugacy class of groups G, by Corollary 5.1.10. The field automorphism acts regularly on the 4-dimensional modules, so $\mathrm{Sz}(q)$ is scalar-normalising in $\mathrm{CSp}_4(q)$. $\qquad\square$

Proposition 5.7.4 *For $q = 3^e$ with e odd, there are exactly two classes of $\mathscr{S}_2^*$-subgroups $\mathrm{R}(q)$ of $\Omega_7(q)$, which are scalar-normalising in $\mathrm{CGO}_7(q)$.*

Proof The irreducible representations of $\mathrm{R}(3^{2\mu+1})$ are all self-dual by Proposition 5.1.12 and the only p-restricted representations of dimension at most 12 have odd dimension, and so are orthoognal.

There is a unique conjugacy class of groups $G \cong R(q)$ in $\mathrm{CGO}_7(q)$, by Corollary 5.1.10. The field automorphism acts regularly on the 7-dimensional modules, so $R(q)$ is scalar-normalising in $\mathrm{CGO}_7(q)$. □

Theorem 5.7.5 *The only candidates for maximal $\mathscr{S}_2^*$-subgroups of the classical groups in dimensions up to 12, other than in extensions of $\mathrm{O}_8^+(q)$ or $\Omega_8^+(q)$, that have composition factor $G_2(q)$, $F_4(q)$, $E_6(q)$, $^2E_6(q)$, $E_7(q)$, $E_8(q)$, $\mathrm{Sz}(2^{2\mu+1})$, $R(3^{2\mu+1})$ or $^2F_4(2^{2\mu+1})$ are those described in Propositions 5.7.1, 5.7.2, 5.7.3 and 5.7.4.*

Proof The minimal degree of a non-trivial representation of any of the groups $F_4(q)$, $E_6(q)$, $^2E_6(q)$, $E_7(q)$, $E_8(q)$, $^2F_4(2^{2\mu+1})$ and their covers in characteristic p is at least 25.

By Table 5.4, the only dimensions we need consider for $G_2(q)$ are 6 (for $p = 2$) and 7 (for p odd), and these are considered in Propositions 5.7.1 and 5.7.2. By Table 5.5, the only dimension we need for $\mathrm{Sz}(2^{2\mu+1})$ is 4, which is dealt with in Proposition 5.7.3. Similarly, by Table 5.5, Proposition 5.7.4 describes the only representations of $R(q)$ in dimension at most 12. □

5.8 Summary of $\mathscr{S}_2^*$-candidates

With the exception of $\Omega = \Omega_8^+(q)$, we have now determined all candidate $\mathscr{S}_2^*$-maximal subgroups of $\overline{\Omega}$ and its almost simple extensions, together with the number c of Ω-classes, and the class stabilisers of these subgroups in the corresponding conformal groups.

In Table 5.6, each row represents a quasi-equivalence class of representations ρ of a simply-connected group of Lie type with non-abelian composition factor S. Column "Group" gives the isomorphism type of $N_\Omega(\mathrm{Im}\,\rho)$, up to the addition of scalars from Ω. Column "Module" describes a module corresponding to one of the representations in the quasi-equivalence class. In Column c we state the number of conjugacy classes in the quasisimple group Ω, whilst Column "Stab" gives the stabiliser of a conjugacy class in the corresponding conformal group.

Theorem 5.8.1 *Let S be an $\mathscr{S}_2^*$-maximal subgroup of a classical group C in dimension at most 12. Then S^∞ is contained in Table 5.6.*

Proof The entries in the table are proved correct in Propositions 5.3.6–5.3.8, 5.4.2, 5.4.5–5.4.8, 5.4.15–5.4.18, 5.4.20, 5.4.21, 5.5.2, 5.5.3, 5.7.1–5.7.4. (The results for $\Omega_8^+(q)$ are also currently included.) The completeness of the tables is documented in Theorems 5.3.9, 5.4.22, 5.5.4, 5.6.1 and 5.7.5. □

Table 5.6: Defining characteristic candidates

Group	Module	Condns on q	Dim	Case	c	Stab
$SL_2(q)$	$S^3(V)$	$p \geqslant 5$	4	**S**	1	$\langle \delta \rangle$
$Sz(q)$	V	$p = 2, q = p^e$	4	**S**	1	1
		$e > 1$ odd				
$L_2(q)$	$S^4(V)$	$p \geqslant 5$	5	**O°**	1	$\langle \delta \rangle$
$SL_3(q)$	$S^2(V)$	$p \geqslant 3$	6	**L**	2	$\langle \delta^2 \rangle$
$SU_3(q)$	$S^2(V)$	$p \geqslant 3$	6	**U**	2	$\langle \delta^2 \rangle$
$SL_2(q)$	$S^5(V)$	$p \geqslant 7$	6	**S**	1	$\langle \delta \rangle$
$G_2(q)$	V	$p = 2 \neq q$	6	**S**	1	1
$L_2(q).2$	$S^6(V)$	$p \geqslant 7$	7	**O°**	2	1
$L_3(q).2$	W (adjoint)	$p = 3$	7	**O°**	2	1
$U_3(q).2$	W (adjoint)	$p = 3$	7	**O°**	2	1
$G_2(q)$	V	$p \geqslant 3$	7	**O°**	2	1
$R(q)$	V	$p = 3, q = p^e$	7	**O°**	2	1
		$e > 1$ odd				
$SL_2(q)$	$S^7(V)$	$p \geqslant 11$	8	**S**	1	$\langle \delta \rangle$
$SL_2(q^3).3$	$V \otimes V^\sigma \otimes V^{\sigma^2}$	$p \geqslant 3$	8	**S**	1	$\langle \delta \rangle$
$L_3(q).3$	W (adjoint)	$q \equiv 1 \pmod 3$	8	**O⁺**	$(q-1,2)^2$	$\langle \gamma \rangle$
$L_3(q)$	W (adjoint)	$q \equiv 2 \pmod 3$	8	**O⁻**	$(q-1,2)$	$\langle \gamma \rangle$
$U_3(q).3$	W (adjoint)	$q \equiv 2 \pmod 3$	8	**O⁺**	$(q-1,2)^2$	$\langle \gamma \rangle$
$U_3(q)$	W (adjoint)	$q \equiv 1 \pmod 3$	8	**O⁻**	$(q-1,2)$	$\langle \gamma \rangle$
$SL_2(q^3).3$	$V \otimes V^\sigma \otimes V^{\sigma^2}$	$p = 2$	8	**O⁺**	2	1
$Sp_6(q)$	spin	$p = 2$	8	**O⁺**	2	1
$2 \cdot \Omega_7(q)$	spin	$p \geqslant 3$	8	**O⁺**	4	$\langle \gamma \rangle$
$(q-1,2) \cdot \Omega_8^- (\sqrt{q})$	spin	$p = 2$	8	**O⁺**	2	1
		$p \geqslant 3$	8	**O⁺**	4	$\langle \gamma \rangle$
$^3D_4(q_0)$	V	$q = q_0^3$	8	**O⁺**	$2(q-1,2)^2$	1
$L_3(q^2).2$	$V \otimes V^\sigma$	$q \equiv 0 \pmod 3$	9	**L**	1	1
$L_3(q^2).S_3$	$V \otimes V^\sigma$	$q \equiv 2 \pmod 3$	9	**L**	1	1
$SL_3(q^2).2$	$V \otimes V^\sigma$	$q \equiv 1 \pmod 9$	9	**L**	3	$\langle \delta^3 \rangle$
$SL_3(q^2).6$	$V \otimes V^\sigma$	$q \equiv 4, 7 \pmod 9$	9	**L**	3	1

Table 5.6: Defining characteristic candidates

Group	Module	Condns on q	Dim	Case	c	Stab
$L_3(q^2).2$	$V \otimes (V^*)^\sigma$	$q \equiv 0 \pmod 3$	9	**U**	1	1
$L_3(q^2).S_3$	$V \otimes (V^*)^\sigma$	$q \equiv 1 \pmod 3$	9	**U**	1	1
$SL_3(q^2).2$	$V \otimes (V^*)^\sigma$	$q \equiv 8 \pmod 9$	9	**U**	3	$\langle \delta^3 \rangle$
$SL_3(q^2).6$	$V \otimes (V^*)^\sigma$	$q \equiv 2, 5 \pmod 9$	9	**U**	3	1
$L_2(q).2$	$S^8(V)$	$p \geqslant 11$	9	**O**$^\circ$	2	1
$L_2(q^2).2$	$S^2(V) \otimes S^2(V^\sigma)$	$p \geqslant 3$	9	**O**$^\circ$	1	$\langle \delta \rangle$
$L_3(q).(q-1,3)$	$S^3(V)$	$p \geqslant 5$	10	**L**	$(q-1,10)$	1
$U_3(q).(q+1,3)$	$S^3(V)$	$p \geqslant 5$	10	**U**	$(q+1,10)$	1
$\frac{(q-1,4)}{2} \cdot SL_4(q) \cdot \frac{(q-1,4)}{2}$	$S^2(V)$	$p \geqslant 3$	10	**L**	$(q-1,5)$	$\langle \delta^5 \rangle$
$\frac{(q+1,4)}{2} \cdot SU_4(q) \cdot \frac{(q+1,4)}{2}$	$S^2(V)$	$p \geqslant 3$	10	**U**	$(q+1,5)$	$\langle \delta^5 \rangle$
$SL_5(q)$	$\Lambda^2(V)$		10	**L**	$(q-1,2)$	$\langle \delta^2 \rangle$
$SU_5(q)$	$\Lambda^2(V)$		10	**U**	$(q+1,2)$	$\langle \delta^2 \rangle$
$SL_2(q)$	$S^9(V)$	$p \geqslant 11$	10	**S**	1	$\langle \delta \rangle$
$S_4(q)$	$S^2(V)$	$q \equiv 1 \pmod 4$	10	**O**$^+$	4	$\langle \delta' \rangle$
$S_4(q)$	$S^2(V)$	$q \equiv 3 \pmod 4$	10	**O**$^-$	4	$\langle \delta' \rangle$
$L_2(q)$	$S^{10}(V)$	$p \geqslant 11$	11	**O**$^\circ$	1	$\langle \delta \rangle$
$SL_2(q)$	$S^{11}(V)$	$p \geqslant 13$	12	**S**	1	$\langle \delta \rangle$
$Sp_4(q)$		$p = 5$	12	**S**	1	$\langle \delta \rangle$

5.9 Determining the effects of duality and field automorphisms

Here we carry out the necessary calculations concerning the action of duality and field automorphisms. The corresponding calculations in the cross characteristic case were introduced in Section 4.6, and we make use of some of the theory developed there. Since we shall be discussing the action of the duality and field automorphisms of a quasisimple classical group Ω on subgroups $G\rho$ which are themselves classical groups, we shall sometimes use subscripts for clarity. For example, ϕ_Ω denotes the field automorphism ϕ of Ω, whilst ϕ_G denotes the induced action of the natural field automorphism of G on $G\rho$.

5.9.1 Cases L and U

We first make some general remarks about examples of the form $U = \mathrm{S}^m(V)$ for $m \in \{2,3\}$, or $U = \Lambda^2(V)$, where V is the natural module for $\mathrm{SL}_d^\pm(q)$ for some d and q, which will be helpful in performing the necessary calculations.

As we explained in Section 5.2, we regard U primarily as a quotient module of the m-th tensor power $V^{\otimes m}$ of V. However, for all characteristics for $U = \Lambda^2(V)$, and for characteristics greater than m for $U = \mathrm{S}^m(V)$, we can identify U with a suitable submodule of $V^{\otimes m}$: see Lemmas 5.2.3 and 5.2.8. When calculating the effect of duality and field automorphisms we shall find it more convenient to work with the submodule version.

Let $(e_1, e_2, \ldots, e_d)$ be the natural basis for V. Order the basis elements of $V^{\otimes m}$ lexicographically. For $g \in \mathrm{SL}_d^\pm(q)$, the matrix representing g in $V^{\otimes m}$ is the m-fold Kronecker product $g^{\otimes m}$ of g. Let A denote an $n \times d^m$ matrix whose rows are the basis vectors for U as a subspace of $V^{\otimes m}$, as given in Definitions 5.2.2 and 5.2.7 (using the isomorphisms from Lemmas 5.2.3 and 5.2.8).

Lemma 5.9.1   *Let $G := \mathrm{SL}_d^\pm(q)$, let V be the natural G-module, and let $\rho : G \to \mathrm{SL}_n^\pm(q^u)$ be the representation with module $U = \mathrm{S}^m(V)$ for $m \in \{2,3\}$, or $U = \Lambda^2(V)$, so that $n = \dim U$. Assume that $p > 3$ if U is a symmetric cube, and p is odd otherwise. Let A denote an $n \times d^m$ matrix as above. Then:*

- (i) *$AA^\mathsf{T}D = I_n$, where $D \in \mathrm{M}_{n \times n}(p)$ has determinant $2^{\binom{d}{2}} \pmod{p}$ when $U = \mathrm{S}^2(V)$, determinant $6^{\binom{d}{3}} \cdot 3^{d(d-1)} \pmod{p}$ when $U = \mathrm{S}^3(V)$, and determinant $2^{-\binom{d}{2}} \pmod{p}$ when $U = \Lambda^2(V)$.*
- (ii) *For $g \in \mathrm{SL}_d^\pm(q)$, the matrix $M(g)$ of the action of g on U is $Ag^{\otimes m}A^\mathsf{T}D$, and furthermore $(Ag^{\otimes m}A^\mathsf{T}D)^{-1} = A(g^{-1})^{\otimes m}A^\mathsf{T}D$.*
- (iii) *If $G = \mathrm{SL}_d(q)$, then the normaliser of $G\rho$ in $\mathrm{Out}\,\mathrm{SL}_n(q)$ contains ϕ_Ω, which induces the outer automorphism ϕ_G of $G\rho$.*
- (iv) *If $G = \mathrm{SL}_d(q)$ and n is even, then the class stabiliser of $G\rho$ in $\mathrm{Out}\,\mathrm{SL}_n(q)$ contains a conjugate of γ_Ω if $\det D$ is a square in $\mathbb{F}_q^\times$, and a conjugate of $\gamma_\Omega\delta_\Omega$ otherwise. In either case, the induced outer automorphism of $G\rho$ is a conjugate of γ_G.*
- (v) *If $G = \mathrm{SU}_d(q)$ then $\mathrm{Im}\,\rho$ consists of isometries of the unitary form with matrix D^{-1}. Futhermore, if n is even, then the class stabiliser of $G\rho$ in $\mathrm{Out}\,\mathrm{SU}_n(q)$ contains a conjugate of ϕ_Ω if $\det D$ is a square in $\mathbb{F}_p^\times$, and a conjugate of $\phi_\Omega\delta_\Omega$ otherwise. In either case, the induced outer automorphism of $G\rho$ is a conjugate of ϕ_G.*

Proof   (i) In our standard bases for $\mathrm{S}^m(V)$ and $\Lambda^2(V)$, each tensor of the form $e_i \otimes e_j$ or $e_i \otimes e_j \otimes e_k$ occurs in precisely one basis vector. The product AA^T has entries the dot product of each row of A with each other row. Thus the only non-zero dot products are those of a row with itself, and D is as given.

(ii) Let $M(g)$ be the matrix of the action of g on U. Then $Ag^{\otimes m} = M(g)A$, and hence (i) gives the first claim. The second now follows from $M(g)^{-1} = M(g^{-1})$.

(iii) Since A and D have entries in $\mathbb{F}_p$, we see by (ii) that the automorphism ϕ_Ω of $\mathrm{SL}_n(q)$ induces ϕ_G. So $G\rho$ is normalised by ϕ_Ω.

(iv) By (ii), again, $M(g^{\gamma_G}) = A(g^{-\mathsf{T}})^{\otimes m}A^\mathsf{T}D$, whereas the image of $M(g)$ under γ_Ω is

$$M(g)^{-\mathsf{T}} = (Ag^{\otimes m}A^\mathsf{T}D)^{-\mathsf{T}} = (A(g^{-1})^{\otimes m}A^\mathsf{T}D)^\mathsf{T} = DA(g^{-\mathsf{T}})^{\otimes m}A^\mathsf{T}.$$

So $M(g^{\gamma_G}) = D^{-1}M(g)^{-\mathsf{T}}D$. In other words, γ_G is induced by γ_Ω followed by conjugation by the diagonal automorphism of Ω induced by conjugation by D. The result now follows from Lemma 4.6.2.

(v) For a matrix g with entries in $\mathbb{F}_{q^2}$, let g^* be the result of applying the map $x \mapsto x^q$ to the entries of g^T. The matrix I_d is our standard form for $\mathrm{GU}_d(q)$, so $g^* = g^{-1}$ for $g \in \mathrm{GU}_d(q)$. Hence $M(g)^* = (Ag^{\otimes m}A^\mathsf{T}D)^* = DA(g^{-1})^{\otimes m}A^\mathsf{T} = D(Ag^{\otimes m}A^\mathsf{T}D)^{-1}D^{-1}$, and so $(Ag^{\otimes m}A^\mathsf{T}D)D^{-1}(Ag^{\otimes m}A^\mathsf{T}D)^* = D^{-1}$. Thus D^{-1} is the matrix of the form stabilised by G.

In the notation of Lemma 4.6.4 we may set $\alpha = \phi_G$ and $\beta = \phi_\Omega$, and then $L = I_n$ and $\lambda = 1$. (The matrix of the fixed unitary form D^{-1} was denoted by B in Lemma 4.6.4.) Using Lemma 4.6.5 we get $\kappa = 1$, so that the class of G is stabilised by ϕ_Ω if and only if $\det D^{-1}$ is a square in $\mathbb{F}_p^\times$, which is equivalent to $\det D$ being a square in $\mathbb{F}_p^\times$. $\square$

We see from Theorem 5.8.1 that $\mathscr{S}_2^*$-maximal candidates exist only in dimensions 6, 9 and 10. As in Section 4.9, in our statements of the results, when we refer to the class stabiliser of an $\mathscr{S}_2^*$ subgroup S of Ω in $\mathrm{Out}\,\Omega$, we are really specifying the stabiliser of the Ω-class of some $\mathrm{Aut}\,\Omega$-conjugate of S.

Proposition 5.9.2   (i) *The class stabiliser of the $\mathscr{S}_2^*$-subgroup $\mathrm{SL}_3(q)$ of $\mathrm{SL}_6(q)$ is $\langle \delta^2, \phi, \gamma \rangle$ if $q \equiv \pm 1 \pmod 8$ and $\langle \delta^2, \phi, \gamma\delta \rangle$ if $q \equiv \pm 3 \pmod 8$.*

(ii) *The class stabiliser of the $\mathscr{S}_2^*$-subgroup $\mathrm{SU}_3(q)$ of $\mathrm{SU}_6(q)$ is $\langle \delta^2, \phi \rangle$ if $p \equiv \pm 1 \pmod 8$ and $\langle \delta^2, \phi\delta \rangle$ if $p \equiv \pm 3 \pmod 8$.*

Proof By Theorem 5.8.1, the group $S \cong \mathrm{SL}_3^\pm(q)$ arises from the module $\mathrm{S}^2(V_3)$ of $\mathrm{L}_3^\pm(q)$, with $p \geqslant 3$, and the class stabiliser of S in the group of conformal automorphisms is $\langle \delta^2 \rangle$. In the notation of Lemma 5.9.1, the determinant $\det D = 2^3$.

In Case **L**, by Lemma 5.9.1 (iii), we may assume that S is normalised by ϕ. Furthermore, by Lemma 5.9.1 (iv), the class stabiliser of S contains an $\mathrm{Out}\,\Omega$-conjugate of γ when 2 is a square in $\mathbb{F}_q^\times$, and an $\mathrm{Out}\,\Omega$-conjugate of $\gamma\delta$ when 2 is non-square in $\mathbb{F}_q^\times$. Since the class stabiliser also contains $\langle \delta^2 \rangle$, it must be $\langle \delta^2, \phi, \gamma \rangle$ or $\langle \delta^2, \phi, \gamma\delta \rangle$ in these two cases.

In Case **U**, by Lemma 5.9.1 the class stabiliser contains an $\mathrm{Out}\,\Omega$-conjugate of ϕ if and only if $\det D$ is a square modulo p, which is the case if and only if $p \equiv \pm 1 \pmod 8$, by Proposition 1.13.7 (ii). □

Recall that our choice of form means that γ is equal to ϕ^e in Case **U**.

Proposition 5.9.3   *Let S be an $\mathscr{S}_2^*$-subgroup of $\mathrm{SL}_9^{\pm}(q)$ with composition factor $\mathrm{L}_3(q^2)$. Then the class stabiliser of S is $\langle \delta^3, \phi, \gamma \rangle$.*

Proof   If $q \equiv 0, 2 \pmod 3$ for $\Omega = \mathrm{SL}_9(q)$, or $q \equiv 0, 1 \pmod 3$ for $\Omega = \mathrm{SU}_9(q)$, then δ_Ω is trivial, and there is a single class of images $(3^{\cdot})\mathrm{L}_3(q^2)$ in Ω. Thus the class stabiliser is $\langle \phi, \gamma \rangle$.

Otherwise, by Theorem 5.8.1 there are three Ω-classes of subgroups G, permuted transitively under the action of δ_Ω. A straightforward calculation reveals that there is a single conjugacy class of subgroups of $\mathrm{Out}\,\Omega$ of index 3 that does not contain δ, with representative $\langle \delta^3, \phi, \gamma \rangle$. □

Proposition 5.9.4   (i) *The class stabiliser of the $\mathscr{S}_2^*$-subgroup $S = \mathrm{PGL}_3(q)$ of $\mathrm{SL}_{10}(q)$ is $\langle \phi, \gamma \rangle$ when $q \equiv \pm 1, \pm 5 \pmod{24}$, and $\langle \phi, \gamma\delta^5 \rangle$ when $q \equiv \pm 7, \pm 11 \pmod{24}$.*

(ii) *The class stabiliser of the $\mathscr{S}_2^*$-subgroup $S = \mathrm{PGU}_3(q)$ of $\mathrm{SU}_{10}(q)$ is $\langle \phi \rangle$ when $p \equiv \pm 1, \pm 5 \pmod{24}$ and $\langle \phi\delta \rangle$ when $p \equiv \pm 7, \pm 11 \pmod{24}$.*

Proof   By Theorem 4.3.3, these subgroups arise from the module $\mathrm{S}^3(V_3)$ of $\mathrm{L}_3^{\pm}(q)$, with $p \geqslant 5$, and the class stabiliser of S in the group of conformal automorphisms is trivial. In the notation of Lemma 5.9.1, the determinant $\det D = 6 \cdot 3^6 \pmod p$.

Consider first $S = \mathrm{PGL}_3(q)$, so that $\Omega = \mathrm{SL}_{10}(q)$. Let X be the stabiliser in $\mathrm{Out}\,\Omega$ of the Ω-class of S. By Lemma 5.9.1 (iii), we may assume that $\phi \in X$. Furthermore, by Lemma 5.9.1 (iv), the group X contains an $\mathrm{Out}\,\Omega$-conjugate of γ when 6 is a square in $\mathbb{F}_q^{\times}$, and (since $\gamma\delta$ and $\gamma\delta^5$ are conjugate in $\mathrm{Out}\,\Omega$) an $\mathrm{Out}\,\Omega$-conjugate of $\gamma\delta^5$ when 6 is non-square in $\mathbb{F}_q^{\times}$. We need to show that X is conjugate in $\mathrm{Out}\,\Omega$ to $\langle \phi, \gamma \rangle$ or $\langle \phi, \gamma\delta^5 \rangle$ in the two cases. Write $\phi = \phi_1\phi_2$, where ϕ_1 has odd order and ϕ_2 has 2-power order. Then, since δ has order 2 or 10, ϕ_1 is central in $\mathrm{Out}\,\Omega$, so all conjugates of X contain ϕ_1. By Sylow's theorem, some conjugate of a Sylow 2-subgroup of X is contained in the abelian self-normalising Sylow 2-subgroup $\langle \delta^5, \phi_2, \gamma \rangle$ of $\mathrm{Aut}\,\Omega$ and, since ϕ_2 is not conjugate to $\phi_2\delta^5$ in $\mathrm{Aut}\,\Omega$, this conjugate of X must contain ϕ_2, and the result follows.

In the unitary case, the class of $\mathrm{SU}_3(q)$ is stabilised by ϕ if and only if $\det D$ is a square modulo p, by Lemma 5.9.1 (iv), and is stabilised by $\phi\delta$ otherwise. This occurs if and only if 6 is a square modulo p. □

The examples arising from the 10-dimensional module $\mathrm{S}^2(V_4)$ of $\mathrm{L}_4^{\pm}(q)$, with $p \geqslant 3$, are similar.

Proposition 5.9.5   (i) *The   class   stabiliser   of   the $\mathscr{S}_2^*$-subgroup* $\frac{(q-1,4)}{2} \cdot \mathrm{SL}_4(q) . \frac{(q-1,4)}{2}$ *of* $\mathrm{SL}_{10}(q)$ *is* $\langle \delta^5, \phi, \gamma \rangle$.

(ii) *The class stabiliser of the $\mathscr{S}_2^*$-subgroup* $\frac{(q+1,4)}{2} \cdot \mathrm{SU}_4(q) . \frac{(q+1,4)}{2}$ *of* $\mathrm{SU}_{10}(q)$ *is* $\langle \delta^5, \phi \rangle$.

Proof In the notation of Lemma 5.9.1, $\det D = 2^6$ is a square. Hence, the result follows as in the previous proposition from Theorem 5.8.1 and Lemma 5.9.1. □

Proposition 5.9.6   (i) *The class stabiliser of the $\mathscr{S}_2^*$-subgroup* $\mathrm{SL}_5(q)$ *of* $\mathrm{SL}_{10}(q)$ *is* $\langle \delta^2, \phi, \gamma \rangle$.

(ii) *The class stabiliser of the $\mathscr{S}_2^*$-subgroup* $\mathrm{SU}_5(q)$ *of* $\mathrm{SU}_{10}(q)$ *is* $\langle \delta^2, \phi \rangle$.

Proof By Theorem 5.8.1, these groups arise from the module $\Lambda^2(V_5)$ of $\mathrm{L}_5^{\pm}(q)$. If q is odd then using Lemma 5.2.3 we get a matrix D of determinant 2^{-10}, a square, and the result follows as before.

Thus we need only consider the case q even. Let A be the matrix defined just before Lemma 5.9.1, and suppose temporarily that A has entries in $\mathbb{Z}$, so that each row contains a 1 and a -1. Let C be the matrix formed from A by replacing the -1 in each row by 0, considered now as a matrix over $\mathbb{F}_q$. Then $AC^{\mathsf{T}} = I_n$, so the matrix of $g \in \mathrm{SL}_d^{\pm}(q)$ in its action on U is $Ag^{\otimes 2}C^{\mathsf{T}}$. It is straightforward to verify that this is equal to $Cg^{\otimes 2}A^{\mathsf{T}}$ for all g, from which it follows that $(Ag^{\otimes 2}C^{\mathsf{T}})^{-\mathsf{T}} = A(g^{-\mathsf{T}})^{\otimes 2}C^{\mathsf{T}}$.

So, in the linear case, γ_G is induced by γ_Ω, and a corresponding argument in the unitary case shows that the image of $\mathrm{SU}_d(q)$ in $\mathrm{GL}_n(q^2)$ consists of isometries of the unitary form with matrix I_n. □

5.9.2 Case S

In this subsection we calculate the class stabiliser in $\mathrm{Out}\,\mathrm{Sp}_n(q)$ of all $\mathrm{Sp}_n(q)$-conjugacy classes of $\mathscr{S}_2^*$-subgroups. Recall from Subsection 1.7.2 that if $(n,p) \neq (4,2)$, then $\mathrm{Out}\,\mathrm{S}_n(q) = \langle \delta, \phi \rangle$ (with δ trivial if $p = 2$), whilst the group $\mathrm{Out}\,\mathrm{S}_4(2^e) = \langle \gamma, \phi \rangle = \langle \gamma \rangle \cong 2e$ for $e > 1$, and $\mathrm{Sp}_4(2)$ is not quasisimple.

Proposition 5.9.7   (i) *The class stabiliser of the $\mathscr{S}_2^*$-subgroup* $\mathrm{Sz}(q)$ *of* $\mathrm{Sp}_4(q)$, *with* $q = 2^{2\mu+1}$ *and* $\mu \geqslant 1$, *is* $\langle \gamma \rangle$.

(ii) *The class stabiliser of all other $\mathscr{S}_2^*$-subgroups of* $\mathrm{S}_n(q)$ *for* $n \leqslant 12$ *is* $\langle \delta, \phi \rangle$.

Proof Suppose first that $n = 4$ and q is even. By Theorem 5.8.1 there is a unique Ω-class of $\mathscr{S}_2$-subgroups, namely $\mathrm{Sz}(q)$ with $q = 2^{2\mu+1}$ and $\mu \geqslant 1$. So this class is stabilised by the full outer automorphism group $\langle \phi, \gamma \rangle$ of $\mathrm{Sp}_4(q)$.

In all other cases, we need only resolve the action of ϕ when $q \neq p$. By Theorem 5.8.1, in all such cases there is a single class of subgroups of Ω, which must therefore be stabilised by ϕ. □

5.9.3 Cases O^ε

In this subsection we calculate the class stabiliser in Out $O_n^\varepsilon(q)$ of all $\Omega_n^\varepsilon(q)$-conjugacy classes of $\mathscr{S}_2^*$-subgroups.

We start with a group that occurs in several dimensions.

Proposition 5.9.8 *The class stabiliser of all $\mathscr{S}_2^*$-subgroups S of $\Omega = \Omega_n(q)$ with composition factor $L_2(q)$, for $n = 7, 9, 11$, contains $\langle \phi \rangle$.*

Proof We have seen already in Section 5.3 that, for the image of the representation corresponding to the module $S^n(V)$ of odd dimension $n + 1$ for $SL_2(q)$, the automorphism ϕ_S is induced by ϕ_Ω. $\qquad \square$

Recall from Subsection 1.7.2 that Out $\Omega_7(q) = \langle \delta, \phi \rangle$.

Proposition 5.9.9 *The class stabiliser of each $\mathscr{S}_2^*$-subgroup of $\Omega_7(q)$ is $\langle \phi \rangle$.*

Proof By Theorem 4.3.3, the groups to consider are $L_2(q).2$ (with $p \geqslant 7$), $L_3(q).2$ (with $p = 3$), $U_3(q).2$ (with $p = 3$), $G_2(q)$ (with $p \geqslant 3$) and $R(q)$ (with $p = 3$ and $e \geqslant 1$ odd). In each case there are two classes, with trivial stabiliser in the conformal group.

Two generators of the image S of the 7-dimensional representation of $G_2(q)$ with q odd are defined in [49, Section 3.6], one of which is stabilised by ϕ while the other is mapped to a power of itself. So the class of S is stabilised by $\langle \phi \rangle$, and ϕ_Ω induces ϕ_S.

It is shown in [64] that, for the appropriate q, the groups $L_2(q).2$, $L_3(q).2$, $U_3(q).2$ and $R(q)$ are all maximal subgroups of $G_2(q)$ that are irreducible in the 7-dimensional representation, and that $\langle \phi \rangle$ stabilises them. $\qquad \square$

We are not calculating the maximal subgroups of $\Omega_8^+(q)$ in this book. Recall from Subsection 1.7.2 that Out $\Omega_8^-(q) = \langle \delta, \gamma, \varphi \rangle$, with δ trivial if $q = p^e$ is even.

Proposition 5.9.10 *The class stabiliser of each $\mathscr{S}_2^*$-subgroup of $\Omega_8^-(q)$ is $\langle \gamma, \varphi \rangle$.*

Proof By Theorem 4.3.3, the $\mathscr{S}_2^*$-subgroups are $L_3(q)$ (with $q \equiv 2 \pmod 3$) and $U_3(q)$ (with $q \equiv 1 \pmod 3$), each acting on their adjoint module. There are $(q - 1, 2)$ classes of these subgroups in $\Omega_8^-(q)$, stabilised by γ, and interchanged by δ when q is odd. So either φ or $\varphi\delta$ lies in the class stabiliser. The result is immediate for even q, so assume from now on that q is odd.

If e is odd, then since both φ^2 and $\gamma = \varphi^e$ are in the class stabiliser, so is φ. This is always the case for $L_3(q)$. It therefore remains to consider $U_3(q)$ with q an even power of an odd prime $p > 3$.

Let $\alpha_1 : U_3(q) \to \Omega_8^-(q)$ and $\alpha_2 : L_3(q^2) \to \Omega_8^+(q^2)$ be the adjoint representations under consideration. Since $SU_3(q) \leqslant GL_3(q^2)$, the group $U_3(q)$ is a subgroup of $L_3(q^2)$; let $\beta_1 : U_3(q) \to L_3(q^2)$ be the natural inclusion map. From Table 2.8 we see that $\Omega_8^-(q)$ is an irreducible subgroup of $\Omega_8^+(q^2)$, such that $\Omega_8^-(q)$ can be written over $\mathbb{F}_q$; let $\beta_2 : \Omega_8^-(q) \to \Omega_8^+(q^2)$ be the natural inclusion map.

As we saw in Subsection 5.4.1, α_1 is defined by first applying β_1, which embeds $U_3(q)$ into $L_3(q^2)$, then applying the adjoint representation α_2 to the image, and then conjugating within $GL_8(q^2)$ (and hence by Lemma 1.8.10 (ii), within $CGO_8^+(q^2)$) to write the result over $\mathbb{F}_q$. So, by taking suitable conjugacy class representatives, we may assume that $\beta_1\alpha_2 = \alpha_1\beta_2$.

Our aim is to prove that φ stabilises the class of $\mathrm{Im}(\alpha_1)$. We shall deduce this by considering the class stabilisers of $\mathrm{Im}(\beta_1)$, $\mathrm{Im}(\beta_2)$ and $\mathrm{Im}(\alpha_2)$. For β_1, it follows directly from our definition of ϕ and choice of form I_n that ϕ on $L_3(q^2)$ stabilises and induce ϕ on $U_3(q) = \mathrm{Im}(\beta_1)$. For α_2, we know from Table 8.50 (which comes from [62]) that the automorphism ϕ of $\Omega_8^+(q^2)$, with our standard form, stabilises the class of $L_3(q^2) = \mathrm{Im}(\alpha_2)$. Notice that these automorphisms ϕ all correspond to the $x \mapsto x^p$ map on matrices. In this final case, the following lemma therefore completes the proof. □

Lemma 5.9.11 *The automorphism ϕ of $\Omega_8^+(q^2)$ stabilises and induces φ on the class of $\mathrm{Im}(\beta_2) \cong \Omega_8^-(q)$.*

Proof Let z and ω be primitive elements of $\mathbb{F}_{q^2}^\times$ and $\mathbb{F}_q^\times$, respectively, with $z^{q+1} = \omega$. Let A be the matrix of our standard form for $\Omega_8^-(q)$, and note that our standard copy of $\Omega_8^-(q)$ is a subgroup of $\Omega_8^+(q^2, A)$. Let $\lambda = z^{(q+1)/2} = \sqrt{\omega}$ and define $g := \mathrm{diag}(\lambda, 1, 1, 1, 1, 1, 1, 1)$. Then $\Omega_8^+(q^2, A)^g = \Omega_8^+(q^2, I_8)$.

Let ϕ_I be the automorphism of $\Omega_8^+(q^2, I_8)$ that acts on matrix entries as $x \mapsto x^p$, and let c_g denote conjugation by g. Then $c_g\phi_I c_g^{-1}$ is an automorphism of $\Omega_8^+(q^2, A)$ that induces ϕ_I. Let $d = \mathrm{diag}(w^{(p-1)/2}, 1, 1, 1, 1, 1, 1, 1)$. Then $c_g\phi_I c_g^{-1} = \phi_I c_g^{\phi_I} c_g^{-1} = \phi_I c_d$. But $\phi_I c_d$ stabilises the class of $\Omega_8^-(q, A)$, and induces the automorphism φ of $\Omega_8^-(q)$ that we defined in Subsection 1.7.1.

It is shown in [6] that, if we define the field automorphism ϕ_B of $\Omega_8^+(q^2, B)$ for a form with matrix B over $\mathbb{F}_p$ by raising matrix entries to the p-th power, then ϕ_B can depend, as an element of $\mathrm{Out}\,\Omega_8^+(q^2)$, on the form B. However, our standard form for $\Omega_8^+(q^2)$ is antidiag $(1, 1, 1, 1, 1, 1, 1, 1)$, so both it and I_8 have square determinants over $\mathbb{F}_p$. Thus it follows from [6, Proof of Proposition 12] that ϕ_I is the same outer automorphism as our standard ϕ. So ϕ induces φ as claimed. □

Recall from Subsection 1.7.2 that $\mathrm{Out}\,\Omega_9(q) = \langle \delta, \phi \rangle$.

Proposition 5.9.12 *The class stabiliser of the $\mathscr{S}_2^*$-subgroup $L_2(q^2).2$ of $\Omega_9(q)$ is $\langle \delta, \phi \rangle$.*

Proof By Theorem 5.8.1 there is a single class of these groups in $\Omega_9(q)$. □

Recall from Subsection 1.7.2 that $\mathrm{Out}\,\Omega_{10}^+(q) = \langle \delta, \gamma, \delta', \phi \rangle$ if $q \equiv 1 \pmod 4$, whilst if $q \equiv 3 \pmod 4$ then $\mathrm{Out}\,\Omega_{10}^-(q) = \langle \delta, \gamma, \delta', \phi \rangle$.

Proposition 5.9.13   *The class stabiliser of the $\mathscr{S}_2^*$-subgroup $\mathrm{S}_4(q)$ of $\Omega_{10}^\pm(q)$ is $\langle \delta', \phi \rangle$.*

Proof By Theorem 5.8.1, the group $G = \mathrm{S}_4(q)$ is a subgroup of $\Omega_{10}^\pm(q)$ when $q \equiv \pm 1 \pmod 4$, respectively, and in both cases there are four classes in Ω, with stabiliser $\langle \delta' \rangle$ in the group of conformal automorphisms.

If $q \equiv -1 \pmod 4$, then $q = p^e$ must be an odd power of p, and so the class stabiliser contains ϕ.

Otherwise, $q \equiv 1 \pmod 4$ and $G \leqslant \Omega_{10}^+(q)$. As noted in the proof of Proposition 5.5.2, the natural symmetric square representation lies in $\Omega_{10}^+(q, F)$, where F is a form matrix with entries in $\mathbb{F}_p$ and $\det F = 16$. With that embedding, the field automorphism ϕ_F (i.e raising matrix entries to the p-th power) of $\Omega_{10}^+(q, F)$ induces ϕ_G on G. Since $\det F$ is a square in $\mathbb{F}_p^\times$, it follows from [6, Proposition 9] that ϕ_F is equal to our standard outer automorphism ϕ of $\Omega_{10}^+(q)$. □

5.10 Containments

To complete the determination of the $\mathscr{S}_2^*$-maximal subgroups, it remains to consider containments between them: recall Definition 2.1.4. We only need consider the relatively small number of cases in Table 5.6 where there is more than one candidate in the same case and dimension.

Lemma 5.10.1   *Let $\Omega = \Omega_7(q)$. Then $\mathrm{G}_2(q)$ and its extensions are the only $\mathscr{S}_2^*$-maximal subgroups of almost simple extensions of Ω.*

Proof Recall from Subsection 1.7.2 that $\mathrm{Out}\,\Omega_7(q) = \langle \delta, \phi \rangle$. By Theorem 5.8.1, the $\mathscr{S}_2^*$-subgroups are $\mathrm{G}_2(q)$, $\mathrm{L}_2(q).2$ ($p \geqslant 7$), $\mathrm{L}_3(q).2$ ($p = 3$), $\mathrm{U}_3(q).2$ ($p = 3$), and $\mathrm{R}(q)$ ($q = 3^e$, e odd). In each case, there are two Ω-classes and a single class in $\mathrm{SO}_7(q)$.

We find from [63] (see our Table 8.41) that $\mathrm{G}_2(q)$ has unique classes of irreducible maximal subgroups isomorphic to each of the other $\mathscr{S}_2^*$-subgroups except for $\mathrm{L}_2(7).2$, and that $\mathrm{G}_2(7)$ has a single class of subgroups $\mathrm{L}_2(7).2$, which are non-maximal. From [84], each of the $\mathscr{S}_2^*$-subgroups has a unique class of irreducible representations in degree 7, up to quasi-equivalence, so the remaining $\mathscr{S}_2^*$-subgroups of $\Omega_7(q)$ are all subgroups of $\mathrm{G}_2(q)$.

Furthermore, Propositions 5.9.8 and 5.9.9 tell us that the full class stabiliser of each of the $\mathscr{S}_2$-candidates is $\langle\phi\rangle$, and it follows that all almost simple extensions of $L_2(q).2$, $L_3(q).2$, $U_3(q).2$ and $R(q)$ are contained in the corresponding almost simple extensions of $G_2(q)$. $\qquad\square$

Lemma 5.10.2 *Let $\Omega = \mathrm{Sp}_8(q)$. Then all $\mathscr{S}_2^*$-subgroups of Ω are $\mathscr{S}_2^*$-maximal.*

Proof From Theorem 5.8.1, the $\mathscr{S}_2^*$-subgroups are $2^{\cdot}L_2(q)$ for $p \geqslant 11$ and $2^{\cdot}L_2(q^3).3$ for q odd. The representation of $2^{\cdot}L_2(q^3).3$ arises from the tensor product of three 2-dimensional modules for $2^{\cdot}L_2(q^3)$ twisted by different powers of the field automorphism, with the result rewritten over $\mathbb{F}_q$. By Corollary 5.3.3, the irreducible representations of $\mathrm{SL}_2(q)$ in dimension 2 have minimal field $\mathbb{F}_q$, so $2^{\cdot}L_2(q)$ can arise only as a $\mathscr{C}_5$-subgroup of $2^{\cdot}L_2(q^3)$. But the restriction of the 8-dimensional module to this subgroup is isomorphic to a tensor product of three copies of the natural module for $2^{\cdot}L_2(q)$ which, by Proposition 5.3.10, is reducible. So there is no containment between the two $\mathscr{S}_2^*$-subgroups. $\qquad\square$

Lemma 5.10.3 *Let $\Omega = \Omega_9(q)$. Then all $\mathscr{S}_2^*$-subgroups of Ω are $\mathscr{S}_2^*$-maximal.*

Proof By Theorem 5.8.1, the $\mathscr{S}_2^*$-subgroups are $L_2(q).2$ for $p \geqslant 11$, and $L_2(q^2).2$ for $p \geqslant 3$. The representation of $L_2(q^2).2$ arises from the tensor product of two 3-dimensional modules for $L_2(q^2)$, and as in the proof of Lemma 5.10.2, the restriction of the 9-dimensional module to its unique subgroup (up to conjugacy) isomorphic to $L_2(q)$ is reducible. $\qquad\square$

Lemma 5.10.4 *Let $\Omega = \mathrm{SL}_{10}^{\pm}(q)$. Then all $\mathscr{S}_2^*$-subgroups of Ω are $\mathscr{S}_2^*$-maximal.*

Proof In Case **L**, by Theorem 5.8.1 there are three $\mathscr{S}_2^*$-subgroups, of which the non-abelian composition factors are $L_3(q)$, $L_4(q)$ and $L_5(q)$. From [84] we find that $L_3(q)$ has no irreducible defining characteristic representations in dimensions 4 and 5, and $L_4(q)$ has no such representations in dimension 5. So $L_3(q)$ arises as a composition factor only of reducible subgroups of $\mathrm{SL}_4(q)$ in its natural representation, and $L_3(q)$ and $L_4(q)$ arise as composition factors only of reducible subgroups of $\mathrm{SL}_5(q)$ in its natural representation.

The $\mathscr{S}_2^*$-subgroup $\mathrm{SL}_4(q)$ is acting via the symmetric square of the natural representation, so that reducible subgroups of $\mathrm{SL}_4(q)$ also act reducibly in the 10-dimensional representation. Similarly, reducible subgroups of $\mathrm{SL}_5(q)$ act reducibly in the 10-dimensional representation. So there are no containments between the $\mathscr{S}_2^*$-subgroups in this case.

Similarly, there are no containments between the $\mathscr{S}_2^*$-candidates in Case **U**, as these have non-abelian composition factors $U_3(q)$, $U_4(q)$ and $U_5(q)$. $\qquad\square$

Proposition 5.10.5 *The only containments between $\mathscr{S}_2^*$-candidate maximals of Ω are in $\Omega_7(q)$. In this case, all almost simple extensions of $L_2(q).2$, $L_3(q).2$*

$(p = 3)$, $U_3(q).2$ $(p = 3)$ and $R(q)$ $(q = 3^e$, e odd$)$ are contained in corresponding extensions of $G_2(q)$.

Proof By Theorem 4.3.3, the only groups Ω which have more than one isomorphism type of $\mathscr{S}_2^*$-maximal subgroup for any fixed q are $\Omega_7(q)$, $Sp_8(q)$, $\Omega_8^+(q)$, $\Omega_9(q)$, and $SL_{10}^{\pm}(q)$. We are not classifying the maximal subgroups of $\Omega_8^+(q)$, and the other possibilities have been considered in Lemmas 5.10.1–5.10.4. □

5.11 Summary of the $\mathscr{S}_2^*$-maximals

As we did for the $\mathscr{S}_1$-maximals in Section 4.9, we conclude the chapter with a summary of the $\mathscr{S}_2^*$-maximals. We use the same conventions as in Section 4.9.

5.11.1 Cases L and U

We remind the reader that in this chapter we treat $SL_2(q)$ as $Sp_2(q)$.

Theorem 5.11.1 *Let G and Ω be as in Convention 4.10.1, with $\Omega = SL_n(q)$ or $SU_n(q)$ for $n = 3, 4, 5, 7, 8, 11$ or 12. Then there are no $\mathscr{S}_2^*$-maximal subgroups of G.*

Proof Follows from Theorem 5.8.1. □

Theorem 5.11.2 *Let G and Ω be as in Convention 4.10.1, with $\Omega = SL_6(q)$ or $SU_6(q)$. Then representatives of the conjugacy classes of $\mathscr{S}_2^*$-maximal subgroups of G are described in the list below.*

Proof Follows from Theorem 5.8.1 and Propositions 5.9.2 and 5.10.5. □

1. $S = SL_3(q)Z < SL_6(q)$ for $p \neq 2$. The class stabiliser is $\langle \delta^2, \phi, \gamma \rangle$ for $q \equiv \pm 1$ (mod 8) and $\langle \delta^2, \phi, \gamma\delta \rangle$ for $q \equiv \pm 3$ (mod 8).

2. $S = SU_3(q)Z < SU_6(q)$ for $p \neq 2$. The class stabiliser is $\langle \delta^2, \phi \rangle$ for $p \equiv \pm 1$ (mod 8) and $\langle \delta^2, \phi\delta \rangle$ for $p \equiv \pm 3$ (mod 8). (Note that the congruences are on p and not on q.)

Theorem 5.11.3 *Let G and Ω be as in Convention 4.10.1, with $\Omega = SL_9(q)$ or $SU_9(q)$. Then representatives of the conjugacy classes of $\mathscr{S}_2^*$-maximal subgroups of G are described in the list below.*

Proof Follows from Theorem 5.8.1 and Proposition 5.9.3. □

1. S has non-abelian composition factor $L_3(q^2)$. We list the possibilities.

(i) $S = L_3(q^2).2 < SL_9(q)$, with $q \equiv 0 \pmod 3$ and class stabiliser $\langle \phi, \gamma \rangle$.
(ii) $S = L_3(q^2).S_3 < SL_9(q)$, with $q \equiv 2 \pmod 3$ and class stabiliser $\langle \phi, \gamma \rangle$.
(iii) $S = SL_3(q^2).2Z < SL_9(q)$, with $q \equiv 1 \pmod 9$ and class stabiliser $\langle \delta^3, \phi, \gamma \rangle$.
(iv) $S = SL_3(q^2).6 < SL_9(q)$, with $q \equiv 4, 7 \pmod 9$ and class stabiliser $\langle \phi, \gamma \rangle$.
(v) $S = L_3(q^2).2 < SU_9(q)$, with $q \equiv 0 \pmod 3$ and class stabiliser $\langle \phi \rangle$.
(vi) $S = L_3(q^2).S_3 < SU_9(q)$, with $q \equiv 1 \pmod 3$ and class stabiliser $\langle \phi \rangle$.
(vii) $S = SL_3(q^2).2Z < SU_9(q)$, with $q \equiv 8 \pmod 9$ and class stabiliser $\langle \delta^3, \phi \rangle$.
(viii) $S = SL_3(q^2).6 < SU_9(q)$, with $q \equiv 2, 5 \pmod 9$ and class stabiliser $\langle \phi \rangle$.

Theorem 5.11.4 *Let G and Ω be as in Convention 4.10.1, with $\Omega = SL_{10}(q)$ or $SU_{10}(q)$. Then representatives of the conjugacy classes of $\mathscr{S}_2^*$-maximal subgroups of G are described in the list below.*

Proof This follows from Theorem 5.8.1 and Propositions 5.9.4, 5.9.5, 5.9.6 and 5.10.5. $\qquad\square$

1. $S = L_3(q).(q-1, 3) \times Z < SL_{10}(q)$, with $p \geqslant 5$. The class stabiliser is $\langle \phi, \gamma \rangle$ for $q \equiv \pm 1, \pm 5 \pmod{24}$ and $\langle \phi, \gamma \delta^5 \rangle$ for $q \equiv \pm 7, \pm 11 \pmod{24}$.

2. $S = U_3(q).(q+1, 3) \times Z < SU_{10}(q)$, with $p \geqslant 5$. The class stabiliser is $\langle \phi \rangle$ for $p \equiv \pm 1, \pm 5 \pmod{24}$ and $\langle \phi \delta \rangle$ for $p \equiv \pm 7, \pm 11 \pmod{24}$.

3. $S = \frac{(q-1,4)}{2} \cdot SL_4(q).\frac{(q-1,4)}{2} Z < SL_{10}(q)$, with $p \geqslant 3$ and class stabiliser $\langle \delta^5, \phi, \gamma \rangle$.

4. $S = \frac{(q+1,4)}{2} \cdot SU_4(q).\frac{(q+1,4)}{2} Z < SU_{10}(q)$, with $p \geqslant 3$ and class stabiliser $\langle \delta^5, \phi \rangle$.

5. $S = SL_5(q)Z < SL_{10}(q)$, with class stabiliser $\langle \delta^2, \phi, \gamma \rangle$.

6. $S = SU_5(q)Z < SU_{10}(q)$, with class stabiliser $\langle \delta^2, \phi \rangle$.

5.11.2 Case S

Theorem 5.11.5 *Let G and Ω be as in Convention 4.10.1, with $\Omega = Sp_2(q)$. Then there are no $\mathscr{S}_2^*$-maximal subgroups of G.*

Proof Follows from Theorem 5.8.1. $\qquad\square$

Theorem 5.11.6 *Let G and Ω be as in Convention 4.10.1, with $\Omega = Sp_4(q)$. Then representatives of the conjugacy classes of $\mathscr{S}_2^*$-maximal subgroups of G are described in the list below.*

Proof Follows from Theorem 5.8.1 and Propositions 5.9.7 and 5.10.5. $\qquad\square$

1. $S = SL_2(q)$, with $p \geqslant 5$ and class stabiliser $\langle \delta, \phi \rangle$.

2. $S = \mathrm{Sz}(2^e)$, with $e > 1$ odd and class stabiliser $\langle \phi, \gamma \rangle = \langle \gamma \rangle$, where γ is the graph automorphism.

Theorem 5.11.7 *Let G and Ω be as in Convention 4.10.1, with $\Omega = \mathrm{Sp}_6(q)$. Then representatives of the conjugacy classes of $\mathscr{S}_2^*$-maximal subgroups of G are described in the list below.*

Proof Follows from Theorem 5.8.1 and Propositions 5.9.7 and 5.10.5. □

1. $S = \mathrm{SL}_2(q)$, with $p \geqslant 7$ and class stabiliser $\langle \delta, \phi \rangle$.
2. $S = \mathrm{G}_2(q)$ with $p = 2$, $q > 2$, and class stabiliser $\langle \phi \rangle$.

Theorem 5.11.8 *Let G and Ω be as in Convention 4.10.1, with $\Omega = \mathrm{Sp}_8(q)$. Then representatives of the conjugacy classes of $\mathscr{S}_2^*$-maximal subgroups of G are described in the list below.*

Proof Follows from Theorem 5.8.1 and Propositions 5.9.7 and 5.10.5. □

1. $S = \mathrm{SL}_2(q)$, with $p \geqslant 11$ and class stabiliser $\langle \delta, \phi \rangle$.
2. $S = \mathrm{SL}_2(q^3).3$, with q odd and class stabiliser $\langle \delta, \phi \rangle$.

Theorem 5.11.9 *Let G and Ω be as in Convention 4.10.1, with $\Omega = \mathrm{Sp}_{10}(q)$. Then representatives of the conjugacy classes of $\mathscr{S}_2^*$-maximal subgroups of G are described in the list below.*

Proof Follows from Theorem 5.8.1 and Proposition 5.9.7. □

1. $S = \mathrm{SL}_2(q)$, with $p \geqslant 11$ and class stabiliser $\langle \delta, \phi \rangle$.

Theorem 5.11.10 *Let G and Ω be as in Convention 4.10.1, with $\Omega = \mathrm{Sp}_{12}(q)$. Then representatives of the conjugacy classes of $\mathscr{S}_2^*$-maximal subgroups of G are described in the list below.*

Proof Follows from Theorem 5.8.1 and Propositions 5.9.7 and 5.10.5. □

1. $S = \mathrm{SL}_2(q)$, with $p \geqslant 13$ and class stabiliser $\langle \delta, \phi \rangle$.
2. $S = \mathrm{Sp}_4(q)$, with $p = 5$ and class stabiliser $\langle \delta, \phi \rangle$.

5.11.3 Cases O^ε

Theorem 5.11.11 *Let G and Ω be as in Convention 4.10.1, with $\Omega = \Omega_7(q)$. Then representatives of the conjugacy classes of $\mathscr{S}_2^*$-maximal subgroups of G are described in the list below.*

Proof Follows from Theorem 5.8.1 and Propositions 5.9.9 and 5.10.5. □

1. $S = \mathrm{G}_2(q)$, with q odd and class stabiliser $\langle \phi \rangle$.

Theorem 5.11.12 *Let G and Ω be as in Convention 4.10.1, with $\Omega = \Omega_8^-(q)$. Then representatives of the conjugacy classes of $\mathscr{S}_2^*$-maximal subgroups of G are described in the list below.*

Proof Follows from Theorem 5.8.1 and Propositions 5.9.10 and 5.10.5.   □

1. $S = \mathrm{L}_3(q)$, with $q \equiv 2 \pmod 3$ and class stabiliser $\langle \gamma, \varphi \rangle$.
2. $S = \mathrm{U}_3(q)$, with $q \equiv 1 \pmod 3$ and class stabiliser $\langle \gamma, \varphi \rangle$.

Theorem 5.11.13 *Let G and Ω be as in Convention 4.10.1, with $\Omega = \Omega_9(q)$. Then representatives of the conjugacy classes of $\mathscr{S}_2^*$-maximal subgroups of G are described in the list below.*

Proof See Theorem 5.8.1 and Propositions 5.9.8, 5.9.12 and 5.10.5.   □

1. $S = \mathrm{L}_2(q).2$, with $p \geqslant 11$ and class stabiliser $\langle \phi \rangle$.
2. $S = \mathrm{L}_2(q^2).2$, with q odd and class stabiliser $\langle \delta, \phi \rangle$.

Theorem 5.11.14 *Let G and Ω be as in Convention 4.10.1, with $\Omega = \Omega_{10}^+(q)$ or $\Omega_{10}^-(q)$. Then representatives of the conjugacy classes of $\mathscr{S}_2^*$-maximal subgroups of G are described in the list below.*

Proof Follows from Theorem 5.8.1 and Proposition 5.9.13.   □

1. $S = 2 \times \mathrm{S}_4(q)$, with p odd. The group $S < \Omega_{10}^+(q)$ when $q \equiv 1 \pmod 4$, and $S < \Omega_{10}^-(q)$ when $q \equiv 3 \pmod 4$. The class stabiliser is $\langle \delta', \phi \rangle$.

Theorem 5.11.15 *Let G and Ω be as in Convention 4.10.1, with $\Omega = \Omega_{11}(q)$. Then representatives of the conjugacy classes of $\mathscr{S}_2^*$-maximal subgroups of G are described in the list below.*

Proof Follows from Theorem 5.8.1 and Proposition 5.9.8.   □

1. $S = \mathrm{L}_2(q)$, with $p \geqslant 11$ and class stabiliser $\langle \delta, \phi \rangle$.

Theorem 5.11.16 *Let G and Ω be as in Convention 4.10.1, with $\Omega = \Omega_{12}^+(q)$ or $\Omega = \Omega_{12}^-(q)$. Then there are no $\mathscr{S}_2^*$-maximal subgroups of G.*

Proof Follows from Theorem 5.8.1.   □

6

Containments involving $\mathscr{S}$-subgroups

6.1 Introduction

Recall Definitions 2.1.2 and 4.1.1 of the geometric, $\mathscr{S}_1$, and $\mathscr{S}_2$-subgroups, and that Aschbacher's theorem (see Theorem 2.1.5) divides the maximal subgroups of most finite classical groups into these three families. In Chapter 3, we determined which groups are maximal among the geometric groups, and in Chapters 4 and 5 we found which groups are $\mathscr{S}_1$-maximal and $\mathscr{S}_2^*$-maximal (see Definition 5.1.15), respectively.

In this chapter we determine all remaining containments between these subgroups, and hence complete the proof of Theorem 2.1.1 for those groups to which Aschbacher's theorem applies.

Definition 6.1.1 A subgroup is $\mathscr{S}^*$-*maximal* if it is maximal among the union of the $\mathscr{S}_1$-maximal and $\mathscr{S}_2^*$-maximal subgroups.

First we shall determine the $\mathscr{S}^*$-maximal subgroups, and then determine the containments between geometric and $\mathscr{S}^*$-maximal subgroups.

Recall the notation introduced in Series 1.1 and 1.2. It is often more convenient to state the containments that arise projectively, that is, as subgroups of $\bar{\Omega}$. As in Chapters 4 and 5, when doing this, if we wish to draw attention to the centre that arises when we lift to Ω, then we put that centre in brackets.

For the convenience of the reader, we state the following standard result: see for example [31, p491].

Lemma 6.1.2 *The order of* $\mathrm{G}_2(q)$ *is* $q^6(q^6 - 1)(q^2 - 1)$.

We remind the reader that, in the case of $\Omega = \mathrm{Sp}_4(2^e)$, we have only determined the subgroups that are maximal amongst the geometric subgroups of those almost simple extensions of Ω that are contained in $\Sigma\mathrm{Sp}_4(2^e)$. We shall deal with maximal subgroups of other almost simple extensions (i.e. those that involve the graph automorphism γ of Ω) in Section 7.2. We saw in Chapters 4 and 5 that the only $\mathscr{S}^*$-subgroup of $\mathrm{Sp}_4(2^e)$ is the $\mathscr{S}_2^*$-maximal $\mathrm{Sz}(2^e)$. So, in this chapter, we shall only determine containments between the candidate maximals of subgroups of $\Sigma\mathrm{Sp}_4(2^e)$ (and in fact there are none).

Although we present proofs of all of our claimed containments, for completeness and self-containedness, we note that there is a wealth of literature on this matter. Amongst other sources, we have checked our results against the following papers when relevant. In [81] the containment of (in our language) $\mathscr{S}_2$-subgroups of $\bar{\Omega}$ in $\mathscr{S}_1$-subgroups is investigated, and explicit lists of containments are given (excluding the possibility that the $\mathscr{S}_1$-subgroup is sporadic). The recent preprint [88] is also concerned with containments of $\mathscr{S}_2$-subgroups in $\mathscr{S}_1$-subgroups. The general theory of containments of $\mathscr{S}_2$-subgroups in other $\mathscr{S}_2$-subgroups is considered in Seitz's monograph [100]. There is an extensive treatment of the problem of which cross characteristic representations of quasisimple group have imprimitive images in the recent preprint [41], which also includes an extensive bibliography on results of this type. In [54] Husen investigates containments of $\mathscr{S}_1$-subgroups A_k or $2\cdot A_k$ in certain $\mathscr{S}_2$-subgroups, for $k \geqslant 10$: some related work (for $k \geqslant 15$ and certain $\mathscr{S}_2$-overgroups) was published in [55]. The paper [89] is also relevant to containments of $\mathscr{S}_1$-subgroups in $\mathscr{S}_2$-subgroups. There are papers by Schaffer [99] and by Cossidente and King [16] on maximality of $\mathscr{S}_2$-subgroups of *twisted tensor product type*, which means that their non-abelian composition factors are of the form $X_n(q^r)$ in $Y_{n^r}(q)$, where X and Y denote classical groups.

We remind the reader that the files of MAGMA calculations that we refer to are available on the webpage http://www.cambridge.org/9780521138604.

6.2 Containments between $\mathscr{S}_1$- and $\mathscr{S}_2^*$-maximal subgroups

Recall Definition 2.1.4 of a containment between two subgroups of a classical group. In this section we determine all containments between the $\mathscr{S}_1$-maximal subgroups, and the $\mathscr{S}_2^*$-maximal subgroups, in dimension up to 12.

6.2.1 Cases L and U

It is convenient to consider dimension 2 under Case **S**. By Theorem 5.11.1, $\mathscr{S}_2^*$-maximals arise only in dimensions 6, 9, and 10. We consider each in turn.

Proposition 6.2.1   (i) *The group* $(3^{\cdot})A_6.2^2$ *is* $\mathscr{S}^*$-*maximal in* $\mathrm{L}_6(p).\langle\gamma\rangle$ *when* $p \equiv 1 \pmod{24}$, *in* $\mathrm{L}_6(p).\langle\gamma\delta\rangle$ *when* $p \equiv 19 \pmod{24}$, *and in* $\mathrm{L}_6(p).\langle\gamma, \delta^3\rangle$ *when* $p \equiv 7$ *or* $13 \pmod{24}$.

(ii) *The group* $(3^{\cdot})A_6.2_2 \cong (3^{\cdot})\mathrm{PGL}_2(9)$ *is* $\mathcal{S}^*$-*maximal in* $\mathrm{L}_6(p).\langle\gamma\rangle$ *when* $p \equiv 7 \pmod{24}$ *and* $p \equiv \pm 2 \pmod 5$, *and in* $\mathrm{L}_6(p).\langle\gamma\delta^3\rangle$ *when* $p \equiv 13$ (mod 24) *and* $p \equiv \pm 2 \pmod 5$.

(iii) *The group* $(3^{\cdot})A_6.2^2$ *is* $\mathcal{S}^*$-*maximal in* $\mathrm{U}_6(p).\langle\gamma\rangle$ *when* $p \equiv 23 \pmod{24}$ *and in* $\mathrm{U}_6(p).\langle\gamma\delta\rangle$ *when* $p \equiv 5 \pmod{24}$ *with* $p > 5$, *and in* $\mathrm{U}_6(p).\langle\gamma, \delta^3\rangle$ *when* $p \equiv 11$ *or* $17 \pmod{24}$.

(iv) *The group* $(3^{\cdot})A_6.2_2 \cong (3^{\cdot})\mathrm{PGL}_2(9)$ *is* $\mathcal{S}^*$-*maximal in* $\mathrm{U}_6(p).\langle\gamma\rangle$ *when* $p \equiv 17 \pmod{24}$ *and* $p \equiv \pm 2 \pmod 5$, *and in* $\mathrm{U}_6(p).\langle\gamma\delta^3\rangle$ *when both* $p \equiv 11 \pmod{24}$ *and* $p \equiv \pm 2 \pmod 5$.

There are no other instances in which extensions of the $\mathcal{S}_1$-*subgroup* $(3^{\cdot})A_6$ *are* $\mathcal{S}^*$-*maximal in almost simple extensions of* $\mathrm{L}_6(q)$ *or* $\mathrm{U}_6(q)$.

All other $\mathcal{S}_1$-*maximal and* $\mathcal{S}_2^*$-*maximal subgroups of all almost simple extensions of* $\mathrm{L}_6(q)$ *and* $\mathrm{U}_6(q)$ *are* $\mathcal{S}^*$-*maximal.*

Proof We first consider the $\mathcal{S}^*$-maximality of the $\mathcal{S}_1$-maximals, and then of the $\mathcal{S}_2^*$-maximals. By Theorem 4.10.5, when q is odd, the $\mathcal{S}_1$-maximal subgroups are extensions of: $3^{\cdot}A_6$ with $q = p \neq 3$; $6^{\cdot}A_6$ with $q = p$ or p^2 and $q \neq 3$; $2^{\cdot}L_2(11)$ with $q = p \neq 3, 11$; $6^{\cdot}A_7$ with $q = p$ or p^2 and $p \neq 3$; $2^{\cdot}M_{12} < \mathrm{SL}_6(3)$; $6^{\cdot}L_3(4)$ with $q = p \neq 3$; and $6_1^{\cdot}U_4(3)$ with $q = p \neq 3$. By Theorem 5.11.2, the $\mathcal{S}_2^*$-maximal subgroups are extensions of $\mathrm{SL}_3(q)$ (Case **L**) and $\mathrm{SU}_3(q)$ (Case **U**), with q odd.

We deal first with the behaviour of $3^{\cdot}A_6$, as decribed in Theorem 4.10.5. The $\mathcal{S}_1$-maximal subgroups of this type are the extensions of the form $(3^{\cdot})A_6.2_2$ and $(3^{\cdot})A_6.2^2$. Suppose first that $p \neq 5$. By Theorem 4.10.2, if $p \equiv 1, 4 \pmod{15}$ then $3^{\cdot}A_6 < \mathrm{SL}_3(p)$, whilst $3^{\cdot}A_6 < \mathrm{SU}_3(p)$ when $p \equiv 11$ or $14 \pmod{15}$. From the character table of $3^{\cdot}A_6$ in [12] and Proposition 5.2.5 we calculate that the 6-dimensional irreducible representation of $3^{\cdot}A_6$ is the symmetric square of the 3-dimensional representations. By Theorem 5.8.1, the same is true for the corresponding representations of $\mathrm{SL}_3(p)$ and $\mathrm{SU}_3(p)$. So there is a containment of $3^{\cdot}A_6$ in $\mathrm{SL}_3(p)$ or $\mathrm{SU}_3(p)$ for these values of p; that is, when $p \equiv \pm 1 \pmod 5$. But when $p \equiv \pm 2 \pmod 5$, there is no such containment, so for these p the $\mathcal{S}_1$-maximal subgroups are $\mathcal{S}^*$-maximal. We therefore assume from now that $p \equiv \pm 1 \pmod 5$.

In Case **L** with $p \equiv 1$ or $4 \pmod{15}$, we saw in Lemma 5.9.1 (iv) and Proposition 5.9.2 that the duality automorphism of $\mathrm{SL}_3(q)$ is induced by duality $(= \gamma)$ of $\mathrm{SL}_6(q)$ when $q \equiv \pm 1 \pmod 8$ and by $\gamma\delta$ when $q \equiv \pm 3 \pmod 8$. From Proposition 4.7.8, we find that the same conditions apply to the induction of the 2_2 automorphism of $3^{\cdot}A_6$. Furthermore, by Theorem 4.7.1 (ii), the duality automorphism of $\mathrm{SL}_3(q)$ induces the 2_2 automorphism of $3^{\cdot}A_6$. So $(3^{\cdot})A_6.2_2$ is not $\mathcal{S}^*$-maximal in $\mathrm{L}_6(p).\langle\gamma\rangle$ or in $\mathrm{L}_6(p).\langle\gamma\delta\rangle$ when $p \equiv 1$ or $4 \pmod{15}$. (Recall that $\langle\gamma\delta\rangle$ and $\langle\gamma\delta^3\rangle$ are conjugate in $\mathrm{Out}\,\mathrm{SL}_6(p)$.) Similar reasoning shows that $(3\dot{})A_6.2_2$ is not $\mathscr{S}^*$-maximal in $U_6(p).\langle\gamma\rangle$ or in $U_6(p).\langle\gamma\delta\rangle$ when $p \equiv 11$ or $14 \pmod{15}$. So, when $p \equiv \pm 1 \pmod 5$, extensions of the form $(3\dot{})A_6.2_2$ are not $\mathscr{S}^*$-maximal.

Continuing our assumption that $p > 5$, we saw in Theorem 4.10.2 that $3\dot{}A_6$ is scalar-normalising in $SL_3(p)$ or $SU_3(p)$. Furthermore, the 6-dimensional representations of $SL_3(p)$ and $SU_3(p)$ are not stabilised by the diagonal automorphism of $SL_6(p)$ or $SU_6(p)$ of order 2, and so $(3\dot{})A_6.2^2$ cannot be contained in any almost simple extension of $L_3(p)$ or $U_3(p)$, and hence is $\mathscr{S}^*$-maximal.

However, by Theorem 4.10.2 the containment of $3\dot{}A_6$ in $SU_3(5)$ extends to a containment $3\dot{}M_{10} < SU_3(5)$. In that case $(3\dot{})A_6.2^2 < U_3(5):2 < U_6(5):\langle\gamma\delta\rangle$, so $(3\dot{})A_6.2^2$ is not $\mathscr{S}^*$-maximal.

Since the image of a faithful 3-dimensional representation of a quasisimple group must have centre of order 1 or 3, and none of the other $\mathscr{S}_1$-groups in the above list of quasisimple groups has centre of order 1 or 3, all other $\mathscr{S}_1$-maximal subgroups are $\mathscr{S}^*$-maximal.

Finally, we consider the $\mathscr{S}^*$-maximality of the $\mathscr{S}_2^*$-maximal subgroups. As noted at the beginning of the proof, there are no $\mathscr{S}_2^*$-maximal subgroups when $p = 2$. When $p = 3$, the only $\mathscr{S}_1$-maximals are extensions of $2\dot{}M_{12}$. Since, by Theorem 1.11.2, the minimum permutation degree of $L_3(3)$ is greater than 12, the group $2\dot{}M_{12}$ has no subgroup isomorphic to $SL_3(3)$. For $p \geqslant 5$, none of the $\mathscr{S}_1$-maximals have order divisible by p^3, whereas $|L_3(q)|$ and $|U_3(q)|$ are divisible by q^3, so the result follows. $\qquad\square$

Proposition 6.2.2 *When $p \equiv 1 \pmod 3$ and $p \equiv \pm 2 \pmod 5$, the $\mathscr{S}_1$-maximal subgroup $3\dot{}A_6.2_3$ of $SL_9(p)$ is contained in the $\mathscr{S}_2^*$-maximal subgroup $SL_3(p^2).6$. When $p \equiv 2 \pmod 3$ and $p \equiv \pm 2 \pmod 5$, the $\mathscr{S}_1$-maximal subgroup $3\dot{}A_6.2_3$ of $SU_9(p)$ is contained in the $\mathscr{S}_2^*$-maximal subgroup $SL_3(p^2).6$. These containments extend to $3\dot{}A_6.2^2 < SL_3(p^2).[12] < SL_9^{\pm}(p).2$.*

All other $\mathscr{S}_1$-maximal and $\mathscr{S}_2^*$-maximal subgroups of all almost simple extensions of $L_9(q)$ or $U_9(q)$ are $\mathscr{S}^*$-maximal.

Proof By Theorem 4.10.8, the $\mathscr{S}_1$-maximals of $SL_9^{\pm}(q)$ are extensions of: $3\dot{}A_6.2_3$ with $q = p \neq 3, 5$; $L_2(19)$ with $q = p \neq 19$; $3\dot{}A_7$ in $SL_9(7)$; and $3\dot{}J_3$ in $SU_9(2)$. By Theorem 5.11.3, the $\mathscr{S}_2^*$-maximals are extensions of $(3\dot{})L_3(q^2)$.

We first consider the possible containments of $3\dot{}A_6.2_3$ in the $\mathscr{S}_2^*$-maximals. When $p \equiv \pm 1 \pmod 5$, we find from Theorem 4.10.2 that the only extension of $(3\dot{})A_6$ that is contained in an almost simple extension of $L_3^{\pm}(q)$ is $(3\dot{})A_6.2_2$, so there are no such containments in that case.

So suppose that $p \equiv \pm 2 \pmod 5$ and recall from Theorem 4.10.8 that $3\dot{}A_6.2_3 < SL_9(p)$ and $SU_9(p)$ when $p \equiv 1$ and $-1 \pmod 3$, respectively. We saw in the proofs of Propositions 5.4.20 and 5.4.21 that the outer automorphisms of order 2 of $SL_3(p^2)$ that are induced by elements of $SL_9^{\pm}(p)$ are (conjugates of) ϕ and $\gamma\phi$, respectively, in these two cases, and we find from Theorem 4.10.2 that these automorphisms do indeed induce the 2_3 automorphism of the $\mathscr{S}_1$-subgroup $3{\cdot}A_6$ of $\mathrm{SL}_3(p^2)$. Furthermore, we find from [12] (or [57] when $p = 2$) that the 2_3 automorphism of $3{\cdot}A_6$ interchanges the two degree 3 characters χ_{14} and χ_{15} of $3{\cdot}A_6$ that are displayed in [12, 57], and that $\chi_{14} \otimes \chi_{15} = \chi_{17}$, which is irreducible of degree 9. So we do have a containment $3{\cdot}A_6.2_3 < \mathrm{SL}_3(p^2).2 < \mathrm{SL}_9^{\pm}(p)$ in this case which, since there are is an odd number (3) of classes of $3{\cdot}A_6$ in $\mathrm{SL}_3(p^2)$, must extend under the action of the γ automorphism of $\mathrm{SL}_9^{\pm}(p)$ to $3{\cdot}A_6.2^2 < \mathrm{SL}_3(p^2).2^2 < \mathrm{SL}_9^{\pm}(p).2$.

We find from Theorem 4.3.3 that none of the $\mathscr{S}_1$-maximals other than $3{\cdot}A_6.2_3$ have 3-dimensional irreducible projective representations in the relevant characteristics, so they cannot be subgroups of the $\mathscr{S}_2$-groups. On the other hand, the $\mathscr{S}_2^*$-maximals have order divisible by q^6, and the only $\mathscr{S}_1$-maximal with this property is $3{\cdot}J_3 < \mathrm{SU}_9(2)$. But by [25], no subgroup of J_3 involves $\mathrm{L}_3(4)$, so the $\mathscr{S}_2^*$-maximals are all $\mathscr{S}^*$-maximal. $\square$

Proposition 6.2.3   *If $2 < p \equiv 1, 2, 4 \pmod 7$, then there are containments* $A_7 < \mathrm{L}_4(q).\frac{(q-1,4)}{2} < \mathrm{L}_{10}(p)$ *and* $A_7.\langle\gamma\rangle < \mathrm{L}_4(q).\frac{(q-1,4)}{2}.\langle\gamma\rangle < \mathrm{L}_{10}(p).\langle\gamma\rangle$. *If* $p \equiv 3, 5, 6 \pmod 7$, *then there are containments* $A_7 < \mathrm{U}_4(q).\frac{(q+1,4)}{2} < \mathrm{U}_{10}(p)$ *and* $A_7.\langle\gamma\rangle < \mathrm{U}_4(q).\frac{(q+1,4)}{2}.\langle\gamma\rangle < \mathrm{U}_{10}(p).\langle\gamma\rangle$.

There is also a containment $\mathrm{L}_3(4).2_2 < \mathrm{U}_4(3).2 < \mathrm{U}_{10}(3)$, *extending to* $\mathrm{L}_3(4).2_2.\langle\gamma\delta\rangle < \mathrm{U}_4(3).2.\langle\gamma\delta\rangle < \mathrm{U}_{10}(3).\langle\gamma\delta\rangle$.

If $p \equiv 1 \pmod 3$, then there are containments $\mathrm{U}_4(2) < \mathrm{L}_5(p) < \mathrm{L}_{10}(p)$ *and* $\mathrm{U}_4(2).\langle\gamma\rangle < \mathrm{L}_5(p).\langle\gamma\rangle < \mathrm{L}_{10}(p).\langle\gamma\rangle$. *If $2 \neq p \equiv 2 \pmod 3$, then there are containments* $\mathrm{U}_4(2) < \mathrm{U}_5(p) < \mathrm{U}_{10}(p)$, *and* $\mathrm{U}_4(2).\langle\gamma\rangle < \mathrm{U}_5(p).\langle\gamma\rangle < \mathrm{U}_{10}(p).\langle\gamma\rangle$.

There are no other containments between $\mathscr{S}_1$-maximal and $\mathscr{S}_2^*$-maximal subgroups of almost simple extensions of $\mathrm{L}_{10}(q)$ or $\mathrm{U}_{10}(q)$.

Proof   By Theorem 4.10.9, the $\mathscr{S}_1$-maximals of $\mathrm{SL}_{10}^{\pm}(q)$ are extensions of: $2{\cdot}\mathrm{L}_2(19)$ with $q = p \neq 2, 19$; A_7 with $q = p \neq 2, 7$; $2{\cdot}M_{12}$ with $q = p \neq 2$; $(2, q - 1){\cdot}M_{22}$ with $q = p \neq 7$; $\mathrm{U}_4(2)$ with $q = p \neq 2, 3$; and $2{\cdot}\mathrm{L}_3(4)$ with $q = p \neq 2, 7$. By Theorem 5.11.4, the $\mathscr{S}_2^*$-maximals are extensions of: $\mathrm{L}_3^{\pm}(q)$ with $p \neq 2, 3$; $\frac{(q\mp1,4)}{2}{\cdot}\mathrm{L}_4^{\pm}(q)$ with $p \neq 2$; and $\mathrm{SL}_5^{\pm}(q)$.

For possible containments of $\mathscr{S}_1$-maximals in $\mathscr{S}_2^*$-maximals, we first use Theorem 4.3.3 to ascertain which of the $\mathscr{S}_1$-maximals have irreducible projective representations of degree 3, 4 or 5 in the relevant characteristic p. We find that: $2{\cdot}A_7$ has two representations of degree 4 for all p; the group $3{\cdot}A_7$ has one of degree 3 in characteristic 5; the group $4_2{\cdot}\mathrm{L}_3(4)$ has two of degree 4 for $p = 3$; the group $\mathrm{U}_4(2)$ has one or two of degree 5 for all p; the group $2{\cdot}\mathrm{U}_4(2)$ has two of degree 4 for all p (faithful for p odd); and there are no other such representations.

By Theorem 5.8.1, the the $\mathscr{S}_2^*$-maximals involving $L_4^\pm(q)$ arise from the symmetric square of the natural representations of $SL_4^\pm(q)$. We find, using Proposition 5.2.5 and the character tables of $2{\cdot}A_7$ (in characteristics other than 2 and 7) and $4_2{\cdot}L_3(4)$ (in characteristic 3), that their 10-dimensional representations also arise as symmetric squares of 4-dimensional representations. This proves the claimed containments of the quasisimple groups $2{\cdot}A_7$ and $4_2{\cdot}L_3(4)$.

By Theorem 5.8.1, in characteristics greater than 3, the $\mathscr{S}_2^*$-maximals involving $L_5^\pm(q)$ arise from the anti-symmetric square of the natural representations of $SL_5^\pm(q)$. By Proposition 5.2.5 and the character table of $U_4(2)$ we find that, in characteristics greater than 3, the two 10-dimensional irreducible representations of $U_4(2)$ are the anti-symmetric squares of the two 5-dimensional representations. The proves the claimed containments of $2{\cdot}U_4(2)$.

We turn now to the containments of the extensions. By Theorem 5.11.4, the conjugacy classes of subgroups $L_4^\pm(q)$ and $L_5^\pm(q)$ of $L_{10}^\pm(q)$ are stabilised by the γ automorphism of $L_{10}^\pm(q)$ which, by Lemma 5.9.1 (iv) and (v), induces the γ automorphism of $L_4^\pm(q)$ or $L_5^\pm(q)$. By Theorems 4.10.3 and 4.10.4 the $\mathscr{S}_1$ subgroups $A_7 < L_4^\pm(q)$ and $U_4(2) < L_5^\pm(q)$ are stabilised by the γ automorphisms of $L_4^\pm(q)$ and $L_5^\pm(q)$. This proves the claimed containments of $A_7.\langle\gamma\rangle$ and $U_4(2).\langle\gamma\rangle$. By Theorem 5.11.4, the full stabiliser of the class of $U_4(3).2$ in $U_{10}(3)$ is $\langle\delta^5, \phi\rangle = \langle\delta, \phi\rangle$ (since $|\delta| = (4, 10) = 2$), which contains $\phi\delta = \gamma\delta$, which induces the outer automorphism $\gamma\delta$ of $U_4(3)$. By Theorem 4.10.3, the automorphism $\gamma\delta$ of $U_4(3)$ stabilises the class of $L_4(3).2 < U_4(3)$, so this proves the claimed containment of $A_7.\langle\gamma\delta\rangle$.

The quasisimple $\mathscr{S}_2^*$-maximal subgroups all have order divisible by q^3, by q^6 if $p \leqslant 3$, and by q^{10} if $p = 2$. None of the quasisimple $\mathscr{S}_1$-subgroups have this property, so the $\mathscr{S}_2^*$-maximals are all $\mathscr{S}^*$-maximal.   □

6.2.2 Case S

Proposition 6.2.4 *There are no containments between $\mathscr{S}_1$-maximal and $\mathscr{S}_2^*$-maximal subgroups of almost simple extensions of $S_2(q)$.*

Proof By Theorem 5.11.5 there are no $\mathscr{S}_2^*$-maximal subgroups.   □

We again remind the reader that, in the case of $\Omega = Sp_4(2^e)$, we shall deal with maximal subgroups of almost simple extensions that involve the graph automorphism γ of Ω in Section 7.2, and that $Sp_4(2)$ is not quasisimple and so is excluded from our classification.

Proposition 6.2.5 *There are containments $(2^{\cdot})L_2(5) < (2^{\cdot})A_6 < S_4(5)$, $(2^{\cdot})L_2(5).2 < (2^{\cdot})A_6.2_1 < S_4(5).2$, and $(2^{\cdot})L_2(7) < (2^{\cdot})A_7 < S_4(7)$. There are no other containments between $\mathscr{S}_1$-maximal and $\mathscr{S}_2^*$-maximal subgroups of* *almost simple extensions of* $S_4(q)$ *that do not involve the exceptional graph auto-morphism of* $S_4(2^e)$. *(So, in particular,* $(2\dot{\ })L_2(7).2$ *is* $\mathscr{S}_2^*$-*maximal in* $S_4(7).2$.)

Proof By Theorem 5.11.6, the $\mathscr{S}_2^*$-maximals of $Sp_4(q)$ are extensions of $SL_2(q)$ with $p \geqslant 5$, and $Sz(p^e)$ with $p = 2$ and e odd. By Theorem 4.10.13, the $\mathscr{S}_1$-maximals are extensions of: $2\dot{\ }A_6$ with $q = p > 3$ and $p \neq 7$; and $2\dot{\ }A_7 < Sp_4(7)$.

By Theorem 4.3.3, none of the $\mathscr{S}_1$-groups have irreducible representations of degree 2 with $p > 3$, and all occur for odd p, so the $\mathscr{S}_1$-maximals are all $\mathscr{S}^*$-maximal.

There are no $\mathscr{S}_1$-maximals with $p = 2$, so $Sz(2^e)$ is $\mathscr{S}^*$-maximal. Since p divides $|L_2(p)|$, the only possible containments of $\mathscr{S}_2^*$-maximal subgroups of Ω in $\mathscr{S}_1$-subgroups are $SL_2(5) < 2\dot{\ }A_6 < Sp_4(5)$ and $SL_2(7) < 2\dot{\ }A_7 < Sp_4(7)$. We find from the character tables for $p = 5$ in [57] of $2\dot{\ }A_6$ and $SL_2(5) \cong 2\dot{\ }A_5$ that one of the degree 4 faithful irreducible representations of $2\dot{\ }A_6$ reduces to an irreducible representation of $2\dot{\ }A_5$, whereas the other reduces to the sum of two degree 2 representations. The two degree 4 representations of $2\dot{\ }A_6$ are interchanged by the 2_2 (and 2_3) automorphism of $2\dot{\ }A_6$, which also interchanges the two classes of maximal subgroups isomorphic to $2\dot{\ }A_5$, so this containment comes from the restriction to one of these two classes. The other containment is similar. Furthermore, by Proposition 4.5.10 (ii) the outer automorphism of $S_4(5)$ induces the 2_1 automorphism of A_6, and $A_6.2_1 \cong S_6$, so the first containment extends to $SL_2(5).2 < 2\dot{\ }A_6.2_1 < Sp_4(5).2$. Conversely, S_7 does not contain $L_2(7).2$. $\qquad\square$

Proposition 6.2.6 *There is a containment* $(2\dot{\ })L_2(5) < (2\dot{\ })L_2(q) < S_6(q)$ *for* $q \equiv \pm 11, \pm 19 \pmod{40}$. *There are containments* $L_2(13) < G_2(4) < S_6(4)$, $L_2(13).2 < G_2(4).2 < S_6(4).2$, $J_2 < G_2(4).2 < S_6(4)$, *and* $J_2.2 < G_2(4).2 < S_6(4).2$. *There are no other containments between* $\mathscr{S}_1$-*maximal and* $\mathscr{S}_2^*$-*maximal subgroups of almost simple extensions of* $S_6(q)$.

Proof By Theorem 5.11.7, the $\mathscr{S}_2^*$-maximals of $Sp_6(q)$ are extensions of $SL_2(q)$ with $p \geqslant 7$ and $G_2(q)$ with $q > p = 2$. By Theorem 4.10.14, the $\mathscr{S}_1$-maximals are extensions of: $2\dot{\ }A_5$ with $q = p \neq 2, 5$; $2\dot{\ }L_3(2)$ with $q = p$ or p^2 and $p \neq 2, 7$; $(2, p-1)\dot{\ }L_2(13)$ with $q = p$ or p^2 and $p \neq 13$; $2\dot{\ }A_7 < Sp_6(9)$; $(2, p-1)\dot{\ }J_2$ with $q = p$ or p^2; and $(2, p-1) \times U_3(3)$ with $q = p \neq 3, 5$.

For possible containments of $\mathscr{S}_1$-maximals in $SL_2(q)$, we first observe from Theorem 4.3.3 that the only $\mathscr{S}_1$-maximal with a faithful representation of degree 2 in characteristic p is $2\dot{\ }A_5$. Since $2\dot{\ }A_5.2$ has no such representation, it is $\mathscr{S}^*$-maximal whenever it is $\mathscr{S}_1$-maximal. By Theorem 4.10.12, when $p \equiv \pm 3, \pm 13 \pmod{40}$, the group $2\dot{\ }A_5$ requires $q = p^2$. So the only possible containments are $2\dot{\ }A_5 < SL_2(p) < Sp_6(p)$ when $p \equiv \pm 11, \pm 19 \pmod{40}$, so assume that p satifies this condition.

In the notation of Section 5.3, we know from Proposition 5.3.10 that, for all $k \geqslant 2$, the module $V_2 \otimes V_k$ for $\mathrm{SL}_2(p)$ has V_{k-1} as a constituent. By definition, $V_k = \mathrm{S}^{k-1}(V_2)$, which is a quotient of $V_2^{\otimes k-1}$, and by Lemma 5.3.1 the module V_k is irreducible for $k \leqslant p$. By induction on k, we see that V_{k+1} is also a constituent of $V_2 \otimes V_k$ for $k < p$. Hence $V_2 \otimes V_k$ has precisely two irreducible constituents, namely V_{k-1} and V_{k+1}, for $2 \leqslant k < p$. Let V_2' be the restriction of V_2 to the subgroup $\mathrm{SL}_2(5)$ of $\mathrm{SL}_2(p)$. This is a 2-dimensional irreducible module in coprime characteristic, which we may assume corresponds to the first row of the character table for $2^{\boldsymbol{\cdot}}\mathrm{A}_5$ in [12]. Denote the module corresponding to the first 3-dimensional character of $2^{\boldsymbol{\cdot}}\mathrm{A}_5$ in [12] by V_3', the faithful 4-dimensional module by V_4', and the 1-, 5- and 6-dimensional modules by V_1', V_5' and V_6', respectively. Then straightforward character calculations (using $\mathrm{b}_5{}^2 = 1 - \mathrm{b}_5$) show that $V_2' \otimes V_k' \cong V_{k-1}' \oplus V_{k+1}'$, for $2 \leqslant k \leqslant 5$. Since $p > 5$, the module $V_2' \otimes V_k'$ therefore has irreducible constituents of the same dimensions as $V_2 \otimes V_k$ for $2 \leqslant k \leqslant 5$, and it follows by induction on k that the restriction of V_k to $\mathrm{SL}_2(5)$ must be isomorphic to V_k' for $2 \leqslant k \leqslant 6$. So the restriction of the 6-dimensional irreducible representation of $\mathrm{SL}_2(p)$ to $\mathrm{SL}_2(5)$ is irreducible, and hence the containments in question are genuine.

The only $\mathscr{S}_1$-maximals with $p = 2$ and $q \geqslant 4$ are $\mathrm{L}_2(13) < \mathrm{Sp}_6(4)$ and $\mathrm{J}_2 < \mathrm{Sp}_6(4)$. We find from [14] (or [12]) that there are unique classes of subgroups $\mathrm{L}_2(13)$ and J_2 in $\mathrm{G}_2(4)$. Since neither $\mathrm{L}_2(13)$ nor J_2 has irreducible representations of degree less than 6 in characteristic 2, these subgroups must be irreducible subgroups of $\mathrm{S}_6(4)$, so there are containments $\mathrm{L}_2(13) < \mathrm{G}_2(4) < \mathrm{Sp}_6(4)$ and $\mathrm{J}_2 < \mathrm{G}_2(4) < \mathrm{Sp}_6(4)$. Since the class stabiliser in each case is $\langle \phi \rangle$, by [12] these extend to $\mathrm{L}_2(13).2 < \mathrm{G}_2(4).2 < \mathrm{Sp}_6(4).2$, $\mathrm{J}_2.2 < \mathrm{G}_2(4).2 < \mathrm{Sp}_6(4).2$.

By Lemma 6.1.2, the group $\mathrm{G}_2(q)$ is larger than any of the $\mathscr{S}_1$-maximals, so is $\mathscr{S}^*$-maximal. The only $\mathscr{S}_1$-maximal with order divisible by q with $p \geqslant 7$ is $\mathrm{U}_3(3)$ with $q = 7$, but $\mathrm{U}_3(3)$ has no subgroup isomorphic to $\mathrm{SL}_2(7)$, so $\mathrm{SL}_2(q)$ and its extensions are also $\mathscr{S}^*$-maximal. $\qquad \square$

Proposition 6.2.7   *There are no containments between $\mathscr{S}_1$-maximal and $\mathscr{S}_2^*$-maximal subgroups of almost simple extensions of $\mathrm{S}_8(q)$.*

Proof   By Theorem 5.11.8, the $\mathscr{S}_2^*$-maximals of $\mathrm{Sp}_8(q)$ are extensions of $\mathrm{SL}_2(q)$ with $p \geqslant 11$ and $\mathrm{SL}_2(q^3).3$ with q odd. By Theorem 4.10.15, the $\mathscr{S}_1$-maximals are extensions of: $2^{\boldsymbol{\cdot}}\mathrm{L}_3(2)$ with $q = p \neq 2, 3, 7$; $2^{\boldsymbol{\cdot}}\mathrm{A}_6$ with $q = p$ or p^2 and $p \neq 2, 3, 5$; $(2, p - 1)^{\boldsymbol{\cdot}}\mathrm{L}_2(17)$ with $q = p$ or p^2 and $p \neq 17$; and $\mathrm{A}_{10}.2 < \mathrm{Sp}_8(2)$.

By Theorem 4.3.3 none of the $\mathscr{S}_1$-maximals have faithful representations of degree 2 in characteristic p. On the other hand, none of the $\mathscr{S}_1$-maximals have order divisible by p when $p \geqslant 11$ or by p^3 when p is odd. The result follows. $\qquad \square$

Proposition 6.2.8   *There are no containments between $\mathscr{S}_1$-maximal and $\mathscr{S}_2^*$-maximal subgroups of almost simple extensions of* $\mathrm{S}_{10}(q)$.

Proof By Theorem 5.11.9, the $\mathscr{S}_2^*$-maximals of $\mathrm{Sp}_{10}(q)$ are extensions of $\mathrm{SL}_2(q)$ with $p \geqslant 13$. By Theorem 4.10.16, the $\mathscr{S}_1$-maximals are extensions of: $2^{\cdot}\mathrm{A}_6$ with $q = p$ or p^2 and $p \neq 2, 3$; $2^{\cdot}\mathrm{L}_2(11)$ with $q = p$ or p^2 and $p \neq 2, 11$; and $2 \times \mathrm{U}_5(2)$ with $q = p \neq 2$.

By Theorem 4.3.3, none of the $\mathscr{S}_1$-maximals have faithful representations of degree 2 in characteristic p. Conversely, none of the $\mathscr{S}_1$-maximals has order divisible by p when $p \geqslant 13$. The result follows.   □

Proposition 6.2.9   *There are no containments between $\mathscr{S}_1$-maximal and $\mathscr{S}_2^*$-maximal subgroups of almost simple extensions of* $\mathrm{S}_{12}(q)$.

Proof By Theorem 5.11.10, the $\mathscr{S}_2^*$-maximals of $\mathrm{Sp}_{12}(q)$ are extensions of $\mathrm{SL}_2(q)$ with $p \geqslant 13$, and $\mathrm{Sp}_4(q)$ with $p = 5$. By Theorem 4.10.17, the $\mathscr{S}_1$-maximals are extensions of: $2^{\cdot}\mathrm{L}_2(11)$ with $q = p$ or p^2 and $p \neq 2, 5, 11$; $2^{\cdot}\mathrm{L}_2(13)$ with $q = p$ or p^3 and $p \neq 2, 7, 13$; $(2, p - 1)^{\cdot}\mathrm{L}_2(25)$ with $q = p \neq 3, 5$; $(2, p - 1)^{\cdot}\mathrm{S}_4(5)$ with $q = p$ or p^2 and $p \neq 5$; $2^{\cdot}\mathrm{G}_2(4)$ with $q = p \neq 2, 3$; $\mathrm{A}_{14}.2 < \mathrm{Sp}_{12}(2)$; and $2^{\cdot}\mathrm{Suz} < \mathrm{Sp}_{12}(3)$

First suppose $p = 5$. The only $\mathscr{S}_1$-maximals are $2^{\cdot}\mathrm{L}_2(13) < \mathrm{Sp}_{12}(5^3)$ and $2^{\cdot}\mathrm{G}_2(4) < \mathrm{Sp}_{12}(5)$. The $\mathscr{S}_1$-subgroups have no faithful representations of degree 4, so are $\mathscr{S}^*$-maximal, whilst the $\mathscr{S}_1$-subgroups do not have order divisible by 5^4, so $\mathrm{Sp}_4(q)$ is $\mathscr{S}^*$-maximal.

We therefore assume that $p \geqslant 13$. By Theorem 4.3.3, the $\mathscr{S}_1$-maximals have no representations of degree 2 in characteristic p, so the $\mathscr{S}_1$-maximals are $\mathscr{S}^*$-maximal. Conversely, the only $\mathscr{S}_1$-maximals with order divisible by q when $p \geqslant 13$ are $2^{\cdot}\mathrm{L}_2(25) < \mathrm{Sp}_{12}(13)$ and $2^{\cdot}\mathrm{G}_2(4) < \mathrm{Sp}_{12}(13)$. Since $2^{\cdot}\mathrm{L}_2(13)$ has no 2-dimensional representation when $p = 5$, the first possibility does not lead to a containment. The group $2^{\cdot}\mathrm{G}_2(4)$ has a subgroup isomorphic to $\mathrm{SL}_2(13)$, but we find from the character values on elements of order 7 of the 12-dimensional representations of $2^{\cdot}\mathrm{G}_2(4)$ and $\mathrm{SL}_2(13)$ as given in [12] that, in dimension 12, this subgroup acts reducibly with two components of degree 6, so it is not a containment of $\mathscr{S}$-subgroups. Thus the $\mathscr{S}_2^*$-maximals are $\mathscr{S}^*$-maximal.   □

6.2.3 Cases O^ε

Proposition 6.2.10   *The $\mathscr{S}_1$-maximal subgroups of $\mathrm{O}_7(q)$ that are extensions of $\mathrm{L}_2(8)$, $\mathrm{L}_2(13)$ and J_1 $(q = 11)$ are contained in the corresponding extensions of the $\mathscr{S}_2^*$-subgroup $\mathrm{G}_2(q)$. There are no other containments between $\mathscr{S}_1$-maximal and $\mathscr{S}_2^*$-maximal subgroups of almost simple extensions of* $\mathrm{O}_7(q)$.

Proof By Theorem 5.11.11, the only $\mathscr{S}_2^*$-maximals of $\Omega_7(q)$ are extensions of

$G_2(q)$, with q odd. By Theorem 4.10.18, the $\mathscr{S}_1$-maximals are extensions of: $L_2(8)$ with $q = p$ or p^3 and $p \neq 2, 3$; $L_2(13)$ with $q = p$ or p^2 and $p \neq 2, 13$; $S_6(2)$ with $q = p \neq 2$; $A_9.2 < \Omega_7(3)$; and $J_1 < \Omega_7(11)$.

By [63] (see Tables 8.41 and 8.42), the groups $L_2(8)$, $L_2(13)$, and J_1 are all subgroups of $G_2(q)$, but $S_6(2)$ and A_9 are not. Furthermore, these three subgroups of $G_2(q)$ are irreducible as subgroups of $\Omega_7(q)$, and correspond to the same representation as the corresponding $\mathscr{S}_1$-maximal subgroup. (We have to be careful with $L_2(8)$ since there are two such $\mathscr{S}_1$-candidates, which we denote $L_2(8)_1$ and $L_2(8)_2$. The $\mathscr{S}_1$-maximal is $L_2(8)_2$, which corresponds to the representation involving the irrationality y_9, and it is clear from the statement of [63, Theorem A] that this is the subgroup of $G_2(q)$ in question.) Since $G_2(q)$ has unique classes of subgroups isomorphic to each of these $\mathscr{S}_1$-maximals, there are also containments $L_2(8).3 < G_2(q).3$ when $q = p^3$ and $L_2(13).2 < G_2(q).2$ when $q = p^2$, so no extension of $L_2(8)$, $L_2(13)$ or J_1 is $\mathscr{S}^*$-maximal.

By Lemma 6.1.2, $|G_2(q)|$ is divisible by q^6, and so is $\mathscr{S}^*$-maximal. $\qquad\square$

We remind the reader that the maximal subgroups of all almost simple extensions of $O_8^+(q)$ were classified by Kleidman in [62].

Proposition 6.2.11 *The $\mathscr{S}^*$-maximal subgroups of extensions of $O_8^-(q)$ are precisely the $\mathscr{S}_2^*$-maximal subgroups.*

Proof By Theorem 5.11.12, the $\mathscr{S}_2^*$-maximals of $\Omega_8^-(q)$ are extensions of $L_3(q)$ with $q \equiv 2 \pmod 3$ and $U_3(q)$ with $q \equiv 1 \pmod 3$. By Theorem 4.10.20, the only $\mathscr{S}_1$-maximals are extensions of $L_3(2)$ with $q = p \equiv \pm2, \pm8, \pm10 \pmod{21}$ and $p \neq 2$.

By Theorem 4.10.2, these are the values of p for which $L_3^{\pm}(p)$ has a subgroup $L_3(2)$. The 8-dimensional module in question for $L_3^{\pm}(q)$ is the adjoint module by Theorem 5.8.1. We see from Definition 5.4.9 that, in characteristic 0, the tensor product $M \otimes M^*$, where M is the natural module of a general linear group, decomposes as the sum of the adjoint module and the trivial module. Now let V be one of the 3-dimensional irreducible modules for $L_3(2)$. We find, from the character table of $L_3(2)$ in [12], that $V \otimes V^*$ decomposes as the sum of the 8-dimensional and the trivial irreducible modules of $L_3(2)$. So there are containments $L_3(2) < L_3^{\pm}(p) < O_8^-(p)$ and $L_3(2).2 < L_3^{\pm}(p).2 < O_8^-(p).\langle\gamma\rangle$. $\qquad\square$

Proposition 6.2.12 *There are containments $L_2(9).2 < A_{10} < O_9(3)$ and $L_2(9).2^2 < A_{10}.2 < O_9(3).2$. There are no other containments between $\mathscr{S}_1$-maximal and $\mathscr{S}_2^*$-maximal subgroups of almost simple extensions of $O_9(q)$.*

Proof By Theorem 5.11.13, the $\mathscr{S}_2^*$-maximal subgroups of $\Omega_9(q)$ are extensions of $L_2(q).2$ with $p \geqslant 11$, and $L_2(q^2).2$ for all odd q. By Theorem 4.10.20, the $\mathscr{S}_1$-maximals are extensions of: $L_2(8)$ with $q = p$ or p^3 and $p \neq 2, 7$; $L_2(17)$ with $q = p$ or p^2 and $p \neq 2, 17$; A_{10} with $q = p \neq 2, 5, 11$; and $A_{11}.2 < O_9(11)$.

None of the $\mathscr{S}_1$-maximals have 2-dimensional irreducible projective representations in characteristic p, so the $\mathscr{S}_1$-maximals are $\mathscr{S}^*$-maximal. Conversely, the only $\mathscr{S}_1$-maximals with order divisible by q^2, or by p when $p \geqslant 11$, are $A_{10} < O_9(3)$ and $A_{11}.2 < O_9(11)$. The group S_{11} does not contain a subgroup $PGL_2(11)$. An easy computer calculation (file `containmentsd9`) demonstrates the claimed containments. $\qquad\qquad\qquad\qquad\qquad\qquad\qquad\qquad\qquad\qquad\qquad$ $\square$

Proposition 6.2.13   *There is a containment $A_6.2_1 < S_4(p) < O_{10}^{\pm}(p)$ when $p \equiv \pm 1 \pmod{12}$. There is a containment $A_6 < S_4(p) < O_{10}^{\pm}(p)$, extending to $A_6.2_1 < S_4(p).2 < O_{10}^{\pm}(p).2$ when $p \equiv \pm 5 \pmod{12}$. There is also a containment $A_7.2 < S_4(7).2 < O_{10}^{-}(7).2$.*

There are no other containments between $\mathscr{S}_1$-maximal and $\mathscr{S}_2^*$-maximal subgroups of almost simple extensions of $O_{10}^{\pm}(q)$. (In particular, when $p \neq 7$, an extension of A_6 is $\mathscr{S}^*$-maximal in $O_{10}^{\pm}(p)$ if and only if it is not contained in $A_6.2_1$ and, when $p = 7$, an extension of A_6 is $\mathscr{S}^*$-maximal in $O_{10}^{-}(p)$ if and only if it is not contained in $A_6.2_1$ or $A_6.2_3$.)

Proof   By Theorem 5.11.14, the $\mathscr{S}_2^*$-maximals of $\Omega_{10}^{\pm}(q)$ are all extensions of $2 \times S_4(q)$ for $p \neq 2$. By Theorem 4.10.21, the $\mathscr{S}_1$-maximals are extensions of: A_6 with $q = p \neq 2, 3$; $L_2(11)$ with $q = p \neq 3, 11$; $A_7 < \Omega_{10}^{-}(7)$; A_{11} with $q = p \neq 2, 3, 11$; $A_{12} < \Omega_{10}^{-}(2)$; $A_{12} < \Omega_{10}^{+}(3)$; $M_{12} < \Omega_{10}^{-}(2)$; $2^{\cdot}M_{22} < \Omega_{10}^{-}(7)$; $2^{\cdot}L_3(4) < \Omega_{10}^{-}(7)$.

By Theorem 4.10.13 the extensions of A_6 and A_7 are subgroups of the corresponding extensions of $S_4(p)$ as claimed, and extensions of A_6 not contained in $A_6.2_1$ are not subgroups of any extension of $S_4(p)$. By Theorem 5.8.1, the 10-dimensional module of $Sp_4(q)$ in question is the symmetric square of the irreducible 4-dimensional representation. By Proposition 5.2.5 and the character tables of $2.A_6$ and $2.A_7$ in characteristic 7, the 10-dimensional representations of A_6 and A_7 are also the symmetric squares of the 4-dimensional representations of $2.A_6$ and $2.A_7$, so these are containments of $\mathscr{S}$-subgroups.

By Theorem 4.3.3, all other $\mathscr{S}_1$-maximals have no irreducible 4-dimensional projective representations in characteristic p, so they are $\mathscr{S}^*$-maximal.

The order of $S_4(p)$ is divisible by p^4, but the only $\mathscr{S}_1$-maximal with this property with p odd is $2 \times A_{12} < \Omega_{10}^{+}(3)$. The group $S_4(3)$ has no faithful permutation representation of degree at most 12, by Theorem 1.11.2, so we are done. $\qquad\qquad\qquad\qquad\qquad\qquad\qquad\qquad\qquad\qquad\qquad\qquad\qquad$ $\square$

Proposition 6.2.14   *There are containments $L_2(11) < A_{12} < O_{11}(11)$ and $L_2(11).2 < A_{12}.2 < O_{11}(11).2$. There are no other containments between $\mathscr{S}_1$-maximal and $\mathscr{S}_2^*$-maximal subgroups of almost simple extensions of $O_{11}(q)$.*

Proof By Theorem 5.11.15, the only $\mathscr{S}_2^*$-maximals of $\Omega_{11}(q)$ are extensions of $L_2(q)$ with $p \geqslant 11$. By Theorem 4.10.22, the $\mathscr{S}_1$-maximals are extensions of: A_{12} with $q = p \neq 2, 3, 13$; $A_{13} < \Omega_{11}(13)$; and $L_3(3).2 < \Omega_{11}(13)$.

By Theorem 4.3.3, none of the $\mathscr{S}_1$-maximals have irreducible projective representations of degree 2, so the $\mathscr{S}_1$-maximals are $\mathscr{S}^*$-maximal.

The $\mathscr{S}_1$-maximals with order divisible by $p \geqslant 11$ are A_{12} with $p = 11$, A_{13} with $p = 13$, and $L_3(3).2$ with $p = 13$. Now $L_2(13)$ has no faithful permutation representation of degree at most 13, so it cannot be contained in A_{13} or in $L_3(3)$. However, we can check from [57] that A_{12} and $L_2(11)$ both have unique irreducible representations of dimension 11 and, since both groups act double-transitively on 12 points, these modules must be the associated deleted permutation modules. Hence the 11-dimensional representation of $L_2(11)$ is the restriction of that of A_{12}, and similarly for $L_2(11).2 < S_{12}$, so the claimed containments exist. $\qquad\square$

Proposition 6.2.15 *There are no containments between $\mathscr{S}_1$-maximal and $\mathscr{S}_2^*$-maximal subgroups of almost simple extensions of $O_{12}^\pm(q)$.*

Proof By Theorem 5.11.16, there are no $\mathscr{S}_2^*$-maximal subgroups of $\Omega_{12}^\pm(q)$. $\quad\square$

6.3 Containments between geometric and $\mathscr{S}^*$-maximal subgroups

Recall Definition 2.1.4 of a containment. In this section we determine all containments between the $\mathscr{S}^*$-maximal subgroups and those subgroups that are maximal amongst the geometric subgroups.

Throughout this section we let Ω be a quasisimple classical group of dimension at most 12, with dimension as in Definition 1.6.20, with A the corresponding group in Column A of Table 1.2. We let G be a group with $\Omega \leqslant G \leqslant A$. As usual, bars denote images modulo $Z(\Omega)$. So $\overline{A} = \operatorname{Aut} \overline{\Omega}$ except when $\Omega = \operatorname{Sp}_4(2^e)$ with $e > 1$, in which case Ω has an additional graph automorphism: see Section 7.2 and recall that $\operatorname{Sp}_4(2)$ is not quasisimple. We let H be maximal amongst the geometric subgroups of G, and let $S < G$ be an $\mathscr{S}^*$-maximal subgroup of G. We define $H_\Omega := H \cap \Omega$ and $S_\Omega := S \cap \Omega$. In the following two subsections, we determine all containments $S < H$ and $H < S$, respectively.

6.3.1 $S < H$

In this subsection, we determine all containments of $\mathscr{S}^*$-maximal subgroups in those subgroups that are maximal amongst the geometric groups. In particular, this subsection completes our classification of the $\mathscr{S}^*$-maximal subgroups of the almost simple classical groups in dimension up to 12, with the exception of extensions of $S_4(2^e)$ that contain a graph automorphism.

Theorem 6.3.1   *With the above notation, the containments $S < H$ are as follows.*

(i) *Dimension 9, Cases* **L** *and* **U**, $\Omega = \mathrm{SL}_9^{\pm}(p)$, $S_\Omega = (p \mp 1, 9)\dot{\,}A_6.2_3$, *and* $H_\Omega = (\mathrm{SL}_3^{\pm}(p) \circ \mathrm{SL}_3^{\pm}(p)).(p \mp 1, 3)^2.2$, *with* $p \equiv \pm 1 \pmod 5$. *These containments extend to* $S_\Omega.\langle\gamma\rangle < H_\Omega.\langle\gamma\rangle < \Omega.\langle\gamma\rangle$.

(ii) *Dimension 12, Cases* **L** *and* **U**. *There are some containments with the group* $S_\Omega = (q \mp 1, 12) \circ 6\dot{\,}A_6$. *When* $q = p \equiv 1, 4 \pmod{15}$ *in Case* **L** *or* $q = p \equiv 11, 14 \pmod{15}$ *in Case* **U**, *there are containments* $S_\Omega < H_\Omega$ *with* $H_\Omega = \mathrm{SL}_2(q)^6.(q \mp 1)^5.S_6$, *and also with* $H_\Omega = \mathrm{SL}_3^{\pm}(q) \times \mathrm{SL}_4^{\pm}(q)$.

   In Case **L** *with* $q = p^2$ *and* $3 < p \equiv 2, 3 \pmod 5$, *there are the same containments with* $G = \Omega$, *and also with* $S = (q - 1, 12) \circ 6\dot{\,}S_6$, *where* $H = H_\Omega.2$ *and* $G = \Omega.2$. *(See Table 8.77 for a precise description of the extension* $G = \Omega.2$, *which depends on* $p \pmod{12}$.)

(iii) *Dimension 12, Case* **U**, $\Omega = \mathrm{SU}_{12}(5)$, $S_\Omega = 6\dot{\,}A_7$, $H_\Omega = \mathrm{SU}_3(q) \times \mathrm{SU}_4(q)$. *This containment extends to* $S_\Omega.\langle\gamma\rangle < H_\Omega.\langle\gamma\rangle < \Omega.\langle\gamma\rangle$.

(iv) *Dimension 12, Case* **O**$^+$, $G = \Omega = \Omega_{12}^+(p)$, $S = 2\dot{\,}M_{12}$, $H = 2^{11}.A_{12}$ *or* $2^{11}.S_{12}$, $q = p \equiv \pm 5, \pm 7, \pm 11 \pmod{24}$.

Hence the tables of maximal $\mathscr{S}$-subgroups in Chapter 8 are correct.

Proof   By definition of Class $\mathscr{S}$, H does not lie in any of the classes $\mathscr{C}_1$, $\mathscr{C}_3$, $\mathscr{C}_5$ or $\mathscr{C}_8$, so we do not consider these as candidate classes for H.

   We prove in Lemma 6.3.2 that if there is a containment $S < H$ with $H \in \mathscr{C}_2$ then the Case is **L**, **U** or **O**$^\varepsilon$. In Proposition 6.3.4 we show that the only containment in Cases **L** or **U** with $H \in \mathscr{C}_2$ is as described in Part (ii) above. We prove in Proposition 6.3.5 that the only containment in Case **O**$^\varepsilon$ with $H \in \mathscr{C}_2$ is as described in Part (iv) above.

   We prove in Proposition 6.3.7 that the only containments in Cases **L** and **U** with $H \in \mathscr{C}_4 \cup \mathscr{C}_7$ are as described in Parts (i), (ii) and (iii) above, and in Proposition 6.3.8 that in Cases **S** and **O**$^\varepsilon$ there are no such containments.

   Finally, we prove in Proposition 6.3.9 that if there is a containment $S < H$ then $H \notin \mathscr{C}_6$.

   The correctness of the tables in Chapter 8 for groups in Class $\mathscr{S}$ can now be deduced as follows. Start by taking the lists of candidate maximals in Sections 4.10 and 5.11. In Cases **L** and **U**, by Theorem 5.11.1, Class $\mathscr{S}_2^*$ is nonempty only for $n = 6, 9, 10$, so now eliminate those groups shown to be non-maximal in Propositions 6.2.1, 6.2.2, and 6.2.3, and then eliminate the groups in Parts (i), (ii), and (iii) above. In Case **S**, it suffices to remove from the lists of candidate maximals those groups described in Propositions 6.2.4 to

6.2.9, as there are no containments $S < H$. Finally, in Case $\mathbf{O}^\varepsilon$, first remove the non-maximal groups described in Propositions 6.2.10 to 6.2.15, and finally the groups in Part (iv) above.   □

H **in Class** $\mathscr{C}_2$. Some of the calculations in this section will involve induced characters [56, Chapter 5]. For a representation ρ with character χ of a subgroup Y of a finite group X, the character χ^X of the induced representation ρ^X of X satisfies

$$\chi^X(g) = \frac{1}{|Y|} \sum_{x \in X} \chi^\circ(xgx^{-1})$$

for $g \in X$, where $\chi^\circ(y)$ is defined to be $\chi(y)$ for $y \in Y$ and 0 for $y \notin Y$. So $\deg \chi^X = |X : Y| \deg \chi$. In particular, if ρ is the trivial representation, then ρ^X is the permutation representation of X acting on the cosets of Y.

   In general, if X is an irreducible imprimitive matrix group acting on V, the subspace W is one of the subspaces in an imprimitive decomposition of V, the group Y is the stabiliser in X of W, and ρ is the representation of Y defined by its action on W, then [56, (5.8)] X acting on V is the image of the induced representation ρ^X. Conversely [56, (5.9)], for any representation ρ of Y acting on W, the image of ρ^X is an imprimitive matrix group with an imprimitive decomposition into $|X : Y|$ subspaces that are isomorphic to W, where Y is the stabiliser of one of those subspaces and ρ is equivalent to the representation defined by the action of Y on that subspace.

   We first produce a candidate list of possible groups S.

Lemma 6.3.2   *Let S and H be as above, with $H \in \mathscr{C}_2$, preserving a decomposition into t subspaces. Then Ω, S and t are one of the following.*

   (i) $\Omega = \mathrm{SL}_6^\pm(q)$, $S_\Omega' = 3^{\cdot}\mathrm{A}_6$, $t = 6$.
  (ii) $\Omega = \mathrm{SL}_6^\pm(q)$, $S_\Omega = 6^{\cdot}\mathrm{A}_6$, $t = 6$.
 (iii) $\Omega = \mathrm{SL}_{12}^\pm(q)$, $S_\Omega' = 6^{\cdot}\mathrm{A}_6$, $t = 6$.
 (iv) $\Omega = \Omega_9(q)$, $S_\Omega = \mathrm{L}_2(8)$, $t = 9$.
  (v) $\Omega = \Omega_{10}^\pm(q)$, $S_\Omega' = \mathrm{A}_6$, $t = 10$.
 (vi) $\Omega = \Omega_{12}^+(q)$, $S_\Omega' = \mathrm{L}_2(11)$ *or* $2^{\cdot}\mathrm{M}_{12}$, $t = 12$.

Proof   Since S^∞ acts irreducibly, it must act transitively on the block system. So we only need to consider cases in which the simple composition factor X of S has a subgroup of index t, with $t|n$. Theorem 1.11.2 identifies the possibilities for S. The possibilities for (X, t) with $t \leqslant 12$, other than (A_t, t), are:

$$(\mathrm{A}_5, 6),\ (\mathrm{L}_2(7), 7),\ (\mathrm{L}_2(7), 8),\ (\mathrm{L}_2(8), 9),\ (\mathrm{A}_5, 10),\ (\mathrm{A}_6, 10),$$
$$(\mathrm{L}_2(11), 11),\ (\mathrm{M}_{11}, 11),\ (\mathrm{A}_5, 12),\ (\mathrm{L}_2(11), 12),\ (\mathrm{M}_{11}, 12),\ (\mathrm{M}_{12}, 12).$$

We go through the $\mathscr{S}^*$-maximals using Sections 4.10, 5.11 and 6.2 to locate the possible pairs (X, t) with $t|n$. Note, in particular, that $t \neq n$ in Case **S** or in $\Omega_8^-(q)$ or $\Omega_{12}^-(q)$ by Table 2.4, and that $S_\Omega = L_2(11)$ with $t = 11$ and $\Omega = \Omega_{11}(11)$ is ruled out by Propositions 6.2.14. $\qquad\square$

The following lemma will be used shortly.

Lemma 6.3.3 *Let N be a group, let $G \leqslant S_n$ be transitive, and let $W = N \wr G$. Then any complement of the base group of W is conjugate in W to a subgroup of $M \wr G$, where M is the image of a homomomorphism $G_1 \to N$ (and G_1 is the stabiliser of 1 with S_n acting on $\{1, \ldots, n\}$).*

Note: It is shown in [48, Corollary 6] that there is a bijection between the set of conjugacy classes of complements of the base group in W and $\mathrm{Hom}(G_1, N)$, and this lemma could probably also be extracted from the proof of that result.

Proof We identify the base group of W with the set N^Ω of functions from $\Omega = \{1, \ldots, n\}$ to N, where W is the semidirect product of N^Ω and G, with the action $f^g(i) = f(i^{g^{-1}})$ for $f \in N^\Omega$, $g \in G$ and $i \in \Omega$. Let $H = G_1$, and let $\{t_i \mid i \in \Omega\}$ be a right transversal of H in G with $t_1 = 1$ and $1^{t_i} = i$ for $i \in \Omega$.

The complements of N^Ω in W have the form $\{g\phi(g) \mid g \in G\}$, where the map $\phi : G \to N^\Omega$ satisfies $\phi(g_1 g_2) = \phi(g_1)^{g_2}\phi(g_2)$ for $g_1, g_2 \in G$. For any homomorphism $\tau : H \to N$, we can define a map $\phi : G \to N^\Omega$ satisfying $\phi(g_1 g_2) = \phi(g_1)^{g_2}\phi(g_2)$, and hence a complement C of N^Ω in W, by letting $\phi(g)(i) = \tau(h)$, where $t_k g = h t_i$ with $h \in H$ and $i = k^g$. Note that $C \leqslant \mathrm{Im}(\tau) \wr G$.

On the other hand, if D is a complement of N^Ω in W, with associated map $\psi : G \to N^\Omega$, then $\psi(g_1 g_2)(1) = \psi(g_1)(1)\,\psi(g_2)(1)$ for all $g_1, g_2 \in H$. That is, the projection onto the first component of N^Ω of the restriction of ψ to H is a homomorphism $\tau : H \to N$. Let C be the complement of N^Ω in W defined from τ, and having associated map ϕ, as described in the preceding paragraph. We shall show that C and D are conjugate in W, thereby proving the lemma.

Define $f \in N^\Omega$ by $f(i) = \psi(t_i^{-1})(1)$ for $i \in \Omega$. Then we calculate that $fg\phi(g)f^{-1} = gf^g\phi(g)f^{-1}$ for all $g \in G$. We shall show that $f^g\phi(g)f^{-1} = \psi(g)$, and hence $fCf^{-1} = D$. For $i \in \Omega$, let $t_k g = h t_i$ with $h \in H$ and $i = k^g$. Then

$$(f^g\phi(g)f^{-1})(i) = f(k)\tau(h)f(i)^{-1} = \psi(t_k^{-1})(1)\,\tau(h)\,\psi(t_i^{-1})(1)^{-1}.$$

Now

$$\psi(t_k^{-1})(1) = \psi(gt_i^{-1}h^{-1})(1) = \psi(gt_i^{-1})^{h^{-1}}(1)\,\psi(h^{-1})(1)$$
$$= \psi(gt_i^{-1})(1)\,\psi(h^{-1})(1),$$

since $h \in H$. Then $\psi(h^{-1})(1) = \tau(h^{-1})$, by definition of τ, so

$$(f^g\phi(g)f^{-1})(i) = \psi(gt_i^{-1})(1)\,\psi(t_i^{-1})(1)^{-1} = \psi(g)^{t_i^{-1}}(1) = \psi(g)(1^{t_i}) = \psi(g)(i),$$

so $f^g\phi(g)f^{-1} = \psi(g)$ as claimed. $\qquad\qquad\qquad\qquad\qquad\qquad\square$

We continue with the notation defined at the beginning of this section.

Proposition 6.3.4 *Let* $\Omega = \mathrm{SL}_n^{\pm}(q)$ *with* $n \leqslant 12$, *let* $H \in \mathscr{C}_2$ *be maximal amongst the geometric subgroups of G, and let S be an $\mathscr{S}^*$-maximal subgroup of G. If $S \leqslant H$ then $n = 12$, as described in Theorem 6.3.1 (ii).*

Proof We consider the three possibilities from Lemma 6.3.2 in turn.

Suppose first that $n = t = 6$, with $S'_\Omega = 3\,{}^{\cdot}A_6$. By Proposition 6.2.1, the only subgroups S with $S'_\Omega = 3\,{}^{\cdot}A_6$ that are $\mathscr{S}^*$-maximal are novelties with the property that $S/\mathrm{Z}(S)$ contains $\mathrm{PGL}_2(9)$. In the $\mathscr{C}_2$-subgroups of type $\mathrm{GL}_1^{\pm}(q) \wr S_6$ in $\mathrm{GL}_6^{\pm}(q)$, the subgroup S_6 is centralised by the automorphisms γ and ϕ, and so no such $\mathscr{C}_2$-subgroup of any almost simple extension of $\mathrm{L}_6(p)$ or $\mathrm{U}_6(p)$ can involve $\mathrm{PGL}_2(9)$ as a section. Hence there are no containments.

Suppose next that $n = t = 6$ again, but with $S_\Omega = 6\,{}^{\cdot}A_6$. As before, the group $H'_\Omega = (q \pm 1)^5.A_6$. By using the facts that $\mathrm{SL}_2(5) < \mathrm{SL}_2(9)$ and 3 does not divide the order of the Schur multiplier of A_5, we find that the subgroups of index 6 in $6\,{}^{\cdot}A_6$ have the structure $3 \times \mathrm{SL}_2(5)$. So, if $S_\Omega = 6\,{}^{\cdot}A_6$ preserved such an imprimitive decomposition, then the stabiliser of a block would be isomorphic to $3 \times \mathrm{SL}_2(5)$. Since $\mathrm{SL}_2(5)$ is perfect, it would have to act trivially on the 1-dimensional subspace that it fixes. But this is impossible, because the central element of order 2 in $\mathrm{SL}_2(5)$ is central in $6\,{}^{\cdot}A_6$ and acts as $-I_6$. So there is no such containment.

Finally, suppose that $S'_\Omega = 6\,{}^{\cdot}A_6$, with $n = 12$ and $t = 6$. So the group $H_\Omega = \mathrm{SL}_2(q)^6.(q \pm 1)^5.S_6$, by Table 2.5. It can be checked by a routine computer calculation or from the character tables in [12] that the 12-dimensional irreducible representations of $6\,{}^{\cdot}A_6$ are induced from 2-dimensional faithful irreducible representations of $3 \times \mathrm{SL}_2(5)$. (To see this observe, from the formula for induced characters given above applied to central elements of $3 \times \mathrm{SL}_2(5)$, that these induced representations are faithful with central elements represented by scalar matrices, so they must be either irreducible or the sum of two faithful irreducibles of degree 6. Since an element of order 5 lies in a unique conjugate of $3 \times \mathrm{SL}_2(5)$ in $6\,{}^{\cdot}A_6$, it follows from the character formula that the induced characters in question take the same values as the characters of $3 \times \mathrm{SL}_2(5)$ on elements of order 5, which is b_5 or $b_5{}^*$, and so the induced characters must be irreducible.) So there is a containment $S_\Omega < H_\Omega$, whenever S is an $\mathscr{S}^*$-maximal subgroup of G.

The automorphisms γ and ϕ of $\mathrm{SL}_{12}^{\pm}(q)$ centralise the permutation subgroup S_6 of H_Ω, and hence no group H for which $H_\Omega = \mathrm{SL}_2(q)^6.(q \pm 1)^5.S_6$ has a subgroup mapping onto $A_6.2_2$ or $A_6.2_3$, so there are no such containments $S < H$ in which S maps onto $A_6.2_2$, $A_6.2_3$ or $A_6.2^2$.

It remains to consider cases in which S maps onto $A_6.2_1 = S_6$, which by Theorem 4.10.11 occur in Case **L** with $q = p^2$ and $p \equiv \pm 2 \pmod 5$. We shall show that there is a containment $6\dot{}S_6 < H = H_\Omega.2 < G = \Omega.2$ in this situation.

The subgroups of $H_\Omega = \mathrm{SL}_2(q)^6.(q \pm 1)^5.S_6$ that are isomorphic to the group $S_\Omega = 6\dot{}A_6$ map onto complements of the base group in $\mathrm{PGL}_2(q) \wr A_6$. So they have conjugates that map into subgroups $A_5 \wr A_6 < \mathrm{PGL}_2(q) \wr A_6$, by Lemma 6.3.3. So we may assume that S_Ω lies in a subgroup K of H_Ω with structure $(3 \times \mathrm{SL}_2(5))^6{:}A_6$. Let K be a fixed subgroup of H_Ω with this structure. To establish the containment $6\dot{}A_6.2_1 < H_\Omega.2 < \Omega.2$, we shall find an outer automorphism of Ω that normalises H_Ω, K, and S_Ω, and extends $S_\Omega = 6\dot{}A_6$ to $6\dot{}S_6$.

We may assume that H_Ω is embedded naturally as a wreath product in the matrix group $\Omega = \mathrm{SL}_{12}(p^2)$, and we assume that the complement of the base group in K isomorphic to A_6 is just the natural complement consisting of the permutation matrices that permute the six 2-dimensional spaces. As we noted earlier, γ and ϕ both centralise the subgroup S_6 of H_Ω consisting of permutation matrices. Furthermore, they both normalise each of the factors $\mathrm{SL}_2(p^2)$ in the base group of H_Ω.

Some conjugate of γ normalises and induces inner automorphisms of each of the six factors $\mathrm{SL}_2(5)$ of the base group of K, whereas γ inverts $O_3(K)$, the largest normal 3-subgroup of K. So, by multiplying γ by an inner automorphism of K, we may assume that it centralises the subgroup $\mathrm{SL}_2(5)^6{:}A_6$ and inverts $O_3(K)$. By Theorem 4.10.12, the map ϕ can be chosen to induce an outer automorphism of the subgroup $\mathrm{SL}_2(5)$ of $\mathrm{SL}_2(q)$, so we may assume that ϕ normalises and induces an outer automorphism of each of the six factors $\mathrm{SL}_2(5)$ of K. Furthermore, ϕ centralises $O_3(K)$ when $p \equiv 1 \pmod 3$ and inverts $O_3(K)$ when $p \equiv 2 \pmod 3$.

An element of $6\dot{}S_6 \setminus 6\dot{}A_6$ inverts the central element of order 3 of $6\dot{}A_6$. If the outer automorphism of Ω that we are trying to construct, which normalises H_Ω, K, and S_Ω, and extends S_Ω to $6\dot{}S_6$, exists, then its action on K can be obtained by multiplying the actions of ϕ (when $p \equiv 2 \pmod 3$) or $\phi\gamma$ (when $p \equiv 1 \pmod 3$) by an automorphism induced by conjugation by an element of $(3 \times \mathrm{SL}_2(5))^6{:}S_6 \setminus K$.

Since we know the induced actions on K of γ and ϕ precisely, it suffices to prove that there exists such an automorphism of K that extends the subgroup $6\dot{}A_6$ to $6\dot{}S_6$. This can be done theoretically, but it is perhaps easier to do it by computer calculation (file `containmentsd12`). It turns out that K has two classes of subgroups with the structure $6\dot{}A_6$, both of which are normalised by the appropriate automorphisms of K that extend them to $6\dot{}S_6$. $\qquad\square$

Proposition 6.3.5 *Let $\Omega = \Omega_n^\varepsilon(q)$ with $n \leqslant 12$, let $H \in \mathscr{C}_2$ be maximal* amongst the geometric subgroups of G, and let S be an $\mathscr{S}^*$-maximal subgroup of G. If $S \leqslant H$ then $\Omega = \Omega_{12}^+(p)$, as described in Theorem 6.3.1 (iv).

Proof We work through the possibilities from Lemma 6.3.2, and show that the only containment is when $n = 12$ and $S_\Omega = 2^{\cdot}M_{12}$.

Let $n = t = 9$, with $S_\Omega = L_2(8)$. It can be checked (see file `charcalc`) that the three quasi-equivalent 9-dimensional irreducible representations of $L_2(8)$ are imprimitive (as subgroups of $SL_9(q)$) and induced from non-trivial 1-dimensional representations of its subgroup K with structure $2^3 {:} 7$. So, in the 9-dimensional representations of S, the group K stabilises and has an element of order 7 acting non-trivially on a 1-dimensional subspace. Hence K does not preserve an orthogonal form on that subspace, and so S_Ω does not preserve the orthogonal decomposition of H_Ω. So there are no such containments.

Let $n = t = 10$, with $S_\Omega' = A_6$. It can be checked (see file `charcalc`) that this representation of A_6 is imprimitive and induced from a 1-dimensional representation of its subgroup K with structure $9{:}4$, with image cyclic of order 4. So, as in the previous case, there are no such containments.

Finally, let $n = t = 12$. If $S_\Omega' = L_2(11)$, then the block stabiliser S_B' is isomorphic to $11{:}5$. If S_B' acted trivially on B, then the module for S_Ω' would be a permutation module, and would not be irreducible. Hence S_B' acts non-trivially on the block, so this action is not orthogonal, and S_Ω' is not contained in H_Ω. For $S_\Omega' = 2^{\cdot}M_{12}$, it can be checked by character table calculations (file `charcalc`), that the 12-dimensional representation of S_Ω' is induced by a 1-dimensional orthogonal representation of its subgroup with structure $2 \times M_{11}$, and so there is a containment $S_\Omega' < H_\Omega$. Note, however, that $M_{12}.2$ is not contained in $A_{12}.2$, so $2^{\cdot}M_{12}.2$ is maximal wherever it occurs. $\square$

H in Class $\mathscr{C}_4$ or $\mathscr{C}_7$. Recall the notation from the beginning of this section, and recall Definition 5.1.15 of the $\mathscr{S}_2^*$-subgroups.

Lemma 6.3.6 *Let $n \leqslant 12$, let S be an $\mathscr{S}^*$-maximal subgroup that is not contained in a member of Class $\mathscr{C}_2$, and let $H \in \mathscr{C}_4 \cup \mathscr{C}_7$. If $S < H$ then Ω and S are one of the following.*

  (i) $\Omega = SL_9^{\pm}(q)$, $S_\Omega' = 3^{\cdot}A_6.2_3$, with $p \neq 3, 5$.
  (ii) $\Omega = SL_{12}^{\pm}(q)$, $S_\Omega' = 6^{\cdot}A_6$ with $p \neq 2, 3, 5$.
  (iii) $\Omega = SU_{12}(5)$, $S_\Omega = 6^{\cdot}A_7$.
  (iv) $\Omega = Sp_6(q)$, $S_\Omega' = 2^{\cdot}A_5$ with $p \neq 2, 5$, and $S_\Omega = SL_2(q)$.
  (v) $\Omega = Sp_8(q)$, $S_\Omega = SL_2(q)$.
  (vi) $\Omega = Sp_{10}(q)$, $S_\Omega = SL_2(q)$.
  (vii) $\Omega = Sp_{12}(q)$, $S_\Omega = SL_2(q)$.
  (viii) $\Omega = \Omega_9(q)$, $S_\Omega = L_2(q)$.

Proof If there is a containment $S_\Omega < H_\Omega$, then S_Ω must have irreducible projective representations in the dimensions of the tensor factors, and at least one of these factors must have dimension 2 or 3. So we can immediately reduce the number of potential containments by recalling that:

(i) the only simple groups with irreducible projective representations of degree 2 over $\mathbb{F}_q$ are A_5 in cross characteristic and $L_2(q_0)$, where q is a power of q_0;

(ii) the only simple groups with irreducible projective representations of degree 3 over $\mathbb{F}_q$ are A_5, A_6 and $L_3(2)$ in cross characteristic, A_7 when $p = 5$, and $L_2(q_0)$, $L_3(q_0)$ and $U_3(q_1)$, where q is a power of q_0 or q_1^2.

Going through the $\mathscr{S}^*$-maximals (using the results stated in Sections 4.10, 5.11 and 6.2), yields the above list. $\qquad\square$

Again, we consider these in turn, and start by identifying corresponding subgroups in Classes $\mathscr{C}_4$ or $\mathscr{C}_7$, if any.

Proposition 6.3.7 *Let $\Omega = \mathrm{SL}_n^\pm(q)$ with $n \leqslant 12$, let $H \in \mathscr{C}_4 \cup \mathscr{C}_7$ be maximal amongst the geometric subgroups of G, and let S be an $\mathscr{S}^*$-maximal subgroup of G. If $S \leqslant H$ then either $\Omega = \mathrm{SL}_9^\pm(p)$, as described in Theorem 6.3.1 (i); or $\Omega = \mathrm{SL}_{12}^\pm(q)$ with $q = p$ or p^2, as described in Theorem 6.3.1 (ii) or (iii).*

Proof Let $n = 9$, with $S_\Omega = 3{\cdot}A_6.2_3$, so that $q = p \neq 3, 5$ by Theorem 4.10.8, and $p \equiv \pm 1 \pmod 5$ by Proposition 6.2.2. By Tables 2.7 and 2.10 the group $H_\Omega = (\mathrm{SL}_3^\pm(p) \circ \mathrm{SL}_3^\pm(p)).(q \mp 1, 3)^2.2 \in \mathscr{C}_7$.

To analyse this case, we need to consider the character table of $3{\cdot}A_6$ in [12]. We also need to know the action of outer automorphisms of $3{\cdot}A_6$ on its conjugacy classes, which can easily be computed directly, but can also be deduced from careful study of the character tables. Specifically, the central 3-element of $3{\cdot}A_6$ is centralised by the 2_3 automorphism and inverted by the 2_1 and 2_2 automorphisms, whereas the two conjugacy classes of 5-elements are fixed by the 2_2 automorphism and interchanged by the 2_1 and 2_3 automorphisms.

There are four faithful irreducible characters of degree 3, two of which are represented by each of the printed characters χ_{14} and χ_{15}. The two represented by the same printed characters are dual to each other and take different values on the central 3-elements. Using the actions on conjugacy classes described above, we see that the 2_2 automorphism interchanges the two characters represented by χ_{14} and the two represented by χ_{15}, whereas the 2_1 and 2_3 automorphism interchange the pairs χ_{14} and χ_{15}, where 2_3 interchanges two characters with the same value on the central 3-element. (We used this fact earlier, in the proof of Proposition 6.2.2.) On the other hand, the two characters of degree 9 represented by χ_{17} are dual, and are fixed by 2_3 and interchanged by 2_1 and by 2_2. Using the fact that the character of a tensor product is the product of the characters of the tensor factors, we see that each of the two 9-dimensional irreducible representations is a tensor product of two of the four 3-dimensional representations that are interchanged by 2_3.

Since $p \equiv \pm 1 \pmod 5$, the group $3\cdot A_6$ is contained in $\mathrm{SL}_3^\pm(p)$ by Theorem 4.10.2. The tensor decomposition of S'_Ω is preserved by the 2_3 automorphism so $S_\Omega = 3\cdot A_6.2_3$ is not maximal. The duality automorphism of Ω fixes H_Ω and induces the duality automorphism of the two tensor factors, which fixes (the class of) S_Ω. So there is a containment $S_\Omega.\langle\gamma\rangle < H_\Omega.\langle\gamma\rangle < \Omega.\langle\gamma\rangle$.

Let $n = 12$, with $S_\Omega = 6\cdot A_6$, and $p \neq 2,3,5$ by Theorem 4.10.11. Class $\mathscr{C}_7$ is empty, so by Table 2.7 the group $H_\Omega = \mathrm{SL}_3^\pm(q) \times \mathrm{SL}_4^\pm(q)$. As can be checked from the character table of $6\cdot A_6$ [12], the four 12-dimensional irreducible representations of $6\cdot A_6$, which are equivalent under its automorphism group, are each a tensor product of a (faithful) 3-dimensional representation of $3\cdot A_6$ and a 4-dimensional representation of $2\cdot A_6$. This results in containments $6\cdot A_6 < \mathrm{SL}_3^\pm(q) \times \mathrm{SL}_4^\pm(q)$ in all characteristics $p > 5$.

By Proposition 4.5.10, the image of the 4-dimensional irreducible representation of $2\cdot A_6$ lies in $\mathrm{Sp}_4(p)$ for all primes $p \neq 2,3$, and no automorphism of $\mathrm{Sp}_4(p)$ normalises and induces the 2_2 or the 2_3 automorphism of $2\cdot A_6$. Any automorphism of $\mathrm{SL}_4^\pm(p)$ or $\mathrm{SL}_4(p^2)$ that normalises the image of $2\cdot A_6$ must lie in $\mathrm{CSp}_4(p)$ by Lemma 1.8.9, and so no such automorphism can induce the 2_2 or the 2_3 automorphism of $2\cdot A_6$. It follows that no extensions of S_Ω of the form $6\cdot A_6.2_2$, $6\cdot A_6.2_3$ and $3\cdot A_6.2^2$ can be contained in Class $\mathscr{C}_4$ groups H, and so all such extensions are maximal in G.

By Theorem 6.3.1 (ii), extensions $6\cdot A_6.2_1$ of S_Ω are contained in extensions of $\mathrm{SL}_3^\pm(q) \times \mathrm{SL}_4^\pm(q)$ that are $\mathscr{C}_4$-subgroups, and we believe this to be true! However, since we saw in Proposition 6.3.4 that extensions of S_Ω of the form $6\cdot A_6.2_1$ are contained in $\mathscr{C}_2$-subgroups, we do not need this containment to prove that such extensions are not maximal, and so we omit the proof.

Finally, let $\Omega = \mathrm{SU}_{12}(5)$, with $S_\Omega = 6\cdot A_7$, as in Theorem 4.10.11. As for $S_\Omega = 6\cdot A_6$, the group H_Ω equals $\mathrm{SU}_3(5) \times \mathrm{SU}_4(5)$, and from the character table of $6\cdot A_7$ in characteristic 5 [57], the two 12-dimensional irreducible representations of $6\cdot A_7$, which are equivalent under $\mathrm{Out}\,A_7$, are each a tensor product of a (faithful) 3-dimensional representation of $3\cdot A_7$ and a 4-dimensional representation of $2\cdot A_7$. This results in the containment $6\cdot A_7 < \mathrm{SU}_3(5) \times \mathrm{SU}_4(5)$. Since the conjugacy classes of the $\mathscr{S}_1$-subgroups $3\cdot A_7 < \mathrm{SU}_3(5)$ and $2\cdot A_7 < \mathrm{SU}_4(5)$ are both normalised by the respective automorphism γ, and the duality automorphism γ of Ω fixes the conjugacy class of H_Ω and induces γ on the two tensor factors, this containment extends to $S_\Omega.\langle\gamma\rangle < H_\Omega.\langle\gamma\rangle < \Omega.\langle\gamma\rangle$. $\square$

Proposition 6.3.8 *Let $\Omega = \mathrm{Sp}_n(q)$ or $\Omega_n^\pm(q)$ with $n \leqslant 12$, let $H \in \mathscr{C}_4 \cup \mathscr{C}_7$ be* *maximal amongst the geometric subgroups of G, and let S be an $\mathscr{S}^*$-maximal subgroup of G. Then $S \not\leqslant H$.*

Proof We continue working through the possibilities given by Lemma 6.3.6.

First, let $\Omega = \mathrm{Sp}_6(q)$, with $S'_\Omega = 2^{\cdot}\mathrm{A}_5$ and $q = p \neq 2, 5$ by Theorem 4.10.14. The full covering group of A_5 is $2^{\cdot}\mathrm{A}_5$, which has two 2-dimensional symplectic representations over $\mathbb{F}_p(\sqrt{5})$ ($p \neq 2, 5$), and two 3-dimensional orthogonal representations over $\mathbb{F}_p(\sqrt{5})$ ($p \neq 2, 5$). It can be observed directly from the character table of $2^{\cdot}\mathrm{A}_5$ that the required 6-dimensional representation of $2^{\cdot}\mathrm{A}_5$ is obtained as $3a \otimes 2b$ or $3b \otimes 2a$ (using the ATLAS ordering).

By Theorem 4.10.14, the group $S_\Omega = 2^{\cdot}\mathrm{A}_5$ is $\mathscr{S}_1$-maximal in $\mathrm{Sp}_6(p)$ when $p \equiv \pm 3 \pmod 8$ (and $p \neq 5$). The group S_Ω is a tensor product over $\mathbb{F}_p$ if and only if 5 is a square in $\mathbb{F}_p^\times$, which is the case if and only if $p \equiv \pm 1 \pmod 5$. Thus if $p \equiv \pm 11, \pm 19 \pmod{40}$ then there is a containment $2^{\cdot}\mathrm{A}_5 < \mathrm{Sp}_2(p) \times \mathrm{SO}_3(p)$. (However, S_Ω is non-maximal for these values of q, by Proposition 6.2.6.) If $p \equiv 3 \pmod 8$ then $2^{\cdot}\mathrm{A}_5$ extends to $2^{\cdot}\mathrm{S}_5^+ < \mathrm{CSp}_6(p)$, whilst if $p \equiv 5 \pmod 8$ then $2^{\cdot}\mathrm{A}_5$ extends to $4.\mathrm{S}_5 = 4 \circ 2^{\cdot}\mathrm{S}_5 < \mathrm{CSp}_6(p)$. Any central extension $Z.2^{\cdot}\mathrm{S}_5^+$ or $Z.4.\mathrm{S}_5$ will contain a normal subgroup $2^{\cdot}\mathrm{A}_5$ together with elements that fuse both 2-dimensional and both 3-dimensional representations of this subgroup, and so the representations of $2^{\cdot}\mathrm{S}_5^+$ and $4.\mathrm{S}_5$ are not tensor products. Thus if $p \equiv \pm 11, \pm 19 \pmod{40}$ then the $\mathscr{S}^*$-maximal subgroup S_5 of $\mathrm{PCSp}_6(p)$ is not contained in any $\mathscr{C}_4$-subgroup. If $p \equiv \pm 1 \pmod 8$ then $S_\Omega = 2^{\cdot}\mathrm{A}_5.2$, which as noted above is not a tensor product, so S is not contained in any $\mathscr{C}_4$-subgroup.

Next, let $\Omega = \mathrm{Sp}_n(q)$ with $n = 6, 8, 10, 12$, and $S_\Omega = \mathrm{SL}_2(q)$: we shall treat these cases simultaneously. By Theorem 5.8.1 these groups S_Ω are acting on modules $\mathrm{S}^{n-1}(V)$: see Section 5.2 for basic properties of $\mathrm{S}^k(V)$. Here, V is the natural module for $\mathrm{SL}_2(q)$, and $p \geqslant 7$ throughout. If $n = 6$ then the group $H_\Omega^\infty = \mathrm{Sp}_2(q) \circ \Omega_3(q)$. If $n = 8$ then $H_\Omega^\infty = \mathrm{Sp}_2(q) \circ \mathrm{Sp}_2(q) \circ \mathrm{Sp}_2(q) \in \mathscr{C}_7$ or $\mathrm{Sp}_2(q) \circ \Omega_4^-(q) \in \mathscr{C}_4$, by Tables 2.7 and 2.10 and Lemma 3.7.6. Similarly, if $n = 10$ then $H_\Omega^\infty = \mathrm{Sp}_2(q) \times \Omega_5(q)$, and if $n = 12$ then $H_\Omega^\infty = \mathrm{Sp}_2(q) \circ \Omega_6^\pm(q)$ or $\Omega_3(q) \times \mathrm{Sp}_4(q)$. These six containments would respectively imply isomorphisms $\mathrm{S}^5(V) \cong V \otimes \mathrm{S}^2(V)$, $\mathrm{S}^7(V) \cong V \otimes V \otimes V$, $\mathrm{S}^7(V) \cong V \otimes \mathrm{S}^3(V)$, $\mathrm{S}^9(V) \cong V \otimes \mathrm{S}^4(V)$, and $\mathrm{S}^{11}(V) \cong V \otimes \mathrm{S}^5(V)$ or $\mathrm{S}^2(V) \otimes \mathrm{S}^3(V)$, each of which is ruled out by Proposition 5.3.10.

Finally, let $\Omega = \Omega_9(q)$, with $S_\Omega = \mathrm{L}_2(q).2$. Then $H_\Omega^\infty = \Omega_3(q)^2 \in \mathscr{C}_7$ by Table 2.10. This containment is ruled out by Proposition 5.3.10, since $\mathrm{S}^2(V) \otimes \mathrm{S}^2(V)$ is reducible and hence not isomorphic to $\mathrm{S}^8(V)$. $\qquad\square$

H in Class $\mathscr{C}_6$. It is straightforward to show that there are no containments of $\mathscr{S}^*$-maximals in $\mathscr{C}_6$-subgroups, completing the proof of Theorem 6.3.1. Recall the notation from the beginning of this section.

Proposition 6.3.9 *Let Ω be a quasisimple classical group with $n \leqslant 12$, let $H \in \mathcal{C}_6$ be maximal amongst the geometric subgroups of G, and let S be an $\mathcal{S}^*$-maximal subgroup of G. Then $S \not\leqslant H$.*

Proof By Definition 2.2.13, the only non-abelian simple composition factors X arising in subgroups in Class $\mathcal{C}_6$ are:

(i) $\Omega = \mathrm{SL}_4^{\pm}(q)$, $X = A_6$.
(ii) $\Omega = \mathrm{Sp}_4(q)$, $X = A_5$.
(iii) $\Omega = \mathrm{SL}_5^{\pm}(q)$, $X = L_2(5)$.
(iv) $\Omega = \mathrm{SL}_7^{\pm}(q)$, $X = L_2(7)$.
(v) $\Omega = \mathrm{SL}_8^{\pm}(q)$, $X = S_6(2)$.
(vi) $\Omega = \mathrm{Sp}_8(q)$, $X = \Omega_6^-(2)$.
(vii) $\Omega = \mathrm{SL}_9^{\pm}(q)$, $X = S_4(3)$.
(viii) $\Omega = \mathrm{SL}_{11}^{\pm}(q)$, $X = L_2(11)$.

By Theorems 4.10.3, 4.10.13, 4.10.4, 4.10.6, 4.10.10, 5.11.1, and 5.11.6 together with Proposition 6.2.5 for $\mathrm{Sp}_4(q)$, in Cases (i), (ii), (iii), (iv) and (viii) above there are no $\mathcal{S}^*$-maximals with simple composition factor contained in X.

In Case (v), it follows from Theorems 4.10.7 and 5.11.1 that all $\mathcal{S}^*$-maximal subgroups have simple composition factor $L_3(4)$. From our analysis of Case **S** in dimension 6 (see Tables 8.28 and 8.29, which we may assume at this point to be correct) we find that no subgroup of $S_6(2)$ involves $L_3(4)$.

In Case (vi), since $\Omega_6^-(2) \cong U_4(2) \cong S_4(3)$, we need only consider the $\mathcal{S}^*$-maximal $S_\Omega' = 2\,{}^{\textstyle\cdot}A_6$, by Theorem 4.10.15, with $H_\Omega' = 2^{1+6}.\Omega_6^-(2)$, by Table 2.9. A computer calculation (file `charcalc`) shows that the restriction of the unique 8-dimensional character of H_Ω' to the unique subgroup isomorphic to $2\,{}^{\textstyle\cdot}A_6$ is rational and reducible, which rules out this containment. (We carried out this calculation using $H < \mathrm{Sp}_8(3)$. We found that the 8-dimensional character of H is rational with Frobenius-Schur indicator -1, and so its reduction modulo p is an irreducible subgroup of $\mathrm{Sp}_8(p)$ for all p not dividing $|H|$, and we can check it directly for $p = 5$. So our conclusion is valid for all odd p.)

For Case (vii), there is an $\mathcal{S}^*$-maximal subgroup with composition factor A_6, and $S_4(3)$ has a subgroup isomorphic to A_6. By Theorem 4.10.8, the group $S_\Omega = (9, q \mp 1){}^{\textstyle\cdot}A_6.2_3$. Tables 8.12 and 8.13, which may be assumed correct at this point, show that $S_4(3)$ has a unique subgroup of shape $A_6.2$, and a computer calculation (file `containmentsd9`) shows that it is isomorphic to $S_6 = A_6.2_1$ rather than to $A_6.2_3$, so there is no containment in this case. $\square$

6.3.2 $H < S$

In this subsection, we determine which of the groups that are maximal amongst the geometric subgroups are in fact maximal. Since we determined the maximal

$\mathcal{S}$-subgroups in Theorem 6.3.1, this subsection completes the proof of correctness of the tables of maximal subgroups of classical groups in Chapter 8, with the exception of the groups with socle $\mathrm{Sp}_4(2^e)$ for $e > 1$ that are not contained in $\Sigma\mathrm{Sp}_4(2^e)$. In particular (again with the possible exception of $\mathrm{Sp}_4(2^e)$), the tables of maximal $\mathcal{S}$-subgroups in Chapter 8 have now been proved correct, so we can use those tables. We continue with the notation from the beginning of this section. Recall Definition 2.1.4 of a containment.

In the case of $\Omega = \mathrm{Sp}_4(2^e)$, in this section we shall only investigate containments $H < S$ for subgroups of $\Sigma\mathrm{Sp}_4(2^e)$. In fact there are no such containments. We shall complete the correctness proofs for Table 8.14 in Section 7.2.

Theorem 6.3.10 *Let H be maximal amongst the geometric subgroups of G, and let S be a maximal $\mathcal{S}$-subgroup of G such that $H \leqslant S$. Then H, G and S are as given below:*

(i) $\Omega = \mathrm{SL}_2(11)$, $S \cong 2 \cdot \mathrm{A}_5$, $H \cong Q_{2(q-1)} \in \mathscr{C}_2$, $G = \Omega$.

(ii) $\Omega = \mathrm{SL}_2(9)$, $S_\Omega \cong 2 \cdot \mathrm{A}_5$, $H_\Omega \cong Q_{2(q+1)} \in \mathscr{C}_3$, $\Omega \leqslant G \leqslant \Omega.\langle\phi\rangle$.

(iii) $\Omega = \mathrm{SL}_2(p)$, $S \cong 2 \cdot \mathrm{A}_5$, $q = p \equiv \pm 11, \pm 19 \pmod{40}$, $H \cong 2^{1+2}_-:3 \cong 2 \cdot \mathrm{A}_4 \in \mathscr{C}_6$, $G = \Omega$.

(iv) $\Omega = \mathrm{SU}_3(3)$, $S_\Omega \cong \mathrm{L}_2(7)$, $H_\Omega \cong 7:3 \in \mathscr{C}_3$, $\Omega \leqslant G \leqslant \Omega.\langle\gamma\rangle$.

(v) $\Omega = \mathrm{SU}_3(5)$, $S_\Omega \cong 3 \cdot \mathrm{A}_7$, $H_\Omega \cong (q+1)^2{:}S_3 \in \mathscr{C}_2$, $\Omega \leqslant G \leqslant \Omega.\langle\gamma\rangle$.

(vi) $\Omega = \mathrm{SU}_3(5)$, $S_\Omega \cong 3 \times \mathrm{L}_2(7)$, $H_\Omega \cong (q^2 - q + 1){:}3 \in \mathscr{C}_3$, $\Omega \leqslant G \leqslant \Omega.\langle\gamma\rangle$.

(vii) $\Omega = \mathrm{SU}_3(5)$, $S_\Omega \cong 3 \cdot \mathrm{A}_7$, $H_\Omega \cong 3 \times \mathrm{SO}_3(5) \in \mathscr{C}_5$, $\Omega \leqslant G \leqslant \Omega.\langle\gamma\rangle$.

(viii) $\Omega = \mathrm{SU}_3(5)$, $S_\Omega \cong 3 \cdot \mathrm{A}_6 \cdot 2_3$, $H_\Omega \cong 3^{1+2}{:}Q_8 \in \mathscr{C}_6$, $\Omega \leqslant G \leqslant \Omega.\langle\gamma\rangle$.

(ix) $\Omega = \mathrm{SU}_5(2)$, $S_\Omega \cong \mathrm{L}_2(11)$, $H_\Omega \cong \frac{q^5+1}{q+1}{:}5 \in \mathscr{C}_3$, $\Omega \leqslant G \leqslant \Omega.\langle\gamma\rangle$.

(x) $\Omega = \mathrm{SU}_6(2)$, $S_\Omega \cong 3 \cdot \mathrm{U}_4(3).2_2$, $H_\Omega \cong 3^5.S_6 \in \mathscr{C}_2$, $\Omega \leqslant G \leqslant \Omega.\langle\gamma\rangle$.

Thus, if H is a geometric subgroup of G that is maximal in G, then H is as listed in the tables in Chapter 8.

In almost all cases, if S is an $\mathcal{S}_1$-subgroup, then we can rule out containments of the form $H < S$ using Lagrange's theorem. In many cases this is because $|H|$ is divisible by a higher power of the defining characteristic p than $|S|$: recall, for example, that by Lemma 2.2.2 (iii) the parabolic subgroups of Ω contain a Sylow p-subgroup of Ω. We shall not go into details, and the "possible containments" of this form considered below are exactly those that do not contradict Lagrange's theorem. Apart from the genuine containments, these can usually be easily eliminated either by direct calculation or by using [12].

Lagrange's theorem can also be used to eliminate many such containments when S is an $\mathcal{S}_2$-subgroup, but this is not generally possible when $H \in \mathscr{C}_5$, or when the order of H does not increase with q, which occurs when $H \in \mathscr{C}_6$ and when $H \in \mathscr{C}_2$ in the orthogonal cases and H has the structure $2^{n-2}.\mathrm{A}_n$ or $2^{n-2}.\mathrm{S}_n$. If H contains a classical group, then such containments can be ruled out by using Theorem 1.11.5 (for $\mathscr{C}_5$) or Theorem 1.11.7 (for $\mathscr{C}_8$). If H contains an alternating group, then these containments can be ruled out by Proposition 1.11.6 and Lemma 2.2.4 (i).

Proposition 6.3.11 *Let $n = 2$ and let $H < G$ be maximal amongst the geometric subgroups of G. If H is not maximal in G then either $q \in \{9, 11\}$ or $q = p \equiv \pm 11, \pm 19 \bmod 40$.*

If $q = 9$, then there is a containment $H \leqslant S < G$ if and only if H is of type $\mathrm{GL}_1(q^2)$, with $G = \mathrm{SL}_2(9)$ or $\mathrm{SL}_2(9).\langle \phi \rangle$, and $S_\Omega \cong 2^{\cdot}\mathrm{A}_5$.

If $q = 11$, then there is a containment $H \leqslant S < G$ if and only if H is of type $\mathrm{GL}_1(q) \wr \mathrm{S}_2$ or type $2^{1+2}.\mathrm{Sp}_2(2)$, with $G = \mathrm{SL}_2(11)$ and $S \cong 2^{\cdot}\mathrm{A}_5$.

If $q = p \equiv \pm 11, \pm 19 \bmod 40$, with $q > 11$, and $H \leqslant S < G$, then H is of type $2^{1+2}.\mathrm{Sp}_2(2)$, with $G = \mathrm{SL}_2(p)$ and $S \cong 2^{\cdot}\mathrm{A}_5$.

Hence $M < G$ is a non-trivial maximal subgroup of G if and only if M is listed in Table 8.1 or 8.2.

Proof Assume that $H \leqslant S < G$, where $S \in \mathscr{S}$. By Table 8.2, the only possible S_Ω is isomorphic to $2^{\cdot}\mathrm{A}_5$, so that $|S_\Omega| = 2^3 \cdot 3 \cdot 5$. The group S only occurs when $q = p \equiv \pm 1 \bmod 10$ or $q = p^2$ with $p \equiv \pm 3 \bmod 10$. $p = 2$ or 5 or Therefore $H \not\in \mathscr{C}_1$ by Lagrange's theorem.

If $H \in \mathscr{C}_2$ with q odd, then $H_\Omega \cong \mathrm{Q}_{2(q-1)}$ by Table 2.4. If $q \leqslant 7$ then Class $\mathscr{S}$ is empty. If $q = 11$ then $\mathrm{Q}_{2(q-1)} < 2^{\cdot}\mathrm{A}_5$, however there are two classes of groups $2^{\cdot}\mathrm{A}_5$ that are interchanged by δ, whilst H_Ω is normalised by δ. For all other values of q for which S arises, this containment is impossible because $2^{\cdot}\mathrm{A}_5$ has no element of order $q - 1$.

If $H \in \mathscr{C}_3$ with q odd, then by Table 2.6 the group $H_\Omega \cong \mathrm{Q}_{2(q+1)}$. If $q = 9$ then $\mathrm{Q}_{2(q+1)} < 2^{\cdot}\mathrm{A}_5$, which extends to $S = S_\Omega.2 < H = H_\Omega.2$ in $G = \mathrm{SL}_2(9).\langle \phi \rangle \cong 2.\mathrm{S}_6$, but not within any other almost simple extensions of $\mathrm{SL}_2(9)$. All other values of q for which S occurs yield impossible containments, because $2^{\cdot}\mathrm{A}_5$ has no element of order $q + 1$.

If $H \in \mathscr{C}_5$ with q odd, then H_Ω is of shape $\mathrm{SL}_2(q_0).(2, r)$ by Table 2.8, where $q = q_0^r$. Thus $H \leqslant S$ is ruled out by Lagrange's theorem for all q.

Finally, consider $H \in \mathscr{C}_6$. By Table 2.9 and Lagrange's theorem, the group $H_\Omega \cong 2_-^{1+2}{:}3 \cong 2^{\cdot}\mathrm{A}_4$ and $q = p \equiv \pm 11, \pm 19 \pmod{40}$. There is a genuine containment $H_\Omega < S_\Omega$, but there are two classes of groups S_Ω that are interchanged by δ, whilst H_Ω is normalised by δ. $\qquad\square$

Proposition 6.3.12 *Let $n = 3$ and let H be maximal amongst the geometric subgroups of G. If H is not maximal in G then $\Omega = \mathrm{SU}_3(q)$ and $q \in \{3, 5\}$.*

If $q = 3$ then there is a containment $H \leqslant S < G$ if and only if H is of type $\mathrm{GU}_1(q^3)$ ($\mathscr{C}_3$), with $\mathrm{SU}_3(3) \trianglelefteq G \leqslant \Gamma\mathrm{U}_3(3)$ and $S_\Omega \cong \mathrm{L}_2(7)$.

If $q = 5$ and $H \leqslant S < G$ then $H_\Omega \leqslant S_\Omega$ is one of: $(q+1)^2{:}\mathrm{S}_3 < 3^{\cdot}\mathrm{A}_7$

($\mathscr{C}_2$), ($q^2 - q + 1$):3 $< 3 \times L_2(7) < 3{\cdot}A_7$ *($\mathscr{C}_3$)*, $3 \times SO_3(5) < 3{\cdot}A_7$ *($\mathscr{C}_5$)*, $3^{1+2}{:}Q_8 < 3{\cdot}A_6{\cdot}2_3$ *($\mathscr{C}_6$)*. *These containments with H in Classes $\mathscr{C}_2$, $\mathscr{C}_3$ and $\mathscr{C}_6$ occur if and only if G is contained in a conjugate of* $SU_3(5).\langle\gamma\rangle$. *Those in Class $\mathscr{C}_5$ occur for all G with* $SU_3(5) \trianglelefteq G \leqslant \Gamma U_3(5)$.

    Hence $M < G$ is a non-trivial maximal subgroup of G if and only if M is listed in Table 8.3, 8.4, 8.5 or 8.6.

Proof   The MAGMA calculations referred to in this proof can all be found in file `containmentsHinS`. Assume that $H \leqslant S < G$, where $S \in \mathscr{S}$. Here, by Tables 8.4 and 8.6, the possibilities for S_Ω are :

- $(q{\pm}1, 3) \times L_2(7)$, of order $(q{\pm}1, 3){\cdot}2^3{\cdot}3{\cdot}7$. This requires $2 \neq q = p \not\equiv 0 \bmod 7$. If $p \equiv 1, 2, 4 \bmod 7$ then $\Omega = SL_3(q)$, otherwise $\Omega = SU_3(q)$.
- $3{\cdot}A_6$, of order $2^3{\cdot}3^3{\cdot}5$. In $SL_3(q)$ this requires $q = p \equiv 1, 4 \bmod 15$ or $q = p^2$, $p \equiv 2, 3 \bmod 5$, $q \neq 9$. In $SU_3(q)$ this requires $q = p \equiv 11, 14 \bmod 15$.
- $3{\cdot}A_6{\cdot}2_3$ (order $2^4 \cdot 3^3 \cdot 5$) and $3{\cdot}A_7$ (order $2^3 \cdot 3^3 \cdot 5 \cdot 7$) in $SU_3(5)$.

    Suppose that $H \in \mathscr{C}_1$: the structure of H is given in Table 2.3. Then $S_\Omega \cong (q \pm 1, 3) \times L_2(7)$ is ruled out by Lagrange's theorem for all q. In $SL_3(4)$ with $H_\Omega \cong GL_2(4)$, the only possible S_Ω is $3{\cdot}A_6$. Although $GL_2(4)$ is isomorphic to a subgroup of $3{\cdot}A_6$, it can be checked by direct computation in MAGMA or from [12] that this containment does not arise when extensions of $GL_2(4)$ are candidates for maximality in extensions of $SL_3(4)$. The only other possibilities not ruled out by Lagrange's theorem are in $SU_3(5)$, with $H_\Omega \cong GU_2(5)$, and $S_\Omega \cong 3{\cdot}A_6{\cdot}2_3$. But note that $GU_2(5)'$ is isomorphic to the perfect group $SL_2(5)$, which does not embed into $3{\cdot}A_6$, so this case does not occur either.

    Next suppose that $H \in \mathscr{C}_2$. In Case **L**, $H_\Omega \cong (q-1)^2{:}S_3$ by Table 2.4, and $q \geqslant 5$ by Proposition 2.3.6, which is ruled out by Lagrange's theorem for all S. In Case **U**, $H_\Omega \cong (q+1)^2{:}S_3$ by Table 2.4, which is ruled out by Lagrange's theorem except when $q = 5$ and $S_\Omega \cong 3{\cdot}A_6{\cdot}2_3$ or $3{\cdot}A_7$. It can be checked in MAGMA that $6^2{:}S_3$ occurs as an imprimitive subgroup of $3{\cdot}A_7$ but not of $3{\cdot}A_6{\cdot}2_3$, and that $3{\cdot}A_7$ has a unique class of subgroups of this order. Now the automorphism γ of $SU_3(5)$ fixes the class of $3{\cdot}A_7$, so the containment extends to $H_\Omega.\langle\gamma\rangle < S_\Omega.\langle\gamma\rangle < SU_3(5).\langle\gamma\rangle$. But the class of H_Ω in $SU_3(5)$ is stabilised by the diagonal automorphism δ of $SU_3(5)$ of order 3, whereas that of S_Ω is not, so if $G \in \{GU_3(5), \Gamma U_3(5)\}$ then H is a novel maximal subgroup of G.

    Next suppose that $H \in \mathscr{C}_3$. In Case **L**, the group $H_\Omega \cong (q^2 + q + 1){:}3$ by Table 2.6, which is ruled out by Lagrange's theorem in all cases. In Case **U**, the group $H_\Omega \cong (q^2 - q + 1){:}3$ by Table 2.6, which is ruled out by Lagrange's theorem except when $S_\Omega \cong (q+1, 3) \times L_2(7)$ with $q = 3$ or 5, or $S_\Omega \cong 3{\cdot}A_7$ when $q = 5$. As in the previous case, it can be checked in MAGMA that these are genuine containments that extend to containments under the action of $\langle\gamma\rangle$. But, again as in the preceding paragraph, in the case $q = 5$ if $G \in \{\mathrm{GU}_3(5), \Gamma\mathrm{U}_3(5)\}$ then H is a novel maximal subgroup of G.

If $H \in \mathscr{C}_5$, then the structure of H is given in Table 2.8. The examples in which H is an extension of $\mathrm{SL}_3(q_0)$ or $\mathrm{SU}_3(q_0)$, where q is a power of q_0, are ruled out by Lagrange's theorem. There remains the possibility $(q+1,3) \times \mathrm{SO}_3(q)$ in Case $\mathbf{U}$, where q is odd and $q \neq 3$ by Proposition 3.2.4. This is ruled out by Lagrange's theorem except when $q = 5$ and $S_\Omega \cong 3\dot{}A_6\dot{}2_3$ or $3\dot{}A_7$. As in earlier cases, it can be checked in MAGMA that $(q+1,3) \times \mathrm{SO}_3(q)$ occurs when $q = 5$ as a $\mathscr{C}_5$-subgroup of $3\dot{}A_7$ but not of $3\dot{}A_6\dot{}2_3$, and that the containments extend under the action of $\langle \gamma \rangle$. But in this case the class of H_Ω is not stabilised by δ, and no extensions of H_Ω are novelties of extensions of G.

Next suppose that $H \in \mathscr{C}_6$. Then by Table 2.9 $H_\Omega \cong 3^{1+2}{:}Q_8$ or $3^{1+2}{:}Q_8.3$, and $S_\Omega \cong (q \pm 1, 3) \times \mathrm{L}_2(7)$ is ruled out by Lagrange's theorem. For $S_\Omega \cong 3\dot{}A_6$, $|S_\Omega|$ is divisible by $|3^{1+2}{:}Q_8|$, but the index would be 5, and $3\dot{}A_6$ has no subgroups of index 5. Once again there remains $\mathrm{SU}_3(5)$, and $S_\Omega \cong 3\dot{}A_6\dot{}2_3$ or $3\dot{}A_7$, which again we investigate in MAGMA. This time it turns out that $3^{1+2}{:}Q_8$ is a subgroup of $3\dot{}A_6\dot{}2_3$ but not of $3\dot{}A_7$. This containment extends to $H_\Omega.\langle \gamma \rangle < S_\Omega.\langle \gamma \rangle \cong 3\dot{}A_6\dot{}2^2$ in $\mathrm{SU}_3(5).\langle \gamma \rangle$ but, since S_Ω is not normalised by the diagonal automorphism δ of $\mathrm{SU}_3(5)$, if $G \in \{\mathrm{GU}_3(5), \Gamma\mathrm{U}_3(5)\}$ then H is a novel maximal subgroup of G.

Finally, suppose that $H \in \mathscr{C}_8$. Then by Table 2.11 the group H_Ω is of shape $\mathrm{SU}_3(q^{1/2}).(q^{1/2}-1,3)$ or shape $\mathrm{SO}_3(q).(q-1,3)$ (with q odd). This is ruled out by Lagrange's theorem except when $q = 4$, $H_\Omega \cong \mathrm{SU}_3(2)$ and $S_\Omega \cong 3\dot{}A_6$. But $\mathrm{SU}_3(2) \cong 3^{1+2}{:}Q_8$ which is not isomorphic to a subgroup of $3\dot{}A_6$. $\square$

Proposition 6.3.13 *Let $n = 4$ and let H be maximal amongst the geometric subgroups of G. If $\mathrm{Sp}_4(2^e) \trianglelefteq G$, then assume that $G < \Sigma\mathrm{Sp}_4(2^e)$. Then H is a maximal subgroup of G. Hence, except for when $\mathrm{Sp}_4(2^e) \trianglelefteq G \not\leq \Sigma\mathrm{Sp}_4(2^e)$, the subgroup M of G is a non-trivial maximal subgroup of G if and only if M is listed in Tables 8.8, 8.9, 8.10, 8.11, 8.12 or 8.13.*

Proof Assume that $H \leqslant S < G$, where $S \in \mathscr{S}$. We first deal with $\mathrm{SL}_4^\pm(q)$. Here, by Tables 8.9 and 8.11, S_Ω is a central extension by a subgroup of order dividing 4 of one of the following: $\mathrm{L}_2(7)$, of order $2^3 \cdot 3 \cdot 7$, with $q = p \neq 2, 3, 7$; A_7, of order $2^3 \cdot 3^2 \cdot 5 \cdot 7$, with $q = p \neq 7$; $\mathrm{U}_4(2)$, of order $2^6 \cdot 3^4 \cdot 5$, with $q = p \neq 2, 3$; $\mathrm{L}_3(4)$, of order $2^6 \cdot 3^2 \cdot 5 \cdot 7$ in $\mathrm{SU}_4(3)$.

For $H \in \mathscr{C}_1$, the only containment allowed by Lagrange's theorem is when $\Omega = \mathrm{SL}_4(2)$, with $H_\Omega \cong \mathrm{GL}_3(q)$ and $S_\Omega \cong A_7$. This is a genuine containment, but $\mathrm{GL}_3(2).2$ is not a subgroup of $A_7.2 = A_8$, so $\mathrm{GL}_3(2).2$ remains a novel maximal subgroup of $\mathrm{SL}_4(2).2$.

For $H \in \mathscr{C}_2$, the possible structures for H are given in Table 2.5, and the situations where H exists are given in Propositions 3.3.2 and 3.3.3. The containments consistent with Lagrange's theorem are:

(i) $H_\Omega \cong (q-1)^3.S_4 \leqslant SL_4(7)$ or $H_\Omega \cong (q+1)^3.S_4 \leqslant SU_4(5)$, and the group $S_\Omega \cong 2^{\cdot}U_4(2)$; and

(ii) $H_\Omega \cong SL_2(q^2).(q-1).2$ in $SU_4(3)$, and $S_\Omega \cong 4_2{}^{\cdot}L_3(4)$.

In Case (i), H_Ω would have index 10 in S_Ω but by Theorem 1.11.2 the group $2^{\cdot}U_4(2)$ has no such subgroup. In Case (ii), H_Ω would have index 28 in S_Ω but by Proposition 6.3.12 the group $4_2{}^{\cdot}L_3(4)$ has no such subgroup.

By Table 2.6 there are no groups H in Class $\mathcal{C}_3$ in $SU_4(q)$ and, in $SL_4(q)$, the only potential containment is with $q = 2$, $H_\Omega \cong SL_2(q^2).(q+1).2$ and $S_\Omega \cong A_7$. But A_7 has no subgroup isomorphic to $SL_2(4).3.2$.

Class $\mathcal{C}_4$ is empty and, for all candidates for $H \in \mathcal{C}_5$, by Table 2.8 the only possible Ω allowed by Lagrange's theorem is $SU_4(3)$, where $H_\Omega \cong SO_4^-(3).4$ and $S_\Omega \cong 4_2{}^{\cdot}L_3(4)$. In that case the index would be 28, and by Proposition 6.3.12 $4_2{}^{\cdot}L_3(4)$ has no such subgroup.

Using Tables 2.9 and 2.11, Lagrange's theorem eliminates all options with $H \in \mathcal{C}_6 \cup \mathcal{C}_8$, and Class $\mathcal{C}_7$ is empty. This completes the arguments for $SL_4^\pm(q)$.

In $Sp_4(q)$ with q odd, by Table 8.13 the possibilities for S_Ω are: $S_\Omega \cong 2^{\cdot}A_6$ of order $2^4 \cdot 3^2 \cdot 5$, with $q = p \equiv \pm 5 \bmod 12$; $2^{\cdot}S_6$ of order $2^5 \cdot 3^2 \cdot 5$, with $q = p \equiv \pm 1 \bmod 12$; $2^{\cdot}A_7$ of order $2^4 \cdot 3^2 \cdot 5 \cdot 7$ when $q = 7$; or $SL_2(q)$ of order $q(q^2 - 1)$ with $p \geqslant 5$. All containments are eliminated immediately by Lagrange's theorem except when H is in Class $\mathcal{C}_5$ or $\mathcal{C}_6$ and $S_\Omega \cong SL_2(q)$. But $SL_2(q)$ has no subgroups of the form $Sp_4(q)$ or $2^{1+4}.A_5$ by Proposition 6.3.11.

In $Sp_4(2^e)$, as we saw in Chapters 4 and 5, the only $\mathcal{S}^*$-maximal is $Sz(2^e)$ in $Sp_4(2^e)$ with e odd and $e > 1$. All possible H have order divisible by 3, but $Sz(2^e)$ does not, so there are no containments in this case. $\quad\square$

Proposition 6.3.14 *Let $n = 5$ and let $H < G$ be maximal amongst the geometric subgroups of G. Then H is a non-maximal subgroup of G if and only if $\Omega = SU_5(2)$ and $H_\Omega \cong \frac{q^5+1}{q+1}:5 < L_2(11)$ (Class $\mathcal{C}_3$).*

Hence $M < G$ is a non-trivial maximal subgroup of G if and only if M is listed in Table 8.18, 8.19, 8.20, or 8.21.

Proof Here by Tables 8.19 and 8.21, the group S_Ω is a central extension by a group of order 1 or 5 of one of: $L_2(11)$ (order $2^2 \cdot 3 \cdot 5 \cdot 11$, with $q = p$); $U_4(2)$ (order $2^6 \cdot 3^4 \cdot 5$, with $q = p$ odd); or M_{11} (order $2^4 \cdot 3^2 \cdot 5 \cdot 11$, with $q = 3$). Since there are no $\mathcal{S}_2^*$-maximals, it is straightforward to use Tables 2.3 to 2.11 to check that the only containment consistent with Lagrange's theorem is $\frac{q^5+1}{q+1}:5 < L_2(11)$, in $SU_5(2)$. As can easily be checked in MAGMA (file `containmentsHinS`) or by Proposition 6.3.11, this is a genuine containment that extends to $S_\Omega.2 < H_\Omega.2$ in $SU_5(2).2$. $\quad\square$

Dimension 6 is more laborious to check, on account of the large number of candidates for S_Ω. To assist with the checking, we start by recording the $\mathscr{S}^*$-maximal subgroups, and making a few general observations.

By Tables 8.25, 8.27 and 8.29, if S is a maximal $\mathscr{S}_1$-subgroup then, for some d dividing 6, S_Ω is one of:

(i) $2 \times 3{\cdot}A_6$, of order $2^4 \cdot 3^3 \cdot 5$, with $q = p \geqslant 7$, in $\mathrm{SL}_6^\pm(q)$;

(ii) $2 \times 3{\cdot}A_6.2_3$, of order $2^5 \cdot 3^3 \cdot 5$, with $q = p \geqslant 5$, in $\mathrm{SL}_6^\pm(q)$;

(iii) $6{\cdot}A_6$, of order $2^4 \cdot 3^3 \cdot 5$, with $q = p$ or $q = p^2$, $p \geqslant 5$, in $\mathrm{SL}_6^\pm(q)$;

(iv) $6{\cdot}A_7$, of order $2^4 \cdot 3^3 \cdot 5 \cdot 7$, with $q = p$ or $q = p^2$, $p \geqslant 5$, in $\mathrm{SL}_6^\pm(q)$;

(v) $d \circ 2{\cdot}L_2(11)$, of order dividing $2^3 \cdot 3^2 \cdot 5 \cdot 11$, with $q = p \geqslant 5$, in $\mathrm{SL}_6^\pm(q)$;

(vi) $6{\cdot}L_3(4)$, of order $2^7 \cdot 3^3 \cdot 5 \cdot 7$, with $q = p \geqslant 7$, in $\mathrm{SL}_6^\pm(q)$;

(vii) $6{\cdot}L_3(4){\cdot}2_1$, of order $2^8 \cdot 3^3 \cdot 5 \cdot 7$, with $q = p \geqslant 5$, in $\mathrm{SL}_6^\pm(q)$;

(viii) $3{\cdot}U_4(3).2_2$, of order $2^8 \cdot 3^7 \cdot 5 \cdot 7$ in $\mathrm{SU}_6(2)$;

(ix) $6{\cdot}U_4(3)$, of order $2^8 \cdot 3^7 \cdot 5 \cdot 7$, with $q = p \geqslant 5$, in $\mathrm{SL}_6^\pm(q)$;

(x) $6{\cdot}U_4(3).2_2$, of order $2^9 \cdot 3^7 \cdot 5 \cdot 7$, with $q = p \geqslant 11$, in $\mathrm{SL}_6^\pm(q)$;

(xi) $2{\cdot}M_{12}$, of order $2^7 \cdot 3^3 \cdot 5 \cdot 11$ in $\mathrm{SL}_6(3)$;

(xii) $3{\cdot}M_{22}$, of order $2^7 \cdot 3^3 \cdot 5 \cdot 7 \cdot 11$ in $\mathrm{SU}_6(2)$;

(xiii) $2{\cdot}A_5$, of order $2^3 \cdot 3 \cdot 5$, with $q = p$ odd, in $\mathrm{Sp}_6(q)$;

(xiv) $2{\cdot}S_5$, of order $2^4 \cdot 3 \cdot 5$, with $q = p \geqslant 7$, in $\mathrm{Sp}_6(q)$;

(xv) $2{\cdot}L_2(7)$, of order $2^4 \cdot 3 \cdot 7$, with $q = p$ or $q = p^2$ odd, in $\mathrm{Sp}_6(q)$;

(xvi) $2{\cdot}L_2(7){\cdot}2$, of order $2^5 \cdot 3 \cdot 7$, with $q = p \geqslant 17$, in $\mathrm{Sp}_6(q)$;

(xvii) $2{\cdot}L_2(13)$, of order $2^3 \cdot 3 \cdot 7 \cdot 13$, with $q = p$ or $q = p^2$, p odd, in $\mathrm{Sp}_6(q)$;

(xviii) $U_3(3).2$, of order $2^6 \cdot 3^3 \cdot 7$ in $\mathrm{Sp}_6(2)$;

(xix) $2 \times U_3(3)$, of order $2^6 \cdot 3^3 \cdot 7$, with $q = p \geqslant 7$, in $\mathrm{Sp}_6(q)$;

(xx) $(2 \times U_3(3)).2$, of order $2^7 \cdot 3^3 \cdot 7$, with $q = p \geqslant 11$, in $\mathrm{Sp}_6(q)$;

(xxi) $2{\cdot}J_2$, of order $2^8 \cdot 3^3 \cdot 5^2 \cdot 7$, with $q = p \geqslant 5$ or $q = p^2 \geqslant 9$, in $\mathrm{Sp}_6(q)$;

(xxii) $2{\cdot}A_7$, of order $2^4 \cdot 3^2 \cdot 5 \cdot 7$ in $\mathrm{Sp}_6(9)$;

By Tables 8.25, 8.27 and 8.29, for some d dividing 6, the maximal $\mathscr{S}_2$-subgroups have S_Ω one of: $d \circ \mathrm{SL}_3(q)$ (Case **L**, q odd), $d \circ \mathrm{SU}_3(q)$ (Case **U**, q odd), $\mathrm{SL}_2(q)$ (Case **S**, $p \geqslant 7$), $G_2(q)$ (Case **S**, $p = 2 < q$). The following lemma is now immediate (using Lemma 6.1.2).

Lemma 6.3.15   *Let $n = 6$ and let $S \in \mathscr{S}^*$. If a higher power of the defining characteristic p divides $|S_\Omega|$ than p itself, then the highest such power of p is:*

(i) q^3 *divides* $|\mathrm{SL}_3(q)|$ *in* $\mathrm{SL}_6(q)$;

(ii) q^3 *divides* $|\mathrm{SU}_3(q)|$ *in* $\mathrm{SU}_6(q)$;

(iii) q *divides* $|\mathrm{SL}_2(q)|$ *in* $\mathrm{Sp}_6(q)$;

(iv) q^6 *divides* $|G_2(q)|$ *in* $\mathrm{Sp}_6(q)$ *($q > 2$ even)*;

(v) 2^8 *divides* $|3{\cdot}U_4(3).2_2|$ *in* $\mathrm{SU}_6(2)$;

(vi) 2^7 *divides* $|3{\cdot}M_{22}|$ *in* $\mathrm{SU}_6(2)$;

(vii) 3^3 *divides* $|2^{\cdot}M_{12}|$ *in* $SL_6(3)$;

(viii) 2^6 *divides* $|U_3(3){:}2|$ *in* $Sp_6(2)$;

(ix) 5^2 *divides* $|2^{\cdot}J_2|$ *in* $Sp_6(5)$;

(x) 3^3 *divides* $|2^{\cdot}J_2|$ *in* $Sp_6(9)$;

(xi) 3^2 *divides* $|2^{\cdot}A_7|$ *in* $Sp_6(9)$.

Proposition 6.3.16 *Let* $\Omega = SL_6^{\pm}(q)$, *and let* $H < G$ *be maximal amongst the geometric subgroups of* G. *Then* H *is a non-maximal subgroup of* G *if and only if* $\Omega = SU_6(2)$, *with* $H_\Omega \cong 3^5.S_6 < S_\Omega \cong 3^{\cdot}U_4(3).2_2$ *(𝒞₂), and* $\Omega \leqslant G \leqslant \Omega.\langle\gamma\rangle$.

Hence $M < G$ *is a non-trivial maximal subgroup of* G *if and only if* M *is listed in Table 8.24, 8.25, 8.26, or 8.27.*

Proof We shall now run through the possibilities for the class of the geometric subgroup H. In most cases, $|H|$ is divisible by p^2, and then the candidates for S_Ω can be found from Lemma 6.3.15.

For $H \in \mathscr{C}_1$, the prime-power q^7 always divides $|H|$ by Table 2.3 (even for the novelties in $SL_6(q)$), which eliminates all possible S_Ω except $3^{\cdot}M_{22}$ and $3^{\cdot}U_4(3).2_2$ in $SU_6(2)$. The only possibility allowed by Lagrange's theorem is then $H_\Omega \cong (SU_4(2) \times SU_2(2)).3$, and $S_\Omega \cong 3^{\cdot}U_4(3).2_2$. But then the index would be 42, and $3^{\cdot}U_4(3).2_2$ has no such subgroup by Proposition 6.3.13.

The possible types for $H \in \mathscr{C}_2$ are given in Table 2.4. First suppose that $H_\Omega \cong (q-1)^5.S_6$ (with $q \geqslant 5$ by Proposition 2.3.6) in $SL_6(q)$ or $H_\Omega \cong (q+1)^5.S_6$ in $SU_6(q)$. Then $|H_\Omega|$ does not divide $|SL_3(q)|$ or $|SU_3(q)|$. For $SU_6(2)$, the only possibility allowed by Lagrange's theorem is $S_\Omega \cong 3^{\cdot}U_4(3).2_2$ and it can be checked by direct computation in $SU_6(2)$ in MAGMA (file `containmentsHinS`) that this is a genuine containment. Also, S_Ω has a unique class of maximal subgroups isomorphic to H_Ω, and so this containment extends to $H_\Omega.2 < S_\Omega.2$. But the stabilisers of the classes of S_Ω and H_Ω in this case are respectively $\langle\gamma\rangle$ and $\langle\gamma,\delta\rangle$, so $S_\Omega.3$ and $S_\Omega.[6]$ are novel maximal subgroups in $\Omega.3$ and $\Omega.[6]$.

For $q \pm 1 = 4$ or $q \pm 1 = 6$, the power of 2 dividing $|H_\Omega|$ is higher than that dividing any other possible $|S_\Omega|$. For $q + 1 = 5$, there is no possible $|S_\Omega|$ divisible by 5^5. For all higher values of q, $|H|$ is larger than any possible $|S|$.

Next suppose that $H \in \mathscr{C}_2$ with $H_\Omega \cong SL_2(q)^3.(q-1)^2.S_3$ in $SL_6(q)$ or $H_\Omega \cong SU_2(q)^3.(q+1)^2{:}S_3$ in $SU_6(q)$. Recall that by Proposition 2.3.6 the group H is not maximal when $q = 2$. Since q^3 divides $|H|$, the only examples to be considered are $S_\Omega \cong 2^{\cdot}M_{12}$ in $SL_6(3)$, and $S_\Omega \cong SL_3(q)$ or $SU_3(q)$. But $|H_\Omega| > |S_\Omega|$ in all of these instances.

For the remaining types in Class $\mathscr{C}_2$, the order of H is divisible by q^6, and if $\Omega \neq SU_6(2)$ then there is no suitable S by Lemma 6.3.15. If $S_\Omega \cong 2^{\cdot}M_{12}$, then H of shape $SU_3(2)^2.3.2$ is ruled out by Lagrange's theorem, and H of shape $SL_3(4).2$ would have index 11, a contradiction [12]. It can be checked by computation in MAGMA (file `containmentsHinS`) that $3 \cdot U_4(3).2_2$ has no subgroups with the same order as $SU_3(2)^2.3.2$ or $SL_3(4).2$.

The possible types for $H \in \mathscr{C}_3$ are given in Table 2.6, and in each case q^3 divides $|H_\Omega|$. We can rule out $\Omega \cong SU_6(2)$ by first noting that 3^4 divides the order of H_Ω, but not the order of $3 \cdot M_{22}$, and then by observing from Tables 8.10 and 8.11 (which have now been proved correct, by Proposition 6.3.13) that $U_4(3)$ has no subgroup involving $SU_2(8)$. For $\Omega \neq SU_6(2)$, the groups $2 \cdot M_{12}$ in $SL_6(3)$ and $SL_3^\pm(q)$ are the only possibilities for S_Ω by Lemma 6.3.15, but the first possibility violates Lagrange's theorem, and for the second $|H_\Omega| > |S_\Omega|$.

The possible types for $H \in \mathscr{C}_4$ are given in Table 2.7. By Proposition 2.3.22 this type is not maximal amongst the geometric subgroups when $q = 2$. For each type, q^4 divides $|H|$, and there is no possible S.

The possible types for $H \in \mathscr{C}_5$ are given in Table 2.8. Lagrange's theorem eliminates all possibilities when $\Omega = SU_6(2)$. Otherwise, since p^6 divides $|H|$ in all cases, the only possible S_Ω are $SL_3^\pm(q)$. This is ruled out by Theorem 1.11.5.

Theorem 1.11.5 similarly rules out all possible $H \in \mathscr{C}_8$ in $SL_6(q)$. $\qquad\square$

Proposition 6.3.17 *Let $\Omega = Sp_6(q)$, and let $H < G$ be maximal amongst the geometric subgroups of G. Then H is a maximal subgroup of G.*

Hence $M < G$ is a non-trivial maximal subgroup of G if and only if M is listed in Table 8.28 or 8.29.

Proof For $H \in \mathscr{C}_1$, the prime-power q^9 divides $|H|$ in all types except for H of type $Sp_2(q) \times Sp_4(q)$, when $|H|$ is divisible by q^5 only. So we must consider this H with $S^\infty \cong G_2(q)$ (q even) and with $q = 2$ and $S_\Omega \cong U_3(3).2$. But $q^4 - 1$ divides $|Sp_4(q)|$, so this is ruled out by Lemma 1.13.3 (ii) and 6.1.2. As for the latter case, 5 divides $|H_\Omega|$ but not $|U_3(3).2|$.

The possible types for $H \in \mathscr{C}_2$ are given in Table 2.4. First suppose that $H_\Omega \cong Sp_2(q)^3.S_3$ (with $q \geqslant 3$ by Proposition 2.3.6), so q^3 divides $|H|$. The only possibility is $S_\Omega \cong G_2(q)$ with q even. If $|H|$ divides $|S|$, then $(q^2 - 1)^3$ divides $(q^6 - 1)(q^2 - 1)$, which is not possible when $q \geqslant 3$. The other possible $H \in \mathscr{C}_2$ is $H_\Omega \cong GL_3(q).2$ with q odd. Again q^3 divides $|H|$ so there is no possible S.

The possible types for $H \in \mathscr{C}_3$ are given in Table 2.6. First suppose that $H_\Omega \cong Sp_2(q^3).3$, so q^3 divides $|H|$. Suppose that $S_\Omega \cong G_2(q)$ with $q > 2$ even. It is proved in [80, Theorem 5.2 (i)] that a proper subgroup L of $G_2(q)$ with $q \neq 2$ has order at most $q^6(q^2 - 1)(q - 1)$ unless $q = 3$ and $L \cong G_2(2)$ or $q = 4$ and $L \cong J_2$. Thus if $q \geqslant 4$ then $|H_\Omega| = 3q^3(q^6 - 1) > q^6(q^2 - 1)(q - 1)$, so there are no containments with $S_\Omega \cong G_2(q)$. For the case $q = 2$, note that $Sp_2(8).3$ would be an index 8 subgroup of $G_2(2) \cong U_3(3){:}2$, contradicting Proposition 6.3.12. The other possibility with $H \in \mathscr{C}_3$ is $H_\Omega \cong GU_3(q).2$ with q odd, but then q^3 divides $|H|$ and there is no possible S.

The possible types for $H \in \mathscr{C}_4$ are given in Table 2.7, where we see that

$H_\Omega \cong \mathrm{SO}_3(q) \times \mathrm{Sp}_2(q)$, so q^2 divides $|H|$, with q odd. The only possibility is $S_\Omega \cong 2 \cdot \mathrm{J}_2$ with $q = 5$, but we can check in [12] that $2 \cdot \mathrm{J}_2$ has no subgroup with this structure.

The possible types for $H \in \mathscr{C}_5$ are given in Table 2.8. Here, $H^\infty \cong \mathrm{Sp}_6(q_0)$, where $q = q_0^r$ for some prime r, so that p^9 divides H. It follows from Theorem 1.11.5 that S_Ω is not $\mathrm{SL}_2(q)$, so the only candidate for S_Ω is $\mathrm{G}_2(q)$ with q even. This is impossible by [66, Proposition 5.2.12 (ii)].

Finally, for $H \in \mathscr{C}_8$, the group $H_\Omega \cong \mathrm{SO}_6^\pm(q)$ with q even, by Table 2.11, so that q^6 divides $|H|$. The only possibility is $S_\Omega \cong \mathrm{G}_2(q)$, and this is again ruled out by [66, Proposition 5.2.12 (ii)]. □

Proposition 6.3.18 *Let $n = 7$ and let $H < G$ be maximal amongst the geometric subgroups of G. Then H is a maximal subgroup of G.*

Hence $M < G$ is a non-trivial maximal subgroup of G if and only if M is listed in Table 8.35, 8.36, 8.37, 8.38, 8.39 or 8.40.

Proof Assume that $H \leqslant S < G$, where $S \in \mathscr{S}$.

We first deal with $\mathrm{SL}_7^\pm(q)$. Here, by Tables 8.36 and 8.38, the group S_Ω is a central extension by a subgroup of order 1 or 7 of $\mathrm{U}_3(3)$, and $q = p > 3$. Thus $|S_\Omega|$ divides $2^5 \cdot 3^3 \cdot 7^2$, and it is straightforward to use Tables 2.3 to 2.11 and Lagrange's theorem to show that there are no possible containments.

In the remainder of the proof we consider $\Omega_7(q)$ (so q is odd). By Table 8.40, the group S_Ω is one of: $\mathrm{S}_6(2)$, of order $2^9 \cdot 3^4 \cdot 5 \cdot 7$, with $q = p$; S_9, of order $2^7 \cdot 3^4 \cdot 5 \cdot 7$, when $q = 3$; $\mathrm{G}_2(q)$, of order $q^6(q^2 - 1)(q^6 - 1)$ by Lemma 6.1.2, for all q.

Consider first the $\mathscr{S}_1$-subgroups. By Proposition 2.3.2 the $\mathscr{C}_1$-subgroup $(\Omega_2^+(3) \times \Omega_5(3)).[4]$ is not maximal amongst the geometric groups. Consulting Tables 2.3 to 2.11, the only containments not ruled out by Lagrange's theorem are when $q = 3$, namely $(\Omega_2^-(q) \times \Omega_5(q)).[4] < \mathrm{S}_6(2)$ with index 7, $(\Omega_3(q) \times \Omega_4^\pm(q)).[4] < \mathrm{S}_6(2)$ with index 105 or 84, or $(\Omega_3(q) \times \Omega_4^-(q)).[4] < \mathrm{S}_9$ with index 21, or $2^6.\mathrm{A}_7 < \mathrm{S}_6(2)$ with index 9. Containments are ruled out in $\mathrm{S}_6(2)$ by Proposition 6.3.17, and in S_9 by considering its maximal subgroups [12].

Next consider $S_\Omega \cong \mathrm{G}_2(q)$. It is a straightforward calculation, using Table 2.3, to show that $H \not\leqslant \mathscr{C}_1$, so only Classes $\mathscr{C}_2$ and $\mathscr{C}_5$ need be considered. For some values of q there are potential containments with $H_\Omega \cong 2^6.\mathrm{A}_7$ or $2^6.\mathrm{S}_7$ (Class $\mathscr{C}_2$) or $\Omega_7(q_0).(r, 2)$ (Class $\mathscr{C}_5$), but these are ruled out by the maximal subgroups of $\mathrm{G}_2(q)$ and $^2\mathrm{G}_2(q)$ given in [64], reproduced here as Tables 8.41 and 8.42. From these tables, the only non-abelian composition factors of maximal subgroups of these groups are $\mathrm{G}_2(q_0)$ and $^2\mathrm{G}_2(q_0)$ for a subfield $\mathbb{F}_{q_0}$ of $\mathbb{F}_q$, $\mathrm{L}_3^\pm(q_0)$, $\mathrm{L}_2(t)$ for various t, $\mathrm{U}_3(3)$ and J_1. By applying Proposition 6.3.12 to $\mathrm{L}_3(q)$, $\mathrm{L}_3(2)$, $\mathrm{U}_3(q)$ and $\mathrm{U}_3(3)$, and then considering the maximal subgroups of

J_1 [12], we see that none of these can involve $\Omega_7(q_0)$, and only $U_3(5^e)$ (which contains $3\dot{\,}A_7$) involves A_7 (note that A_7 is a subgroup of the $\mathscr{C}_2$-subgroup $2^7.A_7$, by Lemma 2.2.4 (i)). But $U_3(5^e)$ has no subgroup $2^6.A_7$. □

Proposition 6.3.19 *Let $n = 8$ and let $H < G$ be maximal amongst the geometric subgroups of G. Then H is a maximal subgroup of G.*

Hence $M < G$ is a non-trivial maximal subgroup of G if and only if M is listed in Table 8.44, 8.45, 8.46, 8.47, 8.48, 8.49, 8.52 or 8.53.

Proof Assume that $H \leqslant S < G$, where $S \in \mathscr{S}$.

We first deal with $SL_8^{\pm}(q)$. Here, by Tables 8.45 and 8.47, the group S_Ω is a central extension by a group of order dividing 8 of $L_3(4)$ or $L_3(4).2_3$, and $q > 3$ is odd. Thus $|S_\Omega|$ divides $2^{10} \cdot 3^2 \cdot 5 \cdot 7$, and it is straightforward to use Tables 2.3 to 2.11 and Lagrange's theorem to eliminate all possible containments.

In $Sp_8(q)$, by Table 8.49, the group S_Ω is a central extension by a subgroup of order 1 or 2 of one of: $L_2(7)$ (order $2^3 \cdot 3 \cdot 7$); $L_2(7).2$; A_6 (order $2^3 \cdot 3^2 \cdot 5$); $A_6.2_2$; $L_2(17)$ (order $2^4 \cdot 3^2 \cdot 17$, with $q = 2$ or q odd); S_{10} (order $2^8 \cdot 3^4 \cdot 5^2 \cdot 7$, with $q = 2$); $L_2(q)$; $L_2(q^3).3$; with p odd unless otherwise indicated. For $q = 2$ the result follows easily from Lagrange's theorem, so assume from now on that q is odd. Then no $|S_\Omega|$ is divisible by a higher power of q than q^3, so $H \notin \mathscr{C}_1 \cup \mathscr{C}_2 \cup \mathscr{C}_3 \cup \mathscr{C}_8$. For $\mathscr{C}_4$ we note from Table 2.7 that q^3 divides $|H_\Omega|$, so $S_\Omega \cong SL_2(q^3)$. By Proposition 3.7.7 the group H is of type $Sp_2(q) \otimes GO_4^-(q)$, so $|H_\Omega|$ is divisible by a Zsigmondy prime for $q^4 - 1$, which does not divide $|S_\Omega|$ by Lemma 1.13.3 (ii). For $\mathscr{C}_5$ we use Theorem 1.11.5. For $\mathscr{C}_6$ we first note that 2^{13} divides $|H_\Omega|$, so $S \in \mathscr{S}_2$. However, by Proposition 6.3.11, for no prime power t does $SL_2(t)$ contain $2^{1+6}.\Omega_6^-(2) \cong 2^{1+6}.S_4(3)$. Finally, for $\mathscr{C}_7$ we note from Table 2.10 that q^3 divides $|H_\Omega|$ and q is odd, so that $S_\Omega \cong SL_2(q^3).3$. But then $(q^2 - 1)^3$ divides $|H_\Omega|$ but not $|S_\Omega|$.

In $\Omega_8^-(q)$, by Table 8.53 the group S_Ω is $L_3^{\pm}(q)$. Classes $\mathscr{C}_2$, $\mathscr{C}_4$, $\mathscr{C}_6$, $\mathscr{C}_7$ and $\mathscr{C}_8$ are all empty. For Classes $\mathscr{C}_1$ and $\mathscr{C}_3$, we use Lagrange's theorem, whilst for Class $\mathscr{C}_5$ we use Theorem 1.11.5. □

Proposition 6.3.20 *Let $n = 9$ and let $H < G$ be maximal amongst the geometric subgroups of G. Then H is a maximal subgroup of G.*

Hence $M < G$ is a non-trivial maximal subgroup of G if and only if M is listed in Table 8.54, 8.55, 8.56, 8.57, 8.58, or 8.59.

Proof Assume that $H \leqslant S < G$, where $S \in \mathscr{S}$.

We first deal with $SL_9^{\pm}(q)$. Here, by Tables 8.55 and 8.57, the group S_Ω is a central extension by a group of order dividing 9 of one of: $L_2(19)$ (order $2^2 \cdot 3^2 \cdot 5 \cdot 19$), $A_6.2_3$ (order $2^4 \cdot 3^2 \cdot 5$), A_7 (order $2^3 \cdot 3^2 \cdot 5 \cdot 7$), J_3 (order $2^7 \cdot 3^5 \cdot 5 \cdot 17 \cdot 19$), $L_3(q^2).2$ or $L_3(q^2).[6]$. The case $q = 2$ is easy, so assume $q > 2$. Then no $\mathscr{S}$-group has order divisible by a higher power of q than q^6. This immediately eliminates

$\mathscr{C}_1$, the decomposition into 3 subspaces in $\mathscr{C}_2$, $\mathscr{C}_3$ and $\mathscr{C}_8$. For the other type of $\mathscr{C}_2$-subgroup, we note that $A_9 \leqslant H$ by Lemma 2.2.4 (i), but is not a subgroup of any of the $\mathscr{S}^*$-subgroups by Lagrange's Theorem and Proposition 1.11.6. Class $\mathscr{C}_4$ is empty. For Classes $\mathscr{C}_5$ and $\mathscr{C}_7$ we use Theorem 1.11.5, whilst for Class $\mathscr{C}_6$ we use Proposition 6.3.12 to see that $L_3(q^2)$ has no subgroups of shape $3^{1+4}.\mathrm{Sp}_4(3)$.

In $\Omega_9(q)$, by Table 8.59 the group S_Ω is one of $L_2(8)$ (order $2^3 \cdot 3^2 \cdot 7$), $L_2(17)$ (order $2^4 \cdot 3^2 \cdot 17$), A_{10} (order $2^7 \cdot 3^4 \cdot 5^2 \cdot 7$), S_{10}, S_{11} (order $2^8 \cdot 3^4 \cdot 5^2 \cdot 7 \cdot 11$), $L_2(q).2$ or $L_2(q^2).2$. The case $q = 3$ is straightforward, so assume that $q \geqslant 5$. Then no $\mathscr{S}$-group is divisible by a higher power of q than q^2. Lagrange's theorem eliminates Class $\mathscr{C}_1$, the $\mathscr{C}_2$-subgroup of type $\mathrm{GO}_3(q) \wr S_3$, and Class $\mathscr{C}_3$. For H^∞ of shape $2^8.A_9$ we argue as in the previous paragraph. Classes $\mathscr{C}_4$, $\mathscr{C}_6$ and $\mathscr{C}_8$ are empty. For Classes $\mathscr{C}_5$ and $\mathscr{C}_7$ we use Lagrange's theorem and Theorem 1.11.5. $\quad\square$

Proposition 6.3.21 *Let $n = 10$ and let $H < G$ be maximal amongst the geometric subgroups of G. Then H is a maximal subgroup of G.*

Hence $M < G$ is a non-trivial maximal subgroup of G if and only if M is listed in Table 8.60, 8.61, 8.62, 8.63, 8.64, 8.65, 8.66, 8.67, 8.68 or 8.69.

Proof Assume that $H \leqslant S < G$, where $S \in \mathscr{S}$.

We first deal with $\mathrm{SL}_{10}^{\pm}(q)$. Here, by Tables 8.61 and 8.63, for some d dividing 10, the group S_Ω is a central extension by a group of order d of one of: $L_2(19)$ (order dividing $2^3 \cdot 3^2 \cdot 5^2 \cdot 19$); M_{12} (order dividing $2^7 \cdot 3^3 \cdot 5^2 \cdot 11$); $M_{12}.2$; M_{22} (order dividing $2^8 \cdot 3^2 \cdot 5^2 \cdot 7 \cdot 11$); $M_{22}.2$; $L_3(4)$ (order dividing $2^7 \cdot 3^2 \cdot 5^2 \cdot 7$); $L_3(4).2_2$; $L_3(q).(q-1,3)$; $L_4(q).\frac{(q-1,4)}{2}$; $L_5(q)$; $U_3(q).(q+1,3)$; $U_4(q).\frac{(q+1,4)}{2}$; $U_5(q)$. Thus the highest power of q to divide $|S_\Omega|$ is q^{10}, which immediately eliminates $\mathscr{C}_1$, all decompositions into two blocks in $\mathscr{C}_2$, the $\mathscr{C}_3$-subgroup preserving a field extension of degree two in $\mathrm{SL}_{10}(q)$, and Class $\mathscr{C}_8$. For the $\mathscr{C}_2$ decomposition into ten blocks, we use Proposition 1.11.6 and Lagrange's theorem. For the $\mathscr{C}_2$ decomposition into five blocks, Lagrange's theorem yields that $S \in \mathscr{S}_2$, and with a bit more work gives a contradiction. The $\mathscr{C}_3$ groups preserving a field extension of degree 5 are divisible both by q^5 and by (in Case **U** the square of) a Zsigmondy prime for $q^{10} - 1$, which is false for all possible S_Ω. Classes $\mathscr{C}_4$, $\mathscr{C}_6$ and $\mathscr{C}_7$ are empty. For Class $\mathscr{C}_5$ we use Lagrange's theorem and Theorem 1.11.5.

In $\mathrm{Sp}_{10}(q)$, by Table 8.65 the prime power q is odd and S_Ω is one of $2\,\dot{}A_6$ (order $2^4 \cdot 3^2 \cdot 5$), $2\,\dot{}A_6.2_2$, $\mathrm{SL}_2(11)$ (order $2^3 \cdot 3 \cdot 5 \cdot 11$), $\mathrm{SL}_2(11).2$, $2 \times U_5(2)$ (order $2^{11} \cdot 3^5 \cdot 5 \cdot 11$), $(2 \times U_5(2)).2$, or $\mathrm{SL}_2(q)$. When $q = 3$, by Lagrange's theorem the only possible containment is $\mathrm{SO}_5(3) \times \mathrm{Sp}_2(3) < 2 \times U_5(2)$, but by Proposition 6.3.14 the group $U_5(2)$ has no such subgroup, so assume that $q \geqslant 5$. Then the highest power of q to divide $|S_\Omega|$ is q, which is sufficient to eliminate Classes $\mathscr{C}_1$, $\mathscr{C}_2$, $\mathscr{C}_3$, $\mathscr{C}_4$ and $\mathscr{C}_8$. For Class $\mathscr{C}_5$ we use Lagrange's theorem and Theorem 1.11.5. Classes $\mathscr{C}_6$ and $\mathscr{C}_7$ are empty.

In $\Omega_{10}^{\pm}(q)$, by Tables 8.67 and 8.69, the group S_Ω is a central extension by a group of order 1 or 2 of one of: A_6 (order $2^3 \cdot 3^2 \cdot 5$), $A_6.2_1$, $L_2(11)$ (order $2^2 \cdot 3 \cdot 5 \cdot 11$), A_7 (order $2^3 \cdot 3^2 \cdot 5 \cdot 7$), A_{11} (order $2^7 \cdot 3^4 \cdot 5^2 \cdot 7 \cdot 11$, with q odd), A_{12} (order $2^9 \cdot 3^5 \cdot 5^2 \cdot 7 \cdot 11$, in $\Omega_{10}^+(3)$ and $\Omega_{10}^-(2)$), M_{12} (order $2^6 \cdot 3^3 \cdot 5 \cdot 11$, with $q = 2$ and Case $\mathbf{O}^-$), M_{22} (order $2^7 \cdot 3^2 \cdot 5 \cdot 7 \cdot 11$, with $q = 7$), $L_3(4)$ (order $2^6 \cdot 3^2 \cdot 5 \cdot 7$, with $q = 7$), or $S_4(q)$ (q odd). In $\Omega_{10}^-(2)$ the only possibility given by Lagrange's theorem is a containment of $(\Omega_4^-(q) \times \Omega_6^+(q)).2$ in $A_{12} < \Omega_{10}^-(2)$, which would be of index 99. However, investigations using MAGMA (file `containmentsHinS`) reveal that A_{12} has no such subgroup. So assume that $\Omega \neq \Omega_{10}^-(2)$. Then the highest power of q to divide $|S_\Omega|$ is q^5, which eliminates Class $\mathscr{C}_1$. For Class $\mathscr{C}_2$, Lagrange's theorem implies that H preserves a decomposition into 5 or 10 subspaces. For $H^\infty \cong 2^9.A_{10}$, Lagrange's theorem implies that the group $S \cong S_4(q)$ (with q odd), but by Lemma 2.2.4 (i), the group H^∞ contains A_{10}, which has no four-dimensional representations by Proposition 1.11.6. For H of type $\mathrm{GO}_2^+(q) \wr S_5$ we note from Proposition 3.9.3 we require $q \geqslant 5$ for H to be maximal amongst the geometric subgroups. Lagrange's theorem then eliminates both groups preserving a decomposition into 5 subspaces, and Class $\mathscr{C}_3$, whilst Theorem 1.11.5 deals with $\mathscr{C}_5$. Classes $\mathscr{C}_4$, $\mathscr{C}_6$, $\mathscr{C}_7$ and $\mathscr{C}_8$ are empty.   □

Proposition 6.3.22 *Let $n = 11$ and let $H < G$ be maximal amongst the geometric subgroups of G. Then H is a maximal subgroup of G.*

Hence $M < G$ is a non-trivial maximal subgroup of G if and only if M is listed in Table 8.70, 8.71, 8.72, 8.73, 8.74 or 8.75.

Proof Assume that $H \leqslant S < G$, where $S \in \mathscr{S}$.

We first deal with $\mathrm{SL}_{11}^{\pm}(q)$. Here, by Tables 8.71 and 8.73, the group S_Ω is a central extension by a subgroup of order 1 or 11 of one of: $U_5(2)$ (order $2^{10} \cdot 3^5 \cdot 5 \cdot 11$), $L_2(23)$ (order $2^3 \cdot 3 \cdot 11 \cdot 23$) or M_{24} (order $2^{10} \cdot 3^3 \cdot 5 \cdot 7 \cdot 11 \cdot 23$). Since there are no $\mathscr{S}_2$-groups, it is straightforward to use Tables 2.3 to 2.11, together with Lagrange's theorem, to see that there are no possible containments.

In $\Omega_{11}(q)$, by Table 8.75 the prime power $q \geqslant 5$ and S_Ω is one of: $L_3(3).2$ (order $2^5 \cdot 3^3 \cdot 13$), A_{12} (order $2^9 \cdot 3^5 \cdot 5^2 \cdot 7 \cdot 11$), S_{12}, A_{13} (order $2^9 \cdot 3^5 \cdot 5^2 \cdot 7 \cdot 11 \cdot 13$) or $L_2(q)$. Lagrange's theorem shows that $H \notin \mathscr{C}_1$. For Class $\mathscr{C}_2$, Lagrange's theorem implies that $S_\Omega = L_2(q)$, but by Proposition 1.11.6 the group A_{11} has no 2-dimensional representation. Classes $\mathscr{C}_3$, $\mathscr{C}_4$, $\mathscr{C}_6$, $\mathscr{C}_7$ and $\mathscr{C}_8$ are empty, whilst Theorem 1.11.5 eliminates Class $\mathscr{C}_5$.   □

Proposition 6.3.23 *Let $n = 12$ and let $H < G$ be maximal amongst the geometric subgroups of G. Then H is a maximal subgroup of G.*

Hence $M < G$ is a non-trivial maximal subgroup of G if and only if M is listed in Table 8.76, 8.77, 8.78, 8.79, 8.80, 8.81, 8.82, 8.83, 8.84, or 8.85.

Proof Assume that $H \leqslant S < G$, where $S \in \mathscr{S}$.

We first deal with $\mathrm{SL}_{12}^{\pm}(q)$. Here, by Tables 8.77 and 8.79, the group S_{Ω} is a central extension by a group of order dividing 12 of one of: A_6 (order dividing $2^5 \cdot 3^3 \cdot 5$); $L_2(23)$ (order dividing $2^5 \cdot 3^2 \cdot 11 \cdot 23$); Suz (order dividing $2^{15} \cdot 3^8 \cdot 5^2 \cdot 7 \cdot 11 \cdot 13$); $L_3(4)$ (order dividing $2^8 \cdot 3^3 \cdot 5 \cdot 7$) with $q = 49$. Thus if $q > 2$ then the highest power of q to divide $|S_{\Omega}|$ is q^2. We note that by Proposition 3.11.2 the groups of type $\mathrm{GL}_1(q) \wr S_{12}$ are non-maximal when $q < 5$, and that by Proposition 3.11.3 the groups of type $\mathrm{GL}_2(q) \wr S_6$ and $\mathrm{GU}_2(q) \wr S_6$ are non-maximal when $q = 2$. Using this, Lagrange's theorem easily eliminates all possibilities for H except H preserving a tensor decomposition into a 3-space and a 4-space, inside $\mathrm{SU}_{12}(2)$. Here there is a potential containment $\mathrm{SU}_3(2) \times \mathrm{SU}_4(2) \leqslant 3\dot{\,}\mathrm{Suz}$ with index 240240, but by [112] (or [12]) the group Suz has no maximal subgroup with index dividing 240240.

In $\mathrm{Sp}_{12}(q)$, by Table 8.81 the group S_{Ω} is one of: $\mathrm{SL}_2(11)$ (order $2^3 \cdot 3 \cdot 5 \cdot 11$), $\mathrm{SL}_2(11).2$, $\mathrm{SL}_2(13)$ (order $2^3 \cdot 3 \cdot 7 \cdot 13$), $\mathrm{SL}_2(13).2$, $\mathrm{SL}_2(25)$ (order $2^4 \cdot 3 \cdot 5^2 \cdot 13$), $L_2(25).2_2$ (with $q = 2$), $\mathrm{Sp}_4(5)$ (order $2^7 \cdot 3^2 \cdot 5^4 \cdot 13$), $S_4(5)$, $2\dot{\,}G_2(4)$ (order $2^{13} \cdot 3^3 \cdot 5^2 \cdot 7 \cdot 13$), $2\dot{\,}G_2(4).2$, $2\dot{\,}\mathrm{Suz}$ (order $2^{14} \cdot 3^7 \cdot 5^2 \cdot 7 \cdot 11 \cdot 13$), S_{14} (order $2^{11} \cdot 3^5 \cdot 5^2 \cdot 7^2 \cdot 11 \cdot 13$), $\mathrm{SL}_2(q)$ or $\mathrm{Sp}_4(q)$. By Lagrange's theorem, Proposition 3.11.3 and Proposition 3.11.8, $q \neq 2$ and if $q = 3$ then the only potential containments are $H_{\Omega} \cong (\mathrm{Sp}_2(3) \circ \mathrm{GO}_6^{\pm}(3)).2$ and $S_{\Omega} \cong 2\dot{\,}\mathrm{Suz}$. However, the indices would be smaller than that of any maximal subgroup of Suz. So we assume that $q > 3$, so the highest power of q dividing $|S_{\Omega}|$ is q^4. Using this, Lagrange's theorem eliminates all possibilities for H in Classes $\mathscr{C}_1$, $\mathscr{C}_2$, $\mathscr{C}_3$, $\mathscr{C}_4$ and $\mathscr{C}_8$. Class $\mathscr{C}_5$ is eliminated by Lagrange's theorem and Theorem 1.11.5.

In $\Omega_{12}^{\pm}(q)$, by Tables 8.83 and 8.85 the group S_{Ω} is a central extension by a group of order at most 2 of one of: $L_2(11)$ (order $2^2 \cdot 3 \cdot 5 \cdot 11$), $L_2(13)$ (order $2^2 \cdot 3 \cdot 7 \cdot 13$), $L_3(3)$ (order $2^4 \cdot 3^3 \cdot 13$), $L_3(3).2$, M_{12} (order $2^6 \cdot 3^3 \cdot 5 \cdot 11$), $M_{12}.2$, A_{13} (order $2^9 \cdot 3^5 \cdot 5^2 \cdot 7 \cdot 11 \cdot 13$), A_{14} (order $2^{10} \cdot 3^5 \cdot 5^2 \cdot 7^2 \cdot 11 \cdot 13$). Thus if $q \geqslant 3$ then p^5 is the maximum power of p to divide $|S_{\Omega}|$. By Proposition 3.11.3 the groups of type $\mathrm{GO}_2^+(q) \wr S_6$ are non-maximal for $q \leqslant 3$, by Proposition 3.11.8 the groups of type $\mathrm{GO}_4^+(q) \otimes \mathrm{GO}_3(q)$ are never maximal, and the groups of type $\mathrm{GO}_4^-(q) \otimes \mathrm{GO}_3(q)$ are non-maximal when $q = 3$. It is now straightforward to use Lagrange's theorem to eliminate all possible containments except for H_{Ω} of shape $\Omega_4^-(8).3 = L_2(64).3$ in $S_{\Omega} = A_{13}$ in $\Omega_{12}^-(2)$. Since $L_2(64)$ has no faithful permutation representation on 13 points, we are done. $\qquad \square$

7

Maximal subgroups of exceptional groups

7.1 Introduction

Aschbacher's theorem does not apply to certain extensions of $S_4(2^i)$ and $O_8^+(q)$; that is, those that involve the exceptional graph automorphism and the triality graph automorphism, respectively. Since the $O_8^+(q)$ case is fully handled in [62], we need not concern ourselves with that. Although Aschbacher's paper [1] does include some results about the extensions of the maximal subgroups of $S_4(2^i)$ in question, we prefer to determine these independently, and we do that in this chapter. We also describe the maximal subgroups of the finite almost simple exceptional groups that have a faithful projective representation in defining characteristic of degree at most 12, namely those with socles $^2B_2(q) = Sz(q)$, $G_2(q)$, $^2G_2(q) = R(q)$ and $^3D_4(q)$. Our principal motivation for doing this is that, if the maximal subgroups of a group G are known, then for many applications it is useful to know also the maximal subgroups of all composition factors of subgroups of G.

Most of the results we need are in the literature, though we have chosen to provide our own proofs in some cases. In general we do so either because only the simple, but not almost simple, groups were treated originally; or because we believe that our proof may be clearer than one already in the literature.

The groups that we shall consider are almost simple extensions of

$$S_4(2^e), \; O_8^+(q), \; Sz(2^e) = {}^2B_2(2^e), \; G_2(q), \; R(3^e) = {}^2G_2(3^e), \; {}^3D_4(q),$$

as it is easy to check that these are the only simple groups of Lie type with faithful projective representations in defining characteristic of degree at most 12 to which Aschbacher's theorem does not apply. Other groups occur in dimension at least 25, and apart from Malle's work [90] on $^2F_4(2^e)$, maximal subgroup classifications are incomplete, although significant information is available.

The maximal subgroups of $Sz(2^e)$, $G_2(q)$, $R(3^e)$, $^3D_4(q)$ are listed in Tables 8.16; 8.30 and 8.41; 8.42; 8.43 and 8.51, respectively. The Aschbacher classes

357

$\mathscr{C}_i$ for these groups have not been formally defined, so the classes listed in the tables should be regarded as being for informal guidance only. Our notation for their outer automorphisms requires some explanation. We shall show in Lemma 7.3.2 that $\operatorname{Out} \operatorname{Sz}(q)$ is generated by the field automorphism ϕ. From [10, Chapter 12], $\operatorname{Out} \operatorname{G}_2(q)$ is generated by the field automorphism ϕ when $p \neq 3$, whereas $\operatorname{Out} \operatorname{G}_2(3^e) \cong \operatorname{C}_{2e}$ is generated by the graph automorphism γ with $\gamma^2 = \phi$. The group $\operatorname{Out} \operatorname{R}(3^e)$ is generated by the field automorphism ϕ [64]. Finally, $\operatorname{Out} {}^3\operatorname{D}_4(q) \cong \operatorname{C}_{3e}$ is generated by the field automorphism ϕ, with $\phi^e = \tau$, the graph automorphism of order 3 [63].

The maximal subgroups of $\operatorname{O}_8^+(q)$ for all q, $\operatorname{G}_2(q)$ for q odd, $\operatorname{R}(3^e)$, and ${}^3\operatorname{D}_4(q)$ for all q, are determined by Kleidman [62, 64, 63], as are the maximal subgroups of all almost simple extensions of these groups. Thus, we consider first $\operatorname{S}_4(2^e)$, then $\operatorname{Sz}(2^e)$, and finally $\operatorname{G}_2(2^e)$.

We recall Definitions 1.3.9 and 1.3.11 of type 1 and type 2 novel maximal subgroup. When working with almost simple extensions of $\operatorname{Sz}(2^e)$ and $\operatorname{G}_2(2^e)$ we need to show that there are no type 2 novelties, and will use the following.

Lemma 7.1.1 *Let $M = \operatorname{N}_G(H)$ be a type 2 novelty of an almost simple group G with socle S, such that $H = \operatorname{N}_S(H) < K < S$, where K is maximal in S. Then:*

(i) $\operatorname{N}_S(N) = H$ *for any non-trivial characteristic subgroup N of H.*

(ii) *There exists $H_0 < K$ such that H and H_0 are conjugate in S but not K.*

(iii) *H has no non-trivial normal Sylow p-subgroups.*

Proof The first part follows from maximality of M and the second from Proposition 1.3.10. If $P \in \operatorname{Syl}_p(H)$ with $P \trianglelefteq H$ then since, by (i), $H = \operatorname{N}_S(P)$, we must have $P \in \operatorname{Syl}_p(S)$. But then Sylow's Theorem in K contradicts (ii). $\square$

7.2 The maximal subgroups of $\operatorname{Sp}_4(2^e)$ and extensions

Let $q = 2^e > 2$ be even. In the earlier chapters of this book, we have determined the maximal subgroups of those almost simple extensions of $\operatorname{Sp}_4(q)$ ($\cong \operatorname{S}_4(q)$) that are subgroups of $\Sigma\operatorname{Sp}_4(q)$, and have determined their stabilisers in $\Sigma\operatorname{Sp}_4(q)$. Recall that $\operatorname{Sp}_4(2)$ is not quasisimple, so is excluded from our classification.

The group $\operatorname{Sp}_4(q)$ has an additional *graph automorphism* γ with $\gamma^2 = \phi$, and it remains to determine the maximal subgroups of almost simple extensions of $\operatorname{Sp}_4(q)$ that are not contained in $\Sigma\operatorname{Sp}_4(q)$: to do so we first need to describe the action of γ on elements of $\operatorname{Sp}_4(q)$. The group $\operatorname{Out} \operatorname{Sp}_4(q) = \langle \gamma \rangle \cong \operatorname{C}_{2e}$, where γ^2 is the field automorphism ϕ, which maps matrix entries x to x^2.

Proposition 7.2.1 *The classes of maximal subgroups of* $\Sigma\mathrm{Sp}_4(2^e)$ *(with* $e > 1$*) are as described in Table 8.14. Each class* C *of maximal subgroups of* $\Sigma\mathrm{Sp}_4(2^e)$ *is stabilised by the field automorphism* ϕ*, and the action of the outer automorphism* γ *of* $\mathrm{Sp}_4(2^e)$ *on* C *is as stated in the table.*

Proof The claims regarding the maximal subgroups of $\Sigma\mathrm{Sp}_4(2^e)$ in class $\mathscr{S}$ follow from Theorem 6.3.1 for subgroups containing $\mathrm{Sz}(2^e)$, and from Proposition 6.3.13 for the other subgroups. The fusion of the classes under γ can be found in [1, Section 14]. □

The aim of this section is to find the remaining (novel) maximals of almost simple groups with socle $\mathrm{Sp}_4(2^e)$ that are not contained in $\Sigma\mathrm{Sp}_4(2^e)$, and hence to prove the correctness of all rows of Table 8.14. Although we could shorten some of our arguments by making use of [1, Theorem 14], it seems worthwhile to present a complete and independent analysis.

For now, let F be any field of characteristic 2, and let the symplectic form for $\mathrm{Sp}_4(F)$ be $\mathrm{antidiag}(1, 1, 1, 1)$. Label the basis for the underlying vector space $V = F^4$ as (e_1, e_2, e_3, e_4). We choose a rather generous generating set for $\mathrm{Sp}_4(F)$ consisting of the following elements:

(i) the permutation matrices corresponding to $\langle (1,2)(3,4), (2,3) \rangle \cong D_8$;
(ii) the diagonal matrices $D(\kappa, \lambda) := \mathrm{diag}(\kappa, \lambda, \lambda^{-1}, \kappa^{-1})$ for $\kappa, \lambda \in F^\times$;
(iii) the unipotent matrices $T(a, b, c, d)$ given below, for $a, b, c, d \in F$.

The matrix $T(a, b, c, d)$ is defined by

$$
T(a, b, c, d) := \begin{bmatrix} 1 & 0 & 0 & 0 \\ a & 1 & 0 & 0 \\ ac+b & c & 1 & 0 \\ d & b & a & 1 \end{bmatrix}.
$$

The Borel subgroup B of $\mathrm{Sp}_4(F)$ is the subgroup generated by all the T-elements and D-elements. We note that the only submodules for B are

$$
0 < \langle e_1 \rangle < \langle e_1, e_2 \rangle < \langle e_1, e_2, e_3 \rangle < V.
$$

7.2.1 The outer automorphisms of $\mathrm{Sp}_4(2^e)$

We start by exploring the action of the graph automorphism γ on some geometries associated with $\mathrm{Sp}_4(2^e)$. Recall Definition 1.4.4 of a perfect field, and that all algebraically closed fields and all finite fields are perfect. There is an analogue of the map γ for $\mathrm{Sp}_4(F)$, where F is any field of characteristic 2.

Consider the action of $\mathrm{Sp}_4(F)$ on $\Lambda^2(V)$ with the basis

$$(e_1 \wedge e_4 + e_2 \wedge e_3, e_1 \wedge e_2, e_1 \wedge e_3, e_2 \wedge e_4, e_3 \wedge e_4, e_1 \wedge e_4).$$

Be warned that this is *not* the basis given in Definition 5.2.2. By Proposition 5.2.4, $\mathrm{Sp}_4(F)$ is a group of isometries of the bilinear form on $\Lambda^2(V)$ with matrix $A = \mathrm{antidiag}(1, 1, 1, 1, 1, 1) + \mathrm{E}_{6,6}$.

We claim that $\Lambda^2(V)$ has submodules

$$W := \langle e_1 \wedge e_4 + e_2 \wedge e_3 \rangle \quad \text{and} \quad U := \langle W, e_1 \wedge e_2, e_1 \wedge e_3, e_2 \wedge e_4, e_3 \wedge e_4 \rangle.$$

This is easily verified for W (by considering the generators specified above), and for U we note that $U = W^{\perp}$ with respect to the form A.

Thus, $g \in \mathrm{Sp}_4(F)$ also has a natural action on U/W, and we denote by g^* the matrix of g in this action, with respect to the following basis of U/W:

$$(e_1 \wedge e_2 + W, \; e_1 \wedge e_3 + W, \; e_2 \wedge e_4 + W, \; e_3 \wedge e_4 + W).$$

Lemma 7.2.2 *Let F be any field of characteristic 2. Then the map $\gamma : g \mapsto g^*$ induces an endomorphism on $\mathrm{Sp}_4(F)$, and acts as follows:*

$$T(a, b, c, d) \mapsto T(c, a^2 c + ab + d, a^2, abc + b^2 + cd), \quad D(\kappa, \lambda) \mapsto D(\kappa\lambda, \kappa\lambda^{-1}),$$

$$(1, 4) \leftrightarrow (1, 3)(2, 4), \quad (2, 3) \leftrightarrow (1, 2)(3, 4), \quad (1, 2, 4, 3) \leftrightarrow (1, 3, 4, 2),$$

with $(1, 4)(2, 3)^{\gamma} = (1, 4)(2, 3)$. Furthermore, $\gamma \notin \Sigma\mathrm{Sp}_4(F)$, and $\gamma^2 = \phi$.

Proof We see from the form matrix A that g^* lies in $\mathrm{Sp}_4(F)$, and so γ induces an endomorphism on $\mathrm{Sp}_4(F)$. The action of γ on the generators is straightforward to calculate, and the action of γ on the traces of the generators shows that $\gamma \notin \Sigma\mathrm{Sp}_4(F)$. We find that γ^2 fixes all the permutations, takes $D(\kappa, \lambda)$ to $D(\kappa^2, \lambda^2)$, and maps $T(a, b, c, d)$ to $T(a^2, b^2, c^2, d^2)$, and hence $\gamma^2 = \phi$. $\square$

From the uniserial action of B on V, we see that the stabiliser $\langle B, (2, 3) \rangle$ of the point $\langle e_1 \rangle$ in $\mathrm{Sp}_4(F)$ acts on V with structure $1 \cdot 2 \cdot 1$, whereas the stabiliser $\langle B, (1, 2)(3, 4) \rangle$ of the totally singular line $\langle e_1, e_2 \rangle$ acts on V with structure $2 \cdot 2$.

Assume for the remainder of this section that F is perfect. Then $B^{\gamma} = B$, so $\langle B, (2, 3) \rangle$ is mapped under γ to $\langle B, (1, 2)(3, 4) \rangle$. Hence γ acts on the union of the sets of totally singular points and lines, and γ swaps these two subsets. Of course, elements of $\Sigma\mathrm{Sp}_4(q)$ stabilise both of these sets.

The *generalised quadrangle* $\mathrm{GQ}(F)$ consists of the (singular) points and totally singular lines of V, with two such objects being incident if and only if one is contained in the other. The associated incidence graph $\Gamma(F)$ is bipartite, with vertex set the union of the sets of (singular) points and totally singular lines of V, and two vertices are joined by an edge if and only if one is contained in the other.

The following result is known, but we include the proof.

Proposition 7.2.3 *Let F be a perfect field of characteristic 2. Then two vertices in $\Gamma(F)$ are at distance at most 4 from each other, and $\mathrm{Aut}\,\mathrm{Sp}_4(F)$ acts distance-transitively on vertices. Furthermore, any pair of vertices at distance 3 from each other are connected by a unique path of length 3.*

Proof Certainly $\mathrm{Sp}_4(F)$ acts transitively on points and on totally singular lines, and these sets are interchanged by γ, so the action is vertex transitive. We see from its generators that the stabiliser in $\mathrm{Sp}_4(F)$ of $\langle e_1 \rangle$, namely $\langle B, (2,3) \rangle$, acts on the totally singular lines containing the point $\langle e_1 \rangle$ via the natural 2-transitive action of $\mathrm{Sp}_2(F) = \mathrm{SL}_2(F)$ on lines, so $\mathrm{Aut}\,\mathrm{Sp}_4(F)$ is transitive on ordered pairs of vertices at distance 1. Considering the pair $(\langle e_1 \rangle, \langle e_1, e_2 \rangle)$, the stabiliser $\langle B, (1,2)(3,4) \rangle$ of $\langle e_1, e_2 \rangle$ is again $\mathrm{Sp}_2(F)$, so $\mathrm{Aut}\,\mathrm{Sp}_4(F)$ is transitive on ordered pairs of vertices at distance 2.

Now consider a non-incident point $\langle f_1 \rangle$ and totally singular line $\langle f_2, f_3 \rangle$. Then clearly $\dim \langle f_1, f_2, f_3 \rangle = 3$, hence $\dim \langle f_1, f_2, f_3 \rangle^\perp = 1$, and therefore $\langle f_1, f_2, f_3 \rangle^\perp < \langle f_2, f_3 \rangle^\perp = \langle f_2, f_3 \rangle$. So if, say, $\langle f_1, f_2, f_3 \rangle^\perp = \langle f_2 \rangle$, then $\langle f_1 \rangle$, $\langle f_1, f_2 \rangle$, $\langle f_2 \rangle$, $\langle f_2, f_3 \rangle$ is the unique path of length 3 connecting the pair, which proves the final statement of the proposition. By choosing (without loss of generality) $f_1 = e_2$, $f_2 = e_1$, we see that the subgroup $\langle T(0,0,c,0) \mid c \in F \rangle$ of $\mathrm{Sp}_4(F)$ stabilises $\langle f_1 \rangle$ and $\langle f_2 \rangle$, and acts transitively on the totally singular lines incident to $\langle f_2 \rangle$ other than $\langle f_1, f_2 \rangle$. It follows that $\mathrm{Sp}_4(F)$ is transitive on such non-incident pairs $(\langle f_1 \rangle, \langle f_2, f_3 \rangle)$ and hence $\mathrm{Aut}\,\mathrm{Sp}_4(F)$ is transitive on ordered pairs of vertices at distance 3.

Finally, for a pair of points $\langle f_1 \rangle$ and $\langle f_2 \rangle$ for which $\langle f_1, f_2 \rangle$ is non-degenerate, there exists $f_3 \in \langle f_1, f_2 \rangle^\perp \setminus \langle f_1, f_2 \rangle$, and so the distance from $\langle f_1 \rangle$ to $\langle f_2 \rangle$ is 4. A similar argument applies to a non-intersecting pair of totally singular lines, so we have now established that any two vertices in $\Gamma(F)$ are at distance at most 4. Let $f_1 = e_1$, then $\langle f_2 \rangle = \langle e_4 + ae_3 + be_2 + ce_1 \rangle$ for some $a, b, c \in F$, and we see directly that the subgroup B of $\mathrm{Sp}_4(F)$ generated by the T-elements stabilises $\langle f_1 \rangle$ and acts transitively on the set of possible $\langle f_2 \rangle$. So $\mathrm{Aut}\,\mathrm{Sp}_4(F)$ is distance transitive, as claimed. $\qquad\square$

7.2.2 The maximal subgroups

Theorem 7.2.4 *Let $\mathrm{Sp}_4(q) \leqslant G \leqslant \mathrm{Aut}\,\mathrm{Sp}_4(q)$ where $q = 2^e > 2$, and suppose that $G \not\leqslant \Sigma\mathrm{Sp}_4(q)$. Then the maximal subgroups of G are as described in Table 8.14.*

Proof Let M be maximal in G with $M \not\geqslant \mathrm{Sp}_4(q)$. Then, by Corollary 1.3.7, $M = \mathrm{N}_G(N)$ for some non-trivial characteristically simple subgroup N of $\mathrm{Sp}_4(q)$. Then $M\mathrm{Sp}_4(q) = G$, so M contains elements swapping totally singular points and lines, and thus N has the same orbit sizes on totally singular points and lines. In Proposition 7.2.5 we classify those M for which N is an elementary abelian 2-group, in Proposition 7.2.7 those for which N is an elementary abelian p-group for p odd, and in Proposition 7.2.6 those for which N is insoluble. $\qquad\qquad\square$

We use the notation of the above proof throughout the following three propositions, and make use of the fact that N has the same orbit sizes on totally singular points and lines.

Proposition 7.2.5 *There is a single class of novel maximal subgroups M of G with minimal normal subgroup N an elementary abelian 2-group. This consists of the stabilisers in G of unordered pairs $\{U, V\}$, where U and V are incident totally singular points and lines of $\mathrm{GQ}(\mathbb{F}_q)$.*

Proof Let the subgroup N of $\mathrm{Sp}_4(q)$ be an elementary abelian 2-group, such that $M = \mathrm{N}_G(N)$ is a novelty and let $H = \mathrm{N}_{\Sigma \mathrm{Sp}_4(q)}(N) = \Sigma \mathrm{Sp}_4(q) \cap M$.

Since N is a 2-group, we have $W := \mathrm{C}_V(N) \neq 0$, and $W \neq V$. The subspace W is H-invariant, and so $W^\perp$ is also H-invariant. So $U := W \cap W^\perp$ is H-invariant and totally singular.

If $U = 0$ then, as an N-module, $V = W \oplus W^\perp$ and, since $W^\perp$ is N-invariant and N consists of elements of order 2, the subspace $\mathrm{C}_{W^\perp}(N) \neq 0$, contradicting the definition of W as $\mathrm{C}_V(N)$. Therefore $U \neq 0$, and so U is a point or line of the generalised quadrangle $\mathrm{GQ}(\mathbb{F}_q)$.

Now $H \cap G$ has index 2 in M, so there exists a $g \in G \setminus \Sigma \mathrm{Sp}_4(q)$ such that $M = (H \cap G) \,\dot{\cup}\, (H \cap G)g$. So U and U^g are different sorts of object in $\mathrm{GQ}(\mathbb{F}_q)$, and thus $U \neq U^g$. However, $H \cap G$ stabilises U, and so $U^M = \{U, U^g\}$, and the pair is stabilised by M. By distance-transitivity of $\Gamma(\mathbb{F}_q)$, there are only two cases to consider: namely when U and U^g are incident, and when they are not.

In the incident case we can assume that $\{U, U^g\} = \{\langle e_1 \rangle, \langle e_1, e_2 \rangle\}$, by Proposition 7.2.3. The stabiliser in $\mathrm{Sp}_4(q)$ of this is the group B (defined earlier), with structure $q^4{:}(q-1)^2$. This subgroup is normalised by γ by Lemma 7.2.2. Now by Proposition 7.2.1, the group B is contained in two maximal $\mathscr{C}_1$-subgroups P_1 and P_2 (the stabilisers of totally singular points and lines) of $\mathrm{Sp}_4(q)$ and, by Lagrange's theorem and the fact that B is uniserial, these are the only such maximals containing B. But $\mathrm{N}_G(P_1), \mathrm{N}_G(P_2) < \Sigma \mathrm{Sp}_4(q)$, so $\mathrm{N}_G(N) = \mathrm{N}_G(B)$ is a novel maximal subgroup of G.

In the non-incident case, by Proposition 7.2.3 we can pick $\{U, U^g\}$ to be $\{\langle e_2 \rangle, \langle e_1, e_3 \rangle\}$. The stabiliser K of this pair in $\mathrm{Sp}_4(q)$ is generated by the diagonal D-elements and the elements $T(0, b, 0, d)$, and hence by Lemma 7.2.2 is normalised by γ. No subgroup of K extends to a novel maximal subgroup since, by Proposition 7.2.3, there is a unique path of length 3 from U to U^g, and this consists of the vertices $\langle e_2, e_1 \rangle$ and $\langle e_1 \rangle$. So any element of $\mathrm{Aut}\,\mathrm{Sp}_4(q)$ stabilising $\{U, U^g\}$ also stabilises $\{\langle e_2, e_1 \rangle, \langle e_1 \rangle\}$. above. (Note that the containment in each other of these full stabilisers in $\mathrm{Sp}_4(q){:}\langle \gamma \rangle$ is proper.) ☐

Proposition 7.2.6 *There are no novel maximal subgroups of G in which a minimal normal subgroup N is insoluble.*

Proof As in the introduction to this section, let V be the natural module of $\mathrm{Sp}_4(q)$. Since N is perfect, at least one composition factor of N in its action on V is at least 2-dimensional.

If N stabilises a 3-dimensional subspace, W say, then N also stabilises the 1-dimensional subspace $W^\perp$. Similarly, if N stabilises a 1-dimensional space W, then N also stabilises the 3-dimensional space $W^\perp$, and $W < W^\perp$. So by self-duality the shape of V as an N-module is either $1 \cdot 2 \cdot 1$ or $2 \oplus 1 \oplus 1$. Both of these are impossible, because N fixes different numbers of totally singular points and lines. So N fixes no singular points, and hence fixes no totally singular lines. Thus either N stabilises a 2-dimensional space, or N acts irreducibly.

If N stabilises a 2-dimensional space, W say, then the space W must be non-degenerate, and $V = W \oplus W^\perp$, with N acting irreducibly on both factors. By Proposition 7.2.1, N is a subgroup either of the imprimitive group $\mathrm{Sp}_2(q) \wr S_2$ or of $\mathrm{SO}_4^+(q)$, and by applying γ to N in the latter case, we may assume that $N < \mathrm{Sp}_2(q) \wr S_2$. Hence, since N is a direct product of isomorphic simple groups and the only non-abelian composition factors of the subgroups of $\mathrm{Sp}_2(q)$ are $\mathrm{Sp}_2(q_0)$ with $q \mid q_0$, we see that N is equal either to $\mathrm{Sp}_2(q_0) \times \mathrm{Sp}_2(q_0)$ or to a diagonal subgroup $\mathrm{Sp}_2(q_0)$ where (in either case) $q_0 \mid q$.

By Proposition 7.2.1, in the first of these cases $N^\gamma = \Omega_4^+(q_0)$, which acts irreducibly on V, so γ cannot normalise N. So suppose that $N \cong \mathrm{Sp}_2(q_0)$. As a diagonal subgroup of $\mathrm{Sp}_2(q_0) \times \mathrm{Sp}_2(q_0)$, N has the form $\{(g, g^\alpha) \mid g \in \mathrm{Sp}_2(q_0)\}$ for some $\alpha \in \mathrm{Aut}(\mathrm{Sp}_2(q_0))$. Recall that, as an $\mathrm{SO}_4^+(q_0)$-module, V decomposes as a tensor product $X \otimes X$ of two copies of the natural module for $\mathrm{Sp}_2(q)$. If $\alpha \in \mathrm{Inn}\,\mathrm{Sp}_2(q_0)$ then, as an N^γ-module, we have $V \cong X \otimes X$ and, since X is self-dual, V has a 1-dimensional invariant subspace (the module U' in Definition 5.4.9). On the other hand, if $\alpha \in \mathrm{Out}(\mathrm{Sp}_2(q_0))$, then α is a field automorphism, and by Theorem 5.1.2 the group N^γ acts irreducibly on V. So again γ cannot normalise N.

The remaining case is when N acts irreducibly on V. If $N \cong \mathrm{Sp}_4(q_0)$ for some $q_0 \mid q$ then $\mathrm{N}_G(N)$ is not a novel maximal subgroup of G. Similarly, if $N \cong \mathrm{Sz}(q_0)$ for some $q_0 \mid q$ then $\mathrm{N}_G(N) \leqslant \mathrm{N}_G(\mathrm{Sz}(q))$ and so $\mathrm{N}_G(N)$ is not a novel maximal subgroup of G. As we shall see in Theorem 7.3.3, the other (local) maximal subgroups of $\mathrm{Sz}(q)$ are all contained in maximal subgroups of $\mathrm{Sp}_4(q)$ other than $\mathrm{Sz}(q)$ itself. So, using Proposition 7.2.1, we can reduce to the case when N is a subgroup of $\mathrm{Sp}_2(q^2){:}2$, $\mathrm{SO}_4^+(q)$, or $\mathrm{SO}_4^-(q)$. If $N < \mathrm{SO}_4^+(q)$, then $N^\gamma < \mathrm{Sp}_2(q) \wr S_2$, which we have already ruled out. Since γ interchanges

$\mathrm{Sp}_2(q^2){:}\mathrm{S}_2$ and $\mathrm{SO}_4^-(q)$, we may assume that $N < \mathrm{Sp}_2(q^2){:}2$, and hence by Tables 8.1 and 8.2 that $N \cong \mathrm{Sp}_2(q_0)$ where $q_0|q^2$. If the index $[\mathbb{F}_{q^2} : \mathbb{F}_{q_0}]$ is even, then $N < \mathrm{Sp}_2(q)$, which acts reducibly (and homogeneously) on V, contrary to assumption. On the other hand, if $[\mathbb{F}_{q^2} : \mathbb{F}_{q_0}]$ is odd, then N acts irreducibly but not absolutely irreducibly on V, whereas $N^\gamma \cong \mathrm{SO}_4^-(\sqrt{q_0})$ acts absolutely irreducibly. So γ does not normalise N.   □

Proposition 7.2.7 *If $q \neq 4$, then there are three classes of novel maximal subgroups of G in which a minimal normal subgroup N is an elementary abelian r-group for an odd prime r. If $q = 4$ then there are two such classes. These are the normalisers in G of subgroups of $\mathrm{Sp}_4(q)$ with the structures C_{q^2+1}, $C_{q-1}{}^2$ with $q \neq 4$, and $C_{q+1}{}^2$.*

Proof By Theorem 1.6.22, the order $|\mathrm{Sp}_4(q)| = q^4(q-1)^2(q+1)^2(q^2+1)$. Since $q = 2^e$ the factors q^4, $(q-1)^2$, $(q+1)^2$ and $q^2 + 1$ are mutually coprime.

We can identify copies of $C_{q-1}{}^2$ and $C_{q+1}{}^2$ within $\mathrm{Sp}_2(q) \wr \mathrm{S}_2$, and a copy of C_{q^2+1} within $\mathrm{Sp}_2(q^2){:}\mathrm{S}_2$. Thus if r is an odd prime then a Sylow r-subgroup, P say, of $\mathrm{Sp}_4(q)$ is a subgroup of one of $C_{q-1}{}^2$, $C_{q+1}{}^2$ or C_{q^2+1}. We consider each of these possibilities.

Suppose first that $r \mid q^2 + 1$. Then $N \leqslant P$ with P cyclic. From the action of the field automorphism of $\mathrm{SL}_2(q^2)$, we see that the cyclic subgroups of order $q^2 + 1$ in $\Sigma\mathrm{L}_2(q^2)$ are self-centralising and have normalisers with structure $C_{q^2+1}{:}4$. So the centraliser and normaliser of N and P in $\mathrm{Sp}_2(q^2){:}2 \cong \Sigma\mathrm{L}_2(q^2)$ are $C := C_{q^2+1}$ and $K := C_{q^2+1}{:}4$ (where the conjugation action is raising to the power q) respectively. As we shall see in Theorem 7.3.3, the group $\mathrm{Sz}(q)$ has no subgroups of order $q^2 + 1$, so $K \not\leqslant \mathrm{Sz}(q)$. By Theorem 1.13.1, there is a prime t dividing $|K|$ that does not divide $2^i - 1$ for any $i < 4e$. Then by Proposition 7.2.1 and Lagrange's theorem, the only maximal subgroups of $\mathrm{Sp}_4(q)$ that could contain K are $L_1 = \mathrm{Sp}_2(q^2){:}2$ and $L_2 = \mathrm{SO}_4^-(q)$. But $\mathrm{N}_G(L_1)$ and $\mathrm{N}_G(L_2)$ are subgroups of $\Sigma\mathrm{Sp}_4(q)$, whereas $\mathrm{N}_G(N)$ is not, because it is the normaliser of a Sylow subgroup of $\mathrm{Sp}_4(q)$). Therefore $\mathrm{N}_G(N)$ is a novel maximal subgroup of G.

Suppose next that $r \mid q - 1$. Then we can take N to consist of diagonal elements in $\mathrm{Sp}_2(q) \wr \mathrm{S}_2$, and note that $N \leqslant C_r{}^2$. Let C be the full diagonal subgroup $C_{q-1}{}^2$. The Sylow r-subgroup P of C satisfies $P \cong C_{r^k}{}^2$ for some $k \geqslant 1$, with $C_{\mathrm{Sp}_2(q) \wr \mathrm{S}_2}(P) = C$ and $K := \mathrm{N}_{\mathrm{Sp}_2(q) \wr \mathrm{S}_2}(P) \cong C_{q-1}{}^2{:}\mathrm{D}_8$. We now divide into two cases, depending on whether $C_{\mathrm{Sp}_4(q)}(N) = C$.

If $C_{\mathrm{Sp}_4(q)}(N) = C$, then $M = \mathrm{N}_G(N) \leqslant \mathrm{N}_G(C)$. Now $P \triangleleft \mathrm{N}_G(C)$, so $P \triangleleft M$ and therefore $K \leqslant M$. By Proposition 7.2.1 and Lagrange's theorem, the only maximal subgroups of $\mathrm{Sp}_4(q)$ that could contain K are $L_1 = \mathrm{Sp}_2(q) \wr \mathrm{S}_2$ and $L_2 = \mathrm{SO}_4^+(q)$, and (if e is even) $\mathrm{Sp}_4(q_0)$ with $q = q_0^2$. Now $\mathrm{N}_G(L_1)$ and $\mathrm{N}_G(L_2)$ are subgroups of $\Sigma\mathrm{Sp}_4(q)$, whereas $\mathrm{N}_G(N)$ is not, so the first two of these containments do not extend to $N_G(N)$. Furthermore, by Proposition 7.2.1 the group $\mathrm{Sp}_4(q_0)$ has subgroups with structure $C_{q_0^2-1}{}^2$ only when $q_0 = 2$, so the containment $K < \mathrm{Sp}_4(q_0)$ is only possible when $q = 4$. Now, since $\mathrm{Sp}_4(2) \cong S_6$ and S_6 contains a unique conjugacy class of subgroups $S_3 \wr S_2$ with the structure $(q-1)^2{:}D_8$, this is a genuine containment, and it extends to $N_G(N) < N_G(\mathrm{Sp}_4(2))$. We conclude that $N_G(N)$ is a novel maximal subgroup of G if and only if $q \neq 4$.

We now assume that $C_{\mathrm{Sp}_4(q)}(N)$ properly contains C, which implies that $N \cong C_r$. As in the previous paragraph, the only maximal subgroups of $\mathrm{Sp}_4(q)$ that could contain C are $\mathrm{Sp}_2(q) \wr S_2$, $\mathrm{SO}_4^+(q)$, and possibly $\mathrm{Sp}_4(q_0)$ with $q = q_0^2$, and then the only possibility for $C_{\mathrm{Sp}_4(q)}(N)$ is a group having $\mathrm{SL}_2(q)$ as minimal normal subgroup, which is ruled out by Proposition 7.2.6.

Lastly, consider the case where $r \mid q + 1$. Fix a subgroup $C \cong C_{q+1}{}^2$ of $\mathrm{Sp}_2(\langle e_1, e_4 \rangle) \times \mathrm{Sp}_2(\langle e_2, e_3 \rangle)$, and let P be a Sylow r-subgroup of C (and $\mathrm{Sp}_4(q)$). As in the previous case, if $C_{\mathrm{Sp}_4(q)}(N) = C$, then $N_G(N) \leqslant N_G(C)$, which normalises P, so $M = N_G(P)$. The normaliser K of C in $\mathrm{Sp}_2(q) \wr S_2$ satisfies $K := N_{\mathrm{Sp}_4(q)}(C) \cong C_{q+1}{}^2{:}D_8$. Once again, by Proposition 7.2.1 and Lagrange's theorem, the only maximal subgroups of $\mathrm{Sp}_4(q)$ that could contain K are $L_1 = \mathrm{Sp}_2(q) \wr S_2$ and $L_2 = \mathrm{SO}_4^+(q)$, but $N_G(L_1), N_G(L_2) < \Sigma\mathrm{Sp}_4(q)$, whereas $N_G(N) \not< \Sigma\mathrm{Sp}_4(q)$, so $N_G(N)$ is a novel maximal subgroup of G.

The remaining case is $C_{\mathrm{Sp}_4(q)}(N) > C$, in which case $N \cong C_r$. We can see from Lemma 7.2.2 that, in $\mathrm{Aut}(\mathrm{Sp}_4(q^2))$, the graph automorphism fixes the naturally embedded subgroup $\mathrm{Sp}_4(q)$ and induces the graph automorphism on that subgroup, so it follows by embedding N into $\mathrm{Sp}_4(q^2)$ and using the above argument for the case when $r \mid q - 1$, that N is not normalised by γ. So we get no further maximals in this case. $\qquad\square$

7.3 The maximal subgroups of Sz(q) and extensions

We start with a brief summary of the definition and properties of the Suzuki groups $\mathrm{Sz}(F)$ in fields F of characteristic 2. We refer the reader to [107] for a more detailed treatment of this material in perfect fields (see Definition 1.4.4). We then move on to the analysis of the maximal subgroups of the almost simple extensions of the groups $\mathrm{Sz}(q)$ over finite fields of characteristic 2, for which we require only properties proved in Suzuki's paper [106].

Let F be a field of characteristic 2, and suppose that there exists a field endomorphism θ of F such that $\theta^2 = \phi$, where $\phi(x) = x^2$. If F is a subfield of $\overline{\mathbb{F}_2}$ then θ, and hence also $\mathrm{Sz}(F)$, exists if and only if F does not contain $\mathbb{F}_4$; in these cases θ is unique. When F is perfect the maps ϕ and θ are automorphisms, and we define the field homomorphism ρ by $\rho : x \mapsto \sqrt{x^\theta}$; thus $\rho = \theta^{-1}$.

Definition 7.3.1 Let F be a field with endomorphism θ as above, and let the endomorphism γ of $\mathrm{Sp}_4(F)$ be as in Lemma 7.2.2. We define the *Suzuki group* $\mathrm{Sz}(F)$, or $\mathrm{Sz}(F, \theta)$ if θ is not unique, to be the set of $g \in \mathrm{Sp}_4(F)$ such that $g^\gamma = g^\theta$.

If ρ exists, then an equivalent definition of $\mathrm{Sz}(F)$ is as the centraliser in $\mathrm{Sp}_4(F)$ of the involution $\gamma\rho$. Note that γ, ϕ, θ and ρ all commute, and that when $F = \mathbb{F}_{2^e}$ is finite, $\mathrm{Sz}(F) = \mathrm{Sz}(2^e)$ exists if and only if e is odd.

Among the generators we wrote down for $\mathrm{Sp}_4(F)$ at the beginning of Section 7.2, the following are in $\mathrm{Sz}(F)$: the permutations 1 and $z := (1,4)(2,3)$; the T-elements

$$\hat{T}(a,b) := T(a, b, a^\theta, a^{2+\theta} + ab + b^\theta),$$

and the D-elements $\tilde{D}(\kappa) := D(\kappa, \kappa^{\theta-1})$.

We now restrict to the case of F a finite field of order 2^e, with $e > 1$ odd.

Lemma 7.3.2 *The group* $\mathrm{Out}\,\mathrm{Sz}(q)$ *is generated by the automorphism* ϕ. *The group* $\mathrm{Sz}(q)$ *acts 2-transitively on a set* $\mathcal{O}$ *of size* $q^2 + 1$, *with point stabiliser a Sylow 2-normaliser of order* $q^2(q-1)$, *where the Sylow 2-subgroup acts regularly on the remaining points. This action on* $\mathcal{O}$ *extends to* $\mathrm{Aut}\,\mathrm{Sz}(q) = \Sigma\mathrm{Sz}(q)$.

Proof The structure of $\mathrm{Out}\,\mathrm{Sz}(q)$, and the action of $\mathrm{Sz}(q)$ on $\mathcal{O}$ are established in [106, Theorem 11, Theorem 7]. The final claim follows from the fact that the Sylow 2-normaliser must be preserved by all outer automorphisms. □

The set of size $q^2 + 1$ is the *Suzuki-Tits ovoid*. In the following proposition, by $A:_q B$, where A and B are cyclic groups, we mean that the conjugation action of some generator of B raises elements of A to their qth powers.

Theorem 7.3.3 *Let* $q > 2$ *be an odd power of 2, and let* $s := \sqrt{2q}$. *Then the group* $\mathrm{Sz}(q)$ *is simple, of order* $q^2(q-1)(q+s+1)(q-s+1)$, *and up to conjugacy the maximal subgroups of* $\mathrm{Sz}(q)$ *are:*

(i) *A Sylow 2-normaliser of order* $q^2(q-1)$, *which is also the stabiliser of a point in* $\mathcal{O}$.

(ii) *Normalisers of cyclic groups of orders* $q-1$, $q+s+1$ *and* $q-s+1$, *with structures* $\mathrm{D}_{2(q-1)}$, $(q+s+1):_q 4$, $(q-s+1):_q 4$, *respectively.*

(iii) *Groups* $\mathrm{Sz}(q_0)$ *with* $q = q_0^r$, *where* $q_0 \neq 2$ *and* r *is prime.*

Proof The (conjugacy classes of) maximal subgroups of the groups $\mathrm{Sz}(q)$ are described in [106, Theorems 9 and 10]. It follows from the analysis of these that the Suzuki groups are simple for $q > 2$. □

Remark 7.3.4 We observe that in [106, Theorem 10], Suzuki appears to assume that the only non-local maximal subgroups have a smaller Suzuki group as minimal normal subgroup N. The odd prime Sylow subgroups of Sz(q) are cyclic by Theorem 7.3.3, so N cannot be a proper power of a non-abelian simple group. Hence we can justify this assumption now by using the result, proved originally by John Thompson but now an easy consequence of the Classification of Finite Simple Groups, that the Suzuki groups are the only finite non-abelian simple groups whose orders are not divisible by 3.

Suzuki does not deal with the maximal subgroups of almost simple extensions of Sz(q), so we do that in the following theorem.

Theorem 7.3.5 *Let* Sz(q) $\leqslant G \leqslant \Sigma$Sz($q$) *with* $q = 2^e$ *and* $e > 1$ *odd.* *Then representatives of the classes of maximal subgroups of* G *are just the normalisers in* G *of representatives of the classes of maximal subgroups of* Sz(q). *In particular, Table 8.16 is correct.*

Proof By Theorem 7.3.3, the classes of maximal subgroups of Sz(q) have distinct structures and orders, so they are all normalised by ϕ, and so their normalisers in G are maximal in G. It therefore follows that there are no type 1 novelties. It remains to prove that there are no type 2 novelties in G. Suppose, by way of contradiction, that M is such a maximal subgroup of G. By Corollary 1.3.7, the group $M = N_G(N)$, where N is a characteristically simple subgroup of M contained in Sz(q).

If N is a 2-group then, since all non-trivial 2-elements of Sz(q) fix a unique point of $\mathcal{O}$ by Lemma 7.3.2, so does N. Therefore $M = N_G(N)$ is contained in, and hence equal to, a point stabiliser in G, a contradiction.

Suppose next that N is a p-group with p odd. Recall from Theorem 7.3.3 that $|$Sz(q)$| = q^2(q-1)(q+s+1)(q-s-1)$ with $s = \sqrt{2q}$, and observe that these factors are mutually coprime. Moreover, by Theorem 7.3.3, the group Sz(q) has cyclic subgroups of orders $q - 1$, $q + s + 1$, $q - s + 1$, of which the normalisers in Sz(q) are maximal in Sz(q). So $|N| = p$ and its normaliser is equal to the normaliser in G of one of these cyclic subgroups, a contradiction.

Finally, suppose that N is insoluble. Then by Remark 7.3.4 and Theorem 7.3.3, we have $N = $ Sz(q_1) with $q = q_1^t$ and $q_1 \neq 2$, and there is a unique conjugacy class of Sz(q_1) for each such q_1, so this class is normalised by ϕ. If t is prime, then N is maximal in Sz(q) and so again M is the normaliser in G of a maximal subgroup of Sz(q). Otherwise, the only maximal subgroups of Sz(q) containing N are Sz(q_0) with q_0 a power of q_1 and $q = q_0^r$ with r prime. It follows (by a straightforward inductive argument on q) that N is self-normalising in Sz(q) and, since Sz(q_0) has a unique class of subgroups isomorphic to Sz(q_1), Lemma 7.1.1 (ii) implies that M cannot be a novel maximal subgroup of G. $\square$

7.4 The maximal subgroups of $G_2(2^e)$ and extensions

Finally, we consider the exceptional group $G_2(q)$ for q even. The outer automorphism group of $G_2(q)$ with $q = 2^e$ is generated by the field automorphism ϕ. Cooperstein [14] finds the maximal subgroups of $G_2(q)$, but not of its almost simple extensions. These are all handled in [2, (17.3)] but the proof there is only sketched, and so we shall include a detailed treatment here. Note that $G_2(2) \cong U_3(3).2$ is not quasisimple, so we are concerned only with the case $e \geqslant 2$. Note also that for q even, $G_2(q) \leqslant Sp_6(q)$ [14, Section 2].

Theorem 7.4.1   *For $q = 2^e$ with $e > 1$, the maximal subgroups of $G_2(q)$ are as described in Table 8.30.*

Proof   The maximal subgroups of $S := G_2(q)$ are described in [14, Theorems 2.3 and 2.4]. $\qquad\square$

In the proofs in this section, we assume that the reader is familiar with the subgroup structure of $SL_2(q)$: see, for example [53, Satz II.8.27] or our Tables 8.1 and 8.2. Note in particular that (since q is even) the insoluble subgroups of $SL_2(q)$ are all isomorphic to $SL_2(2^f)$ with $f|e$, whereas the soluble subgroups are subgroups of dihedral groups of order $2(q \pm 1)$ or groups with the structure $2^e{:}(q-1)$. So all soluble subgroups have non-trivial normal Sylow p-subgroups.

Theorem 7.4.2   *Let $G_2(q) \leqslant G \leqslant \operatorname{Aut} G_2(q)$ with $q = 2^e$ and $e \geqslant 2$. Then representatives of the classes of maximal subgroups of G are just the normalisers in G of representatives of the classes of maximal subgroups of $G_2(q)$. In particular, Table 8.30 is correct.*

Proof   Let $S = G_2(q)$, let V be the natural module for $Sp_6(q)$, which is also the natural module for S, and let X be the set of 1-dimensional subspaces of V. By Theorem 7.4.1, there are two classes of maximal subgroups of S of shape $[q^5]{:}GL_2(q)$, which are the stabilisers of totally singular points and lines of X. Since the subspaces of V of a given dimension are stabilised by ϕ in its semilinear action on V, these two classes are not fused in $\operatorname{Aut} S$ (and can be shown to be non-isomorphic). The other classes of maximal subgroups of S all have distinct isomorphism types, by Theorem 7.4.1. So all classes of maximal subgroups of S are stabilised by ϕ.

It follows immediately from Definition 1.3.9 that there can be no type 1 novelties in any almost simple extension of S. Hence any novelty would necessarily be of type 2, as described in Definition 1.3.11 and Lemma 7.1.1. So let $M := N_G(H)$ be a type 2 novelty, where $H < K < S < G$, and K is maximal in S. We proceed in Lemmas 7.4.3 to 7.4.6 to consider each of the possibilities for K, as given by Theorem 7.4.1, and in each case show that no novelty arises. $\qquad\square$

Lemma 7.4.3 *Let* $S := G_2(q) \leqslant G \leqslant \operatorname{Aut} G_2(q)$. *Then there are no type 2 novel maximal subgroups* M *of* G *with* $H := M \cap S \leqslant K \cong [q^5]{:}\operatorname{GL}_2(q)$.

Proof Let $N := O_2(H)$, the largest normal 2-subgroup of H. Suppose that $N \neq 1$. Then H must contain $O_2(K) = [q^5]$, since otherwise the normaliser of N in $NO_2(K)$ would properly contain N, contradicting Lemma 7.1.1 (i) applied to the characteristic subgroup N of H. If $N = O_2(K)$, then Lemma 7.1.1 (i) gives $H = K$, contrary to assumption. So N has non-trivial intersection with a complement $\operatorname{GL}_2(q)$ of $O_2(K)$ in K and hence N has the form $[q^5]{:}R$, where R is a non-trivial subgroup of a Sylow 2-subgroup of $\operatorname{GL}_2(q)$. But R is abelian, so $R \leqslant H = N_S(N)$ and, since R is normal in all proper subgroups of $\operatorname{GL}_2(q)$ that contain it, we have $R \trianglelefteq H$, contradicting Lemma 7.1.1 (iii).

So $N = 1$, and hence H is isomorphic to a subgroup of $\operatorname{GL}_2(q)$, which is a direct product of its cyclic centre of order $q - 1$ and the simple group $\operatorname{SL}_2(q)$. Suppose that H is perfect, and hence that $H \cong \operatorname{SL}_2(2^f)$ with $f \mid e$. As mentioned earlier, K is the stabiliser of either a point or a singular line in its action on X.

In the first case, $K = P_b$ in the notation of [14]. We claim that H fixes a unique point of X. Suppose not. It is shown in [13, Proof of Lemma 3.1, Method 2] that the orbit lengths of K on X are 1, $q(q+1)$, $q^3(q+1)$, and q^5, where the orbits of $O_2(K)$ within them have lengths 1, q, q^3 and q^5, respectively. So the orbits of K of lengths $q(q+1)$ and $q^3(q+1)$ both consist of $q+1$ orbits of $O_2(K)$, and hence the action of $K/O_2(K) \cong \operatorname{GL}_2(q)$ on each of these sets of $q+1$ orbits of $O_2(K)$ is 2-transitive and has soluble stabiliser. So, since H is insoluble, it cannot fix a point in either of these two orbits of K. The stabiliser of a point in the orbit of K of length q^5 must be a complement of $O_2(K)$ in K and is therefore isomorphic to $\operatorname{GL}_2(q)$. So, if H fixed a point in that orbit, then it would be normalised by the scalars in $\operatorname{GL}_2(q)$, so could not be self-normalising in K. So H fixes a unique point of X, and K is the stabiliser in S of that point. But then any two conjugates of H in S that are contained in K are conjugate in K, contradicting Lemma 7.1.1 (ii).

If, on the other hand, K is the stabiliser in S of a singular line, then $K = P_a$ in the notation of [14, Section 2], which is the normaliser of a long root subgroup Q of K, with $|Q| = q$. It is shown also in [14, Section 2] that Q is central in a Sylow 2-subgroup of K, so Q must be central in the subgroup $[q^5]{:}\operatorname{SL}_2(q)$ of K. Hence, since H is perfect, $H < C_K(Q)$, and so $N_K(H) \neq H$.

So H is not perfect. The subgroup structure of $\operatorname{GL}_2(q)$ and the fact that $N = 1$ implies that there is an odd prime r with $R := O_r(H) \neq 1$. Then, by Sylow's Theorem, R is conjugate in K to a subgroup of a complement $C \cong \operatorname{GL}_2(q)$ of $O_2(K)$ in K, and so we may assume that $R \leqslant C$ and hence (since $H \cap O_2(K) = 1$) that $H = N_K(R) \leqslant C$. Since H is self-normalising in S, and hence in C, it must contain $Z(C)$, which is cyclic of order $q-1$ and, from the subgroup structure of $L_2(q)$, $H/Z(C)$ is either isomorphic to $SL_2(2^f)$ for some $f|e$ or it is dihedral of order $2(q-1)$ or $2(q+1)$. In either case, $Z(C) = Z(H)$ is characteristic in H, and so Lemma 7.1.1 (i) implies that $H = C$. But each complement of $O_2(K)$ in K is a normaliser of a complement in $O_2(K) : Z(C)$, so the complements are all conjugate in K, and hence Lemma 7.1.1 (ii) cannot be fulfilled. $\square$

Lemma 7.4.4   *Let* $S := G_2(q) \leqslant G \leqslant \operatorname{Aut} G_2(q)$. *Then there are no type 2 novel maximal subgroups* M *of* G *with* $H := M \cap S \leqslant K \cong SL_2(q) \times SL_2(q)$.

Proof   Here K is the stabiliser of a non-degenerate 2-dimensional subspace V_2 of V [14, 5.4], and also stabilises the orthogonal complement V_4 of V_2 in V, of dimension 4. Consider the projections of H onto the two direct factors.

Suppose first that both of these projections are soluble (or equivalently that H is soluble). We observed at the beginning of this section that all soluble subgroups of $SL_2(q)$ have non-trivial normal Sylow subgroups. By considering the possible projections onto the two factors, we see that either H has a non-trivial normal Sylow subgroup, contradicting Lemma 7.1.1 (iii), or that H has a characteristic subgroup contained in one of the two direct factors of K, but in that case, by Lemma 7.1.1 (i), H would contain the other factor and so would not be soluble.

If just one of the projections is insoluble, then H has a unique non-abelian normal subgroup isomorphic to $SL_2(2^f)$ with $f|e$, and its centraliser in S contains the other direct factor, again contradicting Lemma 7.1.1 (i). So both projections are insoluble, and either $H \cong SL_2(2^{f_1}) \times SL_2(2^{f_2})$, or H is a diagonal subgroup of K isomorphic to $SL_2(2^f)$, where f_1, f_2 and f all divide e.

It can be verified by computer calculation (file `g22calc`) that the group $I := G_2(2) \cap K \cong SL_2(2) \times SL_2(2)$ acts absolutely irreducibly on both V_2 and V_4, and I acts faithfully on V_4. If $H \cong SL_2(2^{f_1}) \times SL_2(2^{f_2})$, then H contains a conjugate of I, and hence H is absolutely irreducible on V_2 and V_4, and faithful on V_4. So if $H^g = H_0 < K$ with $g \in S$, then g must fix V_2 and V_4, and hence $g \in K$, contradicting Lemma 7.1.1 (ii).

So suppose that $H \cong SL_2(2^f)$ is a diagonal subgroup of K. From the list of maximal subgroups of $Sp_4(2^e)$ (Table 8.14), we see that the action of K on V_4 (which we know to be absolutely irreducible) must arise as the tensor product of two 2-dimensional representations of K (which is the same as the action of $\Omega_4^+(q)$). The restrictions of this representation to the diagonal subgroups of K have corresponding $SL_2(2^f)$-modules $W \otimes W^\sigma$, where W is the natural module for $SL_2(2^f)$ and $\sigma \in \operatorname{Out} SL_2(2^f)$. If $\sigma \neq 1$, then σ is a field automorphism, the action is irreducible by Theorem 5.3.2, and we get the same contradiction as in the case when H is a direct product. If, on the other hand, $\sigma = 1$ then

$W \otimes W^\sigma$ has a 1-dimensional submodule $\langle v \otimes v \rangle$, so H also lies in a maximal subgroup $[q^5]{:}\mathrm{GL}_2(q)$, contradicting Lemma 7.4.3.   □

Lemma 7.4.5 *Let* $S := \mathrm{G}_2(q) \leqslant G \leqslant \mathrm{Aut}\,\mathrm{G}_2(q)$. *Then there are no type 2 novel maximal subgroups* M *of* G *with* $H := M \cap S \leqslant K \cong \mathrm{SL}_3^\pm(q).2$. *Furthermore, if* M *is a type 2 novel maximal subgroup of* G *then* $M \cap S$ *is irreducible.*

Proof First suppose $K \cong \mathrm{SL}_3(q).2$, which is the imprimitive stabiliser of a complementary pair of 3-dimensional totally singular subspaces of V [14, 5.3]. It is immediate from Table 2.4 that $V = V_3 \oplus V_3^*$, where V_3 and V_3^* are dual irreducible $\mathrm{SL}_3(q)$-modules (so they are nonisomorphic, and hence by Lemma 1.8.11 the only $\mathrm{SL}_3(q)$-submodules of V) and K is the extension of $\mathrm{SL}_3(q)$ by the graph (= inverse-transpose) automorphism. The subgroup H is contained in one of the maximal subgroups of K, which can be found in Tables 8.3 and 8.4.

If $H \cap K^\infty$ is reducible (as subgroup of $\mathrm{SL}_3(q)$), then it stabilises a 1-dimensional subspace of at least one of V_3 and V_3^*, and hence H stabilises a 1- or 2-dimensional subspace of V. Hence by [14, Section 5] H is also contained in one of the reducible maximal subgroups of S, and so H has already been eliminated.

If $H \cap K^\infty$ is an imprimitive subgroup of $\mathrm{SL}_3(q)$ then it must stabilise a decomposition into three blocks of dimension 1, and if it is semilinear then its centralising field must have order q^3. So in these cases $H \cap K^\infty$ is contained in a $\mathscr{C}_2$- or $\mathscr{C}_3$-subgroup L of $\mathrm{SL}_3(q)$. From Table 8.3 we see that L is soluble, and has the structure $(q-1)^2{:}S_3$ or $(q^2+q+1){:}C_3$, and the extension $L.2$ of L in $\mathrm{SL}_3(q).2$ has structure $((q-1)^2{:}S_3).2$ or $((q^2+q+1){:}C_3).2$. It is not hard to see that all subgroups of $L.2$ have a non-trivial normal Sylow subgroup, so $H \cap K^\infty$ cannot be imprimitive or semilinear (as subgroup of $\mathrm{SL}_3(q)$) by Lemma 7.1.1 (iii).

We can see now from Tables 8.3, 8.4, 8.5 and 8.6 (recalling that q is even) that the remaining possibilities for $H \cap K^\infty$ are $\mathrm{SL}_3(2^f)(.3)$ with $f \mid e$, $\mathrm{SU}_3(2^f)(.3)$ with $2f \mid e$ and $f > 1$, $3{\cdot}A_6$ with $q = 4$, and a subgroup of the soluble group $\mathrm{SU}_3(2)$. We can check by direct computation (file g22calc) that all subgroups of $\mathrm{SU}_3(2)$ have non-trivial normal Sylow p-subgroups, and so this case is ruled out by Lemma 7.1.1 (iii). In all other cases, V_3 and V_3^* remain irreducible nonisomorphic modules on restriction to H^∞. It follows that if $g \in S$ with $H^g = H_0 < K$, then g must preserve the same system of imprimitivity as K, so $g \in K$, contradicting Lemma 7.1.1 (ii). This completes the elimination of the case $K \cong \mathrm{SL}_3(q).2$.

Since the stabiliser in S of a singular 3-space is contained in $\mathrm{SL}_3(q).2$, this result together with Lemmas 7.4.3 and 7.4.4 eliminate all cases in which H acts reducibly on V, so we can assume from now on that this action is irreducible.

Now consider $K \cong \mathrm{SU}_3(q).2$. From the subgroup structure of $\mathrm{Sp}_6(q)$ (Tables 8.28 and 8.29), we see that K is semilinear, with $\mathrm{SU}_3(q)$ acting absolutely reducibly with centralising field $\mathbb{F}_{q^2}$, and K is the extension of $\mathrm{SU}_3(q)$ by the graph (= inverse-transpose) automorphism. The subgroup H is contained in one of the maximal subgroups of K, which can be found in Tables 8.5 and 8.6. If $H \cap K^\infty$ is reducible (as subgroup of $\mathrm{SU}_3(q)$), then it stabilises a 1-dimensional subspace of the natural module for $\mathrm{SU}_3(q)$ and hence H stabilises a 2-dimensional subspace of V, so H has already been eliminated. The cases when $H \cap K^\infty$ is imprimitive or semilinear (as subgroup of $\mathrm{SU}_3(q)$) are eliminated as in the case $K \cong \mathrm{SL}_3(q).2$ Otherwise $H \cap K^\infty$ is isomorphic to either $\mathrm{SU}_3(2^f)(.3)$ with $f|e$ and $f > 1$, or to a subgroup of $\mathrm{SU}_3(2)$. The second of these possibilities is ruled out by Lemma 7.1.1 (ii) as in the case $K \cong \mathrm{SL}_3(q).2$. Otherwise $H \cap K^\infty$ is itself absolutely reducible with centralising field $\mathbb{F}_{q^2}$. So if $g \in S$ with $H^g = H_0 < K$, then g must normalise the centralising field of K^∞, and hence $g \in K$, contradicting Lemma 7.1.1 (ii).   □

Lemma 7.4.6 *Let $S := \mathrm{G}_2(q) \leqslant G \leqslant \mathrm{Aut}\,\mathrm{G}_2(q)$. Then there are no type 2 novel maximal subgroups M of G with $H := M \cap S \leqslant K$ and $K \cong \mathrm{G}_2(q_0)$, $\mathrm{L}_2(13)$, or J_2.*

Proof Suppose first that $K \cong \mathrm{G}_2(q_0)$ with $q_0 = 2^f$ for some $f|e$. By considering the maximal subgroups of K, we see that the only possibilities for H that have not already been eliminated by Lemmas 7.4.3, 7.4.4, and 7.4.5 are $H \cong \mathrm{G}_2(2^g)$ with $g|f$, or $H \leqslant \mathrm{L}_2(13)$, or $H \leqslant \mathrm{J}_2$. Since K has unique conjugacy classes of primitive absolutely irreducible subgroups isomorphic to $\mathrm{G}_2(2^g)$ and (when f is even) to $\mathrm{L}_2(13)$ and J_2, the group H cannot be isomorphic to any of these groups by Lemma 7.1.1 (ii). We can check by direct computation (file `g22calc`) that all maximal subgroups of $\mathrm{L}_2(13)$ and J_2 are reducible, contradicting Lemma 7.4.5.

Finally, suppose that $K \cong \mathrm{L}_2(13)$ or $K = \mathrm{J}_2$ with $q = 4$. Then, as we just observed, all maximal subgroups of K are reducible, so H is reducible, contradicting Lemma 7.4.5.   □

8

Tables

8.1 Description of the tables

The tables in this chapter list the maximal subgroups of the quasisimple classical groups Ω in dimensions 2–12, as described in Theorem 2.1.1. The tables provide sufficient information to determine the maximal subgroups of all almost simple extensions of $\bar{\Omega} := \Omega/Z(\Omega)$. In addition, there are tables listing the maximal subgroups of those almost simple exceptional groups that arise as subgroups of these classical groups, namely $\mathrm{Sz}(q)$, $\mathrm{G}_2(q)$ (taken from [64] for odd q), $\mathrm{R}(q)$ and $^3\mathrm{D}_4(q)$ (the last two taken from [63]).

The tables are ordered by the dimension of the natural representation, and within that we list first the maximal subgroups of $\mathrm{SL}_n(q)$, then $\mathrm{SU}_n(q)$, then $\mathrm{Sp}_n(q)$ and finally either $\Omega_n^\circ(q)$ or $\Omega_n^+(q)$ and $\Omega_n^-(q)$ (when these groups are quasisimple). Tables for exceptional groups occur immediately after the smallest classical group that contains that family, so for example the groups $\mathrm{Sz}(q)$ are described just after $\mathrm{Sp}_4(q)$, whilst $\mathrm{G}_2(q)$ occurs in dimension 6 for q even and dimension 7 for q odd. Please see Subsection 1.6.3 for a complete description of our notation for the classical groups.

For each family of classical groups (such as $\mathrm{SL}_4(q)$), there are usually two tables, the first listing the subgroups of geometric type and the second listing those in Class $\mathscr{S}$. The two tables have similar but not identical formats.

Let $\mathsf{A} = \mathrm{Aut}\,\bar{\Omega}$. At the top of each table, we provide brief information on $|Z(\Omega)|$ and on the orders of the generators of $\mathrm{Out}\,\bar{\Omega}$. The precise definitions of the outer automorphisms and of their inverse images in A can be found in Subsection 1.7.1, whilst presentations of $\mathrm{Out}\,\bar{\Omega}$ are listed in Subsection 1.7.2. For the exceptional groups, automorphisms are defined in Chapter 7.

The definitions of some of the outer automorphisms can depend on the choice of invariant form in some of the unitary and orthogonal groups [6]. The invariant forms used in this book are defined in Section 1.5.

Each row in the tables describes a representative H of an A-conjugacy class

373

of subgroups of Ω. (We have not kept strictly to this rule in the cases of $\mathrm{Sp}_4(q)$ (q even), and $\Omega_8^+(q)$, as will be explained at the beginning of Tables 8.14 and 8.50, respectively.) Usually, $\bar{H}$ is maximal in $\bar{\Omega}$, but some of the rows define 'novelties', which means that $\bar{H}$ is not maximal in $\bar{\Omega}$ but, for certain subgroups T of $\mathrm{Out}\,\bar{\Omega}$, the group $\bar{\Omega}.T$ has a maximal subgroup K with the structure $\bar{H}.T$ and $K \cap \bar{\Omega} = \bar{H}$. See Definition 1.3.8 and the discussion after it for more information.

Column '$\mathscr{C}_i$'. This is only in the geometric type tables, and specifies the Aschbacher class of H. See Section 2.2 for the Aschbacher classes.

Column 'Subgp'. This describes the structure of H, using the ATLAS conventions [12], which we described in Section 1.2. Note that the '$-$' superscript in examples such as $6^{\cdot}\mathrm{L}_3(4)^{\cdot}2_1^-$ indicates that this is not the group whose character table is displayed in [12], but is the isoclinic variant thereof: see Definition 1.3.3.

Columns 'Notes', 'Nov', 'Conditions on q' . Column 'Notes' occurs only in the geometric type tables, and provides further information, such as restrictions on q, and whether H defines novelties. In the Class $\mathscr{S}$ tables there are two corresponding columns, headed 'Nov' and 'Conditions on q'. We omit the 'Nov' column if it has no entries. See Section 1.3 for a general discussion of novel maximal subgroups. For more information about how to interpret this novelty information see Column 'Stab', below.

Column 'c'. This specifies the number of Ω-conjugacy classes represented by the row of the table: each row represents on A-conjugacy class of subgroups.

Column 'Stab'. This describes the stabiliser S of one Ω-conjugacy class of groups H under the action of $\mathrm{Out}\,\bar{\Omega}$. (So the product of c and the order of the stabiliser should be $|\mathrm{Out}\,\bar{\Omega}|$.) Of course, S is defined only up to conjugacy in $\mathrm{Out}\,\bar{\Omega}$. If the table row is not marked as a novelty then, for any subgroup T of S, the group $\bar{H}$ extends to a maximal subgroup $\bar{H}.T$ of $\bar{\Omega}.T$. If the table row is a novelty, then this is true for some but not all subgroups of S, and we specify those T for which $\bar{H}.T$ is maximal under a 'novelty' entry in the auxiliary table.

Column 'Acts'. In the Class $\mathscr{S}$ tables, this specifies the automorphisms of H induced by the automorphisms of $\bar{\Omega}$ in Column 'Stab'. It is included only when there could be some uncertainty.

The auxiliary table. Some entries are too long to be conveniently included in the main table. When this happens, a symbol such as 'N1' is inserted into the main table, and the auxiliary table then specifies exactly what 'N1' means.

At the time of publication, we know of no errors in these tables, but an errata list has been created at `http://www.cambridge.org/9780521138604`, and we shall keep this up to date. We would be extremely grateful to be informed of any errata.

8.1.1 Examples of use of the tables

Geometric subgroups. Let us first determine the geometric maximal subgroups of $GU_6(2)$, using Table 8.26. We first calculate that the centre of $SU_6(2)$ is $Z(SU_6(3)) = (3,6) = 3 = Z(GU_6(2))$, that $|\delta| = 3$, with $\phi = \gamma$ of order two. From Subsection 1.7.2, $\delta^\phi = \delta^{-1}$, so that $\mathrm{Out}\, U_6(2) \cong S_3$. Note that $GU_6(2) = SU_6(2).\langle\delta\rangle$.

Consider first the reducible groups (Class $\mathscr{C}_1$). In Row 1 we find a class of group of shape $2^{1+8}:(3 \times SU_4(2))$. There is a unique class of such groups ($c = 1$) so their stabiliser is $\mathrm{Out}\, U_6(q) = \langle\delta,\gamma\rangle$. Therefore, the groups in this row extend to maximal subgroups of $GU_6(2)$, of shape $2^{1+8}:(3^2 \times SU_4(2))$. Similarly, Row 2 describes a single class of maximal subgroups of $GU_6(2)$, of shape $2^{4+8}:(GL_2(4) \times GU_2(2))$, and Row 3 lists a class of groups of shape $2^9:GL_3(4)$. These are the three classes of parabolic maximal subgroups. Finally, there are two classes of stabilisers of non-degenerate subspaces, namely the groups $3 \times GU_5(2)$ (Row 4) and $GU_4(2) \times GU_2(2)$ (Row 5).

The first listed imprimitive group (Class $\mathscr{C}_2$) occurs as a novel maximal subgroup (N1) when $q = 2$. By the auxiliary table, the group is maximal under subgroups of its stabiliser that are not contained in a conjugate of $\langle\gamma\rangle$. Now, $\langle\delta\rangle$ is not contained in any conjugate of $\langle\gamma\rangle$ so this group does extend to a maximal subgroup of $GU_6(2)$, of shape $3^6.S_6$. The second listed imprimitive group is non-maximal in all extensions of $SU_6(2)$. The final two rows labelled $\mathscr{C}_2$ give rise to classes of imprimitive maximal subgroups of shape $GU_3(2) \wr S_2$ and $GL_3(4).2$.

Moving on to the semilinear groups (Class $\mathscr{C}_3$), once again we see a novelty. It is also labelled N1, so there is no need to repeat our calculation from the imprimitive case: we know immediately that it will give rise to a maximal subgroup of $GU_6(2)$, of shape $GU_2(8).3$.

There are no maximal tensor product groups (Class $\mathscr{C}_4$), because $q = 2$, and no maximal subfield groups of type $SU_6(q_0)$ (Class $\mathscr{C}_5$), because q is prime. Thus the next class to consider is the subfield groups of type $3 \times Sp_6(2)$. Here, there are three classes of such groups in $SU_6(2)$, and their stabiliser in the outer automorphism group is $\langle\delta^3,\phi\rangle = \langle\phi\rangle$. Since $GU_6(2)$ contains the automorphism δ, which is not contained in $\langle\phi\rangle$, these groups do not extend to maximal subgroups of $GU_6(2)$, instead δ permutes the three classes in $SU_6(2)$.

As a second example, we consider six of the conjugacy classes of subgroups of $P\Omega_8^+(3)$, as described in Table 8.50. The reader should first consult the additional description at the beginning of the table of conjugacy under the triality automorphism. First, we use the information given at the beginning of the table, and the presentation given in Section 1.7.2, to calculate that

$\text{Out}\,\mathrm{P}\Omega_8^+(3) = \langle \delta, \delta', \gamma, \tau \rangle \cong S_4$. There are six classes in $\mathrm{P}\Omega_8^+(3)$ of groups of shape $2^6.A_8$, two in class $\mathscr{C}_2$ and four in class $\mathscr{C}_6$, and these groups are all non-maximal in $\mathrm{P}\Omega_8^+(3)$. The stabiliser of a class is S3, which we see in the auxiliary table is equal to $\langle \gamma, \delta' \rangle \cong 2^2$. Note that with this action, S3 stabilises two of the classes (which can be taken to be H and H^δ, both in $\mathscr{C}_2$), and permutes the remaining four in pairs (which can be taken to be the four groups in $\mathscr{C}_6$). We see in entry N5 of the auxiliary table that the normaliser of H is maximal under subgroups of S3 that are not contained in $\langle \gamma\delta' \rangle$, so the normalisers of H and of H^δ are maximal subgroups of $\mathrm{P}\Omega_8^+(3).\langle \gamma \rangle$ (the elements of $\mathrm{PGO}_8^+(3)$ of spinor norm 1), of $\mathrm{P}\Omega_8^+(3).\langle \delta' \rangle = \mathrm{PSO}_8^+(3)$ and of $\mathrm{P}\Omega_8^+(3).\langle \gamma, \delta' \rangle = \mathrm{PGO}_8^+(3)$, but are otherwise non-maximal.

Examples from Class $\mathscr{S}$. As an example, let us consider the subgroups $H = 6^{\cdot}L_3(4)^{\cdot}2_1^-$ and $H = 6^{\cdot}L_3(4)$ of $\Omega = \mathrm{SL}_6(q)$ and of $\Omega = \mathrm{SU}_6(q)$ that are described in Tables 8.25 (for $\mathrm{SL}_6(q)$) and 8.27 (for $\mathrm{SU}_6(q)$).

From the 'Conditions on q' column, we see that these arise only when $q = p$. The precise values of q for which the two variants of H occur depend on the value of q modulo 24, but observe that they occur only when $p = q \equiv 1 \bmod 6$ in $\mathrm{SL}_6(q)$ and $q \equiv 5 \bmod 6$ in $\mathrm{SU}_6(q)$. This implies that $\mathsf{A} := \text{Out}\,\Omega$ is dihedral of order 12 whenever H occurs.

We are considering two related but non-isomorphic groups H in each of the linear and unitary cases, corresponding to two rows in the tables.

Consider first $H = 6^{\cdot}L_3(4)^{\cdot}2_1^-$, an extension of the quasisimple group $6^{\cdot}L_3(4)$ by its 2_1 automorphism. Each row of the table describes a single A-class of subgroups H of Ω, and we see from Column 'c', that this class splits into six Ω-classes. So the stabiliser in A of each of these Ω-classes of subgroups has order $12/6 = 2$. In the 'Stab' column we find that the stabiliser of one such class is generated by the graph automorphism γ of Ω. So the stabilisers of the other five Ω-classes are conjugates of γ in $\text{Out}\,\Omega$. There is no entry in the 'Nov' column, so these subgroups H are maximal in Ω and the extension $H.\langle \gamma \rangle$ is maximal in $\Omega.\langle \gamma \rangle$. Finally, the 'Acts' column tells us that $H.\langle \gamma \rangle \cong 6^{\cdot}L_3(4)^{\cdot}2^2$.

The second of the two rows that we are considering describes subgroups $H = 6^{\cdot}L_3(4)$. By the 'c' and 'Stab' columns, there are only three Ω-classes of such subgroups, and the stabiliser S of one such class in $\text{Out}\,\Omega$ is the subgroup $\langle \delta^3, \gamma \rangle$ with structure 2^2. The automorphisms of H induced by the generators of S are specified in the 'Acts' column; or rather in entry A3 of the auxiliary table. The entry N5 in the 'Nov' column indicates that H is not maximal in Ω, but that certain extensions $H.T$ are maximal in the corresponding extensions $\Omega.T$. The entry N5 in the auxiliary table then specifies those subgroups T of the class stabiliser S. We see that $H.T$ is maximal when $T = \langle \delta^3 \rangle$, $\langle \gamma\delta^3 \rangle$, and $\langle \delta^3, \gamma \rangle$, but not when $T = 1$ or $\langle \gamma \rangle$.

8.2 The tables

Table 8.1 *The maximal subgroups of* $\mathrm{SL}_2(q)$ $(= \mathrm{Sp}_2(q) \cong \mathrm{SU}_2(q))$ *of geometric type*

$d := |\mathrm{Z}(\mathrm{SL}_2(q))| = (q - 1, 2)$, $|\delta| = d$, $|\phi| = e$, $q = p^e \geqslant 4$.

$\mathscr{C}_i$	Subgp	Notes	c	Stab
$\mathscr{C}_1$	$E_q{:}(q-1)$		1	$\langle \delta, \phi \rangle$
$\mathscr{C}_2$	$Q_{2(q-1)}$	$q \neq 5,7,9,11$; q odd	1	$\langle \delta, \phi \rangle$
		N1 if $q = 7, 11$	1	$\langle \delta \rangle$
		N2 if $q = 9$	1	$\langle \delta, \phi \rangle$
$\mathscr{C}_2$	$D_{2(q-1)}$	q even	1	$\langle \phi \rangle$
$\mathscr{C}_3$	$Q_{2(q+1)}$	$q \neq 7,9$; q odd	1	$\langle \delta, \phi \rangle$
		N1 if $q = 7$	1	$\langle \delta \rangle$
		N2 if $q = 9$	1	$\langle \delta, \phi \rangle$
$\mathscr{C}_3$	$D_{2(q+1)}$	q even	1	$\langle \phi \rangle$
$\mathscr{C}_5$	$\mathrm{SL}_2(q_0).2$	$q = q_0^2$, q odd	2	$\langle \phi \rangle$
$\mathscr{C}_5$	$\mathrm{SL}_2(q_0)$	$q = q_0^r$, q odd, r odd prime	1	$\langle \delta, \phi \rangle$
$\mathscr{C}_5$	$L_2(q_0)$	$q = q_0^r$, q even, $q_0 \neq 2$, r prime	1	$\langle \phi \rangle$
$\mathscr{C}_6$	$2_-^{1+2}.S_3 \cong 2{\cdot}S_4^-$	$q = p \equiv \pm 1 \pmod 8$	2	1
	$2_-^{1+2}{:}3 \cong 2{\cdot}A_4$	$q = p \equiv \pm 3, 5, \pm 13 \pmod{40}$	1	$\langle \delta \rangle$
		N1 if $q = p \equiv \pm 11, \pm 19 \pmod{40}$	1	$\langle \delta \rangle$

N1 Maximal under $\langle \delta \rangle$	N2 Maximal under subgps not contained in $\langle \phi \rangle$

Note: The groups in Classes $\mathscr{C}_2$ and $\mathscr{C}_3$ also lie in $\mathscr{C}_8$, as do the groups in Class $\mathscr{C}_5$ with $r = 2$. The group in Class $\mathscr{C}_5$ with $q = 4$ is maximal, but lies in and is listed under $\mathscr{C}_2$. The Aschbacher classes are a little different when $\mathrm{SL}_2(q)$ is regarded as $\mathrm{SU}_2(q)$ or $\mathrm{Sp}_2(q)$.

Table 8.2 *The maximal subgroups of* $\mathrm{SL}_2(q)(= \mathrm{Sp}_2(q) \cong \mathrm{SU}_2(q))$ *in Class* $\mathscr{S}$

$d := |\mathrm{Z}(\mathrm{SL}_2(q))| = (q - 1, 2)$, $|\delta| = d$, $|\phi| = e$, $q = p^e \geqslant 4$.

Subgp	Conditions on q	c	Stab
$2{\cdot}A_5$	$q = p \equiv \pm 1 \pmod{10}$	2	1
	$q = p^2$, $p \equiv \pm 3 \pmod{10}$	2	$\langle \phi \rangle$

Table 8.3 *The maximal subgroups of* $\mathrm{SL}_3(q)$ *of geometric type*
$d := |\mathrm{Z}(\mathrm{SL}_3(q))| = (q-1,3)$, $|\delta| = d$, $|\phi| = e$, $|\gamma| = 2$, $q = p^e$.

$\mathscr{C}_i$	Subgp	Notes	c	Stab
$\mathscr{C}_1$	$E_q{}^3{:}\mathrm{GL}_2(q)$		2	$\langle \delta, \phi \rangle$
$\mathscr{C}_1$	$E_q{}^{1+2}{:}(q-1)^2$	N1	1	$\langle \delta, \phi, \gamma \rangle$
$\mathscr{C}_1$	$\mathrm{GL}_2(q)$	N1	1	$\langle \delta, \phi, \gamma \rangle$
$\mathscr{C}_2$	$(q-1)^2{:}S_3$	$q \geqslant 5$	1	$\langle \delta, \phi, \gamma \rangle$
$\mathscr{C}_3$	$(q^2+q+1){:}3$	$q \neq 4$	1	$\langle \delta, \phi, \gamma \rangle$
		N2 if $q = 4$	1	$\langle \delta, \phi, \gamma \rangle$
$\mathscr{C}_5$	$\mathrm{SL}_3(q_0).\left(\frac{q-1}{q_0-1},3\right)$	$q = q_0^r$, r prime	$(\frac{q-1}{q_0-1},3)$	$\langle \delta^c, \phi, \gamma \rangle$
$\mathscr{C}_6$	$3^{1+2}_+{:}Q_8.\frac{(q-1,9)}{3}$	$p = q \equiv 1 \bmod 3$	$\frac{(q-1,9)}{3}$	$\langle \delta^c, \gamma \rangle$
$\mathscr{C}_8$	$d \times \mathrm{SO}_3(q)$	q odd	d	$\langle \phi, \gamma \rangle$
$\mathscr{C}_8$	$(q_0-1,3) \times \mathrm{SU}_3(q_0)$	$q = q_0^2$	$(q_0-1,3)$	$\langle \delta^c, \phi, \gamma \rangle$

N1	Maximal under subgroups not contained in $\langle \delta, \phi \rangle$
N2	Maximal under subgroups not contained in $\langle \phi, \gamma \rangle$

Note: The group in Class $\mathscr{C}_2$ with $q = 2$ is non-maximal in $\mathrm{SL}_3(2)$, but extends to a novel maximal subgroup under γ. However, it lies in and is listed under $\mathscr{C}_1$ (type $\mathrm{GL}_2(2)$). The group in Class $\mathscr{C}_2$ with $q = 3$ is maximal, but lies in and is listed under $\mathscr{C}_8$. The group $\mathrm{SU}_3(2)$ in Class $\mathscr{C}_8$ with $q = 4$ is soluble, and is the normaliser of an extraspecial 3-group.

Table 8.4 *The maximal subgroups of* $\mathrm{SL}_3(q)$ *in Class* $\mathscr{S}$
$d := |\mathrm{Z}(\mathrm{SL}_3(q))| = (q-1,3)$, $|\delta| = d$, $|\phi| = e$, $|\gamma| = 2$, $q = p^e$.

Subgp	Conditions on q	c	Stab	Acts
$d \times \mathrm{L}_2(7)$	$q = p \equiv 1, 2, 4 \bmod 7$, $q \neq 2$	d	$\langle \gamma \rangle$	
$3{\cdot}A_6$	$q = p \equiv 1, 4 \bmod 15$	3	$\langle \gamma \rangle$	$\gamma \to 2_2$
	$q = p^2$, $p \equiv 2, 3 \bmod 5$, $p \neq 3$	3	$\langle \phi, \gamma \rangle$	A1

A1	$\gamma \to 2_2$, $\phi \to 2_1$ ($p \equiv 2, 8 \pmod{15}$) or 2_3 ($p \equiv 7, 13 \pmod{15}$)

Table 8.5 *The maximal subgroups of* $SU_3(q)$ *of geometric type*
$d := |Z(SU_3(q))| = (q+1,3)$, $|\delta| = d$, $|\phi| = 2e$, $q = p^e \geqslant 3$.

$\mathscr{C}_i$	Subgp	Notes		c	Stab
$\mathscr{C}_1$	$E_q^{1+2}{:}(q^2-1)$			1	$\langle\delta,\phi\rangle$
$\mathscr{C}_1$	$GU_2(q)$			1	$\langle\delta,\phi\rangle$
$\mathscr{C}_2$	$(q+1)^2{:}S_3$	$q \neq 5$		1	$\langle\delta,\phi\rangle$
		N1 if $q=5$		1	$\langle\delta,\phi\rangle$
$\mathscr{C}_3$	$(q^2-q+1){:}3$	$q \neq 3,5$		1	$\langle\delta,\phi\rangle$
		N1 if $q=5$		1	$\langle\delta,\phi\rangle$
$\mathscr{C}_5$	$SU_3(q_0).\left(\frac{q+1}{q_0+1},3\right)$	$q=q_0^r$, r odd prime		$(\frac{q+1}{q_0+1},3)$	$\langle\delta^c,\phi\rangle$
$\mathscr{C}_5$	$d \times SO_3(q)$	q odd, $q \geqslant 7$		d	$\langle\phi\rangle$
$\mathscr{C}_6$	$3_+^{1+2}{:}Q_8.\frac{(q+1,9)}{3}$	$p = q \equiv 2 \bmod 3$, $q \geqslant 11$		$\frac{(q+1,9)}{3}$	$\langle\delta^c,\phi\rangle$
		N1 if $q=5$		1	$\langle\delta,\phi\rangle$

N1	Maximal under subgroups not contained in $\langle\phi\rangle$

Note: The group $SU_3(2)$ in Class $\mathscr{C}_5$ with $q = 2^r$ is soluble, and is the normaliser of an extraspecial 3-group.

Table 8.6 *The maximal subgroups of* $SU_3(q)$ *in Class* $\mathscr{S}$
In all examples, $q = p \geqslant 3$. So $d := |Z(SU_3(q))| = (q+1,3)$, $|\delta| = d$, $|\phi| = 2$, $\phi = \gamma$.

Subgp	Nov	Conditions on q	c	Stab	Acts
$d \times L_2(7)$		$q = p \equiv 3,5,6 \bmod 7$, $q \neq 5$	d	$\langle\gamma\rangle$	
	N1	$q = 5$	3	$\langle\gamma\rangle$	
$3{\cdot}A_6$		$q = p \equiv 11,14 \bmod 15$	3	$\langle\gamma\rangle$	$\gamma \to 2_2$
$3{\cdot}A_6{\cdot}2_3$		$q = 5$	3	$\langle\gamma\rangle$	
$3{\cdot}A_7$		$q = 5$	3	$\langle\gamma\rangle$	

N1	Maximal under $\langle\gamma\rangle$

Table 8.7 *The maximal subgroups of* $\Omega_3(q)$ $(\cong L_2(q))$

$|Z(\Omega_3(q))| = 1$, $|\delta| = 2$, $|\phi| = e$, $q = p^e \geqslant 5$ odd.

$\mathscr{C}_i$	Subgp	Nov	Conditions on q	c	Stab
$\mathscr{C}_1$	$E_q : \frac{q-1}{2}$			1	$\langle \delta, \phi \rangle$
$\mathscr{C}_1$	D_{q-1}		$q \neq 5, 7, 9, 11$	1	$\langle \delta, \phi \rangle$
		N1	$q = 7, 11$	1	$\langle \delta \rangle$
		N2	$q = 9$	1	$\langle \delta, \phi \rangle$
$\mathscr{C}_1$	D_{q+1}		$q \neq 7, 9$	1	$\langle \delta, \phi \rangle$
		N1	$q = 7$	1	$\langle \delta \rangle$
		N2	$q = 9$	1	$\langle \delta, \phi \rangle$
$\mathscr{C}_2$	$2^2 : S_3 \cong S_4$		$q = p \equiv \pm 1 \pmod 8$	2	1
	$2^2 : 3 \cong A_4$		$q = p \equiv \pm 3, 5, \pm 13 \pmod{40}$	1	$\langle \delta \rangle$
		N1	$q = p \equiv \pm 11, \pm 19 \pmod{40}$	1	$\langle \delta \rangle$
$\mathscr{C}_5$	$\Omega_3(q_0)$		$q = q_0^r$, r odd prime	1	$\langle \delta, \phi \rangle$
$\mathscr{C}_5$	$SO_3(q_0)$		$q = q_0^2$	2	$\langle \phi \rangle$
$\mathscr{S}_1$	A_5		$q = p \equiv \pm 1 \pmod{10}$	2	1
			$q = p^2$, $p \equiv \pm 3 \pmod{10}$	2	$\langle \phi \rangle$

N1	Maximal under $\langle \delta \rangle$
N2	Maximal under subgroups not contained in $\langle \phi \rangle$

Note: The group in Class $\mathscr{C}_5$ with $q = 3^r$ is also imprimitive.

Table 8.8 *The maximal subgroups of* $\mathrm{SL}_4(q)$ *of geometric type*
$d := |Z(\mathrm{SL}_4(q))| = (q-1, 4)$, $|\delta| = d$, $|\phi| = e$, $|\gamma| = 2$, $q = p^e$.

$\mathscr{C}_i$	Subgp	Notes	c	Stab
$\mathscr{C}_1$	${E_q}^3{:}\mathrm{GL}_3(q)$		2	$\langle \delta, \phi \rangle$
$\mathscr{C}_1$	${E_q}^4{:}\mathrm{SL}_2(q)^2{:}(q-1)$		1	$\langle \delta, \phi, \gamma \rangle$
$\mathscr{C}_1$	${E_q}^{1+4}{:}(\mathrm{GL}_2(q) \times (q-1))$	N1	1	$\langle \delta, \phi, \gamma \rangle$
$\mathscr{C}_1$	$\mathrm{GL}_3(q)$	N1	1	$\langle \delta, \phi, \gamma \rangle$
$\mathscr{C}_2$	$(q-1)^3.S_4$	$q \geqslant 7$	1	$\langle \delta, \phi, \gamma \rangle$
		N2 if $q = 5$	1	$\langle \delta, \gamma \rangle$
$\mathscr{C}_2$	$\mathrm{SL}_2(q)^2{:}(q-1).2$	$q \geqslant 4$	1	$\langle \delta, \phi, \gamma \rangle$
		N3 if $q = 3$	1	$\langle \delta, \gamma \rangle$
$\mathscr{C}_3$	$\mathrm{SL}_2(q^2).(q+1).2$		1	$\langle \delta, \phi, \gamma \rangle$
$\mathscr{C}_5$	$\mathrm{SL}_4(q_0).\left[\left(\frac{q-1}{q_0-1}, 4\right)\right]$	$q = q_0^r$, r prime	$\left(\frac{q-1}{q_0-1}, 4\right)$	$\langle \delta^c, \phi, \gamma \rangle$
$\mathscr{C}_6$	$(4 \circ 2^{1+4})^{\textstyle\cdot}S_6$	$p = q \equiv 1 \bmod 8$	4	$\langle \gamma \rangle$
	$(4 \circ 2^{1+4}).A_6$	$p = q \equiv 5 \bmod 8$	2	$\langle \delta^2, \gamma \rangle$
$\mathscr{C}_8$	$\mathrm{SO}_4^+(q).[d]$	q odd	$d/2$	$\langle \delta^c, \phi, \gamma \rangle$
$\mathscr{C}_8$	$\mathrm{SO}_4^-(q).[d]$	q odd	$d/2$	S1
$\mathscr{C}_8$	$\mathrm{Sp}_4(q).(q-1, 2)$		$(q-1, 2)$	$\langle \delta^c, \phi, \gamma \rangle$
$\mathscr{C}_8$	$\mathrm{SU}_4(q_0).(q_0-1, 4)$	$q = q_0^2$	$(q_0-1, 4)$	$\langle \delta^c, \phi, \gamma \rangle$

S1	$\langle \delta^c, \phi\delta^{(p-1)/2}, \gamma\delta \rangle$
N1	Maximal under subgroups not contained in $\langle \delta, \phi \rangle$
N2	Maximal under subgroups not contained in $\langle \delta^2, \gamma \rangle$
N3	Maximal under subgroups not contained in $\langle \gamma \rangle$

Note: The group $\mathrm{SO}_4^+(q).[d]$ in Class $\mathscr{C}_8$ is also tensor induced.

Table 8.9 *The maximal subgroups of* $\mathrm{SL}_4(q)$ *in Class* $\mathscr{S}$
In all examples, $q = p$. So $d := |Z(\mathrm{SL}_4(q))| = (q-1, 4)$, $|\delta| = d$, $|\phi| = 1$, $|\gamma| = 2$.

Subgp	Nov	Conditions on q	c	Stab
$d \circ 2\text{'}\mathrm{L}_2(7)$	N1	$q = p \equiv 1, 2, 4 \bmod 7$, $q \neq 2$	d	S1
A_7		$q = 2$	1	$\langle \gamma \rangle$
$d \circ 2\text{'}\mathrm{A}_7$		$q = p \equiv 1, 2, 4 \bmod 7$, $q \neq 2$	d	$\langle \gamma \rangle$
$d \circ 2\text{'}\mathrm{U}_4(2)$		$q = p \equiv 1 \bmod 6$	d	$\langle \gamma \rangle$

S1	$\langle \gamma \rangle$ ($p \equiv \pm 1 \bmod 8$) or $\langle \delta\gamma \rangle$ ($p \equiv \pm 3 \bmod 8$)	N1	Maximal under S1

Table 8.10 *The maximal subgroups of* $SU_4(q)$ *of geometric type*
$d := |Z(SU_4(q))| = (q+1, 4)$, $|\delta| = d$, $|\phi| = 2e$, $q = p^e$.

$\mathscr{C}_i$	Subgp	Notes	c	Stab
$\mathscr{C}_1$	$E_q^{1+4}{:}SU_2(q){:}(q^2-1)$		1	$\langle \delta, \phi \rangle$
$\mathscr{C}_1$	$E_q^{\,4}{:}SL_2(q^2){:}(q-1)$		1	$\langle \delta, \phi \rangle$
$\mathscr{C}_1$	$GU_3(q)$		1	$\langle \delta, \phi \rangle$
$\mathscr{C}_2$	$(q+1)^3.S_4$	$q \neq 3$	1	$\langle \delta, \phi \rangle$
		N1 if $q = 3$	1	$\langle \delta, \phi \rangle$
$\mathscr{C}_2$	$SU_2(q)^2{:}(q+1).2$	$q \geqslant 3$	1	$\langle \delta, \phi \rangle$
$\mathscr{C}_2$	$SL_2(q^2).(q-1).2$	$q \geqslant 4$	1	$\langle \delta, \phi \rangle$
		N1 if $q = 3$	1	$\langle \delta, \phi \rangle$
$\mathscr{C}_5$	$SU_4(q_0)$	$q = q_0^r,$	1	$\langle \delta, \phi \rangle$
		r odd prime		
$\mathscr{C}_5$	$Sp_4(q).(q+1,2)$		$(q+1,2)$	$\langle \delta^c, \phi \rangle$
$\mathscr{C}_5$	$SO_4^+(q).[d]$	$q \geqslant 5$ odd	$d/2$	$\langle \delta^c, \phi \rangle$
$\mathscr{C}_5$	$SO_4^-(q).[d]$	q odd	$d/2$	$\langle \delta^c, \phi \delta^{(p-1)/2} \rangle$
$\mathscr{C}_6$	$(4 \circ 2^{1+4})^{\textstyle{\cdot}}S_6$	$p = q \equiv 7 \bmod 8$	4	$\langle \phi \rangle$
	$(4 \circ 2^{1+4}).A_6$	$p = q \equiv 3 \bmod 8$	2	$\langle \delta^2, \phi \rangle$

N1   Maximal under subgroups not contained in $\langle \delta^2, \phi \rangle$

Note: The group $SO_4^+(q).[d]$ in Class $\mathscr{C}_5$ is also tensor induced.

Table 8.11 *The maximal subgroups of* $SU_4(q)$ *in Class* $\mathscr{S}$
In all examples, $q = p$. So $d := |Z(SU_4(q))| = (q+1, 4)$, $|\delta| = d$, $|\phi| = 2$, $\phi = \gamma$.

Subgp	Nov	Conditions on q	c	Stab	Acts
$d \circ 2^{\textstyle{\cdot}}L_2(7)$	N1	$q = p \equiv 3, 5, 6 \bmod 7$, $q \neq 3$	d	S1	
$d \circ 2^{\textstyle{\cdot}}A_7$		$q = p \equiv 3, 5, 6 \bmod 7$	d	$\langle \gamma \rangle$	
$4_2{}^{\textstyle{\cdot}}L_3(4)$		$q = 3$	2	$\langle \delta^2, \gamma\delta \rangle$	A1
$d \circ 2^{\textstyle{\cdot}}U_4(2)$		$q = p \equiv 5 \bmod 6$	d	$\langle \gamma \rangle$	

S1	$\langle \gamma \rangle$ ($p \equiv \pm 1 \bmod 8$) or $\langle \gamma\delta \rangle$ ($p \equiv \pm 3 \bmod 8$)
N1	Maximal under S1
A1	$\delta^2 \to 2_2$, $\gamma\delta \to 2_1$ and 2_3 in the two classes

Table 8.12 *The maximal subgroups of* $\mathrm{Sp}_4(q)$ *of geometric type,* q *odd*
$|Z(\mathrm{Sp}_4(q))| = 2$, $|\delta| = 2$, $|\phi| = e$, $q = p^e$ odd.

$\mathscr{C}_i$	Subgp	Notes	c	Stab
$\mathscr{C}_1$	$E_q^{1+2}{:}((q-1) \times \mathrm{Sp}_2(q))$		1	$\langle \delta, \phi \rangle$
$\mathscr{C}_1$	$E_q^{3}{:}\mathrm{GL}_2(q)$		1	$\langle \delta, \phi \rangle$
$\mathscr{C}_2$	$\mathrm{Sp}_2(q)^2{:}2$		1	$\langle \delta, \phi \rangle$
$\mathscr{C}_2$	$\mathrm{GL}_2(q).2$	$q \geqslant 5$	1	$\langle \delta, \phi \rangle$
$\mathscr{C}_3$	$\mathrm{Sp}_2(q^2){:}2$		1	$\langle \delta, \phi \rangle$
$\mathscr{C}_3$	$\mathrm{GU}_2(q).2$	$q \geqslant 5$	1	$\langle \delta, \phi \rangle$
$\mathscr{C}_5$	$\mathrm{Sp}_4(q_0).(2,r)$	$q = q_0^r$, r prime	$(2,r)$	$\langle \delta^c, \phi \rangle$
$\mathscr{C}_6$	$2^{1+4}_-.\mathrm{S}_5$	$q = p \equiv \pm 1 \bmod 8$	2	1
$\mathscr{C}_6$	$2^{1+4}_-.\mathrm{A}_5$	$q = p \equiv \pm 3 \bmod 8$	1	$\langle \delta \rangle$

Table 8.13 *The maximal subgroups of* $\mathrm{Sp}_4(q)$ *in Class* $\mathscr{S}$, q *odd.*
$|Z(\mathrm{Sp}_4(q))| = 2$, $|\delta| = 2$, $|\phi| = e$, $q = p^e$ odd.

Subgp	Nov	Conditions on q	c	Stab	Acts
$2^{\cdot}\mathrm{A}_6$		$q = p \equiv 5,7 \bmod 12$, $q \neq 7$	1	$\langle \delta \rangle$	$\delta \to 2_1$
$2^{\cdot}\mathrm{S}_6$		$q = p \equiv 1,11 \bmod 12$	2	1	
$2^{\cdot}\mathrm{A}_7$		$q = 7$	1	$\langle \delta \rangle$	
$\mathrm{SL}_2(q)$		$p \geqslant 5$, $q > 7$	1	$\langle \delta, \phi \rangle$	
	N1	$q = 7$	1	$\langle \delta \rangle$	

N1	Maximal under $\langle \delta \rangle$

When $p = 2$, the group $\mathrm{S}_4(q)$ has an additional automorphism, the graph automorphism γ, which squares to the field automorphism ϕ. The rows in the following two tables represent the classes of subgroups under Γ, and the fusion under A is given in the final column: note that this departs from our usual convention of each row representing one A-conjugacy class of groups.

Alternative notation for the classes for maximal subgroups of subgroups of $\mathrm{Aut}\,\mathrm{S}_4(2^e)$ not contained in $\mathrm{P\Sigma Sp}_4(2^e)$ is introduced in [1, Section 14]. In the "Classes" column of Table 8.14, the first class is the standard class name from Chapter 2, and the second is the number of the $(\mathrm{Aut}\,\mathrm{S}_4(2^e))$-class as given in [1] but, to avoid a clash with our notation, we have replaced $\mathscr{C}_i$ in [1] by $\mathscr{A}_i$.

Table 8.14 *The maximal subgroups of* $\mathrm{Sp}_4(q) \cong \mathrm{S}_4(q)$, $q \geqslant 4$ *even*

$|Z(\mathrm{Sp}_4(q))| = 1$, $|\delta| = 1$, $|\phi| = e$, $|\gamma| = 2e$, $\gamma^2 = \phi$, $q = 2^e \geqslant 4$.

Classes		Subgp	Nov	Notes	c	Stab	Fus of γ
$\mathscr{C}_1$	—	$E_q{}^3{:}\mathrm{GL}_2(q)$		point stabiliser	1	$\langle\phi\rangle$	line stab
$\mathscr{C}_1$	—	$E_q{}^3{:}\mathrm{GL}_2(q)$		line stabiliser	1	$\langle\phi\rangle$	point stab
—	$\mathscr{A}_1$	$[q^4]{:}C_{q-1}{}^2$	N1		1	$\langle\gamma\rangle$	self
$\mathscr{C}_2$	—	$\mathrm{Sp}_2(q) \wr 2$			1	$\langle\phi\rangle$	$\mathrm{SO}_4^+(q)$
$\mathscr{C}_3$	—	$\mathrm{Sp}_2(q^2){:}2$			1	$\langle\phi\rangle$	$\mathrm{SO}_4^-(q)$
—	$\mathscr{A}_2$	$C_{q-1}{}^2{:}D_8$	N1	$q \neq 4$	1	$\langle\gamma\rangle$	self
—	$\mathscr{A}_2$	$C_{q+1}{}^2{:}D_8$	N1		1	$\langle\gamma\rangle$	self
—	$\mathscr{A}_3$	$C_{q^2+1}{:}4$	N1		1	$\langle\gamma\rangle$	self
$\mathscr{C}_5$	$\mathscr{A}_4$	$\mathrm{Sp}_4(q_0)$		$q = q_0^r$, r prime	1	$\langle\gamma\rangle$	self
$\mathscr{C}_8$	—	$\mathrm{SO}_4^+(q)$			1	$\langle\phi\rangle$	$\mathrm{Sp}_2(q) \wr 2$
$\mathscr{C}_8$	—	$\mathrm{SO}_4^-(q)$			1	$\langle\phi\rangle$	$\mathrm{Sp}_2(q^2){:}2$
$\mathscr{S}_2$	$\mathscr{A}_5$	$\mathrm{Sz}(q)$		$e \geqslant 3$ odd	1	$\langle\gamma\rangle$	self

N1	Maximal under subgroups not contained in $\langle\phi\rangle$

Note: The $\mathscr{C}_8$-subgroup $\mathrm{SO}_4^+(q)$ is also tensor induced.

Table 8.15 *The maximal subgroups of* $\mathrm{Sp}_4(2) \cong \mathrm{S}_4(2) \cong \mathrm{S}_6$

$|Z(\mathrm{Sp}_4(2))| = 1$, $|\delta| = 1$, $|\phi| = 1$, $|\gamma| = 2$.

Classes		Subgp	Nov	c	Stab	Fusion of γ
$\mathscr{C}_1$	—	$\mathrm{S}_4 \times 2$ (point stabiliser)		1	1	line stab
$\mathscr{C}_1$	—	$\mathrm{S}_4 \times 2$ (line stabiliser)		1	1	point stab
—	$\mathscr{A}_1$	$D_8 \times 2$	N1	1	$\langle\gamma\rangle$	self
$\mathscr{C}_3$	—	$\mathrm{Sp}_2(4){:}2 \cong \mathrm{S}_5$		1	1	$\mathrm{SO}_4^-(2)$
$\mathscr{C}_{2/8}$	$\mathscr{A}_2$	$3^2{:}D_8$		1	$\langle\gamma\rangle$	self
—	$\mathscr{A}_{3/5}$	$5{:}4 \cong \mathrm{Sz}(2)$	N1	1	$\langle\gamma\rangle$	self
$\mathscr{C}_8$	—	$\mathrm{SO}_4^-(2) \cong \mathrm{S}_5$		1	1	$\mathrm{Sp}_2(4){:}2$
$\mathscr{T}$	$\mathscr{T}$	A_6		1	$\langle\gamma\rangle$	self

N1	Maximal under $\langle\gamma\rangle$

Note: The A_6-subgroup of $\mathrm{S}_4(2)$ contains the socle of $\mathrm{S}_4(2)$. It is therefore a triviality (denoted as class $\mathscr{T}$), and gives rise to *two* (rather than one) maximal subgroups of $\mathrm{S}_4(2){:}2$ intersecting $\mathrm{S}_4(2)$ in A_6, namely $\mathrm{PGL}_2(9)$ and M_{10}.

Table 8.16 *The maximal subgroups of* $\mathrm{Sz}(q) < \mathrm{Sp}_4(q)$, $q = 2^e$, $e > 1$ *odd.*
$|Z(\mathrm{Sz}(q))| = 1$, $|\phi| = e$. Note that $\mathrm{Sz}(2) \cong \mathrm{F}_{20} \cong 5{:}4$ is soluble.

$\mathscr{C}_i$	Suzuki	Subgp	Notes	c	Stab
$\mathscr{C}_1$	$H(q)$	$\mathrm{E}_q^{1+1}{:}\mathrm{C}_{q-1}$		1	$\langle\phi\rangle$
$\mathscr{C}_2/\mathscr{C}_1$	B_0	$\mathrm{D}_{2(q-1)}$		1	$\langle\phi\rangle$
$\mathscr{C}_3/\mathscr{C}_8$	B_1 or B_2	$(q - \sqrt{2q} + 1){:}4$		1	$\langle\phi\rangle$
$\mathscr{C}_3/\mathscr{C}_8$	B_2 or B_1	$(q + \sqrt{2q} + 1){:}4$		1	$\langle\phi\rangle$
$\mathscr{C}_5$	$G(q_0)$	$\mathrm{Sz}(q_0)$	$q = q_0^r$, r prime, $q_0 \neq 2$	1	$\langle\phi\rangle$

Table 8.17 *The maximal subgroups of* $\Omega_4^-(q)\,(\cong \mathrm{L}_2(q^2))$
$|Z(\Omega_4^-(q))| = 1$, $|\delta| = (q-1, 2)$, $|\gamma| = 2$, $|\varphi| = 2e$, $\varphi^e = \gamma$, $q = p^e$.

$\mathscr{C}_i$	Subgp	Nov	Conditions on q	c	Stab
$\mathscr{C}_1$	$\mathrm{E}_{q^2}{:}\frac{q^2-1}{(q-1,2)}$			1	$\langle\delta,\varphi\rangle$
$\mathscr{C}_1$	$\mathrm{L}_2(q).(q-1,2)$		$q \neq 2$	$(q-1,2)$	$\langle\varphi\rangle$
$\mathscr{C}_1$	$\mathrm{D}_{2(q^2-1)/(q-1,2)}$		$q \neq 3$	1	$\langle\delta,\varphi\rangle$
		N1	$q = 3$	1	$\langle\delta,\gamma\rangle$
$\mathscr{C}_3$	$\mathrm{D}_{2(q^2+1)/(q-1,2)}$		$q \neq 3$	1	$\langle\delta,\varphi\rangle$
		N1	$q = 3$	1	$\langle\delta,\gamma\rangle$
$\mathscr{C}_5$	$\Omega_4^-(q_0) \cong \mathrm{L}_2(q_0^2)$		$q = q_0^r$, r odd prime	1	$\langle\delta,\varphi\rangle$
$\mathscr{S}_1$	A_5		$q = p \equiv \pm 3 \pmod{10}$	2	$\langle\gamma\rangle$

N1	Maximal under subgroups not contained in $\langle\gamma\rangle$

Table 8.18 *The maximal subgroups of* $\mathrm{SL}_5(q)$ *of geometric type*
$d := |Z(\mathrm{SL}_5(q))| = (q-1,5),\ |\delta| = d,\ |\phi| = e,\ |\gamma| = 2,\ q = p^e.$

$\mathscr{C}_i$	Subgp	Notes	c	Stab
$\mathscr{C}_1$	$E_q{}^4{:}\mathrm{GL}_4(q)$		2	$\langle \delta, \phi \rangle$
$\mathscr{C}_1$	$E_q{}^6{:}(\mathrm{SL}_2(q) \times \mathrm{SL}_3(q)){:}(q-1)$		2	$\langle \delta, \phi \rangle$
$\mathscr{C}_1$	$E_q{}^{1+6}{:}(\mathrm{GL}_3(q) \times (q-1))$	N1	1	$\langle \delta, \phi, \gamma \rangle$
$\mathscr{C}_1$	$E_q{}^{4+4}{:}\mathrm{GL}_2(q)^2$	N1	1	$\langle \delta, \phi, \gamma \rangle$
$\mathscr{C}_1$	$\mathrm{GL}_4(q)$	N1	1	$\langle \delta, \phi, \gamma \rangle$
$\mathscr{C}_1$	$(\mathrm{SL}_2(q) \times \mathrm{SL}_3(q)){:}(q-1)$	N1	1	$\langle \delta, \phi, \gamma \rangle$
$\mathscr{C}_2$	$(q-1)^4{:}\mathrm{S}_5$	$q \geqslant 5$	1	$\langle \delta, \phi, \gamma \rangle$
$\mathscr{C}_3$	$\frac{q^5-1}{q-1}{:}5$		1	$\langle \delta, \phi, \gamma \rangle$
$\mathscr{C}_5$	$\mathrm{SL}_5(q_0).\left(\frac{q-1}{q_0-1},5\right)$	$q = q_0^r,\ r$ prime	$\left(\frac{q-1}{q_0-1},5\right)$	$\langle \delta^c, \phi, \gamma \rangle$
$\mathscr{C}_6$	$5_+^{1+2}{:}\mathrm{Sp}_2(5)$	$p = q \equiv 1 \bmod 5$	5	$\langle \gamma \rangle$
$\mathscr{C}_8$	$d \times \mathrm{SO}_5(q)$	q odd	d	$\langle \phi, \gamma \rangle$
$\mathscr{C}_8$	$(q_0-1,5) \times \mathrm{SU}_5(q_0)$	$q = q_0^2$	$(q_0-1,5)$	$\langle \delta^c, \phi, \gamma \rangle$

> N1   Maximal under subgroups not contained in $\langle \delta, \phi \rangle$

Table 8.19 *The maximal subgroups of* $\mathrm{SL}_5(q)$ *in Class* $\mathscr{S}$.

In all examples, $q = p$. So $d := |Z(\mathrm{SL}_5(q))| = (q-1,5),\ |\delta| = d,\ |\phi| = 1,\ |\gamma| = 2.$

Subgp	Nov	Conditions on q	c	Stab
$d \times \mathrm{L}_2(11)$		$q = p \equiv 1,3,4,5,9 \bmod 11,\ q \neq 3$	d	$\langle \gamma \rangle$
	N1	$q = 3$	1	$\langle \gamma \rangle$
M_{11}		$q = 3$	2	1
$d \times \mathrm{U}_4(2)$		$q = p \equiv 1 \bmod 6$	d	$\langle \gamma \rangle$

> N1   Maximal under $\langle \gamma \rangle$

Table 8.20 *The maximal subgroups of* $\mathrm{SU}_5(q)$ *of geometric type*
$d := |Z(\mathrm{SU}_5(q))| = (q+1,5),\ |\delta| = d,\ |\phi| = 2e,\ q = p^e.$

$\mathscr{C}_i$	Subgp	Notes	c	Stab
$\mathscr{C}_1$	$E_q{}^{1+6}{:}\mathrm{SU}_3(q){:}(q^2-1)$		1	$\langle \delta, \phi \rangle$
$\mathscr{C}_1$	$E_q{}^{4+4}{:}\mathrm{GL}_2(q^2)$		1	$\langle \delta, \phi \rangle$
$\mathscr{C}_1$	$\mathrm{GU}_4(q)$		1	$\langle \delta, \phi \rangle$
$\mathscr{C}_1$	$(\mathrm{SU}_3(q) \times \mathrm{SU}_2(q)){:}(q+1)$		1	$\langle \delta, \phi \rangle$
$\mathscr{C}_2$	$(q+1)^4{:}\mathrm{S}_5$		1	$\langle \delta, \phi \rangle$
$\mathscr{C}_3$	$\frac{q^5+1}{q+1}{:}5$	$q \geqslant 3$	1	$\langle \delta, \phi \rangle$
$\mathscr{C}_5$	$\mathrm{SU}_5(q_0).\left(\frac{q+1}{q_0+1},5\right)$	$q = q_0^r,\ r$ odd prime	$\left(\frac{q+1}{q_0+1},5\right)$	$\langle \delta^c, \phi \rangle$
$\mathscr{C}_5$	$d \times \mathrm{SO}_5(q)$	q odd	d	$\langle \phi \rangle$
$\mathscr{C}_6$	$5_+^{1+2}{:}\mathrm{Sp}_2(5)$	$q = p \equiv 4 \bmod 5$, or	5	$\langle \phi \rangle$
		$(q = p^2\ \&\ p \equiv 2,3 \bmod 5)$		

Table 8.21 *The maximal subgroups of* $\mathrm{SU}_5(q)$ *in Class* $\mathscr{S}$
In all examples, $q = p$. So $d := |Z(\mathrm{SU}_5(q))| = (q+1,5),\ |\delta| = d,\ |\phi| = 2,\ \phi = \gamma.$

Subgp	Conditions on q	c	Stab
$d \times \mathrm{L}_2(11)$	$q = p \equiv 2,6,7,8,10 \bmod 11$	d	$\langle \gamma \rangle$
$d \times \mathrm{U}_4(2)$	$q = p \equiv 5 \bmod 6$	d	$\langle \gamma \rangle$

Table 8.22 *The maximal subgroups of* $\Omega_5(q)\ (\cong \mathrm{S}_4(q))$ *of geometric type*
$|Z(\Omega_5(q))| = 1,\ |\delta| = 2,\ |\phi| = e,\ q = p^e$ odd.

$\mathscr{C}_i$	Subgp	Notes	c	Stab
$\mathscr{C}_1$	$E_q{}^3{:}(\frac{q-1}{2} \times \Omega_3(q)).2$		1	$\langle \delta, \phi \rangle$
$\mathscr{C}_1$	$E_q{}^{1+2}{:}\frac{1}{2}\mathrm{GL}_2(q)$		1	$\langle \delta, \phi \rangle$
$\mathscr{C}_1$	$\Omega_4^+(q).2$		1	$\langle \delta, \phi \rangle$
$\mathscr{C}_1$	$\Omega_4^-(q).2$		1	$\langle \delta, \phi \rangle$
$\mathscr{C}_1$	$(\frac{q-1}{2} \times \Omega_3(q)).2^2$	$q \geqslant 5$	1	$\langle \delta, \phi \rangle$
$\mathscr{C}_1$	$(\frac{q+1}{2} \times \Omega_3(q)).2^2$	$q \geqslant 5$	1	$\langle \delta, \phi \rangle$
$\mathscr{C}_2$	$2^4{:}\mathrm{A}_5$	$q = p \equiv 3,5 \bmod 8$	1	$\langle \delta \rangle$
	$2^4{:}\mathrm{S}_5$	$q = p \equiv 1,7 \bmod 8$	2	1
$\mathscr{C}_5$	$\Omega_5(q_0)$	$q = q_0^r,\ r$ odd prime	1	$\langle \delta, \phi \rangle$
$\mathscr{C}_5$	$\mathrm{SO}_5(q_0)$	$q = q_0^2$	2	$\langle \phi \rangle$

Table 8.23 *The maximal subgroups of* $\Omega_5(q)$ $(\cong S_4(q))$ *in Class* $\mathscr{S}$
$|Z(\Omega_5(q))| = 1$, $|\delta| = 2$, $|\phi| = e$, $q = p^e$ *odd.*

Subgp	Nov	Conditions on q	c	Stab	Acts
A_6		$q = p \equiv 5, 7 \bmod 12$, $q \neq 7$	1	$\langle \delta \rangle$	$\delta \to 2_1$
S_6		$q = p \equiv 1, 11 \bmod 12$	2	1	
A_7		$q = 7$	1	$\langle \delta \rangle$	
$L_2(q)$		$p \geqslant 5$, $q > 7$	1	$\langle \delta, \phi \rangle$	
	N1	$q = 7$	1	$\langle \delta \rangle$	

N1	Maximal under $\langle \delta \rangle$

Table 8.24 *The maximal subgroups of* $\mathrm{SL}_6(q)$ *of geometric type*
$d := |Z(\mathrm{SL}_6(q))| = (q - 1, 6)$, $|\delta| = d$, $|\phi| = e$, $|\gamma| = 2$, $q = p^e$.

$\mathscr{C}_i$	Subgp	Notes	c	Stab
$\mathscr{C}_1$	$E_q^{\,5}{:}\mathrm{GL}_5(q)$		2	$\langle \delta, \phi \rangle$
$\mathscr{C}_1$	$E_q^{\,8}{:}(\mathrm{SL}_4(q) \times \mathrm{SL}_2(q)){:}(q - 1)$		2	$\langle \delta, \phi \rangle$
$\mathscr{C}_1$	$E_q^{\,9}{:}(\mathrm{SL}_3(q) \times \mathrm{SL}_3(q)){:}(q - 1)$		1	$\langle \delta, \phi, \gamma \rangle$
$\mathscr{C}_1$	$E_q^{\,1+8}{:}(\mathrm{GL}_4(q) \times (q - 1))$	N1	1	$\langle \delta, \phi, \gamma \rangle$
$\mathscr{C}_1$	$E_q^{\,4+8}{:}\mathrm{SL}_2(q)^3{:}(q - 1)^2$	N1	1	$\langle \delta, \phi, \gamma \rangle$
$\mathscr{C}_1$	$\mathrm{GL}_5(q)$	N1	1	$\langle \delta, \phi, \gamma \rangle$
$\mathscr{C}_1$	$(\mathrm{SL}_4(q) \times \mathrm{SL}_2(q)){:}(q - 1)$	N1	1	$\langle \delta, \phi, \gamma \rangle$
$\mathscr{C}_2$	$(q - 1)^5.S_6$	$q \geqslant 5$	1	$\langle \delta, \phi, \gamma \rangle$
$\mathscr{C}_2$	$\mathrm{SL}_2(q)^3{:}(q - 1)^2.S_3$	$q \geqslant 3$	1	$\langle \delta, \phi, \gamma \rangle$
$\mathscr{C}_2$	$\mathrm{SL}_3(q)^2{:}(q - 1).S_2$		1	$\langle \delta, \phi, \gamma \rangle$
$\mathscr{C}_3$	$\mathrm{SL}_3(q^2).(q + 1).2$		1	$\langle \delta, \phi, \gamma \rangle$
$\mathscr{C}_3$	$\mathrm{SL}_2(q^3).(q^2 + q + 1).3$		1	$\langle \delta, \phi, \gamma \rangle$
$\mathscr{C}_4$	$\mathrm{SL}_2(q) \times \mathrm{SL}_3(q)$	$q \geqslant 3$	1	$\langle \delta, \phi, \gamma \rangle$
$\mathscr{C}_5$	$\mathrm{SL}_6(q_0).\left[\left(\frac{q-1}{q_0-1}, 6\right)\right]$	$q = q_0^r$	$\left(\frac{q-1}{q_0-1}, 6\right)$	$\langle \delta^c, \phi, \gamma \rangle$
		r prime		
$\mathscr{C}_8$	$(q - 1, 3) \times \mathrm{SO}_6^+(q).2$	q odd	$d/2$	$\langle \delta^c, \phi, \gamma \rangle$
$\mathscr{C}_8$	$(q - 1, 3) \times \mathrm{SO}_6^-(q).2$	q odd	$d/2$	S1
$\mathscr{C}_8$	$(q - 1, 3) \times \mathrm{Sp}_6(q)$		$(q - 1, 3)$	$\langle \delta^c, \phi, \gamma \rangle$
$\mathscr{C}_8$	$\mathrm{SU}_6(q_0).(q_0 - 1, 6)$	$q = q_0^2$	$(q_0 - 1, 6)$	$\langle \delta^c, \phi, \gamma \rangle$

S1	$\langle \delta^c, \phi, \gamma \rangle$ $(q \equiv 3 \bmod 4)$ or $\langle \delta^c, \phi\delta^{(p-1)/2}, \gamma\delta^{-1} \rangle$ $(q \equiv 1 \bmod 4)$
N1	Maximal under subgroups not contained in $\langle \delta, \phi \rangle$

Table 8.25 *The maximal subgroups of* $\mathrm{SL}_6(q)$ *in Class* $\mathscr{S}$
$d := |Z(\mathrm{SL}_6(q))| = (q-1,6)$, $|\delta| = d$, $|\phi| = e$, $|\gamma| = 2$, $q = p^e$.

Subgp	Nov	Conditions on q	c	Stab	Acts
$2 \times 3^{\cdot}A_6.2_3$	N1	$q = p \equiv 1 \bmod 24$	6	$\langle\gamma\rangle$	
	N2	$q = p \equiv 19 \bmod 24$	6	$\langle\gamma\delta\rangle$	
$2 \times 3^{\cdot}A_6$	N3	$q = p \equiv 7 \bmod 24$	3	$\langle\delta^3,\gamma\rangle$	A1
	N4	$q = p \equiv 13 \bmod 24$	3	$\langle\delta^3,\gamma\rangle$	A2
$6^{\cdot}A_6$	N1	$q = p \equiv 1,31 \bmod 48$	6	$\langle\gamma\rangle$	$\gamma \to 2_2$
	N2	$q = p \equiv 7,25 \bmod 48$	6	$\langle\gamma\delta\rangle$	$\gamma\delta \to 2_2$
$6^{\cdot}A_6$	N5	$q = p^2, p \equiv 5,11 \bmod 24$	6	$\langle\phi,\gamma\delta^3\rangle$	A3
$6^{\cdot}A_6$	N6	$q = p^2, p \equiv 13,19 \bmod 24$	6	$\langle\phi\gamma,\gamma\delta^3\rangle$	A4
$d \circ 2^{\cdot}L_2(11)$		$q = p \equiv 1,3,4,5,9 \bmod 11, q \neq 3$	d		
		$p \equiv \pm1 \bmod 8$		$\langle\gamma\rangle$	
		$p \equiv \pm3 \bmod 8$		$\langle\gamma\delta\rangle$	
$6^{\cdot}A_7$		$q = p \equiv 1,7 \bmod 24$	12	1	
$6^{\cdot}A_7$		$q = p^2, p \equiv 5,11 \bmod 24$	12	$\langle\phi\rangle$	
$6^{\cdot}A_7$		$q = p^2, p \equiv 13,19 \bmod 24$	12	$\langle\phi\gamma\rangle$	
$6^{\cdot}L_3(4)^{\cdot}2_1^-$		$q = p \equiv 1,19 \bmod 24$	6	$\langle\gamma\rangle$	$\gamma \to 2^2$
$6^{\cdot}L_3(4)$	N7	$q = p \equiv 7,13 \bmod 24$	3	$\langle\delta^3,\gamma\rangle$	A5
$2^{\cdot}M_{12}$		$q = 3$	2	$\langle\gamma\delta\rangle$	
$6_1^{\cdot}U_4(3)^{\cdot}2_2^-$		$q = p \equiv 1 \bmod 12$	6	$\langle\gamma\rangle$	A6
$6_1^{\cdot}U_4(3)$		$q = p \equiv 7 \bmod 12$	3	$\langle\delta^3,\gamma\rangle$	A7
$d \circ \mathrm{SL}_3(q)$		$q \equiv \pm1 \bmod 8$	2	$\langle\delta^2,\phi,\gamma\rangle$	
		$q \equiv \pm3 \bmod 8$	2	$\langle\delta^2,\phi,\gamma\delta\rangle$	

N1 Maximal under $\langle\gamma\rangle$	A1 $\delta^3 \to 2_3, \gamma \to 2_2$
N2 Maximal under $\langle\gamma\delta\rangle$	A2 $\delta^3 \to 2_3, \gamma \to 2_1$
N3 Maximal under $\langle\gamma\rangle$ if $p \equiv \pm2 \bmod 5$, $\langle\delta^3,\gamma\rangle$	A3 $\phi \to 2_1, \gamma\delta^3 \to 2_2$
N4 Maximal under $\langle\gamma\delta^3\rangle$ if $p \equiv \pm2 \bmod 5$, $\langle\delta^3,\gamma\rangle$	A4 $\phi\gamma \to 2_1, \gamma\delta^3 \to 2_2$
N5 Maximal under subgroups not contained in $\langle\phi\rangle$	A5 $\delta^3 \to 2_1, \gamma \to 2_2$
N6 Maximal under subgroups not contained in $\langle\phi\gamma\rangle$	A6 $\gamma \to (2^2)_{122}$
N7 Maximal under subgroups not contained in $\langle\gamma\rangle$	A7 $\delta^3 \to 2_2, \gamma \to 2_1$

Table 8.26 *The maximal subgroups of* $SU_6(q)$ *of geometric type*
$d := |Z(SU_6(q))| = (q+1,6)$, $|\delta| = d$, $|\phi| = 2e$, $\phi^e = \gamma$, $q = p^e$.

$\mathscr{C}_i$	Subgp	Notes	c	Stab
$\mathscr{C}_1$	$E_q^{1+8}{:}SU_4(q){:}(q^2-1)$		1	$\langle\delta,\phi\rangle$
$\mathscr{C}_1$	$E_q^{4+8}{:}(SL_2(q^2)\times SU_2(q)){:}(q^2-1)$		1	$\langle\delta,\phi\rangle$
$\mathscr{C}_1$	$E_q^{9}{:}SL_3(q^2){:}(q-1)$		1	$\langle\delta,\phi\rangle$
$\mathscr{C}_1$	$GU_5(q)$		1	$\langle\delta,\phi\rangle$
$\mathscr{C}_1$	$(SU_4(q)\times SU_2(q)){:}(q+1)$		1	$\langle\delta,\phi\rangle$
$\mathscr{C}_2$	$(q+1)^5.S_6$	$q \geqslant 3$	1	$\langle\delta,\phi\rangle$
		N1 if $q=2$	1	$\langle\delta,\gamma\rangle$
$\mathscr{C}_2$	$SU_2(q)^3{:}(q+1)^2.S_3$	$q \geqslant 3$	1	$\langle\delta,\phi\rangle$
$\mathscr{C}_2$	$SU_3(q)^2{:}(q+1).S_2$		1	$\langle\delta,\phi\rangle$
$\mathscr{C}_2$	$SL_3(q^2).(q-1).2$		1	$\langle\delta,\phi\rangle$
$\mathscr{C}_3$	$SU_2(q^3).(q^2-q+1).3$	$q \geqslant 3$	1	$\langle\delta,\phi\rangle$
		N1 if $q=2$	1	$\langle\delta,\gamma\rangle$
$\mathscr{C}_4$	$SU_2(q)\times SU_3(q)$	$q \geqslant 3$	1	$\langle\delta,\phi\rangle$
$\mathscr{C}_5$	$SU_6(q_0).\left[\left(\frac{q+1}{q_0+1},6\right)\right]$	$q = q_0^r$,	$\left(\frac{q+1}{q_0+1},6\right)$	$\langle\delta^c,\phi\rangle$
		r odd prime		
$\mathscr{C}_5$	$(q+1,3)\times Sp_6(q)$		$(q+1,3)$	$\langle\delta^c,\phi\rangle$
$\mathscr{C}_5$	$(q+1,3)\times SO_6^+(q).2$	q odd	$(q+1,3)$	S1
$\mathscr{C}_5$	$(q+1,3)\times SO_6^-(q).2$	q odd	$(q+1,3)$	S2

S1	$\langle\delta^c,\phi\rangle$ $(q \equiv 1 \bmod 4)$ or $\langle\delta^c,\phi\delta^{(p-1)/2}\rangle$ $(q \equiv 3 \bmod 4)$
S2	$\langle\delta^c,\phi\delta^{(p-1)/2}\rangle$ $(q \equiv 1 \bmod 4)$ or $\langle\delta^c,\phi\rangle$ $(q \equiv 3 \bmod 4)$
N1	Maximal under subgroups not contained in $\langle\gamma\rangle$

Table 8.27 *The maximal subgroups of* $\mathrm{SU}_6(q)$ *in Class* $\mathscr{S}$
$d := |Z(\mathrm{SU}_6(q))| = (q+1, 6)$, $|\delta| = d$, $|\phi| = 2e$, $\phi^e = \gamma$, $q = p^e$.

Subgp	Nov	Conditions on q	c	Stab	Acts
$2 \times 3^{\cdot}\mathrm{A}_6$	N1	$q = p \equiv 11 \bmod 24$	3	$\langle \delta^3, \gamma \rangle$	A1
	N2	$q = p \equiv 17 \bmod 24$	3	$\langle \delta^3, \gamma \rangle$	A2
$2 \times 3^{\cdot}\mathrm{A}_6.2_3$	N3	$q = p \equiv 5 \bmod 24$, $q \neq 5$	6	$\langle \gamma\delta \rangle$	
	N4	$q = p \equiv 23 \bmod 24$	6	$\langle \gamma \rangle$	
$6^{\cdot}\mathrm{A}_6$	N4	$q = p \equiv 17, 47 \bmod 48$	6	$\langle \gamma \rangle$	$\gamma \to 2_2$
	N3	$q = p \equiv 23, 41 \bmod 48$	6	$\langle \gamma\delta \rangle$	$\gamma\delta \to 2_2$
$d \circ 2^{\cdot}\mathrm{L}_2(11)$		$q = p \equiv 2, 6, 7, 8, 10 \bmod 11$, $q \neq 2$	d		
		$p \equiv \pm 1 \bmod 8$		$\langle \gamma \rangle$	
		$p \equiv \pm 3 \bmod 8$		$\langle \gamma\delta \rangle$	
$6^{\cdot}\mathrm{A}_7$		$q = p \equiv 17, 23 \bmod 24$	12	1	
$6^{\cdot}\mathrm{L}_3(4)$	N5	$q = p \equiv 11, 17 \bmod 24$	3	$\langle \delta^3, \gamma \rangle$	A3
$6^{\cdot}\mathrm{L}_3(4)^{\cdot}2_1^-$		$q = p \equiv 5, 23 \bmod 24$	6	$\langle \gamma \rangle$	$\gamma \to 2^2$
$3^{\cdot}\mathrm{M}_{22}$		$q = 2$	3	$\langle \gamma \rangle$	
$3_1^{\cdot}\mathrm{U}_4(3):2_2$		$q = 2$	3	$\langle \gamma \rangle$	A4
$6_1^{\cdot}\mathrm{U}_4(3)$		$q = p \equiv 5 \bmod 12$	3	$\langle \delta^3, \gamma \rangle$	A5
$6_1^{\cdot}\mathrm{U}_4(3)^{\cdot}2_2^-$		$q = p \equiv 11 \bmod 12$	6	$\langle \gamma \rangle$	A4
$d \circ \mathrm{SU}_3(q)$		$p \equiv \pm 1 \bmod 8$	2	$\langle \delta^2, \phi \rangle$	
		$p \equiv \pm 3 \bmod 8$	2	$\langle \delta^2, \phi\delta \rangle$	

N1	Maximal under $\langle \gamma\delta^3 \rangle$ if $p \equiv \pm 2 \bmod 5$, $\langle \delta^3, \gamma \rangle$	A1	$\delta^3 \to 2_3, \gamma \to 2_1$
N2	Maximal under $\langle \gamma \rangle$ if $p \equiv \pm 2 \bmod 5$, $\langle \delta^3, \gamma \rangle$	A2	$\delta^3 \to 2_3, \gamma \to 2_2$
N3	Maximal under $\langle \gamma\delta \rangle$	A3	$\delta^3 \to 2_1, \gamma \to 2_2$
N4	Maximal under $\langle \gamma \rangle$	A4	$\gamma \to (2^2)_{122}$
N5	Maximal under subgroups not contained in $\langle \gamma \rangle$	A5	$\delta^3 \to 2_2, \gamma \to 2_1$

Table 8.28 *The maximal subgroups of* $\mathrm{Sp}_6(q)$ *of geometric type*
$d := |Z(\mathrm{Sp}_6(q))| = (q-1, 2)$, $|\delta| = d$, $|\phi| = e$, $q = p^e$.

$\mathscr{C}_i$	Subgp	Notes	c	Stab
$\mathscr{C}_1$	$\mathrm{E}_q^{1+4}:((q-1) \times \mathrm{Sp}_4(q))$	q odd	1	$\langle \delta, \phi \rangle$
$\mathscr{C}_1$	$\mathrm{E}_q^5:((q-1) \times \mathrm{Sp}_4(q))$	q even	1	$\langle \phi \rangle$
$\mathscr{C}_1$	$\mathrm{E}_q^{3+4}:(\mathrm{GL}_2(q) \times \mathrm{Sp}_2(q))$		1	$\langle \delta, \phi \rangle$
$\mathscr{C}_1$	$\mathrm{E}_q^6:\mathrm{GL}_3(q)$		1	$\langle \delta, \phi \rangle$
$\mathscr{C}_1$	$\mathrm{Sp}_2(q) \times \mathrm{Sp}_4(q)$		1	$\langle \delta, \phi \rangle$
$\mathscr{C}_2$	$\mathrm{Sp}_2(q)^3:\mathrm{S}_3$	$q \geqslant 3$	1	$\langle \delta, \phi \rangle$
$\mathscr{C}_2$	$\mathrm{GL}_3(q).2$	q odd	1	$\langle \delta, \phi \rangle$
$\mathscr{C}_3$	$\mathrm{Sp}_2(q^3):3$		1	$\langle \delta, \phi \rangle$
$\mathscr{C}_3$	$\mathrm{GU}_3(q).2$	q odd	1	$\langle \delta, \phi \rangle$
$\mathscr{C}_4$	$\mathrm{Sp}_2(q) \circ \mathrm{GO}_3(q)$	$q \geqslant 5$ odd	1	$\langle \delta, \phi \rangle$
$\mathscr{C}_5$	$\mathrm{Sp}_6(q_0).(d, r)$	$q = q_0^r$, r prime	(d, r)	$\langle \delta^c, \phi \rangle$
$\mathscr{C}_8$	$\mathrm{SO}_6^+(q)$	q even	1	$\langle \phi \rangle$
$\mathscr{C}_8$	$\mathrm{SO}_6^-(q)$	q even	1	$\langle \phi \rangle$

Table 8.29 *The maximal subgroups of* $\mathrm{Sp}_6(q)$ *in Class* $\mathscr{S}$

$d := |Z(\mathrm{Sp}_6(q))| = (q-1,2)$, $|\delta| = d$, $|\phi| = e$, $q = p^e$.

Subgp	Nov	Conditions on q	c	Stab
$2\dot{\,}A_5$		$q = p \equiv \pm 3, \pm 13 \bmod 40$	1	$\langle\delta\rangle$
	N1	$q = p \equiv \pm 11, \pm 19 \bmod 40$	1	$\langle\delta\rangle$
$2\dot{\,}S_5^-$		$q = p \equiv \pm 1 \bmod 8$	2	1
$2\dot{\,}L_2(7)\dot{\,}2^+$		$q = p \equiv \pm 1 \bmod 16$	2	1
$2\dot{\,}L_2(7)\dot{\,}2^+$		$q = p \equiv \pm 1 \bmod 16$	2	1
$2\dot{\,}L_2(7)$		$q = p \equiv \pm 7 \bmod 16, q \neq 7$	1	$\langle\delta\rangle$
$2\dot{\,}L_2(7)$		$q = p \equiv \pm 7 \bmod 16, q \neq 7$	1	$\langle\delta\rangle$
$2\dot{\,}L_2(7)$		$q = p^2, p \equiv \pm 3, \pm 5 \bmod 16, p \neq 3$	2	$\langle\delta\rangle$
	N1	$q = 9$	2	$\langle\delta\rangle$
$2\dot{\,}L_2(13)$		$q = p \equiv \pm 1, \pm 3, \pm 4 \bmod 13$	2	1
$2\dot{\,}L_2(13)$		$q = p^2, p \equiv \pm 2, \pm 5, \pm 6 \bmod 13, p \neq 2$	2	$\langle\phi\rangle$
$2\dot{\,}A_7$		$q = 9$	2	$\langle\phi\rangle$
$U_3(3){:}2$		$q = 2$	1	1
$2 \times U_3(3)$		$q = p \equiv \pm 7, \pm 17 \bmod 60$	1	$\langle\delta\rangle$
	N1	$q = p \equiv \pm 19, \pm 29 \bmod 60$	1	$\langle\delta\rangle$
$(2 \times U_3(3)).2$		$q = p \equiv \pm 1 \bmod 12$	2	1
$2\dot{\,}J_2$		$q = p \equiv \pm 1 \bmod 5$	2	1
$2\dot{\,}J_2$		$q = 5$	1	$\langle\delta\rangle$
$2\dot{\,}J_2$		$q = p^2, p \equiv \pm 2 \bmod 5, p \neq 2$	2	$\langle\phi\rangle$
$2\dot{\,}L_2(q)$		$p \geqslant 7$	1	$\langle\delta, \phi\rangle$
$G_2(q)$		$p = 2 \neq q$	1	$\langle\phi\rangle$

N1	Maximal under $\langle\delta\rangle$

Table 8.30 *The maximal subgroups of* $G_2(q) < \mathrm{Sp}_6(q)$, $q = 2^e$, $e > 1$.

$|Z(G_2(q))| = 1$, $|\phi| = e$. Note that $G_2(2) \cong U_3(3).2$.

This table is taken from [14]. It is proved in [2] that there are no novel maximal subgroups, but see also Section 7.4. The Aschbacher classes specified have no formal definitions, and are just intended to give a rough idea of the nature of the subgroups.

$\mathcal{C}_i$	Subgp	Notes	c	Stab
$\mathscr{C}_1$	$[q^5]{:}GL_2(q)$		1	$\langle\phi\rangle$
$\mathscr{C}_1$	$[q^5]{:}GL_2(q)$		1	$\langle\phi\rangle$
$\mathscr{C}_1$	$SL_2(q) \times SL_2(q)$		1	$\langle\phi\rangle$
$\mathscr{C}_2$	$SL_3(q).2$		1	$\langle\phi\rangle$
$\mathscr{C}_3$	$SU_3(q).2$		1	$\langle\phi\rangle$
$\mathscr{C}_5$	$G_2(q_0)$	$q = q_0^r$, r prime	1	$\langle\phi\rangle$
$\mathscr{S}_1$	$L_2(13)$	$q = 4$	1	$\langle\phi\rangle$
$\mathscr{S}_1$	J_2	$q = 4$	1	$\langle\phi\rangle$

Table 8.31 *The maximal subgroups of $\Omega_6^+(q)$ ($\cong SL_4(q)/\langle -I_4\rangle$) of geometric type.*

$q = 2^e$: $|Z(\Omega_6^+(q))| = 1$, $|\delta| = 1$, $|\gamma| = 2$, $|\delta'| = 1$, $|\phi| = e$.

$q = p^e \equiv 1 \bmod 4$: $|Z(\Omega_6^+(q))| = 2$, $|\delta| = 4$, $|\gamma| = 2$, $\delta^2 = \delta'$, $|\phi| = e$.

$q = p^e \equiv 3 \bmod 4$: $|Z(\Omega_6^+(q))| = 1$, $|\delta| = 2$, $|\gamma| = 2$, $|\delta'| = 1$, $|\phi| = e$.

$\mathscr{C}_i$	Subgp	Notes	c	Stab
$\mathscr{C}_1$	$E_q^{\,4}{:}(\frac{q-1}{(q-1,2)} \times \Omega_4^+(q)).(q-1,2)$		1	$\langle\delta,\gamma,\phi\rangle$
$\mathscr{C}_1$	$E_q^{\,1+4}{:}\frac{1}{(q-1,2)}(GL_2(q) \times (q-1))$	N1	1	$\langle\delta,\gamma,\phi\rangle$
$\mathscr{C}_1$	$E_q^{\,3}{:}\frac{1}{(q-1,2)}GL_3(q)$		2	$\langle\delta,\phi\rangle$
$\mathscr{C}_1$	$\Omega_5(q).2$	q odd	2	S1
$\mathscr{C}_1$	$\Omega_4^+(q){:}(q-1){:}2$	$q \geqslant 4$	1	$\langle\delta,\gamma,\phi\rangle$
		N2 if $q = 3$	1	$\langle\delta,\gamma\rangle$
$\mathscr{C}_1$	$\Omega_4^-(q).(q+1).2$		1	$\langle\delta,\gamma,\phi\rangle$
$\mathscr{C}_1$	$Sp_4(q)$	q even	1	$\langle\gamma,\phi\rangle$
$\mathscr{C}_2$	$2^5{\cdot}S_6$	$p = q \equiv 1 \bmod 8$	4	$\langle\gamma\rangle$
$\mathscr{C}_2$	$2^5{:}A_6$	$p = q \equiv 5 \bmod 8$	2	$\langle\delta',\gamma\rangle$
$\mathscr{C}_2$	$\frac{1}{2(q-1,2)}GO_2^+(q)^3.S_3$	$q \geqslant 7$	1	$\langle\delta,\gamma,\phi\rangle$
		N3 if $q = 5$	1	$\langle\delta,\gamma\rangle$
$\mathscr{C}_2$	$\frac{1}{4}GO_3(q)^2.S_2$	$q \equiv 1 \bmod 4$	2	$\langle\gamma,\delta',\phi\rangle$
$\mathscr{C}_2$	$SO_3(q)^2$	$q \equiv 3 \bmod 4$	1	$\langle\delta,\gamma,\phi\rangle$
$\mathscr{C}_2$	$\frac{1}{(q-1,2)}GL_3(q)$	N1	1	$\langle\delta,\gamma,\phi\rangle$
$\mathscr{C}_3$	$\frac{(q-1,4)}{2} \times \Omega_3(q^2).2$	$q \equiv 1 \bmod 4$	2	S2
		$q \equiv 3 \bmod 4$	1	$\langle\delta,\gamma,\phi\rangle$
$\mathscr{C}_5$	$\Omega_6^+(q_0)$	$q = q_0^r$, r prime, r odd or q even	1	$\langle\delta,\gamma,\phi\rangle$
$\mathscr{C}_5$	$SO_6^+(q_0)$	$q = q_0^2$, $q_0 \equiv 1 \bmod 4$	2	$\langle\gamma,\delta',\phi\rangle$
$\mathscr{C}_5$	$SO_6^+(q_0).2$	$q = q_0^2$, $q_0 \equiv 3 \bmod 4$	4	$\langle\gamma,\phi\rangle$
$\mathscr{C}_5$	$\Omega_6^-(q_0)$	$q = q_0^2$, q even	1	$\langle\gamma,\phi\rangle$
$\mathscr{C}_5$	$SO_6^-(q_0).2$	$q = q_0^2$, $q_0 \equiv 1 \bmod 4$	4	$\langle\gamma,\phi\rangle$
$\mathscr{C}_5$	$SO_6^-(q_0)$	$q = q_0^2$, $q_0 \equiv 3 \bmod 4$	2	$\langle\gamma,\delta',\phi\rangle$

S1	$\langle\gamma,\delta',\phi\rangle$ ($q \equiv 1 \bmod 4$) or $\langle\gamma,\phi\rangle$ ($q \equiv 3 \bmod 4$)
S2	$\langle\gamma\delta,\delta',\phi\rangle$ ($p \equiv 1 \bmod 4$) or $\langle\gamma\delta,\delta',\gamma\phi\rangle$ ($p \equiv 3 \bmod 4$)
N1	Maximal under subgps not contained in $\langle\delta,\phi\rangle$
N2	Maximal under subgps not contained in $\langle\gamma\rangle$
N3	Maximal under subgps not contained in $\langle\gamma,\delta'\rangle$

Table 8.32 *The maximal subgroups of $\Omega_6^+(q)$ $(\cong \mathrm{SL}_4(q)/\langle -\mathrm{I}_4 \rangle)$ in Class $\mathscr{S}$*

Note that $q = p$ in all examples, so $|\phi| = 1$.

$q = 2$: $|Z(\Omega_6^+(q))| = 1$, $|\gamma| = 2$.

$q = p$ odd: $d := |Z(\Omega_6^+(q))| = \frac{(q-1,4)}{2}$, $|\delta| = 2d$, $|\gamma| = 2$, $\delta^2 = \delta'$.

Subgp	Nov	Conditions on q	c	Stab
$d \times \mathrm{L}_2(7)$	N1	$q = p \equiv 1, 2, 4 \bmod 7$, $q \neq 2$	$(q-1,4)$	S1
A_7		$q = 2$	1	$\langle \gamma \rangle$
$d \times \mathrm{A}_7$		$q = p \equiv 1, 2, 4 \bmod 7$, $q \neq 2$	$(q-1,4)$	$\langle \gamma \rangle$
$d \times \mathrm{U}_4(2)$		$q = p \equiv 1 \bmod 6$	$(q-1,4)$	$\langle \gamma \rangle$

S1 $\quad \langle \gamma \rangle$ ($p \equiv \pm 1 \bmod 8$) or $\langle \delta\gamma \rangle$ ($p \equiv \pm 3 \bmod 8$)	N1 $\quad$ Maximal under S1

Table 8.33 *The maximal subgroups of $\Omega_6^-(q)$ $(\cong \mathrm{SU}_4(q)/\langle -\mathrm{I}_4 \rangle)$ of geometric type*

$q = 2^e$: $|Z(\Omega_6^-(q))| = 1$, $|\delta| = 1$, $|\gamma| = 2$, $|\delta'| = 1$, $|\varphi| = 2e$, $\varphi^e = \gamma$.

$q = p^e \equiv 1 \bmod 4$: $|Z(\Omega_6^-(q))| = 1$, $|\delta| = 2$, $|\gamma| = 2$, $|\delta'| = 1$, $|\varphi| = 2e$, $\varphi^e = \gamma$.

$q = p^e \equiv 3 \bmod 4$: $|Z(\Omega_6^-(q))| = 2$, $|\delta| = 4$, $|\gamma| = 2$, $\delta^2 = \delta'$, $|\phi| = e$.

$\mathscr{C}_i$	Subgp	Notes	c	Stab
$\mathscr{C}_1$	$E_q^4 {:} (\frac{q-1}{(q-1,2)} \times \Omega_4^-(q)).(q-1,2)$		1	S1
$\mathscr{C}_1$	$E_q^{1+4} {:} \frac{1}{(q-1,2)} (\mathrm{GL}_2(q) \times (q+1))$		1	S1
$\mathscr{C}_1$	$\Omega_5(q).2$	q odd	2	S2
$\mathscr{C}_1$	$\Omega_4^+(q).(q+1).2$	$q \geqslant 3$	1	S1
$\mathscr{C}_1$	$\Omega_4^-(q).(q-1).2$	$q \geqslant 4$	1	S1
		N1 if $q = 3$	1	$\langle \delta, \gamma \rangle$
$\mathscr{C}_1$	$\mathrm{Sp}_4(q)$	q even	1	$\langle \varphi \rangle$
$\mathscr{C}_2$	$2^5 {\cdot} \mathrm{S}_6$	$p = q \equiv 7 \bmod 8$	4	$\langle \gamma \rangle$
$\mathscr{C}_2$	$2^5 {:} \mathrm{A}_6$	$p = q \equiv 3 \bmod 8$	2	$\langle \gamma, \delta' \rangle$
$\mathscr{C}_2$	$\frac{1}{2(q-1,2)} \mathrm{GO}_2^-(q)^3 {.} \mathrm{S}_3$	$q \neq 3$	1	S1
		N1 if $q = 3$	1	$\langle \delta, \gamma \rangle$
$\mathscr{C}_2$	$\frac{1}{4} \mathrm{GO}_3(q)^2 {.} \mathrm{S}_2$	$q \equiv 3 \bmod 4$, $q \neq 3$	2	$\langle \gamma, \delta', \phi \rangle$
$\mathscr{C}_2$	$\mathrm{SO}_3(q)^2$	$q \equiv 1 \bmod 4$	1	$\langle \delta, \varphi \rangle$
$\mathscr{C}_3$	$\Omega_3(q^2).2$	$q \equiv 1 \bmod 4$	1	$\langle \delta, \varphi \rangle$
$\mathscr{C}_3$	$2 \times \Omega_3(q^2).2$	$q \equiv 3 \bmod 4$	2	$\langle \delta\gamma, \delta', \phi \rangle$
$\mathscr{C}_3$	$\frac{1}{(q-1,2)} \mathrm{GU}_3(q)$		1	S1
$\mathscr{C}_5$	$\Omega_6^-(q_0)$	$q = q_0^r$, r odd prime	1	S1

S1 $\quad \langle \delta, \varphi \rangle$ ($q \equiv 1 \bmod 4$) or	S2	$\langle \varphi \rangle$ ($q \equiv 1 \bmod 4$) or	
$\quad \langle \delta, \gamma, \phi \rangle$ ($q \equiv 3 \bmod 4$)		$\langle \gamma, \delta', \phi \rangle$ ($q \equiv 3 \bmod 4$)	
$\quad$ or $\langle \varphi \rangle$ (q even)	N1	Maximal under subgps not contained in S2	

Table 8.34 *The maximal subgroups of* $\Omega_6^-(q)$ $(\cong SU_4(q)/\langle -I_4\rangle)$ *in Class* $\mathscr{S}$.
Note that $q = p$ is odd in all examples, so $|\phi| = 1$ and $\varphi = \gamma$.
$q = p$ odd: $d := |Z(\Omega_6^-(q))| = \frac{(q+1,4)}{2}$, $|\delta| = 2d$, $|\gamma| = 2$, $\delta^2 = \delta'$.

Subgp	Nov	Conditions on q	c	Stab	Acts
$d \times L_2(7)$	N1	$q = p \equiv 3, 5, 6 \bmod 7$, $q \neq 3$	$(q+1,4)$	S1	
$d \times A_7$		$q = p \equiv 3, 5, 6 \bmod 7$	$(q+1,4)$	$\langle \gamma \rangle$	
$2{\cdot}L_3(4)$		$q = 3$	2	$\langle \delta\gamma, \delta' \rangle$	A1
$d \times U_4(2)$		$q = p \equiv 5 \bmod 6$	$(q+1,4)$	$\langle \gamma \rangle$	

S1	$\langle \gamma \rangle$ $(p \equiv \pm 1 \bmod 8)$ or $\langle \delta\gamma \rangle$ $(p \equiv \pm 3 \bmod 8)$	N1	Maximal under S1
A1	$\delta' \to 2_2$, $\delta\gamma \to 2_1$ and 2_3 in the two classes		

Table 8.35 *The maximal subgroups of* $SL_7(q)$ *of geometric type*
$d := |Z(SL_7(q))| = (q-1, 7)$, $|\delta| = d$, $|\phi| = e$, $|\gamma| = 2$, $q = p^e$.

$\mathscr{C}_i$	Subgp	Notes	c	Stab
$\mathscr{C}_1$	$E_q{}^6{:}GL_6(q)$		2	$\langle \delta, \phi \rangle$
$\mathscr{C}_1$	$E_q{}^{10}{:}(SL_5(q) \times SL_2(q)){:}(q-1)$		2	$\langle \delta, \phi \rangle$
$\mathscr{C}_1$	$E_q{}^{12}{:}(SL_4(q) \times SL_3(q)){:}(q-1)$		2	$\langle \delta, \phi \rangle$
$\mathscr{C}_1$	$E_q{}^{1+10}{:}(GL_5(q) \times (q-1))$	N1	1	$\langle \delta, \phi, \gamma \rangle$
$\mathscr{C}_1$	$E_q{}^{4+12}{:}(SL_2(q)^2 \times SL_3(q)){:}(q-1)^2$	N1	1	$\langle \delta, \phi, \gamma \rangle$
$\mathscr{C}_1$	$E_q{}^{9+6}{:}GL_3(q)^2$	N1	1	$\langle \delta, \phi, \gamma \rangle$
$\mathscr{C}_1$	$GL_6(q)$	N1	1	$\langle \delta, \phi, \gamma \rangle$
$\mathscr{C}_1$	$(SL_5(q) \times SL_2(q)){:}(q-1)$	N1	1	$\langle \delta, \phi, \gamma \rangle$
$\mathscr{C}_1$	$(SL_4(q) \times SL_3(q)){:}(q-1)$	N1	1	$\langle \delta, \phi, \gamma \rangle$
$\mathscr{C}_2$	$(q-1)^6{:}S_7$	$q \geqslant 5$	1	$\langle \delta, \phi, \gamma \rangle$
$\mathscr{C}_3$	$\frac{q^7-1}{q-1}{:}7$		1	$\langle \delta, \phi, \gamma \rangle$
$\mathscr{C}_5$	$SL_7(q_0) . \left(\frac{q-1}{q_0-1}, 7 \right)$	$q = q_0^r$, r prime	$\left(\frac{q-1}{q_0-1}, 7 \right)$	$\langle \delta^c, \phi, \gamma \rangle$
$\mathscr{C}_6$	$7_+^{1+2}{:}Sp_2(7)$	$q = p \equiv 1 \bmod 7$ or $(q = p^3$ & $p \equiv 2, 4 \bmod 7)$	7	$\langle \phi, \gamma \rangle$
$\mathscr{C}_8$	$d \times SO_7(q)$	q odd	d	$\langle \phi, \gamma \rangle$
$\mathscr{C}_8$	$(q_0-1, 7) \times SU_7(q_0)$	$q = q_0^2$	$(q_0-1, 7)$	$\langle \delta^c, \phi, \gamma \rangle$

N1	Maximal under subgroups not contained in $\langle \delta, \phi \rangle$

Table 8.36 *The maximal subgroups of* $SL_7(q)$ *in Class* $\mathscr{S}$
In all examples, $q = p$. So $d := |Z(SL_7(q))| = (q-1, 7)$, $|\delta| = d$, $|\phi| = 1$, $|\gamma| = 2$.

Subgp	Conditions on q	c	Stab
$d \times U_3(3)$	$q = p \equiv 1 \bmod 4$	d	$\langle \gamma \rangle$

Table 8.37 *The maximal subgroups of* $\mathrm{SU}_7(q)$ *of geometric type*
$d := |Z(\mathrm{SU}_7(q))| = (q+1,7)$, $|\delta| = d$, $|\phi| = 2e$, $q = p^e$.

$\mathscr{C}_i$	Subgp	Notes	c	Stab
$\mathscr{C}_1$	$\mathrm{E}_q^{1+10}{:}\mathrm{SU}_5(q){:}(q^2-1)$		1	$\langle \delta, \phi \rangle$
$\mathscr{C}_1$	$\mathrm{E}_q^{4+12}{:}(\mathrm{SL}_2(q^2) \times \mathrm{SU}_3(q)){:}(q^2-1)$		1	$\langle \delta, \phi \rangle$
$\mathscr{C}_1$	$\mathrm{E}_q^{9+6}{:}\mathrm{GL}_3(q^2)$		1	$\langle \delta, \phi \rangle$
$\mathscr{C}_1$	$\mathrm{GU}_6(q)$		1	$\langle \delta, \phi \rangle$
$\mathscr{C}_1$	$(\mathrm{SU}_5(q) \times \mathrm{SU}_2(q)){:}(q+1)$		1	$\langle \delta, \phi \rangle$
$\mathscr{C}_1$	$(\mathrm{SU}_4(q) \times \mathrm{SU}_3(q)){:}(q+1)$		1	$\langle \delta, \phi \rangle$
$\mathscr{C}_2$	$(q+1)^6{:}\mathrm{S}_7$		1	$\langle \delta, \phi \rangle$
$\mathscr{C}_3$	$\frac{q^7+1}{q+1}{:}7$		1	$\langle \delta, \phi \rangle$
$\mathscr{C}_5$	$\mathrm{SU}_7(q_0).\left(\frac{q+1}{q_0+1},7\right)$	$q_0^r = q$, r odd prime	$\left(\frac{q+1}{q_0+1},7\right)$	$\langle \delta^c, \phi \rangle$
$\mathscr{C}_5$	$d \times \mathrm{SO}_7(q)$	q odd	d	$\langle \phi \rangle$
$\mathscr{C}_6$	$7_+^{1+2}{:}\mathrm{Sp}_2(7)$	$q = p \equiv 6 \bmod 7$, or $(q = p^3$ & $p \equiv 3, 5 \bmod 7)$	7	$\langle \phi \rangle$

Table 8.38 *The maximal subgroups of* $\mathrm{SU}_7(q)$ *in Class* $\mathscr{S}$
In all examples, $q = p$. So $d := |Z(\mathrm{SU}_7(q))| = (q+1,7)$, $|\delta| = d$, $|\phi| = 2$, $\phi = \gamma$.

Subgp	Conditions on q	c	Stab
$d \times \mathrm{U}_3(3)$	$q = p \equiv 3 \bmod 4$, $q \neq 3$	d	$\langle \gamma \rangle$

Table 8.39 *The maximal subgroups of* $\Omega_7(q)$ *of geometric type*
$|Z(\Omega_7(q))| = 1$, $|\delta| = 2$, $|\phi| = e$, $q = p^e$ odd.

$\mathscr{C}_i$	Subgp	Notes	c	Stab
$\mathscr{C}_1$	$\mathrm{E}_q^5{:}(\frac{q-1}{2} \times \Omega_5(q)).2$		1	$\langle \delta, \phi \rangle$
$\mathscr{C}_1$	$\mathrm{E}_q^{1+6}{:}(\frac{1}{2}\mathrm{GL}_2(q) \times \Omega_3(q)).2$		1	$\langle \delta, \phi \rangle$
$\mathscr{C}_1$	$\mathrm{E}_q^{3+3}{:}\frac{1}{2}\mathrm{GL}_3(q)$		1	$\langle \delta, \phi \rangle$
$\mathscr{C}_1$	$\Omega_6^+(q).2$		1	$\langle \delta, \phi \rangle$
$\mathscr{C}_1$	$\Omega_6^-(q).2$		1	$\langle \delta, \phi \rangle$
$\mathscr{C}_1$	$(\Omega_2^+(q) \times \Omega_5(q)).2^2$	$q \geqslant 5$	1	$\langle \delta, \phi \rangle$
$\mathscr{C}_1$	$(\Omega_2^-(q) \times \Omega_5(q)).2^2$		1	$\langle \delta, \phi \rangle$
$\mathscr{C}_1$	$(\Omega_3(q) \times \Omega_4^+(q)).2^2$		1	$\langle \delta, \phi \rangle$
$\mathscr{C}_1$	$(\Omega_3(q) \times \Omega_4^-(q)).2^2$		1	$\langle \delta, \phi \rangle$
$\mathscr{C}_2$	$2^6{:}\mathrm{A}_7$	$p = q \equiv \pm 3 \bmod 8$	1	$\langle \delta \rangle$
$\mathscr{C}_2$	$2^6{:}\mathrm{S}_7$	$p = q \equiv \pm 1 \bmod 8$	2	1
$\mathscr{C}_5$	$\Omega_7(q_0)$	$q = q_0^r$, r odd prime	1	$\langle \delta, \phi \rangle$
$\mathscr{C}_5$	$\mathrm{SO}_7(q_0)$	$q = q_0^2$	2	$\langle \phi \rangle$

Table 8.40 *The maximal subgroups of* $\Omega_7(q)$ *in Class* $\mathscr{S}$
$|Z(\Omega_7(q))| = 1$, $|\delta| = 2$, $|\phi| = e$, $q = p^e$ odd.

Subgp	Conditions on q	c	Stab
S_9	$q = 3$	2	1
$S_6(2)$	$q = p$	2	1
$G_2(q)$		2	$\langle \phi \rangle$

Table 8.41 *The maximal subgroups of* $G_2(q) < \Omega_7(q)$, $q = p^e$, $p \geqslant 5$.
$|Z(G_2(q))| = 1$, $|\phi| = e$.

This and the following two tables are taken from [64]. The specified Aschbacher classes have no formal definitions, and are just intended to give a rough idea of the nature of the subgroups.

$\mathscr{C}_i$	Subgp	Notes	c	Stab
$\mathscr{C}_1$	$E_q{}^{2+1+2}{:}GL_2(q)$		1	$\langle \phi \rangle$
$\mathscr{C}_1$	$E_q{}^{1+4}{:}GL_2(q)$		1	$\langle \phi \rangle$
$\mathscr{C}_1$	$(SL_2(q) \circ SL_2(q)).2$		1	$\langle \phi \rangle$
$\mathscr{C}_1$	$SL_3(q){:}2$		1	$\langle \phi \rangle$
$\mathscr{C}_1$	$SU_3(q){:}2$		1	$\langle \phi \rangle$
$\mathscr{C}_2$	$2^{3\text{'}}L_3(2)$	$q = p$	1	$\langle \phi \rangle$
$\mathscr{C}_5$	$G_2(q_0)$	$q = q_0^r$, r prime	1	$\langle \phi \rangle$
$\mathscr{S}_1$	$L_2(13)$	$q = p \equiv 1, 3, 4, 9, 10, 12 \pmod{13}$	1	1
		$q = p^2$, $p \equiv 2, 5, 6, 7, 8, 11 \pmod{13}$	1	$\langle \phi \rangle$
$\mathscr{S}_1$	$L_2(8)$	$q = p \equiv 1, 8 \pmod{9}$	1	1
		$q = p^3$, $p \equiv 2, 4, 5, 7 \pmod{9}$	1	$\langle \phi \rangle$
$\mathscr{S}_1$	$U_3(3){:}2$	$q = p \geqslant 5$	1	1
$\mathscr{S}_1$	J_1	$q = 11$	1	1
$\mathscr{S}_2$	$PGL_2(q)$	$p \geqslant 7$, $q \geqslant 11$	1	$\langle \phi \rangle$

Table 8.42 *The maximal subgroups of* $\mathrm{G}_2(q) < \Omega_7(q)$, $q = 3^e$.

$|Z(\mathrm{G}_2(q))| = 1$, $|\gamma| = 2e$, $\gamma^2 = \phi$.

For some of the groups in this table, a single A-class ($A = \mathrm{Aut}\,\Omega$) splits into two Ω-classes, which are interchanged by γ. As in Table 8.50 below, we have split these A-classes over two table rows. The value of c for these classes is obtained by adding the values of c specified in the two rows representing this A-class.

$\mathscr{C}_i$	$\mathscr{C}_i$	Subgp	Notes	c	Stab
$\mathscr{C}_1$		$[q^5]{:}\mathrm{GL}_2(q)$		1	$\langle\phi\rangle$
	$\mathscr{C}_1$	$[q^5]{:}\mathrm{GL}_2(q)$		1	$\langle\phi\rangle$
$\mathscr{C}_1$		$[q^6]{:}(q-1)^2$	N1	1	$\langle\gamma\rangle$
$\mathscr{C}_1$		$(\mathrm{SL}_2(q) \circ \mathrm{SL}_2(q)).2$		1	$\langle\gamma\rangle$
$\mathscr{C}_1$		$\mathrm{SL}_3(q){:}2$		1	$\langle\phi\rangle$
	$\mathscr{S}_2$	$\mathrm{SL}_3(q){:}2$		1	$\langle\phi\rangle$
$\mathscr{C}_1$		$\mathrm{SU}_3(q){:}2$		1	$\langle\phi\rangle$
	$\mathscr{S}_2$	$\mathrm{SU}_3(q){:}2$		1	$\langle\phi\rangle$
$\mathscr{C}_1$		$(q-1)^2.\mathrm{D}_{12}$	$q \geqslant 9$, N1	1	$\langle\gamma\rangle$
$\mathscr{C}_1$		$(q+1)^2.\mathrm{D}_{12}$	$q \geqslant 9$, N1	1	$\langle\gamma\rangle$
$\mathscr{C}_1$		$(q^2+q+1).6$	$q \geqslant 9$, N1	1	$\langle\gamma\rangle$
$\mathscr{C}_1$		$(q^2-q+1).6$	$q \geqslant 9$, N1	1	$\langle\gamma\rangle$
$\mathscr{C}_2$		$2^3{\cdot}\mathrm{L}_3(2)$	$q = 3$	1	$\langle\gamma\rangle$
$\mathscr{C}_5$		$\mathrm{G}_2(q_0)$	$q = q_0^r$, r prime	1	$\langle\gamma\rangle$
$\mathscr{S}_1$		$\mathrm{L}_2(13)$	$q = 3$	1	$\langle\gamma\rangle$
$\mathscr{S}_2$		$\mathrm{R}(q) = {}^2\mathrm{G}_2(q)$	e odd	1	$\langle\gamma\rangle$

N1	Maximal under subgroups not contained in $\langle\phi\rangle$

Table 8.43 *The maximal subgroups of* $\mathrm{R}(q) = {}^2\mathrm{G}_2(q) < \mathrm{G}_2(q)$

$|Z(\mathrm{R}(q))| = 1$, $|\phi| = e$, $q = 3^e$, $e > 1$ odd. Note: $\mathrm{R}(3) \cong \mathrm{L}_2(8){:}3$.

Class	Subgp	Notes	c	Stab
$\mathscr{C}_1$	$\mathrm{E}_q{}^{1+1+1}{:}(q-1)$		1	$\langle\phi\rangle$
$\mathscr{C}_1$	$2 \times \mathrm{L}_2(q)$		1	$\langle\phi\rangle$
$\mathscr{C}_1$	$(2^2 \times \mathrm{D}_{\frac{q+1}{2}}){:}3$		1	$\langle\phi\rangle$
$\mathscr{C}_3$	$(q - \sqrt{3q} + 1){:}6$		1	$\langle\phi\rangle$
$\mathscr{C}_3$	$(q + \sqrt{3q} + 1){:}6$		1	$\langle\phi\rangle$
$\mathscr{C}_5$	$\mathrm{R}(q_0)$	$q = q_0^r$, r prime	1	$\langle\phi\rangle$

Table 8.44 *The maximal subgroups of* $\mathrm{SL}_8(q)$ *of geometric type*
$d := |\mathrm{Z}(\mathrm{SL}_8(q))| = (q-1,8)$, $|\delta| = d$, $|\phi| = e$, $|\gamma| = 2$, $q = p^e$.

$\mathscr{C}_i$	Subgp	Notes	c	Stab
$\mathscr{C}_1$	$E_q{}^7{:}\mathrm{GL}_7(q)$		2	$\langle\delta,\phi\rangle$
$\mathscr{C}_1$	$E_q{}^{12}{:}(\mathrm{SL}_6(q) \times \mathrm{SL}_2(q)){:}(q-1)$		2	$\langle\delta,\phi\rangle$
$\mathscr{C}_1$	$E_q{}^{15}{:}(\mathrm{SL}_5(q) \times \mathrm{SL}_3(q)){:}(q-1)$		2	$\langle\delta,\phi\rangle$
$\mathscr{C}_1$	$E_q{}^{16}{:}(\mathrm{SL}_4(q) \times \mathrm{SL}_4(q)){:}(q-1)$		1	$\langle\delta,\phi,\gamma\rangle$
$\mathscr{C}_1$	$E_q{}^{1+12}{:}(\mathrm{GL}_6(q) \times (q-1))$	N1	1	$\langle\delta,\phi,\gamma\rangle$
$\mathscr{C}_1$	$E_q{}^{4+16}{:}(\mathrm{SL}_2(q)^2 \times \mathrm{SL}_4(q)){:}(q-1)^2$	N1	1	$\langle\delta,\phi,\gamma\rangle$
$\mathscr{C}_1$	$E_q{}^{9+12}{:}(\mathrm{SL}_3(q)^2 \times \mathrm{SL}_2(q)){:}(q-1)^2$	N1	1	$\langle\delta,\phi,\gamma\rangle$
$\mathscr{C}_1$	$\mathrm{GL}_7(q)$	N1	1	$\langle\delta,\phi,\gamma\rangle$
$\mathscr{C}_1$	$(\mathrm{SL}_6(q) \times \mathrm{SL}_2(q)){:}(q-1)$	N1	1	$\langle\delta,\phi,\gamma\rangle$
$\mathscr{C}_1$	$(\mathrm{SL}_5(q) \times \mathrm{SL}_3(q)){:}(q-1)$	N1	1	$\langle\delta,\phi,\gamma\rangle$
$\mathscr{C}_2$	$(q-1)^7.\mathrm{S}_8$	$q \geqslant 5$	1	$\langle\delta,\phi,\gamma\rangle$
$\mathscr{C}_2$	$\mathrm{SL}_2(q)^4{:}(q-1)^3.\mathrm{S}_4$	$q \geqslant 3$	1	$\langle\delta,\phi,\gamma\rangle$
$\mathscr{C}_2$	$\mathrm{SL}_4(q)^2{:}(q-1).\mathrm{S}_2$		1	$\langle\delta,\phi,\gamma\rangle$
$\mathscr{C}_3$	$(((q-1,4)(q+1))\circ\mathrm{SL}_4(q^2)).\frac{(q^2-1,4)}{(q-1,4)}.2$		1	$\langle\delta,\phi,\gamma\rangle$
$\mathscr{C}_4$	$(\mathrm{SL}_2(q)\circ\mathrm{SL}_4(q)).(q-1,2)^2$	$q \geqslant 3$	$(q-1,2)$	$\langle\delta^c,\phi,\gamma\rangle$
$\mathscr{C}_5$	$\mathrm{SL}_8(q_0).\left[\left(\frac{q-1}{q_0-1},8\right)\right]$	$q = q_0^r$, r prime	$\left(\frac{q-1}{q_0-1},8\right)$	$\langle\delta^c,\phi,\gamma\rangle$
$\mathscr{C}_6$	$(d\circ 2^{1+6})^{\cdot}\mathrm{Sp}_6(2)$	$q = p$, $q \equiv 1 \bmod 4$	d	$\langle\gamma\rangle$
$\mathscr{C}_8$	$\mathrm{SO}_8^+(q).[d]$	q odd	$d/2$	$\langle\delta^c,\phi,\gamma\rangle$
$\mathscr{C}_8$	$\mathrm{SO}_8^-(q).[d]$	q odd	$d/2$	S1
$\mathscr{C}_8$	$\mathrm{Sp}_8(q).[(q-1,4)]$		$(q-1,4)$	$\langle\delta^c,\phi,\gamma\rangle$
$\mathscr{C}_8$	$\mathrm{SU}_8(q_0).(q_0-1,8)$	$q = q_0^2$	$(q_0-1,8)$	$\langle\delta^c,\phi,\gamma\rangle$

S1 $\langle\delta^c,\phi\delta^{(p-1)/2},\gamma\delta^{-1}\rangle$	N1 Maximal under subgroups not contained in $\langle\delta,\phi\rangle$

Table 8.45 *The maximal subgroups of* $\mathrm{SL}_8(q)$ *in Class* $\mathscr{S}$
$d := |\mathrm{Z}(\mathrm{SL}_8(q))| = (q-1,8)$, $|\delta| = d$, $|\phi| = e$, $|\gamma| = 2$, $q = p^e$.

Subgp	Conditions on q	c	Stab	Acts
$4_1{}^{\cdot}\mathrm{L}_3(4)$	$q = 5$	2	$\langle\delta^2,\gamma\rangle$	A1
$d \circ 4_1{}^{\cdot}\mathrm{L}_3(4)$	$q = p \equiv 9, 21, 29, 41, 61, 69 \bmod 80$	d	$\langle\delta^{c/2}\rangle$	A2
$8 \circ 4_1{}^{\cdot}\mathrm{L}_3(4).2_3$	$q = p \equiv 1, 49 \bmod 80$	16	1	
$8 \circ 4_1{}^{\cdot}\mathrm{L}_3(4)$	$q = p^2, p \equiv \pm3, \pm13, \pm27, \pm37 \bmod 80$	8	S1	A3
$8 \circ 4_1{}^{\cdot}\mathrm{L}_3(4).2_3$	$q = p^2, p \equiv \pm7, \pm17, \pm23, \pm33 \bmod 80$	16	S2	

S1 $\langle\delta^4,\phi\rangle$ $(p \equiv 3 \bmod 4)$ or $\langle\delta^4,\phi\gamma\rangle$ $(p \equiv 1 \bmod 4)$	A1 $\delta^2 \to 2_3$, $\gamma \to 2_1$ and 2_2 in the two classes
	A2 $\delta^{c/2} \to 2_3$
S2 $\langle\phi\rangle$ $(p \equiv 7 \bmod 8)$ or $\langle\phi\gamma\rangle$ $(p \equiv 1 \bmod 8)$	A3 $\delta^4 \to 2_3$, ϕ or $\phi\gamma \to 2_1$ or 2_2

Table 8.46 *The maximal subgroups of* $\mathrm{SU}_8(q)$ *of geometric type*
$d := |\mathrm{Z}(\mathrm{SU}_8(q))| = (q+1, 8)$, $|\delta| = d$, $|\phi| = 2e$, $q = p^e$.

$\mathscr{C}_i$	Subgp	Notes	c	Stab
$\mathscr{C}_1$	$E_q{}^{1+12}{:}\mathrm{SU}_6(q){:}(q^2-1)$		1	$\langle \delta, \phi \rangle$
$\mathscr{C}_1$	$E_q{}^{4+16}{:}(\mathrm{SL}_2(q^2) \times \mathrm{SU}_4(q)){:}(q^2-1)$		1	$\langle \delta, \phi \rangle$
$\mathscr{C}_1$	$E_q{}^{9+12}{:}(\mathrm{SL}_3(q^2) \times \mathrm{SU}_2(q)){:}(q^2-1)$		1	$\langle \delta, \phi \rangle$
$\mathscr{C}_1$	$E_q{}^{16}{:}\mathrm{SL}_4(q^2){:}(q-1)$		1	$\langle \delta, \phi \rangle$
$\mathscr{C}_1$	$\mathrm{GU}_7(q)$		1	$\langle \delta, \phi \rangle$
$\mathscr{C}_1$	$(\mathrm{SU}_6(q) \times \mathrm{SU}_2(q)){:}(q+1)$		1	$\langle \delta, \phi \rangle$
$\mathscr{C}_1$	$(\mathrm{SU}_5(q) \times \mathrm{SU}_3(q)){:}(q+1)$		1	$\langle \delta, \phi \rangle$
$\mathscr{C}_2$	$(q+1)^7.\mathrm{S}_8$		1	$\langle \delta, \phi \rangle$
$\mathscr{C}_2$	$\mathrm{SU}_2(q)^4{:}(q+1)^3.\mathrm{S}_4$	$q \geqslant 3$	1	$\langle \delta, \phi \rangle$
$\mathscr{C}_2$	$\mathrm{SU}_4(q)^2{:}(q+1).\mathrm{S}_2$		1	$\langle \delta, \phi \rangle$
$\mathscr{C}_2$	$\mathrm{SL}_4(q^2).(q-1).2$		1	$\langle \delta, \phi \rangle$
$\mathscr{C}_4$	$(\mathrm{SU}_2(q) \circ \mathrm{SU}_4(q)).(q+1,2)^2$	$q \geqslant 3$	$(q+1,2)$	$\langle \delta^c, \phi \rangle$
$\mathscr{C}_5$	$\mathrm{SU}_8(q_0)$	$q_0^r = q$, r odd prime	1	$\langle \delta, \phi \rangle$
$\mathscr{C}_5$	$\mathrm{Sp}_8(q).[(q+1,4)]$		$(q+1,4)$	$\langle \delta^c, \phi \rangle$
$\mathscr{C}_5$	$\mathrm{SO}_8^+(q).[d]$	q odd	$d/2$	$\langle \delta^c, \phi \rangle$
$\mathscr{C}_5$	$\mathrm{SO}_8^-(q).[d]$	q odd	$d/2$	S1
$\mathscr{C}_6$	$(d \circ 2^{1+6})\dot{\ }\mathrm{Sp}_6(2)$	$q = p \equiv 3 \bmod 4$	d	$\langle \phi \rangle$

S1	$\langle \delta^c, \phi \delta^{(p-1)/2} \rangle$

Table 8.47 *The maximal subgroups of* $\mathrm{SU}_8(q)$ *in Class* $\mathscr{S}$
In all examples, $q = p$. So $d := |\mathrm{Z}(\mathrm{SU}_8(q))| = (q+1, 8)$, $|\delta| = d$, $|\phi| = 2$, $\phi = \gamma$.

Subgp	Conditions on q	c	Stab	Acts
$d \circ 4_1\dot{\ }\mathrm{L}_3(4)$	$q = p \equiv 11, 19, 39, 51, 59, 71 \bmod 80$	d	$\langle \delta^{c/2} \rangle$	$\delta^{c/2} \to 2_3$
$8 \circ 4_1\dot{\ }\mathrm{L}_3(4).2_3$	$q = p \equiv 31, 79 \bmod 80$	16	1	

Table 8.48 *The maximal subgroups of* $\mathrm{Sp}_8(q)$ *of geometric type*
$d := |Z(\mathrm{Sp}_8(q))| = (q-1,2)$, $|\delta| = d$, $|\phi| = e$, $q = p^e$.

$\mathscr{C}_i$	Subgp	Notes	c	Stab
$\mathscr{C}_1$	$E_q^{1+6}:((q-1) \times \mathrm{Sp}_6(q))$	q odd	1	$\langle \delta, \phi \rangle$
$\mathscr{C}_1$	$E_q^{7}:((q-1) \times \mathrm{Sp}_6(q))$	q even	1	$\langle \phi \rangle$
$\mathscr{C}_1$	$E_q^{3+8}:(\mathrm{GL}_2(q) \times \mathrm{Sp}_4(q))$		1	$\langle \delta, \phi \rangle$
$\mathscr{C}_1$	$E_q^{6+6}:(\mathrm{GL}_3(q) \times \mathrm{Sp}_2(q))$		1	$\langle \delta, \phi \rangle$
$\mathscr{C}_1$	$E_q^{10}:\mathrm{GL}_4(q)$		1	$\langle \delta, \phi \rangle$
$\mathscr{C}_1$	$\mathrm{Sp}_6(q) \times \mathrm{Sp}_2(q)$		1	$\langle \delta, \phi \rangle$
$\mathscr{C}_2$	$\mathrm{Sp}_2(q)^4{:}S_4$	$q \geqslant 3$	1	$\langle \delta, \phi \rangle$
$\mathscr{C}_2$	$\mathrm{Sp}_4(q)^2{:}S_2$		1	$\langle \delta, \phi \rangle$
$\mathscr{C}_2$	$\mathrm{GL}_4(q).2$	q odd	1	$\langle \delta, \phi \rangle$
$\mathscr{C}_3$	$\mathrm{Sp}_4(q^2){:}2$		1	$\langle \delta, \phi \rangle$
$\mathscr{C}_3$	$\mathrm{GU}_4(q).2$	q odd	1	$\langle \delta, \phi \rangle$
$\mathscr{C}_4$	$(\mathrm{Sp}_2(q) \circ \mathrm{GO}_4^-(q)).2$	q odd	1	$\langle \delta, \phi \rangle$
$\mathscr{C}_5$	$\mathrm{Sp}_8(q_0).(d,r)$	$q = q_0^r$, r prime	(d,r)	$\langle \delta^c, \phi \rangle$
$\mathscr{C}_6$	$2_-^{1+6}{\cdot}\mathrm{SO}_6^-(2)$	$q = p \equiv \pm 1 \bmod 8$	2	1
$\mathscr{C}_6$	$2_-^{1+0}{\cdot}\Omega_6^-(2)$	$q = p \equiv \pm 3 \bmod 8$	1	$\langle \delta \rangle$
$\mathscr{C}_7$	$(\mathrm{Sp}_2(q) \circ \mathrm{Sp}_2(q) \circ \mathrm{Sp}_2(q)).2^2.S_3$	$q \geqslant 5$ odd	1	$\langle \delta, \phi \rangle$
$\mathscr{C}_8$	$\mathrm{SO}_8^+(q)$	q even	1	$\langle \phi \rangle$
$\mathscr{C}_8$	$\mathrm{SO}_8^-(q)$	q even	1	$\langle \phi \rangle$

Table 8.49 *The maximal subgroups of* $\mathrm{Sp}_8(q)$ *in Class* $\mathscr{S}$
$d := |Z(\mathrm{Sp}_8(q))| = (q-1,2)$, $|\delta| = d$, $|\phi| = e$, $q = p^e$.

Subgp	Conditions on q	c	Stab	Acts
$2^{\cdot}L_2(7)$	$q = p \equiv \pm 5 \bmod 12$, $q \neq 7$	1	$\langle \delta \rangle$	
$2^{\cdot}L_2(7).2$	$q = p \equiv \pm 1 \bmod 12$	2	1	
$2^{\cdot}A_6$	$q = p \equiv \pm 9 \bmod 20$	1	$\langle \delta \rangle$	$\delta \to 2_2$
$2^{\cdot}A_6.2_2$	$q = p \equiv \pm 1 \bmod 20$	2	1	
$2^{\cdot}A_6$	$q = p^2$, $p \equiv \pm 2 \bmod 5$, $p \neq 2, 3$	1	$\langle \delta, \phi \rangle$	$\delta \to 2_2$, $\phi \to 2_1$
$L_2(17)$	$q = 2$	1	1	
$2^{\cdot}L_2(17)$	$q = p \equiv \pm 1, \pm 2, \pm 4, \pm 8 \bmod 17$, $q \neq 2$	2	1	
$2^{\cdot}L_2(17)$	$q = p^2$, $p \equiv \pm 3, \pm 5, \pm 6, \pm 7 \bmod 17$	2	$\langle \phi \rangle$	
S_{10}	$q = 2$	1	1	
$2^{\cdot}L_2(q)$	$p \geqslant 11$	1	$\langle \delta, \phi \rangle$	
$2^{\cdot}L_2(q^3).3$	q odd	1	$\langle \delta, \phi \rangle$	

Table 8.50 *The maximal subgroups of* $\Omega_8^+(q)$
$d := |Z(\Omega_8^+(q))| = (q-1,2), |\delta| = d, |\gamma| = 2, |\delta'| = d, |\tau| = 3, |\phi| = e, q = p^e.$

This table is taken from [62]. The entries occur in the same order as in [62], but we have usually compressed several lines there to a single line here. In this table, we have departed from our rule that each A-class $(A = \text{Aut}\,\overline{\Omega})$ of subgroups is described by a single row of the table. There are are some cases in which representatives of a single A-class lie in two different Aschbacher classes (or in two different types within the same class). We have then split the entry for the A-class over two rows, and specified the second Aschbacher class in the second column of the second row. The number c for the complete A-class is the sum of the entries for c in the two rows. We have not repeated restrictions on q in the second of the two rows, since these are the same as in the first row, and the stabiliser of a representative in the second row (column Stab) can be taken to be the conjugate under the triality automorphism τ of the stabiliser in the first row: only half of this stabiliser lies in $\text{C}\Gamma\text{O}_8^+(q)$.

For convenience, we note the following identifications between our notation for outer automorphisms and the notation in [62]:

$$\gamma = (1\ 2), \ \tau = (1\ 2\ 3), \ \delta' = (1\ 2)(3\ 4), \ \delta = (1\ 3)(2\ 4), \ \langle\delta,\delta'\rangle = V_4.$$

$\mathscr{C}_i$	$\mathscr{C}_i$	Subgp	Notes	c	Stab
$\mathscr{C}_1$		$E_q{}^6:(\frac{q-1}{d} \times \Omega_6^+(q)).d$		1	S1
	$\mathscr{C}_1$	$E_q{}^6:\frac{1}{d}GL_4(q)$		2	
$\mathscr{C}_1$		$[q^{11}]:[\frac{q-1}{d}]^2.\frac{1}{d}GL_2(q).d^2$	N1	1	S2
$\mathscr{C}_1$		$E_q{}^{1+8}:(\frac{1}{d}GL_2(q)\times\Omega_4^+(q)).d$		1	S2
$\mathscr{C}_1$		$E_q{}^{3+6}:(\frac{1}{d}GL_3(q) \times \frac{q-1}{d}).d$	N2	1	S1
	$\mathscr{C}_1$	$E_q{}^{3+6}:(\frac{1}{d}GL_3(q) \times \frac{q-1}{d}).d$		2	
$\mathscr{C}_1$		$2 \times \Omega_7(q)$	q odd	2	S3
	$\mathscr{S}$	$2{\cdot}\Omega_7(q)$		4	
$\mathscr{C}_1$		$Sp_6(q)$	q even	1	S3
	$\mathscr{S}$	$Sp_6(q)$		2	
$\mathscr{C}_1$		$d \times G_2(q)$	N1	d^2	$\langle\gamma,\tau,\phi\rangle$
$\mathscr{C}_1$		$(\Omega_2^+(q) \times \Omega_6^+(q)).[2d]$	$q \geqslant 4$; N3 if $q = 3$	1	S1
	$\mathscr{C}_2$	$SL_4(q).\frac{q-1}{d}.2$		2	
$\mathscr{C}_1$		$(\Omega_2^+(q) \times \frac{1}{d}GL_3(q)).[2d]$	$q \geqslant 3$, N4	1	S2

$\mathscr{C}_i$	$\mathscr{C}_i$	Subgp	Notes	c	Stab
$\mathscr{C}_1$		$(\Omega_2^-(q) \times \Omega_6^-(q)).[2d]$		1	S1
	$\mathscr{C}_3$	$\mathrm{SU}_4(q).\frac{q+1}{d}.2$		2	
$\mathscr{C}_1$		$(\Omega_2^-(q) \times \frac{1}{d}\mathrm{GU}_3(q)).[2d]$	N1	1	S2
$\mathscr{C}_1$		$(\Omega_3(q) \times \Omega_5(q)).[4]$	q odd	2	S3
	$\mathscr{C}_4$	$(\mathrm{Sp}_2(q) \circ \mathrm{Sp}_4(q)).2$		4	
$\mathscr{C}_2$		$2^7{:}\mathrm{A}_8$	$q=p\equiv \pm 3 \bmod 8$, N5	2	S3
	$\mathscr{C}_6$	$2^{1+6}_+.\mathrm{A}_8$		4	
$\mathscr{C}_2$		$2^7{\cdot}\mathrm{S}_8$	$q = p \equiv \pm 1 \bmod 8$	4	$\langle \gamma \rangle$
	$\mathscr{C}_6$	$2^{1+6}_+{\cdot}\mathrm{S}_8$		8	
$\mathscr{C}_2$		$2^4.2^6.\mathrm{L}_3(2)$	$q = p$ odd, N1	4	$\langle \gamma, \tau \rangle$
$\mathscr{C}_2$		$\Omega_2^+(q)^4.(2d)^3.\mathrm{S}_4$	$q \geqslant 7$; N6 if $q = 5$	1	S2
$\mathscr{C}_2$		$\Omega_2^-(q)^4.(2d)^3.\mathrm{S}_4$	$q \neq 3$; N6 if $q = 3$	1	S2
$\mathscr{C}_2$		$\Omega_4^+(q)^2.[2d].\mathrm{S}_2$	$q \geqslant 3$	1	S2
$\mathscr{C}_2$		$\Omega_4^-(q)^2.[2d].\mathrm{S}_2$		1	S1
	$\mathscr{C}_3$	$\Omega_4^+(q^2).[4]$		2	
$\mathscr{C}_2$		$(\mathrm{D}_{2(q^2+1)/d})^2.[2d].\mathrm{S}_2$	N4	1	S2
$\mathscr{C}_5$		$\Omega_8^+(q_0)$	$q = q_0^r$, r prime	1	S2
			r odd or q even		
$\mathscr{C}_5$		$\mathrm{SO}_8^+(q_0).2$	$q = q_0^2$, q odd	4	$\langle \gamma, \tau, \phi \rangle$
$\mathscr{C}_5$		$d \times \Omega_8^-(q_0)$	$q = q_0^2$	d	S3
	$\mathscr{S}$	$d{\cdot}\Omega_8^-(q_0)$		$2d$	
	$\mathscr{S}$	$d \times \mathrm{L}_3(q).3$	$q \equiv 1 \pmod 3$	d^2	$\langle \gamma, \tau, \phi \rangle$
	$\mathscr{S}$	$d \times \mathrm{U}_3(q).3$	$q \equiv 2 \pmod 3$, $q \neq 2$	d^2	$\langle \gamma, \tau, \phi \rangle$
	$\mathscr{S}$	$d \times {}^3\mathrm{D}_4(q_0)$	$q = q_0^3$	$2d^2$	$\langle \tau, \phi \rangle$
	$\mathscr{S}$	$2{\cdot}\Omega_8^+(2)$	$q = p$ odd	4	$\langle \gamma, \tau \rangle$
	$\mathscr{S}$	$2{\cdot}\mathrm{Sz}(8)$	$q = 5$	8	$\langle \tau \rangle$
	$\mathscr{S}$	A_9	$q = 2$	3	$\langle \gamma \rangle$
	$\mathscr{S}$	$2 \times \mathrm{A}_{10}$	$q = 5$	4	$\langle \gamma \rangle$
	$\mathscr{S}$	$2{\cdot}\mathrm{A}_{10}$	$q = 5$	8	

S1	$\langle \delta, \gamma, \delta', \phi \rangle$ (q odd), $\langle \gamma, \phi \rangle$ (q even)
S2	$\langle \delta, \gamma, \delta', \tau, \phi \rangle$ (q odd) $\langle \gamma, \tau, \phi \rangle$ (q even)
S3	$\langle \gamma, \delta', \phi \rangle$ (q odd) $\langle \gamma, \phi \rangle$ (q even)
N1	Maximal under subgroups whose order mod $\langle \phi \rangle$ is a multiple of 3
N2	Maximal under subgroups not contained in $\langle \delta, \delta', \phi \rangle$ (q odd), $\langle \phi \rangle$ (q even)
N3	Maximal under subgroups not contained in $\langle \gamma, \delta' \rangle$
N4	Maximal under subgroups whose order mod $\langle \phi \rangle$ is a multiple of 3 if $q \neq 3$, or under subgroups containing $\langle \delta, \delta', \tau \rangle$ if $q = 3$
N5	Maximal under subgroups not contained in $\langle \gamma\delta' \rangle$
N6	Maximal under subgroups not contained in a conj of S3 or of $\langle \gamma, \tau \rangle$

Table 8.51 *The maximal subgroups of* ${}^3\mathrm{D}_4(q) < \Omega_8^+(q^3)$.

$|Z({}^3\mathrm{D}_4(q))| = 1$, $|\phi| = 3e$, $\phi^e = \tau$, $q = p^e$.

This table is taken from [63]. The specified Aschbacher classes have no formal definitions, and are just intended to give a rough idea of the nature of the subgroups.

$\mathscr{C}_i$	Subgp	Notes	c	Stab
$\mathscr{C}_1$	$\mathrm{E}_q{}^{1+8}{:}((q-1) \circ \mathrm{SL}_2(q^3)).(q-1,2)$		1	$\langle \phi \rangle$
$\mathscr{C}_1$	$[q^{11}]{:}((q^3-1) \circ \mathrm{SL}_2(q)).(q-1,2)$		1	$\langle \phi \rangle$
$\mathscr{C}_1$	$\mathrm{G}_2(q)$		1	$\langle \phi \rangle$
$\mathscr{C}_1$	$(\mathrm{SL}_2(q^3) \circ \mathrm{SL}_2(q))).(q-1,2)$		1	$\langle \phi \rangle$
$\mathscr{C}_1$	$((q^2+q+1) \circ \mathrm{SL}_3(q)).(q^2+q+1,3).2$		1	$\langle \phi \rangle$
$\mathscr{C}_1$	$((q^2-q+1) \circ \mathrm{SU}_3(q)).(q^2-q+1,3).2$		1	$\langle \phi \rangle$
$\mathscr{C}_1$	$(q^4-q^2+1).4$		1	$\langle \phi \rangle$
$\mathscr{C}_2$	$(q^2+q+1)^2.\mathrm{SL}_2(3)$		1	$\langle \phi \rangle$
$\mathscr{C}_2$	$(q^2-q+1)^2.\mathrm{SL}_2(3)$		1	$\langle \phi \rangle$
$\mathscr{C}_5$	${}^3\mathrm{D}_4(q_0)$	$q = q_0^r$, $3 \neq r$ prime	1	$\langle \phi \rangle$
$\mathscr{S}_2$	$\mathrm{PGL}_3(q)$	$q \equiv 1 \pmod 3$	1	$\langle \phi \rangle$
$\mathscr{S}_2$	$\mathrm{PGU}_3(q)$	$q \equiv 2 \pmod 3$, $q \neq 2$	1	$\langle \phi \rangle$

Table 8.52 *The maximal subgroups of* $\Omega_8^-(q)$ *of geometric type*

$|Z(\Omega_8^-(q))| = 1$, $|\delta| = (q-1,2)$, $|\gamma| = 2$, $|\varphi| = 2e$, $\varphi^e = \gamma$, $q = p^e$.

$\mathscr{C}_i$	Subgp	Notes	c	Stab
$\mathscr{C}_1$	$E_q{}^6 : (\frac{q-1}{(q-1,2)} \times \Omega_6^-(q)).(q-1,2)$		1	$\langle \delta, \varphi \rangle$
$\mathscr{C}_1$	$E_q{}^{1+8} : (\frac{1}{(q-1,2)}GL_2(q) \times \Omega_4^-(q)).(q-1,2)$		1	$\langle \delta, \varphi \rangle$
$\mathscr{C}_1$	$E_q{}^{3+6} : (\frac{1}{(q-1,2)}GL_3(q) \times \Omega_2^-(q)).(q-1,2)$		1	$\langle \delta, \varphi \rangle$
$\mathscr{C}_1$	$\Omega_7(q).2$	q odd	2	$\langle \varphi \rangle$
$\mathscr{C}_1$	$(\Omega_2^+(q) \times \Omega_6^-(q)).2^{(q-1,2)}$	$q \geqslant 4$	1	$\langle \delta, \varphi \rangle$
		N1 if $q = 3$	1	$\langle \delta, \gamma \rangle$
$\mathscr{C}_1$	$(\Omega_2^-(q) \times \Omega_6^+(q)).2^{(q-1,2)}$		1	$\langle \delta, \varphi \rangle$
$\mathscr{C}_1$	$(\Omega_3(q) \times \Omega_5(q)).2^2$	q odd	2	$\langle \varphi \rangle$
$\mathscr{C}_1$	$(\Omega_4^+(q) \times \Omega_4^-(q)).2^{(q-1,2)}$		1	$\langle \delta, \varphi \rangle$
$\mathscr{C}_1$	$Sp_6(q)$	q even	1	$\langle \varphi \rangle$
$\mathscr{C}_3$	$\Omega_4^-(q^2).2$		1	$\langle \delta, \varphi \rangle$
$\mathscr{C}_5$	$\Omega_8^-(q_0)$	$q = q_0^r$, r odd prime	1	$\langle \delta, \varphi \rangle$

N1	Maximal under subgroups not contained in $\langle \gamma \rangle$

Table 8.53 *The maximal subgroups of* $\Omega_8^-(q)$ *in Class* $\mathscr{S}$

$|Z(\Omega_8^-(q))| = 1$, $|\delta| = (q-1,2)$, $|\gamma| = 2$, $|\varphi| = 2e$, $\varphi^e = \gamma$, $q = p^e$.

Subgp	Conditions on q	c	Stab
$L_3(q)$	$q \equiv 2 \bmod 3$	$(q-1,2)$	$\langle \gamma, \varphi \rangle$
$U_3(q)$	$q \equiv 1 \bmod 3$	$(q-1,2)$	$\langle \gamma, \varphi \rangle$

Table 8.54 *The maximal subgroups of* $\mathrm{SL}_9(q)$ *of geometric type*
$$d := |Z(\mathrm{SL}_9(q))| = (q-1,9),\ |\delta| = d,\ |\phi| = e,\ |\gamma| = 2,\ q = p^e.$$

$\mathscr{C}_i$	Subgp	Notes	c	Stab
$\mathscr{C}_1$	$E_q{}^8{:}\mathrm{GL}_8(q)$		2	$\langle\delta,\phi\rangle$
$\mathscr{C}_1$	$E_q{}^{14}{:}(\mathrm{SL}_7(q) \times \mathrm{SL}_2(q)){:}(q-1)$		2	$\langle\delta,\phi\rangle$
$\mathscr{C}_1$	$E_q{}^{18}{:}(\mathrm{SL}_6(q) \times \mathrm{SL}_3(q)){:}(q-1)$		2	$\langle\delta,\phi\rangle$
$\mathscr{C}_1$	$E_q{}^{20}{:}(\mathrm{SL}_5(q) \times \mathrm{SL}_4(q)){:}(q-1)$		2	$\langle\delta,\phi\rangle$
$\mathscr{C}_1$	$E_q{}^{1+14}{:}(\mathrm{GL}_7(q) \times (q-1))$	N1	1	$\langle\delta,\phi,\gamma\rangle$
$\mathscr{C}_1$	$E_q{}^{4+20}{:}(\mathrm{SL}_2(q)^2 \times \mathrm{SL}_5(q)){:}(q-1)^2$	N1	1	$\langle\delta,\phi,\gamma\rangle$
$\mathscr{C}_1$	$E_q{}^{9+18}{:}\mathrm{SL}_3(q)^3{:}(q-1)^2$	N1	1	$\langle\delta,\phi,\gamma\rangle$
$\mathscr{C}_1$	$E_q{}^{16+8}{:}\mathrm{GL}_4(q)^2$	N1	1	$\langle\delta,\phi,\gamma\rangle$
$\mathscr{C}_1$	$\mathrm{GL}_8(q)$	N1	1	$\langle\delta,\phi,\gamma\rangle$
$\mathscr{C}_1$	$(\mathrm{SL}_7(q) \times \mathrm{SL}_2(q)){:}(q-1)$	N1	1	$\langle\delta,\phi,\gamma\rangle$
$\mathscr{C}_1$	$(\mathrm{SL}_6(q) \times \mathrm{SL}_3(q)){:}(q-1)$	N1	1	$\langle\delta,\phi,\gamma\rangle$
$\mathscr{C}_1$	$(\mathrm{SL}_5(q) \times \mathrm{SL}_4(q)){:}(q-1)$	N1	1	$\langle\delta,\phi,\gamma\rangle$
$\mathscr{C}_2$	$(q-1)^8{:}S_9$	$q \geqslant 5$	1	$\langle\delta,\phi,\gamma\rangle$
$\mathscr{C}_2$	$\mathrm{SL}_3(q)^3{:}(q-1)^2.S_3$		1	$\langle\delta,\phi,\gamma\rangle$
$\mathscr{C}_3$	$(((q-1,3)(q^2+q+1))\circ\mathrm{SL}_3(q^3)).3$		1	$\langle\delta,\phi,\gamma\rangle$
$\mathscr{C}_5$	$\mathrm{SL}_9(q_0).\left[\left(\frac{q-1}{q_0-1},9\right)\right]$	$q = q_0^r$, r prime	$\left(\frac{q-1}{q_0-1},9\right)$	$\langle\delta^c,\phi,\gamma\rangle$
$\mathscr{C}_6$	$(d \circ 3_+^{1+4}){:}\mathrm{Sp}_4(3)$	$q=p\equiv 1 \bmod 3$	d	$\langle\gamma\rangle$
$\mathscr{C}_7$	$(\mathrm{SL}_3(q) \circ \mathrm{SL}_3(q)).(q-1,3)^2.2$		$(q-1,3)$	$\langle\delta^c,\phi,\gamma\rangle$
$\mathscr{C}_8$	$d \times \mathrm{SO}_9(q)$	q odd	d	$\langle\phi,\gamma\rangle$
$\mathscr{C}_8$	$(q_0-1,9) \times \mathrm{SU}_9(q_0)$	$q = q_0^2$	$(q_0-1,9)$	$\langle\delta^c,\phi,\gamma\rangle$

N1	Maximal under subgroups not contained in $\langle\delta,\phi\rangle$

Table 8.55 *The maximal subgroups of* $\mathrm{SL}_9(q)$ *in Class* $\mathscr{S}$
$$d := |Z(\mathrm{SL}_9(q))| = (q-1,9),\ |\delta| = d,\ |\phi| = e,\ |\gamma| = 2,\ q = p^e.$$

Subgp	Conditions on q	c	Stab
$3{\cdot}A_7$	$q = 7$	3	$\langle\gamma\rangle$
$d \times \mathrm{L}_2(19)$	$q = p \equiv 1,4,5,6,7,9,11,16,17 \bmod 19$	d	$\langle\gamma\rangle$
$\mathrm{L}_3(q^2).2$	$q \equiv 0 \bmod 3$	1	$\langle\phi,\gamma\rangle$
$\mathrm{L}_3(q^2).S_3$	$q \equiv 2 \bmod 3$	1	$\langle\phi,\gamma\rangle$
$9 \circ \mathrm{SL}_3(q^2).2$	$q \equiv 1 \bmod 9$	3	$\langle\delta^3,\phi,\gamma\rangle$
$\mathrm{SL}_3(q^2).6$	$q \equiv 4,7 \bmod 9$	3	$\langle\phi,\gamma\rangle$

Notes: The extension of $\mathrm{L}_3(q^2)$ of degree 2 is from the automorphism $x \mapsto x^q$ of $\mathbb{F}_{q^2}$. The duality automorphism of $\mathrm{L}_3(q^2)$ is induced by γ.

Table 8.56 *The maximal subgroups of* $\mathrm{SU}_9(q)$ *of geometric type*

$$d := |Z(\mathrm{SU}_9(q))| = (q+1, 9),\ |\delta| = d,\ |\phi| = 2e,\ q = p^e.$$

$\mathscr{C}_i$	Subgp	Notes	c	Stab
$\mathscr{C}_1$	$E_q{}^{1+14}{:}\mathrm{SU}_7(q){:}(q^2-1)$		1	$\langle \delta, \phi \rangle$
$\mathscr{C}_1$	$E_q{}^{4+20}{:}(\mathrm{SL}_2(q^2) \times \mathrm{SU}_5(q)){:}(q^2-1)$		1	$\langle \delta, \phi \rangle$
$\mathscr{C}_1$	$E_q{}^{9+18}{:}(\mathrm{SL}_3(q^2) \times \mathrm{SU}_3(q)){:}(q^2-1)$		1	$\langle \delta, \phi \rangle$
$\mathscr{C}_1$	$E_q{}^{16+8}{:}\mathrm{GL}_4(q^2)$		1	$\langle \delta, \phi \rangle$
$\mathscr{C}_1$	$\mathrm{GU}_8(q)$		1	$\langle \delta, \phi \rangle$
$\mathscr{C}_1$	$(\mathrm{SU}_7(q) \times \mathrm{SU}_2(q)){:}(q+1)$		1	$\langle \delta, \phi \rangle$
$\mathscr{C}_1$	$(\mathrm{SU}_6(q) \times \mathrm{SU}_3(q)){:}(q+1)$		1	$\langle \delta, \phi \rangle$
$\mathscr{C}_1$	$(\mathrm{SU}_5(q) \times \mathrm{SU}_4(q)){:}(q+1)$		1	$\langle \delta, \phi \rangle$
$\mathscr{C}_2$	$(q+1)^8{:}\mathrm{S}_9$		1	$\langle \delta, \phi \rangle$
$\mathscr{C}_2$	$\mathrm{SU}_3(q)^3{:}(q+1)^2.\mathrm{S}_3$		1	$\langle \delta, \phi \rangle$
$\mathscr{C}_3$	$(((q+1,3)(q^2-q+1)) \circ \mathrm{SU}_3(q^3)).3$		1	$\langle \delta, \phi \rangle$
$\mathscr{C}_5$	$\mathrm{SU}_9(q_0).\left[\left(\frac{q+1}{q_0+1}, 9\right)\right]$	$q_0^r = q,$ r odd prime	$\left(\frac{q+1}{q_0+1}, 9\right)$	$\langle \delta^c, \phi \rangle$
$\mathscr{C}_5$	$d \times \mathrm{SO}_9(q)$	q odd	d	$\langle \phi \rangle$
$\mathscr{C}_6$	$(d \circ 3_+^{1+4}){:}\mathrm{Sp}_4(3)$	$q = p \equiv 2 \bmod 3$	d	$\langle \phi \rangle$
$\mathscr{C}_7$	$(\mathrm{SU}_3(q) \circ \mathrm{SU}_3(q)).(q+1,3)^2.2$	$q \geqslant 3$	$(q+1,3)$	$\langle \delta^c, \phi \rangle$

Table 8.57 *The maximal subgroups of* $\mathrm{SU}_9(q)$ *in Class* $\mathscr{S}$

$$d := |Z(\mathrm{SU}_9(q))| = (q+1, 9),\ |\delta| = d,\ |\phi| = 2e,\ \phi^e = \gamma,\ q = p^e.$$

Subgp	Nov	Conditions on q	c	Stab
$d \times \mathrm{L}_2(19)$		$q = p \equiv 2, 3, 8, 10, 12, 13, 14, 15, 18 \bmod 19$ $q \neq 2$	d	$\langle \gamma \rangle$
$3 \times \mathrm{L}_2(19)$	N1	$q = 2$	3	$\langle \gamma \rangle$
$3{}^{\cdot}\mathrm{J}_3$		$q = 2$	3	$\langle \gamma \rangle$
$\mathrm{L}_3(q^2).2$		$q \equiv 0 \bmod 3$	1	$\langle \phi \rangle$
$\mathrm{L}_3(q^2).\mathrm{S}_3$		$q \equiv 1 \bmod 3$	1	$\langle \phi \rangle$
$9 \circ \mathrm{SL}_3(q^2).2$		$q \equiv 8 \bmod 9$	3	$\langle \delta^3, \phi \rangle$
$\mathrm{SL}_3(q^2).6$		$q \equiv 2, 5 \bmod 9$	3	$\langle \phi \rangle$

N1	Maximal under $\langle \gamma \rangle$

Notes: The extension of $\mathrm{L}_3(q^2)$ of degree 2 is from the product of the duality automorphism and the automorphism $x \mapsto x^q$ of $\mathbb{F}_{q^2}$. The field automorphism of $\mathrm{L}_3(q^2)$ is induced by ϕ.

Table 8.58 *The maximal subgroups of $\Omega_9(q)$ of geometric type*
$|Z(\Omega_9(q))| = 1$, $|\delta| = 2$, $|\phi| = e$, $q = p^e$ odd.

$\mathscr{C}_i$	Subgp	Notes	c	Stab
$\mathscr{C}_1$	$E_q^{\;7}{:}(\frac{q-1}{2} \times \Omega_7(q)).2$		1	$\langle \delta, \phi \rangle$
$\mathscr{C}_1$	$E_q^{\;1+10}{:}(\frac{1}{2}\mathrm{GL}_2(q) \times \Omega_5(q)).2$		1	$\langle \delta, \phi \rangle$
$\mathscr{C}_1$	$E_q^{\;3+9}{:}(\frac{1}{2}\mathrm{GL}_3(q) \times \Omega_3(q)).2$		1	$\langle \delta, \phi \rangle$
$\mathscr{C}_1$	$E_q^{\;6+4}{:}\frac{1}{2}\mathrm{GL}_4(q)$		1	$\langle \delta, \phi \rangle$
$\mathscr{C}_1$	$\Omega_8^+(q).2$		1	$\langle \delta, \phi \rangle$
$\mathscr{C}_1$	$\Omega_8^-(q).2$		1	$\langle \delta, \phi \rangle$
$\mathscr{C}_1$	$(\Omega_2^+(q) \times \Omega_7(q)).2^2$	$q \geqslant 5$	1	$\langle \delta, \phi \rangle$
$\mathscr{C}_1$	$(\Omega_2^-(q) \times \Omega_7(q)).2^2$		1	$\langle \delta, \phi \rangle$
$\mathscr{C}_1$	$(\Omega_3(q) \times \Omega_6^+(q)).2^2$		1	$\langle \delta, \phi \rangle$
$\mathscr{C}_1$	$(\Omega_3(q) \times \Omega_6^-(q)).2^2$		1	$\langle \delta, \phi \rangle$
$\mathscr{C}_1$	$(\Omega_4^+(q) \times \Omega_5(q)).2^2$		1	$\langle \delta, \phi \rangle$
$\mathscr{C}_1$	$(\Omega_4^-(q) \times \Omega_5(q)).2^2$		1	$\langle \delta, \phi \rangle$
$\mathscr{C}_2$	$2^8{:}A_9$	$p = q \equiv \pm 3 \bmod 8$	1	$\langle \delta \rangle$
$\mathscr{C}_2$	$2^8{:}S_9$	$p = q \equiv \pm 1 \bmod 8$	2	1
$\mathscr{C}_2$	$\Omega_3(q)^3.2^4.S_3$	$q \geqslant 5$	1	$\langle \delta, \phi \rangle$
$\mathscr{C}_3$	$\Omega_3(q^3).3$		1	$\langle \delta, \phi \rangle$
$\mathscr{C}_5$	$\Omega_9(q_0)$	$q = q_0^r$, r odd prime	1	$\langle \delta, \phi \rangle$
$\mathscr{C}_5$	$SO_9(q_0)$	$q = q_0^2$	2	$\langle \phi \rangle$
$\mathscr{C}_7$	$\Omega_3(q)^2.[4]$	$q \geqslant 5$	1	$\langle \delta, \phi \rangle$

Table 8.59 *The maximal subgroups of $\Omega_9(q)$ in Class $\mathscr{S}$*
$|Z(\Omega_9(q))| = 1$, $|\delta| = 2$, $|\phi| = e$, $q = p^e$ odd.

Subgp	Conditions on q	c	Stab
$L_2(8)$	$q = p \equiv \pm 1 \bmod 7$	2	1
$L_2(8)$	$q = p^3$, $p \equiv \pm 2, \pm 3 \bmod 7$	2	$\langle \phi \rangle$
$L_2(17)$	$q = p \equiv \pm 1, \pm 2, \pm 4, \pm 8 \bmod 17$	2	1
$L_2(17)$	$q = p^2$, $p \equiv \pm 3, \pm 5, \pm 6, \pm 7 \bmod 17$	2	$\langle \phi \rangle$
A_{10}	$q = p \equiv \pm 2 \bmod 5$	1	$\langle \delta \rangle$
S_{10}	$q = p \equiv \pm 1 \bmod 5$, $q \neq 11$	2	1
S_{11}	$q = 11$	2	1
$L_2(q).2$	$p \geqslant 11$	2	$\langle \phi \rangle$
$L_2(q^2).2$	$q \neq 3$	1	$\langle \delta, \phi \rangle$

Note: The extension of $L_2(q^2)$ of degree 2 is from the automorphism $x \mapsto x^q$ of $\mathbb{F}_{q^2}$ when $q \equiv \pm 1 \pmod 8$ and from the product of the diagonal automorphism and the field automorphism when $q \equiv \pm 3 \pmod 8$.

Table 8.60 *The maximal subgroups of* $\mathrm{SL}_{10}(q)$ *of geometric type*
$d := |Z(\mathrm{SL}_{10}(q))| = (q-1, 10)$, $|\delta| = d$, $|\phi| = e$, $|\gamma| = 2$, $q = p^e$.

$\mathscr{C}_i$	Subgp	Notes	c	Stab
$\mathscr{C}_1$	$E_q{}^9{:}\mathrm{GL}_9(q)$		2	$\langle \delta, \phi \rangle$
$\mathscr{C}_1$	$E_q{}^{16}{:}(\mathrm{SL}_8(q) \times \mathrm{SL}_2(q)){:}(q-1)$		2	$\langle \delta, \phi \rangle$
$\mathscr{C}_1$	$E_q{}^{21}{:}(\mathrm{SL}_7(q) \times \mathrm{SL}_3(q)){:}(q-1)$		2	$\langle \delta, \phi \rangle$
$\mathscr{C}_1$	$E_q{}^{24}{:}(\mathrm{SL}_6(q) \times \mathrm{SL}_4(q)){:}(q-1)$		2	$\langle \delta, \phi \rangle$
$\mathscr{C}_1$	$E_q{}^{25}{:}(\mathrm{SL}_5(q) \times \mathrm{SL}_5(q)){:}(q-1)$		1	$\langle \delta, \phi, \gamma \rangle$
$\mathscr{C}_1$	$E_q{}^{1+16}{:}(\mathrm{GL}_8(q) \times (q-1))$	N1	1	$\langle \delta, \phi, \gamma \rangle$
$\mathscr{C}_1$	$E_q{}^{4+24}{:}(\mathrm{SL}_2(q)^2 \times \mathrm{SL}_6(q)){:}(q-1)^2$	N1	1	$\langle \delta, \phi, \gamma \rangle$
$\mathscr{C}_1$	$E_q{}^{9+24}{:}(\mathrm{SL}_3(q)^2 \times \mathrm{SL}_4(q)){:}(q-1)^2$	N1	1	$\langle \delta, \phi, \gamma \rangle$
$\mathscr{C}_1$	$E_q{}^{16+16}{:}(\mathrm{SL}_4(q)^2 \times \mathrm{SL}_2(q)){:}(q-1)^2$	N1	1	$\langle \delta, \phi, \gamma \rangle$
$\mathscr{C}_1$	$\mathrm{GL}_9(q)$	N1	1	$\langle \delta, \phi, \gamma \rangle$
$\mathscr{C}_1$	$(\mathrm{SL}_8(q) \times \mathrm{SL}_2(q)){:}(q-1)$	N1	1	$\langle \delta, \phi, \gamma \rangle$
$\mathscr{C}_1$	$(\mathrm{SL}_7(q) \times \mathrm{SL}_3(q)){:}(q-1)$	N1	1	$\langle \delta, \phi, \gamma \rangle$
$\mathscr{C}_1$	$(\mathrm{SL}_6(q) \times \mathrm{SL}_4(q)){:}(q-1)$	N1	1	$\langle \delta, \phi, \gamma \rangle$
$\mathscr{C}_2$	$(q-1)^9.\mathrm{S}_{10}$	$q \geqslant 5$	1	$\langle \delta, \phi, \gamma \rangle$
$\mathscr{C}_2$	$\mathrm{SL}_2(q)^5{:}(q-1)^4.\mathrm{S}_5$	$q \geqslant 3$	1	$\langle \delta, \phi, \gamma \rangle$
$\mathscr{C}_2$	$\mathrm{SL}_5(q)^2{:}(q-1).\mathrm{S}_2$		1	$\langle \delta, \phi, \gamma \rangle$
$\mathscr{C}_3$	$(\frac{q^5-1}{q-1} \times \mathrm{SL}_2(q^5)).5$		1	$\langle \delta, \phi, \gamma \rangle$
$\mathscr{C}_3$	$(((q-1,5)(q+1)){\circ}\mathrm{SL}_5(q^2)).(q+1,5).2$		1	$\langle \delta, \phi, \gamma \rangle$
$\mathscr{C}_4$	$\mathrm{SL}_2(q) \times \mathrm{SL}_5(q)$	$q \geqslant 3$	1	$\langle \delta, \phi, \gamma \rangle$
$\mathscr{C}_5$	$\mathrm{SL}_{10}(q_0).\left[\left(\frac{q-1}{q_0-1}, 10\right)\right]$	$q = q_0^r$,	$\left(\frac{q-1}{q_0-1}, 10\right)$	$\langle \delta^c, \phi, \gamma \rangle$
		r prime		
$\mathscr{C}_8$	$(q-1,5) \times \mathrm{SO}_{10}^+(q).2$	q odd	$d/2$	$\langle \delta^c, \phi, \gamma \rangle$
$\mathscr{C}_8$	$(q-1,5) \times \mathrm{SO}_{10}^-(q).2$	q odd	$d/2$	S1
$\mathscr{C}_8$	$(q-1,5) \times \mathrm{Sp}_{10}(q)$		$(q-1,5)$	$\langle \delta^c, \phi, \gamma \rangle$
$\mathscr{C}_8$	$\mathrm{SU}_{10}(q_0).(q_0-1,10)$	$q = q_0^2$	$(q_0-1,10)$	$\langle \delta^c, \phi, \gamma \rangle$

S1	$\langle \delta^c, \phi, \gamma \rangle$ $(q \equiv 3 \bmod 4)$ or $\langle \delta^c, \phi\delta^{(p-1)/2}, \gamma\delta^{-1} \rangle$ $(q \equiv 1 \bmod 4)$
N1	Maximal under subgroups not contained in $\langle \delta, \phi \rangle$

Table 8.61 *The maximal subgroups of* $\mathrm{SL}_{10}(q)$ *in Class* $\mathscr{S}$
$d := |Z(\mathrm{SL}_{10}(q))| = (q-1, 10)$, $|\delta| = d$, $|\phi| = e$, $|\gamma| = 2$, $q = p^e$.

Subgp	Nov	Conditions on q	c	Stab	Acts
$d \circ 2^{\cdot}\mathrm{L}_2(19)$		$q=p\equiv 1, 4, 5, 6, 7, 9,$	d	S1	
		$11, 16, 17 \bmod 19$			
$d \circ 2^{\cdot}\mathrm{L}_3(4)$	N1	$q=p\equiv 11, 15, 23 \bmod 28$	$d/2$	$\langle \delta^5, \gamma \rangle$	A1
$d \circ 2^{\cdot}\mathrm{L}_3(4).2_2$	N2	$q=p\equiv 1, 9, 25 \bmod 28$	d	S2	
$d \circ 2^{\cdot}\mathrm{M}_{12}$		$q=p\equiv 3 \bmod 8$	d	$\langle \delta^5 \rangle$	
$d \circ 2^{\cdot}\mathrm{M}_{12}.2$		$q=p\equiv 1 \bmod 8$	$2d$	1	
$\mathrm{M}_{22}.2$		$q = 2$	2	1	
$d \circ 2^{\cdot}\mathrm{M}_{22}$		$q=p\equiv 11, 15, 23 \bmod 28$	d	$\langle \delta^5 \rangle$	
$d \circ 2^{\cdot}\mathrm{M}_{22}.2$		$q=p\equiv 1, 9, 25 \bmod 28$	$2d$	1	
$d \times \mathrm{L}_3(q).(q-1, 3)$		$p \geqslant 5$	d	S3	A2
$d \circ \frac{(q-1,4)}{2}{\cdot}\mathrm{L}_4(q).\frac{(q-1,4)}{2}$		$p \geqslant 3$	$d/2$	$\langle \delta^5, \phi, \gamma \rangle$	A3
$d \circ \mathrm{SL}_5(q)$			$(q-1, 2)$	$\langle \delta^2, \phi, \gamma \rangle$	A4

S1	$\langle \gamma \rangle$ ($p \equiv \pm 1 \bmod 8$) or $\langle \gamma\delta \rangle$ ($p \equiv \pm 3 \bmod 8$)
S2	$\langle \gamma \rangle$ ($p \equiv 1 \bmod 8$) or $\langle \gamma\delta \rangle$ ($p \equiv 5 \bmod 8$)
S3	$\langle \phi, \gamma \rangle$ ($q \equiv \pm 1, \pm 5 \bmod 24$) or $\langle \phi, \gamma\delta^5 \rangle$ ($q \equiv \pm 7, \pm 11 \bmod 24$)
N1	Maximal under subgroups not contained in $\langle \delta^5 \rangle$
N2	Maximal under S2
A1	$\delta^5 \to 2_2$, $\gamma \to 2_1$ ($p \equiv 3 \bmod 8$) or $2_1 2_2$ ($p \equiv 1 \bmod 8$)
A2	γ_Ω or $\gamma_\Omega \delta_\Omega^5 \to \gamma_S$, $\phi_\Omega \to \phi_S$
A3	$\delta_\Omega^5 \to \delta_S$, $\gamma_\Omega \to \gamma_S$, $\phi_\Omega \to \phi_S$
A4	$\delta_\Omega^2 \to \delta_S$, $\gamma_\Omega \to \gamma_S$, $\phi_\Omega \to \phi_S$

Table 8.62 *The maximal subgroups of* $\mathrm{SU}_{10}(q)$ *of geometric type*
$d := |Z(\mathrm{SU}_{10}(q))| = (q+1,10)$, $|\delta| = d$, $|\phi| = 2e$, $q = p^e$.

$\mathscr{C}_i$	Subgp	Notes	c	Stab
$\mathscr{C}_1$	$E_q{}^{1+16}{:}\mathrm{SU}_8(q){:}(q^2-1)$		1	$\langle \delta, \phi \rangle$
$\mathscr{C}_1$	$E_q{}^{4+24}{:}(\mathrm{SL}_2(q^2) \times \mathrm{SU}_6(q)){:}(q^2-1)$		1	$\langle \delta, \phi \rangle$
$\mathscr{C}_1$	$E_q{}^{9+24}{:}(\mathrm{SL}_3(q^2) \times \mathrm{SU}_4(q)){:}(q^2-1)$		1	$\langle \delta, \phi \rangle$
$\mathscr{C}_1$	$E_q{}^{16+16}{:}(\mathrm{SL}_4(q^2) \times \mathrm{SU}_2(q)){:}(q^2-1)$		1	$\langle \delta, \phi \rangle$
$\mathscr{C}_1$	$E_q{}^{25}{:}\mathrm{SL}_5(q^2){:}(q-1)$		1	$\langle \delta, \phi \rangle$
$\mathscr{C}_1$	$\mathrm{GU}_9(q)$		1	$\langle \delta, \phi \rangle$
$\mathscr{C}_1$	$(\mathrm{SU}_8(q) \times \mathrm{SU}_2(q)){:}(q+1)$		1	$\langle \delta, \phi \rangle$
$\mathscr{C}_1$	$(\mathrm{SU}_7(q) \times \mathrm{SU}_3(q)){:}(q+1)$		1	$\langle \delta, \phi \rangle$
$\mathscr{C}_1$	$(\mathrm{SU}_6(q) \times \mathrm{SU}_4(q)){:}(q+1)$		1	$\langle \delta, \phi \rangle$
$\mathscr{C}_2$	$(q+1)^9.\mathrm{S}_{10}$		1	$\langle \delta, \phi \rangle$
$\mathscr{C}_2$	$\mathrm{SU}_2(q)^5{:}(q+1)^4.\mathrm{S}_5$	$q \geqslant 3$	1	$\langle \delta, \phi \rangle$
$\mathscr{C}_2$	$\mathrm{SU}_5(q)^2{:}(q+1).\mathrm{S}_2$		1	$\langle \delta, \phi \rangle$
$\mathscr{C}_2$	$\mathrm{SL}_5(q^2).(q-1).2$		1	$\langle \delta, \phi \rangle$
$\mathscr{C}_3$	$(\frac{q^5+1}{q+1} \times \mathrm{SU}_2(q^5)).5$		1	$\langle \delta, \phi \rangle$
$\mathscr{C}_4$	$\mathrm{SU}_2(q) \times \mathrm{SU}_5(q)$	$q \geqslant 3$	1	$\langle \delta, \phi \rangle$
$\mathscr{C}_5$	$\mathrm{SU}_{10}(q_0).\left[\left(\frac{q+1}{q_0+1},10\right)\right]$	$q_0^r = q$, r odd prime	$\left(\frac{q+1}{q_0+1},10\right)$	$\langle \delta^c, \phi \rangle$
$\mathscr{C}_5$	$(q+1,5) \times \mathrm{Sp}_{10}(q)$		$(q+1,5)$	$\langle \delta^c, \phi \rangle$
$\mathscr{C}_5$	$(q+1,5) \times \mathrm{SO}_{10}^+(q).2$	q odd	$(q+1,5)$	S1
$\mathscr{C}_5$	$(q+1,5) \times \mathrm{SO}_{10}^-(q).2$	q odd	$(q+1,5)$	S2

S1	$\langle \delta^c, \phi \rangle$ ($q \equiv 1 \bmod 4$) or $\langle \delta^c, \phi\delta^{(p-1)/2} \rangle$ ($q \equiv 3 \bmod 4$)
S2	$\langle \delta^c, \phi\delta^{(p-1)/2} \rangle$ ($q \equiv 1 \bmod 4$) or $\langle \delta^c, \phi \rangle$ ($q \equiv 3 \bmod 4$)

Table 8.63 *The maximal subgroups of* $\mathrm{SU}_{10}(q)$ *in Class* $\mathscr{S}$
$d := |\mathrm{Z}(\mathrm{SU}_{10}(q))| = (q+1, 10),\ |\delta| = d,\ |\phi| = 2e,\ \phi^e = \gamma,\ q = p^e.$

Subgp	Nov	Conditions on q	c	Stab	Acts
$d \circ 2^{\cdot}\mathrm{L}_2(19)$		$q = p \equiv 2, 3, 8, 10, 12, 13,$	d	S1	
		$14, 15, 18 \bmod 19,\ q \neq 2$			
$d \circ 2^{\cdot}\mathrm{L}_3(4)$	N1	$q = p \equiv 5, 13, 17 \bmod 28$	$d/2$	$\langle \delta^5, \gamma \rangle$	A1
$d \circ 2^{\cdot}\mathrm{L}_3(4).2_2$	N2	$q = p \equiv 3, 19, 27 \bmod 28,$	d	S2	
		$q \neq 3$			
$d \circ 2^{\cdot}\mathrm{M}_{12}$		$q = p \equiv 5 \bmod 8$	d	$\langle \delta^5 \rangle$	
$d \circ 2^{\cdot}\mathrm{M}_{12}.2$		$q = p \equiv 7 \bmod 8$	$2d$	1	
$d \circ 2^{\cdot}\mathrm{M}_{22}$		$q = p \equiv 5, 13, 17 \bmod 28$	d	$\langle \delta^5 \rangle$	
$d \circ 2^{\cdot}\mathrm{M}_{22}.2$		$q = p \equiv 3, 19, 27 \bmod 28$	$2d$	1	
$d \times \mathrm{U}_3(q).(q+1, 3)$		$p \geqslant 5$	d	S3	A2
$d \circ \frac{(q+1,4)}{2} {}^{\cdot}\mathrm{U}_4(q).\frac{(q+1,4)}{2}$		$p \geqslant 3$	$d/2$	$\langle \delta^5, \phi \rangle$	A3
$d \circ \mathrm{SU}_5(q)$			$(q+1, 2)$	$\langle \delta^2, \phi \rangle$	A4

S1	$\langle \gamma \rangle$ $(p \equiv \pm 1 \bmod 8)$ or $\langle \gamma\delta \rangle$ $(p \equiv \pm 3 \bmod 8)$
S2	$\langle \gamma \rangle$ $(p \equiv 7 \bmod 8)$ or $\langle \gamma\delta \rangle$ $(p \equiv 3 \bmod 8)$
S3	$\langle \phi \rangle$ $(p \equiv \pm 1, \pm 5 \bmod 24)$ or $\langle \phi\delta \rangle$ $(p \equiv \pm 7, \pm 11 \bmod 24)$
N1	Maximal under subgroups not contained in $\langle \delta^5 \rangle$
N2	Maximal under S2
A1	$\delta^5 \to 2_2,\ \gamma \to 2_1$ $(p \equiv 5 \bmod 8)$ or $2_1 2_2$ $(p \equiv 7 \bmod 8)$
A2	ϕ_Ω or $\phi_\Omega \delta_\Omega \to \phi_S$
A3	$\delta_\Omega^5 \to \delta_S,\ \phi_\Omega \to \phi_S$
A4	$\delta_\Omega^2 \to \delta_S,\ \phi_\Omega \to \phi_S$

Table 8.64 *The maximal subgroups of* $\mathrm{Sp}_{10}(q)$ *of geometric type*
$d := |Z(\mathrm{Sp}_{10}(q))| = (q-1, 2),\ |\delta| = d,\ |\phi| = e,\ q = p^e.$

$\mathscr{C}_i$	Subgp	Notes	c	Stab
$\mathscr{C}_1$	$E_q^{1+8}{:}((q-1) \times \mathrm{Sp}_8(q))$	q odd	1	$\langle \delta, \phi \rangle$
$\mathscr{C}_1$	$E_q^{9}{:}((q-1) \times \mathrm{Sp}_8(q))$	q even	1	$\langle \phi \rangle$
$\mathscr{C}_1$	$E_q^{3+12}{:}(\mathrm{GL}_2(q) \times \mathrm{Sp}_6(q))$		1	$\langle \delta, \phi \rangle$
$\mathscr{C}_1$	$E_q^{6+12}{:}(\mathrm{GL}_3(q) \times \mathrm{Sp}_4(q))$		1	$\langle \delta, \phi \rangle$
$\mathscr{C}_1$	$E_q^{10+8}{:}(\mathrm{GL}_4(q) \times \mathrm{Sp}_2(q))$		1	$\langle \delta, \phi \rangle$
$\mathscr{C}_1$	$E_q^{15}{:}\mathrm{GL}_5(q)$		1	$\langle \delta, \phi \rangle$
$\mathscr{C}_1$	$\mathrm{Sp}_8(q) \times \mathrm{Sp}_2(q)$		1	$\langle \delta, \phi \rangle$
$\mathscr{C}_1$	$\mathrm{Sp}_6(q) \times \mathrm{Sp}_4(q)$		1	$\langle \delta, \phi \rangle$
$\mathscr{C}_2$	$\mathrm{Sp}_2(q)^5{:}\mathrm{S}_5$	$q \geqslant 3$	1	$\langle \delta, \phi \rangle$
$\mathscr{C}_2$	$\mathrm{GL}_5(q).2$	q odd	1	$\langle \delta, \phi \rangle$
$\mathscr{C}_3$	$\mathrm{Sp}_2(q^5){:}5$		1	$\langle \delta, \phi \rangle$
$\mathscr{C}_3$	$\mathrm{GU}_5(q).2$	q odd	1	$\langle \delta, \phi \rangle$
$\mathscr{C}_4$	$\mathrm{Sp}_2(q) \circ \mathrm{GO}_5(q)$	q odd	1	$\langle \delta, \phi \rangle$
$\mathscr{C}_5$	$\mathrm{Sp}_{10}(q_0).(d,r)$	$q = q_0^r,\ r$ prime	(d,r)	$\langle \delta^c, \phi \rangle$
$\mathscr{C}_8$	$\mathrm{SO}_{10}^+(q)$	q even	1	$\langle \phi \rangle$
$\mathscr{C}_8$	$\mathrm{SO}_{10}^-(q)$	q even	1	$\langle \phi \rangle$

Table 8.65 *The maximal subgroups of* $\mathrm{Sp}_{10}(q)$ *in Class* $\mathscr{S}$
$d := |Z(\mathrm{Sp}_{10}(q))| = (q-1, 2),\ |\delta| = d,\ |\phi| = e,\ q = p^e.$

Subgp	Conditions on q	c	Stab	Acts
$2{\cdot}\mathrm{A}_6$	$q = p \equiv \pm 7 \bmod 16$	1	$\langle \delta \rangle$	
$2{\cdot}\mathrm{A}_6.2_2$	$q = p \equiv \pm 1 \bmod 16$	2	1	
$2{\cdot}\mathrm{A}_6$	$q = p^2,\ p \equiv \pm 3 \bmod 8,\ p \neq 3$	1	$\langle \delta, \phi \rangle$	$\delta \to 2_2,\ \phi \to 2_1$
$2{\cdot}\mathrm{L}_2(11)$	$q = p \equiv \pm 3 \bmod 8,\ q \neq 11$	1	$\langle \delta \rangle$	
$2{\cdot}\mathrm{L}_2(11).2$	$q = p \equiv \pm 1 \bmod 8$	2	1	
$2{\cdot}\mathrm{L}_2(11)$	$q = p \equiv \pm 11 \bmod 24,\ q \neq 11$	1	$\langle \delta \rangle$	
$2{\cdot}\mathrm{L}_2(11)$	$q = p \equiv \pm 11 \bmod 24,\ q \neq 11$	1	$\langle \delta \rangle$	
$2{\cdot}\mathrm{L}_2(11).2$	$q = p \equiv \pm 1 \bmod 24$	2	1	
$2{\cdot}\mathrm{L}_2(11).2$	$q = p \equiv \pm 1 \bmod 24$	2	1	
$2{\cdot}\mathrm{L}_2(11).2$	$q = p^2,\ p \equiv \pm 5 \bmod 12$	4	1	
$2 \times \mathrm{U}_5(2)$	$q = p \equiv \pm 3 \bmod 8$	1	$\langle \delta \rangle$	
$(2 \times \mathrm{U}_5(2)).2$	$q = p \equiv \pm 1 \bmod 8$	2	1	
$2{\cdot}\mathrm{L}_2(q)$	$p \geqslant 11$	1	$\langle \delta, \phi \rangle$	

Table 8.66 *The maximal subgroups of* $\Omega_{10}^+(q)$ *of geometric type*

$q = 2^e$: $|Z(\Omega_{10}^+(q))| = 1$, $|\delta| = 1$, $|\gamma| = 2$, $|\delta'| = 1$, $|\phi| = e$.
$q = p^e \equiv 1 \bmod 4$: $|Z(\Omega_{10}^+(q))| = 2$, $|\delta| = 4$, $|\gamma| = 2$, $\delta^2 = \delta'$, $|\phi| = e$.
$q = p^e \equiv 3 \bmod 4$: $|Z(\Omega_{10}^+(q))| = 1$, $|\delta| = 2$, $|\gamma| = 2$, $|\delta'| = 1$, $|\phi| = e$.

$\mathscr{C}_i$	Subgp	Notes	c	Stab
$\mathscr{C}_1$	$E_q{}^8 : (\frac{q-1}{(q-1,2)} \times \Omega_8^+(q)).(q-1,2)$		1	$\langle\delta,\gamma,\phi\rangle$
$\mathscr{C}_1$	$E_q{}^{1+12} : (\frac{1}{(q-1,2)}GL_2(q) \times \Omega_6^+(q)).(q-1,2)$		1	$\langle\delta,\gamma,\phi\rangle$
$\mathscr{C}_1$	$E_q{}^{3+12} : (\frac{1}{(q-1,2)}GL_3(q) \times \Omega_4^+(q)).(q-1,2)$		1	$\langle\delta,\gamma,\phi\rangle$
$\mathscr{C}_1$	$E_q{}^{6+8} : (\frac{1}{(q-1,2)}GL_4(q) \times \Omega_2^+(q)).(q-1,2)$	N1	1	$\langle\delta,\gamma,\phi\rangle$
$\mathscr{C}_1$	$E_q{}^{10} : \frac{1}{(q-1,2)}GL_5(q)$		2	$\langle\delta,\phi\rangle$
$\mathscr{C}_1$	$\Omega_9(q).2$	q odd	2	S1
$\mathscr{C}_1$	$(\Omega_2^+(q) \times \Omega_8^+(q)).2^{(q-1,2)}$	$q \geqslant 4$; N2 if $q = 3$	1	$\langle\delta,\gamma,\phi\rangle$
$\mathscr{C}_1$	$(\Omega_2^-(q) \times \Omega_8^-(q)).2^{(q-1,2)}$		1	$\langle\delta,\gamma,\phi\rangle$
$\mathscr{C}_1$	$(\Omega_3(q) \times \Omega_7(q)).2^2$	q odd	2	S1
$\mathscr{C}_1$	$(\Omega_4^+(q) \times \Omega_6^+(q)).2^{(q-1,2)}$		1	$\langle\delta,\gamma,\phi\rangle$
$\mathscr{C}_1$	$(\Omega_4^-(q) \times \Omega_6^-(q)).2^{(q-1,2)}$		1	$\langle\delta,\gamma,\phi\rangle$
$\mathscr{C}_1$	$Sp_8(q)$	q even	1	$\langle\gamma,\phi\rangle$
$\mathscr{C}_2$	$2^9 : A_{10}$	$p = q \equiv 5 \bmod 8$	2	$\langle\gamma,\delta'\rangle$
$\mathscr{C}_2$	$2^9 \cdot S_{10}$	$p = q \equiv 1 \bmod 8$	4	$\langle\gamma\rangle$
$\mathscr{C}_2$	$\Omega_2^+(q)^5.2^{4(2,q-1)}.S_5$	$q \geqslant 7$	1	$\langle\delta,\gamma,\phi\rangle$
		N3 if $q = 5$	1	$\langle\delta,\gamma\rangle$
$\mathscr{C}_2$	$\Omega_5(q)^2.2^2.S_2$	$q \equiv 1 \bmod 4$	2	$\langle\gamma,\delta',\phi\rangle$
$\mathscr{C}_2$	$SL_5(q).\frac{q-1}{(q-1,2)}$	N1	1	$\langle\delta,\gamma,\phi\rangle$
$\mathscr{C}_2$	$SO_5(q)^2$	$q \equiv 3 \bmod 4$	1	$\langle\delta,\gamma,\phi\rangle$
$\mathscr{C}_3$	$\Omega_5(q^2).2$	$q \equiv 3 \bmod 4$	1	$\langle\delta,\gamma,\phi\rangle$
$\mathscr{C}_3$	$2 \times \Omega_5(q^2).2$	$q \equiv 1 \bmod 4$	2	S2
$\mathscr{C}_5$	$\Omega_{10}^+(q_0)$	$q = q_0^r$, r prime,	1	$\langle\delta,\gamma,\phi\rangle$
		r odd or q even		
$\mathscr{C}_5$	$SO_{10}^+(q_0)$	$q = q_0^2$, $q_0 \equiv 1 \bmod 4$	2	$\langle\gamma,\delta',\phi\rangle$
$\mathscr{C}_5$	$SO_{10}^+(q_0).2$	$q = q_0^2$, $q_0 \equiv 3 \bmod 4$	4	$\langle\gamma,\phi\rangle$
$\mathscr{C}_5$	$\Omega_{10}^-(q_0)$	$q = q_0^2$, q even	1	$\langle\gamma,\phi\rangle$
$\mathscr{C}_5$	$SO_{10}^-(q_0).2$	$q = q_0^2$, $q_0 \equiv 1 \bmod 4$	4	$\langle\gamma,\phi\rangle$
$\mathscr{C}_5$	$SO_{10}^-(q_0)$	$q = q_0^2$, $q_0 \equiv 3 \bmod 4$	2	$\langle\gamma,\delta',\phi\rangle$

S1	$\langle\gamma,\delta',\phi\rangle$ $(q \equiv 1 \bmod 4)$ or $\langle\gamma,\phi\rangle$ $(q \equiv 3 \bmod 4)$
S2	$\langle\gamma\delta,\delta',\phi\rangle$ $(p \equiv 1 \bmod 4)$ or $\langle\gamma\delta,\delta',\gamma\phi\rangle$ $(p \equiv 3 \bmod 4)$
N1	Maximal under subgroups not contained in $\langle\delta,\phi\rangle$
N2	Maximal under subgroups not contained in $\langle\gamma\rangle$
N3	Maximal under subgroups not contained in $\langle\gamma,\delta'\rangle$

Table 8.67 *The maximal subgroups of* $\Omega_{10}^+(q)$ *in Class* $\mathscr{S}$

$q = 2^e$: $|Z(\Omega_{10}^+(q))| = 1$, $|\delta| = 1$, $|\gamma| = 2$, $|\phi| = e$.
$q = p^e \equiv 1 \bmod 4 : d := |Z(\Omega_{10}^+(q))| = 2$, $|\delta| = 4$, $|\gamma| = 2$, $\delta^2 = \delta'$, $|\phi| = e$.
$q = p^e \equiv 3 \bmod 4 : d := |Z(\Omega_{10}^+(q))| = 1$, $|\delta| = 2$, $|\gamma| = 2$, $|\delta'| = 1$, $|\phi| = e$.

Subgp	Nov	Conditions on q	c	Stab	Acts
$2 \times A_6$	N1	$q = p \equiv 5 \bmod 12$	2	S1	A1
$2 \times A_6.2_1$	N2	$q = p \equiv 1 \bmod 12$	4	S2	
$d \times L_2(11)$		$q = p \equiv 1,3,4,5,9 \bmod 11$, $q \neq 3$	$(q{-}1,4)$	$\langle \gamma \rangle$	
$d \times L_2(11)$	N3	$q = p \equiv 1,3,4,5,9 \bmod 11$, $q \neq 3$	$(q{-}1,4)$	S3	
$d \times A_{11}$		$q = p \equiv 1,3,4,5,9 \bmod 11$, $q \neq 3$	$(q{-}1,4)$	$\langle \gamma \rangle$	
A_{12}		$q = 3$	2	$\langle \gamma \rangle$	
$2 \times S_4(q)$		$q \equiv 1 \bmod 4$	4	$\langle \delta', \phi \rangle$	

S1	$\langle \delta', \gamma \rangle$ $(p \equiv 17 \bmod 24)$ or $\langle \delta', \delta\gamma \rangle$ $(p \equiv 5 \bmod 24)$
S2	$\langle \gamma \rangle$ $(p \equiv 1 \bmod 24)$ or $\langle \delta\gamma \rangle$ $(p \equiv 13 \bmod 24)$
S3	$\langle \gamma \rangle$ $(p \equiv \pm 1 \bmod 12)$ or $\langle \delta\gamma \rangle$ $(p \equiv \pm 5 \bmod 12)$
N1	Maximal under subgroups not contained in $\langle \delta' \rangle$
N2	Maximal under S2
N3	Maximal under S3
A1	$\delta' \to 2_1$, γ or $\delta\gamma \to 2_2$ in one of the two classes and 2_3 in the other

Table 8.68 *The maximal subgroups of $\Omega_{10}^-(q)$ of geometric type*

$q = 2^e$: $|Z(\Omega_{10}^-(q))| = 1$, $|\delta| = 1$, $|\gamma| = 2$, $|\delta'| = 1$, $|\varphi| = 2e$, $\varphi^e = \gamma$.

$q = p^e \equiv 1 \bmod 4$: $|Z(\Omega_{10}^-(q))| = 1$, $|\delta| = 2$, $|\gamma| = 2$, $|\delta'| = 1$, $|\varphi| = 2e$, $\varphi^e = \gamma$.

$q = p^e \equiv 3 \bmod 4$: $|Z(\Omega_{10}^-(q))| = 2$, $|\delta| = 4$, $|\gamma| = 2$, $\delta^2 = \delta'$, $|\phi| = e$.

$\mathscr{C}_i$	Subgp	Notes	c	Stab
$\mathscr{C}_1$	$E_q^{\,8}{:}(\frac{q-1}{(q-1,2)} \times \Omega_8^-(q)).(q-1,2)$		1	S1
$\mathscr{C}_1$	$E_q^{1+12}{:}(\frac{1}{(q-1,2)}GL_2(q) \times \Omega_6^-(q)).(q-1,2)$		1	S1
$\mathscr{C}_1$	$E_q^{3+12}{:}(\frac{1}{(q-1,2)}GL_3(q) \times \Omega_4^-(q)).(q-1,2)$		1	S1
$\mathscr{C}_1$	$E_q^{6+8}{:}(\frac{1}{(q-1,2)}GL_4(q) \times \Omega_2^-(q)).(q-1,2)$		1	S1
$\mathscr{C}_1$	$\Omega_9(q).2$	q odd	2	S2
$\mathscr{C}_1$	$(\Omega_2^+(q) \times \Omega_8^-(q)).2^{(q-1,2)}$	$q \geqslant 4$	1	S1
		N1 if $q = 3$	1	$\langle \delta, \gamma \rangle$
$\mathscr{C}_1$	$(\Omega_2^-(q) \times \Omega_8^+(q)).2^{(q-1,2)}$		1	S1
$\mathscr{C}_1$	$(\Omega_3(q) \times \Omega_7(q)).2^2$	q odd	2	S2
$\mathscr{C}_1$	$(\Omega_4^+(q) \times \Omega_6^-(q)).2^{(q-1,2)}$		1	S1
$\mathscr{C}_1$	$(\Omega_4^-(q) \times \Omega_6^+(q)).2^{(q-1,2)}$		1	S1
$\mathscr{C}_1$	$Sp_8(q)$	q even	1	$\langle \varphi \rangle$
$\mathscr{C}_2$	$2^9{:}A_{10}$	$p = q \equiv 3 \bmod 8$	2	$\langle \gamma, \delta' \rangle$
$\mathscr{C}_2$	$2^9{\cdot}S_{10}$	$p = q \equiv 7 \bmod 8$	4	$\langle \gamma \rangle$
$\mathscr{C}_2$	$\Omega_2^-(q)^5.2^{4(2,q-1)}.S_5$	$q \neq 3$	1	S1
		N1 if $q = 3$	1	S1
$\mathscr{C}_2$	$\Omega_5(q)^2.2^2.S_2$	$q \equiv 3 \bmod 4$	2	$\langle \gamma, \delta', \phi \rangle$
$\mathscr{C}_2$	$SO_5(q)^2$	$q \equiv 1 \bmod 4$	1	$\langle \delta, \varphi \rangle$
$\mathscr{C}_3$	$\Omega_5(q^2).2$	$q \equiv 1 \bmod 4$	1	$\langle \delta, \varphi \rangle$
$\mathscr{C}_3$	$2 \times \Omega_5(q^2).2$	$q \equiv 3 \bmod 4$	2	$\langle \delta\gamma, \delta', \phi \rangle$
$\mathscr{C}_3$	$(\frac{q+1}{(q+1,2)} \circ SU_5(q)).(q+1,5)$		1	S1
$\mathscr{C}_5$	$\Omega_{10}^-(q_0)$	$q = q_0^r$,	1	S1
		r odd prime		

S1	$\langle \delta, \varphi \rangle$ ($q \equiv 1 \bmod 4$) or $\langle \delta, \gamma, \phi \rangle$ ($q \equiv 3 \bmod 4$) or $\langle \varphi \rangle$ (q even)
S2	$\langle \varphi \rangle$ ($q \equiv 1 \bmod 4$) or $\langle \gamma, \delta', \phi \rangle$ ($q \equiv 3 \bmod 4$)
N1	Maximal under subgroups not contained in S2

Table 8.69 *The maximal subgroups of* $\Omega_{10}^-(q)$ *in Class* $\mathscr{S}$

$q = 2^e$: $d := |Z(\Omega_{10}^-(q))| = 1$, $|\delta| = 1$, $|\gamma| = 2$, $|\delta'| = 1$, $|\varphi| = 2e$, $\varphi^e = \gamma$.

$q = p^e \equiv 1 \bmod 4$: $d := |Z(\Omega_{10}^-(q))| = 1$, $|\delta| = 2$, $|\gamma| = 2$, $|\delta'| = 1$, $|\varphi| = 2e$, $\varphi^e = \gamma$.

$q = p^e \equiv 3 \bmod 4$: $d := |Z(\Omega_{10}^-(q))| = 2$, $|\delta| = 4$, $|\gamma| = 2$, $\delta^2 = \delta'$, $|\phi| = e$.

Subgp	Nov	Conditions on q	c	Stab	Acts
$2 \times A_6$	N1	$q = p \equiv 7 \bmod 12$, $q \neq 7$	2	S1	A1
	N2	$q = 7$	2	S1	A1
$2 \times A_6.2_1$	N3	$q = p \equiv 11 \bmod 12$	4	S2	
$d \times L_2(11)$		$q = p \equiv 2, 6, 7, 8, 10 \bmod 11$, $q \neq 2, 7$	$(q+1, 4)$	$\langle\gamma\rangle$	
$d \times L_2(11)$	N4	$q = p \equiv 2, 6, 7, 8, 10 \bmod 11$, $q \neq 2$	$(q+1, 4)$	S3	
$2^{\cdot}L_3(4)$	N5	$q = 7$	2	$\langle\gamma, \delta'\rangle$	A2
M_{12}	N6	$q = 2$	1	$\langle\gamma\rangle$	
$2^{\cdot}M_{22}$		$q = 7$	4	$\langle\gamma\rangle$	
$d \times A_{11}$		$q = p \equiv 2, 6, 7, 8, 10 \bmod 11$, $q \neq 2$	$(q+1, 4)$	$\langle\gamma\rangle$	
A_{12}		$q = 2$	1	$\langle\gamma\rangle$	
$2 \times S_4(q)$		$q \equiv 3 \bmod 4$	4	$\langle\delta', \phi\rangle$	

S1	$\langle\gamma, \delta'\rangle$ ($p \equiv 7 \bmod 24$) or $\langle\delta\gamma, \delta'\rangle$ ($p \equiv 19 \bmod 24$)
S2	$\langle\gamma\rangle$ ($p \equiv 23 \bmod 24$) or $\langle\delta\gamma\rangle$ ($p \equiv 11 \bmod 24$)
S3	$\langle\gamma\rangle$ ($p \equiv \pm 1, 2 \bmod 12$) or $\langle\delta\gamma\rangle$ ($p \equiv \pm 5 \bmod 12$)
N1	Maximal under subgroups not contained in $\langle\delta'\rangle$
N2	Extensions of $2 \times A_6$ that are not contained in $2 \times A_6.2_1$ or $2 \times A_6.2_3$ are maximal
N3	Maximal under S2
N4	Maximal under S3
N5	Extensions of $2^{\cdot}L_3(4)$ that are not contained in $2^{\cdot}L_3(4).2_2$ are maximal
N6	Maximal under $\langle\gamma\rangle$
A1	$\delta' \to 2_1$, γ or $\delta\gamma \to 2_2$ in one of the two classes and 2_3 in the other
A2	$\delta' \to 2_3$, $\gamma \to 2_1$ and 2_2 in the two classes

Table 8.70 *The maximal subgroups of* $\mathrm{SL}_{11}(q)$ *of geometric type*
$d := |\mathrm{Z}(\mathrm{SL}_{11}(q))| = (q-1, 11)$, $|\delta| = d$, $|\phi| = e$, $|\gamma| = 2$, $q = p^e$.

$\mathscr{C}_i$	Subgp	Notes	c	Stab
$\mathscr{C}_1$	$E_q{}^{10}{:}\mathrm{GL}_{10}(q)$		2	$\langle \delta, \phi \rangle$
$\mathscr{C}_1$	$E_q{}^{18}{:}(\mathrm{SL}_9(q) \times \mathrm{SL}_2(q)){:}(q-1)$		2	$\langle \delta, \phi \rangle$
$\mathscr{C}_1$	$E_q{}^{24}{:}(\mathrm{SL}_8(q) \times \mathrm{SL}_3(q)){:}(q-1)$		2	$\langle \delta, \phi \rangle$
$\mathscr{C}_1$	$E_q{}^{28}{:}(\mathrm{SL}_7(q) \times \mathrm{SL}_4(q)){:}(q-1)$		2	$\langle \delta, \phi \rangle$
$\mathscr{C}_1$	$E_q{}^{30}{:}(\mathrm{SL}_6(q) \times \mathrm{SL}_5(q)){:}(q-1)$		2	$\langle \delta, \phi \rangle$
$\mathscr{C}_1$	$E_q{}^{1+18}{:}(\mathrm{GL}_9(q) \times (q-1))$	N1	1	$\langle \delta, \phi, \gamma \rangle$
$\mathscr{C}_1$	$E_q{}^{4+28}{:}(\mathrm{SL}_2(q)^2 \times \mathrm{SL}_7(q)){:}(q-1)^2$	N1	1	$\langle \delta, \phi, \gamma \rangle$
$\mathscr{C}_1$	$E_q{}^{9+30}{:}(\mathrm{SL}_3(q)^2 \times \mathrm{SL}_5(q)){:}(q-1)^2$	N1	1	$\langle \delta, \phi, \gamma \rangle$
$\mathscr{C}_1$	$E_q{}^{16+24}{:}(\mathrm{SL}_4(q)^2 \times \mathrm{SL}_3(q)){:}(q-1)^2$	N1	1	$\langle \delta, \phi, \gamma \rangle$
$\mathscr{C}_1$	$E_q{}^{25+10}{:}\mathrm{GL}_5(q)^2$	N1	1	$\langle \delta, \phi, \gamma \rangle$
$\mathscr{C}_1$	$\mathrm{GL}_{10}(q)$	N1	1	$\langle \delta, \phi, \gamma \rangle$
$\mathscr{C}_1$	$(\mathrm{SL}_9(q) \times \mathrm{SL}_2(q)){:}(q-1)$	N1	1	$\langle \delta, \phi, \gamma \rangle$
$\mathscr{C}_1$	$(\mathrm{SL}_8(q) \times \mathrm{SL}_3(q)){:}(q-1)$	N1	1	$\langle \delta, \phi, \gamma \rangle$
$\mathscr{C}_1$	$(\mathrm{SL}_7(q) \times \mathrm{SL}_4(q)){:}(q-1)$	N1	1	$\langle \delta, \phi, \gamma \rangle$
$\mathscr{C}_1$	$(\mathrm{SL}_6(q) \times \mathrm{SL}_5(q)){:}(q-1)$	N1	1	$\langle \delta, \phi, \gamma \rangle$
$\mathscr{C}_2$	$(q-1)^{10}{:}S_{11}$	$q \geqslant 5$	1	$\langle \delta, \phi, \gamma \rangle$
$\mathscr{C}_3$	$\frac{q^{11}-1}{q-1}{:}11$		1	$\langle \delta, \phi, \gamma \rangle$
$\mathscr{C}_5$	$\mathrm{SL}_{11}(q_0) . \left(\frac{q-1}{q_0-1}, 11 \right)$	$q = q_0^r$, r prime	$\left(\frac{q-1}{q_0-1}, 11 \right)$	$\langle \delta^c, \phi, \gamma \rangle$
$\mathscr{C}_6$	$11^{1+2}_+{:}\mathrm{Sp}_2(11)$	$q = p \equiv 1 \bmod 11$ or ($q = p^5$ & $p \equiv 3,4,5,9 \bmod 11$)	11	$\langle \gamma \rangle$
$\mathscr{C}_8$	$d \times \mathrm{SO}_{11}(q)$	q odd	d	$\langle \phi, \gamma \rangle$
$\mathscr{C}_8$	$(q_0 - 1, 11) \times \mathrm{SU}_{11}(q_0)$	$q = q_0^2$	$(q_0 - 1, 11)$	$\langle \delta^c, \phi, \gamma \rangle$

> N1   Maximal under subgroups not contained in $\langle \delta, \phi \rangle$

Table 8.71 *The maximal subgroups of* $\mathrm{SL}_{11}(q)$ *in Class* $\mathscr{S}$

In all examples, $q = p$. So $d := |\mathrm{Z}(\mathrm{SL}_{11}(q))| = (q-1, 11)$, $|\delta| = d$, $|\phi| = 1$, $|\gamma| = 2$.

Subgp	Nov	Conditions on q	c	Stab
$L_2(23)$	N1	$q = 2$	1	$\langle \gamma \rangle$
$d \times L_2(23)$		$q = p \equiv 1,2,3,4,6,8,9,12,13,16,18 \bmod 23$, $q \neq 2$	d	$\langle \gamma \rangle$
$d \times U_5(2)$		$q = p \equiv 1 \bmod 3$	d	$\langle \gamma \rangle$
M_{24}		$q = 2$	2	1

> N1   Maximal under $\langle \gamma \rangle$

Table 8.72 *The maximal subgroups of* $\mathrm{SU}_{11}(q)$ *of geometric type*
$d := |\mathrm{Z}(\mathrm{SU}_{11}(q))| = (q+1,11)$, $|\delta| = d$, $|\phi| = 2e$, $q = p^e$.

$\mathscr{C}_i$	Subgp	Notes	c	Stab
$\mathscr{C}_1$	$\mathrm{E}_q{}^{1+18}{:}\mathrm{SU}_9(q){:}(q^2-1)$		1	$\langle\delta,\phi\rangle$
$\mathscr{C}_1$	$\mathrm{E}_q{}^{4+28}{:}(\mathrm{SL}_2(q^2)\times\mathrm{SU}_7(q)){:}(q^2-1)$		1	$\langle\delta,\phi\rangle$
$\mathscr{C}_1$	$\mathrm{E}_q{}^{9+30}{:}(\mathrm{SL}_3(q^2)\times\mathrm{SU}_5(q)){:}(q^2-1)$		1	$\langle\delta,\phi\rangle$
$\mathscr{C}_1$	$\mathrm{E}_q{}^{16+24}{:}(\mathrm{SL}_4(q^2)\times\mathrm{SU}_3(q)){:}(q^2-1)$		1	$\langle\delta,\phi\rangle$
$\mathscr{C}_1$	$\mathrm{E}_q{}^{25+10}{:}\mathrm{GL}_5(q^2)$		1	$\langle\delta,\phi\rangle$
$\mathscr{C}_1$	$\mathrm{GU}_{10}(q)$		1	$\langle\delta,\phi\rangle$
$\mathscr{C}_1$	$(\mathrm{SU}_9(q)\times\mathrm{SU}_2(q)){:}(q+1)$		1	$\langle\delta,\phi\rangle$
$\mathscr{C}_1$	$(\mathrm{SU}_8(q)\times\mathrm{SU}_3(q)){:}(q+1)$		1	$\langle\delta,\phi\rangle$
$\mathscr{C}_1$	$(\mathrm{SU}_7(q)\times\mathrm{SU}_4(q)){:}(q+1)$		1	$\langle\delta,\phi\rangle$
$\mathscr{C}_1$	$(\mathrm{SU}_6(q)\times\mathrm{SU}_5(q)){:}(q+1)$		1	$\langle\delta,\phi\rangle$
$\mathscr{C}_2$	$(q+1)^{10}{:}\mathrm{S}_{11}$		1	$\langle\delta,\phi\rangle$
$\mathscr{C}_3$	$\frac{q^{11}+1}{q+1}{:}11$		1	$\langle\delta,\phi\rangle$
$\mathscr{C}_5$	$\mathrm{SU}_{11}(q_0).\left(\frac{q+1}{q_0+1},11\right)$	$q_0^r = q$, r odd prime	$(\frac{q+1}{q_0+1},11)$	$\langle\delta^c,\phi\rangle$
$\mathscr{C}_5$	$d\times\mathrm{SO}_{11}(q)$	q odd	d	$\langle\phi\rangle$
$\mathscr{C}_6$	$11_+^{1+2}{:}\mathrm{Sp}_2(11)$	$q = p \equiv 10 \bmod 11$, or $(q = p^5$ & $p \equiv 2,6,7,8 \bmod 11)$	11	$\langle\phi\rangle$

Table 8.73 *The maximal subgroups of* $\mathrm{SU}_{11}(q)$ *in Class* $\mathscr{S}$

In all examples, $q = p$. So $d := |\mathrm{Z}(\mathrm{SU}_{11}(q))| = (q+1,11)$, $|\delta| = d$, $|\phi| = 2$, $\phi = \gamma$.

Subgp	Conditions on q	c	Stab
$d\times\mathrm{L}_2(23)$	$q = p \equiv 5,7,10,11,14,15,17,19,20,21,22 \bmod 23$	d	$\langle\gamma\rangle$
$d\times\mathrm{U}_5(2)$	$q = p \equiv 2 \bmod 3$, $q \neq 2$	d	$\langle\gamma\rangle$

Table 8.74 *The maximal subgroups of* $\Omega_{11}(q)$ *of geometric type*
$|Z(\Omega_{11}(q))| = 1$, $|\delta| = 2$, $|\phi| = e$, $q = p^e$ odd.

$\mathscr{C}_i$	Subgp	Notes	c	Stab
$\mathscr{C}_1$	$E_q{}^9 : (\frac{q-1}{2} \times \Omega_9(q)).2$		1	$\langle \delta, \phi \rangle$
$\mathscr{C}_1$	$E_q{}^{1+14} : (\frac{1}{2}\mathrm{GL}_2(q) \times \Omega_7(q)).2$		1	$\langle \delta, \phi \rangle$
$\mathscr{C}_1$	$E_q{}^{3+15} : (\frac{1}{2}\mathrm{GL}_3(q) \times \Omega_5(q)).2$		1	$\langle \delta, \phi \rangle$
$\mathscr{C}_1$	$E_q{}^{6+12} : (\frac{1}{2}\mathrm{GL}_4(q) \times \Omega_3(q)).2$		1	$\langle \delta, \phi \rangle$
$\mathscr{C}_1$	$E_q{}^{10+5} : \frac{1}{2}\mathrm{GL}_5(q)$		1	$\langle \delta, \phi \rangle$
$\mathscr{C}_1$	$\Omega_{10}^+(q).2$		1	$\langle \delta, \phi \rangle$
$\mathscr{C}_1$	$\Omega_{10}^-(q).2$		1	$\langle \delta, \phi \rangle$
$\mathscr{C}_1$	$(\Omega_2^+(q) \times \Omega_9(q)).2^2$	$q \geqslant 5$	1	$\langle \delta, \phi \rangle$
$\mathscr{C}_1$	$(\Omega_2^-(q) \times \Omega_9(q)).2^2$		1	$\langle \delta, \phi \rangle$
$\mathscr{C}_1$	$(\Omega_3(q) \times \Omega_8^+(q)).2^2$		1	$\langle \delta, \phi \rangle$
$\mathscr{C}_1$	$(\Omega_3(q) \times \Omega_8^-(q)).2^2$		1	$\langle \delta, \phi \rangle$
$\mathscr{C}_1$	$(\Omega_4^+(q) \times \Omega_7(q)).2^2$		1	$\langle \delta, \phi \rangle$
$\mathscr{C}_1$	$(\Omega_4^-(q) \times \Omega_7(q)).2^2$		1	$\langle \delta, \phi \rangle$
$\mathscr{C}_1$	$(\Omega_5(q) \times \Omega_6^+(q)).2^2$		1	$\langle \delta, \phi \rangle$
$\mathscr{C}_1$	$(\Omega_5(q) \times \Omega_6^-(q)).2^2$		1	$\langle \delta, \phi \rangle$
$\mathscr{C}_2$	$2^{10} : A_{11}$	$p = q \equiv \pm 3 \bmod 8$	1	$\langle \delta \rangle$
$\mathscr{C}_2$	$2^{10} : S_{11}$	$p = q \equiv \pm 1 \bmod 8$	2	1
$\mathscr{C}_5$	$\Omega_{11}(q_0)$	$q = q_0^r$, r odd prime	1	$\langle \delta, \phi \rangle$
$\mathscr{C}_5$	$\mathrm{SO}_{11}(q_0)$	$q = q_0^2$	2	$\langle \phi \rangle$

Table 8.75 *The maximal subgroups of* $\Omega_{11}(q)$ *in Class* $\mathscr{S}$
$|Z(\Omega_{11}(q))| = 1$, $|\delta| = 2$, $|\phi| = e$, $q = p^e$ odd.

Subgp	Conditions on q	c	Stab
$L_3(3).2$	$q = 13$	2	1
A_{12}	$q = p \equiv \pm 7, \pm 11 \bmod 24$, $q \neq 13$	1	$\langle \delta \rangle$
S_{12}	$q = p \equiv \pm 1, \pm 5 \bmod 24$	2	1
A_{13}	$q = 13$	1	$\langle \delta \rangle$
$L_2(q)$	$p \geqslant 11, q \neq 11$	1	$\langle \delta, \phi \rangle$

Table 8.76 *The maximal subgroups of* $\mathrm{SL}_{12}(q)$ *of geometric type*
$d := |\mathrm{Z}(\mathrm{SL}_{12}(q))| = (q-1, 12)$, $|\delta| = d$, $|\phi| = e$, $|\gamma| = 2$, $q = p^e$.

$\mathscr{C}_i$	Subgp	Notes	c	Stab
$\mathscr{C}_1$	$\mathrm{E}_q^{\ 11}{:}\mathrm{GL}_{11}(q)$		2	$\langle\delta,\phi\rangle$
$\mathscr{C}_1$	$\mathrm{E}_q^{\ 20}{:}(\mathrm{SL}_{10}(q)\times\mathrm{SL}_2(q)){:}(q-1)$		2	$\langle\delta,\phi\rangle$
$\mathscr{C}_1$	$\mathrm{E}_q^{\ 27}{:}(\mathrm{SL}_9(q)\times\mathrm{SL}_3(q)){:}(q-1)$		2	$\langle\delta,\phi\rangle$
$\mathscr{C}_1$	$\mathrm{E}_q^{\ 32}{:}(\mathrm{SL}_8(q)\times\mathrm{SL}_4(q)){:}(q-1)$		2	$\langle\delta,\phi\rangle$
$\mathscr{C}_1$	$\mathrm{E}_q^{\ 35}{:}(\mathrm{SL}_7(q)\times\mathrm{SL}_5(q)){:}(q-1)$		2	$\langle\delta,\phi\rangle$
$\mathscr{C}_1$	$\mathrm{E}_q^{\ 36}{:}(\mathrm{SL}_6(q)\times\mathrm{SL}_6(q)){:}(q-1)$		1	$\langle\delta,\phi,\gamma\rangle$
$\mathscr{C}_1$	$\mathrm{E}_q^{\ 1+20}{:}(\mathrm{GL}_{10}(q)\times(q-1))$	N1	1	$\langle\delta,\phi,\gamma\rangle$
$\mathscr{C}_1$	$\mathrm{E}_q^{\ 4+32}{:}(\mathrm{SL}_2(q)^2\times\mathrm{SL}_8(q)){:}(q-1)^2$	N1	1	$\langle\delta,\phi,\gamma\rangle$
$\mathscr{C}_1$	$\mathrm{E}_q^{\ 9+36}{:}(\mathrm{SL}_3(q)^2\times\mathrm{SL}_6(q)){:}(q-1)^2$	N1	1	$\langle\delta,\phi,\gamma\rangle$
$\mathscr{C}_1$	$\mathrm{E}_q^{\ 16+32}{:}\mathrm{SL}_4(q)^3{:}(q-1)^2$	N1	1	$\langle\delta,\phi,\gamma\rangle$
$\mathscr{C}_1$	$\mathrm{E}_q^{\ 25+20}{:}(\mathrm{SL}_5(q)^2\times\mathrm{SL}_2(q)){:}(q-1)^2$	N1	1	$\langle\delta,\phi,\gamma\rangle$
$\mathscr{C}_1$	$\mathrm{GL}_{11}(q)$	N1	1	$\langle\delta,\phi,\gamma\rangle$
$\mathscr{C}_1$	$(\mathrm{SL}_{10}(q)\times\mathrm{SL}_2(q)){:}(q-1)$	N1	1	$\langle\delta,\phi,\gamma\rangle$
$\mathscr{C}_1$	$(\mathrm{SL}_9(q)\times\mathrm{SL}_3(q)){:}(q-1)$	N1	1	$\langle\delta,\phi,\gamma\rangle$
$\mathscr{C}_1$	$(\mathrm{SL}_8(q)\times\mathrm{SL}_4(q)){:}(q-1)$	N1	1	$\langle\delta,\phi,\gamma\rangle$
$\mathscr{C}_1$	$(\mathrm{SL}_7(q)\times\mathrm{SL}_5(q)){:}(q-1)$	N1	1	$\langle\delta,\phi,\gamma\rangle$
$\mathscr{C}_2$	$(q-1)^{11}.\mathrm{S}_{12}$	$q\geqslant 5$	1	$\langle\delta,\phi,\gamma\rangle$
$\mathscr{C}_2$	$\mathrm{SL}_2(q)^6{:}(q-1)^5.\mathrm{S}_6$	$q\geqslant 3$	1	$\langle\delta,\phi,\gamma\rangle$
$\mathscr{C}_2$	$\mathrm{SL}_3(q)^4{:}(q-1)^3.\mathrm{S}_4$		1	$\langle\delta,\phi,\gamma\rangle$
$\mathscr{C}_2$	$\mathrm{SL}_4(q)^3{:}(q-1)^2.\mathrm{S}_3$		1	$\langle\delta,\phi,\gamma\rangle$
$\mathscr{C}_2$	$\mathrm{SL}_6(q)^2{:}(q-1).\mathrm{S}_2$		1	$\langle\delta,\phi,\gamma\rangle$
$\mathscr{C}_3$	$(((q-1,6)(q+1))\circ\mathrm{SL}_6(q^2)).(q+1,3).2$		1	$\langle\delta,\phi,\gamma\rangle$
$\mathscr{C}_3$	$(((q-1,4)(q^2+q+1))\circ\mathrm{SL}_4(q^3)).3$		1	$\langle\delta,\phi,\gamma\rangle$
$\mathscr{C}_4$	$(\mathrm{SL}_2(q)\circ\mathrm{SL}_6(q)).(q-1,2)^2$	$q\geqslant 3$	$(q-1,2)$	$\langle\delta^c,\phi,\gamma\rangle$
$\mathscr{C}_4$	$\mathrm{SL}_3(q)\times\mathrm{SL}_4(q)$		1	$\langle\delta,\phi,\gamma\rangle$
$\mathscr{C}_5$	$\mathrm{SL}_{12}(q_0).\left[\left(\frac{q-1}{q_0-1},12\right)\right]$	$q=q_0^r$, r prime	$\left(\frac{q-1}{q_0-1},12\right)$	$\langle\delta^c,\phi,\gamma\rangle$
$\mathscr{C}_8$	$\mathrm{SO}_{12}^+(q).[d]$	q odd	$d/2$	$\langle\delta^c,\phi,\gamma\rangle$
$\mathscr{C}_8$	$\mathrm{SO}_{12}^-(q).[d]$	q odd	$d/2$	S1
$\mathscr{C}_8$	$\mathrm{Sp}_{12}(q).[(q-1,6)]$		$(q-1,6)$	$\langle\delta^c,\phi,\gamma\rangle$
$\mathscr{C}_8$	$\mathrm{SU}_{12}(q_0).[(q_0-1,12)]$	$q=q_0^2$	$(q_0-1,12)$	$\langle\delta^c,\phi,\gamma\rangle$

S1	$\langle\delta^c,\phi\delta^{(p-1)/2},\gamma\delta^{-1}\rangle$
N1	Maximal under subgroups not contained in $\langle\delta,\phi\rangle$

Table 8.77 *The maximal subgroups of* $\mathrm{SL}_{12}(q)$ *in Class* $\mathscr{S}$

$d := |\mathrm{Z}(\mathrm{SL}_{12}(q))| = (q-1,12)$, $|\delta| = d$, $|\phi| = e$, $|\gamma| = 2$, $q = p^e$.

Subgp	Nov	Conditions on q	c	Stab	Acts
$d \circ 6\dot{\,}A_6$	N1	$q = p \equiv 1,4 \bmod 15$	d	S1	A1
$12 \circ 6\dot{\,}A_6$	N2	$q = p^2$, $p \equiv 2,3 \bmod 5$, $p \neq 2,3$	d	S2	A2
$d \circ 2\dot{\,}L_2(23)$		$q = p \equiv 1,2,3,4,6,8,9,12,13,$	d	S1	
		$\phantom{q = p \equiv{}} 16, 18 \bmod 23$, $p \neq 2$			
$12_2\dot{\,}L_3(4)$		$q = 49$	12	$\langle \phi, \gamma \rangle$	A3
$d \circ 6\dot{\,}\mathrm{Suz}$		$q = p \equiv 1 \bmod 3$	d	$\langle \gamma \rangle$	

S1	$\langle \gamma \rangle$ $(p \equiv \pm 1 \bmod 8)$ or $\langle \gamma \delta \rangle$ $(p \equiv \pm 3 \bmod 8)$
S2	$\langle \phi, \gamma \rangle$ $(p \equiv \pm 5 \bmod 12)$ or $\langle \phi \delta^6, \gamma \rangle$ $(p \equiv \pm 1 \bmod 12)$
N1	Maximal under S1
N2	Maximal under subgroups of S2 that do not lie in $\langle \gamma \phi \delta^6 \rangle$ $(p \equiv 1 \bmod 12)$,
	$\langle \phi \rangle$ $(p \equiv 5 \bmod 12)$, $\langle \gamma \phi \rangle$ $(p \equiv 7 \bmod 12)$, $\langle \phi \delta^6 \rangle$ $(p \equiv 11 \bmod 12)$.
	Equivalently, extensions of $6\dot{\,}A_6$ not contained in $6\dot{\,}A_6.2_1$ are maximal.
	Also, when $p = 7$, $q = 49$, the extension of $6\dot{\,}A_6$ by $\langle \phi \rangle$ with structure
	$6\dot{\,}A_6.2_3$ is not maximal.
A1	γ or $\gamma \delta \to 2_2$
A2	$\gamma \to 2_2$, ϕ or $\phi \delta^6 \to 2_1$ $(p \equiv 5, 11 \bmod 12)$ or 2_3 $(p \equiv 1, 7 \bmod 12)$
A3	$\gamma \to 2_3$, $\phi \to 2_1$

Table 8.78 *The maximal subgroups of* $\mathrm{SU}_{12}(q)$ *of geometric type*
$$d := |Z(\mathrm{SU}_{12}(q))| = (q+1, 12), \ |\delta| = d, \ |\phi| = 2e, \ q = p^e.$$

$\mathscr{C}_i$	Subgp	Notes	c	Stab
$\mathscr{C}_1$	$E_q{}^{1+20}{:}\mathrm{SU}_{10}(q){:}(q^2-1)$		1	$\langle \delta, \phi \rangle$
$\mathscr{C}_1$	$E_q{}^{4+32}{:}(\mathrm{SL}_2(q^2) \times \mathrm{SU}_8(q)){:}(q^2-1)$		1	$\langle \delta, \phi \rangle$
$\mathscr{C}_1$	$E_q{}^{9+36}{:}(\mathrm{SL}_3(q^2) \times \mathrm{SU}_6(q)){:}(q^2-1)$		1	$\langle \delta, \phi \rangle$
$\mathscr{C}_1$	$E_q{}^{16+32}{:}(\mathrm{SL}_4(q^2) \times \mathrm{SU}_4(q)){:}(q^2-1)$		1	$\langle \delta, \phi \rangle$
$\mathscr{C}_1$	$E_q{}^{25+20}{:}(\mathrm{SL}_5(q^2) \times \mathrm{SU}_2(q)){:}(q^2-1)$		1	$\langle \delta, \phi \rangle$
$\mathscr{C}_1$	$E_q{}^{36}{:}\mathrm{SL}_6(q^2){:}(q-1)$		1	$\langle \delta, \phi \rangle$
$\mathscr{C}_1$	$\mathrm{GU}_{11}(q)$		1	$\langle \delta, \phi \rangle$
$\mathscr{C}_1$	$(\mathrm{SU}_{10}(q) \times \mathrm{SU}_2(q)){:}(q+1)$		1	$\langle \delta, \phi \rangle$
$\mathscr{C}_1$	$(\mathrm{SU}_9(q) \times \mathrm{SU}_3(q)){:}(q+1)$		1	$\langle \delta, \phi \rangle$
$\mathscr{C}_1$	$(\mathrm{SU}_8(q) \times \mathrm{SU}_4(q)){:}(q+1)$		1	$\langle \delta, \phi \rangle$
$\mathscr{C}_1$	$(\mathrm{SU}_7(q) \times \mathrm{SU}_5(q)){:}(q+1)$		1	$\langle \delta, \phi \rangle$
$\mathscr{C}_2$	$(q+1)^{11}.\mathrm{S}_{12}$		1	$\langle \delta, \phi \rangle$
$\mathscr{C}_2$	$\mathrm{SU}_2(q)^6{:}(q+1)^5.\mathrm{S}_6$	$q \geqslant 3$	1	$\langle \delta, \phi \rangle$
$\mathscr{C}_2$	$\mathrm{SU}_3(q)^4{:}(q+1)^3.\mathrm{S}_4$		1	$\langle \delta, \phi \rangle$
$\mathscr{C}_2$	$\mathrm{SU}_4(q)^3{:}(q+1)^2.\mathrm{S}_3$		1	$\langle \delta, \phi \rangle$
$\mathscr{C}_2$	$\mathrm{SU}_6(q)^2{:}(q+1).\mathrm{S}_2$		1	$\langle \delta, \phi \rangle$
$\mathscr{C}_2$	$\mathrm{SL}_6(q^2).(q-1).2$		1	$\langle \delta, \phi \rangle$
$\mathscr{C}_3$	$(((q+1,4)(q^2-q+1)) \circ \mathrm{SU}_4(q^3)).3$		1	$\langle \delta, \phi \rangle$
$\mathscr{C}_4$	$(\mathrm{SU}_2(q) \circ \mathrm{SU}_6(q)).(q+1,2)^2$	$q \geqslant 3$	$(q+1,2)$	$\langle \delta^c, \phi \rangle$
$\mathscr{C}_4$	$\mathrm{SU}_3(q) \times \mathrm{SU}_4(q)$		1	$\langle \delta, \phi \rangle$
$\mathscr{C}_5$	$\mathrm{SU}_{12}(q_0).\left[\left(\frac{q+1}{q_0+1}, 12\right)\right]$	$q_0{}^r = q,$	$\left(\frac{q+1}{q_0+1}, 12\right)$	$\langle \delta^c, \phi \rangle$
		r odd prime		
$\mathscr{C}_5$	$\mathrm{Sp}_{12}(q).[(q+1,6)]$		$(q+1,6)$	$\langle \delta^c, \phi \rangle$
$\mathscr{C}_5$	$\mathrm{SO}_{12}^+(q).[d]$	q odd	$d/2$	$\langle \delta^c, \phi \rangle$
$\mathscr{C}_5$	$\mathrm{SO}_{12}^-(q).[d]$	q odd	$d/2$	S1

S1	$\langle \delta^c, \phi \delta^{(p-1)/2} \rangle$

Table 8.79 *The maximal subgroups of* $\mathrm{SU}_{12}(q)$ *in Class* $\mathscr{S}$

In all examples, $q = p$. So $d := |Z(\mathrm{SU}_{12}(q))| = (q+1, 12)$, $|\delta| = d$, $|\phi| = 2$, $\phi = \gamma$.

Subgp	Nov	Conditions on q	c	Stab	Acts
$d \circ 6{\cdot}\mathrm{A}_6$	N1	$q = p \equiv 11, 14 \bmod 15$	d	S1	γ or $\gamma\delta \to 2_2$
$d \circ 2{\cdot}\mathrm{L}_2(23)$		$q = p \equiv 5, 7, 10, 11, 14, 15, 17,$	d	S1	
		$19, 20, 21, 22 \bmod 23$			
$3{\cdot}\mathrm{Suz}$		$q = 2$	d	$\langle \gamma \rangle$	
$d \circ 6{\cdot}\mathrm{Suz}$		$q = p \equiv 2 \bmod 3, \ q \neq 2$	d	$\langle \gamma \rangle$	

S1	$\langle \gamma \rangle$ ($p \equiv \pm 1 \bmod 8$) or $\langle \gamma\delta \rangle$ ($p \equiv \pm 3 \bmod 8$)	N1	Maximal under S1

Table 8.80 *The maximal subgroups of* $\mathrm{Sp}_{12}(q)$ *of geometric type*
$d := |\mathrm{Z}(\mathrm{Sp}_{12}(q))| = (q-1,2), |\delta| = d, |\phi| = e, q = p^e.$

$\mathscr{C}_i$	Subgp	Notes	c	Stab
$\mathscr{C}_1$	$\mathrm{E}_q^{1+10}{:}((q-1) \times \mathrm{Sp}_{10}(q))$	q odd	1	$\langle \delta, \phi \rangle$
$\mathscr{C}_1$	$\mathrm{E}_q^{11}{:}((q-1) \times \mathrm{Sp}_{10}(q))$	q even	1	$\langle \phi \rangle$
$\mathscr{C}_1$	$\mathrm{E}_q^{3+16}{:}(\mathrm{GL}_2(q) \times \mathrm{Sp}_8(q))$		1	$\langle \delta, \phi \rangle$
$\mathscr{C}_1$	$\mathrm{E}_q^{6+18}{:}(\mathrm{GL}_3(q) \times \mathrm{Sp}_6(q))$		1	$\langle \delta, \phi \rangle$
$\mathscr{C}_1$	$\mathrm{E}_q^{10+16}{:}(\mathrm{GL}_4(q) \times \mathrm{Sp}_4(q))$		1	$\langle \delta, \phi \rangle$
$\mathscr{C}_1$	$\mathrm{E}_q^{15+10}{:}(\mathrm{GL}_5(q) \times \mathrm{Sp}_2(q))$		1	$\langle \delta, \phi \rangle$
$\mathscr{C}_1$	$\mathrm{E}_q^{21}{:}\mathrm{GL}_6(q)$		1	$\langle \delta, \phi \rangle$
$\mathscr{C}_1$	$\mathrm{Sp}_{10}(q) \times \mathrm{Sp}_2(q)$		1	$\langle \delta, \phi \rangle$
$\mathscr{C}_1$	$\mathrm{Sp}_8(q) \times \mathrm{Sp}_4(q)$		1	$\langle \delta, \phi \rangle$
$\mathscr{C}_2$	$\mathrm{Sp}_2(q)^6{:}\mathrm{S}_6$	$q \geqslant 3$	1	$\langle \delta, \phi \rangle$
$\mathscr{C}_2$	$\mathrm{Sp}_4(q)^3{:}\mathrm{S}_3$		1	$\langle \delta, \phi \rangle$
$\mathscr{C}_2$	$\mathrm{Sp}_6(q)^2{:}\mathrm{S}_2$		1	$\langle \delta, \phi \rangle$
$\mathscr{C}_2$	$\mathrm{GL}_6(q).2$	q odd	1	$\langle \delta, \phi \rangle$
$\mathscr{C}_3$	$\mathrm{Sp}_6(q^2){:}2$		1	$\langle \delta, \phi \rangle$
$\mathscr{C}_3$	$\mathrm{Sp}_4(q^3){:}3$		1	$\langle \delta, \phi \rangle$
$\mathscr{C}_3$	$\mathrm{GU}_6(q).2$	q odd	1	$\langle \delta, \phi \rangle$
$\mathscr{C}_4$	$(\mathrm{Sp}_2(q) \circ \mathrm{GO}_6^+(q)).2$	q odd	1	$\langle \delta, \phi \rangle$
$\mathscr{C}_4$	$(\mathrm{Sp}_2(q) \circ \mathrm{GO}_6^-(q)).2$	q odd	1	$\langle \delta, \phi \rangle$
$\mathscr{C}_4$	$\mathrm{Sp}_4(q) \circ \mathrm{GO}_3(q)$	$q \geqslant 5$ odd	1	$\langle \delta, \phi \rangle$
$\mathscr{C}_5$	$\mathrm{Sp}_{12}(q_0).(d,r)$	$q = q_0^r$, r prime	(d,r)	$\langle \delta^c, \phi \rangle$
$\mathscr{C}_8$	$\mathrm{SO}_{12}^+(q)$	q even	1	$\langle \phi \rangle$
$\mathscr{C}_8$	$\mathrm{SO}_{12}^-(q)$	q even	1	$\langle \phi \rangle$

Table 8.81 *The maximal subgroups of* $\mathrm{Sp}_{12}(q)$ *in Class* $\mathscr{S}$
$d := |Z(\mathrm{Sp}_{12}(q))| = (q-1,2)$, $|\delta| = d$, $|\phi| = e$, $q = p^e$.

Subgp	Conditions on q	c	Stab	Acts
$2^{\cdot}\mathrm{L}_2(11)$	$q = p \equiv \pm 9 \bmod 20$, $q \neq 11$	1	$\langle \delta \rangle$	
$2^{\cdot}\mathrm{L}_2(11)$	$q = p \equiv \pm 9 \bmod 20$, $q \neq 11$	1	$\langle \delta \rangle$	
$2^{\cdot}\mathrm{L}_2(11).2$	$q = p \equiv \pm 1 \bmod 20$	2	1	
$2^{\cdot}\mathrm{L}_2(11).2$	$q = p \equiv \pm 1 \bmod 20$	2	1	
$2^{\cdot}\mathrm{L}_2(11)$	$q = p^2$, $p \equiv \pm 2 \bmod 5$, $p \neq 2$	2	$\langle \delta \rangle$	
$2^{\cdot}\mathrm{L}_2(13)$	$q = p \equiv \pm 13 \bmod 28$, $q \neq 13$	1	$\langle \delta \rangle$	
$2^{\cdot}\mathrm{L}_2(13)$	$q = p \equiv \pm 13 \bmod 28$, $q \neq 13$	1	$\langle \delta \rangle$	
$2^{\cdot}\mathrm{L}_2(13)$	$q = p \equiv \pm 13 \bmod 28$, $q \neq 13$	1	$\langle \delta \rangle$	
$2^{\cdot}\mathrm{L}_2(13).2$	$q = p \equiv \pm 1 \bmod 28$	2	1	
$2^{\cdot}\mathrm{L}_2(13).2$	$q = p \equiv \pm 1 \bmod 28$	2	1	
$2^{\cdot}\mathrm{L}_2(13).2$	$q = p \equiv \pm 1 \bmod 28$	2	1	
$2^{\cdot}\mathrm{L}_2(13)$	$q = p^3$, $p \equiv \pm 2, \pm 3 \bmod 7$, $p \neq 2$	3	$\langle \delta \rangle$	
$2^{\cdot}\mathrm{L}_2(25)$	$q = p \equiv \pm 2 \bmod 5$, $q \neq 2, 3$	1	$\langle \delta \rangle$	$\delta \to 2_2$
$\mathrm{L}_2(25).2_2$	$q = 2$	1	1	
$2^{\cdot}\mathrm{S}_4(5)$	$q = p \equiv \pm 1 \bmod 5$	2	1	
$2^{\cdot}\mathrm{S}_4(5)$	$q = p^2$, $p \equiv \pm 2 \bmod 5$, $p \neq 2$	2	$\langle \phi \rangle$	
$\mathrm{S}_4(5)$	$q = 4$	1	$\langle \phi \rangle$	
$2^{\cdot}\mathrm{G}_2(4)$	$q = p \equiv \pm 3 \bmod 8$, $q \neq 3$	1	$\langle \delta \rangle$	
$2^{\cdot}\mathrm{G}_2(4).2$	$q = p \equiv \pm 1 \bmod 8$	2	1	
S_{14}	$q = 2$	1	1	
$2^{\cdot}\mathrm{Suz}$	$q = 3$	1	$\langle \delta \rangle$	
$2^{\cdot}\mathrm{L}_2(q)$	$p \geqslant 13$	1	$\langle \delta, \phi \rangle$	
$2^{\cdot}\mathrm{S}_4(q)$	$p = 5$	1	$\langle \delta, \phi \rangle$	

Table 8.82 *The maximal subgroups of $\Omega_{12}^+(q)$ of geometric type*
$d := |Z(\Omega_{12}^+(q))| = (q-1,2)$, $|\delta| = d$, $|\gamma| = 2$, $|\delta'| = d$, $|\phi| = e$, $q = p^e$.

$\mathscr{C}_i$	Subgp	Notes	c	Stab
$\mathscr{C}_1$	$E_q{}^{10}:(\frac{q-1}{(q-1,2)} \times \Omega_{10}^+(q)).d$		1	$\langle \delta, \gamma, \delta', \phi \rangle$
$\mathscr{C}_1$	$E_q{}^{1+16}:(\frac{1}{(q-1,2)}GL_2(q) \times \Omega_8^+(q)).d$		1	$\langle \delta, \gamma, \delta', \phi \rangle$
$\mathscr{C}_1$	$E_q{}^{3+18}:(\frac{1}{(q-1,2)}GL_3(q) \times \Omega_6^+(q)).d$		1	$\langle \delta, \gamma, \delta', \phi \rangle$
$\mathscr{C}_1$	$E_q{}^{6+16}:(\frac{1}{(q-1,2)}GL_4(q) \times \Omega_4^+(q)).d$		1	$\langle \delta, \gamma, \delta', \phi \rangle$
$\mathscr{C}_1$	$E_q{}^{10+10}:(\frac{1}{(q-1,2)}GL_5(q) \times \Omega_2^+(q)).d$	N1	1	$\langle \delta, \gamma, \delta', \phi \rangle$
$\mathscr{C}_1$	$E_q{}^{15}:\frac{1}{(q-1,2)}GL_6(q)$		2	$\langle \delta, \delta', \phi \rangle$
$\mathscr{C}_1$	$\Omega_{11}(q).2$	q odd	2	$\langle \gamma, \delta', \phi \rangle$
$\mathscr{C}_1$	$(\Omega_2^+(q) \times \Omega_{10}^+(q)).2^d$	$q \geqslant 4$; N2 if $q = 3$	1	$\langle \delta, \gamma, \delta', \phi \rangle$
$\mathscr{C}_1$	$(\Omega_2^-(q) \times \Omega_{10}^-(q)).2^d$		1	$\langle \delta, \gamma, \delta', \phi \rangle$
$\mathscr{C}_1$	$(\Omega_3(q) \times \Omega_9(q)).2^2$	q odd	2	$\langle \gamma, \delta', \phi \rangle$
$\mathscr{C}_1$	$(\Omega_4^+(q) \times \Omega_8^+(q)).2^d$		1	$\langle \delta, \gamma, \delta', \phi \rangle$
$\mathscr{C}_1$	$(\Omega_4^-(q) \times \Omega_8^-(q)).2^d$		1	$\langle \delta, \gamma, \delta', \phi \rangle$
$\mathscr{C}_1$	$(\Omega_5(q) \times \Omega_7(q)).2^2$	q odd	2	$\langle \gamma, \delta', \phi \rangle$
$\mathscr{C}_1$	$Sp_{10}(q)$	q even	1	$\langle \gamma, \phi \rangle$
$\mathscr{C}_2$	$2^{11}{}^{\cdot}S_{12}$	$p = q \equiv \pm 1 \bmod 8$	4	$\langle \gamma \rangle$
$\mathscr{C}_2$	$2^{11}:A_{12}$	$p = q \equiv \pm 3 \bmod 8$	2	$\langle \gamma, \delta' \rangle$
$\mathscr{C}_2$	$\Omega_2^+(q)^6.2^{5d}.S_6$	$q \geqslant 7$; N2 if $q = 5$	1	$\langle \delta, \gamma, \delta', \phi \rangle$
$\mathscr{C}_2$	$\Omega_2^-(q)^6.2^{5d}.S_6$	$q \neq 3$; N2 if $q = 3$	1	$\langle \delta, \gamma, \delta', \phi \rangle$
$\mathscr{C}_2$	$\Omega_3(q)^4.2^6.S_4$	$q \geqslant 5$ odd	2	$\langle \gamma, \delta', \phi \rangle$
$\mathscr{C}_2$	$\Omega_4^+(q)^3.2^{2d}.S_3$	$q \geqslant 3$	1	$\langle \delta, \gamma, \delta', \phi \rangle$
$\mathscr{C}_2$	$\Omega_6^+(q)^2.2^d.S_2$		1	$\langle \delta, \gamma, \delta', \phi \rangle$
$\mathscr{C}_2$	$\Omega_6^-(q)^2.2^d.S_2$		1	$\langle \delta, \gamma, \delta', \phi \rangle$
$\mathscr{C}_2$	$SL_6(q).\frac{(q-1)}{d}.2$		2	$\langle \delta, \delta', \phi \rangle$
$\mathscr{C}_3$	$\Omega_6^+(q^2).[4]$		2	$\langle \delta, \delta', \phi \rangle$
$\mathscr{C}_3$	$\Omega_4^+(q^3).3$		1	$\langle \delta, \gamma, \delta', \phi \rangle$
$\mathscr{C}_3$	$((q+1) \circ SU_6(q)).[2(q+1,3)]$		2	$\langle \delta, \delta', \phi \rangle$
$\mathscr{C}_4$	$Sp_2(q) \circ Sp_6(q)$	$q \geqslant 3$	2	$\langle \delta, \delta', \phi \rangle$
$\mathscr{C}_4$	$\Omega_4^+(q) \times SO_3(q)$	$q \geqslant 5$, q odd, N1	1	$\langle \delta, \gamma, \delta', \phi \rangle$
$\mathscr{C}_5$	$\Omega_{12}^+(q_0)$	$q = q_0^r$, r prime, r odd or q even	1	$\langle \delta, \gamma, \delta', \phi \rangle$
$\mathscr{C}_5$	$SO_{12}^+(q_0).2$	$q = q_0^2$, q odd	4	$\langle \gamma, \phi \rangle$
$\mathscr{C}_5$	$\Omega_{12}^-(q_0)$	$q = q_0^2$, q even	1	$\langle \gamma, \phi \rangle$
$\mathscr{C}_5$	$SO_{12}^-(q_0)$	$q = q_0^2$, q odd	2	$\langle \gamma, \delta', \phi \rangle$

N1	Maximal under subgroups not contained in $\langle \delta, \delta', \phi \rangle$
N2	Maximal under subgroups not contained in $\langle \gamma, \delta' \rangle$

Table 8.83 *The maximal subgroups of $\Omega_{12}^+(q)$ in Class $\mathscr{S}$*

$d := |Z(\Omega_{12}^+(q))| = (q-1, 2)$, $|\delta| = d$, $|\gamma| = 2$, $|\delta'| = d$, $|\phi| = e$, $q = p^e$.

Subgp	Nov	Conditions on q	c	Stab
$2 \times L_2(11)$		$q = p \equiv \pm 1, \pm 16, \pm 19, \pm 24, \pm 26 \bmod 55$	4	$\langle\gamma\rangle$
$2 \times L_2(11)$		$q = p \equiv \pm 1, \pm 16, \pm 19, \pm 24, \pm 26 \bmod 55$	4	$\langle\gamma\rangle$
$2 \times L_2(13)$		$q = p \equiv \pm 1 \bmod 7$, $p \equiv \pm 1, \pm 3, \pm 4 \bmod 13$	4	$\langle\gamma\rangle$
$2 \times L_2(13)$		$q = p \equiv \pm 1 \bmod 7$, $p \equiv \pm 1, \pm 3, \pm 4 \bmod 13$	4	$\langle\gamma\rangle$
$2 \times L_2(13)$		$q = p \equiv \pm 1 \bmod 7$, $p \equiv \pm 1, \pm 3, \pm 4 \bmod 13$	4	$\langle\gamma\rangle$
$2 \times L_2(13)$		$q = p^3$, $p \equiv \pm 3, \pm 4, \pm 9, \pm 10, \pm 12, \pm 16, \pm 17,$ $\pm 23, \pm 25, \pm 30, \pm 38, \pm 40 \bmod 91$	12	$\langle\gamma\rangle$
$2 \times L_3(3)$	N1	$q = p \equiv \pm 1, \pm 3, \pm 4 \bmod 13$, $q \neq 3$, C1	4	S1
$2 \times L_3(3).2$		$q = p \equiv \pm 1, \pm 3, \pm 4 \bmod 13$, $q \neq 3$, C2	8	1
$2{\cdot}M_{12}$	N2	$q = p \equiv \pm 5, \pm 7, \pm 11 \bmod 24$	4	S2
$2{\cdot}M_{12}.2$		$q = p \equiv \pm 1 \bmod 24$	8	1
$2 \times A_{13}$		$q = p \equiv \pm 1, \pm 3, \pm 4 \bmod 13$	4	$\langle\gamma\rangle$

S1	$\langle\delta'\rangle$ ($p \equiv \pm 1 \bmod 12$) or $\langle\delta\rangle$ ($p \equiv \pm 5 \bmod 12$)
S2	$\langle\delta'\rangle$ ($p \equiv \pm 11 \bmod 24$) or $\langle\delta\rangle$ ($p \equiv \pm 5 \bmod 12$)
N1	Maximal under S1
N2	Maximal under S2
C1	$p \equiv \pm 5 \bmod 12$ or $x^4 - 10x^2 + 13$ has no roots in $\mathbb{F}_p$
C2	$p \equiv \pm 1 \bmod 12$ and $x^4 - 10x^2 + 13$ has four roots in $\mathbb{F}_p$

Table 8.84 *The maximal subgroups of* $\Omega_{12}^-(q)$ *of geometric type*

$|Z(\Omega_{12}^-(q))| = 1$, $|\delta| = (q-1,2)$, $|\gamma| = 2$, $|\varphi| = 2e$, $\varphi^e = \gamma$, $q = p^e$.

$\mathscr{C}_i$	Subgp	Notes	c	Stab
$\mathscr{C}_1$	$E_q{}^{10}{:}(\frac{q-1}{(q-1,2)} \times \Omega_{10}^-(q)).(q-1,2)$		1	$\langle \delta, \varphi \rangle$
$\mathscr{C}_1$	$E_q{}^{1+16}{:}(\frac{1}{(q-1,2)}\mathrm{GL}_2(q) \times \Omega_8^-(q)).(q-1,2)$		1	$\langle \delta, \varphi \rangle$
$\mathscr{C}_1$	$E_q{}^{3+18}{:}(\frac{1}{(q-1,2)}\mathrm{GL}_3(q) \times \Omega_6^-(q)).(q-1,2)$		1	$\langle \delta, \varphi \rangle$
$\mathscr{C}_1$	$E_q{}^{6+16}{:}(\frac{1}{(q-1,2)}\mathrm{GL}_4(q) \times \Omega_4^-(q)).(q-1,2)$		1	$\langle \delta, \varphi \rangle$
$\mathscr{C}_1$	$E_q{}^{10+10}{:}(\frac{1}{(q-1,2)}\mathrm{GL}_5(q) \times \Omega_2^-(q)).(q-1,2)$		1	$\langle \delta, \varphi \rangle$
$\mathscr{C}_1$	$\Omega_{11}(q).2$	q odd	2	$\langle \varphi \rangle$
$\mathscr{C}_1$	$(\Omega_2^+(q) \times \Omega_{10}^-(q)).2^{(q-1,2)}$	$q \geqslant 4$; N1 if $q = 3$	1	$\langle \delta, \varphi \rangle$
$\mathscr{C}_1$	$(\Omega_2^-(q) \times \Omega_{10}^+(q)).2^{(q-1,2)}$		1	$\langle \delta, \varphi \rangle$
$\mathscr{C}_1$	$(\Omega_3(q) \times \Omega_9(q)).2^2$	q odd	2	$\langle \varphi \rangle$
$\mathscr{C}_1$	$(\Omega_4^+(q) \times \Omega_8^-(q)).2^{(q-1,2)}$		1	$\langle \delta, \varphi \rangle$
$\mathscr{C}_1$	$(\Omega_4^-(q) \times \Omega_8^+(q)).2^{(q-1,2)}$		1	$\langle \delta, \varphi \rangle$
$\mathscr{C}_1$	$(\Omega_5(q) \times \Omega_7(q)).2^2$	q odd	2	$\langle \varphi \rangle$
$\mathscr{C}_1$	$(\Omega_6^+(q) \times \Omega_6^-(q)).2^{(q-1,2)}$		1	$\langle \delta, \varphi \rangle$
$\mathscr{C}_1$	$\mathrm{Sp}_{10}(q)$	q even	1	$\langle \delta, \varphi \rangle$
$\mathscr{C}_2$	$\Omega_4^-(q)^3.2^{2(q-1,2)}.S_3$		1	$\langle \delta, \varphi \rangle$
$\mathscr{C}_3$	$\Omega_6^-(q^2).2$		1	$\langle \delta, \varphi \rangle$
$\mathscr{C}_3$	$\Omega_4^-(q^3).3$		1	$\langle \delta, \varphi \rangle$
$\mathscr{C}_4$	$\Omega_4^-(q) \times \mathrm{SO}_3(q)$	$q \geqslant 5$ odd	1	$\langle \delta, \varphi \rangle$
$\mathscr{C}_5$	$\Omega_{12}^-(q_0)$	$q = q_0^r$, r odd prime	1	$\langle \delta, \varphi \rangle$

N1   Maximal under subgroups not contained in $\langle \varphi \rangle$

Table 8.85 *The maximal subgroups of* $\Omega_{12}^-(q)$ *in Class* $\mathscr{S}$

$|Z(\Omega_{12}^-(q))| = 1$, $|\delta| = (q-1,2)$, $|\gamma| = 2$, $|\varphi| = 2e$, $\varphi^e = \gamma$, $q = p^e$.

Subgp	Nov	Conditions on q	c	Stab
$L_2(11)$		$q = p \equiv \pm4, \pm6, \pm9, \pm14, \pm21 \bmod 55$	2	$\langle \gamma \rangle$
$L_2(11)$		$q = p \equiv \pm4, \pm6, \pm9, \pm14, \pm21 \bmod 55$	2	$\langle \gamma \rangle$
$L_2(11)$		$q = p^2$, $p \equiv \pm2 \bmod 5$	$2(q+1,2)$	$\langle \gamma \rangle$
$L_2(13)$		$q = p \equiv \pm1 \bmod 7$, $p \equiv \pm2, \pm5, \pm6 \bmod 13$	2	$\langle \gamma \rangle$
$L_2(13)$		$q = p \equiv \pm1 \bmod 7$, $p \equiv \pm2, \pm5, \pm6 \bmod 13$	2	$\langle \gamma \rangle$
$L_2(13)$		$q = p \equiv \pm1 \bmod 7$, $p \equiv \pm2, \pm5, \pm6 \bmod 13$	2	$\langle \gamma \rangle$
$L_2(13)$		$q = p^3$, $p \equiv \pm2, \pm5, \pm11, \pm18, \pm19, \pm24, \pm31$ $\pm32, \pm33, \pm37, \pm44, \pm45 \bmod 91$	$3(q+1,2)$	$\langle \gamma \rangle$
$L_3(3)$	N1	$q = p \equiv \pm2, \pm5, \pm6 \bmod 13$, $p \equiv \pm5 \bmod 12$	2	$\langle \delta \rangle$
$L_3(3).2$		$q = p \equiv \pm2, \pm5, \pm6 \bmod 13$, $p \equiv \pm1, 2 \bmod 12$	$2(q+1,2)$	1
A_{13}		$q = p \equiv \pm2, \pm5, \pm6 \bmod 13$, $q \neq 7$	$(q+1,2)$	$\langle \gamma \rangle$
A_{14}		$q = 7$	2	$\langle \gamma \rangle$

N1   Maximal under $\langle \delta \rangle$

References

[1] M. Aschbacher. On the maximal subgroups of the finite classical groups. *Invent. Math.* **76** (1984), 469–514.

[2] M. Aschbacher. Chevalley groups of type G_2 as the group of a trilinear form. *J. Algebra* **109** (1987), 193–259.

[3] D. M. Bloom. The subgroups of $PSL(3, q)$ for odd q. *Trans. Amer. Math. Soc.* **127** (1967), 150–178.

[4] R. Brauer and C. Nesbitt. On the modular characters of groups. *Ann. of Math. (2)* **42** (1941), 556–590.

[5] W. Bosma, J. Cannon and C. Playoust. The Magma algebra system. I. The user language. *J. Symbolic Comput.* **24** (1997), 235–265.

[6] J. N. Bray, D. F. Holt and C. M. Roney-Dougal. Certain classical groups are not well-defined. *J. Group Theory* **12** (2009), 171–180.

[7] J. Brundan and A. Kleschev. Lower bounds for the degrees of representations of irreducible Brauer characters of finite general linear groups. *J. Algebra* **223** (2000), 615–629.

[8] Peter J. Cameron. Permutation Groups. London Math. Soc. Student Texts, 45. *Cambridge University Press, Cambridge*, 1999.

[9] J. J. Cannon and D. F. Holt. Computing maximal subgroups of finite groups. *J. Symbolic Comput.* **37** (2004), 589–609.

[10] Roger W. Carter. Simple Groups of Lie Type. *John Wiley and Sons, London–New York–Sydney*, 1972.

[11] P. M. Cohn. Basic Algebra: Groups, Rings and Fields. *Springer-Verlag, London*, 2003.

[12] J. H. Conway, R. T. Curtis, S. P. Norton, R. A. Parker and R. A. Wilson. An ATLAS of Finite Groups. *Clarendon Press, Oxford*, 1985; reprinted with corrections 2003.

[13] B. N. Cooperstein. The geometry of root subgroups in exceptional groups. I. *Geom. Dedicata* **8** (1979), 317–338.

[14] B. N. Cooperstein. Maximal subgroups of $G_2(2^n)$. *J. Algebra* **70** (1981), 23–36.

[15] A. Cossidente and O. H. King. Maximal subgroups of finite orthogonal groups stabilizing spreads of lines. *Comm. Algebra* **34** (2006), 4291–4309.

[16] A. Cossidente and O. H. King. On twisted tensor product group embeddings and the spin representation of symplectic groups. *Adv. Geom.* **7** (2007), 55–64.

[17] A. Cossidente and A. Siciliano. On some maximal subgroups in Aschbacher's class $\mathscr{C}_5$. *Linear Algebra Appl.* **403** (2005), 285–290.

[18] H. J. Coutts, M. R. Quick and C. M. Roney-Dougal. The primitive groups of degree less than 4096. *Comm. Algebra* **39** (2011), 3526–3546.

[19] Charles W. Curtis and Irving Reiner. Representation Theory of Finite Groups and Associative Algebras. Reprint of the 1962 original. Wiley Classics Library. *John Wiley & Sons, Inc., New York*, 1988.

[20] Charles W. Curtis and Irving Reiner. Methods of Representation Theory. Vol. II. With applications to finite groups and orders. *John Wiley & Sons, Inc., New York*, 1987.

[21] M. R. Darafsheh. Maximal subgroups of the group $GL_6(2)$. *Bull. Malaysian Math. Soc. (2)* **7** (1984), 49–55.

[22] Leonard Eugene Dickson. Linear groups, with an exposition of the Galois field theory. *Teubner, Leipzig*, 1901 (Dover reprint 1958).

[23] L. Di Martino and A. Wagner. The irreducible subgroups of $PSL(V_5, q)$, where q is odd. *Resultate Math.* **2** (1979), 54–61.

[24] L. Finkelstein and A. Rudvalis. Maximal subgroups of the Hall-Janko-Wales group. *J. Algebra* **24** (1973), 486–493.

[25] L. Finkelstein and A. Rudvalis. The maximal subgroups of Janko's simple group of order $50,232,960$. *J. Algebra* **30** (1974), 122–143.

[26] D. E. Flesner. Finite symplectic geometry in dimension four and characteristic two. *Illinois J. Math.* **19** (1975), 41-47.

[27] D. E. Flesner. Maximal subgroups of $PSp_4(2^n)$ containing central elations or noncentered skew elations. *Illinois J. Math.* **19** (1975), 247-268.

[28] William Fulton and Joe Harris. Representation Theory. Graduate Texts in Mathematics 129, *Springer–Verlag, New York*, 1991.

[29] The GAP Group, GAP – Groups, Algorithms, and Programming, Version 4.5.7; 2012. <http://www.gap-system.org>

[30] Nick Gill. Polar spaces and embeddings of classical groups. *New Zealand J. Math.* **36** (2007), 175–184.

[31] Daniel Gorenstein. Finite Groups. *Harper and Row, New York–London*, 1968.

[32] Daniel Gorenstein. The classification of finite simple groups, Number 3, Part I, Chapter A, Almost simple K-groups. Mathematical Surveys and Monographs, 40.3. *American Mathematical Society, Providence, RI*, 1998.

[33] Ronald L. Graham, Donald E. Knuth and Oren Patashnik. Concrete mathematics. A foundation for computer science. Second edition. *Addison-Wesley Publishing Company, Reading, MA*, 1994.

[34] R. M. Guralnick, K. Magaard, J. Saxl and P. H. Tiep. Cross characteristic representations of symplectic and unitary groups. *J. Algebra* **257** (2002), 291–347. Addendum: *J. Algebra* **299** (2006), 443–446.

[35] Robert Guralnick and Gunter Malle. Products of conjugacy classes and fixed point spaces. *J. Amer. Math. Soc.* **25** (2012), 77–121.

[36] R. Guralnick, T. Penttila, C. E. Praeger and J. Saxl. Linear groups with orders having certain large prime divisors. *Proc. London Math. Soc.* **78** (1999), 167–214.

[37] R. M. Guralnick and P. H. Tiep. Low-dimensional representations of special linear groups in cross characteristics. *Proc. London Math. Soc.* **78** (1999), 116–138.

[38] R. M. Guralnick and P. H. Tiep. Cross characteristic representations of even characteristic symplectic groups *Trans. Amer. Math. Soc.* **356** (2004), 4969–5023.

[39] K. Harada and H. Yamaki. The irreducible subgroups of $GL_n(2)$ with $n \leq 6$. *C. R. Math. Rep. Acad. Sci. Canada*, **1** (1978/79), 75–78.

[40] R. W. Hartley. Determination of the ternary collineation groups whose coefficients lie in the $GF(2^n)$. *Ann. of Math.* **27** (1925/6), 140–158.

[41] G. Hiß, W. J. Husen and K. Magaard. Imprimitive irreducible modules for finite quasisimple groups. <http://arxiv.org/abs/1211.6350>.

[42] G. Hiß and G. Malle. Low-dimensional representations of quasi-simple groups. *LMS J. Comput. Math.* **4** (2001), 22–63. Corrigenda: *LMS J. Comput. Math.* **5** (2002), 95–126.

[43] G. Hiß and G. Malle. Low-dimensional representations of special unitary groups. *J. Algebra* **236** (2001), 747–767.

[44] D. F. Holt and S. Rees. Testing modules for irreducibility. *J. Austral. Math. Soc. Ser. A* **57** (1994), 1–16.

[45] D. F. Holt and C. M. Roney-Dougal. Constructing maximal subgroups of classical groups. *LMS J. Comput. Math.* **8** (2005), 46–79.

[46] D. F. Holt and C. M. Roney-Dougal. Constructing maximal subgroups of orthogonal groups. *LMS J. Comput. Math.* **13** (2010), 164–191.

[47] C. Hoffman. Cross characteristic projective representations for some classical groups. *J. Algebra* **229** (2000), 666–677.

[48] C. H. Houghton. Wreath products of groupoids. *J. London Math. Soc. (2)* **10** (1975), 179–188.

[49] R. B. Howlett, L. J. Rylands and D. E. Taylor. Matrix generators for exceptional groups of Lie type. *J. Symbolic Comput.* **31** (2001), 429–445.

[50] James E. Humphreys. Linear Algebraic Groups. Graduate Texts in Mathematics, 21. *Springer-Verlag, Berlin*, 1975.

[51] J. E. Humphreys. Modular representations of finite groups of Lie type. In *Finite Simple Groups II, Durham 1978*. Ed. M. J. Collins, *Academic Press, London* (1980), 259–290.

[52] B. Huppert. Singer-Zyklen in klassischen Gruppen. *Math. Z.* **117** (1970), 141–150.

[53] B. Huppert. Endliche Gruppen I. *Springer-Verlag, Berlin*, 1967.

[54] W. J. Husen. Maximal embeddings of alternating groups in the classical groups. Ph.D. Thesis, Wayne State University, 1997.

[55] W. J. Husen. Irreducible modules for classical and alternating groups. *J. Algebra* **226** (2000), 977–989.

[56] I. Martin Isaacs. Character theory of finite groups. Pure Appl. Math., 69. *Academic Press, New York–London*, 1976.

[57] Christoph Jansen, Klaus Lux, Richard Parker and Robert Wilson. An ATLAS of Brauer characters. *The Clarendon Press, Oxford University Press, New York*, 1995.

[58] Jens Carsten Jantzen. Representations of Algebraic Groups. Pure Appl. Math., 131. *Academic Press, Boston MA*, 1987.

[59] W. M. Kantor and R. A. Liebler. The rank 3 permutation representations of the finite classical groups. *Trans. Amer. Math. Soc.* **271** (1982), 1-71.

[60] O. H. King. The subgroup structure of finite classical groups in terms of geometric configurations. In *Surveys in combinatorics, 2005.* Ed. B. S. Webb, London Math. Soc. Lecture Note Ser., 327. *Cambridge University Press, Cambridge,* 2005, 29–56

[61] P. B. Kleidman. The maximal subgroups of the low-dimensional classical groups. PhD Thesis, University of Cambridge, 1987.

[62] P. B. Kleidman. The maximal subgroups of the finite 8-dimensional orthogonal groups $P\Omega_8^+(q)$ and of their automorphism groups. *J. Algebra* **110** (1987), 173–242.

[63] P. B. Kleidman. The maximal subgroups of the Steinberg triality groups $^3D_4(q)$ and of their automorphism groups. *J. Algebra* **115** (1988), 182–199.

[64] P. B. Kleidman. The maximal subgroups of the Chevalley groups $G_2(q)$ with q odd, the Ree groups $^2G_2(q)$, and their automorphism groups. *J. Algebra* **117** (1988), 30–71.

[65] P. B. Kleidman and M. W. Liebeck. A survey of the maximal subgroups of the finite simple groups. *Geom. Dedicata* **25** (1988), 375–389.

[66] Peter Kleidman and Martin Liebeck. The subgroup structure of the finite classical groups. London Math. Soc. Lecture Note Ser., 129. *Cambridge University Press, Cambridge*, 1990.

[67] A. S. Kondrat'ev. Irreducible subgroups of the group GL(7, 2). *Mat. Zametki* **37** (1985), 317–321, 460.

[68] A. S. Kondrat'ev. Linear groups of small degree over a field of order 2 (Russian). *Algebra i Logika* **25** (1986), 544–565.

[69] A. S. Kondrat'ev. Irreducible subgroups of the group GL(9, 2). *Mat. Zametki* **39** (1986), 320–329, 460.

[70] A. S. Kondrat'ev. The irreducible subgroups of the group $GL_8(2)$. *Comm. Algebra* **15** (1987), 1039–1093.

[71] A. S. Kondrat'ev. Finite linear groups of degree 6. *Algebra i Logika* **28** (1989), 181–206, 245.

[72] A. S. Kondratiev. Finite linear groups of small degree. In *The Atlas of Finite Groups: Ten Years On (Birmingham, 1995).* London Math. Soc. Lecture Note Ser., 249. *Cambridge University Press, Cambridge*, 1990, 139–148.

[73] A. S. Kondrat'ev. Finite linear groups of small degree. II. *Comm. Algebra* **29** (2001), 4103–4123.

[74] Serge Lang. Algebraic Number Theory. *Addison-Wesley, Reading, Mass.–London–Don Mills, Ont.*, 1970.

[75] V. Landazuri and G. M. Seitz. On the minimal degrees of projective representations of the finite Chevalley groups. *J. Algebra* **32** (1974), 418–443.

[76] V. M. Levchuk and Ya. N. Nuzhin. Structure of Ree groups. *Algebra i Logika* **24** (1985), 26–41.

[77] M. W. Liebeck. On the orders of maximal subgroups of the finite classical groups. *Proc. London Math. Soc. (3)* **50** (1985), 426–446.

[78] M. W. Liebeck, C. E. Praeger and J. Saxl. A classification of the maximal subgroups of the finite alternating and symmetric groups. *J. Algebra* **111** (1987), 365–383.

[79] M. W. Liebeck, C. E. Praeger and J. Saxl. On the O'Nan–Scott theorem for finite primitive permutation groups. *J. Austral. Math. Soc. Ser. A* **44** (1988), 389–396.

[80] M. W. Liebeck and J. Saxl. Primitive permutation groups containing an element of large prime order. *J. London Math. Soc.* **31** (1985), 237–249.

[81] M. W. Liebeck, J. Saxl and G. M. Seitz. On the overgroups of irreducible subgroups of the finite classical groups. *Proc. London Math. Soc.* **55** (1987), 507–537.

[82] M. W. Liebeck and G. M. Seitz. On the subgroup structure of classical groups. *Invent. Math.* **134** (1998), 427–453.

[83] M. W. Liebeck and G. M. Seitz. A survey of maximal subgroups of exceptional groups of Lie type. In *Groups, combinatorics & geometry (Durham, 2001).* World Sci. Publ., River Edge, NJ, (2003), 139–146.

[84] F. Lübeck. Small degree representations of finite Chevalley groups in defining characteristic. *LMS J. Comput. Math.* **4** (2001), 135–169.

[85] F. Lübeck. Tables of Weight Multiplicities,
    <http://www.math.rwth-aachen.de/~Frank.Luebeck/chev/WMSmall/>

[86] F. Lübeck. Conway polynomials for finite fields.
    <http://www.math.rwth-aachen.de/~Frank.Luebeck/data/ConwayPol/>

[87] Klaus Lux and Herbert Pahlings. Representations of groups: a computational approach. Cambridge studies in advanced mathematics, 124. *Cambridge University Press, Cambridge*, 2010.

[88] K. Magaard, G. Roehrle and D. Testerman. On the irreducibility of symmetrizations of cross-characteristic representations of finite classical groups.
    <http://arxiv.org/abs/1201.2057>.

[89] K. Magaard, G. Malle and P. H. Tiep. Irreducibility of tensor squares, symmetric squares and alternating squares. *Pacific J. Math.* **202** (2002),379-427.

[90] G. Malle. The maximal subgroups of $^2F_4(q^2)$. *J. Algebra* **139** (1991), 52–69.

[91] Gunter Malle and Donna Testerman. Linear algebraic groups and finite groups of Lie type. Cambridge studies in advanced mathematics, 133. *Cambridge University Press, Cambridge*, 2011.

[92] H. H. Mitchell. Determination of the ordinary and modular ternary linear groups. *Trans. Amer. Math. Soc.* **12** (1911), 207–242.

[93] H. H. Mitchell. The subgroups of the quaternary abelian linear group. *Trans. Amer. Math. Soc.* **15** (1914), 379–396.

[94] E. H. Moore. The subgroups of the generalized finite modular group. *Dicennial publications of the University of Chicago* **9** (1904), 141–190.

[95] B. Mwene. On the subgroups of the group $PSL_4(2^m)$. *J. Algebra* **41** (1976), 79–107.

[96] B. Mwene. On some subgroups of the group $PSL_4(q)$, q odd. *Geom. Dedicata* **12** (1982), 189–199.

[97] W. Nickel. Endliche Körper in dem gruppentheoretischen Programmsystem GAP. Diplomarbeit, RWTH, Aachen, 1988.

[98] C. M. Roney-Dougal. The primitive groups of degree less than 2500. *J. Algebra* **292** (2005), 154–183.

[99] M. Schaffer. Twisted tensor product subgroups of finite classical groups. *Comm. Algebra* **27** (1999), 5097–5166.

[100] Gary M. Seitz. The maximal subgroups of classical algebraic groups. *Mem. Amer. Math. Soc.* **67** (1987).

[101] G. M. Seitz and A. E. Zalesskii. On the minimal degrees of projective representations of the finite Chevalley groups. II. *J. Algebra* **158** (1993), 233–243.

[102] C. C. Sims. Computational methods in the study of permutation groups. In *Computational Problems in Abstract Algebra (Proc. Conf., Oxford, 1967)*. *Pergamon, Oxford* (1970), 169–183.

[103] A. K. Steel. Construction of ordinary irreducible representations of finite groups. PhD Thesis, University of Sydney,2012.

[104] R. Steinberg. Representations of algebraic groups. *Nagoya Math. J.* **22** (1963), 33–56.

[105] Robert Steinberg. Lectures on Chevalley Groups. *Yale University Mathematics Department*, 1968.

[106] M. Suzuki. On a class of doubly transitive groups. *Ann. of Math. (2)* **75** (1962), 105–145.

[107] J. Tits. Ovoïdes et groupes de Suzuki. *Arch. Math.* **13** (1962), 187–198.

[108] Donald E. Taylor. The geometry of the classical groups. *Heldermann Verlag, Berlin*, 1992.

[109] P. H. Tiep and A. E. Zalesskii. Some aspects of finite linear groups: a survey. *J. Math. Sci. (New York)* **100** (2000), 1893-1914.

[110] A. Wagner. The subgroups of PSL$(5, 2^a)$. *Resultate Math.* **1** (1978), 207–226.

[111] R. A. Wilson, P. G. Walsh, J. Tripp, I. A. I. Suleiman, R. A. Parker, S. P. Norton, S. J. Nickerson, S. A. Linton, J. N. Bray and R. A. Abbott. ATLAS of Finite Group Representations. <http://brauer.maths.qmul.ac.uk/Atlas/v3/>

[112] R. A. Wilson. The complex Leech lattice and maximal subgroups of the Suzuki groups. *J. Algebra* **84** (1983), 151–188.

[113] R. A. Wilson. Maximal subgroups of automorphism groups of simple groups. *J. London Math. Soc. (2)* **32** (1985), 460–466.

[114] Robert A. Wilson. The Finite Simple Groups. Graduate Texts in Mathematics, 251. *Springer-Verlag London, Ltd., London*, 2009.

[115] A. Wiman. Bestimmung aller Untergruppen einer doppelt unendlichen Reihe von einfachen Gruppen. *Stockh. Akad. Bihang* **25** (1899), 1–47.

[116] A. E. Zalesskii. Classification of the finite linear groups of degrees 4 and 5 over a field of characteristic 2. *Dokl. Akad. Nauk BSSR* **21** (1977), 389–392, 475.

[117] A. E. Zalesskii and I. D. Suprunenko. Classification of finite irreducible linear groups of rank 4 over a field of characteristic $p > 5$. *Vestsi Akad. Navuk BSSR Ser. Fz.-Mat. Navuk* **138** (1978), 9-15.

[118] K. Zsigmondy. Zur Theorie der Potenzreste. *Monatsh. für Math. u. Phys.* **3** (1892), 265–284.

Index of Definitions

Printed in the United States
by Baker & Taylor Publisher Services